Space structures 5

Edited by
G A R Parke and P Disney

Volume 1

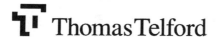 Thomas Telford

This volume contains the proceedings of the Fifth International Conference on Space Structures, held at the University of Surrey, Guildford, UK on 19–21 August 2002

Published for the conference organizers by
Thomas Telford Publishing, Thomas Telford Ltd, 1 Heron Quay, London E14 4JD.
URL: http://www.thomastelford.com

Distributors for Thomas Telford books are
USA: ASCE Press, 1801 Alexander Bell Drive, Reston, VA 20191-4400, USA
Japan: Maruzen Co. Ltd, Book Department, 3–10 Nihonbashi 2-chome, Chuo-ku, Tokyo 103
Australia: DA Books and Journals, 648 Whitehorse Road, Mitcham 3132, Victoria

First published 2002

A catalogue record for this book is available from the British Library

ISBN: 0 7277 3173 4

Printed and bound in Great Britain by MPG Books, Bodmin, Cornwall

Preface

These two volumes on Space Structures contain the latest work of leading Engineers and Architects submitted for presentation at the Fifth International Conference on Space Structures. The Conference was held at the University of Surrey, Guildford, UK in August 2002. The volumes contain a collection of one hundred and seventy papers, providing a wealth of up-to-date information on the analysis and design of space structures ranging from skeletal towers, grids and domes through to tension and tensegrity structures.

The Editors are indebted to the many engineers, architects and academics who have produced faultless manuscripts for publication. Also, the organisation of the Conference was helped greatly by the members of the International Organising Committee who have given limitless advice and encouragement. All of their support was essential to the smooth running of the Conference and to the valuable interchange of ideas and exciting new concepts presented during the event.

The Editors of the two volumes are staff at the Space Structures Research Centre of the University of Surrey which was founded in 1964 by Professor Z. S. Makowski. The Research Centre, under the initial directorship of Professor Makowski, followed by Professor H Nooshin, and recently by Dr G A R Parke, has now been the host to a total of five International Conferences on Space Structures - the first of which was held in London in 1965.

International organising committee

Z S Makowski (Chairman)
J F Abel (USA)
F G A Albermani
 (Australia)
S C Baird (UK)
M R Barnes (UK)
B S Benjamin (USA)
M Burt (Israel)
J W Butterworth (New
 Zealand)
Y K Cheung (Hong Kong)
J C Chilton (UK)
I Collins (USA)
N Dianat (Iran)
M G T Dickson (UK)
M Eekhout (The
 Netherlands)
A I El-Sheikh (UK)
F Escrig (Spain)
J F Gabriel (USA)
C Gantes (Greece)
K Ghavami (Brazil)
V Gioncu (Romania)
K K Gupta (USA)
A Hanaor (Israel)
P Huybers (The
 Netherlands)

K Imai (Japan)
K Ishii (Japan)
K Ishikawa (Japan)
Y Isono (Japan)
S Kato (Japan)
A Kaveh (Iran)
M Kawaguchi (Japan)
H Klimke (Germany)
L A Kubik (UK)
I Kubodera (Japan)
A S K Kwan (UK)
H Lalvani (USA)
T T Lan (China)
W J Lewis (UK)
K Linkwitz (Germany)
S Maalek (Iran)
M Majowiecki (Italy)
R B Malla (USA)
C Marsh (Canada)
F Matsushita (Japan)
S J Medwadowski (USA)
J L Meek (Australia)
H Moghaddam (Iran)
M Mollaert (Belgium)
R Motro (France)
I Mungan (Turkey)
E Murtha-Smith (USA)

A H Noble (South Africa)
J B Obrebski (Poland)
M Papadrakakis (Greece)
S Pellegrino (UK)
S Rajasekaran (India)
G S Ramaswamy (India)
E Ramm (Germany)
J Rebielak (Poland)
M Saidani (UK)
M Saitoh (Japan)
J Schlaich (Germany)
L C Schmidt (Australia)
S Z Shen (China)
V V Shugaev (Russia)
B W Smith (UK)
A Sollazzo (Italy)
N K Srivastava (Canada)
N Subramanian (India)
R Sundaram (India)
T Tarnai (Hungary)
T Wester (Denmark)
D T Wright (Canada)
H L Zhao (China)
A Zingoni (South Africa)

Executive committee
Dr G A R Parke (Chairman)
Professor H Nooshin
Dr P Disney

Sponsors

**THE INSTITUTION OF
CIVIL ENGINEERS**

**The Institution of
Structural Engineers**

Contents

Construction of three-strut tension systems

K. KAWAGUCHI and Z-Y. LU
University of Tokyo, Tokyo, Japan

INTRODUCTION

Tensegrity structures have been appealing to many architects and researchers with their peculiar appearance. However even a very typical model have never been applied for structural elements of buildings. Space grids assembled with tensegrity elements have been proposed by some researchers but never been constructed for practical use [5]. Cable domes and hyper tensegrity domes have already been constructed in large scales. However they are rather cable nets with struts in rigid compression rings and are somewhat different from the general image of the tensegrities. One of the authors have proposed and constructed "truss structure stabilized by cable tension". This system was found as a result of the exploration between space trusses and tensegrities [6].

In the paper the results of a numerical study, which was carried out for the pre-design of a structure, on structural behavior of three-strut tension systems are briefly discussed. Then the full-scale test and the construction of a pair of three-strut tension systems as supporting structures for a membrane roof, especially their tension introduction process, is reported.

THREE-STRUT TENSION SYSTEM

We firstly compare the structural behavior of three variations of the three-strut tension system, shown in fig.1, by means of numerical analysis. Since we had planned to use three-strut tension system to support a membrane roof the numerical models are in a special trapezoidal proportion. Model 1, fig.1(a), is a trapezoidal version of the most well known three-strut tension system, sometimes called as a "simplex", a triangular prism module of three-strut tension system. This model has six joints, three struts and nine tension members. Model 2, fig.1(b), is a variation of model 1. Two triangles, at the top and the bottom, are replaced by three-pronged tension members. New two joints appear at the center of upper and bottom triangles, where three tension members meet together. The numbers of struts and tension members remain same. Since only two tension members gather at each end of struts and they are arranged in one plane the joint details may become simpler than model 1. Model 3, fig.1(c), is also a variation of model 1 and has other three more tension members.

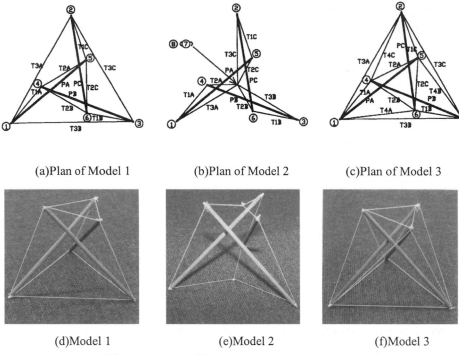

(a)Plan of Model 1   (b)Plan of Model 2   (c)Plan of Model 3

(d)Model 1   (e)Model 2   (f)Model 3

Figure 1. Trapezoidal three-strut tension systems

Table 1. Member properties for 3 models

Model 1			
Member	Area (cm²)	Initial Force (kN)	Yielding Stress(MPa)
T1	4.9	78.40	274
T2	4.9	59.31	274
T3	4.9	25.42	274
P	40.4	-93.58	323
Model 2			
Member	Area	Initial Force	Yield Stress
T1	4.9	78.40	274
T2	4.9	102.72	274
T3	4.9	44.02	274
P	40.4	-93.58	323
Model 3			
Member	Area	Initial Force	Yield Stress
T1	4.9	77.91	274
T2	4.9	62.07	274
T3	4.9	91.22	274
T4	4.9	26.61	274
P	40.4	-116.47	323
Young's modulus : 210 Gpa Poisson's Ratio : 0.3			

Table 2. Joint coordinates for the 3 models

Model 1			
Joint No.	x (cm)	y(cm)	z(cm)
1	-606.2	-350.0	0.00
2	0.00	700.0	0.00
3	606.2	-350.0	0.00
4	-300.0	0.00	800.0
5	150.0	259.8	800.0
6	150.0	-259.8	800.0
Model 2			
Joint No.	x (cm)	y(cm)	z(cm)
1	-606.2	-350.0	0.00
2	0.00	700.0	0.00
3	606.2	-350.0	0.00
4	-300.0	0.00	800.0
5	150.0	259.8	800.0
6	150.0	-259.8	800.0
7	0.00	0.00	0.00
8	0.00	0.00	0.00
Model 3			
Joint No.	x (cm)	y(cm)	z(cm)
1	-606.3	-350.0	0.00
2	0.00	700.1	0.00
3	606.3	-350.0	0.00
4	-299.3	0.00	790.6
5	168.1	212.7	790.6
6	131.2	-248.6	790.6

These three models have the common dimension. The height is 800cm, the side length of the triangle at the bottom is 1212cm and the side length of the triangle at the top is 520cm.

R.Motro carried out a research including loading test for a triangular prism model, similar to the model 1, and reported its great flexibility and geometrical nonlinearity [2]. Model 1 is statically and kinematically indeterminate structure of order one. Its inextensional displacement mode is twisting motion between the top and the bottom triangles and is stiffened only by introduction of initial tension.

Model 2 has same number of members as model 1 while it has another two joints. Therefore it is supposed to have more kinematic indeterminacy than model 1. Further since one strut and two tension members are arranged in one plane at the end of each strut, it is easily imagined that stiffness for out of plane motions is supposed to be low. The selfstress mode for model 2 is easily derived from the equilibrium conditions at each joint.

Model 3 is statically indeterminate and kinematically determinate structure since it has three more members than model 1. For the practical application a certain statically indeterminacy is preferable just in case of member failure. The number of tension members meeting at the end of each strut is now four, which makes joint design more complicated than other models.

NUMERICAL ANALYSIS
In this section the results of geometrically nonlinear analyses for models shown in fig.1 with parameters indicated in table 1 and 2 are shown. Joints 4,5 and 6 were loaded vertically and equally and joints 1, 2 and 3 are constrained in z-direction and freely supported in xy-plain. The initial axial forces in T1 members are set to be equal for all models.

Load-displacement relationships in z-direction at joint 4 under vertical load are plotted in fig.2. Each calculation was terminated when one of the members reached yielding stress level. Slackening of tension members was also considered.

Initial behavior of models 1 and 2 are similar. Model 1 gradually shows nonlinear behavior and T1 members reach yielding level about 100kN. T3 members of model 2 reach yielding level around 30 kN, which is much earlier than model 1.

Model 3 exhibits very high rigidity until about 50kN where T4 members are slackened. This high rigidity is due to the action of T4 members as compressive members until their slackening. After the slackening of these members the load-displacement relationship of model 3 is similar to those of model 1 and 2 and finally T3 members yield at around 150kN. The high rigidity of the model 3 is remarkable and different from the other two models.

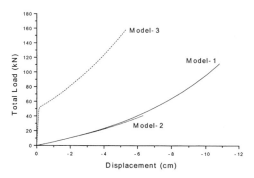

Figure 2. Load-displacement relationship

According to the results of eigenvalue analyses [7] model 1 and 2 have "zero" eigenvalue in the elastic stiffness. Their stiffness is provided only by geometric stiffness under the existence of initial stress while the stiffness of model 3 is sufficiently provided by the elastic stiffness until the member T4s' slackening.

Table 3. Initial Joint Coordinate of the Full-Scale Model

Joint No.	Designed joint coordinate (cm)			Measured joint coordinate (cm)		
	x-direction	y-direction	z-direction	x-direction	y-direction	z-direction
1	-1528.3	0.0	300.0	-1528.3	0.0	300.0
2	-886.8	-374.8	400.0	-887.3	-374.2	400.2
3	-886.8	374.8	400.0	-887.9	374.6	400.5
4	-1180.5	-170.9	879.6	-1182.8	-171.2	879.1
5	-1326.7	85.4	856.8	-1325.4	87.6	857.3
6	-1034.3	85.4	902.4	-1033.2	83.0	902.2

Table 4. Member Properties and Designed Axial Force

Member No.	Member Properties				Axial Force (kN)		
	Diameter (mm)	Sectional Area (cm²)	Length (cm)	Yielding stress (Mpa)	Step 1	Step 2	Step 3
T1A	32	8.04	598.8	440	29.4	40.0	156.6
T1B	32	8.04	598.8	440	29.4	39.3	155.8
T1C	32	8.04	598.8	440	29.4	40.8	157.3
T2A	28	6.16	296.2	440	21.3	30.6	121.3
T2B	28	6.16	296.2	440	21.3	31.1	121.7
T2C	28	6.16	296.2	440	21.3	31.7	122.4
T3A	25	8.04	750	440	8.2	13.7	49.5
T3B	25	8.04	750	440	8.2	14.2	49.9
T3C	25	8.04	750	440	8.2	13.2	48.9
T4A	25	4.91	681.3	440			18.5
T4B	25	4.91	681.3	440			18.5
T4C	25	4.91	681.3	440			18.5
PA	216.3x8.2	53.61	783.4	323	-38.5	-56.8	-230.5
PB	216.3x8.2	53.61	783.4	323	-38.5	-58.8	-232.5
PC	216.3x8.2	53.61	783.4	323	-38.5	-57.8	-231.5

TENSION INTRODUCTION TEST

According to the results of numerical analyses, we found that model 3 had high rigidity and high strength, which was enough for the practical use. Before going to the real design and construction we had an opportunity to carry out full scale assembling test.

Tension Introduction

For the realization of designed structural performance of tensegrities, introduction of initial tension is one of the most important issues. Therefore the tension introducing process should be carefully planned.

Prior to the construction, we had an opportunity to carry out a full-scale and tried to ensure our tension introduction scheme. Since model 1 in fig.1 has just one state of selfstress the initial prestressed state can be easily realized by shortening any of the tension members. On the other hand, model 3 has three independent selfstress states [7]. In order to realize the designed initial stress states the selfstress states should be appropriately adjusted. Further if we desire to introduce the tension through manual operation of turnbuckles inserted in the tension members, without any electric or hydraulic devices, we should consider the limit of the torque that can be induced by human power at the construction site.

In our scheme we firstly prepare model 1 with certain initial tension and then set three more members, T4's, and shorten them with low torque by long lengths. The objective tension was 155kN in T1 members and 18kN in T4 members.
The tension introduction scheme was as follows:

    Step 1 : Support the joints 4, 5 and 6, vertically. Assemble model 1 and introduce
           30kN into T1 members by shortening their lengths simultaneously.
    Step 2: Remove the supports.
    Step 3 : Set T4 members and introduce 155kN into T1 members by shortening T4
           members simultaneously until T4 members' tension become 18kN.

Wire strain gauges were mounted on the all of the members and their strain was monitored at real time so that the readings can be immediately fed back to the tension introduction process.

Test Results of the tension introduction test

Before start the test one of the T1 members was shortened during model 1 state and we confirmed that satisfactory tension distribution could be realized by just one member operation. In order to avoid uneven member length change we operated the three T1 members simultaneously. Even after the setting of T4 members tension introduction process was not difficult than we expected because of the effective feedback scheme from the real time reading of the member strain.

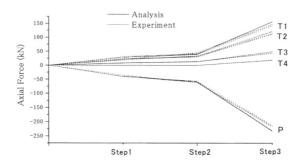

Figure 3. The results of the test: Tension distribution

(a) General view of the test   (b) Tension introduction

Figure 4. Full-scale tension introduction test

The results of tension introduction test are plotted in fig.3. The test results agreed very well with the designed quantities. In order to increase the tension in T1 members from 30kN to 155kN, T4 members were shortened by about 8 cm.

CONSTRUCTION

Gaining confidence from the results of a number of numerical analyses and a full-scale tension introducing test we designed a pair of supporting structures for a membrane roof employing three-strut tension systems of model 3 (figs.5 and 6). One post member was added for each tension-strut system, hung from three joints of top triangle by tension rods, to push up the membrane roof (fig.5). Joint at the bottom end of the post is flexibly connected to the tension rods so that the large displacement of the membrane roof can be absorbed by the inclining action of the post. The bigger tension-strut system was called tensegrity A and the other was called tensegrity B (fig.6). The height and the side length of the bottom triangle of the tensegrity A are 9m and 12m, respectively, and 6m and 9m for the tensegrity B.

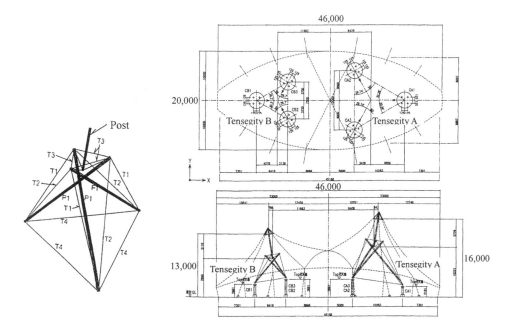

Figure 5. The supporting structure   Figure 6. Plan and section of the structure

Tension Introduction

After the foundation was prepared two strut tension systems were assembled one by one. Joints of the top triangle were set on the jacks on the support columns and members were assembled in the form of model 1 in fig.1. Three T1, T5 for tensegrity B, members were equally shortened until the 42kN self-equilibrated axial force was introduced to them. Then the jacks were removed since the structure could stand by itself. Three more rods, T2's for the tensegrity A and T6's for the tensegrity B, were then set in and the system had the form of model 3 in fig.1. These new tension rods were carefully and equally shortened until their axial force reach 30kN and the tension introduction process was completed. It should be noted that the structure experiences large deformation during the tension introduction scheme. The joint details should be designed that they can accommodate to this large deformation.

All the tension introduction works were manually done just by human power (fig.4). A pair of wire strain gauges was mounted on the every member, the tension rod and the compression pipe, and its strain was always monitored at real time at the site. The readings were immediately fed back to the tension introduction work. Members for the tensegrity A was newly prepared while the members used for the preceding tension introduction test were diverted to the tensegrity B.

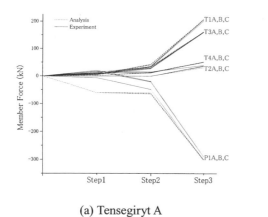
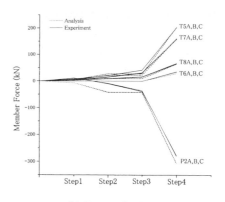

(a) Tensegiryt A   (b) Tensegrity B

Figure 7. The measured axial forces during the tension introducion work

(a) Tension introduction work   (b) Real time measuring at the site

(c) Two three-strut tension systems

Figure 8. Supporting structures under the construction

Results of the tension introduction at the site

The measured axial forces during tension introduction work are shown in fig.7. Just like the results of the preceding test in fig.3 the results agreed with the designed quantities well enough to expect the designed structural performance to the structures.

For tensegrity A the shortened length of the tension rods T2 during the operation from the model 1 stage to the model 3 stage was about 17cm and the displacements of the joints of the top triangle were about 16cm in twisting motion. For tensegrity B the the tension rods T2 were shortened during the same operation by about 14cm and the displacements of the joints of the top triangle were about 15cm also in twisting motion. These long distance changes of the rods enable us the manual work for the tension introduction although a big amount of pretension must be introduced in the structure.

CONCLUSION

A number of numerical analyses and a full-scale tension introduction test were carried out before the design and construction of the tension strut systems. A part of the numerical analyses was shown and the validity of the model 3 in the application to practical use was discussed in this paper. The remarkable improvement in structural performance of model 3, comparing to the model 1, is attributed to the effective reaction in compression of pretensioned members. The proposed tension introduction scheme was successful and effective so that just manual works were enough to complete the tension introduction process.

(a) Tensegrity B supporting   (b) Interior view   (c) Exterior view
the membrane roof

Figure 9. General view of the completed structure

REFERENCES

1. Calladine, C.R., Buckminster Fuller's "Tensegrity" structures and Clerk Maxwell's rules for the construction of stiff frames, *Internationa Journal of Solids Structures*, 1978, Vol.14, pp161-172.

2. Motro, R., Forms and forces in tensegrity systems, *Proceedigs of the Third International Conference on Space Structures*, 1984, pp283-288.

3. Tanaka, H. and Hangai, Y., Rigid Body Displacement and Stablization Conditions of Unstable Truss Structures, *Proceedigs of the IASS Symposium on Membrane Structures and Space Frames*, ed. By Heki, 1986, pp55-62.

4. Motro, R., Tensegrity Systems – Past and Future, *Proceedigs of the IASS International Symposium '97*, 1997, pp69-79.

5. Hanaor, A. and Liao, M.K., Double-Layer Tensegrity Grids: Static Load Response I: Analytical Study and II: Experimental Study, *ASCE-Journal of Structural Engineering*, 1991, Vol.117, No.6, pp1660-1684.

6. Kawaguchi, K., Oda, K. and Hangai, y., Experiments and Construction of Truss Structure Stabilized by Cable Tension, *Proceedigs of the IASS International Symposium '97*, 1997, pp421-429.

7. Ken'ichi Kawaguchi and Lu Zhen-Yu, A Study of Trapezoidal Three-Strut Tension System, Proceedings of IASS2001, International Symposium on Theory, Design and Realization of Shell and Spatial Structures, edited by H.Kunieda, CD-ROM, TP142, 8pages, 2001.10.

The Concept of Structural Depth as Applied to Certain Bar-tendon Assemblies

ARIEL HANAOR, National Building Research Institute, The Technion, Israel Institute of Technology, Haifa, Israel

STRUCTURAL EFFICIENCY AND STRUCTURAL DEPTH

Definitions and Scope

Structural efficiency

Structural efficiency is a yardstick used, sometimes, for evaluating and comparing structural systems, and particularly lightweight spatial systems. It is of primarily aesthetic-philosophical significance, although in certain applications, such as in outer space or in structures of extreme spans, it can assume primary technical and economical significance. In the context of this paper, the *Structural efficiency ratio* of a given structure is defined as the ratio of the load bearing capacity of the structure to its weight. The applied load, usually gravity load, includes the live load and the load arising from any non-load bearing components, but excluding the weight of the structural elements (addition of unity to the efficiency ratio will allow for the inclusion of self-weight).

The structural efficiency ratio is thus a measure of the lightness of a structure – the higher the ratio the lighter the structure. However, care should be exercised when applying it for comparison between structures or structural systems. When comparing structures of similar systems but differing load-bearing capacities (or design loads), it should be borne in mind that this parameter is not linearly proportional to the load. For instance, when members of a bar structure, such as a truss, are designed efficiently, the cross-section area, and therefore the weight, of tensile members is roughly proportional to the applied load. On the other hand, the weight of compressive members whose design is governed by elastic buckling, is roughly proportional to the cubic root of the load (since for a tube of constant wall thickness both cross-section area and radius of gyration are approximately proportional to the diameter). Thus, the structural efficiency ratios of structures of similar type and geometry tend to be higher the more heavily the structure is loaded, even though the actual weight is larger.

Another effect, which tends to distort the interpretation of efficiency ratio as a measure of comparison between structures or structural systems, is the material effect. It is obvious that a structure made of aluminium, for instance, would be lighter than the same structure, subjected to the same load, but made of steel. When considering systems containing components made of different grades of the same material (e.g. mild steel and high-strength steel), the material effect also plays a role. This is particularly relevant, as in the present study, when comparing structural systems consisting primarily of bars made of, for instance, mild-steel tubes, with systems that contain a large proportion of high-strength cables. The efficiency ratio, as defined above, thus tends to confound the material effect with the purely structural effect, and, depending on the purpose of the comparison, the material effect should be factored out or taken into consideration.

Structural depth
Structural depth at a cross-section through the structure, is defined as the lever arm of the resultant internal force couple at the cross-section, balancing the overturning moment produced by the external load on a free body bound by the cross section in question [ref. 1]. The structural depth, thus, varies over the structure, but for most structures of the type considered here, it is the structural depth at a particular cross-section – the one governing the design – that is of interest. For instance, for simply supported spanning structures, the critical cross-section is usually at mid-span. Clearly, there is some relation, albeit subtle, between structural depth and structural efficiency, since the lever arm of the internal force couple determines, to a large extent, the magnitude of internal forces in members in the critical cross-section, and thus their dimensions. Varying the structural depth, however, affects the dimensions of the structure as a whole, and thus the effect on overall weight is complex. As is well known, the optimal structural depth for a given structural system varies within relatively narrow bounds (see for instance ref. 2). The primary significance of the concept of structural depth is, arguably, not its influence on structural efficiency, but as a tool for understanding structural behaviour at the conceptual level, and perhaps also at the rudimentary analytical level. It is thus particularly valuable for the assessment of geometrically complex structural systems of the type considered in this paper.

A Survey of the Structural Efficiency of some Systems
In this section some data encountered in the literature is presented with a focus on structural efficiency. The main purpose for presenting this data is for comparison with data recently reported in the literature for some double-layer bar-tendon systems [refs. 3, 4], bearing in mind the limitations of structural efficiency as a yardstick for the comparison, as mentioned above.

Double-layer Grids
Figure 1 presents some data gleaned from the literature and processed to represent efficiency ratio of some double-layer grid structures made of steel. As mentioned above, due to load differences, it is difficult to compare systems. What is more, load is not given for all systems surveyed. For instance, ref. 6 does not provide load data, and it was assumed that the imposed load for all structures reported there is 120 kg/m^2, as in ref. 2. In an attempt to adjust the efficiency ratio for a standard load, it was assumed that the weight of tensile members is ⅓ of the total weight and that the weight of tensile members is proportional to the load and the weight of compression members is proportional to the cubic root of the load. The total weight was then adjusted to an imposed load of 100 kg/m^2. It should be noted also that structures related to refs. 5 and 6 are actually built structures, whereas structures of refs. 2-4 are based only on theoretical design, with ref. 2 incorporating an optimization procedure (although for cost, not weight).

Domes
Figure 2 presents efficiency data for some braced domes, obtained from the literature. The solid lines represent design data for triangular grid domes with steel bars employing the Triodetic connector, as presented in ref. 7. The strong influence of the load on the efficiency ratio is apparent, particularly for single-layer domes. Also presented are the adjusted values for a load of 100 kg/m^2, assuming efficiency ratio to load relation of cubic root, although the actual relation derived from the data is even weaker. Also presented (by a solid triangle) is

the data for a dome of this type actually built (in Israel by the Geometrica company). This data seems to fit with the design data given in ref. 7. Data is given also for two large span lamella domes (Lousiana Superdome and Houston Astrodome – ref. 8), and for domes constructed in China as given in ref. 6. Since the source does not give load data, a load of 120 kg/m^2 was assumed in Fig. 2. In view of the strong dependence of the efficiency ratio on the load, this latter data is in doubt.

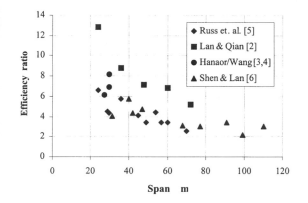

Figure 1.

Efficiency ratio vs. span of double-layer space trusses, adjusted for imposed load of 100 kg/m^2.

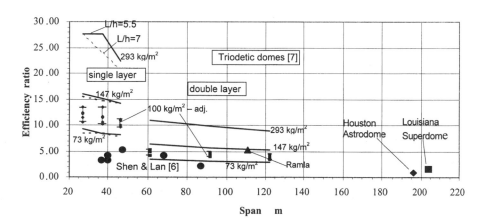

Figure2. Efficiency ratios of braced domes

Cable roofs
Fig. 3 presents some data for various types of cable roofs, including saddle shaped cable nets; circular layouts comprising radial cables, external compression ring and internal tension ring ("bicycle wheel"); as well as the well-known Georgia cable dome, which is the largest span structure of this type (oval shape with span of 186/233.5 m – refs. 8,9). The majority of the structures are from ref. 6, which does not provide load data, but as the weight of cable roofs is governed by tensile elements (assuming the compression ring and boundary members are excluded), the efficiency ratio is not strongly dependent on the load. It is interesting to note that the structural efficiency of large span cable roofs is not very high, contrary to common perception. See also structural depth analysis below.

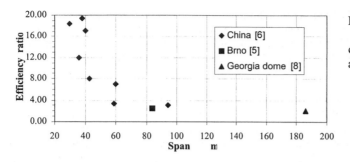

Figure 3.
Structural efficiency of constructed cable roofs and domes.

Bar-tendon assemblies

Fig. 4 presents the structural efficiency of some bar-tendon double-layer grids proposed in ref. 3 and more extensively in ref. 4. In this paper the term *bar* indicates a rectilinear member capable of sustaining compression and tension. A *tendon* is a rectilinear member capable of sustaining tension only. All the configurations presented in Fig. 4 are composed of basic prestressable units termed *simplexes*. When the simplexes are tensegrity units, bars in the unit itself are not in contact, but their connection to form double-layer grids can generate grids in which bars are not in contact – "pure" tensegrity grids, or grids in which bars are in contact. In both cases bars are inclined and there are no continuous chords made of bars. The other type of grids, termed *cable-strut* grids by Wang [ref. 4] is composed of units in which bars are already in contact in the basic simplex. When these units are joined to form a double-layer grid, the grid contains continuous lines of bars forming chords in the upper part of the grid.

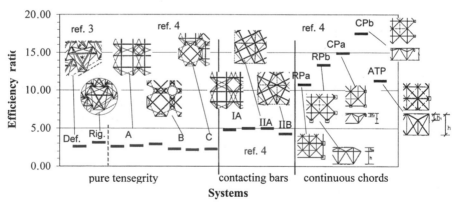

Figure 4. Structural efficiency of designed bar-tendon double-layer grids (refs. 3,4), adjusted for imposed load of 100 kg/m². Span = 27-30 m.

All the grids investigated have a span of 27-30 m, and thus there is no information on the variation of structural efficiency with span. The grids in reference 4 were all designed for imposed design load of 100 kg/m². The grids in ref. 3 were designed for a load of 195 kg/m² and the data in fig. 4, was adjusted for a load of 100 kg/m² based on the actual weight ratio of cables and bars and on the assumption that bar weights are proportional to the cubic root of the load. When comparing efficiency ratios of this type of structures with those of bar structures (such as double-layer grids or braced domes), it should be borne in mind that the design strength of cables is 2.5-3 times that of the bars. The weight of cables ranges from ca. 15% in tensegrity and ATP grids (cf. Fig. 4) to 20-25% in continuous chord grids (RP, CP in Fig. 4). Correspondingly, 22%-37% of the efficiency ratio is due to material, rather than structural effect.

STRUCTURAL DEPTH ANALYSIS APPLIED TO BAR-TENDON STRUCTURES

The evaluation of structural depth in bar structures of the spanning type is straightforward. For instance, the structural depth of truss structures, whether planar or spatial, is the distance between the chords (at the cross-section considered). Similarly, the structural depth of braced domes under symmetrical loading is the overall height of the dome. Understanding and evaluating structural depth in bar-tendon assemblies may, however, be more subtle and may occasionally lead to unexpected results. In general, the structures considered in this paper are all simply supported around their boundary. They possess a top chord and a bottom chord consisting of tendons, which may or may not be prestressed, and web members that include bars in various configurations. Some of the systems contain also an intermediate chord composed of bars. At first thought it may seem that when the chords are prestressed, the structural depth is the distance between the chords, since the compressive component of the internal couple can be accommodated by a reduction of the prestress in the corresponding chord. At the ultimate state, however, the critical tendons of the compressive chord slacken and the compressive component has to be borne by bar members. In fact, the prestress itself can only be maintained through compression in bar members, so that prestress is not a viable means for improving structural efficiency (although it can be used to increase stiffness). The structural depth analysis that follows, which is essentially at the conceptual level, employs as basic models planar analogues of the spatial configurations, for the sake of simplicity and clarity.

Cable roofs and cable domes

Radial cable roofs
The planar model for a type of radial cable roof is presented in Fig. 5. This is a simple structural system. Under gravity loads, the top cables slacken and the structural depth of the system is the sag of the bottom cable. The compressive component of the internal couple is supported by the horizontal intermediate chord in the planar model. In the actual structure the compression ring beam performs this function. Under uplift load, the role of the two families of cables is reversed. The relative sags of the top and bottom families of cables can be adjusted in accordance with the relative magnitude of gravity and uplift loads. The relative position of the two caable families can be also shifted, to reduce the overall depth of the roof. In this case, each family has its own compression ring (or chord, in the planar model).

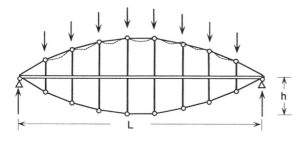

Figure 5.
Planar model for a type of radial cable roof.

Cable domes
The term *cable dome* is an oxymoron (although it is an improvement on the term *tensegrity domes*, which is occasionally employed). Domes are essentially compressive structures (under gravity loads) whereas cable structures are purely tensile. Indeed, as the following analysis demonstrates, this system is not a dome at all, in the structural sense, but a type of cable net roof. Figure 6 presents a simplified planar cable truss model, which possesses the main structural features of this type of structure, even though it is geometrically different, in that its top chord is straight, rather than curved. In effect, the entire load on internal panels is transmitted via the diagonal tendons to the vertical struts adjacent to the supports, where it is supported on the bottom chord as a simple cable. The compressive force component is supported by the intermediate bar chord, which in a spatial framework would be provided by a ring beam. Similarly, the bottom tendon chord would be replaced in the spatial configuration by a series of cable hoops passing through the corresponding nodes. The structural depth of this structure is clearly the sag of the bottom chord.

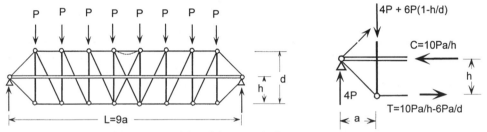

Figure 6. A planar cable truss model and force analysis.

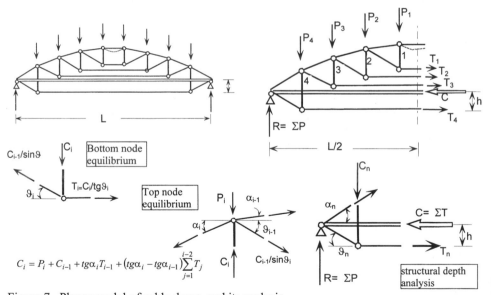

Figure 7. Planar model of cable dome, and its analysis.

Figure 7 presents the planar analogue of a cable dome. As can be seen, the load transfer system is similar to that of the cable truss of Fig. 6. In fact, the domical shape causes the forces in the bars and chords to increase, compared with the flat configuration of Fig. 6. This is the penalty for forcing the radial cables to assume a shape contrary to the natural catenary shape. This explains, perhaps, the shallow profile of the largest cable dome constructed to date – the Georgia Dome – Figure 8. On the other hand, the structural depth under uplift load (which may be the dominant load) is the full rise of the dome from the ring beam, where the radial cables (the top chord) act in a natural way.

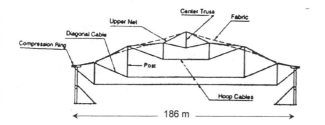

figure 8.
Cross-section of the Georgia Dome. From Ref. 8.

The simple analysis presented in Figs, 6 and 7 is a first order analysis. The structures discussed possess low stiffness and are subject to large deformations, requiring a non-linear analysis. However, the deformations tend to increase structural depth and the first-order analysis can be conservatively employed for preliminary design and check of computer analysis.

Double-layer bar-tendon grids
The structures examined in this section are those featuring in Figure 4, namely double-layer tensegrity grids with contacting and with non-contacting bars, and grids proposed by Wang (ref. 4) that contain continuous chords comprised of bars.

Tensegrity grids

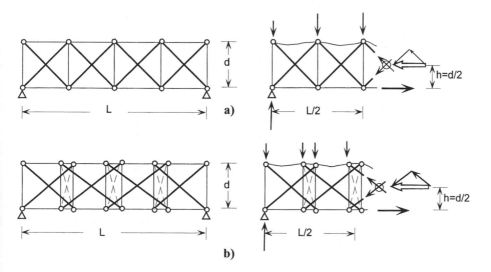

Figure 9. Planar model for double-layer tensegrity grids: a) with contacting bars; b) pure tensegrity.

Figure 9 presents the planar analogue models for the two types of tensegrity grids. As can be seen from the structural depth analysis shown in the drawing, the two types of structures are similar in this respect, although the indirect connection of bars through tendons may cause significant reduction in stiffness, as observed in ref. 4. This effect may also cause significant differences in force distribution, even in grids of similar layouts (such as, for instance, the pure tensegrity grid denoted as type A in Fig. 4 and the grid with contacting bars denoted as type IA). Such differences will show up only in the essentially non-linear analysis and cannot be accounted for by the simplified analysis presented here.

The structural depth analysis clearly demonstrates the three sources of the relatively low structural efficiency of grids constructed from tensegrity modules, as compared to grids with continuous bar chords [ref. 4]. These sources are: a) A reduction in structural depth (half overall depth as compared to full depth in double-layer grids constructed of bars); b) Inclination of the bars relative to the direction of the compressive component of the internal force couple; c) Increased length of the bars relative to the panel length. Figure 10 presents a cross-section of a double-layer tensegrity grid with bars in contact (grid type IA in Fig. 4), together with the application of structural depth analysis to this type of structure. A crude computation, assuming that the load on the two-way square grid is distributed equally in the two directions and uniformly along panels in each direction, yields a result for the compressive force in bars at mid-span, which is in close agreement with data presented in ref. 4 (ca. 9 tonne). On the other hand the analysis fails to account for the much higher bar forces (av. 20 tonne) reported in ref. 4 for the pure tensegrity grid of similar layout (type A in Fig. 4). The main reason for the discrepancy is that the indirect transfer of forces between non-contacting bars causes the tendons in the top chord to remain in tension. This, in turn, causes the tensile force resultant to lift above the bottom chord, resulting in further reduction of the effective structural depth. Naturally, the simplified planar model cannot account for this effect.

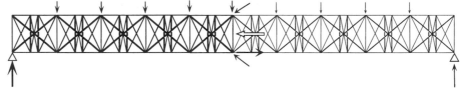

Figure 10. Structural depth analysis applied to the cross-section of a double layer tensegrity grid (ref. 4).

The inclined bars in the panels of the cable truss model (and in tensegrity grids) serve a dual role as providing both the vertical force component needed to balance the shear force at a cross-section, and the compressive component of the internal couple. Accordingly, the maximum bar force may occur near the support, where shear governs (for short spans), or at mid-span, where flexure governs (for longer spans). A simple calculation shows that for the planar truss shown in Fig. 9a, and assuming panels of equal lengths, the "break even" point (i.e. when the maximum bar force at mid-span is equal to the bar force near the support) occurs when the span contains eight panels, regardless of dimensions.

Double-layer tensegrity grids can be implemented as curves surfaces, to produce tensegrity domes or shells. Figure 11 presents the planar analogue to such a dome, which is an arch. As can be observed, the structural depth analysis is quite different to that of Figure 9. The structural depth is the rise of the arch (dome) to its mid-surface. In the simplified analysis, the tendon chords have only a secondary role (for instance, in case the arch is not the funicular shape for the applied load), and the compressive force is sustained by the bars alone. The inclination of the bars to the mid-surface still reduces their effectiveness, compared with arches or domes containing continuous bar surfaces, but improved structural efficiency can be achieved relative to the flat surface grids due to the improved structural depth.

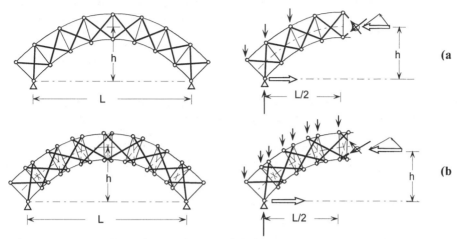

Figure 11. Tensegrity arches as planar models for tensegrity domes: a) with bars in contact; b) pure tensegrity.

Bar-tendon grids propsed by Wang (ref. 4)

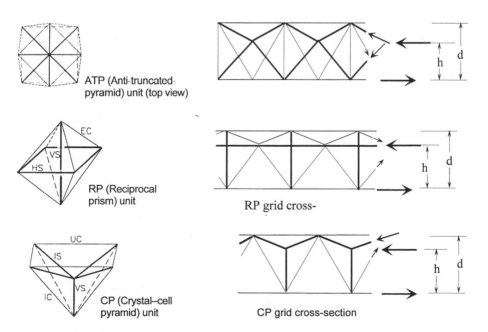

Figure 12. Bar-tendon grids with continuous bar chord, to ref. 4: simplex units and structural depth analysis.

The bar-tendon grids presented in Figure 12 were proposed by Wang in ref. 4 (and termed *Cable-strut* grids by the author). These configurations are based on simplex units that contain bars in contact. These structures are conceptually simple for structural depth analysis, as they possess continuous bar chords serving for the compressive component of the internal couple, similarly to conventional bar structures, although in some configurations (CP, ATP in Fig. 12) the cords do not form straight lines. The principle of these structures, apart for their construction from individual *simplex* units, is to replace tensile members with tendons and to reduce the lengths of compressive bars, thus achieving high structural efficiency. It should be noted, however, that configurations presented in ref. 4 are designed for gravity loads. In the presence of uplift their structural depth is substantially reduced, and in the case of CP grids vanishes. In the presence of substantial uplift, some optimization of the relative structural depths for gravity and uplift loads can be performed, but it is doubtful if the result would be an improved structural efficiency compared to conventional double-chord bar grids (at least when material efficiency is factored out).

REFERENCES

1. HANAOR A, Principles of Structures, Blackwell Science, 1998.

2. LAN TT, QIAN R, A Study on the Optimum Design of Space Trusses – Optimal Geometrical Configuration and Selection of Type, Shells, Membranes and Space Frames, Proc. IASS Symp. Osaka, Japan, Sept. 1986, Elsevier, Amsterdam, V. 3, pp 191-198.

3. HANAOR A, Geometrically Rigid Double-Layer Tensegrity Grids, Int. J. Space Structures, V9, N4, 1994' pp 227-238.

4. WANG BB, From Tensegrity Grids to Cable-Strut Grids, Int. J. Space Structures, V16, N4, pp 279-314.

5. RUSS R, HODER F, DOHNAL J, Prefabricated Steel Space Structures designed by Chemoprojekt Prague – CSSR, Proc. 3rd Intnl. conf. on Space Structures, Uni. of Surrey, Sept. 1984, Elsevier Applied Science Publ., London, pp. 42-52.

6. SHEN SZ and LAN TT, A Review of the Development of Spatial Structures in China, Appendix: Important Spatial Structures in China, Int. J. Space Structures, Special Issue on Space Structures in China, Shen S.Z. and Lan T.T, guest editors, V 16, N 3, 2001, pp 157-172.

7. MAKOWSKI ZS, A History of the Development of Domes and a Review of Recent Achievements World-wide, Analysis Design and Construction of Braced Domes, Makowski Z.S., Editor, Nichols Publ. Co., New York, 1984, pp 1-85, Tables 1.2a, 1/2b, 1.3, pp 81-82.

8. Internet site: http://www.columbia.edu/cu/gsapp/BT/DOMES/ .

9. LEVY M, The Innovation of Lightness, IASS Bulletin, V35, N2, August 1994, pp 77-84.

A study of tensegrity cable domes

F. FU
Beijing Institute of Architectural Design and Research, Beijing, China
T. T. LAN
Chinese Academy of Building Research, Beijing, China

INTRODUCTION

In the process of designing long span space structures, how to reduce the self-weight of the structure and consequently the cost of the building is the key issue. Among different types of structures, the 'Tensegrity System', that is a self-equilibrium system composed of continuous prestressed cables and individual compression bars, is one of the most promising solutions. The concept of 'Tensegrity' was first conceived by B. Fuller, which reflected his idea of 'nature relies on continuous tension to embrace islanded compression elements'. Unfortunately, his 'tensegrity dome' has never been executed in engineering project. It was D.H. Geiger, who made use of Fuller's thought and designed an innovative structure 'cable dome'. It has been successfully realized in the circular roof structures of Gymnastic and Fencing Arenas for the Seoul Olympic Games in 1986. For the 1992 Atlanta Olympic Games, M. Levy further improved the layout of the cable dome and built the Georgia Dome in quasi-elliptical shape. Among eight cable domes of tensegrity system built to date, there exit different variations of the network geometry of the dome. For designers, it is interesting to know the relative pros and cons of different layout of the network, which will influence precisely the weight and the cost of the structure.

Since most long span sports halls are non-circular in plan and usually have the configuration in oval shape, the Georgia Dome is taken as the prototype for the cable dome. Some other structural types with different layouts are presented as examples for comparison.

STRUCTURAL BEHAVIOUR OF TENSEGRITY DOME

In order to design different types of tensegrity domes in oval plans, its static behavior is first investigated by analyzing the force and deformation in the structure. Fig.1 shows the structural layout of the Georgia Dome. It is noticed that instead of using the radial cable-and-strut trusses as Geiger designed in his cable domes, Levy preferred the triangulated geometry for the network. In Fig.1, a, b, c, d denotes the ridge cable in each layer; e, f, g, h denotes the diagonal cable; l, m, n denotes the tension hoop cable, while P1, P2, P3, P4 denotes the compression strut, all ascending from the bottom to top. The following materials are used for calculation: 5mm high strength tensile wires for cables with a tensile strength of $1670 N/mm^2$ and Q345 circular tubes for struts with a yielding strength of $345 N/mm^2$ Uniform superimposed load of 0.6 kN/m^2 is applied to the top surface of the dome.

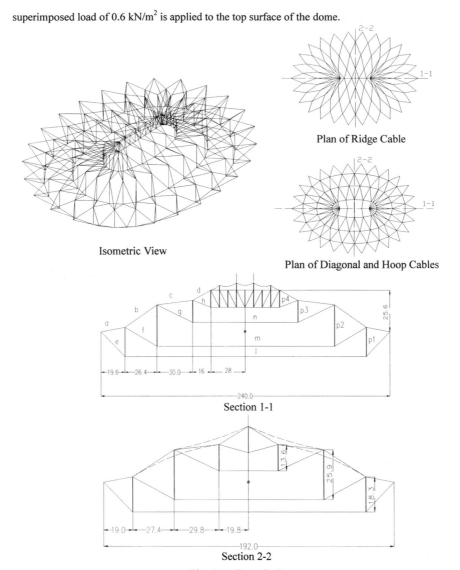

Fig. 1.   Georgia Dome---prototype.

As a result of design, the following sectional areas are used. For ridge cables on a,b,c,d layers, the areas are 85.4cm^2,46.2cm^2,21.8cm^2 and 21.8cm^2 respectively. For diagonal cables on e,f,g,h layers, the areas are 47.4cm^2,56cm^2,23.9cm^2 and 23.9cm^2 respectively. For tension hoop cables on l,m,n layers , the areas are 189.7cm^2,267.8cm^2 and 75.6cm^2 respectively. For struts P1,P2,P3,P4, the areas are 270.8cm^2,270.8cm^2,115cm^2 and 400cm^2 respectively.

Using a nonlinear finite element analysis program, it was found that the variation of member forces and deformation of a tensegrity dome depicts certain special features. The variation of forces with loading is basically linear. When the load increases, each type of the cable responds in different manner. The forces in ridge cables decrease; with those in inner and top layer decrease more rapidly than the outer ones. The variation of forces in diagonal cables depends on its position. For outer and lower layer, it increases, while for inner and upper layer, it decreases. Forces in tension hoop cables increase with the load, the rate of increasing is much larger in the bottom layer.

It is interesting to note the failure mode of the dome. When the load keeps increasing, the forces in the ridge cables on d layer as well as the diagonal cables on h layer all decrease. When the load attains certain value, the force in one of the ridge cables, usually in the central section of the dome, will decrease to zero and thus the cable becomes slack.. However, the forces in hoop cables and part of the diagonal cables are still increasing. The structure can still maintain its bearing capacity, but the deformation is increasing significantly. If the load increases further until one of the diagonal cables on h layer also becomes slack, then failure occurs to the whole structure. The failure criteria of a tensegrity dome are neither the breaking of the tension hoops and diagonal cables, nor the buckling of the struts. It is the slackening of the ridge cable and diagonal cable in the central section of the dome that determines the bearing strength. An efficient way to increase the bearing capacity is to increase the prestressing forces in the ridge and diagonal cables on the inner and upper layer of the central section of the dome.

STRUCTURAL TYPES
Taking the Georgia Dome as the prototype, several structural schemes of cable domes of Tensegrity system are analyzed and designed. For the purpose of comparison, all schemes are designed in a quasi-elliptical plan with a longitudinal span of 240 m,

Structural Type 2
The circular cable dome designed by Geiger demonstrates some significant advantages. The ways of forming networks in wedge shape of the cable dome is simpler than the triangulated networks. The number of cable elements is less, and hence less weight. As there are less cables connecting at a node, the construction of a joint is more or less easy. The advantages of such network are shown on circular domes, since the construction of the joints in each layer is the same, resulting a minimum type of joints. However, the stiffness of cable dome is smaller comparing with the triangulated dome system, especially in the horizontal direction. There are no links between the top chords joints in the circumferential direction of the cable dome. For triangulated networks, all the top chord joints are connected by the ridge cables, thus a greater horizontal stiffness can be obtained.

In structural type 2 (Fig. 2), networks of triangulated dome and cable dome are used simultaneously so as to utilize their respective advantages. The quasi-elliptical configuration is composed of two semi-circles at both ends where networks of cable dome are used, and a rectangular central section using triangulated networks. From the studies before, it was found that the weakest position in an elliptical tensegrity dome is in the central section, where the slackening of ridge cables always occurs and the displacements are larger. The concept of the design of this structural type is to strengthen the central section of the dome. Under the action of the vertical load, the semi-circles at both ends tend to pull away the central section. It is required a stiffer central section to resist the horizontal displacements, thus triangulated networks with strong horizontal stiffness are used. For the section of two semi-circles at both ends, networks of cable dome are used, so as to simply the construction of connecting joints and to lessen the types of the joint. Furthermore, the top chord in the central section tends to resist compression, in order to increase the load bearing capacity; circumferential bars are established in the top chord.

Under vertical uniform load, the sectional areas of structural member are taken as follows. For ridge cables - 100cm^2, for diagonal cables - 60 cm^2, for tension hoop cables on n layer - 80cm^2, on other layers - 200cm^2, for top chord bars - 115cm^2, for strut P3 - 96.7cm^2, other struts - 213.8cm^2.

Structural type 3
As can be seen from Fig. 3, this structural type is similar to type 2, except the central section is all constructed by rigid bars instead of cables. This structural scheme can be imagined as a cable dome to be divided into two halves and connected by a tranverse truss system. The central truss acts as a center tension ring in a cable dome.

The calculated sectional area for ridge cables is 100cm^2, for diagonal cables - 80cm^2, for tension hoop cables - 200cm^2. The areas of all bar elements in the central truss are 203.4cm^2.

Structural type 4
This structural scheme has already been constructed as a realized project of Marine Midland Arena in Buffalo, U.S.A. with an overall plan size of 140m x 103m. The 'tension braced dome roof', as the designer named, was conceived on the concept of the combination of a tensegrity cable dome with a braced barrel dome. The main objective was to develop a structural roof system that combined the lightweight, clear span advantage of a tensegrity dome with the clear and simple load path of a single-layer braced dome. Since all members in this type are rigid bar elements without any flexible cables, in a strict sense, this structural type does not fall into the category of tensegrity system. It is presented here as one of the structural types for the purpose of comparison.

To follow the configuration of the prototype, the structural scheme is designed to consist a semi-circular sector at both ends with a plan radius of 90m. A 60m long barrel-shaped central section connects the two half-domed ends. The resulting roof is composed of 24 radial trusses, the central truss, three circumferential rings and the braced ring along the perimeter. (Fig. 4) All radial trusses are connected between the central truss and braced ring. Diagonal bracings are established between the top chords of the radial trusses in the central section to provide stability of the structural system.

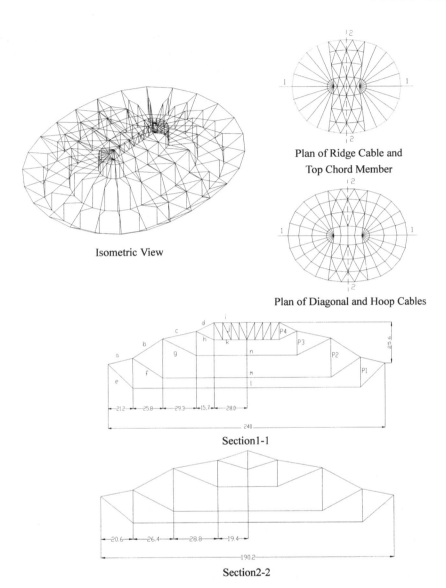

Plan of Ridge Cable and
Top Chord Member

Plan of Diagonal and Hoop Cables

Section1-1

Section2-2

Fig. 2.   Structural Type 2.

Analysis and design of the roof structure was performed by the computer program SAP91. The primary load paths are compression in the central truss, compression and flexure in the trusses of central section, and dome-type compression in the radial trusses at the ends.   Wide flange milled H-sections are used for the members.   The sectional areas are designed as follows.   For top chord members, ridge member - 404.3cm^2, other members - 197.5cm^2, for member in the central truss - 195.1cm^2, for all bottom chord, diagonal and vertical strut members - 105cm^2.

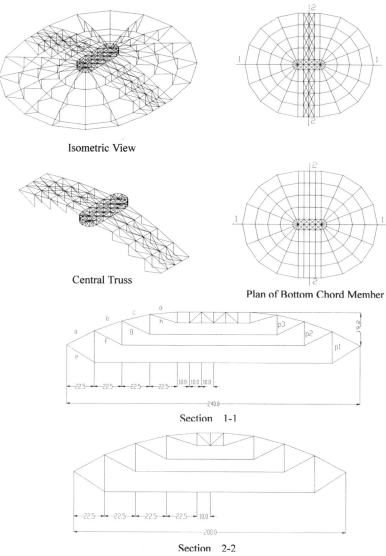

Isometric View

Central Truss

Plan of Bottom Chord Member

Section 1-1

Section 2-2

Fig. 3. Structural Type 3

Calculating results show that such type of tension braced dome has a comparatively great stiffness and a small displacement. With the increase of span, the compressive forces in the top chord members will also increase significantly. At the same time, the calculating length of the member will increase, resulting a large slenderness ratio. This will inevitably cause buckling in the compression members. Therefore it is necessary to make the network closer or to add more bracings; otherwise the sectional areas of the compression members should be increased. Both will increase the weight of the structure for long span roofs.

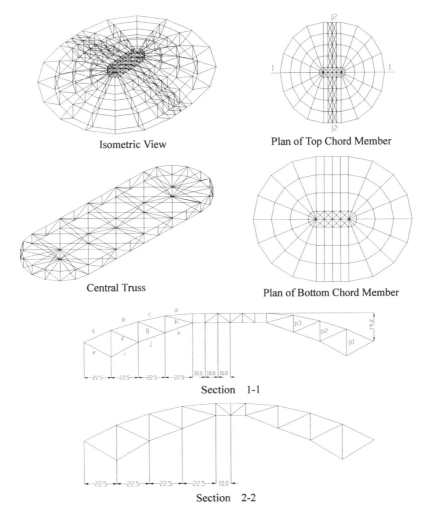

Isometric View

Plan of Top Chord Member

Central Truss

Plan of Bottom Chord Member

Section   1-1

Section   2-2

Fig. 4.   Structural Type 4.

Structural type 5

A structural system named 'Truss Structure Stabilized by Cable Tension' (TSC) was developed by Hangai et al.   TSC is an assemblage of the 'Unit Structure', which is composed of four truss members connected by pin joints and stabilized by using a central post and eight connecting cables.   The cables connect the upper and lower ends of the post to four joints of the unstable truss structure.   Such unit structure has no inextentional displacement, but a self-equilibrated stress system.   Domes with different configuration, such as cylindrical and domical shape, can be formed by assembling these unit structures.   The constitution and structural behavior of unit structures have been studied both theoretically and experimentally. An experimental building of cylindrical TSC in the size of 13.5m x 22.5m was constructed in the Institute of Industrial Science, University of Tokyo.

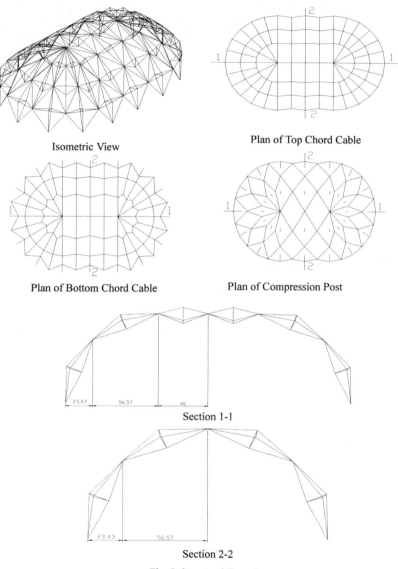

Isometric View

Plan of Top Chord Cable

Plan of Bottom Chord Cable

Plan of Compression Post

Section 1-1

Section 2-2

Fig. 5. Structural Type 5.

For structural type 5, a tensegrity dome is formed by adopting the TSC system. The quasi-elliptical plan can again be composed of two semi-circles at both ends with a radius of 80m connected by a central rectangular section of 80m long. The structural layout is shown in Fig. 5. The calculated sectional area for top chord cables is 40cm^2, for bottom chord cables is 60cm^2, and the area of the compression posts varies from 153.8cm^2 to 442.2cm^2.

The calculating results show that this structural type has a small displacement comparing with other types of tensegrity dome, and hence a great stiffness.   When the span is large, the forces in the compression posts will also be large, which will induce buckling in the compression members.   The same problem emerges as in structural type 4.   Since the dome is assembled from many unit structures, a large amount of bar elements will cause the increasing of the self-weight of the structure.

TYPE COMPARISON

The four structural types that have been investigated are compared with the prototype – Georgia Dome in terms of nominal steel weight and maximum vertical displacement in Table 1.   As the cost of the cables is approximately twice the cost of the steel sections, the weight of the cable is multiplied by 2 and then added to the weight of the steel sections to give the nominal steel weight of the corresponding structural type.   This will reflect the cost of the structure in a more objective sense.

Table 1.   Comparison of structural types.

Structural type	Steel wt (kg/m$^{2)}$	Nominal steel. wt (kg/m^2)	Max. vertical displacement (mm)
Georgia dome	23.3	37.8	706
2	20.5	33.4	846
3	31.5	42.2	407
4	50.1	50.1	46
5	47.2	59.5	182

From Table 1, it can be seen that structural types 2 and 3 all demonstrate a low steel consumption.   For type 2, it is even lower than that of Georgia Dome, but it is less stiff. However, the maximum vertical displacements of type 2 and Georgia Dome, around 1/227 to 1/272 of the shorter span, are acceptable.   This proves that the concept of designing quasi-elliptical domes, i.e. to strengthen the central section and to simplify the semi-circular sectors, is effective.   Both structural types 4 and 5 have a strong stiffness in the sense of small vertical displacements, but they have to bear the shortcoming of high steel weight.

CONCLUSIONS

A comprehensive study on the structural types of cable domes of tensegrity system is presented.   A tensegrity dome in quasi-elliptical plan with a longitudinal span of 240m, similar to Georgia Dome, is taken as the prototype.   Four additional types with different layouts are designed and used as examples for comparison.   The main conclusions are obtained as follows.

1. The structural behavior of a tensegrity dome is different from conventional dome structures.   The failure mode is characterized by the slackening of the ridge and diagonal cables in the central section of the dome.
2. The weakest position in an elliptical tensegrity dome is in the central section.   The

concept of design is to strengthen this part of the dome, especially the horizontal stiffness.

3. For long span roofs around 200m, either the networks of cable dome in wedge shape or the triangulated network is appropriate. Each type of network geometry has its advantages and disadvantages. The combination of these two will provide a satisfactory solution.

4. The 'tensioned braced domed roof' constructed with all steel sections and the tensegrity dome assembled from unit structures like TSC system are not recommended for domes of span over 200m. The steel consumption is relatively high.

REFERENCES

1. MOTRO R, Tensegrity Systems: The State of the Art, International Journal of Space Structures, Vol.7, No.2, 1992, pp75-81.

2. YAMAGUCHI I, OKADA K, et al., A Study on the Mechanism and Structural Behavior of Cable Dome, Proceedings of the International Colloquium on Space Structures for Sports Buildings, Beijing, 1987, pp534-549.

3. GEIGER D, STEFANIUK A, CHEN D, The Design and Construction of Two Cable Domes for the Korean Olympics, Shells, Membranes and Space Frames, Proceedings of IASS Symposium, Osaka, 1986, Vol.2, pp265-272.

4. CAMPELL D M, CHEN D, Effects of Spatial Triangulation on the Behavior of "Tensegrity" Domes, Spatial, Lattice and Tension Structures, Proceedings of the IASS – ASCE International Symposium 1994, pp652-663.

5. LEVY M, Hypar-Tensegrity Dome, Proceedings of International Symposium on Sports Architecture, Beijing, 1989, pp157-162.

6. LEVY M, Floating Fabric over Georgia Dome, Civil Engineering ASCE, Nov. 1991, pp34-37.

7. TERRY W R, STORM G A, et al., Building Tension in Buffalo, Civil Engineering ASCE, May 1997, pp40-43.

8. HANGAI Y, KAWAGUCHI K-I, ODA K, Self-equilibrated Stress System and Structural. Behavior of Truss Structures Stabilized by Cable Tension, International Journal of Space Structures, Vol.7, No.2, 1992, pp91-99.

Tensegrity systems selfstress state implementation methodology

J. AVERSENG
Laboratoire de Mécanique et Génie Civil, Université Montpellier II, France
M. N. KAZI-AOUAL
Laboratoire de Mécanique et Génie Civil, Université Montpellier II, France
B. CROSNIER
Laboratoire de Mécanique et Génie Civil, Université Montpellier II, France

INTRODUCTION

Tensegrity systems are a sub class of reticulate space systems. Their development began at the end of the forties, with the sculptures of Kenneth Snelson [Ref 1] but also in works of Richard Buckminster Fuller [Ref 2] and David Georges Emmerich [Ref 3].

These systems contain a discontinuous set of compressed bars, inside a continuous set of tensioned cables, and they are stabilized by a selfstress state. They bring as interesting characteristics a great rigidity for a weak weight.

The shape of these systems, coming out at the origin from the inspiration of sculptors such as Johansen en 1921 or, later, Snelson in 1948 [Ref 1], is since that time the result of a form-finding process, which couples geometry and selfstress state, with different methods [Refs 4-5-6].

The selfstress state is the internal stress state established during assembling and which ensures the auto-balance of the system, in the absence of any external action. We know that it governs also its behaviour so its level and distribution should be correctly adapted to the requirements of the designer. That is why it is necessary to control its implementation with precision.

We propose in this paper an implementation method for the selfstress state and we present its application to a real system (1:1 scale).

SELFSTRESS

Using the force density method [Refs 4-6], we can write the equilibrium under a simple matrix form :

$$[A]\{q\} = \{f\} \qquad (1)$$

with :

- $[A]$: the equilibrium matrix of the system which is dependent on its design geometry and its relational structure.

- $\{q\}$: density force vector where each component, corresponding to an element, is

defined by $q_i = \dfrac{T_i}{l_i^0}$

      T_i : normal force(N)

      l_i^0 : length at design initial geometry (selfstress but no external load)

- $\{f\}$ nodal external forces vector

With this formulation, we see that the allowable selfstress states are in a subset of the null space of the equilibrium matrix, in which the stress states are compatible with the unilateral behaviour of elements. We can obtain, with numerical computation, a family (q^0) of independent vectors from this compatible subset [Ref 7]. We show that this selfstress base contains a lot of local states (which concern only a few elements). As a consequence, the efforts inside a local state elements are linked, so a modification in one of them has an impact on the others. Active cables are those elements, chosen in order to act on the selfstress state of the system.

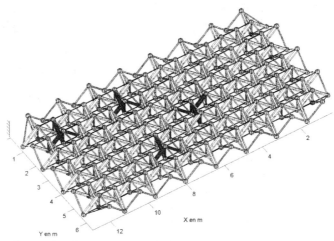

Fig. 1. Four local states in a large "woven" double layer grid.

The selfstress state, designed for a system in a given geometry, imposes the elements manufacturing length l_i^{lib}. Indeed, we assume the elements' behaviour to be linear elastic, so :

$$T_i = E_i.A_i \frac{l_i - l_i^{lib}}{l_i^{lib}} \quad \Leftrightarrow \quad l_i^{lib} = \frac{l_i}{1 + \dfrac{T_i}{E_i.A_i}} \tag{2}$$

with : - T_i : normal effort (positive in traction) in N
- E_i, A_i : respectively the Young's modulus and cross section area
- l_i, l_i^{lib} : respectively current and manufacturing length

Assembling the system in an ideal way should bring it to the target state E^0 (T^0 and l^0). But when using materials of high rigidity, the difference between the manufacturing length l^{lib} and the current length in the selfstress state l^0 is too small, about the order of the manufacturing margins of the elements: selfstress state is sensitive to these tolerances [Ref 8]. We will have to adjust efforts inside elements after assembling.

THE MODEL

The structure studied in this paper is what we call "mini-grid". It's a square double layer grid 3.2m*3.2m*0.8m, the architectural principle of which is based upon weaving [Ref 9]. It is a juxtaposition of "module expanders" (two V been linked around a vertical cable). Built in the Spring 2000, it is the prototype of a larger double layer grid of 12.8*6.4*0.8m based on the same principle, assembled in Autumn 2000.

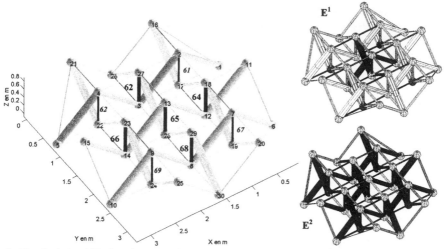

Fig. 2. The "mini-grid", the chosen active cables and the two selfstress state.

This structure accepts two selfstress states : a local one E^1 upon the central expander and a global one E^2 (see Figure 2).

Real elements are of three different types :

- cables in up and down layer : ϕ8 mm, 19 wires, Young's modulus=100000 MPa.

- rods as vertical (ϕ10 mm) and peripheral (ϕ8 mm) tensioned, Young's modulus=270000 MPa

- tubes as compressed elements : ϕ33.7 mm, thickness 2.9 mm, Young's modulus=210000 MPa.

Nodes were specifically designed and make the connection between elements :

- by obstacle with bars and cables (sleeve)

- by screwing with rods. In consequence, those elements are the only ones that we can adjust during the assembling.

We choose five active cables among the nine vertical rods to act upon the selfstress state : the central and the four adjacent ones (see Figure 2). This choice allows a homogeneous action on the structure.

Assembling, adjustment

We proceeded as follows : we began by assembling the layers of cables, then we screwed in the vertical rods between the two layers, and we plugged in the bars. Finally, we screwed in the peripheral rods.

The technology used gives us some freedom of adjustment upon vertical and peripheral rods, screwed in the nodes. A rotation of these elements around their axis allows us to move closer or to take away the end nodes. This is a way of adjusting virtually their manufacturing length.

At the beginning, when there's no stress, the distance between those nodes should be set to the theoretical manufacturing length, given by the target selfstress state and geometry according to relation 2. So we first adjusted the non-active elements. Then, we finished adjusting active cables.

During this last step, the system goes from a null stress state, shapeless, to a selfstress state without transition. So, it is impossible for us to set it to the transitional state, shape defined but with a null stress state. Stresses inside the structure after assembling are indefinite.

OUR STRATEGY : TO CORRECT A REAL STATE

In order to avoid inaccuracies of manufacturing margins and assembling, we suggest rectifying the imperfect state present inside the structure after assembly (Figure 3).

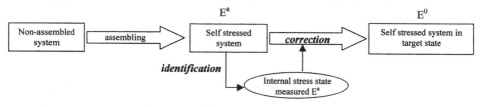

Fig. 3. Implementing a seflstress state E^0.

This correction is based upon the knowledge of the current stress state E^a. Our goal is to make it converge toward the required state E^0. Although one could think of several identification-correction cycles, we plan to get E^0 quickly, in one step.

MEASURE OF A CURRENT STATE E^a

To identify the current state E^a, we look for its co-ordinates in the selfstress base of the system so we have its representation according to the base vectors. To do this, we need some elements' tension values. Although two conditions are necessary in theory, we will use the tensions of the nine vertical cables.

Dynamic measures

To measure a tension, we used a "pulse-test" type method. The response of every cable was recorded by a micro-accelerometer fixed in the middle of the element. The post analysis gave us the first modal frequency f_i^0 of the cable.

Then, we deducted the value of the tension inside from an experimental law, previously established on the same type of element, with the same edge conditions [Ref 10].

Identification

We use an implementation of the simplex method to minimise the f function defined by:

$$f = \max_i \left| E_{,i}(\alpha_1, \alpha_2) - E_{,i}^m \right| \tag{3}$$

with: $E_{,i}(\alpha_1, \alpha_2) = \alpha_1 * E_{,i}^1 + \alpha_2 * E_{,i}^2$
(α_1, α_2): identified state's co-ordinates in (E^1, E^2)
E_i^m : component i of the measured state

This minimisation process results in a compatible state, the components of which are the closest to measures. Current geometry, in which the system is during identification, is a little different from theoretical target geometry, used to calculate the selfstress base (E^1, E^2), because of the non-adjustable elements' manufacturing margins. We noticed an average difference lower than 2.5%.

Table 1. Measures of the nine vertical cables and identification of current state E^a.

elements nb.	Measured T (N)	Identified (E^a) T (N)	difference %	Goal (E^0) T (N)
61	9392	9614	2.31%	12000
62	10946	11097	1.36%	14400
63	9837	9614	-2.32%	12000
64	11213	11097	-1.04%	14400
65	14285	14063	-1.58%	19200
66	11079	11097	0.16%	14400
67	9837	9614	-2.32%	12000
68	10880	11097	1.96%	14400
69	9645	9614	-0.32%	12000

CORRECTION

We now know which modification to do to the current selfstress state in order to have the system in the target selfstress state. Simply, it is the difference $\Delta E = E^0 - E^a$.

This correction is to be achieved by acting successively on active cables. So we have to know their influence over the internal selfstress state.

We define the influence dE_j of the active cable j over selfstress state as the state's variation consecutive to a unit modification of its manufacturing length. Indeed, for the rods, this can be artificially done by screwing them in the nodes.

$$dE_j = \frac{\partial E}{dl_j^{lib}} \tag{4}$$

Applied to the set of active cables, we can build a family (dE_j) of influence vectors that we call influence base.

We can easily show that these influences are mainly stochastic and lightly geometric. Indeed, a shortening of only 1mm can lead to a 3 kN normal effort change. With the hypothesis of small deformations, it is then possible to linearly combine influence vectors in order to write the total state variation dE as:

$$dE = \sum_j c_j.dE_j \tag{5}$$

Actually, the influences will be calculated from differences $\Delta E^{unit/j}$ between two stress states after unit modifications $\Delta l_j^{lib/unit}$ on the cable j, so the notation becomes:

$$\Delta E = \sum_j c_j.\Delta E_j \quad \text{with} \quad \Delta E_j = \frac{\Delta E^{unit/j}}{\Delta l_j^{lib/unit}} \tag{6}$$

With this decomposition of the ΔE vector in the calculated influences base (ΔE_j), we get the c_j coefficients and then the variations to be made to active cables.

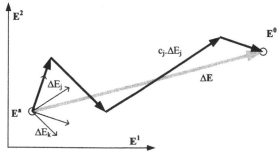

Fig. 4. Graphical representation of the linear combination of influences ΔE_j.

Numerical tool

We used a numerical model of the system in order to quantify unit influences and to simulate the tensioning process. This code computes the equilibrium of a self-stressed system under given disturbances, by an iterative Newton-Raphson method. Elements are modelled as bars (no flexion) with a linear elastic material. It takes into account, on the one hand, the unilateral behaviour of cables, by eliminating them from the calculation when they are in compression, and on the other hand, the effects of the second order, integrated during the equilibrium computation iterations.

In our particular case, there is no external load, we start from the target geometry and we disturb the system by modifying chosen active cables' manufacturing length l^{lib}. Every iteration, internal efforts are calculated according to relation 2 and are used to update the geometrical terms of the rigidity matrix of the system.

The derivative of expression 2:

$$\frac{dT_i}{dl_i^{lib}} = \frac{1}{l_i^{lib}}\left[E_i.A_i\left(\frac{dl_i}{dl_i^{lib}} - 1 \right) - T_i \right]$$ (7)

shows the non linear effect of a change of an element's manufacturing length on its normal effort. This effect is also a function of the effort T. This demonstrates that the choice of unit modifications $\Delta l_i^{lib/unit}$ for the computed influences has an important impact in the accuracy of our linear combination. Actually, it should be in the same range as the real variation to be found. There is also an impact of the level of the stress state: the higher the level is, the more significant are the influences.

Manufacturing length modifications

We chose to compute the active cables influences ΔE_j from the target state E^0, simulating the same unit l_j^{lib} elongation of 1mm in every active cables.

Determination of c_j coefficients followed the same method used for the current state identification. We searched the minimum of the g function defined by :

$$g = \max_i \left| E_{,i}^m + \Delta E_{,i} - E_{,i}^{0} \right|$$ (8)

with : $-\Delta E_{,i} = \sum_j c_j . \Delta E_{j,i}$

- E^m is the measured state

The product of c_j coefficients by the unit 1mm length modification gave us the variations Δl^{lib} to apply to the system.

EXPERIMENT

Technologically, it isn't possible to control the manufacturing length of a stressed element. We can only measure it's current length, with a ±0.01mm display precision, thanks to a specific tool designed from a digital calliper. The accuracy of this device is more rationally around ±0.03mm.

In order to get the distances between nodes from manufacturing length variations Δl^{lib}, we computed a simulation of the process and we noted geometric changes Δl^{app}. We began by a shortening of the central cable 65 which has the greatest influence.

Table 2. Manufacturing length variations Δl^{lib}, apparent variations Δl^{app}.

elements nb.	Δl^{lib} (mm)	Δl^{app} (mm)
65	-0.57	-0.51
62	-0.6298	-0.60
64	-0.631	-0.60
66	-0.631	-0.60
68	-0.626	-0.60

The Δl^{app} correction were then applied in the same order to the real elements.

FINAL STATE
We measured the normal effort in the nine vertical cables according to the same procedure as during the identification of the current state E^a. Table 3 shows final tensions T^{fin} and a comparison with the target values T^0.

Table 3. Final measures T^{fin} and comparison with target tensions T^0

elements	T^{fin} (N) measure	T^0 (N) design	difference
61	12162	12000	1.35%
62	14651	14400	1.74%
63	12369	12000	3.08%
64	15019	14400	4.30%
65	19787	19200	3.06%
66	14871	14400	3.27%
67	12508	12000	4.23%
68	14431	14400	0.22%
69	12369	12000	3.08%

These final values are close to the target ones, with a mean difference inferior to 5%. The goal state E^0 is set with a reasonably good precision, in spite of several imperfections introduced by the technology employed and adjustment process.

CONCLUSION
To linearly combine the active cable influence seems to give good results with this particular structure, thanks to its high rigidity which validates the small disturbance hypothesis. Indeed, geometric variations are very small : 0.6mm for 800mm long elements. This ensures the non-linear geometrical effects to be insignificant.
The methodology demonstrated on this small tensegrity system, which has only two selfstress states, has to be generalised to more complex systems with more selfstress states.

REFERENCES
1. SNELSON K., Tensegrity mast, Shelter Publications, 1973.
2. FULLER R.B., Tensile Integrity Structures, US Patent 3.063.572, 1962.
3. EMMERICH D. G., Structures tendues et autotendantes, Ecole d'architecture de Paris La Vilette, 1988.
4. SHECK H.J., The force density method for form-finding and computation of networks, Computer Method in applied mechanics and Engineering 3, 1974, pp 115-134.
5. BARNES M.R., Applications of dynamic relaxation to the design and analysis of cable, membrane and pneumatic structures, 2nd International Conference on Space Structures, Guilford, 1975.

6. PELLEGRINO S. and CALLADINE C. R., Matrix analysis of statically and kinematically indeterminate frameworks, International Journal of Solids and Structures, vol 22, 1986.

7. QUIRANT J., KAZI AOUAL M. N., LAPORTE R., Tensegrity systems : the application of linear programming in search of compatible selfstress states, submitted to International Journal of Solids and Structures, 2002.

8. QUIRANT, J., KÉBICHE, K., et KAZI AOUAL, M. N., Systèmes de tenségrité : Comportement et sensibilité de modules simples, Revue Française de Génie Civil, Vol 4, 2000, pp 429-442.

9. RADUCANU V., MOTRO R ., Composition structurale de systèmes de tenségrité, Société de Tissage et Enduction Ferrari, Brevet n° 0104 822, 16 mars 2001.

10. CROSNIER B., CEVAËR F., Stratégie de mise en précontrainte dans les systèmes de tenségrité et contrôle, Colloque Lagrange, Rome, 2001.

A novel portable and collapsible tensegrity unit for the rapid assembly of tensegrity networks

K. A. LIAPI
Architectural Engineering Program, Department of Civil Engineering, University of Texas at Austin, USA

ABSTRACT

Pre-assembly of building components can significantly reduce on site construction time and is particularly appropriate for emergency or temporary structures as well as for construction on sites with extreme weather condition. Pre-assembly can be an efficient method for the construction of double layer tensegrity structures. Double layer tensegrity structures are self-stressed cable networks that consist of two parallel layers of cable nets with bars confined between them. Such networks can find applications in building design and construction and typically result from the assembly of interconnected tensegrity units. This paper focuses on the conceptual design of a novel tensegrity unit that can be used for the rapid assembly of double layer tensegrity structures. The new unit is a lightweight, portable and collapsible tensegrity structure of simple prismatic geometry that allows for two distinct functional configurations. In the deployed state, the unit is rigid and can be attached to other deployed units to form a large tensegrity structure, whereas, when collapsed for transportation and storage, the unit presents no loose bars or cables. Functionality requirements and geometric considerations that led to the design of the novel unit are presented. The conceptual design of the new unit has been tested experimentally with the construction of a full-scale model of a collapsible tensegrity module of square base. The unit has been used for the assembly of a sixteen unit tensegrity structure.

BACKGROUND

There are many and varied building needs and site conditions that dictate the use of structures that can be rapidly assembled on site, and/or partially or fully pre-assembled before being transported to site. Examples of structures that can significantly benefit from speed in construction range from emergency medical stations, moving theatres or exhibition spaces, temporary storage spaces, to military shelters and bridges. Transportability, ease of assembly, and at times, easy and rapid disassembly, are significant considerations in the design or selection of a rapidly assembled structure. Categorization of rapidly assembled structures is usually based on the method or mechanism by which building members or parts are attached to each other. Hinged, pinned, clamped and sliding are some examples of assembly methods (Bulson 91). Combining modular construction with other methods of rapid assembly can further reduce on site construction time. The term modular assembly is most often used to address large scale units, and it usually refers to functional modules like manufactured homes, school classroom units etc. In this paper the term modular refers to pre-assembled structural units which, when attached to each other, can form a large structure in a short time.

Pre-assembly is hypothesized to significantly reduce the time needed for the construction of double layer tensegrity structures. Tensegrity structures can be defined as spatial lattices consisting of a continuous network of cables maintained in tension by isolated compression bars. The type of tensegrity structures that are most appropriate for applications in building design are the double layer tensegrity. Double layer tensegrity networks, like other forms of double layer grids, typically occur from a regular arrangement of three-dimensional structural units. In this regard a modular construction of tensegrity structures, where each unit can become a separate building module, is directly related to the geometric conception of tensegrity structures.

A potential challenge that pre-assembly construction of three dimensional modules presents is that individual components can be bulky, and when transportation of the modules is to be considered, the total volume of modules to be brought to site is equal to the volume of the structure. A method for assembling tensegrity building structures from collapsible modules that can be transported to site in a stowed configuration, is a new approach to the construction of double layer tensegrity structures, currently under investigation by the author and the subject matter of an upcoming publication. The development of an efficient collapsible unit is the necessary first step in the formulation of this method.

In the following sections the functional and geometric requirements that led to the conception and design of a novel collapsible tensegrity unit to be used in the construction of double layer tensegrity networks, are presented.

CONCEPTUAL DESIGN OF THE NOVEL UNIT

A tensegrity structure, as mentioned earlier, is a self-stressed equilibrium network in which a continuum of tension members (e.g. cables) interacts with a discontinuous system of compression members (e.g. bars) to provide the structure structural integrity. Releasing tension causes the structure to partially or fully collapse. The size of the structure is thus reduced or minimized. Inversely, a collapsed tensgrity structure in theory can be deployed by applying tension to appropriate tension members. Two main methods for deploying tensegrity structures have been identified in relevant literature: by releasing tension on one or more tension members or by reducing the length of one or more compression members. A combination of both methods has also been suggested (Hanaor 97, Bourdebala et al. 97). For structures of significant size, issues such as tangling of tension members, and collision between members have been identified as difficult problems to solve.

A series of experiments have been conducted by the author to determine differences between various methods for deploying/ collapsing units. The results have been taken into account in the formulation of the overall method for the erection of the structure and the features of the collapsible units. This paper focuses on the functionality requirements and the geometric features of the collapsible units only.

Funcionality requirements

The overall conception of the method for assembling double layer tensegrity structures from detachable and collapsible tensegrity modules has suggested several functionality requirements that the unit should fulfill. These can be summarized as follows:

- A collapsible unit allows for two distinct functional configurations:

1. A deployed configuration, in which the unit may be attached to other units to form a tensegrity structure.

2. A collapsed configuration, in which the tension members remain attached to compression members, so that they are no loose ends.

- The size, shape and weight of the collapsed modules should be such that one person can carry at least one of them at a time. A preferred shape of a collapsed unit is therefore one that collapses as a bundle in an upright position. This requirements sets limitations to the size of individual units as well

- The steps in the deployment/ collapse of the unit, including the number of joints that require manual control, are kept to the minimum. A main objective has been to minimize the steps in the process and the total manual effort required in the erection/collapse and maintenance of the unit.

Geometric configuration of the novel unit and technical considerations

Functionality requirements set by the overall conception of the method for assembling double layer tensegrity structures from detachable and collapsible tensegrity modules have been considered in the design of the new unit.

Double-layer tensegrity structures that are considered here typically occur from the assembly of simple tensegrity units such as tensegrity prisms or truncated pyramids.(Figure 1).

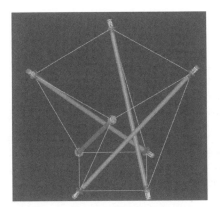 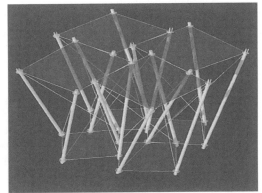

Fig. 1. Schematic plans of an *n*-sided tensegrity unit.

A tensegrity prism is a skew prism, or more precisely an anti-prism, formed by cables along the edges of the prism and with bars along the diagonals of the side faces in a consistent right handed or left-handed sense. In the prestressed state, when all cables are in tension, the two parallel bases of the prism are rotated relative to each other by an angle that is dictated by the requirement for stability of the shape. The angle between the two base polygons of the tensegrity prism is unique for each polygon of *n* sides and is given by the formula:

$$\alpha = 90° - \frac{180°}{n} \quad \text{(Kenner, 76).}$$

According to this, for a tensegrity unit of a triangular base the angle is 30 degrees, for a tensegrity unit of a square base it is 45 degrees, etc. (Figure 2).

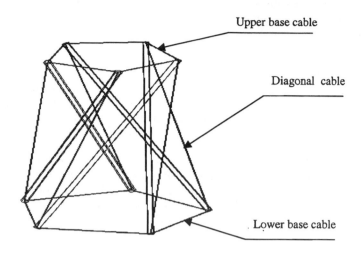

Fig. 2. Schematic plans of an *n*-sided tensegrity unit.

The same rules apply for tensegrity units in which the two parallel bases are similar but of different size. The closest regular polyhedron to this type of tensegrity unit is the truncated pyramid.

Tensegrity units of prismatic or pyramidal form consist of (a) two polygonal bases of *n* cables each, that is 2*n* base-cables in total, (b) *n* side-cables and (c) *n* diagonal struts (bars) that hold the two polygonal cable bases apart. Thus, a tensegrity unit typically consists of 3*n* tensioned members (cables) and *n* compressive members (bars), where *n* is the number of sides of the base polygon (Figure 2).

The basic feature of the geometric configuration of tensegrity units that has been taken into account in the design of the novel unit is that the longest linear components in any unit are the bars which are the diagonal connections between the two base cables and which fall within the solid angle defined by base and side cables.

Preliminary experiments conducted by the author have indicated that releasing or elongating the side cables of the unit until their length becomes equal to, or longer than the length of the bars, forces the unit to collapse in the upright position. In this instance, the simple geometry of the triangle formed by a side cable, a compression element and a base cable dictates that side-cable length + base-cable length > bar length. This observation suggests that by replacing each side-cable and one of its adjacent base-cables with a continuous cable, the length of which is equal to the total length of the two component cables, a new configuration of a tensegrity unit is defined.

The new configuration in its general form is composed of a) $2n$ base-cables, n on each base polygon, and n continuous-cables, and b) n bars. The length of each continuous cable is equal to the sum of the lengths of the base-cable and the diagonal-cable. Thus, is in the new configuration n tension members form one base polygon (Figure 3). In this configuration cables can have fixed lengths, so that the ends of the tension members are not free when the unit is either in the collapsed or in the deployed configuration, or during the passage from one state to the other. The compressive members on the other hand can have adjustable lengths.

According to the above, in any tensegrity unit of $2n$-sided polygonal bases there are always n continuous cables that, if properly controlled, will allow the unit to fold in an upright position, with the bars standing parallel to each other. In addition, the functionality requirement that one person be able to hold the unit in the upright position implies that the joints of the upper base are the most convenient for manual control.

A significant feature of this novel unit is that it retains its topological description in all configurations. In addition, all cables retain their size and are independent of changes in the geometric configuration of the unit. This same concept and collapse method can apply to units of any geometry, as long as bars are confined between two polygonal bases, which do not have to be identical, similar, regular, or parallel to each other.

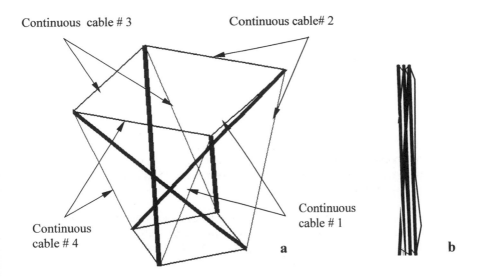

Continuous cable # 3

Continuous cable# 2

Continuous cable # 1

Continuous cable # 4

a

b

Fig. 3. Schematic plans of a collapsible tensegrity unit of square base.
a Deployed configuration.
b. Collapsed configuration.

The design of the joints has been instrumental in the deployment and collapse of the unit and the assembly of the tensegrity network. In its general form each tensegrity unit with n –sided polygonal cable bases has $2n$ nodes. At each node three cables and one bar meet. The overall concept of the structure, and the more specific concept of the deployable tensegrity unit have set the requirements for the design of the joints at the nodes.

In very general terms, the main requirements that were taken into account in the design of the joints was that they should allow for an easy manual snap of the cable in the pre-stressed configuration and for the attachment of one unit to the other. A technical requirement that was of critical importance for the design of the novel unit was that the joint, , in the deployed configuration, should allow each continuous-cable to perform, geometrically and structurally, as two independent cables: a base-cable and a side-cable.

The requirements for the joints have been taken into account in the design and the construction of a typical deployable tensegrity unit of square base as illustrated in figure 4. A 16-unit structure has also been constructed and is shown in Figures 5 and 6.

Fig. 4. Tensegrity unit in the deployed configuration.

Fig. 5. View of a tensegrity structure composed of sixteen units of square base.

Fig. 6. Front view of a sixteen unit tensegrity structure.

CONCLUSIONS

Pre-assembly and modular construction can be an effective method for the rapid assembly of double layer tensegrity networks. The development of a new collapsible unit is an integral part of the method. The features of the novel unit are shown to fulfill the functionality requirements introduced by the proposed method.

The deployment and collapse of the tensegrity unit does not require on site assembly of units from separate compressive and tension members. Unlike other "cable" deployment methods reported in literature, the proposed one does not require loosening cable ends for deploying or collapsing, or for pre-stressing the unit. Loose cables are prone to entanglement and pre-tensioning cables is usually a tedious process.

Many areas worthy of exploration remain to be investigated in follow-up work. Special consideration will be given to the optimized design of joints and the attachment of covering.

REFERENCES

1. BOURDEBALA, M., MOTRO, M., Folding Tensegrity Systems, Proceedings of the International Colloquium on Structural Morphology-Towards the New Millennium, August 1997, Nottingham, pp.115-122.
2. HANAOR A., "Tensegrity: Theory and Application," Beyond the Cube: The Architecture of Space Frames and Polyhedra, edited by J. Francois Gabriel, John Wiley and Sons, Inc., 1997, pp. 385-408.
3. HANAOR, A., "Aspects of Design of Double –Layer Tensegrity Domes," International Journal of Space Structures, Vol. 7, No 2, 1992, pp. 101-113.
4. KENNER, H., Geodesic Math and How to Use it, University of California press, 1976.

How does a metal cobweb behave?

I MUTOH
Dept. of Architecture and Civil Engineering, Gifu National College of Technology, JAPAN

INTRODUCTION
The patterns and forms found in natural systems have been used to create an image for artefacts: *e.g.* arts and architecture. Since topology and geometry of these patterns and forms are often as a basis for the human recognition of actual living system and also they form the basis for the methods and the process of creation, we have been enforced by observation of them to inspire something. Among these natural objects, a classic and a particular interest configuration in nature is the spider's web.

Since the colloquium IL 6 in 1973 SFB 64, Ref 1 was published, a spider's web construction and their forms has been one of the interesting artefacts to be realized as cable nets structures. The nets of orb webs are adjusted to a special way of catching prey, and all the different web types should be related to basic structural action. The orb (called *Garden cross spider* and *Feather-legged spider*) web is close to being an ideally engineered plane pure lattice structure with pre-tensioned straight and pin - or adhesive jointed bars, as stated in Ref 2.

On the other hand, some spiders called *Filmy dome spider, grass spider* form three-dimensional webs made a dense forest of sticky mesh over bushes or hedges, see Ref 3. Like the canopy landing in a tree: the strength and elasticity of branches, will be placed. Some analogical competence can be extent to a special spatial structure, illustrated as in *Fig 1*. Recently, the awareness of possibility of an orb web to make a key for form finding starting block is raised, see Refs 4 – 6. This paper is also concerned with one of applications of web building strategy to form a metal structure. However the idea should be evident to get an initial structural model and to illustrate how to make the initial form and to assess how the web will behave.

Fig. 1. Grass spider's web (left) and Toyota Stadium (right).

Space Structures 5, Thomas Telford, London, 2002

TOOLS (PREVIEW)
It has been developed that a shape of natural object is implicated its structural characteristic through fractal dimension, and as well as descriptive features of artefacts (skyline of landscape and of envelope of cities) to be implicated. Basically, since 1975 to 1982, the works by *Mandelbrot* has continued in a field of fractal geometry and, defined as a fractal is by definition a set for which the *Hausdorff-Besicovitch* dimension strictly exceeds the topological dimension. Since then, many artful pictures in computer graphics has been reported, e.g. see SIGRGRAPH, on the basis of *IFS* (Iterated Function System) algorithm by searching for a "fitting" attractor. In order to do so, it is said that to cover the original object with locally *affined* images of itself is needed.

On the other hand, an approach for unification of computer graphic algorithm may be based on system for "string re-writing" scheme. This system was often called "L-system": a mathematical formalism which was proposed by *Lindenmayer* in 1968 as a foundation for an axiomatic theory of development. At later time Smith 1984 proposed *L-system* as a tool for synthesizing realistic images of plants, and pointed out the relationship between *L-system* and the concept of fractals, as well as *Mandelbrot* doing so.

The author has investigated preliminary into how to apply the technique of L-system to "structural morphogenesis" in conjunction with a concept of A-life (see, Ref 7) From a point of view for a technique of A-life, *L-system* and cellular automata are representative algorithm for development, and genetic algorithm may be dominant tool to evolve and emerge the object. Again, a A-life technique has also a function of recursively generated objects: "reproduction" function after Dawkins' bio-morph. All the direction to study above may get closer to natural things, instead to artefacts.

Here, we present one of ideas for computational morphogenesis (artefacts called, roof-framework like spatial structures) on the basis of *L-system* and fractal geometry. According to Hangai's message that the spatial structures are one of resistant structures with configuration form, form consists with geometric shape and internal (structural) system. Also the spatial structures: structural system of space frames is governed by arrangement (mesh- pattern) of constituent members. One is a method of subdivision over continuum surface, the another is a method of accumulation of base unit. The former is called "stability degradation method", and the later is called "stability increasing method" (Ref 8). However the paper will discuss in view of conceptual fractal world: such a category will be disappeared. Why?

FUNDAMENTALS
Conceptual descriptions as a reticulated (latticed) spatial structure composed of joints and members, a kind of branching system is one of basic arrangement methods: (1) subdivision and/or (2) accumulation. In particular, in order to figure a specific shape the fundamental (discrete) unit has to be generated.

An *L-system* previously stated is a parallel rewriting system operating on branching structures, represented as bracketed strings of modules. Matching pairs of square brackets enclose branches. Simulation like set-up of genetic algorithm, begins with an initial string called the axiom, and proceeds in sequence of discrete derivation steps. In each step, rewriting rules (production rules) replace all modules in the predecessor string by successor modules simultaneously in parallel action. The applicability of a production depends on a predecessor's context, values of parameters, and on random factors (ref. 5). An example for a production rules as shown below.

$$id : lc < pred > rc: cond \ succ: prob \hspace{3cm} (1)$$

Where *id* is the production identifier (label), *lc, pred,* and *rc* are the left context, the strict predecessor, and the right context; *cond* in the condition, *succ* is the successor, and *prob* is the probability of production application.

In this paper, among many families of *L-systems,* the simplest class of *L-systems,* those which are deterministic and context-free, called DOL-systems is used. Consider strings built of two letters **a** and **b**, which may occur many times in a string. Each letter is associated with a rewriting rule. The rewriting process starts from a distinguished string called the axiom. Formal definitions (mathematically) describing *DOL-systems* and their operation are given in elsewhere.

To construct an *L-system* which captures given structure (fundamental structural unit) or sequence of structures representing a developmental process: called "inference problem". As noted, a fractal graphic procedure for basic figure like Koch islands and curves, the initiator corresponds to the axiom and the generator corresponds to the production successor. The predecessor represents a single edge. The *L-systems* specified in this way can be perceived as coding for Koch constructions. Here Koch curve is characterized by three remarkable properties: there s no tangent defined at any of its points, it is locally self-similar on each scale, and the total length of the curve is infinite. Fractals show a self-similar structure.

On the other hand, a method for generation of objects using formal languages, like *L-systems,* iterative geometric constructions (IGC) is available especially for highly computerized image generator system. It begins from simple rules, geometric production rules, stepwise, then more complex rules can be built. The element of the geometric construction is (1) base element, (2) initiator, (3) generator and the procedure iteration algorithm. In any way, any procedure and algorithm illustrated above, from basic seed (gene, axiom, base element or fundamental structural unit) is first defined, then some algorithm (rules) drives seed to develop (genotype: chromosome) recursively or iterative way, with graphical objects (phenotype). Here you can see a way of A-life world. Recursively generated objects, lead to general approach to built-in genotype/phenotype system *Langton* illustrate the notion of recursively generated objects by means of : *L-systems,* cellular automata, and computer animation. Tab 1 shows the relationships between fractal and others.

Table. 1. Algorithmic relationships.

A-life : Emergence by genetic algorithm & development by L systems (in this paper; branching growth bottom up manner as categorized "accumulation" method)	*L-system*: parallel string rewriting grammar with recursively generated objects illustration by parameters. Close to natural objects through strict artefact generated by simple algorithm, and to implicate pure natural behaviours.	
Fractal: artificial self-immorality is related to configuration algorithm for L-systems.	What is a kind of bridge over some? Fractal mechanics??	*Structural morphology*: computerized Morphogenesis??

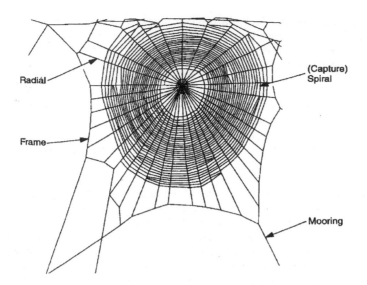

Fig. 2. *Araneeus diadematus* web: pre-tension in threads: mooring, frame and radii in ratio of 10 : 7 : 1 (for example).

Orb webs, as shown in Fig 2, which have been investigated into how spiders to build web, *e.g.* Refs 13 and 14, might be important to control angles formed by adjacent threads, distances and directions moved, and the forces and strains in the threads. A spider orb web comes close to a visco-elastic least weight structure stabilized by the tension of the threads. Also the spider exactly adjusts the web mechanics to the particular environmental conditions, as mentioned in Refs 3 and15.

However the measurements of thread geometry, or web pattern, showed that statistics of (1) radii and central angle elements were not so regular, (2) spiral elements less regularity, (3) spiral area varies widely, and (4) mesh size or density of the threads in spiral area is just adjusted due to saving of labor. As known the Euler dimension of orb web is 2, plane structure and it of grass spiders' double skin web may be 3, three-dimensional structure. The observations indicated that the geometry of web pattern is rather irregular but thought of its fractal nature!?

If the geometry, pattern, be in fractal nature, then the pre-tension in threads and response to disturbances may be in a manner of *fractal-like mechanics*. In this paper the aim to be investigated and discussed is this presumption into fractal characteristic for initial form generation and the behavior of modified orb webs.

POSSIBILITY
From Ref 9, Otto stated that asking on a daily basis why we are dabbling in Biology:

Every material object has the ability of conveying forces. We are studying the capability of convey forces independent of form, material and type of load and, in all objects of organic and in-organic nature, technology and art.

We are investigating the process which characterizes forms in all fields. These processes in in-organic nature differ greatly from those in organic nature and in technology. The forming process of organic nature is characterized by evolution, by accident and selection. It is essentially free of human manipulation.

As shown Fig 2, a cob web consists of viscous string (often called *viscous thread*) which a spider makes first diagonal strings with boundary supports attached to tip or branch of trees or other and he walks around with generating strings with one stoke (under radiation pattern of *radial threads*): not jointed each segment of strings but glued to diagonal strings. Moreover you can see natural cobwebs in a manner as : example pictures adopted as shown in Fig 1.

Figure 1 (the left) shows the lower safety web and the upper strong trapping web over tree-like supports. The right in Fig 1, you can see the dense mesh-like positive curvature with a circus tent over pole-like supports. What are you thinking about? If imagine that web under self-weight hang down with natural surface, then you can enforce the web in flat reticulated plate due to pre-stress and finally, if you want, almost twice the same displacement amplitude in a hanging-down state is applied to lift-up overall web net at the same time element thread turned out to be much stiffer in both axial and bending rigidity. That is a kind of reticulated shell with positive curvature !

PROCEDURES

The initial shape will be generated first non-fractal object on the basis of "Force Density " method and "Measure Potential" function. The results are compared to a dimension, between fractal dimension and Euclidean. The next, its shape will be modified by means of fractal geometry algorithm in conjunction with *L-system* algorithm. And the analytical investigation will be carried out by the same method. In addition, the generated artificial cobweb will be applied to carry out an eigen-value analysis in order to get buckling mode shapes. Finally, the mode of both fundamental elastic deformations and eigen-modes will be compared with each other based on the fractal dimension.

Form-finding by using density methods was proposed by Schek which is devoted to the shape-finding of general cable networks. Another configuration method from geometric potential was introduced by Butterworth which is gave an approach to a regular subdivision of a surface as arrangement of the set of node coordinates by moving on the surface until equilibrium is achieved. One idea for from-finding of various types was presented by Ijima and Obiya which is the application of the tangent stiffness method to the analysis of the equilibrium forms consisting of basic elements with some defined function of element potential.

All the above can be treated with the assumption to investigate mentioned before:
(1) the force density, stress density and strain density will be in fractal nature, e.g. distribution function of power law.
(2) the potential function will be in fractal nature, e.g. distribution function of power law.
Briefly followings illustrate how to relate the distribution with fractal mechanics.

Geometric potential concept: (original notations)
The geometric potential G is defined as below.

$G = \sum\limits_{i \neq j} m_i\, m_j\, d_{ij}^{-q}$ with q is a positive integer defining the power law of repulsion between

nodes, d_{ij} means inter-distance between node i to j and m_i, m_j mean the weight for intensity of each node's interference. Then an effect of power law is if higher, more distant nodes will have very little effect and the major contribution to G will come from the interference of a node with its nearest neighbors only. Namely more uniform dense mesh will be created. When the power law with non-integer q is used then what kind of arrangement will be appeared ?

Measure potential function of element: (original notations)
Element potential function P is defined by using a measurement A such as the element length or area as $P=f(A)$. Then the element force N is derived by derivative operation to this function by the vector consisting of the element measurements. Finally, the equilibrium equation is obtained as the relationship between nodal force vector and the element force through the transforming matrix. For example, in one-dimensional element, the element potential function is $P=Cl^n$, and the axial force $N=nCl^{n-1}$. If n=1 then the axial force is constant regardless of the element length. If n>1, the non-stress length of element is zero. If n=2, the tangent stiffness equation becomes linear regarding the nodal displacement. Therefore, the lengths of all elements are more uniform than the form derived from n=1, and also the mechanical property is the same between the net form with n=2 and the isotropic tension form. Here again what happened to calculate by using the exponent n with non-integer, fractal dimension.

Force density method:
The force density method is based on that the force on the end of a linear element can be represented by the product of the element force and a unit vector in the direction of the element. Therefore the problems can be formulated in which the node coordinates, shape, are easily found by solving a system of linear equations. By using the vector *w* consisted of pre-specified spatial distance in form w_k, force densities q is defined as q=s/w =(force)/(stressed length) ratio. This method is the special extension of the "updated reference strategy" to cable nets * using a homologous mapping concept. Another notations for the deformation gradient F which is determined as l/L with stressed length and original length L. Again the ratio of the pre-scribed tension force to stressed length means the force density which can be interpreted as a second Piola-Kirchhoff force (Ref *).Then we put forward the force density into fractal distribution of member length, individual element however behaves due to its own density. Thus a designer using this method can specify the force density as well as the element force even with pre-stressed member. The mechanics of overall configuration may be of fractal nature or else, it is interesting to investigate!?

Supplements:
The shape energy concept, Ref **: a static energy on the basis of products of force and distance, *work*, which is allocated to an equilibrium shape, and attributed to the statistical distribution of external forces. Also the both quantities will be fractal, then fractal mechanics formalism can be described.

The diffusion of the stresses concept, Ref ***: the diffusion of stresses is the attitude of a structure to use the highest number of elements to carry loads. The maximum diffusion shows a constant distribution of stresses and a probability density function nil everywhere except for one value of the element force, the same for all the elements.

EXAMPLES

We adopted to carry out numerical analysis by means of "force density method": (1) cobwebs having only extensional rigidity under self-weight hanging with boundary, (2) cobwebs raised up by an amount initial displacement of 2 times initial configuration. As stated by Thompson D'arcy, spiral objects in general can be described with mathematical equation even circles as below:

$$x = a\theta\cos\theta \qquad\qquad x = e^{a\theta}\cos\theta \qquad\qquad y = a\sin\theta$$
$$y = a\theta\sin\theta \qquad\qquad y = e^{a\theta}\sin\theta \qquad\qquad x = a\cos\theta$$

Archimedes spiral   Logarithmic spiral   Concentric circle

Fundamental study for such spirals of web spiral (capture, viscid thread) as shown in Fig 2 was illustrated as shown in Fig 3.

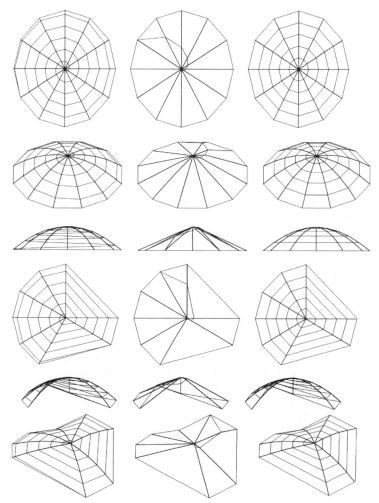

Fig. 3. Cobweb generation (example; bottom figure means some variation).

Description and how does metal cobwebs modeled behave in view of linear elastic and/or of linear eigen-value problem, are to be studied. The scope of this paper was from basic concept/idea to be adopted a tool like fractal dimension to a measurement of surface roughness for both initial configuration due to a characteristics of fractal nature and eigen-mode deformation patterns through an equilibrium condition at bifurcation state at the Conference to be appear.

CLOSURE

The Author has discussed a view of the comparison between degree of fractal dimension of the initial shape as a pattern of initial geometric imperfection due to self stressed reticulated metal cobwebs (as implicated form finding scheme) and bifurcation modes. The outcome to be obtained may be a measure for appropriate distribution of actual or artificial mode of initial geometric imperfection: how does its distribution can be of similarity against eigen-mode qualitatively.

REFERENCES (with *Supplements*)

1 PETERS von H M, Spiders and Their Webs, IL 6 SFB 64 (Otto F), 1974, pp. 28-40.

2 WESTER T, Some Remarkable Statical Qualities of Structures in Nature, Structural Morphology Towards the New Millennium, Nottingham, UK, 1997, pp.12-21.

3 VOLLRATH F, Spider Webs & Silks, Scientific American, 266, 1992, pp.52-58.

4 WESTER T, The Nature of Structural Morphology and Some Interdisciplinary Examples, Spatial Lattice and Tension Structures, ASCE, 1994, pp.1000-1009.

5 FUJU G & XILAIANG L, The Proposal Study on Spatial Structure and Bionic Engineering, IASS Symposium 2001, Nagoya, 2001, TP132 (on CD-ROM), 8 pages.

6 TETIOR A, Basic Principles of Creation of Natural Spatial Structures and Their Use in Building, IASS Symposium 2001, Nagoya, 2001, TP131 (on CD-ROM), 8 pages.

7 MUTOH I & KATO S, Structural morphogenesis by a concept of Alife, Conceptual Design of Structures(IASS Symposium) Oct.,1996 (Stuttgart), Vol. I, 323-330.

8 HANGAI Y, Metal Space Structures-Form, System and Performance, Current and Emerging Technologies of Shell and Spatial Structures, IASS Madrid, 1997, pp. 63-76.

9 OTTO F & RASCH B, Finding Form Towards an Architecture of the Minimal, Edition Axel Menges, 1995, 239 pages.

10 SCHEK H J, The Force Densities Method for Form Finding and Computation of General Networks, Computer Methods in Applied Mechanics and Engineering, 3, 1974, pp. 115-134.

11 IJIMA K & OBIYA H, Form-finding of Single Layer Structure by Measure Potential Function of Element, Structural Morphology Towards the New Millennium, Nottingham, UK, 1997, pp. 249-256.

12 BUTTERWORTH J W, Structural Configurations from Geometric Potential, Proceedings of the 3rd International Conference on Space Structures, Ed. H Nooshin, Elsevier Applied Science Publishers, London, 1984, pp. 84-87.

13 REED C F, WITT P N & JONES R L, The Measuring Function of the First Legs of *Araneus Diadematus* CL., Behaviour, 25, 1965, pp.98-119.

14 PETERS von H M, Uber das Kreuzspinnennetz und seine Probleme, Die Naturwissenschaften, Heft 47, Nov 1939, 778-786.

15 LIN L H & SOBEK W, Structural Hierarchy in Spider Webs and Spiderweb-type Systems. The Structural Engineer, vol. 76, no 4, Feb 1998, pp.59-64.

* BLETZINGER K-U, Form Finding of Tensile Structures by the Updated Reference Strategy, Structural Morphology Towards the New Millennium, Nottingham, UK, 1997, pp. 68-75.

**SEMENETZ L, Short Hand Design Methods for Tensioned Membrane Structures by the Shape Energy Concept, Space Structures for Sports Buildings, Proc of ICSB (Beijing), 1987, 582-591.
***CLEMENTE P & D'Apuzzo M, Structural Optimization Indices, Spatial Structures: Heritage, Present and Future, Atlanta, ASCE, 1994, pp. 39-46.

Tensarch : A tensegrity double layer grid prototype

R. MOTRO
School of Architecture Languedoc Roussillon and University of Sciences Montpellier II, France

INTRODUCTION

Between 1998 and 2000, a project of a tensegrity double layer grid has been realised with the financial help of the industrial firm Ferrari. The resulting prototype is a plane double layer grid, rectangular in plane and covering an area of more than 80 square meters, with an own weight of 12 kgf per square meter.

The challenge was to demonstrate that beyond an apparent complexity it was possible to build tensegrity systems in the same way as other space structures. Several studies were undertaken and they concerned : conceptual design, sizing for usual conditions, study of sensitivity to manufacturing imperfections, node design, selfstress implementation policy, and finally realisation.

It is obvious that nothing could have been done without a collective work associating many people at the different stages of the project. The grid was completed at the end of year 2000. We understood that we reached our goal when after many years of questions on tensegrity feasibility, a new question arose "how much does it cost ?". Of course the price of the prototype could not constitute a serious basis, since factors have not been considered like optimisation of sizing, or choice of appropriate lighter material such as composites. Nevertheless with this communication, since the prototype itself is not the only result, we want to share simultaneously what we realised and what we learned during this project.

This communication is submitted under the responsibility of the project's leader, but it is again a collective work which implied people from two research laboratories ("Laboratoire de Mécanique et Génie Civil", University Montpellier II, team "Structures Légères pour l'Architecture", school of architecture Languedoc Roussillon), and also three departments of the Universitary Institute of Technology of Nîmes ("Génie civil", "Génie Mécanique et Productique", and "Sciences et Génie des Matériaux"). Last but not least, nothing would have been possible without help of our industrial partner Ferrari. Main actors of this project are : J. Averseng, F. Cevaër, B. Crosnier, G. Fras, A. Grandjean, M.N. Kazi-Aoual, O. Noël, V. Raducanu, R. Sanchez Sandoval, J. Tuset.

MAIN STEPS OF THE PROJECT

As it is impossible to describe in details each step of the project we give information in form of tables in which the main features are illustrated by figures. The whole project is described in Ref 1, and the structural composition was patented (Ref 2).

We began the work by an initial thought about the design process : avoiding the usual agglomeration of cells (Fig. 1), and we developed the concept of extender (Fig. 2). This led us to several structural compositions, whose some possible networks are illustrated by Table 2. All are

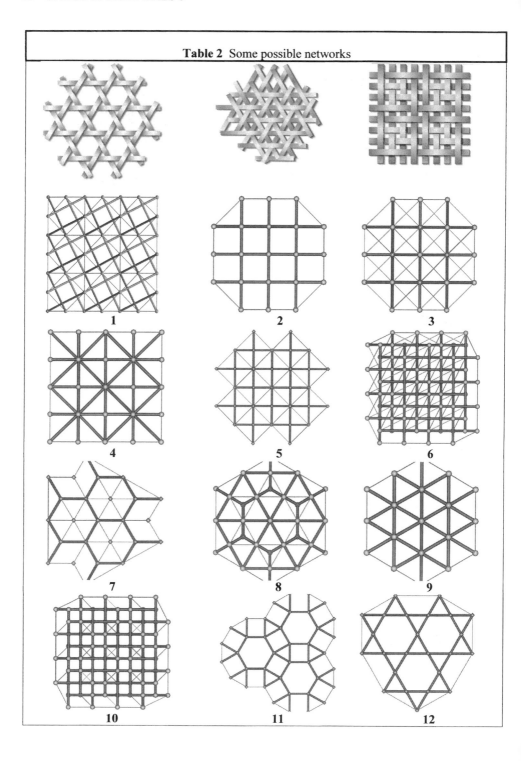

Table 2 Some possible networks

Table 3 Practical and theoretical studies

Fig. 5. Node for prototype at 1/5 scale.

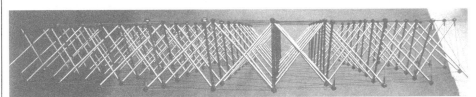

Fig. 6. Grid model.

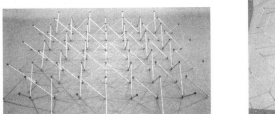

Fig. 7. Grid model.

Fig. 8. Deflection study (augmented vertical scale).

Fig. 9. Selsftress study.

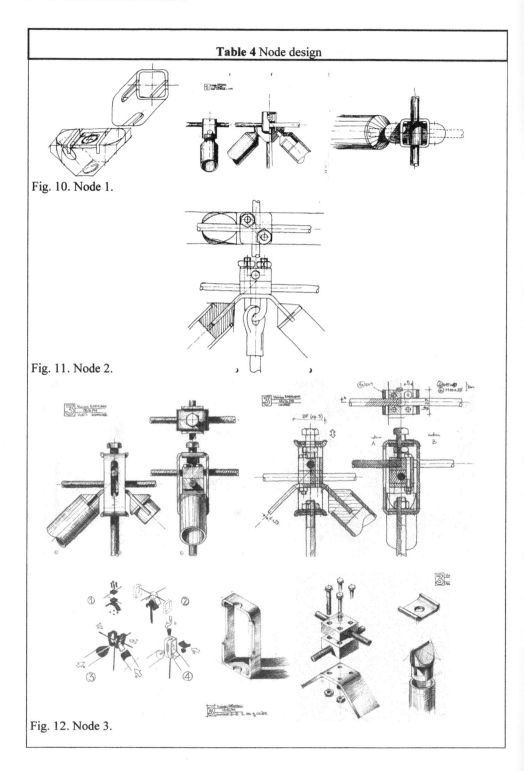

Table 4 Node design

Fig. 10. Node 1.

Fig. 11. Node 2.

Fig. 12. Node 3.

Table 5 Industrial nodes and cables

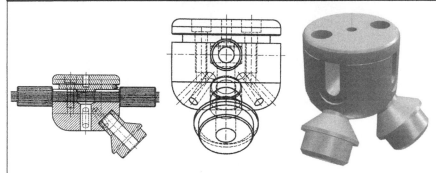

Fig. 13. Node design.

Fig. 14. Cables for so called "mini-grid."

Fig. 15. From design to realization.

Table 6 Grid construction

Fig. 16. Layers disposition.

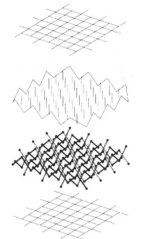

Fig. 17. Grid composition and strut introduction.

Fig. 18. Resulting grid.

Table 7 Geometry and selfstress adjustments

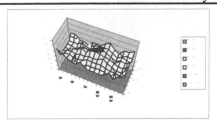 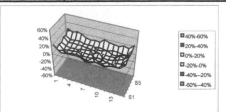

Fig. 19. Geometry and selfstress disturbances.

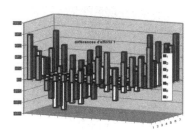 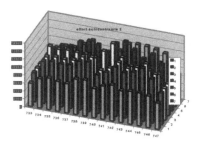

Fig. 20. Selfstress regulation.

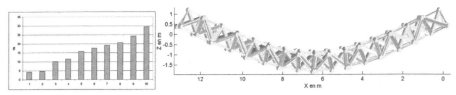

Fig. 21. Theoretical dynamic study.

 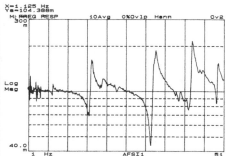

Fig. 22. Experimental dynamic study (with a resulting transfer function).

REFERENCES
1. RADUCANU V., MOTRO R. Système constructif et architecture : cas des systèmes de tenségrité. Contrat SLA/LMGC/Ferrari. Rapport Final . Septembre 2001
2. RADUCANU V., MOTRO R., Composition structurale de systèmes de tenségrité, Société de Tissage et Enduction Serge Ferrari, déposé le 16 Mars 2001 sous le numéro 0104 822.
3. MOTRO R., RADUCANU V., Tensegrity Systems, submitted to International Journal of Space Structures.

FMEM for form-finding and loading analysis of membrane structures

HONG-TAO BAI, and QI-LIN ZHANG
Department of Civil Engineering, University of Tongji, China

INTRODUCTION

In the widely used approach of design and construction of membrane structures, an initial shape of equilibrium is first found for the given boundary conditions and initial stress ratios in two principal directions by using dynamic relaxation method[1][2][3]], density methods[4][5][6] and finite element method[7]. The finite element method was firstly introduced in form-finding by E. Haug and G. H. Powell[7] in 1970 and its applicability and validity were exhibited. From then on, many methods were proposed to find the shape of membrane structures based on the finite element method.

This paper presents a study of finite membrane element method (FMEM) for form-finding and loading analysis of membrane structures. Based on large deformation theory, the geometrical nonlinear membrane element formulas are deduced to analyze membrane structures. The minimal surfaces and the equilibrium surfaces can be effectively formed by the FMEM derived in this paper. Loading analysis can be studied by the same method. A new treatment is adopted to avoid producing press stress or wrinkling elements in the membrane structures. Finally, a numerical example is carefully made and listed.

FINITE MEMBRANE ELEMENT METHOD

Basic Hypotheses

Three basic hypotheses are applied in finite membrane element method:
1. The membrane is modeled as linear orthotropic material.
2. The element has large deformation and small strain.
3. Ignoring bending stiffness of the membrane material.

Definition And Transform of Coordinate Systems

Defining Some Coordinate Systems

Let $O - X_1 X_2 X_3$, $o - x_1 x_2 x_3$ and $o' - x_1' x_2' x_3'$ denote global coordinate system, local coordinate system and inertia coordinate system, respectively, as shown in Figure 1.

A membrane is discretized by using the triangular finite element modal with constant strain.

The global coordinates of node 1, node 2 and node 3 are written as:

$$\{X_e\} = \{X_1^1 X_2^1 X_3^1 X_1^2 X_2^2 X_3^2 X_1^3 X_2^3 X_3^3\} \tag{1}$$

The local coordinates of node 1, node 2 and node 3 are written as:

$$\{x_e\} = \{x_1^1 x_2^1 x_3^1 x_1^2 x_2^2 x_3^2 x_1^3 x_2^3 x_3^3\} \tag{2}$$

The positive local coordinate direction $\overrightarrow{ox_1}$ is defined to be parallel to the vector $\overrightarrow{12}$; $\overrightarrow{ox_2}$ direction is in the plane of 123 and points to node 3.

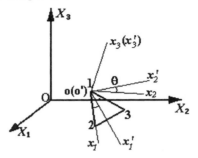

Fig. 1. Definition of coordinate systems.

The membrane is modeled as a linear orthotropic material. The principal direction of warp and weft are assumed to coincide with the direction of the inertial coordinate system. The elastic modulus in the principal directions are denoted by E_1 and E_2, where $E_1 > E_2$, $\overrightarrow{ox_1}$ corresponding to E_1 and $\overrightarrow{ox_2'}$ corresponding to E_2. There is angle θ between axis $\overrightarrow{ox_1'}$ and axis $\overrightarrow{ox_1}$.

Transform Matrix Between Local And Global Coordinate Systems

Axis $\overrightarrow{ox_1}$ is:

$$\overrightarrow{ox_1} = \{l_1 m_1 n_1\} \tag{3}$$

$$\begin{cases} l_1 = (X_1^2 - X_1^1)/L_{12} \\ m_1 = (X_2^2 - X_2^1)/L_{12} \\ n_1 = (X_3^2 - X_3^1)/L_{12} \end{cases}$$

similarly,

$$\overrightarrow{13} = \{(X_1^3 - X_1^1)/L_{13}, (X_2^3 - X_2^1)/L_{13}, (X_3^3 - X_3^1)/L_{13}\}$$

$$\overrightarrow{ox_3} = \overrightarrow{12} \times \overrightarrow{13} / \left| \overrightarrow{12} \times \overrightarrow{13} \right| = \{l_3 m_3 n_3\} \tag{4}$$

$$\overrightarrow{ox_2} = \overrightarrow{ox_3} \times \overrightarrow{ox_1} = \{l_2 m_2 n_2\} \qquad (5)$$

And then the transform matrix of the displacement vector of three nodes is:

$$[t_c] = \begin{bmatrix} l_1 & m_1 & n_1 \\ l_2 & m_2 & n_2 \\ l_3 & m_3 & n_3 \end{bmatrix} \qquad (6)$$

The relation between the deformation vector in local coordinate system and that in global coordinate system is derived as:

$$\{u_e\} = [T_c]\{U_e\} \qquad (7)$$

where,

$$\{u_e\} = \{u_1^1 u_1^2 u_1^3 u_2^1 u_2^2 u_2^3 u_3^1 u_3^2 u_3^3\} \qquad (8)$$

$$\{U_e\} = \{U_1^1 U_1^2 U_1^3 U_2^1 U_2^2 U_2^3 U_3^1 U_3^2 U_3^3\} \qquad (9)$$

$$[T_c] = \begin{bmatrix} [t_c] & 0 & 0 \\ 0 & [t_c] & 0 \\ 0 & 0 & [t_c] \end{bmatrix} \qquad (10)$$

Transform Matrix Between Inertia And Local Coordinate System

According to right hand coordinate system, axis $\overrightarrow{ox_1}$ rotates angle θ around axis x_3 and

gets in axis $\overrightarrow{o'x_1'}$, as shown in Figure 1.

Transform matrix between nodal displacement vector in inertia coordinate system and that of local coordinate system is written as:

$$[t_1] = \begin{bmatrix} \cos\theta & \sin\theta & 0 \\ -\sin\theta & \cos\theta & 0 \\ 0 & 0 & 1 \end{bmatrix} \qquad (11)$$

$$[T_1] = \begin{bmatrix} [t_1] & 0 & 0 \\ 0 & [t_1] & 0 \\ 0 & 0 & [t_1] \end{bmatrix} \qquad (12)$$

$$\{u_e'\} = [T_1]\{u_e\} \qquad (13)$$

where

$$\{u_e'\} = \begin{bmatrix} u_1^{1'} & u_2^{1'} & u_3^{1'} & u_1^{2'} & u_2^{2'} & u_3^{2'} & u_1^{3'} & u_2^{3'} & u_3^{3'} \end{bmatrix}^T \qquad (14)$$

$$\{u_e\} = \begin{bmatrix} u_1^1 & u_2^1 & u_3^1 & u_1^2 & u_2^2 & u_3^2 & u_1^3 & u_2^3 & u_3^3 \end{bmatrix}^T \qquad (15)$$

Displacement Functions

The displacements of one arbitrary point in one plane triangle element can be written as:

$$u_i = C_{i1} + C_{i2}x_1 + C_{i3}x_2 \qquad\qquad i = 1,2,3 \qquad\qquad (16)$$

Then we can get the displacements of node k from Eq. (16)

$$u_i^k = C_{i1} + C_{i2}x_1^k + C_{i3}x_2^k \qquad\qquad i = 1,2,3 \qquad\qquad (17)$$

To solve the equations to get the coefficients C_{i1}, C_{i2}, C_{i3}, then the displacement functions can be written as:

$$u_i = C_{i1} + C_{i2}x_1 + C_{i3}x_2 = \begin{bmatrix} 1 & x_1 & x_2 \end{bmatrix} \begin{Bmatrix} C_{i1} \\ C_{i2} \\ C_{i3} \end{Bmatrix}$$

$$= \frac{1}{2A} \begin{bmatrix} N_1 N_2 N_3 \end{bmatrix} \begin{Bmatrix} u_i^1 \\ u_i^2 \\ u_i^3 \end{Bmatrix} \qquad\qquad (18)$$

In the expression, A denotes the area of the triangle membrane element, and

$$A = \frac{1}{2} \begin{vmatrix} 1 & x_i & y_i \\ 1 & x_j & y_j \\ 1 & x_k & y_k \end{vmatrix}$$

N_k is an expression relevant to local coordinates of three nodes of one triangle element .

$$N_k = A_{k1} + A_{k2}x_1 + A_{k3}x_2$$

Then Eq.(18) can be written as:

$$\begin{Bmatrix} u_1 \\ u_2 \\ u_3 \end{Bmatrix} = [N]\{u_e\} \qquad\qquad (19)$$

$$[N] = \frac{1}{2A} \begin{bmatrix} N_1 & 0 & 0 & N_2 & 0 & 0 & N_3 & 0 & 0 \\ 0 & N_1 & 0 & 0 & N_2 & 0 & 0 & N_3 & 0 \\ 0 & 0 & N_1 & 0 & 0 & N_2 & 0 & 0 & N_3 \end{bmatrix}$$

Geometry Conditions

The strains of one arbitrary point in one element can be denoted as:

$$2e_{ij} = \left(\frac{\partial u_j}{\partial x_i} + \frac{\partial u_i}{\partial x_j} + \frac{\partial u_k}{\partial x_i} \frac{\partial u_k}{\partial x_j} \right) \qquad\qquad (20)$$

in $t \to t + \Delta t$ increment step,

$$2\Delta e_{ij} = 2\Delta e_{ij}^{\ L} + 2\Delta e_{ij}^{\ N} \tag{21}$$

$$2\Delta e_{ij}^{\ L} = \left(\frac{\partial \Delta u_j}{\partial x_i} + \frac{\partial \Delta u_i}{\partial x_j} \right)$$

$$2\Delta e_{ij}^{\ N} = \frac{\partial \Delta u_k}{\partial x_i} \frac{\partial \Delta u_k}{\partial x_j}$$

Substitution of Eq.(26) into Eq.(21) yields:

$$\{\Delta \varepsilon\} = ([B_L] + [B_{NL}])\{\Delta u_e\} \tag{22}$$

where, $\{\Delta \varepsilon\} = \{\Delta \varepsilon_{11} \Delta \varepsilon_{22} \Delta \varepsilon_{33} \gamma_{12} \gamma_{13} \gamma_{23}\}$; $[B_L]$ is a matrix with constant coefficients; $[B_{NL}]$ is a non-linear matrix relevant to displacements.

Stress-Strain Relationship
Do not consider the material non-linearity, and the stress-strain relationship of triangle membrane element in local coordinate is written as:

$$\{\Delta \sigma\} = [D]\{\Delta \varepsilon\} \tag{23}$$

where, $[D]$ is the elastic matrix of triangle membrane element; $\{\Delta \sigma\}$ is the increment of stresses in Δt time. $\{\Delta \varepsilon\}$ is the increment of strains in Δt time.

$$\{\Delta \sigma\} = [\Delta \sigma_{11} \quad \Delta \sigma_{22} \quad \Delta \sigma_{33} \quad \Delta \tau_{12} \quad \Delta \tau_{13} \quad \Delta \tau_{23}]^T$$

$$\{\Delta \varepsilon\} = [\Delta \varepsilon_{11} \quad \Delta \varepsilon_{22} \quad \Delta \varepsilon_{33} \quad \gamma_{12} \quad \gamma_{13} \quad \gamma_{23}]^T$$

Stress-strain relationship in inertia coordinate system is written as:

$$\{\Delta \sigma'\} = [D']\{\Delta \varepsilon'\} \tag{24}$$

where, $[D']$ is elastic matrix in inertia coordinate system, and for the linear orthotropic material, it can be written as:

$$[D'] = \begin{bmatrix} d_{11} & d_{12} & 0 & 0 & 0 & 0 \\ d_{21} & d_{22} & 0 & 0 & 0 & 0 \\ 0 & 0 & 0 & 0 & 0 & 0 \\ 0 & 0 & 0 & d_{44} & 0 & 0 \\ 0 & 0 & 0 & 0 & 0 & 0 \\ 0 & 0 & 0 & 0 & 0 & 0 \end{bmatrix} \tag{25}$$

where,

$$d_{11} = \frac{E_1}{1-\mu_{12}\mu_{21}}, d_{22} = \frac{E_2}{1-\mu_{12}\mu_{21}}$$
$$d_{12} = d_{21} = \frac{\mu_{21}E_1}{1-\mu_{12}\mu_{21}} = \frac{\mu_{12}E_2}{1-\mu_{12}\mu_{21}}, \quad d_{44} = G_{12}$$

where, E_1 and E_2 are the elastic modulus in the principal directions; μ_1 and μ_2 are the Poisson's ratios in two orthogonal directions.

And by deducing, we can get the elastic matrix of a triangle membrane element in coordinate system is written as:

$$[D] = [H_\sigma][D'][H_\sigma]^T \tag{26}$$

where, $[H_\sigma]$ is the rotation matrix of stress. Specially, for the triangle membrane element, the rotation matrix of stresses can be written as:

$$[H_\sigma] = \begin{bmatrix} \cos^2\theta & \sin^2\theta & 0 & 2\cos\theta\sin\theta & 0 & 0 \\ \sin^2\theta & \cos^2\theta & 0 & -2\cos\theta\sin\theta & 0 & 0 \\ 0 & 0 & 1 & 0 & 0 & 0 \\ -\cos\theta\sin\theta & -\cos\theta\sin\theta & 0 & \cos^2\theta-\sin^2\theta & 0 & 0 \\ 0 & 0 & 0 & 0 & 1 & 0 \\ 0 & 0 & 0 & 0 & 0 & 1 \end{bmatrix} \tag{27}$$

Equilibrium Equations
Element Equilibrium Equations In Local Coordinate System
According to the principle of virtual work, the equation in $t+\Delta t$ time can be written as:

$$\int\{\delta\Delta\varepsilon\}^T\{\sigma\}dv = \int\{\delta\Delta u\}^T\{\overline{p}\}dv + \int\{\delta\Delta u\}^T\{\overline{q}\}ds \tag{28}$$

where, $\{\overline{p}\}$ and $\{\overline{q}\}$ are the volume loads and area loads on the membrane element.

Substitution of Eq. (22) and Eq.(23) into Eq.(28) yields:

$$[k_e]\{\Delta u_e\} = \{p_e\} - \{f_e\} \tag{29}$$

where,

$$[k_e] = [k_E] + [k_G] + [k_\varepsilon] \tag{30}$$

$$[k_E] = \int[B_L]^T[D][B_L]dv \tag{31}$$

$$[k_G] = \int[G]^T[M][G]dv \tag{32}$$

$$[k_\varepsilon] = \int\left[\frac{1}{2}[B_L]^T[D][B_{NL}] + [B_{NL}]^T[D][B_L] + \frac{1}{2}[B_{NL}]^T[D][B_{NL}]\right]dv \tag{33}$$

$$\{p_e\} = \int [N]^T \{\bar{p}\} dv + \int [N]^T \{\bar{q}\} ds \qquad (34)$$

$$\{f_e\} = \int [B_L]^T \{\sigma\} dv \qquad (35)$$

Element Equilibrium Equations And Structure Equilibrium Equations In Global Coordinate System

The transform relations of nodal load vector and nodal displacement vector between in local coordinate system and in global coordinate system can be written as:

$$\{p_e\} = [T_c]\{P_e\} \qquad (36)$$

$$\{f_e\} = [T_c]\{F_e\} \qquad (37)$$

$$\{u_e\} = [T_c]\{U_e\} \qquad (38)$$

Substitution of Eq.(36) ~(38) into Eq.(29) yields:

$$[K_e]\{\Delta U_e\} = \{P_e\} - \{F_e\} \qquad (39)$$

$$[K_e] = [T_c]^T [k_e][T_c] \qquad (40)$$

Eq.(40) is assembled into the total structure to drive the equilibrium equation as:

$$[K]\{\Delta U\} = \{P\} - \{F\} \qquad (41)$$

where, let $[K], \{\Delta U\}, \{P\}$ and $\{F\}$ denote the total stiffness matrix of the structures, the total displacement vector, the total nodal load vector and the equivalent nodal force vector produced by initial stresses in global coordinate system. And Eq.(41) is a non-linear finite equation, as can be solved by such methods as Newton-Raphson method, Updated Newton-Raphson method and so forth.

Treatment Adopted To Avoid Wrinkling Elements In Iterative Process

For the reasons that membrane material can not bear press force, when the pull force in one direction fading away, one membrane element will lose its ability to bear loads and produces wrinkles in this direction. When only very few elements are wrinkling elements, the total structures can keep their ability to bear loads, but their beauty will be influenced.

In the iterative process of form-finding and loading analysis, every iterative step must be checked to see whether some elements produce wrinkles. In every element, according to the three stresses σ_x, σ_y and τ_{xy}, the maximal stress σ_1 and the minimal stress σ_2 can be solved, where $\sigma_1 \geq \sigma_2$. According to the different conditions of the maximal stress and the minimal stress to define the corresponding stress-strain relationship in inertia coordinate system. Then substitute of it into Eq.(26) to consider the possible wrinkling element.

1. when $\sigma_1 \geq \sigma_2 > 0$, do not produce wrinkling elements.

2. when $0 \geq \sigma_1 \geq \sigma_2$, the element exits work. The relation matrix $[D']$ in Eq.(25) can be written as:

$$[D']=\begin{bmatrix} \varepsilon & 0 & 0 & 0 & 0 & 0 \\ 0 & \varepsilon & 0 & 0 & 0 & 0 \\ 0 & 0 & 0 & 0 & 0 & 0 \\ 0 & 0 & 0 & \varepsilon & 0 & 0 \\ 0 & 0 & 0 & 0 & 0 & 0 \\ 0 & 0 & 0 & 0 & 0 & 0 \end{bmatrix} \qquad (42)$$

where, ε is an very small number specified by the author, in the article, $\varepsilon = 1 \times 10^{-8}$.

3. when $\sigma_1 > 0$, $\sigma_2 \leq 0$, the element can only bear the pull force in σ_1 direction and the relation matrix $[D']$ in main stress coordinate system can be written as:

$$[D_1]=\begin{bmatrix} d_{11} & 0 & 0 & 0 & 0 & 0 \\ 0 & 0 & 0 & 0 & 0 & 0 \\ 0 & 0 & 0 & 0 & 0 & 0 \\ 0 & 0 & 0 & 0 & 0 & 0 \\ 0 & 0 & 0 & 0 & 0 & 0 \\ 0 & 0 & 0 & 0 & 0 & 0 \end{bmatrix} \qquad (43)$$

We assume main stress coordinate can arrive in inertia coordinate system by contra rotating angle α, Eq.(26) yields:

$$[D']=[H_\sigma][D_1][H_\sigma]^T \qquad (44)$$

$$[H_\sigma]=\begin{bmatrix} \cos^2\alpha & \sin^2\alpha & 0 & 2\cos\alpha\sin\alpha & 0 & 0 \\ \sin^2\alpha & \cos^2\alpha & 0 & -2\cos\alpha\sin\alpha & 0 & 0 \\ 0 & 0 & 1 & 0 & 0 & 0 \\ -\cos\alpha\sin\alpha & \cos\alpha\sin\alpha & 0 & \cos^2\alpha-\sin^2\alpha & 0 & 0 \\ 0 & 0 & 0 & 0 & 1 & 0 \\ 0 & 0 & 0 & 0 & 0 & 1 \end{bmatrix} \qquad (45)$$

EXAMPLE
A membrane structure with leaf shape
Figure 2 is the boundary dimension of a leaf. Figure 3a, b are its initial elevation and perspective drawing. The thickness of its material is 0.9mm and its equal stress is $2000\,kN/m^2$, and let us adopt FMEM to find its minimal shape. Figure 3c is its elevation of having minimal shape and Figure 3d is the perspective drawing of minimum shape calculated by FMEM.

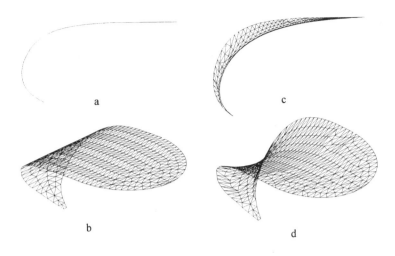

a

c

b

d

Fig. 2. Boundary dimension of a leaf (unit: mm).

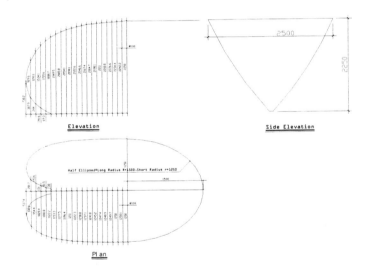

Fig. 3. Initial geomtry shape and the minimum form found by FMEM.

REFERENCES

1. A. S. Day, An Introduction to Dynamic Relaxation, The Engineer, Jan.29, 1965:218-221.

2. M. R. Barnes, Form Finding and Analysis of Tension Structures by Dynamic Relaxation, International Journal of Space structures, Vol. 14, No.2, 1999:89-104.

3. B. H. V. Topping, The Application of Dynamic Relaxation to the Design of Modular Space Structures, Ph.D. Thesis, The City University, London, 1978.

4. B. Maurin, R. Motro, Investigation of Minimal Forms with density methods, Journal of Int. Association for Shell and Spatial Structures: IASS, Vol.38 (1997) n.125.

5. H. J. Schek, The Force Density Method for Form-Finding and Computation of General Metwork, Comp. Meth. in Appl. Mech. And Engr. Vol.3, 1974:115-134.

6. L. Grundig, Minimal Surface for Finding Forms of structural Membrane, Computer & Structural, Vol.30 NO.3,1998: 679-683.

7. R. H. Haber: Computer-Aided Cable Reinforced Membrane structures, Ph.D. Dissertation, Connell University, 1980.

Keeping pace with progress: from reinforced concrete shells to membrane structures

V. GIONCU

Politechnica University Timisoara, Romania

PROGRESS IN SPACE STRUCTURES

In the latter part of the last century much progress has been observed in the art and science of architecture and engineering. The understanding of materials, fabrication and construction and finally, the manner in which structural systems behave has become significantly deeper. The process was accelerated in the last few decades by the phenomenal growth in development of computers, which truly make possible previously inaccessible projects.

A crucial step, perhaps the most important one, is the deeper understanding of role of space structures in the modern architecture. The visible progress in development of space structures is extremely significant emphasizing continuous achievement of creative designers, engineers and architects alike. Over the last period engineers have been called upon to built larger and larger structures, all of them with less material and less cost. So, the best engineers matured under the discipline of extreme economy and their ideas developed under competitive cost control. Architectural concepts of aesthetics are changed all the time, leading to the introduction of new structural systems.

During the last decades designers agree that, for large span structures, conventional planar beam and truss solutions are uneconomical. So they looked to the innovative solutions, as the space structures. The advantages of these structures have been known for many years. The second half of XX century could be characterized as the era of space structures, being the result of cooperation between architects and engineers or the perfect harmony between image and technology. Architects started to experiment with new shapes and some years ago **reinforced concrete (RC) shells** were in fashion. The visual beauty of these structures appealed to architects.

However, the reinforced concrete is no the ideal structural material. Some progressive engineers realized the limitation of RC shells (long construction time, elaborate scaffolding and framework) and turned their attention to a new structure type, the **steel space structures**. For many years the engineers appreciated their ability to cover large spans with minimum weight, but the difficulty of complex analysis of these systems originally contributed to their limited use. The introduction of electronic computers changed this situation and produced an important impact in development of steel space structures. In this area the progress was almost unbelievable and the use of light-weight structures greatly change the engineering approach to design.

After this progress in space structure design, a new stage is created by the rapid improvement in structural textiles. This lead within the last decade to the dramatic development of a new structural system, the **membrane structures**. Rarely attempted three decades ago, this new form is now accepted by the progressive designers as a best solution to cover large span buildings. But the membrane systems are kinematically undetermined and they can be rigid only by using the pre-stressing and some rigid elements. Therefore, during the last decade a new technique has been developed, the **hybrid membrane structures**, by assembling a flexible membrane fabric with a rigid steel structure.

Keeping pace with the progress, some space structures were designed by a team from Timisoara Technical University and Building Research Institute. The author wish to thank their colleagues from this team, dr. Nicolae Balut, Dorin Porumb and Nicolae Rennon. To exemplify this progress, in the present paper two space structures are present, showing the way from the reinforced concrete shells to the membrane structures.

RC SHELL FOR ORSOVA CATHOLIC CHURCH

The erection of a church has represented always the most important event in the life of a town. The church was the result of the commune efforts of natives and artisans, leaded by a "chief master", usually an architect. From this co-operation it was risen a monumental masterpiece, which became the central attraction of the town. The success of construction will depend on the good co-operation between two personalities: architect and structural engineer. There is no other clearer example than the erection of a Catholic Church where the co-operation architect-engineer to be more evident. Giving up the old dogma, classical of roman, gothic or baroque architecture, and the frame in the rhythm of society development, is the characteristic of Catholic Church spirit. This open spirit has permitted the erection of some churches with modern architecture as an expression of improvement the beauty, useful and economic notions. On no other domain of architecture the results where so amazing. With no doubt this feeling to imagine forms and to confer structural consistence to them, represents the most clearly co-operation of artistic synthesis and rigorously analysis.

As result of Iron Gates hydro-electrical station construction on Danube, the entire Orsova town had to rebuild on other place. The reconstruction included buildings and social endowments. Among these ones it was the Catholic Church. The complex contains three objects, the Church (1), the house of priest (2) and the belfry (3) (Fig. 1 a). In the followings only the church, the most interesting part from structural point of view, will be presented. After the analysis of some volume variants, the architect decided that the most adequate, with also an ecclesiastical philosophy, is the parallelepiped. From this form are cut up some portions so that to obtain a volume as a sarcophagus, with the Christian's sign, pointing on the cross on the lid (Fig. 1 b). Another aspect imposed by the architect was to underline the idea of a massive block outside and the Cross lightning inside. So, the ceiling and the lateral walls are full, the open surface towards exterior being only the cross on the top of the church.

The roof form represents the intersection of four curved surfaces, which must be ruled surfaces in order to be covered with wood. Therefore, there were examined three surface types (Fig.2): (i) hypars referred to rectilinear generators; (ii) hypars with curved edges; (iii) conoids. By analysis of these three solutions it was found that the conoid surfaces are simple enough for execution and it was decided for this solution for the roof. (Fig. 3)

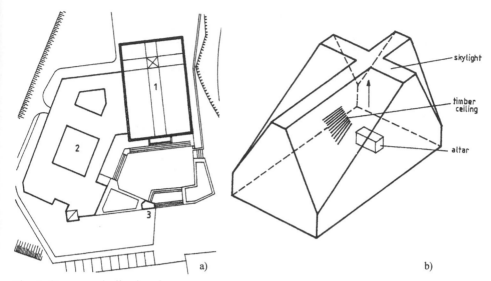

Fig. 1. Orsova catholic church.

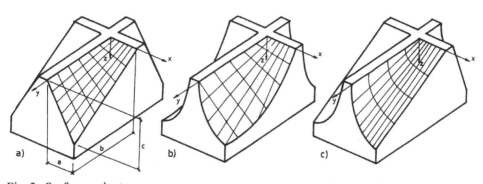

Fig. 2. Surface variants.

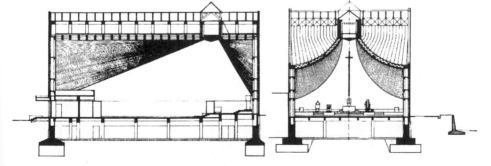

Fig. 3. Transversal and longitudinal sections.

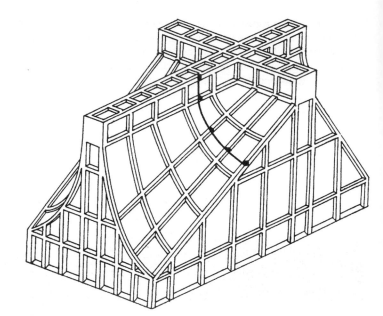

Fig. 4. Church
structure.

Fig. 4 shows the Church structure composed by the four shells form of roof, the cross space frames which support the roof and the lateral walls. One can observe that the entire building forms a space structure where all elements work together. Today, using the computers, the space analysis of this structure does not represent anymore a problem. But when this structure was designed, during 1973-74, we had not such possibilities and for design there were used some approximate methods. So, for shells, the differential equations of membrane theory were integrated with complete edges (edges able to take horizontal forces). Due to the inequality, as dimensions, of the two adjacent shells, high horizontal forces were obtained, which cannot be over take by the transversal beams. For this reason, there were introduced some transversal ribs (Fig. 5a), which leaded to a most uniform distribution of these horizontal forces. The role of these ribs is also to take over some asymmetrical loads from wind and snow, hard to be checked for this unusual form of roof. The supporting structure of the roof is the space cross frames (Fig. 5b). As for the shells the difficult problem was that of frame asymmetry, which introduced in the transversal beams important moments of torsion.

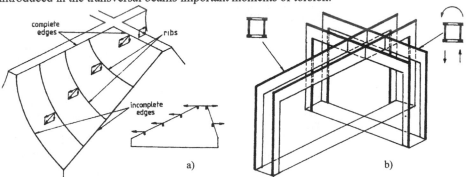

Fig. 5. Structure analysis.

Inside and outside aspects of the church can be seen in Fig. 6. It must be mention that the erection of this Church was made during a hard period of our country from political point of view, when I could not publish anything about it. This is the reason why only today, after already 25 years after finishing I have the opportunity to present it. The architect of this masterpiece was my dear late colleague and friend Hans Fakelmann.

Fig. 6. Church aspects.

TENSIONED HYBRID MEMBRANE FOR TIMISOARA MARKET

This structure represents my clear progress from the RC shells to the new modern structures. This project had to solve the rehabilitation and conversion of an ancient marked place and a covered platform as a permanent market with private commerce for fruits, vegetables and flowers (Fig.7).

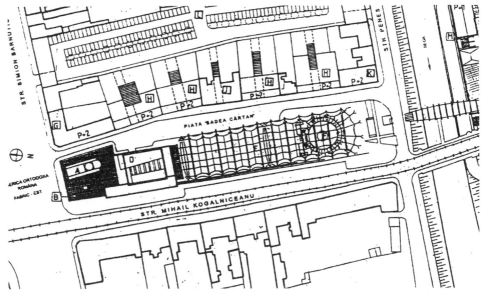

Fig. 7. The Timisoara market.

After an analysis of other structural solutions, decision to use a tensioned membrane was assumed. This structure type gives for architects and engineers the possibility to create new exciting solutions, in which the architectural expression and structure form are in perfect concordance. Because the textile material is very light-weight, the solution is very efficient and economic. But there are also some disadvantages, as complex surface formation, large deformations, intricate anchorage, etc., which give many problems in design. A very good solution is the tensioned hybrid membranes, obtained by the combination of a flexible membrane with a rigid supporting steel structure. Therefore, this was the reason for which to cover the market this solution was selected. The steel structure is composed by transversal frames (laced arches, tubular built-up columns and external pre-stressed ties), longitudinal stiffening systems (transversal pre-stressed cables and longitudinal pre-stressed ties) and end tension system (inclined columns and ties). Intermediary tension frames were placed at the half of building and at extremity, given the possibility of a future market extension (Fig. 8). Two membrane types, the middle one in the form of saddle hypars, and the end one in the free shape (Fig. 9) compose the market roof.

The design of structure follows some very different steps in comparison with the first space structure, showing the progress in structural deign. The first step was the form-finding for the membrane surface, using the Easy software package. The second step was the calculation of the individual components such as membranes, cables and fittings, considering the external forces and the cable pre-stressing forces. The third step was the cutting pattern of membrane, based on the surface geodesic lines of deformed surfaces. A computer program was used for stress analysis of steel structure, performed for dead, snow (uniformly distributed and agglomerations), wind (pressure and suction) and seismic loads and pre-stressing of external and longitudinal ties. A special attention was paid for the case of partial or total loss of pre-stresses due to an accident.

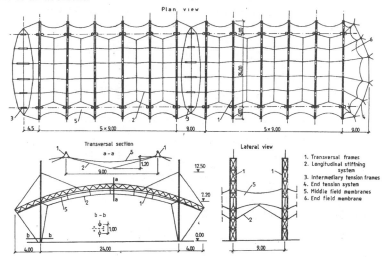

1. Transversal frames
2. Longitudinal stiffning system
3. Intermediary tension frames
4. End tension system
5. Middle field membranes
6. End field membrane

Fig. 8. Structure configuration.

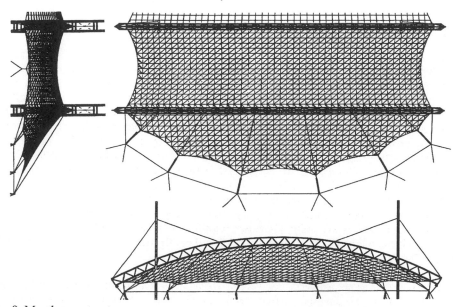

Fig. 9. Membrane structure.

The steel structure received the European Convention for Constructional Steelwork, ECCS, Steel Design Award in 1997 (Fig. 10).

Fig. 10. ECCS Steel–
Design Award in
1997.

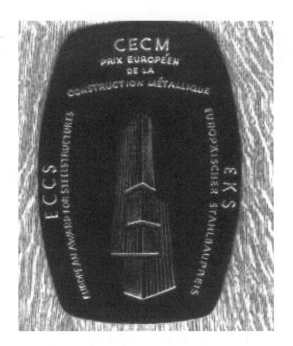

Fig. 11a.
Market aspects.

Fig. 11b.

Fig. 11c.

Some aspects of the market can be seen in the Figs. 11(a, b and c). The design and erection of this structure were the result of a co-operation of an international complex team from Romania, Germany and Austria, emphasizing a new step in construction philosophy in my Country. Another important progress must be noticed. If for the Church the architect was one of my colleagues, for the market the architects were my former students, the youngs Ioan Andreescu and Vlad Gaivoronschi, marking the changing of the generations.

ADAPTENT: A Generating System for Cable Net Structures

Mollaert M., Department of Architecture, Vrije Universiteit Brussel, Belgium, and
Hebbelinck S., Department of Architecture, Vrije Universiteit Brussel, Belgium

INTRODUCTION

In the last decennia public events have increased considerably in both number and size. More and more outdoor venues are used for festivals, fairs, exhibitions etc. Temporary shelter has to be provided for people and equipment alike. Tensile structures are very appropriate for this task for they are lightweight, easy to handle and transport, relatively fast erected from prefabricated parts and easily disassembled.

Two possibilities are on offer for the event organiser: *standardised "tents"* and *customised fabric structures*. Both types of tensile structures have distinct characteristics, advantages and disadvantages. Whereas the standardised "tents" provide readily available rental solutions in a systematic way, built of standard elements taken from the tent-provider's stock, they offer rather low creative freedom as far as design is concerned. They are based upon repetition of fixed plan forms, usually squares, rectangles and hexagons. They do lack the "free-form" character typical for customised fabric structures.

According to the principles of tensile architecture creative designers can make beautiful and non-conventional, eye-catching structures. Such attractive structures can add greatly to the appearance of an event. But custom designed and manufactured structures are expensive and in view of their mostly short live span, the investment is in many cases not justifiable.

A four-year research project, sponsored by IWT, looked deeper into the discrepancy between custom fabric structures and standardised "tents". A generating system for cable net structures was developed and its feasibility was evaluated.

BASICS OF THE PROJECT

Combination of two design approaches

The idea was put forward whether it wouldn't be possible to combine the advantages of the standardised tenting systems (Figure 1) -standardised elements, immediate delivery, expandability, re-use and re-configuration of elements- with the specific features of custom tensile structures (Figure 2) -freedom of design, non-conventional shapes, attractive appearance?

Figure 1. Typical rental tent.

Figure 2. BMW Pavilion for the IAA'95 [6].

Space Structures 5, Thomas Telford, London, 2002

Research priorities

Tensile structures mainly consist of four distinct, ever reoccurring parts: *supporting structure, anchorage, pretension devices* and *tensile surface*. It is a typical feature of tensile architecture that all these parts work together to establish the whole structure's stability.

It is however the tensile surface that poses the most specific problems in view of standardisation and systematisation. *The tensile surface of the structures to be generated therefore got prior attention.*

To generate a stable (roof) surface able to withstand external loads, stiffness has to be given to the surface in a mechanical way. This is done by giving the surface a double curved shape -like a horse saddle- and by prestressing it. For each design the engineer has to calculate the correct shape and prestress level in order to deliver a safe and sound tensile roof.

Once the form found, the roof surface is manufactured welding together pre-calculated flat patterns into the desired double curved shape. This action is irrevocable and the tensile surface thus has a fixed form. Another design with another shape requires other patterns. Manufactured membrane covers are thus non-adaptable.

Another way to make tensile roofs is by prestressing nets made of cables and cladding them afterwards (Figure 3). The Olympic stadium roof in Munich (Figure 4) is a well-known example of this technique. A cable net basically adapts to double curvature by changing the angles of its meshes. To give the net the required double curved shape, all cables in the net as well as its edge cables have to be given precalculated lengths. Correct positioning of the connection of the net-cables to the edge-cables is very important too. Most of the time cable nets are constructed from steel cables clamped and bolted together. Again a fixed, non-adaptable shape is the result. To enlarge a cable net, meshes have to be added or the meshes should be re-distributed and the constructing cables lengthened. This last idea was further studied and led to the ADAPTENT concept. It was chosen to *focus at first upon the structural core of the tensile surface of the structures to be generated.*

Figure 3. Membrane cover hung under pretensioned cable net structure; IL Stuttgart, 1966 by Frei Otto [1]

Figure 4. Cable net clad with transparent panels on neoprene pedestals, Olympic stadium roof, Munich, by Frei Otto and J. Schlaich, 1972 [2]

BASIC CONCEPT OF THE NEW GENERATING SYSTEM

Nets with sliding nodes

In order to make a net that can be enlarged or reduced by redistribution of the current meshes, two specific features have to be present: *the cables that build up the net have to be able to change length* and *the nodes in the net have to be able to change position*.

When the nodal clamps of a traditional cable net are removed, an assembly of two series of cables lying orthogonal on top of each other remains. When this net is prestressed into a double curved shape, every "node" will be formed by a system of two cables tensioned one over the other in space. In such a system, the contact-point of the two cables -from now on called "node"- will have a unique equilibrium position in space, depending on the geometry of the cables and the amount of pretension induced in each of them. Geometrically this equilibrium is characterised by the fact that each cable is situated in the bisector plane of the other cable and that the force in each cable is constant. Changing the boundary conditions of the two-cable system and/or the levels of prestress will result in another shape, generated using the same elements. A special feature of this equilibrium is that, when composed into a network, the path followed by each cable coincides with a geodesic line on the generated double curved surface. The cables thus materialise the shortest path between their start and end point on the surface they generate (Figure 6). This particular equilibrium shape can be simulated using the Force Density method [3] in an iterative process.

Figure 5. A net calculated with a constant force density in all the links.	Figure 6. A net calculated with a constant force: the cables follow geodesic lines on the surface.

Adaptable net-components
Connecting together precalculated patterns produces double curved membrane surfaces. In order to enable differently sized and shaped double curved surfaces to be generated using the same components two special features have to be provided: *the components have to be adaptable to the different "patterns"* (Figure 8) *needed to build up the different double curved surfaces* and *the components have to be connected together in a completely reversible way* (Figure 9, Figure 17).

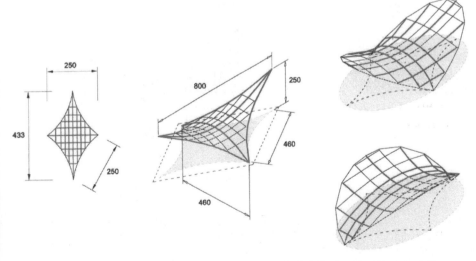

Figure 7. Minimal flat configuration of D1 [5]	Figure 8. Maximal symmetrical saddle shaped configurations of net-component D1 [5]

Applying the concept of "sliding node nets", components having these two features can be developed. These components will be nets with a predefined basic shape -that can be turned into a hypar- that have non-fixed nodes and in which the length of the tensile elements that build up the net can be adapted. These net-components can then be given different double curved shapes, within a predefined range, by adaptation of lengths and redistribution of nodes. Thus they can be configured into the different patterns needed to build up a larger net and connected together using a reversible connection method.

Each net-component will thus form a re-configurable and re-usable "cutting pattern" in the overall cable net surface.

Design process

This new way to construct pretensioned cable net roofs follows a design process very similar to that used to conceive a conventional tensile structure.

At first the boundary conditions of the structure have to be defined. For the ADAPTENT system a software tool called ADSUPSOFT has been developed to help the designer in this task. This tool enables the user to compose the outline of his ADAPTENT structure putting together several net-components and altering their shapes. The computer will take into account all limits and possibilities of the system and will prevent the designer to make shape alterations impossible within the limits of the system. This means that the designer can always be sure that his ADAPTENT design made using the ADSUPSOFT tool is feasible.

Secondly the structure's equilibrium shape is defined in a numeric form-finding process. For an ADAPTENT structure this equilibrium is calculated through an iterative process using the Force Density based calculation program EASY [4] (Figure 9).

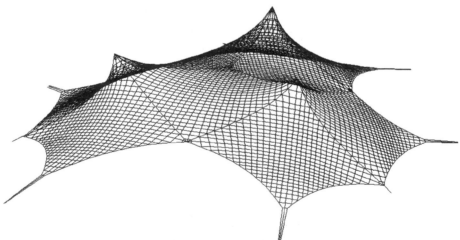

Figure 9. Equilibrium shape of an ADAPTENT structure consisting of 9 net-components.

Next step is to analyse the behaviour of the proposed equilibrium shape under external loads like for any other tensile roof.

Once the equilibrium shape and prestress levels are approved, in the traditional process the cutting patterns of the roof surface are calculated. For the ADAPTENT structure, the major difference to the traditional process is to be found in this step. For the ADAPTENT structure data to configure each net-component of the structure is calculated.

In a last step before transportation and erection, the precalculated roof surface is manufactured. In the traditional process this is done in an irrevocable way by welding together the calculated cutting patters. For the ADAPTENT roof surface the required net-components are taken from the stock, configured according to the calculated data and connected together using reversible techniques.

At the end of the structure's live span, the biggest difference between the two procedures occurs. Whereas the traditional tensile roof surface is thrown away or stored to gather dust, the ADAPTENT tensile surface is disassembled into its different net-components. All elements and components are then stored, waiting to be used in a new configuration later on.

FROM CONCEPT TO PHYSICAL GENERATING SYSTEM

Demands to comply with
In order to develop an actual generating system for tensile structures, the basic concept has to be materialised. The aims set for the project require that:
- Tensile structures, in different shapes and sizes, using their tensile elements as fully fledged structural parts, should be produced,
- Only a limited set of elements should be used for this, it should be easy to manipulate, store and transport them,
- Elements should be completely reusable and components re-configurable,
- Large as well as small free spans are to be generated, referring to the spans of existing rental tents.

The specific demands to make the "sliding node nets" work are to be summarized as follows:
- The "nodes" of the nets may not be fixed,
- The friction coefficient between the two series of cables building up the nets has to be as low as possible,
- The length of the cables building up the nets have to be adjustable,
- Internal cables have to be fixed at the edges of the net,
- Tensile edges and rigid -straight or arched- edges have to be possible,
- Edges have to be adaptable in length,
- All connections have to be reversible.

Elements, sub-components and components
Taking into account these demands, the concept has been materialised in a hierarchical way. 24 different *elements* -the most basic parts in the hierarchy that cannot be disassembled in a non-destructive way- build up 4 different types of *sub-components*: net sub-components, net-to-edge-connection sub-components, tensile edge sub-components and corner sub-components.

Out of the first two sub-components, the adaptable net-components are composed. Together with the remaining sub-components several net-components can be connected together and composed into a net surface to be prestressed into a precalculated shape.

The way the concept has been materialised in the scope of this work is rather exemplary. The main aim set for this work was to proof the value of the concept and as such the possibility to develop an actual generating system for tensile structures that offer a lot more possibilities and freedom of design than the existing standardised tenting systems. The presented materialisation therefore is certainly not the most aesthetically pleasing one and has a number of elements that are over-dimensioned in order to be compatible with other elements. Refinement of the materialisation is certainly needed at a later stage (Figure 10, Figure 11)

Basic elements for the presented materialisation are polyester belts used in the lifting and hoisting industry. They are strong, easy to handle, readily available, inexpensive, show rather low elongation at break and have a whole series of accessories readily available. From these accessories compression and friction buckles have been used to make the length of the belts adaptable. A nodal element of two aluminium plates is introduced to compose the belts into coherent net sub-components. The belts run through slits in the aluminium plates. When tension in the belts is raised however, the nodal elements get stuck between the two belts, thus inducing friction and more or less fixing the nodes into their equilibrium positions. This ensures extra stability to the net when loaded.

Chains and their accessories have been chosen to materialise the tensile edges of the ADAPTENT nets. Although they are rather heavy, they are very flexible and easy to extend and shorten. It is also very easy to connect the net-component to it by fixing shackles into the chain links.

Figure 10. The ADAPTENT net-component: polyester belts, aluminium sliding node elements... [5]

Figure 11. ...compression buckles, steel rings and shackles connected to chains [5]

Adaptable corner sub-components consisting of reconfigurable steel plate elements have been designed to connect the nets to their supports and to pretension them.

Dimensions and limits of the net-components

The concept described asks for some basic net-components that can be given different (saddle) shapes within a predefined range. Preshaped net-components are then to be combined into larger structures.

To establish the dimensions and lengths of the complete net-components and the individual belts, a diamond shape with 60° and 120° angles and a square have been studied, because of their special combination properties. The dimensions and range of adaptability of the net-components (Figure 12) has been defined in accordance to the most common dimensions to be found in the existing standardised tenting systems.

Using the different net-components, structures can be made consisting of four pointed saddle shapes with free spans ranging from 3.54m to 32.00m. These saddle shaped "patterns" can be symmetrical, semi-symmetrical or completely asymmetrical as long as some restrictions for the net-component used are respected (Figure 13):

- The diagonal belts (D1, D2) do not exceed their maximal predefined length and can not be shorter than their predetermined minimal length,
- The diagonals remain orthogonal in plan,
- In plan view, the distance (L1) from each corner point to the intersection point of the diagonals does not exceed its predefined maximum,
- The edge length (L2) does not exceed its predefined maximum.

These design rules have been implemented in the aforementioned ADSUPSOFT design tool that accompanies the ADAPTENT system and are checked each time the designer alters the shape of a net-component in a project.

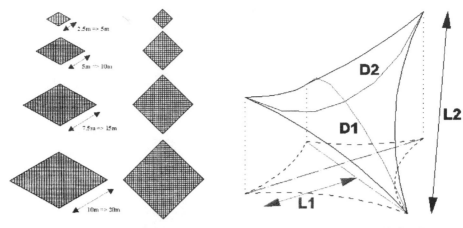

Figure 12. The net-components [5]   Figure 13. Restrictions on specific lengths [5]

Other design rules are to be observed when designing an ADAPTENT structure:
- The net surface has to be double curved and pretensioned with about 2.5kN/belt,
- Tensile edges have to be curved,
- Connection of two net-components with different orientation is to be done using an internal chain edge,
- The diagonal belts have to be connected to the supports separately from the edges,
- When tensile forces of over 100kN are to be expected at the edges, the rotating corner plates have to be doubled,
- For compositions of large net-components, heavier chain edges are to be applied.

PROTOTYPE TESTING

Aims of the experiments
To examine whether the theoretical concept and its materialisation actually worked as predicted, a basic net-component was built and put to the test in the laboratory of civil engineering at the Vrije Universiteit Brussel. Three main topics had to be studied:
- The verification of the "sliding node nets" concept and the presented materialisation,
- The accuracy of the computer simulations and construction data generated as output,
- The methodology to use and work with the system.

Test procedure
To find reliable answers to the different questions enclosed within these three topics, the prototype net was configured and re-configured into five different shapes; two completely symmetrical ones, two with only one axis of symmetry and a totally asymmetrical one (Figure 14). Forces in the structures have been measured using dynamometers at the fixed high points and at five belts in the net and by calibrated manometers recording the pressure in the hydraulic cylinders that were used at the anchor points to pretension the nets. Furthermore the geometry was checked, measuring the position of the different nodes according to the level of the laboratory floor. All these measurements were done when the structure was prestressed applying the precalculated tension forces at the low anchor points and when the prestressed structure had been loaded with a 0.5kN/m² uniformly distributed load. All experimental results were then compared to the numerical results of the computer simulations.

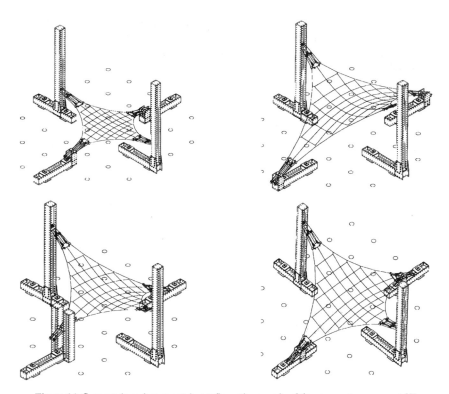

Figure 14. Symmetric and asymmetric configurations made of the same net-component [5]

Experimental findings
About the working of the concept and the performance of the nets could be concluded that:
- The designed ADAPTENT nets can be configured in different shapes and pretensioned accordingly into stable structures,
- The sliding nodes of the nets take on a position that deviates little from the theoretical equilibrium position,
- The deformation under load of the prestressed nets is low and the position of the sliding nodes is guaranteed when sufficient prestress is induced.

About the accuracy and reliability of the computer simulations it was concluded that:
- The shape of the prestressed structure corresponds to the simulated shape,
- When the actual net is prestressed applying the pre-calculated tension forces at the anchorage points, the average level of prestress induced in the belts forming the net, when accurately set, is about 30% higher then the pre-calculated prestress,
- The sagging of the net under external load corresponds with the simulated deflection under that load.

About the methodology to use and work with the system it has been concluded that:
- Setting carefully the pre-calculated belt lengths and edge-net connection positions results in the correct "patterning" for the designed structure, the inevitable inaccuracies in the configuration of the actual net have no important negative effect,
- With the construction data file provided by the ADSUPSOFT software, the net components are easily configured into the desired shape,
- Setting the length of the belts is done easily with a scale printed onto the belts,

- Re-configuration does not ask for complete dismantling of the net component, the compression buckles are to be moved along the belt into their new position,
- The weight of the corner sub-components and the chain edges hinders the manipulation of the nets to certain extend but has no effect on the prestressed structure.

Building a prototype ADAPTENT net also showed that having focused upon the structural core of the tensile surface of the structures to be generated had been a good choice. The problem of fixed cutting patterns has been overcome. But in order to test a prototype net-component, the other features of a tensile structure had to be provided too. This proved to be quite easy. An *adaptable supporting structure* was built using the Meccano-like system available at the laboratory of civil engineering designed to build up experimental set ups. The configured prototype nets were *anchored* to this structure without a problem and an appropriate way to *pretension* them was found in using hydraulic cylinders that could be taken out of the set up once it was tensioned to be applied elsewhere, thus fitting very well into a concept of compatibility and re-use.

EVALUATING THE OUTCOME OF THE PROJECT

Figure 15. Symmetric configuration [5]

Figure 16. Asymmetric configuration [5]

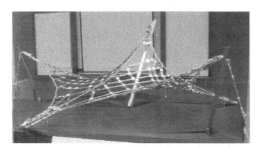

Figure 17. Presentation model with 2 net-components [5]

Figure 18. Prototype covered with foil [5]

Covering the nets

The prototype structures proved the value and workability of the ADAPTENT generating system in its present state, being a system that generates, in a reversible way, the structural core of the tensile surfaces of differently sized and shaped tensile structures.

In order to serve as temporary shelters, the ADAPTENT nets need to be covered in a continuous, waterproof way however. This goes for traditional cable net structures too, but the ADAPTENT nets show the particularity, and difficulty, of having: an *adaptable shape* and *differently shaped and sized meshes*.

The first feature makes the traditional covering with membrane sheets difficult for the sheets will ask for patterning, so the initial problem of cutting patterns reoccurs. The second feature on the other hand makes the commonly used cladding with tiles rather difficult because the position of the points of connection -usually the nodes- is irregular and different for each configuration. Some other possibilities have therefore been studied. The most interesting of these methods is the covering of the prestressed nets with PE foil from the packaging industry. It cannot be re-used but the waste material can be recycled completely. The low cost of the foil justifies its temporary character.

Either *wrapping* or *shrinking* is possible. The *wrapping* of a large structure can be a difficult task to accomplish. One should be aware that the wrapping always occurs in the direction of the concave curve on top or underneath the surface. The latest developments in the industry have revealed a new type of PE *shrink foil* that is stronger than the one used for packaging and that comes in large sheets or on large roll. It has already been used to cover scaffolding or to pack large irregular objects. The main advantage of it is that it can be shrunk using simple manual heating devices. This option has not been tested yet but definitely needs further attention.

Application of the "sliding node nets" concept
The problem of covering can be a drawback for the further development of the ADAPTENT system into a fully-fledged generating system for tensile shelters. Tensile structures are however not the only application field of structural nets. A number of different fields in which nets are frequently used can also benefit greatly from the newly introduced adaptability and re-usability. The developed and approved "sliding node nets" concept is maybe even more appropriate in other applications than the original tensile shelters application. Some application fields that could benefit from the new concept are:
  - Gauze covered temporary shading structures,
  - Adaptable safety nets for personal safety or to fix large irregular loads onto transportation vehicles,
  - Re-usable and re-configurable "moulds" for laminated double curved panels or adaptable formwork for double curved shells.

FINAL REMARK
Does this project that aimed at a new generating system for a wide range of "free-form" tensile roofs lead to a concept that has far wider application possibilities in different fields and will it be used in applications outside the original field of tensile architecture? Time and the creative minds of the different readers of this paper will tell.

REFERENCES
1. OTTO F. and RASCH B., Gestalt Finden, Axel Menges, 1995.
2. VANDENBERG M., Cable Nets, Academy Editions, 1998.
3. SCHEK H.-J., The force density method for formfinding and computation of general networks. Computer methods in applied mechanics and engineering. Vol. 3, No. 1, North-Holland Publishing Company. 1974.
4. GRÜNDIG L. and MONCRIEFF E., Form Finding of Textile Structures. Studiedag Textiel-strukturen, 25/05/1993, Vrije Universiteit Brussel.
5. HEBBELINCK S., ADAPTENT, A generating system for temporary, adaptable and reusable nets and tensile structures. Phd. Vrije Universiteit Brussel, 2002.
6. Exhibition Pavilion in Frankfurt, DETAIL Serie 1996, 8, pp. 1240-1244.

Tensegrity : the state of the art

R. MOTRO
School of Architecture Languedoc Roussillon and University of Sciences Montpellier II,
France

INTRODUCTION
Ten years ago, the International Journal of Space Structures published a special issue under the simple title "Tensegrity systems" (Ref.1). Since this time and more precisely since among five years, an increase of interest for Tensegrity can be observed in various places in the world. The aim of this paper is to report on the main advances concerning Tensegrity, and also to evoke open questions.
Apart the "tensegrity fashion", which has no meaningful interest, several real projects closely related to tensegrity, have been now completed.
On theoretical viewpoint several studies have been published by several authors : identification of selfstress states and infinitesimal mechanisms, studies on tensegrity masts, analytical and numerical prestressability characterization, dynamic behaviour, morphological possibilities, constitute the main themes of these studies.

DEFINITION
The "Double X" as reference
If we consider the patent delivered to Kenneth Snelson, we find the fundamental "X-shape" as a basic piece of his design process. If we neglect the fact that the two struts(Fig. 1). are not lying in the same plane because of their thickness, the forces, which were mentioned by Snelson, correspond to a self stress, which implies a compression in the two struts and associated tensions in the peripheral cables. If one of the cables is suppressed the associated effect can be described by two outward pushing forces at nodes "1" and "2" The X shape acts like an *active equivalent compression member* if the distance between nodes 1 and 2 has to be constant. This effect is clearly related to the selfstress of the initial X-shape : all the developments by Snelson after his initial two sculptures are based on this fundamental idea. Mechanically speaking the components which appeared in these two first sculptures, that we described in another conference (Ref.2) are equivalent to these shapes. The only modification on the components of initial sculptures, is to rotate the two upper members relatively to the two lower one constituting so a plane X-shape

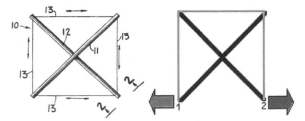

Fig. 1 "X-shape" as *active equivalent compression member.*

Space Structures 5, Thomas Telford, London, 2002

It is then possible to realise the "Double X" by inserting a rhombus of cables between two "X-shapes"; this rhombus is in equilibrium under the active strut effect of the two "X-shapes", characterised by forces "H" : upper one is equivalent to lower black arrows, lower one to upper black arrows. This set of four forces is in equilibrium. Under this only set of forces resulting from self-stress, the rhombus could be plane. It is not because of gravitational effects : the weight of the upper X can not be stabilised by a straight horizontal equivalent cable. The effect of gravitational forces is represented by the vertical forces (Ref.2). Some extra cables are added in order to prevent out of plane instability.

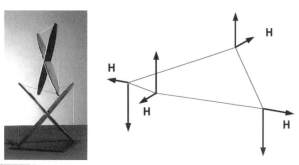

Fig. 2. "Double X" and rhombus equilibrium.

Tensegrity masts

It is necessary to enhance the link between the first works by Snelson ("X-shape" and "Double X") and the tensegrity masts that have been developed during the following monthes by Snelson himself and also by Fuller. If we refer to one of Fuller's books (Ref.3), we find page 165 a figure showing Fuller with "Tensegrity mast made by Kenneth Snelson, 1949" (Fig. 3 **A**), and in the following page "University of Oregon Tensegrity Mast" (Fig. 3 **B**). The later seems to have been realised under Fuller's direction, but both of them contain similar compressed assemblies which are obviously derived from Snelson work (Fig. 3 **C**).

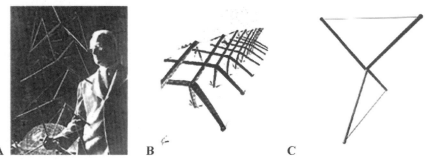

A   B   C

Fig. 3 Tensegrity masts and compressed component.

Even if many people think that tensegrity systems contain only disclosed struts, these examples show that Fuller and Snelson themselves developed tensegrity systems with compressed subassemblies.

Extended definition for a tensegrity "state"

It is well known that the word tensegrity was an invention of R. B. Fuller. This word corresponds to a concept enounced by Fuller "*as islands of compression in a sea of tension*". This is a poetic expression of a structural principle. I submitted (Ref.4) a first definition based on structures described by Emmerich, Fuller and Snelson in their patents: the three give explanations on structures containing struts and cables. It is why I called this definition "historical definition". If previous definitions given also by people like A. Pugh (Ref. 5) were closely related to struts and cables tensegrity systems, this structural principle can be applied for other purposes and even for fields of interest that are not in the scope of architecture and construction. Some recent investigations in biology use this structural principle. See for instance D. Ingber works on endothelial cells (Ref.6).

It is why I suggested the following extended definition for the tensegrity **state**:
"A tensegrity state is a stable self equilibrated state of a system containing a discontinuous set of compressed components inside a continuum of tensioned components"

The expression **"self equilibrated state"** expresses the initial mechanical state of the system, before any loading, even gravitational one. The system has to be in a self-equilibrium state, which could be equivalent to a self stress state, with any self stress level. Furthermore this equilibrium is stable.

The shape of the **"compressed component"** is not prescribed to be linear, surface or volume type: it can be a strut, a cable, a piece of membrane, or an air volume… It can be a combination of one or several of elementary components assembled in a component of higher order. The matter of the component is not prescribed: air, steel, composite… Even if a component is very complex in term of shape, or if it is the result of an assembly of identifiable elementary components, the condition is that its matter has to be compressed at any point.

Expressions **"discontinuous set"** and **"continuum"** are closely related to words "*islands*" and "*sea*" used by Buckminster Fuller. Each compressed component constitutes an "*island*"; when a system is defined, it is necessary to identify each compressed component. If there is more than one, these components have do be disconnected. If we used graph theory their graph would be disconnected. Systems with only one compressed component constitute a specific case and can be also considered in the scope of tensegrity systems. A controversy could be engaged for tensioned components since the corresponding set has to be continuous and consequently the whole tensioned components could be defined as a higher order component. There are several "seas" in our earth, but it seems that using the expression **"continuum of tensioned components"** is adapted to our objective and does not causes any controversy. It is always possible to identify components of lower order inside the continuum.

"**Inside**" is a key word in the definition, since it will allow to separate two kinds of structural design: the first one, which is a part of our constructive culture and based on compression as the sustaining solicitation, and an opposite one based on tension as fundamental "support". In order to know if "*islands*" of compression are, or not, inside an "*sea of tension*" it is necessary to establish a clear definition of the limit, of the frontier between the inside and the outside of the system.

A direct consequence of this definition is that all the points on this envelope belong to the continuum of tensioned components. All the actions lines lying on the boundary surface between the outside and the inside of the system are tension lines. Of course this is the crucial point of tensegrity systems definition, since the self-equilibrium is based on tension and this is not usual in the history of constructions. This fact is certainly at the basis of the frequently evoked "*surprise*" and "*fascination*" for tensegrity systems.

As example we may quote the case of the double layer grid of Fig. 4, there are five compressed components (four of them are identical and contain three struts); in this case some tensioned components of the continuum are inside the envelope. Our recent studies demonstrate that tensegrity systems with tensioned components inside the continuum are stiffer than systems for which the continuum is homeomorphic to a sphere (for simply connex systems).

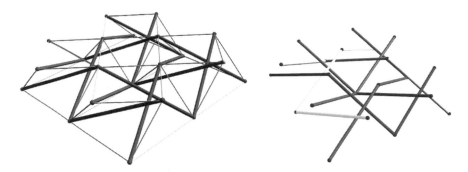

Fig. 4. Double layer grid and its compressed components.

FORMS AND FORCES
The case of initially stressed systems
As initially stressed systems, tensegrity systems satisfy form and force requirements : their geometry must be compatible with at least one self-stress state. Designers need to reach simultaneously a double goal characterised by geometric and mechanical features. Geometry is visible, but nobody saw a state of stress, even if everybody uses it to establish, for instance, static conditions. Form-finding procedures have to give a correct answer to the following problem : find a geometry compatible with a self-stress state.

Designing tensegrity systems
It can be claimed that two main methods are available. They can be respectively qualified as "form controlled" or "force controlled" methods. The first one is illustrated by the work that has been achieved by sculptors and mainly by Kenneth Snelson. The objective is to develop tensegrity systems without any criteria about regularity of components and moreover regardless of generalisation potentiality with mechanical characters. The stability is ensured by a heuristic method based on experimentation and "trial an errors" process. This gave very impressive results. David Georges Emmerich worked also in this design mode. The world of polyhedric geometry was the basis of his researches (Ref. 7).

The second form-finding method was developed in order to ensure the mechanical requirements by a theoretically modelled form-finding process. As it can be anticipated, these models have to be simultaneously aware of geometry and prestressability to be successful. This kind of method can give precious results, but it can fail. It must also be underlined that, if in this last method, resulting shapes are very regular; they have not the richness of the heuristic way. Secondly it appeared that "mechanically" based solutions require very long developments. Consequently a **mixed** process was approached in our research team : a general principle is defined so as to generate a tensegrity system, the results of its application are then tested according to prestressability criterion. We chose this way for designing some new tensegrity grids like the prototype that we describe in another paper of this conference. The basic idea is described in Fig. 5.

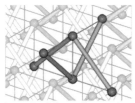

Fig. 5. Extender principle and associated cubic stitch.

ONGOING RESEARCHES
Prestressability

Among present available studies those devoted to selfstress constitute the main part of the corresponding literature. Apart experimental design, theoretical methods for designing systems, which allow the implementation of at least one state of selfstress, can be classified in two main processes, numerical and analytical.

Analytical models

Analytical models can only be developed for small systems, since the difficulty is to choose a few number of parameters for describing the system. Available studies concern mainly masts, and are sometimes related to controlled structures. The constitution of these masts (Fig. 6) results from assemblies of simple polygonal tensegrity prisms with extra cables. The basic idea was of course found in Snelson's work. His "Needle tower" is an impressive sculpture which caused fascination of many people. Its structural composition results from assembly of several layers of simplex such as upper triangle of one simplex be joined with lower one of a second simplex creating an hexagonal out of plane polygon of cables. Three extra cables are then added and allow an "interpenetration" of the two cells.

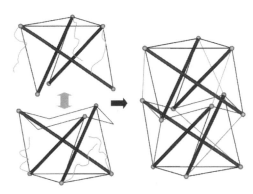

Fig. 6. Needle Tower and how to build a tensegrity mast.

Cornel Sultan solved analytically the general prestressability conditions for several cases (Ref. 8). "The methodology uses symbolic and numeric computation...For certain classes of structures, particular motions are investigated and simpler dynamic equations are derived. These equations are next used for a simple, efficient, tendon control reconfiguration procedure. For certain classes of tensegrity structures linear parametric dynamical models are also developed". A similar problem was addressed recently by Michelleti (Ref. 9) who established the indeterminacy condition for tensegrity towers in order to find their configuration. In the Lagrange colloquium, Williamson et al. submitted equilibrium conditions for tensegrity (Ref. 10). Their paper is very theoretical and would need applications to know if their conditions can be simply applied.

Numerical models

In their "Review of Form-Finding Methods for Tensegrity Structures" (Ref. 11), A.G. Tibert and S. Pellegrino concluded "that the kinematical methods are best suited to obtaining only configurations **details** of structures that are already known, the force density method is best suited to searching new configurations, but affords no control over the lengths of the elements of the structure. The reduced coordinates method offers a greater control of elements lengths, but requires more extensive symbolic manipulations". The work that Vassart did in his thesis(Ref. 12) showed that it is possible to have also a control of some node coordinates when using force density method. Some suraboundant coordinates allow to have a control of geometry besides the selfstress control.

If numerical methods like force density are useful for designing some others are necessary when any structure has to be checked relatively to selfstress implementation possibilities. They are based on the analysis of the kernel of equilibrium matrix. This study is only a first step, which allows to have a selfstress basis. Choice of appropriate combination and of selfstress level are also subject of studies by Kazi-Aoual et al (Ref. 13). Finding selfstress states is not sufficient since their stability has to be checked. Vassart established energetic criteria for this purpose (Ref. 14).

Mechanical behavior

Since tensegrity systems are mainly characterized by selfstress, their mechanical behavior has principally be studied in relation with this associated characteristic. Even if many static analysis models are available, emphasis has to be put on dynamic studies : they allow of course, to know more on vibrational aspects, but simultaneously they can be used for other purpose like selfstress level adjustment (Refs.15-16), and structural control (Ref. 17).

The number of papers related to mechanical behavior is now increasing. It is useful to quote the work achieved by Kono et al (Ref. 18) since it is associated with an interesting experimental work. B.B. Wang published also many papers (see for instance Refs. 19-20). These last authors present a study on cable strut systems. I think that the specificities of tensegrity systems give other properties than the simple one resulting from an assembly of cables and struts.

Deployable tensegrities

According to their nature, tensegrity systems offer new possibilities for designing deployable systems. A. Hanaor opened the way, and M. Bouderbala developed some ideas concerning double layer grids and masts (Refs. 21 to 23). Some problems related to strut-strut contact and dynamic effects of unilateral rigidity of cables are actually addressed by C. Lesaux (Ref. 24). Moreover these properties are used for spatial applications : an example will be presented in the next section. It has been developed by G. Tibert in cooperation with S. Pellegrino (Ref. 25).

Conclusion

It is not possible to give an exhaustive analysis of the numerous publications that been edited in recent years. Interested people may have a look to reference lists included in the publications in the reference list of this paper and they certainly find many other information. I devoted a major part of this section to "prestressability" of tensegrity systems, since it is certainly the key point for studying these structures when dealing with formfinding, with selfstress level adjustment, sizing and mechanical behaviour.

APPLICATIONS

If tensegrity systems have been for a too long time limited to specific applications in the fields of sculpture, of mathematical curiosities, they became a very new structural principle whose applications are found in different other fields. I only quote for memory actual studies by D. Ingber et al in biology since it is the most surprising application of tensegrity. Some examples of application are given in this section and give a more precise idea of the potentialities of these structures.

Experimental realisations

Some 1/1 scale prototypes have been built for experimental validation of structural theories. Two of them have been quoted in this text (Refs. 18 and 26).

Fig. 7. **A** Tensarch double layer grid- **B** Kono's project.

Adaptative systems

If in recent years some theoretical studies have been developed on active control of tensegrity (see Ref.17), a major improvement must be noticed in experimental work realised at Federal Polytechnic School of Lausanne under the direction of Prof. I. Smith.(Ref.27). The total assembly contains five cells designed initially on Pedretti cell model.

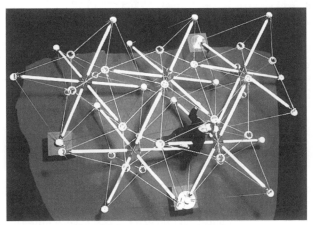

Fig. 8. Experimental tensegrity grid for active control study.

12. VASSART N., Recherche de forme et stabilité des systèmes réticulés spatiaux autocontraints : applications aux systèmes de tenségrité. Thèse de doctorat. Université Montpellier II, France, 1997.

13. KAZI-AOUAL M. N., QUIRANT J. and LAPORTE R., L'autocontrainte dans les systèmes de tenségrité, Colloque Lagrange. « Tenségrité : analyse et projets », Rome, 6-8.05.01. Actes à paraître dans la Revue Française de Génie Civil.

14. VASSART N., LAPORTE R.., MOTRO R., Determination of mechanism's order for kinematically and statically indetermined systems, International Journal of Solids and Structures 37, 28(2000), pp 3807-3839.

15. CROSNIER B., LE ROHELLEC F. Dynamic identification of physical parameters with subspace state models, Proceedings to be published by Soringer Verlag for the Colloquium Lagrangianum, Taormina, Italie, 2000.

16. AVERSENG J., KAZI AOUAL M.N., CROSNIER B., Strategy for selfstress implementation in tensegrity systems, 5° International Conference on Space Structures, Guildford, 2002

17. DJOUADI S., MOTRO R., PONS J.C., CROSNIER B., Active control of tensegrity systems, Journal of Aerospace Division (ASCE), vol. 11, 2, paper n°16880, 1998, pp. 37-44.

18. KONO Y., CHOONG K.K., SHIMADA T., KUNIEDA H., "An experimental investigation of a type of double layer tensegrity grids" , Journal of IASS, Vol. 41, n°131. 2000

19. WANG B. B., YAN YUN Li, Definition of tensegrity systems. Can dispute be settled?, Proceedings of LSA98 "Lightweight structures in architecture engineering and construction",. Edited by Richard Hough & Robert Melchers, ISBN 0 9586065 0 1, Vol.2., 1998, pp.713-719.

20. WANG B. B., YAN YUN Li, From Tensegrity Grids to Cable-strut Grids, International Journal of Space Structures, Vol. 16 N°4, 2001,pp 279-314.

21. BOUDERBALA M., Systèmes spatiaux pliables/dépliables : cas des systèmes de tensegrité, Thèse de doctorat, Université de Montpellier II, 1998.

22. BOUDERBALA M., MOTRO R., Folding Tensegrity Systems, Solid Mechanics and its Applications. IUTAM-IASS Symposium on Deployable Structures (sept 1998): Theory and Applications. S. Pellegrino and S.D. Guest (Eds). Luwer Academic Publishers. ISBN 0-7923-6516-X. 2000 , pp. 27-36.

23. MOTRO R. , BOUDERBALA M., LE SAUX C., CEVAER F., Foldable Tensegrities , Chapter 11 « Deployable structures ». CISM Courses and Lectures N0 412. Edited by S. Pellegrino. 2001. ISBN3-211-83685-3. CISM Springer Wien New York, Pp. 199-237.

24. LE SAUX C., BOUDERBALA M., CEVAER F., MOTRO R., Strut-strut contact in numerical modeling of tensegrity systems folding, 40th Anniversary Congress of IASS : Shell and spatial structures : from recent past to the next millennium. Septembre 1999.D1-D10 Retractable & Deployable Structures.

25. TIBERT G., Deployable Tensegrity Structures For Space Applications, Doctoral Thesis, Stockholm, Royal Institute of Technology, Department of Mechanics, 2002.

26. MOTRO R. (report by), Tensarch project, 5° International Conference on Space Structures, Guildford, 2002.

27. DOMER B., FEST E., "Active control of Tensegrity", Proceedings of Colloquium Lagrangianum Strutture tensegrity: analisi e progetti, Rome, 2001.to be published in "Revue Française de Génie Civil"

28. PEDRETTI M., Smart Tensegrity Structures For The Swiss Expo 2001, Proceedings of LSA98 "Lightweight structures in architecture engineering and construction",. Edited by Richard Hough & Robert Melchers, ISBN 0 9586065 0 1, Vol.2, 1998, pp.684-691

Symmetry of quasi-spherical polyhedra and simplexes in tensegrity systems

Z. BIENIEK
Faculty of Civil and Environmental Engineering, Rzeszów University of Technology, Poland

INTRODUCTION

Throughout recorded history man experimented with designs and materials towards the goal of doing more structures with less. Polyhedra have been a part of the fabric of mathematics for two thousand years and have been the inspiration for contributions to many branches of the subject. Structural forms based on polyhedra have subsequently become popular in architecture. In practice, the term 'space structure' is used to refer to a number of families of structures that include grids, barrel vaults, domes, towers, cable nets, membrane systems, foldable assemblies and tensegrity forms, Refs 10-20. Space structures cover an enormous range of shapes. A sphere, all regular and semi-regular polyhedra, the prisms and the rhombic polyhedra appear in this context to be members of a large family of forms that have great potentials for application in building. These solids have a form that is so perfect, that they exert a great attraction to both artists, scientists and engineers.

A sphere represents the least amount of material surface area possible to enclose a given volume of space. When bisected, the half sphere becomes one of the most efficient shapes known to enclose a given floor area. A 'dome' is a structural system that consists of one or more layers of elements that are arched in all directions. The surface of a dome may be a part of a single surface such as a sphere or a paraboloid, or it may consist of a patchwork of different surfaces, Refs. 2, 6, 8.

The element of common definition of Platonic and Archimedean polyhedra is: "All vertices of a polyhedron lie on one circumscribed sphere". The volume of this sphere is therefore larger than that of the corresponding polyhedron. Such polyhedron can be considered as the first degree of accuracy for approximation of sphere. This paper pays attention to a family of forms related to these so called 'quasi-surface polyhedra'. The author proposed new alternative method for approximation of spherical surface, Refs 1-8. From the geometric point of view, the vertices of polyhedron does not require lie on one circumscribe sphere. This leads to new, economic and spectacular kinds of subdivision with relatively small numbers of different member lengths.

CUBIC SYMMETRY TYPES AND ASSUMPTIONS OF METHOD

The quasi-spherical polyhedra considered in this work have the same axes of rotational symmetry as a cube. There are three axes of 4-fold rotational symmetry which join the centres of opposite faces; four axes of 3-fold rotational symmetry which join diagonally opposite vertices; and six 2-fold rotation axes joining the midpoints of opposite edges. The system of reflection symmetries is equally rich (see Figure 1). There are three mirror planes, each of which contains two of the 4-fold axes, and which together form a system of mutually

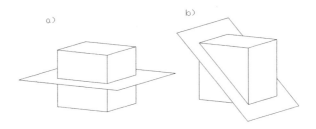

Fig. 1.

The axes of rotational symmetry of highest degree in this system are **4**-fold. The most convenient approach in this regard is to establish a standard co-ordinate system and to use it for all configuration that are based on a cube. The x, y and z-axes of Cartesian co-ordinate system are simultaneously three rotation axes, indicated as **4**, for the unit cubic lattice. The unit vector indicated as $\bar{v}$, see the left part of Figure 2. All nodes of lattice can be group in the cutting planes μ_{zi} (like in μ_{xi} or μ_{yi}). Each one of these cutting planes, z = i; i = 0, ±1, ±2, ..., ±n, is perpendicular to the z-axis and contains the set of points of cubic lattice. The sets of such nodes give the square plane lattices. If the planes are perpendicular to the axes indicated as **3**, in this case the corresponding nodes formed the triangular plane lattices. The cutting planes perpendicular to the axes indicated as **2** contained the points of rectangular plane lattices. In this context a quasi-spherical polyhedra can be generated by three variants, Ref. 7.

The standard system shown in Figure 2 used as the basis for formulation of the transformations necessary for the creation of models representing polyhedric configuration. The origin **O** of Cartesian co-ordinate system [x,y,z] is concentric with the sphere on which a polyhedron is based. Radius of sphere R can be more than or equal to zero. Intersections of hemisphere $\hat{\omega}$ and cutting planes μ_{zi} give the parallels of latitude $\hat{o}_{zi}$. The left part of Figure 3 shows obtained configuration. Each of these lines can be the circle or the point of contact $\hat{\omega}$ and μ_{zi} (the pole of hemisphere $\hat{\omega}$).

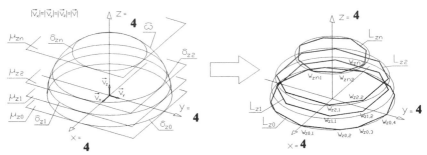

Fig. 2.

Each one of the parallels of latitude $\hat{o}_{zi}$ can be altered into the broken line L_{zi}; where i = 0, ±1, ±2, ..., ±n. A few of broken lines L_{zi} are shown in the right part of Figure 2. The circle $\hat{o}_{xi}$ can be altered into the broken line L_{xi}, etc. All vertices W_{zij} of the broken lines L_{zi} are a members of the unit cubic lattice; where $j = 1, 2,$. In order to determine broken lines L_{zi} it is necessary to have a means of finding the vertices W_{zij}. The shape of circle $\hat{o}_{zi}$ can be approximated in many ways. The author proposed three criteria for approximation of the parallels of latitude $\hat{o}_{zi}$, Refs 1-4. This is demonstrated in Figure 3. Each one of these broken lines L_{zi} has four reflection planes marked as α, β, γ and δ.

(i)   The first criterion for approximation is shown in Figure 3a. Each point of broken line L_{zi}^{I} lie outside or is a member of circle $\hat{o}_{zi}$.

I – index of the first criterion for approximation,

(ii)   The second criterion is shown in Figure 3b. Each point of broken line L_{zi}^{II} lie inside or is a member of circle $\hat{o}_{zi}$.

II – index of the second criterion for approximation.

(iii)   In the configuration of Figure 3c, the points of broken line L_{zi}^{III} can be lie both inside and outside the parallel of latitude $\hat{o}_{zi}$. It is the third criterion for approximation.

III – index of the third criterion for approximation.

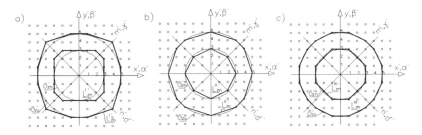

Fig. 3.

Are three general formulas for evaluation of the faces of quasi-spherical polyhedron:
1. The set of all vertices of quasi-spherical polyhedron is contained in the set of points W_{zij}
   (like is contained in the set of points W_{xij} or W_{yij}).
2. The quasi-spherical polyhedron is convex.
3. The volume of quasi-spherical polyhedron must be maximum.

For example, if the surface $\hat{\omega}$ is chosen to be sphere with radius R = 2, then the results will be configurations that of Figs 4, 5 and 6. Figure 4 illustrates a stepwise process of the production of quasi-spherical polyhedron, based on the first criterion for approximation. Figures 4a and 4b shows the set of broken lines L_{zi}^{I} ; where i = 0, ± 1, ± 2

All faces of polyhedron are formed by the coplanar sets of points marked as W_{zij}^{I}. These polygons are placed in specific arrangement and position. This can be done as demonstrated in Fig. 4c. Figure 5 shows a stepwise process based on the second criterion for approximation. Figure 6 illustrates a such process based on the third criterion for approximation.

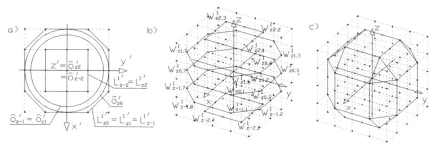

Fig. 4.

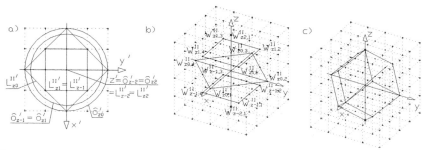

Fig. 5.

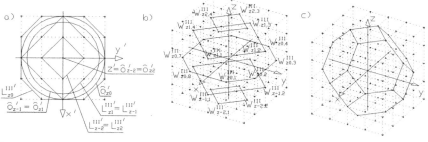

Fig. 6.

SYMMETRY OF QUASI-SPHERICAL POLYHEDRA

All the quasi-spherical polyhedra are highly symmetrical and the nature of this symmetry is the same in every case, Refs 5, 7. Examples of the geometrical characteristics of such structures are shown in Figures 7a-c. If the surface $\hat{\omega}$ is chosen to be hemisphere with radius R = 5 then result will be the following configurations based on three criteria for approximation.

The symmetries of a polyhedron are rules which describe the relationship between its various parts, Ref 9. A quasi-spherical polyhedron is characterised by the fact that its shape can be constructed with forty eight congruent figures (but of course require the aid of a mirror). The fact that a symmetry can be applied to a quasi-spherical polyhedra provides an alternative method of describing the appearance. In addition to the same modularity of the quasi-spherical polyhedra can be easily demonstrated of Figures 4c, 5c and 6c. This remarkable symmetry is the same in all quasi-spherical polyhedra even though they have such a diversity of forms.

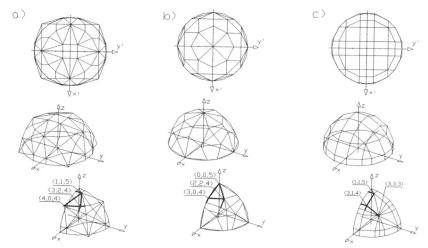

Fig. 7.

QUASI-SPHERICAL DOMES

Single layer domes and domical surface can be formed on the basis of quasi-spherical polyhedra, Refs 2, 3, 6, 8. The surface may consist of a 'patchwork' of different quasi-spherical polyhedra that are connected together along their edges, see Figure 8.

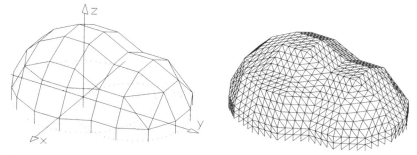

Fig. 8.

A double-layer quasi-spherical dome consists of two (nominally) parallel layers. The creation of double-layer polyhedral configuration can be carried out by:
- designing the two different polyhedra onto two concentric spheres, as shown in Figure 9a
- designing the two similar polyhedra onto two concentric spheres, see Figure 9b
- designing the two identical polyhedra and extending one of these, see Figure 9c.

Fig. 9.

SPACE TRIANGULATION AND CLOSE PACKING

The triangle is the only polygon that is stable (rigid) by virtue of its geometry. A triangular arrangement of structural members reduces non-axial forces on the truss to a set of axial forces in the member. Triangulation is the technique for organizing a structure or system in triangular form. Triangulation imparts strength to structures even before the physical characteristics of their material components are taken into account. As a design, it provides an inherent advantage: guarantee of structural rigidity, based on the natural stability of the triangle. It imparts this advantage regardless of scale.

Tetrahedron(triangular pyramid) is a solid figure bounded by four triangular faces. A regular tetrahedron has four congruent equilateral triangles as its faces. The regular tetrahedron, minimally structured, is the strongest of solids, being most able to resist external forces from all directions. It has the greatest surface area for volume of all Platonic polyhedra. The regular tetrahedron is described as having **2, 3, 3**-fold symmetry, Ref. 9.

The rhombic dodecahedron has twelve identical rhombic faces (see Figures 5c and 10d). It can be visualized as the solid that results when the edges of a cube are sufficiently beveled at 45 degrees, as shown in Figure 10.

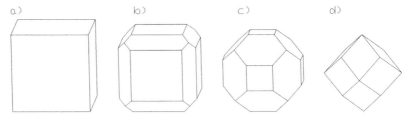

Fig. 10.

It is one of very few symmetrical solids that pack to fill space, two others being the cube and truncated octahedron (see Figures 6c and 10c). Like the cube, it has three 4-fold axes of symmetry, four 3-fold axes, and six 2-fold axes. When viewed along any of its 4-fold axes it appears square in profile, however along any of its 3-fold axes it appears hexagonal (same as the cube), as shown in Figure 11.

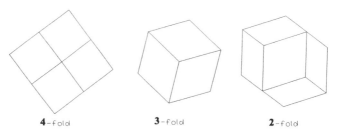

Fig. 11.

Is it possible to subdivide a space into congruent and disjoint tetrahedra ? The task is not possible if we specify the regular tetrahedra. The three-dimensional space can divide into congruent and disjoint a cubes. Following we can easily divide the cube into six congruent and disjoint pyramids, each having one face of the cube as a base, and the center of the cube as a vertex. They are not tetrahedra, though, since they have five faces (the base, plus four slanting sides) instead of four. However, can bisect each of these pyramids along a plane which contains its vertex and a diagonal of its base. That subdivides the cube into twelve congruent and disjoint tetrahedra, as shown in Figure 12a.

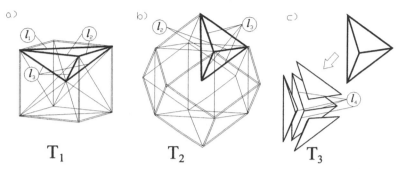

Fig. 12.

They are not regular solids. Each of these tetrahedra has two edges l_2 which are the edges of cube simultaneously.

l_2 = given edge length of cube, ($l_2 = |\bar{v}| = 1$).

Each of these tetrahedra has three edges l_3 which length is equal $\frac{\sqrt{3}}{2}$ and one edge $l_1 = \sqrt{2}$ respectively. Each a single tetrahedron has only one the plane of symmetry. The plane of symmetry is perpendicular to the edge indicated as l_1.

The three-dimensional space can divide into congruent and disjoint a rhombic dodecahedra. Following we can easily divide the rhombic dodecahedra into twelve congruent and disjoint pyramids, each having one face of the rhombic dodecahedra as a base, and the center of the rhombic dodecahedra as a vertex. They are not tetrahedra, though, since they have five faces(the base, plus four slanting sides) instead of four. However, can bisect each of these pyramids along a plane which contains its vertex and short diagonal of its base. That subdivides the rhombic dodecahedra into twenty four congruent and disjoint tetrahedra, as shown in Figure 12b.
They are not regular solids. Each of these tetrahedra has two perpendicular opposite edges l_2.
l_2 = given edge length of cube = given short diagonal length of rhombic faces, ($l_2 = |\bar{v}| = 1$).

Each of these tetrahedra has four edges l_3 which length is equal $\frac{\sqrt{3}}{2}$. Each a single tetrahedron has only one 2-fold axis of symmetry through the centers of opposite edges l_2. The tetrahedron has four congruent isosceles triangles as its faces.

Following we can divide such a tetrahedron into four congruent and disjoint smaller tetrahedra, each having one face of the great tetrahedron as a base, and the center of the great tetrahedron as a vertex. As shown in Figure 12c, all new edges l_4 of small tetrahedra are equal length ($l_4 = \frac{\sqrt{5}}{4}$).

Polyhedra lend themselves to be put together in tight packings. That makes them suitable as the basic configuration for space frames.

TENSEGRITY FRAMEWORK

A 'tensegrity framework' is an ordered finite collection of points in Euclidean space, called a configuration, with certain pairs of these points, called 'cables' constrained not to get further apart; and certain pairs of these points, called 'bars' or 'struts' constrained not to get closer together.

The word 'tensegrity' was coined by R.B. Fuller. It is a contraction of 'tensional integrity'. The term 'tensegrity' is not well defined. Fuller's definition implies a network consisting of tension members (cables) and compression members(bars and struts), in which the cable network is continuous (hence 'tensional integrity'), and the bar-strut system presumably is not. Bars are straight members of fixed length and can sustain either compression or tension. Struts are straight members with a lower bound on length. They cannot contract but they can extend indefinitely in a 'telescoping' fashion and therefore cannot support tension. Cables are straight members with an upper bound on length. They cannot extend but can contract freely and therefore cannot sustain compression. This definition covers the whole range of pin-jointed structures, including trusses and cable networks.

The simplest three-dimensional tensegrity objects are sometimes termed 'simplexes'. Figure 13 shows some relatively simple configurations, but any configuration can be constructed as a tensegrity from a range tetrahedra T_1, T_2 and T_3 (Figures 12 a, b and c, respectively).

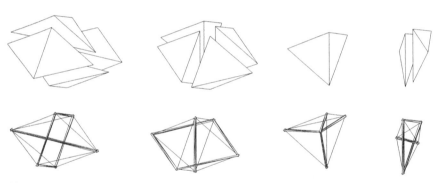

Fig. 13.

CONCLUSIONS

Structural forms based on polyhedra are popular with engineers and architects. However, the difficulty of data generation for these polyhedric configurations has been a barrier to advancement in this area. This work can helped to evolve procedures based on the concepts of 'formex algebra' to facilitate the generation of polyhedric forms and to broaden the boundaries of knowledge in this field. As a consequence of this work, polyhedric forms of many kinds can be generated with easy for practical applications. In viev on their peculiar features, both positive and negative, it is expected that such structures will find useful, replacing more familiar structural systems.

REFERENCES

1. BIENIEK Z, Koncepcja jednowarstwowych kratownic quasi-powierzchniowych generowanych na sferze, ZN PRz, Seria: Budownictwo i Inżynieria Środowiska, Rzeszów 1999, pp. 55-72.

2. BIENIEK Z, Some comments on geometry of one-layer and quasi-surface trusses made on sphere or rotational ellipsoid, VI Seminarium „Geometria i komputer", Wisła 1999 pp. 12-13.

3. BIENIEK Z, Kilka uwag o geometrycznej strukturze kratownic quasi-sferycznych, Konferencja o geometrii, Wydawnictwo PCz, Częstochowa 1999, pp. 62-70.

4. BIENIEK Z, Geometria jednowarstwowych kratownic quasi-powierzchniowych generowanych na elipsoidzie obrotowej, ZN PŚl, Seria: Geometria i Grafika Inżynierska, Gliwice 2000, pp. 45-54.

5. BIENIEK Z, Parametry geometryczne wybranej grupy wielościanów quasi-sferycznych, IVKonferencja Naukowo-Techniczna „Aktualne problemy naukowo badawcze budownictwa", Olsztyn-Łańsk 2000, pp. 49-57.

6. BIENIEK Z, Examples of the quasi-surface polyhedrons application in forming of the engineery object, III Seminarium „Geometria i grafika inżynierska w nauczaniu współczesnego inżyniera", Wisła2000, pp. 13-15.

7. BIENIEK Z, The variants of method quasi-surface polyhedrons generation, Proceedings of 7th Seminar „Geometry and computer", Wisła 2001, pp. 10-12.

8. BIENIEK Z, Structural qualities of quasi-spherical polyhedra, Lightweight Structures in Civil Engineering, Proc. of the Local Seminar of IASS Polish Chapter, Micro-Publisher Jan B. Obrębski Wydawnictwo Naukowe, Warsaw-Wrocław 2001, pp. 9-12.

9. CROMWELL P.R, Polyhedra, Cambridge University Press, 1997

10. EMMERICH D.G, Self-Tensioning Spherical Structures: Single and Double Layer Spheroids, International Journal of Space Structures (Special Issue on Geodesic Forms), T. Tarnai, ed., Vol. 5, No. 3/4, 1990, pp. 353-374.

11. FULINSKI J, Geometria kratownic powierzchniowych, Prace Naukowe Wrocławskiego Towarzystwa Naukowego, Seria B, Nr 178, Wrocław 1973.

12. HANAOR A, Tensegrity: Theory and Application, J.F. Gabriel, editor: Beyond the cube. The architecture of space frames and polyhedra, John Wiley & Sons, Inc., New York 1997.

13. HUYBERS P, Polyhedra for building structures: why and how?, International Conference on Lightweight Structures in Civil Engineering, Warsaw 1995.

14. HUYBERS P, Polyhedroids, 4th International Symposium on "Structural Morphology – Bridge between Civil Engineering and Architecture", Delft, The Netherlands,17-19 August, 2000, pp.100-107

15. NOOSHIN H, DISNEY P.L, and CHAMPION O.C, Computer-Aided Processing of Polyhedric Configurations, J.F. Gabriel, editor: Beyond the cube. The architecture of space frames and polyhedra, John Wiley & Sons, Inc., New York 1997.

16. RĘBIELAK J, Struktury przestrzenne o dużych rozpiętościach, Wydawnictwo Politechniki Wrocławskiej, Wrocław, 1992.

17. RĘBIELAK J, Proposals of shaping of surface tension-strut girders by means of simple form of tension-strut trusses, Lightweight Structures in Civil Engineering, Proc. of the Local Seminar of IASS Polish Chapter, Micro-Publisher Jan B. Obrębski Wydawnictwo Naukowe, Warsaw-Wrocław 2001, pp. 59-60.

18. RĘBIELAK J, Proposition of shaping of linear tension-strut girders, Lightweight Structures in Civil Engineering, Proc. of the Local Seminar of IASS Polish Chapter, Micro-Publisher Jan B. Obrębski Wydawnictwo Naukowe, Warsaw-Wrocław 2001, pp. 74-77.

19. TARNAI T, editor: Spherical grid structures. Geometric essays on geodesic domes, Hungarian Institute for Building Science, Budapest 1987.

20. WANG B, Simplexes in tensegrity systems, Journal of International Association for Shell and Spatial Structures: IASS, Vol. 40, n. 129, 1999

Deflections of Wood Domes: Deficiencies in American Codes

BEZALEEL S. BENJAMIN, School of Architecture, University of Kansas, USA

INTRODUCTION
Stiff-jointed wood domes, with ribs and rings and without diagonals, depend for their stability on the rigid connections between the ribs and rings at the joints of the dome (Ref 1). These joints are made by the use of a variety of steel connectors that require the use of steel bolts, lag bolts, split rings or shear connectors (Ref 2). When load is applied to the dome, these steel connectors dig into the wood, resulting in an internal, localized crushing of the wood fibers. The National Design Specification (NDS) for Wood Construction (Ref 3) recognizes this and even defines the zones, for each mode of failure, in which such crushing of the wood fibers can be expected to occur. In two places, the NDS while discussing the various modes of failure, states: "Mode II represents pivoting of the fastener at the shear plane of a single shear connection with limited localized crushing of the wood fibers near the face(s) of the wood members, as may occur in bolted connections with oversize bolt holes" and again in the same paragraph, "Mode IV represents fastener yield in bending at two plastic hinge points per shear plane, with limited localized crushing of wood fibers near the shear plane(s)." In all wood structures, this results in small additional deformations at the joints that cannot be predicted by the theory, simply because the theoretical calculations do not take this localized crushing of the fibers into account.

In a wood dome, with very many joints, these deformations become cumulative. Since the base of the dome cannot move, all deformations at every one of the joints translates upwards towards the apex of the dome, becoming additive to the very small elastic deflections of the dome as predicted by the theoretical analysis of the dome structure.

EXPERIMENTAL WORK
In order to understand the nature of the problem, experimental work was carried out on two model domes, 1524 mm (60 inch) in diameter, made of basswood, with members 10 mm (3/8 inch) square. The geometry of the dome is shown in Figure 1. An experimental model dome is shown in Figure 2. The domes were loaded with two loading conditions; a point load at the apex of the dome and loads at the top ring level. Both models were identical in form and materials, except that two different joint configurations were used. In Model 1, sheet aluminum connectors were stapled into the members, using a single staple at the centroidal axis, as shown in Figure 3. In Model 2, however, steel pins were used to connect the sheet aluminum at both upper and lower edges, with a lever arm distance between them, for more positive transfer of moments between members. This type of connection is shown in Figure 4. Deflections under the two loading conditions, at several target points, were determined using a Kern E-1 theodolite.

THEORETICAL DEFLECTIONS OF THE DOMES AS PREDICTED BY NDS

For bending deflections of wood members the NDS states: "If deflection due to bending is a factor in design, it shall be calculated by standard methods of engineering mechanics using

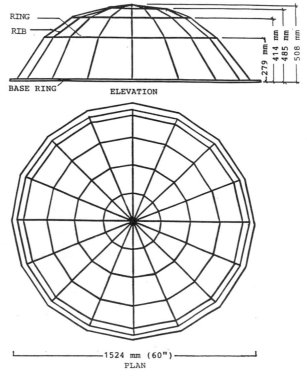

Figure 1. The geometry of the model dome

Figure 2. Experimental model dome, 1524 mm (60 inch) in diameter

Figure 3. Aluminum connector with steel staples

Figure 4. Aluminum connector with steel pins

the allowable modulus of elasticity, E', from this Specification" (Ref 3). The value of E, the modulus of elasticity of the wood species is modified by several factors such as 'wet service factor', 'temperature factor', 'incising factor' and 'buckling stiffness factor'. There is no mention, whatsoever, of a 'connection factor', that could take into account the localized crushing of the wood fibers that result in deformations that are far in excess of those calculated by "standard methods of engineering mechanics".

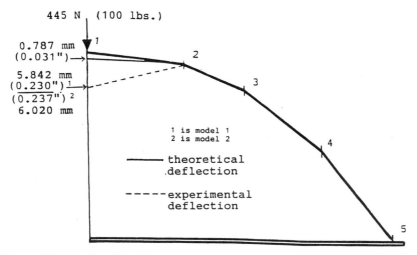

Figure 5. Deflection of the dome: Load at the apex

The theoretical analysis of the model domes was carried out using IMAGES-3D, a computer program out of Berkeley, California, that was capable of handling 1000 members, 1000 joints and 1000 plates, and including a combination of linear members and plates. The analysis was checked for accuracy by the author's students, using other computer software programs, with precisely the same results. The theoretical load for the computer analysis was assumed to be 445 N (100 lbs) at the crown and a total of 712 N (160 lbs) at the top ring level.

COMPARISON OF THEORETICAL AND EXPERIMENTAL DEFLECTIONS
A comparison of the theoretical and experimental deflections, is shown in Figures 5 and 6. With the load at the crown, as shown in Figure 5, there is very little difference between the actual crown deflections of the two models, with the different joint connections. However, both of these actual deflections are more than 7 times the theoretical deflection as predicted by "standard methods of engineering mechanics". The deflections at other points are too small to be graphically represented.

With the load at the top ring level, there is a difference in the behavior of the two models, as shown in Figure 6. But even with steel pins at both upper and lower edges, the actual deflections are still more than 3 ½ times the theoretically predicted deflection. Once again, the deflections at other points are very small and cannot be graphically represented.

CONCLUSIONS
The experimental work clearly shows that the localized crushing of wood fibers in joints of wood domes using steel connectors, cause deflections of the dome that are far in excess of

the theoretically predicted elastic deflections of the dome. This crushing is not easily visible because it is hidden behind the steel plates in the body of the joint. Though the actual total deflections are still small, because of the doubly-curved shape of the dome structure, care must be taken in those situations in which deflection-sensitive, nonstructural finish materials are applied to the dome. While the tests were carried out only under vertical static loading, the application of cyclic wind loading with uplift conditions could cause localized wood fiber crushing on both sides of the metal connector, making the problem even more acute.

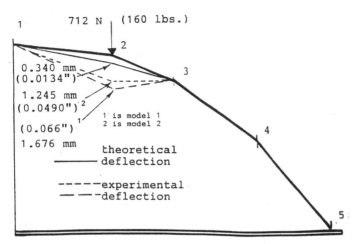

Figure 6. Deflection of the dome: Load at the first ring level

In order to take this localized crushing of the wood fibers into account, the author suggests that the NDS (or other codes) allow for the increased deformations resulting from such crushing by the provision of a "connection factor" of a value less than one. This connection factor would reduce the allowable E-value for the wood in question. The connection factor itself will depend on a number of variables such as the density of the wood, diameter and numbers of steel bolts, and the nature of the connection whether in tension, compression, bending or shear. While such a connection factor at each joint could modify the E-value of the wood, for purposes of theoretical calculation, at an individual connection, there would still need to be some overall modification that would be necessary depending on the type of structure itself. For instance, in the case of the model domes considered in this paper, there are a total of 65 joints. For load at the crown of the dome, a connection factor of 0.9 applied at each joint would cause an increase in deformation of 1/0.9 or 11% at each joint. When multiplied by 65 joints, the theoretical deflection of the dome at the apex would be increased by 7.15. However, this would apply only to domes, where the deformations at the individual joints are cumulative leading to the largest deflection at the crown of the dome.

ACKNOWLEDGEMENTS
I would like to acknowledge my thanks to my students, who did specific work, over a number of semesters, on this project. These include Julia Mathias, Curtis Geise, Travis Green, David Meier, Brian Karpen, Bret Dante, and Jenny Burke.

REFERENCES
1.   Benjamin B S, Structures for Architects, 2nd Ed., Van Nostrand Reinhold, New York, 1984, pp 352.

2.   Wood Reference Handbook, Canadian Wood Council, Ottawa, Canada, 1991, pp 264.

3.   National Design Specification for Wood Construction, American Forest and Paper Assoc. and American Wood Council, Washington D.C., 1997, pp164, pp 20.

Design, engineering, production and realisation of spatial glass structures for 'Digital Baroque' architecture

M. EEKHOUT
Faculty of Architecture, Delft University of Technology, the Netherlands

INTRODUCTION

The second half of the twentieth century saw the development of a number of spatial and systemized lightweight structures. Shell structures, space frames, tensile structures, cable net structures, pneumatic structures, folded plate structures and 'tensegrity' structures. Most of these structures were developed by dedicated pioneers: Felix Candela, Frei Otto, Max Mengeringhausen, Richard Buckminster Fuller, Zygmunt Makowski, Walter Bird, Peter Rice et al [Ref 1]. They designed, analyzed and built impressive amounts of new concepts. The shared basic idea was to minimize the amount of consumed material. To achieve this, extensive intellectual investments in man-hours for design and engineering were necessary. Computer analysis programs assisted the accurate analysis of the complex component geometries of these three-dimensional though, in our current view, highly regular 3D structures. Thanks to the further development of accurate analysis programs, based on non-linear structural behaviour, these 3D structures can now be designed by structural engineers all over the world. They reached a status of acknowledged and mature technology. In the eighties, Peter Rice introduced the intricate use of structural glass in buildings, also based on regularity and systemization.

Meanwhile, the ratio between material and labour has changed dramatically. No longer, the amount of material for cost effective structures counts, but rather the total of invested manpower. The post-war adagio of 'minimal material' became an intellectual objective to architects and structural engineers only. It is not an item for the building industry as a whole anymore. This had a great influence on the choice of building technologies in recent decades. Of course, it differs in every country. While wealth increased, buildings were realized in a more elaborate fashion than the minimalist results of the Modern Movement.

Nevertheless, the development of architecture will be more capricious. The Modern Movement with her worldwide replacement of local building technology by concrete technology was caught up by other approaches of following generations of architects, who wanted to express themselves with innovating built results as compared to previous generations. The subsequent changes in architectural sub-styles from the Sixties onward, following Modernism, i.e. Structuralism and Postmodernism may be regarded as variations on the Modern Movement. Style is not just a way of building, but a total cultural embedment in all the Arts and in the entire society. It is too early to predict if the current computer generation, introduced to fast moving images, derived from flashing videos, television zapping and computer games, will lead to an inherent different behaviour of a new type of architects, currently growing up with all these new technologies and hence with a new style.

To build means to freeze the dynamics of this culture in immobile materials. Actually, the growing complexity in society is increasingly expressed in buildings. Archigram started this expression with its flexibility in built forms, generated by mobility in function [Ref 2]. Deconstructivism followed with a different philosophical approach to life and it went on with Free Form or Digital Baroque architecture. Time will tell if it will develop as a style or as a variation on the 'Mother of all Contemporary Architecture': Modern Architecture. The way of designing Free Form buildings is much more dynamic and it does not follow the rules of the last generation anymore. The growing complexity of the building process and the building itself as a result, show a loss of interest for regular 3D-structures amongst architects, putting an end to a pioneering era. Several causes may be responsible for this:

- The higher building budgets in the last decades (compared to the post-war era);
- The conversion from production to consumption dominated building industry;
- The generation of digitized architects seeking their own identity;
- The development of computer programs for designing and engineering.

Initially, computers were used for the calculation of complexities, and for the support of standardization in design and production as a components collection. Later, a certain degree of systematization with pre-designs for systems and post-designs for applications was introduced. The last step was a high degree of special and spatial designs without the repetition of pre-designed components. This led to special designs from the composing components to the entire artefact of the building.

Architects grew weary of regular systemized structures and building components, which were not designed and developed by them, and showed too clear a mark of the developing designers. They now try to develop their own building technical design concepts, specific elements, components and details that fit in the building design at large. By lack of design experience in this field, these technical designs are usually governed or overdone with merely aesthetical considerations, enlarging the gap between architects and structural engineers.

Meanwhile, as a result of the fading popularity of regular and systemized 3D structures as ready-made products and marketable structural systems, producers of 3D structures left their standard and system assortments of structural products behind and became highly specialized contractors for spatial structures, using much structural glass in the last decade. Of course, they kept their expertise, insight and vision, but obtained a more flexible engineering and production machine potential than ever before. They changed their discipline and fixed organizational routings for project-based collaborative designing, engineering and adaptable machine productions that were available in a wide network of sub-contractors, each of them specialized in a different niche.

The newest trend in the designs of Free Form or Digital Baroque buildings is non-linear, non-repetitive and in their conceptual stage only derived from clay-modelled sculptures, either by making concepts in actual clay or by modelling and generating them in a similar way on the computer. Nowadays, computer designing and rendering programs, i.e. 3D design 'Maya', are capable of juggling and of generating all sorts of geometrical forms, including the ones without any regularity in their geometric patterns. In the conceptual design phase, architects usually do not aim at geometrical repetitive forms and systemized structural schemes, or their behaviour. Like artists, they rather design a very new building with a mega-surprise for the whole world. This is the Digital Baroque building design environment of this decade.

Initially, structural engineers feel quite paralyzed when they have to develop a load bearing structure in the contours of these geometrical forms in order to materialize the structural concept of the envelope of the building. The same applies to building technical engineers, who elaborate these designs with more precision up to the level of shop drawings. The question is how this 'computer supported sculpturalism' can be reconciliated with sound structural design and industrial prefabrication principles. This must be done in such a way that a proper balance is found for revitalizing the excellent and extensive experiences of the twentieth century 3D lightweight structures. Preferably, this should already take place in the conceptual stage, so that both existing know-how and experience would be activated and the budgeting of these buildings becomes less surprising. The relation between principle and application is at stake here. Principles were conquered and secured by pioneers and later by scientists, while architects do as they like in both surprising and pleasing the society at the same time with their created applications, acting as composers but sometimes with the elitism of prima ballerinas.

RELATION BETWEEN ARCHITECTURE & BUILDING TECHNOLOGY

Gaudi developed a suitable building technology to realize his buildings. He extended his influence from architecture to building technology and to the material technology of his Free Form load bearing stone and brick structures. The Gaudi technology still puzzles scientists today, who try to analyze the unwritten regularities to complete the cathedral of the Sagrada Familia. Mark Burry, working on the engineering of the completion of this masterpiece from his office in Australia [ref 4], may serve as an example.

Intermediate fields of technology, i.e. Building Applications, could only contain a different composition of materials, elements and components, without introducing a new approach in these fields, to form the designed building. Some Digital Baroque designs can be built with contemporary materials in a contemporary production and building method. Others need adaptations from current production technologies. There are several possible combinations of conventional and new materials, components, structural schemes, productions, geometries and integrations. Some of the clearest combinations are:

- Conventional materials with conventional components, productions for conventional structural schemes and geometries with conventional integrations;
- Conventional materials with conventional productions for new geometries;
- Conventional materials with new geometries and integrations;
- New materials with new components and productions for new geometries and integrations;
- Conventional materials with new components and productions acting with new structural schemes, caused by new geometries and integrations.

In building technical respect, Digital Baroque designs are primarily material compositions with a new and unconventional geometry, of which architects hope that the spatial composition will be the one and only derivation in the building cycle. However, a complicated geometry requires complicated geometrical surveying, both in the design and engineering phase, and in the phase of the individual productions of building parts and their composition and integration on the building site. Therefore, this is mainly a computer supported logistic affair, both for the main architect and for the main contractor. Furthermore, the interest in structural complexity has changed into geometrical complexity, which is essential for conferences as this fifth International Conference of Space Structures.

The very nature of many of the curved forms of building elements causes them to be produced in an alternative way. It could be done by casting of free material into a complex element form. It could also be done by the transformation of economical commercial plate materials into 2,5D or 3D forms. The 2,5 D element form can be developed from a flat plane, but for the formation of a 3D panel, more rigorous formation techniques with regard to temperature and pressure are necessary, i.e. explosion transformation of aluminium panels and hot mould transformation of glass panels. The geometrical definition and fixation of these 3D elements will greatly complicate the engineering of these elements, but so will the production and fitting of the panel collection that belongs to one building part. Furthermore, there is the joining of the different building parts, engineered and produced by different parties in the building industry: the 'Building Seam'. Therefore, the decomposition of a geometrically complex building requires an optimal description in the form of a 3D computer model. In this respect, it is of great importance to remember the hierarchy of building products, as published in *POPO* [Ref 4], so that materials and commercial materials will not be confused with elements and components. After thirty years of development, Free Form architecture is back on the scene, but not like the early shell structures of Candela cum suis. Free Forms do not seem to have limitations in geometry.

DIGITAL BAROQUE AFTER O'GEHRY
Out of the blue, in 1997 the Guggenheim museum in Bilbao opened, a perfectionist American design, blended with the Spanish way of building. The opening of this museum, designed by Frank O'Gehry, amazed the world. It really boosted the 'Digital Baroque' or 'blob' era. O'Gehry designs his buildings in clay, like sculptures. He measures his most satisfying models electronically and feeds them into a geometrical computer program. To this purpose, O'Gehry's office in Santa Monica uses the French Dassault-based program Catia, which was developed for aeronautical engineering. When done, the building geometry is fixed and the building is engineered and tendered as a total package. The subscribing chief contractors must find sub-contractors who are willing to engineer, produce and build the building parts exactly as O'Gehry designed them. Sub-contractors have to buy, and work with, the Catia program in order to detail the rough geometry as put down in the main design. From this 3D Catia model, the construction and composition of all elements and components of each different building part, taken in hand by each sub-contractor, is derived and fixed, particularly when these elements and components have to be prefabricated.

Other young architects may be more flexible with the overall geometry. (See Asymptote, Floriade pavilion). The project architect must fix these non-rectangular geometries. The structural analysis has to exactly follow the whims of the architectural form. For the detailing elements and components, the usual tolerances from engineering, production of elements and their positioning, will have to be compensated. There is no room for deviating geometries in one building part, since this part will never fit in the other building parts. This type of building designs dictates a very close cooperation between the participating engineering, producing and building parties. Largely due to the obligatory prefabricated character of this architecture, much more than the so-called high tech architecture of the eighties, the term should be 'High precision engineering and production'.

HIGHER DEGREE OF COLLABORATION
A new type of Free Form geometry, involving all building parts of the building design, automatically leads to a very accurate cooperation between the building team partners, much more frequent and intense than ever before. For most of the architects concerned, it takes a number of projects to agree with this concept, before they change their usual distance from the production and phases of building, and start working towards an integrated approach of all building team parties concerned. The building team must be defined as the sum of all

participating designers, advisors, chief contractor, building managers, sub-contractors and specialized producers involved in the project.

Four major stages with different characteristics can be defined:

- The design of the building and its components;
- The engineering of the building parts;
- The production of elements and components;
- The building on site and the installation of prefabricated components.

Each of the four stages has its own characteristics, design considerations and the assuring quality of the building as the final product, which will be a composition of different building parts, produced and installed on the building site by different building team partners. The phase of the building design and its components will be the broad domain of the architect and his/her advisors. Digital Baroque designs tend to systemize standard products and have building systems become special. The final product, the building as product, is extremely special. The demand for special components will increase, because the special geometry of the building influences the form and position of each composing element and component.

The design phase has to result in a 3D CAD parent model of the building, drawn up by the architect. S/he must integrate the principle connections, basic component sizes, and their connections in their model, since this model will be the start of the engineering for the various building team members. As an assembly of centre lines, the model will not be sufficient. The architect has to incorporate in his/her design all relevant data of all different components of the different building parts. An individual building team member must work out each building part. The information this model holds, must be as broad as possible. The 3D parent model will not be used for tendering purposes, since it is not understandable to other professionals than CAD engineers, i.e. quantity surveyors. For the time being, drawings will remain the carriers of information in the tendering process.

Engineering of the building parts

Various building team parties are involved in engineering their own production. All these engineering activities must be based on the 3D parent CAD model. This certified model is the basis for the engineering of the entire building. This model will provide two directions to work from: each part separately or in collaboration. Only the latter mode has a good chance to survive. Each party works on the 3D CAD parent model and is allowed 'slot time' (as in air traffic coordination). During the start, the situation is set. Details and modifications of elements and components can be fed in. The whole process has to be worked through. The final situation will be fixed and communicated to all building parties. After the proper closing off the slot time of one party, the next party is allowed their slot time. Simultaneous work by more than one contractor is not allowed, as it will be confusing and may cause possible legal problems. As the result of different projects with different programs, the staff of the engineering department in one production company could be working with different programs. This will result in mistakes and confusion. Therefore, an appeal is made for a universal 3D computer program with rough descriptions, structural analyses and component descriptions. All involved building team members should use such a program.

When all participants contributed their building parts-directed engineering, a geometrical check on a regular basis is necessary. Neglecting this will result in great problems for production and installation and hence, too much effort will be wasted. Liability is also at stake here. Three building parties are capable to check this: the architect, the contractor and the geodetic surveyor. Of course, each option has its advantages and disadvantages.

Production of elements and components
Both the data from the overall 3D parent model or the individual overall CAD model or drawings, must result in the making of individual element drawings, in the form of shop or production drawings. They are made to feed the production. This will be done directly in CAD/CAM for the cutting, drilling, punching and machine operations. It may also be done indirectly by manual machine activities, i.e. welding and bending operations, the casting of steel joints and the assembly of elements into components, hot dip galvanisation, painting or coating afterwards and protection for the transport to the building site.

Building on site and installation of prefabricated components
The engineering part of site activities is the installation of assembly and building drawings that indicate the identification of the transported components and their location by XYZ co-ordinates of the prominent click points. These click points will have to be established, based on the characteristic geometric points of the 3D CAD parent model. The fixation of these click points during the progress on site, is a service the main contractor provides. Because of the complexity of the geometry and the absence of straight and orthogonal lines, which can easily be fixed by craftsmen (water level, plummet and their mechanical refinements), Digital Baroque architecture definitely needs this new service of a geodetic surveyor to realize these complex buildings.

After completion of the work of a sub-contractor, the surveyor will compare the respective exterior click points with the theoretical ones and their tolerances. This is done to prepare and inform the next sub-contractor, who will have to start building from these data. This sub-contractor can only compensate certain tolerances, since his production is already completed when he will start on the building site. In the near future, the discipline of prefabrication and industrialization, and the installation of subsequent trades on the site, will have to switch to a discipline of industrialized complex building geometries. The building process of Digital Baroque designs is often approached with the same attitude as the traditional building process is, whereby most of the building parts are produced on the building site and the mistakes of the earlier trades are compensated by the later ones. Dealing with Digital Baroque in this traditional way leads to longer legal struggles than the building time itself, to disappointments and even bankruptcy of the weaker parties. Tendering documents should contain the followed procedures and relations, but unfortunately seldom do.

CASE 1: D. G. BANK, BERLIN, GERMANY, OCTATUBE DESIGN ALTERNATIVE
The design of Frank O'Gehry called for a triangular network in the shape of the body of a whale, constructed of stainless steel solid square rods, in triangulated form and covered with double and triple glazing panels. The joints in the shape of fingers, all have six fingers in different vertical and horizontal directions. An extremely difficult job, since all joints were different, all bars had different lengths and connection angles, and all panels were different in size.

The alternative design of Octatube consisted of hollow spherical cast joints and tubular CHS members, a reminder of the space frame era. The joints could only be drilled in the exact direction. The length of the tubes would form the desired spatial envelope. The drawing illustrates the geometry. However, the tendering process resulted in a contract for Gartner in the original design. The building was completed in the year 2000. The accuracy, high degree of workmanship and the finishing of materials required incredibly high precision, much like Swiss clockwork.

Fig. 1. Internal view of skylight [ref glas 6/99].

CASE 2: TOWN HALL, ALPHEN, THE NETHERLANDS

This design of the Dutch architect Erick van Egeraat and ABT engineers is the first major example of Digital Baroque architecture in the Netherlands. The author was involved, as principal of Octatube, in the engineering, production and installation of the frameless glazing façades. This building has a façade of frameless glass panels, fully screened with graphical motives of trees, leaves and flowers in a rather ad hoc fashion. The panels are supported by elliptical façade mullions, 75 x 150 and 110 x 220 up to 20 metres height, spaced at approximately 1.8 metres, with glass support joints in-between. The high yield, slender hot-rolled elliptical mullions are visually excellent in freestanding use of frameless glazing. Their use in Quattro façades, either vertically or horizontally and suspended from the roof, is standard. The glass panels, approximately 850 pieces, are all unique in shape and print design. The glass panels have been screened on surface 2 and have a low E-coating on side 3. Most of the panels are 10.12.10 double glazed units in fully tempered clear glass panels. The additional insulated roof panels have laminated lower panels 6.6. All panels are fully tempered.

In the highly irregular geometry, parts of the façade are conical upward and downward, cylindrical, spherical, anti-clastical and only some parts are straight. Because of the geometrical differences between the lining of the façade mullions and glass panels, the columns are positioned in varying angles to the glass panels. The glass connectors are irregular: more than 70 different types were engineered. Not one of the 90 mullions is equal to the other. In the anti-clastical surface (roughly 10 x 10 metres), the rectangular glass panels are twisted and made into two triangular panels. The elliptical mullions have up to nine bends in their longitudinal axis, which are cut and welded on jigs in the factory, directly from the engineering drawings. They all fitted.

Because of the random shapes of the intersecting bays, approximately 500 glass panels are installed at the 3D curved back of the building, all of them in model form (i.e. non-rectangular). The design called for a twisted row of glass panels. In the first six months of the development and engineering phase, a timber window factory tried to develop glass windows and suitable details to that purpose. After they gave up, Octatube came with a simple, but fitting solution. The idea was to get rid of the window frames, and use only double glass panels, composed of two panes of fully tempered glass, laminated in panels under angles less than 80°. The individual glass panels were to be slightly warped. The maximum size of 900mm. width and 1800mm. length had to warp for 40 mm. perpendicular to its surface. This was done by cold transformation. Tests in the Octatube laboratory showed that this was feasible. Static analysis of the tensions by bending showed that only 10 to 20% of the maximum tensions were used in bending. The stresses in the sealant were acceptably low (20% of the working stresses) and the sealant manufacturer gave his usual guarantee. It was an innovation that cold twisted insulated glass was used as a solution much simpler than the original glazed timber window frame solution.

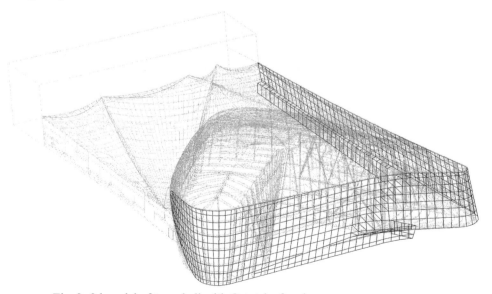

Fig. 2. 3d model of town hall with Octatube facades.

Yet, this type of Digital Baroque architecture required the utmost of the engineering department: thrice the time consumption of a regular project, including many problems with the matching of other building parts. An intensive cooperation was required, involving all building parties. The town hall matured in June 2002.

CASE 3: FLORIADE PAVILION, HAARLEMMERMEER, THE NETHERLANDS
The winning design of the Asymptote Architects competition originally contained a building volume in an arbitrary form of a Blob, with two sloped surfaces of glass. In a later planning stage, this glass roof was partly replaced by aluminium panels. Over both roof surfaces water is running down continuously, as a sign of the watery culture of the Netherlands. This project has three technical innovations.

The most remarkable is the suspended 3D glass water basin of approximately 5 x 12 metres, containing approximately 1200 mm of water in the top of the building. The development target was the realization of the glass panels in 2D, 2,5D and 3D glass panels. These panels were sized 1 x 1.4 metres and point supported. The target of a smooth basin was reached by an experimental route of an initially dual thermal transformation into a 3D shape, subsequent (certified) chemical treatment, lamination of the chemically treated prestressed panels by liquid epoxy, testing and comparison with the theoretically calculated final results and installation. However, due to break risks, the replacement time was estimated at two months: too long for the six months duration of the Floriade. For this reason, the transition to flat 12.12 panels within a polygonal assembly was made.

The second innovation consisted of 2 x 2m² laminated, fully tempered and laminated glass panels, cold bent on site, based upon four corners, with two points on each curved side pushing outward. The bending stresses consumed 60% of the working stresses.

Fig. 3. 3d presentation views of Floriade Pavilion.

The third innovation consisted of 3D curved aluminium panels on top of the building volume. They were prepared from CAD machined polystyrene blocks, smoothened with epoxy resin and glass fibre weave and cast with reinforced fibre concrete. After curing, they formed the positive mould on top of which a 5 mm. thick aluminium panel was vacuum placed, with 300 mm. of water on top and a TNT explosion pressure to press the panel in the mould shape.

After cutting to exact size on a wooden jig, the edges were welded and the 2 x 7 panels were smoothened and coated by spraying them. The 10 mm. wide sealant seams made the system watertight. The Japanese approach to aim for the duplicate project 'half the time, half the effort and half the money' will succeed this initial experimental procedure.

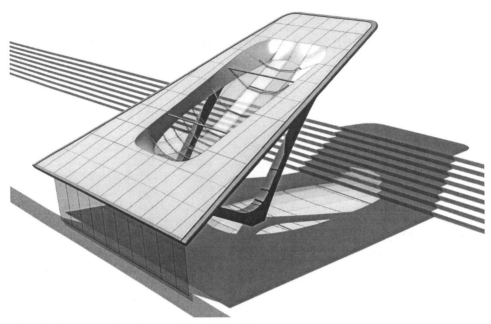

Fig. 4. 3d presentation views of Floriade Pavilion.

CONCLUSION
Architects mainly generate the new Digital Baroque designs for buildings with their arbitrary and non-rectilinear form, from sculptural considerations. All lessons from past decades, where systemized spatial structures and economical building industrialization and their regularities were developed, the do's and don'ts were learned, seem forgotten. The load bearing structures of these buildings, disguised as artefacts, must be developed later than the initial design. The structural engineer rarely gets a position in the front row of the design process. The structural and cladding components of these buildings require an enormous effort in collaborative design and engineering. It should be recommended that, at least in the design phase, the concept of the building technical composition would be developed simultaneously with the architectural concept. Both the design and engineering phase and the production and realization phase, need a definite extremely high degree of collaboration between all the involved capable building parties, to reach the goal of a successful fluent design building, which will be a success to all parties.

REFERENCES
1. EEKHOUT M, Architecture in Space Structures, 010 Publishers, Rotterdam, 1989.
2. COOK P, Archigram, Revised Edition, Princeton Architectural Press, New York, 1999.
3. GOMEZ, COLL, MELERO, BURRY, La Sagrada Familia De Gaudi al Cad, 1996.
4. EEKHOUT M, POPO, Procesorganisatie voor productontwikkeling, Delft University Press, 1997.

Buckling behaviour of compression elements in bamboo space structure

K. GHAVAMI
Pontificia Universidade Catolica do Rio de Janeiro, Brazil
L. E. MOREIRA
Universidade Federal de Minas Gerais, Belo Horizonte, Brazil

INTRODUCTION

During the last two decades bamboo structures have passed from an intuitive structural application to a controlled one (see Figs. 1 and 2). An attempt has been made to determine to which extent it becomes viable to control the structural behaviour of bamboo in its natural state, conforming to the type of technical requirement and environmental conditions [1,2]. It is a fact that the intuitive application of bamboo in construction, during millenniums, was and still is extremely useful and without great risks for the users within the limits already established. This attributes to the mechanical control, the task to make it possible to establish a safety index for bamboo constructions, making the structure more economical and making some type of requirements feasible, which so far had been avoided, exactly for involving a greater risk and demanding greater knowledge of the mechanical behaviour.

From the structural mechanics point of view, bamboo, mainly in order to counteract wind load and the own weight, acquired several natural geometries, which turns it a most optimum structure to the requirement of deflection-compression. Among others the following properties of the bamboo culm are mentioned:
-   a conical form along the culm,
-   an approximately circular transversal section in most bamboo species,
-   a hollow form in most species, which reduces its weight,
-   non-uniformly distributed diaphragms along the culm
-   functionally graded composite material in radial direction of the bamboo's cross-section

To draw on this raw material, trying to integrate it into the contemporary universe and establishing its form of utilization is as well returning it to the sovereignty of nature and Earth and creating a sustainable technology. Thus, with the objective to be able to utilize bamboo, as it is produced by nature, in different civil engineering constructions, such as space structures, geodesics, bridges and cable stayed structures in general, we present part of our studies in this paper [4,5]. This article is related to the evaluations of natural deviations along the longitudinal axis and their relevance to the ultimate load limit, when the bamboo culms are subjected to axial compression. The obtained data then is used to determine the buckling behaviour of the compression elements used in the fabrication of the bamboo space structure as shown in Fig.3.

a-Bamboo bridge in Germany   *b-Bamboo bridge in Latin America*

Fig. 1.   Modern bamboo bridges.

a-↑ *ZERI pavilion in Hanover Expo2000*

b-Bamboo rural construction in Latin America →

Fig. 2.   Modern bamboo structures.

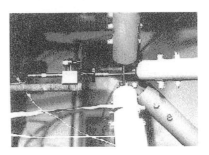

a- **Space bamboo structure during testing**   **b**– *Detail of the joint*

Fig. 3.   Double layer bamboo space structure during test.

MATERIALS AND METHODS

The critical buckling load F_E, for a long and straight column, deduced by Euler, $F_E = \dfrac{\pi^2 EI}{l^2}$, corresponds to a theoretical load with the assumption that the material is homogeneous, isotropic and obeys the Hooke law with a constant transversal section and perfect hinges at the extremities. In the experiments the Euler load only will be effective in a perfectly straight slender column along the longitudinal axis. However, real columns are not perfectly straight. Therefore, to establish the critical buckling load and failure mode of a bamboo column, its density gradient of the transversal section, in radial direction and from the inside outwards, as shown in Fig. 4a, in addition to the initial imperfections of the axis δ_0 should be found.

The study of the behaviour of bamboo under compression leaves us then with some doubts:
- How to measure δ_0 with precision?
- The geometrical approximation of the transversal section as a circular ring with a constant section throughout the element leads mathematically to the necessary results? Where can one assume the transversal section to be constant?
- Does the density gradient of the transversal section, in radial direction and from the inside outwards, seen in Fig. 4a, affect the results?
- How does the failure of the elements occur?

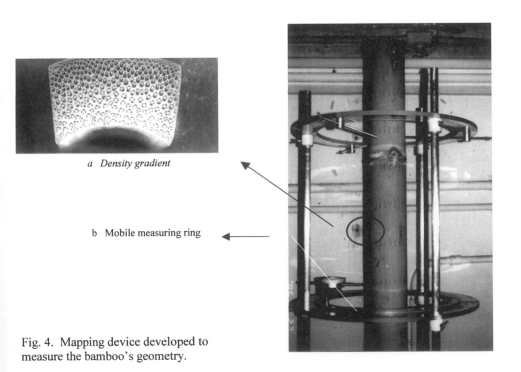

a *Density gradient*

b Mobile measuring ring

Fig. 4. Mapping device developed to measure the bamboo's geometry.

Density Gradient

The fibres of the *schlerenquima* the part of bamboo's cross-section with high strength, visible as black points in Fig. 4a, increase their concentration when approaching the external surface. This causes the transversal section to have a density gradient in the radial direction, from the inside to outwards of the thickness, which interferes with the moment of inertia I of the section. From microstructural studies [3,4], the gradient function one considers for the radial density, shown in Fig. 4a, is assumed to have a constant density of 75% of the thickness of the internal wall, ρ_i, and a higher density close to 25% of the external thickness, ρ_e. In this way it is assumed that the physical inertia, I_f, for a transversal section larger than the geometrical inertia, I_g, is expressed by equation 1.

$$I_f = I_{gi} + I_{ge}k_1 \tag{1}$$

where $k_1 = \dfrac{\rho_e}{\rho_i}$, I_{gi} is the geometrical inertia of the internal part ($\approx$ 75 % of wall thickness t) and I_{ge} the geometrical inertia of the external part.

Mapping of bamboo

To measure with precision the circumference of the bamboo culm a special mapping devise consisting of three aluminum bars, fixed to a wooden circular base plate and an acrylic ring at the upper end was designed. A second ring is fixed to the upper extremity of the bamboo and a third ring is the mobile measuring ring as shown in Fig. 4b.

The mobile ring, consisting of an external ring, is fixed to aluminum bars and an internal ring, which contains a dial gauge with the precision of 1/1000mm. This internal ring can turn 360^0 in steps of 15^0. Knowing the distance from the end of the pointer of the dial gauge to the center of the ring, it is possible to reproduce the whole circumference of the bamboo at a determined height.

The mapping of the element's axis was obtained by projecting the centroid of each circumference on a plane normal to an imaginary axis which links the circumferential centroids of the two bamboo ends, as shown in Fig. 5. Each single mapped bamboo presented a description of the axis different from the others.

BUCKLING

To eliminate the initial eccentricity of the applied load application and to create perfect hinge ends a specially designed hinged device as shown in Fig. 6b was constructed. To the effective length l_0 of the bamboo culm 5 *cm* of the threaded bar end must be added. The lateral displacements δ at the mid length of the element and the mapping of the culm centroid in the $\theta^{'}$ direction was carried out by means of LVDT. The longitudinal and vertical surface strains were measured using L type electrical strain gauges.

Anéis			
0	51.33	5133	
100	51.97	5197	
200	52.56	5256	
300	52.4	5240	
400	52.48	5248	
450	53.5	5248	
500	52.48	5219	
600	52.19	5239	
700	52.39	5227	
800	52.27	5223	

Fig. 5.   Projection of the centroids.

a) General View of Test Set-up

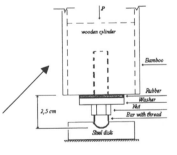

b) Hinge Boundary Conditions of the Column

Fig. 6.   Experimental buckling set – up.

TEST RESULTS
Geometrical description of the mapping
11 bamboo culms of 2m length and 10cm diameter of the specie *Dendrocalamus giganteus* were mapped and then their buckling behaviour subjected to axial load was studied. In Fig. 5, the extreme circumferences are coincidental, numbers $1 \equiv 19$ and represent the straight imaginary axis of bamboo. It was possible to detect the section most distant from the axis, obtaining therefore the maximum imperfection δ_0 and the orientation θ· of this maximum imperfection, assuming the preferential direction of the arching of the bamboo. For this reason the LVDT is fixed in this direction.

Figures 7 to 11 show, respectively, the type of variation along the element, the average radius R, the average thickness of the wall t, the average area A, as well as the geometrical inertias I_g and physical I_f, of the transversal sections, calculated for each circumferential ring along the axis of the column at a distance of 10 *cm* from each other. It can be observed that there are peaks in all curves close to the regions of the diaphragms. Further it can be seen that the direction of the radius increase coincides with the direction of the increase of the wall thickness, which is as well the direction top-base of the bamboo.

Density gradient
Bamboo columns of 2 meter length each, were divided in cylindrical segments of 15 *cm* height and in each of these segments $k_1 = \rho_e / \rho_i$ was determined and from $k_2 = t_e / t_i$ an average value for all the bamboo was calculated afterwards.

For bamboo with an average density equal to $566 \pm 33 \, \frac{kg}{m^3}$, considered to be low, one finds the ratio $k_1 = 1{,}95 \pm 0{,}18$ for $k_2 = 0{,}28 \pm 004$; now for bamboo with an average density of $883 \pm 57 \, \frac{kg}{m^3}$, one finds $k_1 = 1{,}35 \pm 0{,}09$ for $k_2 = 0{,}27 \pm 0{,}05$.

Consequently, for bamboo with a smaller density, the density gradient causes an average of an increase of inertia for the transversal section, described by ratio I_f / I_g close to 30 %. For bamboo with a higher density this ratio falls to approximately 10%.

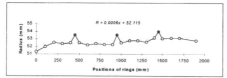

Fig. 7. Average radii along the longitudinal axis.

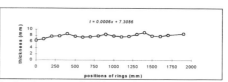

Fig. 8. Average wall thickness along longitudinal axis.

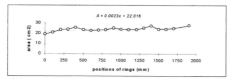

Fig. 9. Average area along longitudinal axis.

Fig. 10. Geometrical inertia along the longitudinal axis.

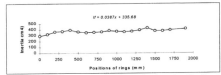

Figure 11 Physical inertia along the longitudinal axis

Results of buckling tests

Table 1 shows the final geometrical characteristics for the studied elements, assuming as the mean value of the characteristics of the extreme transversal sections, which is approximately equal to the arithmetic average of the values obtained from the 19 transversal sections. For each element the elasticity modulus E was determined. The Euler load F_E was obtained using the Southwell diagram and $P_{\lim}$ was obtained from the test.

Table 1: Geometry of specimens and test results.

Test sample	R_e (mm)	t (mm)	I_g (cm⁴)	I_f (cm⁴)	F_E (kN)	$P_{\lim}$ (kN)
5	52,3	7,3	264	303	147,1	46,6
7	45,5	6,2	148	191	60,3	30,7
8	46,0	5,0	128	163	55,0	28,0
9	37,0	4,0	55	76	23,4	16,8
10	42,0	5,1	100	129	30,5	18,2
11	43,5	4,6	99	127	40,8	24,2
12	38,5	6,1	86	109	39,4	16,8
16	49,0	7,2	208	266	78,7	32,6
17	46,0	6,5	161	208	52,6	19,6
18	42,5	6,7	129	167	55,6	19,6
15	52,5	7,9	286	371	95,3	55,9

The instrumentation and the scheme of the buckling test are presented in Fig. 6. Fig. 12 shows the curves P x δ_t for the tested bamboo after the total mapping, where δ_t is the total lateral displacement or the sum of the maximum imperfection δ_0 with the deflections measured at the moment of the experiment, δ. By plotting the values $\delta/P \times P$, where P is the applied load, one obtains the Southwell diagram for the bamboo (shown in Fig. 13), a classic procedure to study the buckling behaviour, independent of the material. The typical failure type of the element occurs by progressing squashing of the fibres in the concave region subjected to larger compression stress. The longitudinal extensometer situated in the most compressed zone, measures the maximum longitudinal strain.

ANALYSIS AND DISCUSSION OF RESULTS

Theoretically, for a non-prismatic column, the form of the elastic line $y(x)$ must be given by a superposition of sinusoidal half waves, Eq. (2):

$$y = \alpha \left(\frac{\delta_0}{1-\alpha} \sin \frac{\pi x}{l} + \frac{\delta_1}{2^2 - \alpha} \sin \frac{2\pi x}{l} + ... \right) \tag{2}$$

	CP05	CP07	CP08	CP09	CP10	CP11
3.3				0		
3.4						
4				4.5		
4.5						
5				8		
5.5						
6				10.7		
6.5						
7				12.7		
7.53	0			13.5		
8	8.2			14.05		
8.2						0
8.5	16.4			14.6		
9	24.1			15		3.5
9.3	28.5		0	15.2		
9.5	31.36		0.3	15.4		
9.8	35.6		1.6	15.6		
10			2.5	15.7		6.9

Fig. 12. Curves $P \times \delta_t$

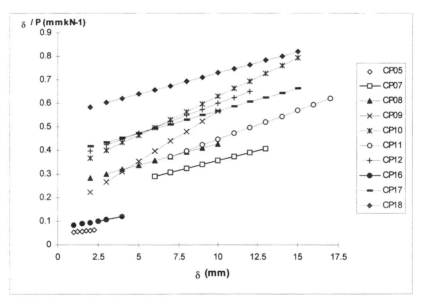

Fig. 13. Southwell diagram.

As $\alpha = P / F_E$ is always smaller than 1 and approaches the unit while P approaches the buckling load of Euler F_E, the first term in this series is usually predominant over the others and for this it can be considered that

$$y(x) = \alpha \left(\frac{\delta_0}{1-\alpha} \right) \text{sen} \left(\frac{\pi x}{l_0} \right)$$

(3)

For $x = l_0/2$ there is a lateral displacement given by

$$\delta = \frac{\delta_0}{\dfrac{F_E}{P} - 1}$$

(4)

By adding δ_0 to the above expression one gets

$$\delta_t = \frac{\delta_0}{1 - \dfrac{P}{F_E}}$$

(5)

Rearranging equation 3 one can write

$$\frac{F_E}{P} \delta - \delta = \delta_0$$

(6)

Therefore, when plotting $\dfrac{\delta}{P} \times \delta$ one obtains a straight line with an inclination of $\tan \theta = 1/F_E$ which crosses the axis of the abscissa in $\delta = \delta_0'$. So, the Southwell diagram, Fig. 13 provides the axial buckling load of Euler obtained from the test, as well as the initial imperfection of the element's axis. The comparison between the imperfections obtained through the mapping δ_0, and the results δ_0' of the Southwell diagram can be noted in Table 2.

The difference between the values δ_0 and δ_0' can be explained by the influence of some parameters of difficult control, such as:
-   The presence of small eccentricities in the load application.
-   The Southwell Diagram being developed for a prismatic bar and the dimensions are considered at $l/2$. Bamboo has an variable inertia and the δ_0 measured in the mapping not always occurs in the center of the bamboo.
It is important to point out that the obtained results of the Southwell Diagram represent a global response of the system. By obtaining the buckling load of Euler $F_E = \dfrac{1}{tg\theta}$, the bending rigidity $(EI)_s$ of the system then can be expressed through equation 7.

$$(EI)_s = \frac{F_E l_0^2}{\pi^2}$$

(7)

As the value of E_{exp} was determined experimentally for each element, the mean value of the moment of inertia of bamboo can be obtained by $I_s = \dfrac{(EI)_s}{E_{exp}}$. The comparison of these results with those moments of inertia results obtained by the mapping and density measurements is given also in Table 2.

Table 2: Comparison of mapping results and Southwell diagram.

Cp	5	7	8	9	10	11	12	15	16	17	18
$\delta o(mm)$	7,6	9,3	16,5	7,0	11,1	8,2	14,3	7,7	3,7	19,5	30,0
$\delta'o(mm)$	7,0	11,6	13,5	3,3	9,2	8,3	13,6	7,0	5,5	19,9	30,5
$If(cm4)$	303	191	163	76	129	127	109	354	266	208	167
$Is(cm4)$	304	170	168	65	136	120	124	373	269	196	153

It can be seen that the density gradient really affects the results, since I_s is closer to I_f than to I_g. The measuring of the maximum strain through electrical strain gauges has shown that the failure of the elements occurs mainly by progressive squashing of the fibres in the concave region of the element followed by the local buckling of the bamboo wall. Therefore, one can assume that the maximum compressive stress responsible for the local failure of long bamboo subjected to buckling would be the failure stress σ_R in short compression tests.

The stress-strain curve $(\sigma \times \varepsilon)$ for bamboo segments of diameter D and of height *2D*, subjected to uniform compression, is presented in Fig. 14.

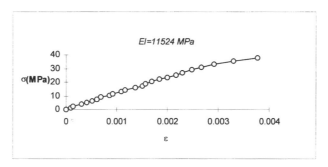

Fig. 14. Stress × strain curve for bamboo in compression.

Nevertheless, a more profound study of the measured strain in real time, during the buckling test, in the highly compressed zone in longitudinal direction should be realized. Then the experimental results should be compared with the ultimate stress limit σ_p of equation 8, corresponding to the ultimate limit load P_{lim}. It is not clear yet if P_{lim} also corresponds to an instability of the composite material itself. At the moment of the local buckling, the displacement of the bamboo wall was observed as well as the squashing of the fibres. It is not clear which occurred first. Thus, both ultimate limit states must be examined for the analysis of bamboo under axial compression. The ultimate limit state of strength guarantees the integrity of the column in compression, expressed by equation 8, in which the influence of the axial imperfection is expressed by $\delta_0 / (1 - P/F_E)$.

$$\frac{P}{A} + \frac{P\delta_0 D}{2I_f\left(1 - \dfrac{P}{F_E}\right)} < \sigma_p \qquad (8)$$

Under the calculated ultimate limit load, P_{lim}, bamboo would have a theoretical lateral deflection given by equation 9.

$$\delta_t = \frac{\delta_0}{1 - \frac{P_{\lim}}{F_E}} \tag{9}$$

The strain limits and the maximum displacements of a structural element known as limit state of serviceability, are mainly associated with the useful life of the materials, aesthetic factors and the comfort of the user. In the case of bamboo, where δ_0 is always present, various test specimens reached maximum loading with the bamboo quite arched, reaching tensile stress in the convex side of the element, mainly in those element with relatively high δ_0.

For the project design purpose the gain of inertia I_f can be considered negligible and to work with $I = I_g$, $I = \pi \overline{R}^3 t$ in favor of safety and solving the problem limiting the maximum lateral deflection. As $\delta_t = n\delta_0$, equation 9 can be rearranged in equation 10

$$P_{\lim}^{\cdot} = \frac{\pi^3 E \overline{R}^3 t (n-1)}{l_0^2 n} \tag{10}$$

where $P_{\lim}^{\cdot}$ represents both the ultimate limit states of strength and service. When an ultimate limit state, due to buckling, is reached, the structural element fails; but if the service limit state is reached there is no failure risk. Therefore, the ultimate buckling limit state must always be great than the service limit state. Therefore, when $P_{\lim}$ represents the ultimate limit load for the serviceability limit, one can write $P_{\lim} = \gamma P_{\lim}^{\cdot}$, where $\gamma > 1$ is estimated for each loading. Consequently, for $P_{\lim}^{\cdot}$ from equation 10, $P_{\lim} \times A$ can be evaluated through equation 11, where $A = 2\pi \overline{R} t$. It is important to register that the proposed sequence of the evaluation of these equations is a function of high initial imperfection of the bamboo axis. However, nothing prevents that the ultimate limit state of strength is evaluated first and then the ultimate service limit state.

$$\frac{P_{\lim}}{2\pi \overline{R} t} + \frac{P_{\lim} n \delta_o R_e}{\pi \overline{R}^3 t} < \sigma_p \tag{11}$$

According to the mapping results, the natural form of columns can be selected with a tolerance of $\delta_0 = \frac{l_0}{150}$, where l_0 is the buckling length.   This value is twice the initial imperfection for wooden strips. As well one can create data for each bamboo specie, also for 3 distinct regions of the bamboo culm i.e. base, center and top parts, in a selection process with industrial characteristics, the parameter $\lambda_D = D_i / D_e = R_i / R_e$, which represents the relation of internal diameter or radius to external diameter or radius. If $\overline{R} = \frac{R_i + R_e}{2}$, the average radius of the center of the element, one can rewrite the equation 10 as:

$$P_{\lim}^{\cdot} = \frac{\pi^3 E R_e^4 (1 - \lambda_D^2)(1 + \lambda_D)^2}{8 l_0^2} \left(\frac{n-1}{n} \right) \tag{12}$$

In the same way the equation 11 can be rewritten as:

$$\frac{P_{\lim} \left[R_e (1 + \lambda_D)^2 + 8n\delta_o \right]}{\pi R_e^3 (1 - \lambda_D^2)(1 + \lambda_D)^2} < \sigma_p \tag{13}$$

CONCLUSION

Through the mapping of the entire bamboo, it was possible establish its buckling behaviour due to the number of relevant variables to the phenomenon. The adopted procedure allowed to determine the buckling plane described by θ, the imperfections characterized by δ_0 as well as the chaotic distribution of the centroids.

Bamboo tubes in perfect state have behaved as a Euler column with a moment of inertia I_f increased by the density gradient of the fibres in the radial direction. The consideration of the bamboo as a prismatic tube with a constant moment of inertia equal to the mean of the moment of inertia of the ends of the element   provide good results for these segments whose maximum mean slenderness is 70.

The ultimate limit load of the tested elements $P_{\lim}$ were defined by the local instability of the bamboo wall close to the center of the element, in the concave side of the bamboo, when the measured strain and estimated stress were both close to ε_p and σ_p, respectively, and not close to ε_r or σ_r given by the stress strain curve of the compression test. In the exact moment of the failure, measured strains have increased indefinitely, resulting in the squashing of the fibres and the displacement of bamboo wall between two adjacent nodes. It was clear that the compression stress that corresponds to this phenomenon is the limit of proportionality σ_p in uniform compression but it was not clear if this level of stress   may also correspond to an instability of the bamboo material itself.

The proposed equations were fitted to the test results and thus are correct to evaluate the buckling load of whole bamboo culms. It is also important to remember that these equations are working with nominal values of the loads and resistance of bamboo without any coefficient of safety factors. Therefore, for the design purpose, it is necessary to consider the suitable safety factors, recommended by norms, and $I = I_g$ can be used instead of I_f, in favor of safety.

ACKNOWLEDGEMENTS

The authors express their appreciation for the financial support received from CNPq, CAPES and FAPERJ.

REFERENCES

1.   GHAVAMI, K. and MARINHO, A.B. *Determinação das propriedades dos bambus das spécies : Moso, Matake, Guadua tagora e Dendrocalamus giganteus para utilização na Engenharia.* Technical Report, PUC-Rio, Rio de Janeiro, 2001, (Portuguese).

2.   GHAVAMI, K., *Application of Bamboo as a Low-cost Construction Material*, Proc. of Int. Bamboo Workshop, Cochin, India, 1988, pp.270-279.

3.   MOREIRA, L.E., *Aspectos Singulares das Estruturas de Bambu – Flambagem e Conexões*, Doctorate Thesis, PUC/Rio, Rio de Janeiro, RJ, 1998 (Portuguese).

4.   GHAVAMI, K. and MOREIRA, L.E., *Double Layer Bamboo Space Structure*, Space Structure IV, Surrey, Guildford, Pub. Thomas Telford, London, Sept. 1993, Vol.1, pp. 573-581.

5.   CULZONI, R. A . M., *Características dos Bambus e sua utilização como material alternativo no concreto*, MSc Thesis, PUC/Rio, Rio de Janeiro, RJ, 1986 (Portuguese).

Lightweight Glazing Structures – 5 Recent Examples

S J GIBLETT Senior Engineer, Connell Mott MacDonald, London,
J F WEBB, Principal, Connell Mott MacDonald, Sydney,
O MARTIN, Principal, Connell Mott MacDonald, Sydney,
K KAYVANI, Associate, Connell Mott MacDonald, Sydney, and
N McCLELLAND, Senior Engineer, Connell Mott MacDonald, Sydney.

INTRODUCTION

The desire of architects and clients to produce highly transparent structures of long spans has led to the development of many interesting systems for the support of Glazing. In these structures, the structural engineer often takes a substantial role in developing the form and hence the aesthetics. This is evidenced in case studies of five recent projects presented in this paper. They all demonstrate close collaboration between architect and structural engineer to determine the form and details and with specialist subcontractor to achieve high quality execution.

GLASS BOX ENTRY FOYER –363 GEORGE STREET

363 George St is a prestige 38 storey office development in Sydney completed in 2000. The building is constructed partially above three heritage buildings, and partially above a very constrained basement. The office tower starts and level 8, some 30 metres above George St, and there is a very large void space below this level. The office tower lobby is formed by an 11m high glass structure, which has a very unusual structural form (Plate 1).

Plate 1

Design Brief

The architect desired a highly rectilinear form for the enclosure. His preference was to have a glass box suspended clear of the structure with strong horizontal rather than vertical lines. The obvious solution would have been to provide a series of steel frames stiffened by opposing catenaries on the wall and by a single sagging catenary on the roof. The walls and roof of the structure could be held on spacers inside the main structural frame.

Structural System

Although the architect was comfortable with the catenary solution described above, the curved catenary shapes departed from his desired rectilinear form. Connell Mott MacDonald devised several alternatives to produce a rectilinear form and minimise the visual weight of the structure. The solution chosen involved a "Vierendeel cable truss", shown in Figure 1. This form has not been used previously to our knowledge. It gains its strength and stiffness from three different actions:

- Elongation of the rod, due to its composite action with the column and the resulting varying rotations of the arms, which are connected to the column. For inward load, this tends to occur in the central rod elements, and for outward load, in the top and bottom elements as shown in Figure 1.
- Beam action of the column.
- Resistance provided by the taut rod acting as a catenary when deformed in either direction.

This form achieved the architect's desires in a way not previously contemplated. It also gave a consistent geometry at each node, reducing the number of different components to be fabricated by the contractor. This would not be the case with conventional catenary solutions which have a different geometry where the draping catenary crosses each support point.

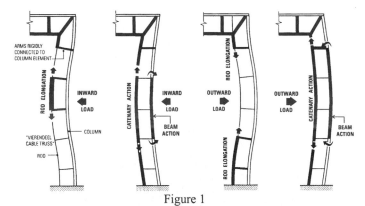

Figure 1

Integration of Sprinklers

Aside from structurally challenging solutions, the design had to be integrated with all of the services.

Sprinkler pipes were integrated within the structure, with the heads fed through a casting connected to a diagonal hollow rod in the corner of the structure. Issues such as how to hide the plumbing for the sprinklers and the cables for lighting took many months of refinement.

For example, the rods through which the sprinklers are fed are so small (25 mm diameter) that it would be impossible to know that a pipe runs inside them (Refer plate 2). Electrical services are also integrated inside the structure. These details were developed in a process involving members of the design team pushing the standard use of materials to achieve what the design architect, Richard Johnson, calls a "crafted design'.

Plate 2 - Corner Detail showing integrated sprinkler pipe

RAILWAY SQUARE

Design Brief
The project involved the redevelopment of the existing Railway Square in Sydney into a vibrant and dynamic transport interchange shown in Plate 3.

Plate 3 – Railway Square

The overall architectural concept is one that communicates the idea of shelter, arrival and gateway, which is reflected by its prominent location at the intersection of four major roads, and its proximity to Sydney Central Railway station in the Sydney CBD.

The architectural statement reflects structural honesty and artistry.

Structural Systems

The architect's ideas for the shelters were broken down into two principal forms - the main bus canopy and the bus shelters. The structural form for both canopies is similar.

The canopies are totally glazed and the supporting structure has been engineered to have minimal impact on the glazing. The design is also characterised by the obvious lack of bracing (Refer to Plate 4).

Plate 4 – End view of Canopies

The solution developed by the architect and structural engineer consists of a series of cantilever blades supporting glazed panels. The cantilever blades are supported by a torsional element, which in turn is supported by a series of inclined columns. The overall overturning forces are resolved into bending moments and axial forces (both tension and compression). This approach to forces leads to a minimisation of member sizes compared to other approaches.

The dominant design load case for the blades is downward load (dead and wind) for both strength and serviceability, and overall movement prediction and limitation was critical to the success of the design.

Extensive computer studies were carried out to determine overall frame displacements and forces by both non-linear and finite element analysis techniques. The blades themselves were ultimately designed by buckling methods to achieve the minimum structure. It also included provision for partial collapse of the glazing panels, which are bolted to the blades. The blades taper parabolically from the tip to the support, reflecting strongly the non-linear increase in bending moment. The slenderness of the cantilever blades is less than half that of conventional cantilever design.

In recent times the appearance of a number of glazed structures has been devalued by the presence of bracing and large numbers of glazing bars. In the end bays of each canopy, glazed panels are rigidly bolted to the blade sections so that horizontal forces are transferred to the steel structure by diaphragm action. The result is the total elimination of cross bracing, enhancing greatly the overall appearance of the canopies (Plate 4).

Fabrication
Fabrication of the structure, tolerances and their relationship to the final geometrical layout was absolutely crucial. Computer studies indicated that the slightest lack of fit or fabrication errors would lead to considerable mis-alignment of the cantilever blades. Expected movement diagrams were included on the drawings for the builder's information. The documentation also reinforced the precision of tolerance and fit between various components.

Once various elements had been fabricated, trial assemblies were carried out by the Builder. As a result, changes were made to a number of bolted connections, both bolt size and configuration to minimise slip in the final structure.

60 CASTLEREAGH STREET

Design Brief
The design brief from the architect was to produce a transparent and unique feature glass wall for the main entrances to the 60 Castlereagh Street building. The builder and curtain wall subcontractor had the added restraint of a limited budget for these elements.

Plate 5 – 60 Castlereagh St

The location of the existing support structure, and the comparatively moderate width of the openings, meant that a single, tensioned horizontal support cable was a viable structural solution. To our knowledge, a single tensioned cable system has not previously been used in Australia to support a glass wall. Transparency and economy were achieved, as there was minimal support structure to the glazing.

Structural Systems

The main load case on the large area of glass is from wind pressure. The out-of-plane wind loads are resisted by a series of tensioned horizontal cables.

In a number of recent projects, transparency has been 'achieved' using twin tensioned bowstring trusses. While each bowstring truss is designed to resist either positive or negative wind loads, a straight cable resists wind loads from both directions.

A planar cable system is more flexible than a bowstring system as it does not have the same depth or catenary shape. Accordingly, the required pre-tension in a horizontal cable system is higher than for a bowstring system. Deflections are also higher, a major design consideration for the glass and glazing connections. Given the location of the glass walls above the entry doors, large deflections can be tolerated.

With the required high pre-tensions in the horizontal cables, the support structure at each end of the cables needs to be significant. The existing structural layout of the building included concrete columns adjacent to the glass walls. These provided the support.

A geometric, non-linear analysis of the cables was carried out, and multiple load cases were considered. These included wind load, thermal load, support deflections (elastic and creep), cable pre-tension and cable relaxation. The strength of the system was determined by the load case of a cold, windy day just after construction (maximum wind load, negative thermal load, no support creep and no loss of cable pre-tension). The serviceability of the system was determined by the load case of a hot, windy day 30 years after construction (maximum wind load, positive thermal load, support creep and cable pre-tension relaxation).

Vertical hanger rods support the dead load of the glazing and the door head beam. They are located close to the plane of the glass and are aligned with the glass joints to minimise visual impact. The overall structural depth is less than 200 mm and the rods are 32 mm in diameter. The comparable depth of a bowstring truss would be approximately 500 mm.

The glass used on the feature walls is 12 mm toughened. As stress concentrations can occur in glass at patch fitting connections, toughened glass was selected for strength. A glass module width of 1280 mm was selected to suit existing architectural modules and to enable the use of locally available 12 mm glass. A wider module would have meant that disproportionately more expensive, imported, 15 mm toughened glass would be required.

A critical element of the glass design is the patch fitting. For large displacement support structures, a 'rigid' patch fitting connection results in excessive glass stresses. Articulated ball joint patch fittings are traditionally used for engineered glass systems. However, these fittings can be expensive. A less expensive, 'semi-rigid' patch fitting connection, based upon relatively thick, soft silicone pads, was used on the project. This was carefully analysed using a brick element FEA model. The local stresses in the glass at the patch fitting were found to be only marginally higher than for an articulated ball joint patch fitting. This detail has been used on several projects.

Materials

Innovative material selection of the main horizontal cable was made by the specialist glazing subcontractor. To minimise the cable size, a high strength material is required. High strength stainless steel rods are expensive and delivery lead times can be a problem with certain sizes.

A standard 25 mm diameter, high strength, stressing bar was used. This was clad in a light gauge 32 mm diameter stainless steel welded tube. The strength of the inexpensive standard rod and the aesthetics of the inexpensive stainless steel tube were combined to achieve the desired result in a cost-effective manner. The same 32 mm diameter stainless steel welded tube was used for the dead load hanger rod. While the diameter used was larger than required for strength, it was acceptable aesthetically and allowed for easy erection. Vertical tolerance adjustment was easily accommodated as the thin-walled, stainless steel could be site drilled at the node connections.

Plate 6 – 60 Castlereagh Street Detail

Cost
The entire system is very cost effective in relation to the impressive architectural statement that it makes.

Cost savings over conventional bowstring trusses were made in:
- Less material used than a in bowstring system
- Half the number of cables
- Less components
- Less component geometries (types)
- Selection of standard, locally available material for the tensioned cables and components
- Use of fabricated glazing nodes rather than cast nodes
- Use of 'semi-rigid' patch fittings constructed of readily available materials
- Use of load indicators and conventional tools for tensioning
- Use of load indicators that remain insitu to reduce monitoring costs.

RUNDLE MALL CANOPY

The prominent feature of the recent upgrade of Adelaide's Rundle Mall is the sculpture-like, glass canopy at the intersection of Rundle Mall with Gawler Place (Plate 7). This highly three-dimensional structure is set on four tree-like columns branching out to support four circular roof quadrants, which are sloped towards a common focal point. The "spider web" appearance of the canopy is maintained by using very slender steel fins supported by circular blade-type girders. A tensile fabric shade structure is also installed to the underside of the glazed canopy.

The many loading combinations, including dead, live and wind loads as well as tensile forces imposed by the shade structure, necessitated extensive computer analyses to resolve the final geometry and to determine minimum member sizes. The selection of member sizes was a balance between producing elements with least visual impact on one hand, and controlling stresses, deflections and movements on the other.

Accurate assessment of the structural response to the imposed loads was obtained using advanced finite element analysis of a typical quarter of the canopy. This was required since the lower tier code design rules could not justify the adequacy of the members having the (architecturally) desired level of slenderness. This special analysis was necessary to determine accurate flexural and lateral-torsional buckling capacities of the fins, circular girders and tree branches.

Plate 7 – Rundle Mall Canopy

CIRCULAR QUAY FERRY TERMINAL

The Circular Quay ferry terminal was one of many Sydney facilities upgraded for the Olympics. Passenger comfort on the wharves has always been affected by strong winds. The screens were required to provide wind shelter but let some air pass for comfort during the summer. In additional they were required to be highly transparent to maintain the superb view to the Opera House, Kirribilli and Harbour Bridge.

Structural Systems

The vertical mullion elements were twin cables formed into a truss Bay simple crank at midspan. Glazing was arranged on opposite faces of this cable truss to provide the desired substantial blockage to the wind. With the major load to the cable truss occurring at midspan, the truss shape approximated the funicular shape.

To resolve the forces from these trusses into the existing structure a cable stayed horizontal beam was used at the top. The truss was anchored to the existing wharf structure at the base.

The result is a very lightweight screens which achieve the design brief in a very attractive and cost effective manner.

ACKNOWLEDGMENTS

Project	Architect	Client
363 George St	Denton Corker Marshall	Australian Growth Properties
Railway Square	Noel Bell Ridley Smith	Dept of Public Works & Services
60 Castlereagh Street	Scott Carver	Prudential Portfolio Managers
Rundle Mall Canopy	Noel Bell Ridley Smith	Dept of Public Works & Services
Circular Quay Screens	Noel Bell Ridley Smith	Dept of Public Works & Services

REFERENCES

1. WEBB J, MARTIN O, KAYVANI K, McCLELLAND N. Lightweight Glazing Structures – 5 Recent. The Australasian Structural Engineering Conference, Gold Coast, Australia, 2001.

Development of the KT-wood space truss system with round timber as a new structural material

K. IMAI, Osaka University, Osaka, Japan
Y. FUJITA, Fujita Architect Office, Osaka, Japan
T. FURUKAWA, Osaka University, Osaka, Japan
K. WAKIYAMA, Osaka Sangyo University, Osaka, Japan
S. TSUJIOKA, Fukui University of Technology, Fukui, Japan
M. FUJIMOTO, Graduate School of Engineering, Osaka City University, Osaka, Japan
M. INADA, Osaka University, Osaka, Japan
A. TAKINO, Osaka University, Osaka, Japan
M. YOSHINAGA, Kawatetsu Civil Co. Ltd., Kobe, Japan

INTRODUCTION
This paper presents the development of the KT-Wood Space Truss System (KT-W) utilizing round timber processed from young trees culled from forests. As a significant percentage of forestation problems in Japan can be traced to young trees not being properly or selectively culled, their effective utilization contributes to maintaining a forest in prime condition. This results from their added value giving an economic incentive to cull them and thereby contributes the proper forest maintenance and the protection of the forest environment.
The purpose of the development is to utilize effectively the medium and low quality timbers (Japanese cedar and Japanese cypress: conifer trees) as value added material. The feature of the joint system is the main point to use such material.

To the KT-Series (KT-I~III) that is patented in many countries around the world and been applied to over 350 structures in Japan, the KT-W is a new addition. KT-I~III and W have mutual compatibility in the installation of their respective parts.
The joint system is designed in order to endure more severe loads and/or the detrimental effects of environmental conditions than ever before and to fit well to the esthetic requirement such as graceful appearance. In order to realize these architectural requirements, the system has following special features: 1) ease of fabrication without the need of skilled labor (no welding process), 2) high reliability factor when joints under static and cyclic loading, 3) ease of assembly at the construction site, 4) exceptional resistance to corrosion of metal part, 5) mixed installation of steel and wood structures because of the full compatibility of KT-series, and 6) ease of replacement of decayed members.

The paper reports the outline of the system, full size material tests (tensile, compression and pull-out test of lag screw) for constructing wooden space truss system.

OUTLINE OF THE SYSTEM
Joint System
KT-W joint consists of KT-truss joint assembly, joint cone, end disk at the end of round timber and lag screws in order to anchor the end disk. As an initial clamping force to the lag

screw can be introduced, looseness of the joint that will be generally serious defect of the wooden structures can be easily eliminated. In case of the shear bolt joint that is most commonly used, it is not easy to eliminate such defect. This idea is similar to "pull bolt joint (so called Hiki-bolt)" that is the traditional Japanese carpenter's technique. Another important

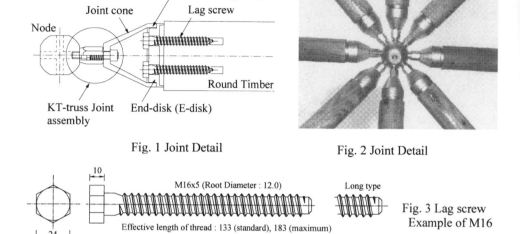

Fig. 1 Joint Detail

Fig. 2 Joint Detail

Fig. 3 Lag screw
Example of M16

advantage is that any bending moment doesn't occur at the joint, because of no joint eccentricity. This feature is especially important for the weak material in joining such as wood. The joint cone is the interconnecting part between KT-truss joint assembly and end disk. Joint cone is screwed into the end disk and glued (see Fig. 1,2). Specially designed lag screw that pull out resistance is strengthened is applied for the joint system (See Fig. 3). Fully mechanical joint realizes no need of skilled labor for fabrication and installation. All metal wares are elegantly powder coated in general and perform a graceful appearance. The powder coating endures preferably more than 15 years even in the outdoors in our experience.

The system consists of the following components.
1) A threaded spherical node (50-235 mm in diameter).
2) KT-truss joint assembly (high strength bolt / M8x1.25~M42x3). (allowable stress for permanent use is 14.2~445.9KN).
3) Round timber, 50-300mm in diameter. (allowable stress for permanent use is 12.5-450.3KN for cypress and 10.6 – 381.0KN for cedar in tension. 16.3 – 588.9KN for cypress and 14.5 – 519.5KN for cedar in compression.).
4) Lag screw, M12x4-36x6 (standard & long type).

Installation of Members
During the assembly of the space truss, the threaded end of the bolt is pushed into the hexagonal sleeve and then the hexagonal sleeve is turned at the threaded hole of the node (see Fig. 4). In this manner, assisted by the pushing device, which provides a pushing action, the bolt enters the node hole smoothly.

The installation of members is, therefore, simplified assuming that the distance between adjacent nodes accurately reflects the correct measurement of the final assemblage. This mechanism, which eliminates the necessity to consider the order in which members are assembled, drastically simplifies dismounting in the event that disassembled bars occur. Especially the ease of the replacement of decayed members is inevitable in wooden structures. The members of traditional systems cannot be assembled in this manner due to the fact that the bolt must project slightly from the hollow sleeve in order to be screwed into the node hole. Because of full compatibility of installation in the KT-I~III and W, the mixed use of wooden

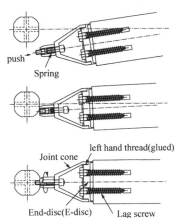

push
Spring
left hand thread(glued)
Joint cone
End-disc(E-disc) Lag screw

Fig. 4 Installation process

Fig. 5 Citizen's Workshop for construction of elliptic dome

and steel structures is possible.

And the ease of fabrication without the need of skilled labors (no welding process and ease of installation) realizes the economical production and construction everywhere.

Fig. 5 shows the construction by the citizen's workshop including children at the forest park. The ease of the installation, dismount and replacement is clarified by the workshop.

The range of the system is shown in Table 1 in combining the truss bolt and lag screw size with timber diameter.

Table 1 Range of the system : cypress (extracted)

(KN)

Medium quality (cypress)			High tensile bolt of KT-Truss					
timber diameter (mm)	Allowable stress for permanent load		M8	M10	M12	M16	M20	up to M24 M30 M36 M42
	tension	compression	14.21(P.L.) 21.32(T.L.)	22.54(P.L.) 33.81(T.L.)	32.73(P.L.) 49.10(T.L.)	60.96(P.L.) 91.44(T.L.)	95.16(P.L.) 142.74(T.L.)	
φ50	12.51	13.47	1-M12 7.61					
			1-M12 9.83			(P.L.) Allowable stress for Permanent Load		
φ60	18.01	19.40	1-M12 7.61	1-M12 7.61		(T.L.) Allowable stress for Temporary Load		
			1-M12 9.83	2-M12 15.22				
			1-M16 13.10	1-M16 13.10				
				1-M16 18.03				
φ80	32.02	34.48		1-M16 13.10	1-M16 13.10			
				1-M16 18.03	1-M16 18.03			
				2-M16 26.21	2-M16 26.21			
				1-M20 22.17	1-M20 28.32			
φ100	50.03	53.88				2-M16 36.06	2-M16 36.06	
						1-M20 22.17	1-M20 22.17	
						1-M20 28.32	1-M20 28.32	
							2-M20 56.65	
						1-M24 33.99	1-M24 33.99	
							1-M24 41.38	
φ120	72.04	77.58				2-M16 36.06		
						1-M20 22.17	1-M20 28.32	
						1-M20 28.32	2-M20 56.65	
						1-M24 33.99	1-M24 41.38	
φ150	112.57	121.23					2-M20 44.33	2-M20 44.33
							2-M20 56.65	4-M20 88.67
							1-M24 41.38	2-M24 82.76
							1-M30 72.41	1-M30 72.41
								3-M20 66.50
								4-M20 88.67
φ180	162.10	174.57						2-M24 82.76
								1-M30 72.41
								2-M24 67.98
								3-M24 101.97
								2-M24 82.76
φ210	220.63	237.60	long type screw					1-M30 72.41

Legend

number	Lag screw size	Pull out stress
	2-M16	26.21
	2-M16	36.06
long type screw	2-M16	

up to φ240, 270 & 300

Allowable compression and tension stress of timber are based on the medium quality cypress by AIJ code (Japanese code) so far. Pull out strength is based on the author's test in the succeeding section. Pull out strength is possibly the ruling factor of the joint. Designer can choose the suitable number of lag screws in order to make the structure economical regarding the magnitude of the tensile stress of member. By the full size tensile test, three times of the AIJ code is obtained. In the future, such result that is quite different from the code should be evaluated based on sufficient number of data.

TEST PROCEDURE AND RESULTS OF THE FULL SIZE TEST OF TIMBERS

Specimens were made from Japanese cedar and cypress those are relatively young and low quality trees. 24 pieces of full size round timbers were tested under tensile loading and 24 pieces under compressive loading. 73 pull out specimens of lag screw were tested under tensile loading simulating the joint. Fig. 6 shows the tensile and pull out test. This kind of tests hardly tried so

Fig. 6 Test set-up
(Left : pull out   Right : tensile)

far because of the difficulty of loading. Specially designed testing jig was successfully developed.

Fig. 7 shows typical failure modes under tensile loading. Fig. 9 and 10 show typical load-deformation relations.

In tension, the specimens fractured brittle without any plastic deformation. The maximum strength is more than from two to three times compared with the minimum specified material strength of high grade material by AIJ code (solid lines in Fig. 9). The results by this study are quite different from the small specimens without defects based on JIS (Z 2101) that specifies AIJ codes.

Fig. 7 Failure mode (tensile test)

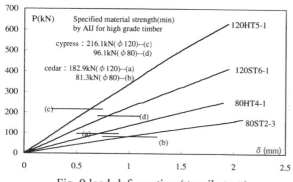

Fig. 9 load-deformation (tensile test)

Fig. 8 Failure mode
(compression test)

This shows the possibility that the round timbers with so called medium and low quality can be applied widely equally to high grade material. In case of cypress timber with narrow annual rings, strength/weight ratio is greater or equal to ultra-duralumin. Surface defect doesn't affect seriously to the strength of the timber. And also it is clarified that the damping factor of the wooden space frame is considerably high compared with steel member by authors' another study in this conference. By this analysis, wood has a possibility as a new structural material for the space frame.

In compression, specimens fractured by the fiber buckling (Fig. 8). The maximum stress is grater than the minimum specified material strength of medium grade material by AIJ code roughly. The results show the ductile deformation after yielding.

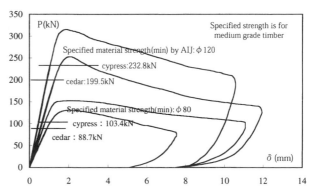

Fig. 10 load-deformation (compression test)

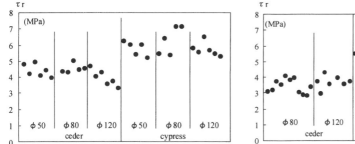
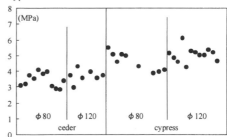

Fig. 11 Pull out strength of lag screw (left :single bolt, right :double bolts)

Fig. 11 shows the pull out unit shear strength (τ r) per surface area of lag screw. There is a tendency that τ r of single bolt is greater than double bolts. This results from unequal pull out strength because of unequal tightening of bolt. By improving tightening method, strength will be higher.

CONCLUSIONS

The wooden space truss system that utilizes effectively the so-called medium and low quality timbers with small and medium size is successfully developed. And the effectiveness of combining the previously existing steel space truss system to the wooden system is clarified.

The possibility of the wood as a new structural material for the space frame was clarified. The potentialities of the wooden space truss were also shown in this study.

Acknowledgement

The authors express their gratitude to the forester (manager) of Hyogo Prefecture Mr. Shigeru Okada for his beneficial and hospital suggestions.

REFERENCES
1. Imai K., Morita T., Yamaoka Y., Wakiyama K. and Tsujioka.S. : The KT Space Truss System", FOURTH INTERNATIONAL CONFERENCE ON SPACE STRUCTURES, vol. 2, GULDFORD, 1993, pp.1374-1382.
2. Mengeringhausen Max :Raumfachwerke aus Stäben und Knoten, Bauverlag, Berlin, 1975
3. Nooshin H.: Third International Conference on Space Structures, Elsevier Applied Science Publishers, London, 1984.
4. Imai K., Wakiyama K., Tsujioka.S., Fujimoto M., Hideki W., Masanobu I. and Morita T. : Development of the KT-Wood Space Truss System, IABSE Conference Lahti, Finland, August, 2001
5. Imai K, Fujita Y, Wakiyama K, Tsujioka S, Fujimoto M, Watanabe H, Inada M, and Morita T: FURTHER DEVELOPMENT OF THE KT-WOOD SPACE TRUSS SYSTEM, Proc. of IASS Int. Symposium 2001, Nagoya, Oct. 9–13, TPNo.157, 2001.10
6. Furukawa T., Imai K., Fujimoto M, Komedani N., Inoue R., Okamoto K. and Fujita Y.:Dynamic Loading Experiment of Wooden Single Layer Two-Way Grid Cylindrical Shell Roof, FIFTH INTERNATIONAL CONFERENCE ON SPACE STRUCTURES, GULDFORD, 2002, will be issued

The structural engineering of the Downland gridshell

R. HARRIS and O. KELLY
Buro Happold, Bath, UK

ABSTRACT
The Downland Gridshell, at the Weald and Downland Museum in Sussex, England, forms the roof of the new Archive Store and Workshop building.

The prime benefit of the timber gridshell is the simplicity of the construction sequence: a regular grid of slender timber laths is laid out flat; at each intersection point the members are connected by a special connector; finally the grid is shaped so that it takes up a doubly-curved form. It is able to do this because of the flexibility of the laths and the rotational freedom at the gridshell intersections. After erection of the lattice, bracing is introduced which triangulates the square grid, providing shear strength.

The complete form of the Downland Gridshell structure is a triple hourglass, 12 – 15 metres wide and 50 metres long, 7m to 10m high. It is the double curvature of the shell that generates the geometric stiffness and is fundamental to its structural action in resisting asymmetric loads. Extensive structural analysis is necessary to find the shape that can be formed from the original flat pattern. Issues addressed in this paper include the formfinding, specification of the timber laths, the development of an appropriate formation technique and methods to monitor and assist the formation process.

The gridshell construction was completed in August 2001. The building was completed in April 2002.

INTRODUCTION
The location and reason for building
At the Weald & Downland Open Air Museum, in Sussex, England more than 45 historic buildings from South East England have been rescued and rebuilt. The Museum wanted a new building that would stand as a testament to architectural and building techniques of the early 21st Century. The resulting structure incorporates a wide range of carpentry disciplines and skills, but, most significantly, the Downland Gridshell is the first structure to be built in the UK using techniques similar to those used for the Mannheim Gridshell in 1975.

The building commissioned by the Museum is a two-storey structure. The upper storey is a Workshop. It is the Workshop roof that is a timber gridshell. In principle, the construction sequence, of the gridshell is simple: a regular grid of slender timber laths is laid out flat and is subsequently shaped into a doubly curved form. It is possible to do this because of the flexibility of the laths and the rotational freedom at the gridshell intersections. This is in contrast to a steel or concrete gridshell that is erected in its doubly curved shape, with the difficulty of prefabricating hundreds or thousands of individual, different nodes. After forming the shell, shear stiffness is provided by adding cross bracing to the doubly curved grid of timber laths.

Precedents for the building
The first double-layered timber gridshell was constructed for the Mannheim Bundesgartenschau[1] in 1975. The Mannheim gridshell was constructed by laying out the flat lattice on the ground and then pushing it up using spreader beams and scaffolding towers at strategic locations. In transforming the flat lattice to the final shape there was a significant component of lateral movement related to the gain in height. To accommodate this the scaffolding towers were moved using forklifts. The building was has been very successful and remains as one of the finest buildings of the 20th century, despite its original two-year design life. The process of pushing up on discrete areas during erection did concentrate stresses within particular areas, the consequence of which was a notable number of breakages of laths and finger joints. It also required operatives to work at height and under the temporarily supported grid; a scenario that would not be permitted under today's CDM regulations.

More recently Buro Happold have designed two single-layer gridshells, one in timber, the other in cardboard. The first were the gridshell sculptures at the Earth Centre in Doncaster constructed in 1998, Figure 1. These were small timber lattice structures composed of a single layer of thin timber laths. A crane was used to lift the lattice into position and then it was manipulated with struts and ties from a flat grid to the final shape. An important lesson learnt on this project was that the nodes need to be extremely loose to enable rotation during formation.

Fig. 1. The gridshell at the Earth Centre in Doncaster.

The second recent project was the Japanese Pavilion at the Hanover Expo 2000. This building was similar in shape to the Downland Gridshell and composed of cardboard tubes, Figure 2. The erection used the modular scaffold system PERI-UP, shown in Figure 3. The flat mat was laid out on a low-level scaffold bed and was pushed up into position using proprietary PERI jack known as a MULITPROP. Due to the similarity of the shape of this shell with the Downland Gridshell, a great deal was learnt from the experience of erecting the Hanover building.

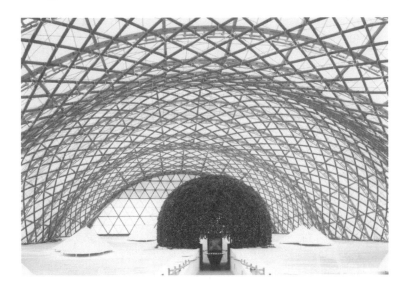

Fig. 2. The Japanese Pavilion, Hanover Expo 2000.

Fig. 3. Erection of Hanover Gridshell using PERI scaffolding.

Leading on from this experience, three of the most significant decisions for Downland Gridshell were the level at which the laying out platform was to be set, the sequence to achieve the triple bulb hourglass shape, and the use the PERI scaffolding system.

THE BUILDING
Background
The building design has succeeded through combining the skills of the Architect, Edward Cullinan Architects, the Engineer, Buro Happold, and the Carpenter, Green Oak Carpentry Company. A large part of the finance for the building comes from the Heritage Lottery Fund. They have been particularly supportive in providing sufficient funds at the early stage of the project to develop the design prior to firming up a tender price. This enabled the construction of a structurally accurate physical model (see section 3 below) and the early appointment of the Carpenter to provide input to the design, particularly through the development of construction methods.

Fig. 4. North elevation (picture: Edward Cullinan Architects).

The Archive Store
The building is dug into the chalk hillside, gaining the environmental benefit of the thermal mass of the ground in moderating the environment for the archive store (Fig 5). The roof of the archive store forms the floor of the workshop. It is constructed with glued laminated beams and solid timber flooring. It cantilevers over the reinforced masonry walls of the archive store, providing the support to the gridshell.

Fig. 5. Cross section (picture: Edward Cullinan Architects).

The Gridshell Roof
The workshop roof is a doubly curved shell made from 50mm wide x 35mm thick oak laths in four layers (Figure 6). It is a double hourglass shape, 48 metres long, 16 metres across at its widest points and 11 metres wide at the waists. The internal height varies between 7 metres and 10 metres. The lath spacing is 500mm in areas of high load, increasing to 1000mmm over substantial areas of the shell. At scheme design stage, the spacing was

500mm all over the shell; the detailed computer analysis (see section *3.3*) enabled the forces and stresses to be examined carefully and lead to the increased spacing in more lightly loaded areas. This gave significant cost and timesavings and was welcomed by the Architect as it emphasises the structural form.

Fig. 6. Elevation of gridshell model.

The weight of the timber structure (gridshell mat and diagonal bracing) is approximately 6 tonnes (7.5kg/m2). It is clad with an undulating flat roof at the crown, a strip of clerestory windows along each side and, beneath these cedar boards. Fully clad, the weight of the roof increases to 20 tonnes. Shear blocks are screwed into place between layers 1 and 3, and between layers 2 and 4. In this way, parallel lines of laths act compositely; the size and position of the shear blocks is arranged to suit the forces derived from the analysis.

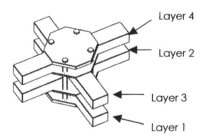

Fig. 7. A nodal connection.

The node connector (Figure 7) was developed to suit the particular requirements of timber gridshell formation. It consists of three plates, the centre one having pins to locate the grid geometry of layers 2 and 3 and the outer ones loosely holding layers 1 and 4 in place during the formation of the shell. Two of the four bolts locating the plates are used to connect the diagonal bracing bolted in place to provide shear stiffness after forming the shell. The connector, which has proved to be very successful, has been patented.

STRUCTURAL MODELLING AND ANALYSIS
Computer and physical modelling have been carried out side by side throughout the design development. The difficulty of this process explains why such an apparently attractive method of construction has been rarely used since the success of the Mannheim Gridshell in 1975.

Physical Modelling
Physical modelling of the structure has been a central element of the design process. A good physical representation of the building was found to be essential to enable effective discussion of concepts. Initially, scheme models were made at small scale, using wire mesh. These led to the construction of a 1:30 model using strips of wood to model the laths. The construction of this model was very instructive in developing a feel for the likely behaviour of the structure during the erection process.

The geometry of the physical model was used to determine the boundary conditions for the computer form-finding model. It also served as a presentation model, which was of great use in persuading the funding bodies of the viability of the scheme. Lastly, a model was made with weights hung down from steel mesh to model the correct self-weight bending characteristics of the laths. This has been used to explore the formation sequence of the shell.

Computer Form-finding
The process of form finding for gridshells is not straightforward, and involves numerous interactions to develop a smooth form. The shape depends upon the current length and position of the sections of lath along the boundary. For a square starting grid there is a relationship between the form-finding model and the construction of the grid, thus the accuracy of the modelling is important.

The shape of the gridshell was obtained using a modified dynamic relaxation method. Dynamic relaxation is an interative process that modifies an initial approximation to the desired shape by monitoring the kinetic energy of the model as it is made to oscillate. Such programmes generally examine the oscillation of a pure catenary to generate the final shape. However, the Weald and Downland Museum gridshell is not a purely funicular shape: a hanging chain model could not form the saddles.

To generate the form it was necessary to modify the software to make allowance for the bending stiffness of the timber laths. Longitudinal and transverse timber rib-laths, fixed to the nodes once the shell is formed, provide shear resistance and lock-in the shape. The waisting along the building gives it strength and stiffness against asymmetric loads. The output from the analysis includes details of the final curvature of all the structural members.

Structural Analysis
The model geometry of the final shape of the formfinding model was loaded into elastic analysis software (STAAD Pro). The first stage of the analysis was to check for adequate factor of safety against buckling failure. Two methods were used. Using the dynamic relaxation method, second order analysis was carried out to determine the loads at which buckling instability occurs. Using the STAAD model, P-Δ analysis was carried out and the deflected shape was compared with the non-linear analysis under the same loads. This confirmed that the elastic P-Δ analysis gave a good prediction of the actual behaviour and the method described in the German timber code (Ref 2) clause 9.6 was used to show that under

working loads the behaviour is elastic, with an adequate factor of safety against buckling (equation 1).

Deflection due to 3 x Working Load	$\leq$	4.5	DIN 1052.1 – 9.6
Deflection due to 2 x Working Load			

$$\frac{\text{Deflection due to 3 x Working Load}}{\text{Deflection due to 2 x Working Load}} = \frac{155.24}{103.49} = 1.5 < 4.5$$

At 1 x (DL + LL + WL), the maximum deflection is 49mm.

The STAAD elastic analysis provided loads for all the structural elements and joints. Detailed stress checks were made in accordance with the Eurocode EC5 (ENV 1995-1-1), Ref 3. The timber grade used was D30.

As the shell is a double layer, the full stiffness required to resist loads is only developed once the shear blocks have been added and the longitudinal/transverse laths have been bolted into position. At this stage the Multiprops can be completely removed and the scaffold used as an access platform for the construction of the cladding.

TIMBER
Selection of Timber
A number of timber species were considered for use in the Gridshell. These included larch, douglas fir, chestnut and oak. They were selected for the following reasons:

- They are all naturally durable, creating the possibility to omit the timber treatment
- Oak and chestnut are the most common materials used in the Museum's collection of buildings
- They are all readily available from sustainable sources in the UK.

Tests carried out on laths at Bath University indicated that the performance of the oak exceeded that of the other species. Whilst it was stiffer than the other timbers tested, needing a larger force to achieve a given curvature, it had a considerably higher bending strength, achieving a smaller bending radius prior to failure. In addition the failure mode was not sudden; there was a degree of plasticity at failure.

Short Grain
A significant problem noted was the variability of the bending strength of oak due to "short grain". The straightness of the grain along the length of the log is a function of the way in which the tree grows. This varies from species to species and, whilst there are trees that grow with a more spiral pattern, oak does not grow with the straightness of grain of a timber such as Hemlock (chosen for the Mannheim gridshell for this reason). In small sections, the problem is considerably increased.

To overcome the problem the defects were cut out and finger joints were used to join the short lengths together. Thus laths of the required length, of a consistently high quality, were produced from normal grade timber.

Timber Grade

The gridshell was analysed and designed in accordance with the Eurocode 5: Design of timber Structures (Ref 3), using a timber grade of D30. Eurocode 5 uses load and material factors in the design; timber grade D30 has a characteristic bending strength of 30N/mm². To meet the design requirements, the 5-percentile characteristic bending strength had to equal to or be in excess of this value. Preliminary testing had proved that solid oak laths had the required strength and could easily achieve the 6m radius of curvature required by the lattice. Thus the solid oaks were satisfactory and attention turned to the specification of the joints.

Lath production

The gridshell lattice required 6000 linear metres of lath; considering that individual pieces of graded lath averaged 0.6m in length this represented 10,000 finger joints. Using specialist machines, this work was completed in three shifts. Although the timber had to be transported to the specialist machine the total weight was only 6 tonnes. Such a small quantity has little difficulty being transported in one load and so does not represent a significant transport problem either in terms of cost or sustainable design.

The advantage of the above approach to produce the 'improved' oak laths is that the quality of the material is maximised very quickly and cheaply with minimum wastage. Figure 8 shows a typical finger joint after the laths had been planed down to 50mm x 35mm section. The finger joint is almost indistinguishable within the lath; the low visual impact is one of the advantages of this jointing method.

Fig. 8. A finger joint.

Specification
The specification stipulated:

- Maximum slope of grain on either face 1:10
- No dead knots. No live knots. Small clusters of pin knots allowed provided that they do not form more than 20% of the width of any one face.
- No shakes or splits
- No sapwood (Heartwood of oak is naturally durable and resistant to infestation but its sapwood is not)

Using the finger jointing technique, this specification could be achieved with almost any source of oak. The problem is that the lower the quality, the greater the number of defects and the larger the amount cut out and discarded. This adds significantly to the cost of the final product. It is necessary to find the balance between low cost poor quality timber with a high rejection rate and higher cost source material with a lower rejection rate. For this reason, suppliers suggested that the timber should be sourced in Normandy.

Site Jointing
The next stage in the process was to join the 6m lengths of 'improved' timber to produce continuous laths up to 37m long for the lattice laths and 50m long for the longitudinal rib laths. This work was carried out on site under the protection of a polytunnel. The 6m lengths were joined using scarf joints with a slope 1 in 7. The slope of 1 in 7 gives the scarf joints a glue-line area the same as that for the finger joints. Figure 9 shows the construction of a typical scarf joint. There is an interesting contrast in two jointing methods used; the finger joints are the latest timber joining technology whereas the scarf joint has been used for centuries.

Fig. 9. Construction of scarf joint.

Joint performance
Of the 10,000 joints in the structure there were approximately 145 breakages during forming. Almost all of these were failures of the finger joints. The cause of these failures included:

- Pinching of the lattice on the scaffold support
- Tight curvature
- Tension build up because of relative sliding between the two layers being restricted
- Dry joints

The low failure rate vindicates the design test and construction methods. The simple repair technique consisted of introducing solid blocking at he point of failure.

STRUCTURAL DETAILS
The diagonal bracing runs longitudinally in the lower area of the shell to support the Western Red Cedar weatherboarding and transversely at the top of the shell to support the "ribbon roof", which is a tradition joist and plywood "flat" roof, undulating in elevation. Where the shell is supported by the cantilever glued laminated beams forming the floor of the workshop (Figure 6), the laths are bolted between two 25mm thick curved plywood sheets. These are bolted to the ends of the glued laminated cantilever beams of the workshop floor with steel brackets.

The wind loads on the gable wall are resisted by a green oak frame spanning vertically between the workshop floor and the shell, the horizontal reaction at the roof being carried down to the workshop floor by the diagonal bracing laths, in the plane of the shell. The vertical and transverse loads from the end of the shell are carried through the green oak frame of the gable wall into the masonry walls of the archive store.

CONSTRUCTION
The benefit of timber gridshells becomes apparent in the construction stage. Scaffolding supplied by the German company PERI was used in the construction of the Japanese Pavilion where it had proved to be very suitable and effective. For the Downland gridshell, PERI were able to bring their experience to bear in deciding the best procedure for formation. Full details of the construction sequence have been described by the authors in a previous paper (Ref 4).

Layout of Flat Mat
The lattice was laid out as a flat mat composed of squares with a 1m edge length; the resulting lattice mat, which would be 47m long x 25m wide if laid out orthogonally, was stretched longitudinally to achieve the 50m length of the completed structure.

Scaffolding System
The scaffolding supplied by PERI was the PERI-UP system. Its main advantage stems from the fact that it is a modular system coupled with several innovative, patented accessories that endow it with the flexibility demanded in gridshell construction.

An innovation was the choice of layout level. Instead of laying out the mat at or near the floor level, as had been done on previous gridshells, a birdcage of scaffolding with a plan area the size of the layout mat was built to a height of 7m above the workshop floor.

Fig. 10. The completed layout of the flat lattice mat.

By laying out the mat at the height of the valleys, the lattice could be lowered into position. By starting at an elevated level the formation process for the Downland Gridshell harnessed gravity in achieving the desired shape.

The completed shape is presented in Figure 11.

Fig. 11. Final shape.

CONCLUSIONS
The Downland gridshell is the first double layer timber gridshell in the UK and only the fifth worldwide. Despite being an efficient and environmentally sustainable structural form there seems to be a reluctance to adopt double layer timber gridshells. A possible reason for this is the difficulties encountered in forming the double curvature of the complete shape from an initial flat mat of laths.

The success of the Downland gridshell project is the result of close co-operation of the design and construction teams. In this spirit of teamwork many of the problems that arose were overcome by constructive input from all concerned.

The starting point for the formation process was to lay out the flat lattice on a scaffolding platform at the height of the valleys in the final shape, thereby taking full advantage of gravity throughout formation.

Observation was seen as the key control. Problems and potential problems were most effectively isolated by continually observing the lattice and it behaviour.

Health and Safety was paramount on the site. There was a debriefing session every morning providing the opportunity to review and discuss safety. The result was no reportable accidents nor near accidents.

Client – Weald and Downland Open Air Museum

Architect – Edward Cullinan Architects

Engineer – Buro Happold

Carpenter – Green Oak Carpentry Company

Cost Consultants – Boxall Sayer

Timber Testing – University of Bath

Main Contractor – EA Chiverton

Scaffold Design & Supply – PERI Formwork and Scaffolding

Figure 12. Interior of the Final Building
(picture: Edward Cullinan Architects)

REFERENCES

1. HAPPOLD E., LIDDELL W. I., 'Timber Lattice Roof of the Mannheim Bundesgartenshau', The Structural Engineer, March 1975.
2. *DIN* 1052 (Holzbau), Clause 9.6.
3. Eurocode EC5. 'Design of Timber Structures'. DD ENV 1995-1-1:1994.
4. KELLY O.J., HARRIS R.J.L., DICKSON M.G.T., ROWE J.A. "Construction of the Downland Gridshell" The Structural Engineer, Vol 79 No. 17, 4 September 2001.

Wooden poles for larger structural applications

P. HUYBERS
Delft University of Technology, The Netherlands

ABSTRACT

Roundwood has always been a very common building material. It was and is available all over the world, it is cheap and has often excellent structural qualities. It is a source of material that is self-replenishing and it is therefore understandable that it is still very popular in many developing countries (Ref 2). But it has also gained renewed interest in more industrialized countries, because thin roundwood makes up almost half the production of wood and it is nowadays greatly under-utilized. The research Group Building Technology of the Civil Engineering Department of the Delft Technological University, has been carrying out since many years experiments to investigate the potential of thin roundwood for structural applications and it had the opportunity to apply this knowledge in a number of actually realized structures, such as space frames (1986 to 1990), a little dome structure of 6 meter height and width and a 27 meter high tower structure (1995). In January 1996 the international co-operation project CT95-0091 was initiated in the context of the European FAIR-2 Programme. This project was called 'Round Small Diameter Timber For Construction' and its task was to develop design criteria for building constructions of roundwood poles with diameters in principle not exceeding 150 mm. This forms a category of wood that makes up a considerably large proportion of the yield but which has up to now little economic value for the industry. Institutes from five different European countries took part in this project: Finland as the initiator, together with Austria, England, France and The Netherlands (Ref 3). As an outcome of this project a demonstration structure in the form of an observation tower of 10 meters height has been built (1999). This paper will report on some of the results of the mentioned investigations.

STRENGTH DATA OF ROUNDWOOD

The material properties of different kinds of roundwood have been thoroughly investigated on large series of test specimens in the laboratories of the partners involved. The next two tables give a review of the most relevant data that were found (Ref 5, page 6].

Table 1. Compression strength of the different wood species.

Species (country)	Sample number	Compression strength [N/mm^2]
Scots pine (FIN + UK)	250	24
Spruce (FIN)	150	27
Larch (NL)	58	38
Douglas (FIN)	190	26
Sitka Spruce (UK)	100	20

Table 2. Bending characteristics of various kinds of locally available roundwood (EN384).

Species (country)	Sample number	Mean density [kg/m³]	Charact. Density	Bending Strength [N/mm²]	E-modulus In bending [KN/m²]	Strength Class
Scots pine (FIN + UK)	250	492	427	37	12.9	C30
Spruce (FIN)	200	434	323	35	12.9	C30
Spruce (AUS)	143	451	360	38	12.9	C24
Larch (NL)	178	580	509	63	14.3	>C40
Douglas (FIN)	180	442	365	37	11.1	C24
Sitka Spruce (UK)	100	478	392	39	15.8	C30

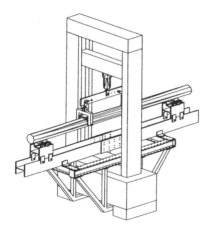

Fig. 1. Set-up of laboratory bending test. (Courtesy Peter de Vries)

Fig. 2. Drying stack for round wood. (Ref 1)

STRENGTH OF CONNECTIONS

The high values for larch wood are very prominent in these tables. The strength of a connection detail can be calculated according the dimensioning criteria that were developed in the course of the roundwood project. Rules for connections were laid down in a so-called Guideline Document (Ref 5). As is said there: *"The engineered guidelines provide the design and the calculation rules (related to the European code EC5, "Design of Timber Structures") with which the dimensioning of round wood structures and connections can be carried out."*

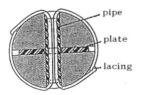

pipe

plate

lacing

Figs. 3 and 4. Cross-section and isometric sketch of connection in round wood structure.

Table 3. Comparison of test results under laboratory conditions with yield theory at a moisture content of 12%.

Test series	Wood type	Number of tests	Pole diam. mm	Number of dowels	Diam.of dowels mm	Ave. test values kN	Failure mode 1 kN	Failure mode 2 kN	Failure mode 3 kN
Pl	Scots P.	9	100	2	17	71.1	91.1	55.5	69.2
P2	Scots P.	16	100	2	17	68.9	91.1	51.4	60.3
L2	Larch	8	120	2	21.3	124.6	154.5	94.9	119.0
L6	Larch	10	200	4	21.3	281.3	466.7	230.6	222.6
RVS-4	Larch	16	120	2	17	75.3	133.9	63.2	72.9
Galv-4	Larch	15	120	2	17	71.8	108.6	60.8	70.8
P3	Scots P.	4	100	2	17	58.4	84.6	48.7	58.1
L3	Larch	8	120	2	21.3	107.4	153.9	94.7	118.8
Galv-2	Larch	5	120	2	17	69.4	111.2	61.7	71.0
L4	Larch	8	150	2	16	152.0	140.7	90.3	116.1
L5	Larch	6	120	2	16	135.9	114.4	85.5	116.1
L7	Larch	10	200	4	16	311.1	378.1	205.8	232.3

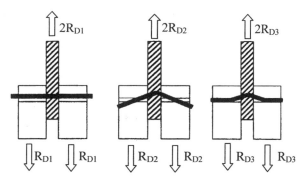

Fig. 5. Three different possible failure modes of doweled timber connections are considered.

$2R_{D1}$: Failure on embedding strength.

$2R_{D2}$: Failure of the dowel with one plastic hinge + embedding strength.

$2R_{D3}$: Failure of the dowel with three plastic hinges + embedding strength.

STRUCTURAL APPLICATIONS

Square offset space frames

A number of structures were built since 1986 on the principle of a space frame (Ref 4). They had overall dimensions of 10.8 x 16.2 m, 8.1 x 16.2 m , 8.1 x 18.9 m in plan, at a height of 3.8 to 6.0 m. The roof frames were all composed of poles with a nominal diameter of 10 cm. The columns are generally thicker and are of 15 cm diameter. The poles meet each other against circular or octagonal steel nodes and are interconnected with bolts (see Figure 4). This connection method facilitated the construction of relatively large span load-bearing frames for buildings, made of roundwood members with diameters of generally not more than about 10 cm. A few of these structures have been realized during the recent years in the Netherlands and in England. Their structural behaviour is promising; they tend to be strong and stiff. In most of these cases the wood was debarked only and required no other machining apart from the provisions at their ends, that were necessary for the connections.

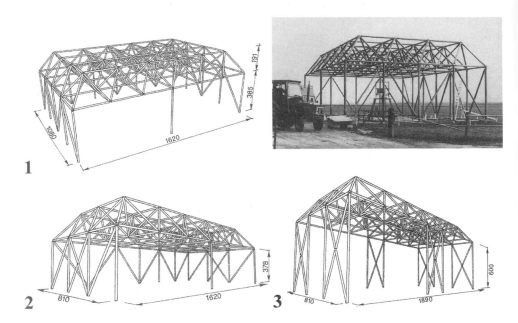

Fig. 6. Realized roundwood space structures in The Netherlands (1 and 2) and in Winchester, England (3).

Fig. 7. Detail view of structure in Lelystad.

Diagonal, nodeless space frame

A particular roundwood space frame has been built in 1990 in Rotterdam. This consists of two individual parts in a diagonal arrangement with respect to the groundfloor plan. The larger part forms the roof frame of a closed building and had been clad later on. The other part is free standing (Ref 7).

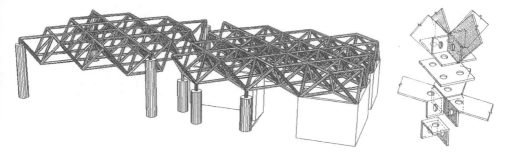

Figs. 8 and 9. Roundwood structure on a children's play ground in Rotterdam with scketch of standard detail.

Figs. 10 and 11. Roundwood structure in Rotterdam.

Figs. 12 and 13. Standard detail of the building.

The interconnection differs from the previous cases. The plates in the ends of the poles are provided with welded-on pieces of angle steel. They fit together in a special way, so that in fact no extra node elements are needed. Diagonal arrangements are easy to be realized with this solution.

Observation tower

In April 1995 the construction of an observation tower with a total height of 27 meters took place. It is composed of 12, 15 and 20 cm thick poles. It has connecting nodes, that basically consist of 4 identical steel elements, made out of standard angle steel. They fit together and allow a maximum number of 18 meeting poles.

Figs. 14 to 16. Sketch and photographs of observation tower and a structural detail.

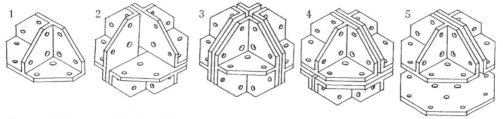

Fig. 17. Different combinations of basic node element.

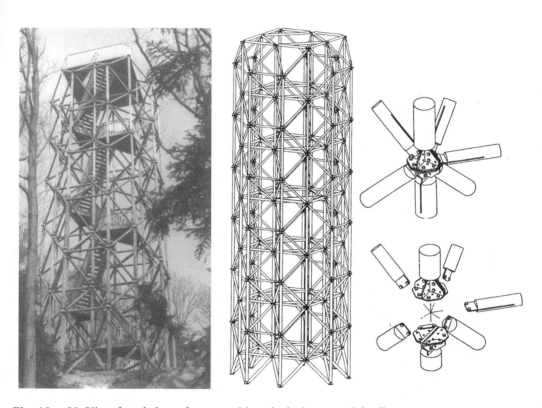

Figs 18 to 20. View from below of tower and its principal structural detail.

In this case the plates in the poles are 10 mm thick and they are fixed to the poles with bolts and then secured with wire lacings. The tower has a staircase and is accessible to the public. Consulting office De Bondt acted for this project as the supervisor for the construction (Ref 3).

Dome structures

A little temporary domical structure with a diameter of 6 m and a total height of also 6 m, composed of 10 cm thick cylindrical poles had been built in the early beginning of the structural round wood research (1984). This dome was used several times as a pavilion for exhibition purposes. Later it had been provided with a tent structure, suspended from the nodal points, and from 1999 to 2000 it was exposed at a building site in Kootwijk.

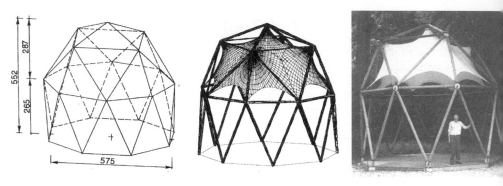

Figs. 21 to 23. Dome structure with PVC coated fabric tent, suspended from the nodes.

The design of a larger dome had been worked out with a span of 25 m. It appeared from the statical analysis, that poles of not more than 12 cm diameter would be sufficient. For the standard detail a somewhat different solution has been suggested: a hollow steel sphere against which the poles were connected with a central bolt. For this purpose, the plate in the end of the pole would be provided with a cylindrical part with an internal threaded hole.

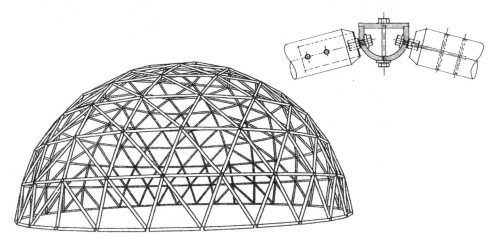

Fig. 24. Plan of 25 meter diameter dome structure of roundwood.

Watch tower

In June 1999 a tower structure of thin round wood poles has been completed in Kootwijk, The Netherlands (Ref 8). It is used as a watchtower and it is situated in a dune landscape. This tower structure was one of the main outcomes of the Dutch contribution to the international co-operation project FAIR-2. This tower has the shape of three stacked octagonal antiprisms. The height of each antiprism is 3 m. The edge length of the top and the bottom octagon is 2.5 m. The two middle octagons have the smaller edge length of 1.91 m. so that a narrower 'waist' is formed at half height, giving the impression of a sand-glass. This shape was considered as to form a characteristic landmark in the surroundings. The octagon edges and the inclined struts were made of 140 mm thick poles. The central mast fulfils a supporting and binding function and has a diameter of 200 mm. The stairs were made of steel plate with timber clad steps. The building has been constructed of debarked, non-calibrated larch wood. The occurring tapering was accepted if not greater than 10 mm/m. In all cases the interconnection of the struts takes place via steel plates in the ends of the wooden poles. The poles have a slit, in which the plate is embedded. In corresponding holes a steel dowel is put through pole and plate, two in each end of the thinner poles and four in those of the central mast. The dowel is fixed in the hole by nails and around the two opposite holes a 25 mm wide stainless steel band is wrapped and subsequently sealed. This band is prevented from shifting by two wood screws. The final appearance of such a connection is very neat, because nothing but the band straps is visible.

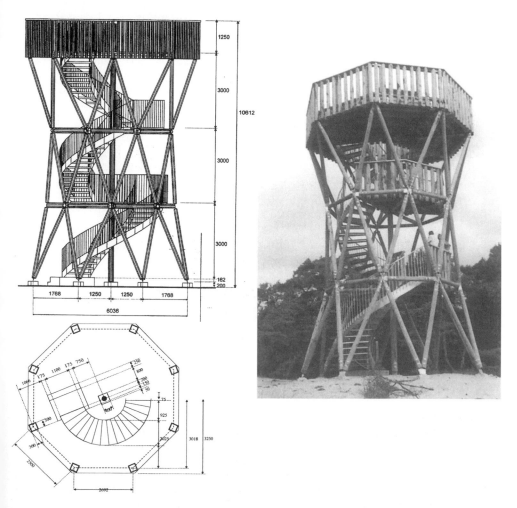

Figs. 25 and 26. Sketch and photograph of the tower structure at Kootwijk.

During the evaluation process after the completion of the tower, it was concluded, that the tower as a whole had a very good structural and visual impact. The connection detail appeared however to be quite complex and it was suggested to use in following cases the hollow sphere of Fig. 24. This might form a more practical and economic solution.

Fig. 27. Typical connection detail in mantle of structure in Kootwijk.

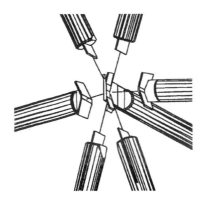

Fig. 28. Sketch of typical connection detail.

Fig. 29. Detail photograph during official opening of watch tower, showing the mantle detail.

ACKNOWLEDGEMENTS

Special mention must be made of those, who had an essential contribution to the realization of the above structures: Gerrit van der Ende, Caspar Groot, Rogier Houtman, Jaap Lanser, Sier van der Reijken, Erik Sluis, Peter de Vries (all from TU-Delft), Rico Golstein (Bureau De Bondt, structural advisor of the two tower structures), Peter Mulder (construction of tower in Apeldoorn), Erik Klein-Lebbink (Staatsbosbeheer, construction of the watch tower in Kootwijk). Thanks to their joint scientific approach many valuable results were gained, that may help exploit a source of material that so far has been very much underestimated.

REFERENCES
1. JAYANETTI, D.L., Timber pole construction, Intermediate Technology Publications, 1990.
2. RANTA-MAUNUS, A., Editor, Round small-diameter timber for construction, Final report of project FAIR-CT-95-0091, VTT Publications 383, Espoo, 1999, 191 pp.
3. RANTA-MAUNUS, A., Co-ordinator, Round small-diameter timber for construction, Progress Report 1998.
4. HUYBERS, P., The structural application of thin roundwood poles in building, Final report of project FAIR-CT-95-0091, Task 2A, Sept. 1996, 122 pp.
5. ADJANOHOUN G., *et al*, Design guidelines for engineered structures and connections, using small diameter round timber. FAIR-CT-95-0091.
6. HUYBERS, P., A node-less space frame system, LSA'98 conference on Lightweight Structures in Architecture, Engineering and Construction, 5-9/10, 1998, Sydney, Australia, p. 187-195.
7. HUYBERS, P., The materialisation of space frames, 40th Anniversary Congress of the IASS, Shells and spatial structures: from the recent past to the next millennium, Madrid, 20-24 September, 1999, p. F33-40.
8. HUYBERS, P., An antiprismatic tower structure, IASS/BLS Conference, 29/5-2/6/2000, Istanbul, p. 557-566.

The r oof s tructure " Expodach" at the World Exhibition Hanover 2000

J. NATTERER
Engineer Bois Consult Natterer, Germany
N. BURGER
IEZ Natterer GmbH, Germany
A. MÜLLER
Engineer Bois Consult Natterer, Germany

ABSTRACT
For the central meeting area of the EXPO 2000 a wide spaced roof construction of 10 square "umbrellas" was constructed serving as the main event location. Each "umbrella" covers an area of 40 m x 40 m and is about 26 m high. It consists of four double curved shell surfaces, having been constructed as partly glue-laminated timber ribbed shells. The shells hang over about 26 m and hang on four cantilevers. In the middle of the "umbrella" a big steel structure transfers the forces onto a t ower c onstruction. T he " umbrellas" a re c onnected a t t he o uter bending edges of the shells and the ends of the cantilevers. The structural components were mostly manufactured in the plant and in an exhibition hall near the site. Big cranes assembled the completely pre-manufactured construction parts step by step.

Fig. 1. General view of "Expodach".

THE IDEA

The concept behind the 16,000 m^2 roof without enveloping walls is a weather protection for the Expo 2000 main event location. The timber construction engineer Prof. Julius Natterer in Lausanne executed the idea – "the roof as the basic form of weather protection " – by the architect Prof. Thomas Herzog, Munich. The planning and execution demanded an extraordinary effort of everybody involved – planners, experts, construction worker and the client. The planning, preparing and manufacturing of the different construction parts were executed simultaneously and requested a significant coordination effort.

The technically and architecturally highly demanding and innovative timber construction is the aspects of the main spirit of the EXPO 2000 symbolizing in an outstanding manner – "Human-Being, Nature, Technique". The building demonstrates the possibilities of using timber for constructions in all varieties – round timber, sawn timber, boards, glued timber and laminated veneer lumber. The choice of the respective means of construction and of the material for each of the construction parts depended on the specific load-carrying capacity, the requested standards and the different material qualities.

THE MAIN STRUCTURE

Each single "umbrella" consists of different elements: 4 shells, 4 cantilevers, the central steel structure and the tower construction.

The shell areas show a double contrary curvature and carry the loads forward to the edge girders and the steel structure. This is made possible by the use of the shell capacity as well as the bending capacity.

The more than 19 m long cantilever carries two different weights: firstly the single force from the outer girder at the very end, secondly the continuous load of the shell itself. The lower bent beam follows the curvature of the shell's edge and is combined with the upper straight beam in the last third of the girder. From outside to inside the height of the girder increases, according to the loads.

The tower construction, consisting of 4 columns and triangular timber frames planked with laminated veneer lumber, carries all vertical and horizontal forces to the foundations. A steel joint connects the central steel construction with the head of the abutments. The frames transfer the horizontal forces of the wind and the geometrical imperfections.

Steel-elements at the lower end of the pole transfer the forces to the foundation. The foundation consists of 4 vertical drilling poles 10 to 15 m long. The heads of the poles are connected with a reinforced concrete wreath to balance the different horizontal forces.

Fig. 2. View of one umbrella.

The maximum numerical deformations, including the deformation of the tower and the steel structure, are 13 cm at the end of the cantilever and 36 cm at the unsupported end of the shell, due to snow and wind loads. Deformations of 17 cm and 50 cm respectively would be the result, if the "umbrellas" were not connected. A coupling at the unsupported end of the ribbed shell and the end of the cantilever was made to deal with this difference in deformation. In addition these couplings prevent the tower from dynamic torsion effects.

The whole construction is covered with a synthetic skin, levitating 5 cm above the timber structure. The fixation was made by cables running in the direction of the longitudinal ribs of the shell. The membranous tensions resulting from the loads and the necessary pretension are carried forward from the edge of the shells to the ribs.

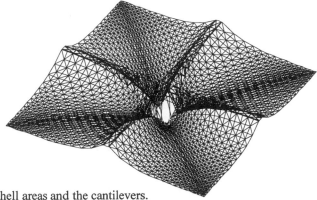

Fig. 3. Computer model of the shell areas and the cantilevers.

LOADS AND MODELS

The main loads of the roof are its self-weight, snow and wind. Other influences had to be considered such as geometric imperfections, dynamic loads and reactions resulting from the couplings. The wind loads and the accumulation of snow were simulated in a wind tunnel. Also dynamic effects caused by wind-induced oscillations were investigated.

The results were the opposite of what could have been expected. Unlike canopy roofs the main wind load is directed downwards, causing an additional load to the structure. This reaction is due to the special form comparable to a wing turned around. It creates negative pressure underneath the roof and pulls down the shells.

Snow and wind are causing an additional tension stress in the membrane since the covering membrane doesn't touch the structure. This tension remains almost constant, if a certain load is reached and the membrane is touching the shell.

The size of the complex structure made it necessary to develop different systems and computer models of some parts of the structure to calculate the forces and to design the elements. The structure is mainly composed of one-directional elements. Therefore all computer models were also made up of bars. The most important models were: A whole "umbrella" without the tower and the central steel structure, but containing four shells and cantilevers. The central steel structure and the tower were separate systems. All in all about 150 different steps of the systems were developed and calculated.

The model of the "umbrella" contained about 2500 node points and 9000 bars which had to be defined in position, geometry, material, the type of bar and its endings. The connections to the central steel structure were defined as fixed supports. The compliance of the connections was considered by reducing the stiffness of the elements.

Due to tight schedule the calculation of the tower construction had to be given priority. The loads of the "umbrella" were transformed into equal forces appearing at the top of the tower. Therefore the load of the foundation was known at a very early stage of the planning.

The conditions of the single models were checked at all intersections on the basis of equilibrium and compatibility to control the forces and the deformations. This was absolutely necessary, since the system is highly undetermined causing complex interactions between the shells and the other parts of the structure.

THE RIBBED SHELL

The ribbed shell is a construction of perpendicular crossing ribs built up of stacked planks. One shell covers an area of about 19 x 19 m. The vertical distance between the lowest and the highest point is 6,0 m. Each shell weighs 36 t.

The direction and distances of the ribs are given by the forces existing in the shell. They are made of 8-10 board layers with a cross-section of 30/160 mm each. At the intersection of the ribs the layers (made of boards S 10 and S 13) run through alternately. In areas of high stress the board layers consist of laminated veneer lumber produced in the complete length of the rib. Every other board is a filling board. A bolt is placed in the middle of each intersection.

The layers are screwed together, so the connection is compliant. In areas of high stress (e.g. towards the center of the "umbrella" and along the main diagonal), as well as at the connection of the ribs to the edge girders, the layers are glued together. In order to maintain the excellent dynamic capacity of a screwed construction, the glued areas were reduced to a necessary minimum. The ribs are connected to the girders with bolts and metal plates.

To obtain a sufficient stability of the ribbed shell, 2 additional layers of boards are located on top of the ribs, placed at an angle of 45° towards the ribs. The boards with cross sections of 24/100 mm are placed at a distance of 10 cm from each other. Because of the curvature and the high forces in the inner area a shell composed of six glued and thin layers of veneer lumber is used instead of the board layers.

A construction with curved edge girders was innovative and had to be taken in to account, when choosing the geometry of the shell. Therefore the formally

Fig. 4. Edge of a shell at the cantilever. Connection of the ribs.

known "HP-shells" resulting in straight edges over a squared ground plan were not applicable in this case.

The coordinates of the shell are prescribed by a mathematical function. The coordinates of the four corners of the shell are fixed points that determine the form. The vertical spread of the structure, meaning the difference of height of these four points, has an enormous influence on the resulting forces and also on the deformation and load capacity of the shell.

The axes of the ribs were defined by the geodetic lines using the mathematical form of the shell. The distances between the single ribs were made according to the forces appearing, varying from 38 cm in the area of the main diagonal to 160 cm in the area of the highest points of the shell.

THE CANTILEVER

The cantilever has a length of 19 m, a width of 3 m with a maximum height of 7 m. It weighs around 22 t and consists of two inclined parts of variable height.

The straight upper beams consist of glued laminated timber (BS 14) with a dimension of 22/100 cm. The curved lower beams are made of glulam as well and curved according to the edge of the ribbed shell.

The height of the lower beam is variable with a cross section of 22/110 cm up to 22/145 cm. Towards the central steel structure, the upper and lower beams are connected by a double plate of laminated veneer lumber (2 x 33 mm) that are completely glued together, pressured by screws. The cantilever is stabilized by diagonals forming a "K" (called K-diagonals) between the two lower beams. To balance the horizontal forces of each single ribbed shell, horizontal steel bars run perpendicular to the cantilever at the level of the K-diagonals.

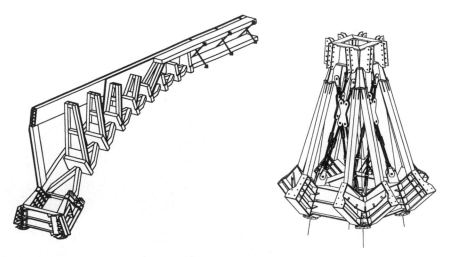

Fig. 5. Isometry of the inner parts of the cantilever.

Fig. 6. Isometry of the central steel structure.

The outer end of the cantilever is made like a box girder of glulam. Because of its low height the main load is a momentum caused by the suspended edge girders of the shell. The unequal load of the shells causes an additional torsional moment in the cantilever. The different parts of the box girder are connected with a special system called "Bertsche-VA-Connection", using steel-shear-connectors. After putting the box together, the cavities are filled with a volume-neutral cement mortar. This ensures a complete closing of the gaps and cavities and gives a high stiffness to various intersections.

THE CENTRAL STEEL STRUCTURE

The forces of the four shells and the cantilevers are concentrated in the central steel structure and passed on to the tower. This steel structure has a base of 5.5 x 5.5 m and a height of 7 m. It weighs 32 t and is composed of an upper and lower ring, where the main beams are fixed. The two rings are connected by eight steel columns and stabilized with prestressed steel bars to minimize deformations. All steel elements have got the same geometrical axis as the tower and the cantilever.

Eight welded steel elements are attached to the lower ring to support the lower beams of the cantilever. The inner part of the shell is also held by an especially formed steel element. A disk spring separates this part of the shell from the steel structure to obtain an equal distribution of the forces (up to 2500 kN compression) in this part of the shell.

THE TOWER

The tower is composed of four columns and stabilized by triangular timber frames. The columns are 16 m long and have a diameter of 68 cm to 74 cm at the lower end and of 95 to 110 cm at the upper end. Each one is made of a single up to 200 year old silver fir, which has been cut along the longitudinal direction into two half sections to speed up the natural drying and to simplify the connections. The half sections are tight together with two dowels and small timber pieces, placed every 50 to 75 cm with a remaining gap of 63 mm between the half sections.

The trees were selected visually considering different criteria such as number and diameter of the branches, spiral grain and the diameter at the lower end. The diameter at the height of 20 m was measured with a theodolite. The trees were also tested by ultrasound to find possible decay in the trunk.

Fig. 7: View of the tower.

Because of their enormous dimensions, it wasn't possible to dry the trees artificially. Therefore they were stocked on a sunny and windy place and prepared for further treatment. They were pealed with highly pressured water and cut in the longitudinal direction into two half sections.

The moisture content of the trunks was up to 200 % MC after cutting the trees. During construction seven months later the moisture content was still more than the equilibrium moisture content of approx. 16 % to 18 %. Therefore the allowable axial stresses had to be reduced by 1/3 and the stiffness by 1/6 as recommended by the German design code. The construction and the gap between the half sections will ensure a sufficient drying process during the next two to four years.

The distance between the four columns is decreasing towards the top. The forces are higher in the upper parts of the columns, so the top of the trunks with smaller diameters is located at the lower end of the columns.

The connectors of the columns had to satisfy high standards concerning their aptitude to the changing humidity of the timber as well as their stiffness and the execution of the connections. Therefore the Bertsche-BVD-connector was finally chosen for the joint at the bottom and at the top of the columns. This product is composed of a long, profiled steel element put in a drilled hole in direction of the wood's fibers. Twenty-four dowels are placed perpendicular to this steel element to keep it in place. All gaps are finally filled with a volume-neutral mortar. This procedure guarantees a stiff connection, which is able to transmit heavy loads of more than 300 kN per connector. Tests have proven the high performance this special case.

The stabilization of the tower was made of a timber frame construction by diagonals of glulam planked with laminated veneer lumber (LVL) of 33 mm thickness. These frames are continuously connected with the columns by bolts and dowels. To compensate the shrinkage of the cross-section all bolts have disk springs to maintain the tension in the bolts.

All connectors in the tower, except for the BVD-joints, are compliant to minimize additional stress in the structure due to shrinkage during the period of drying. The bolts of the dowels in the columns can still be reached and have to be tightened once the equilibrium moisture content is reached.

The horizontal distortion of the triangular frames is prevented by horizontal crosses of steel bars on three levels and additional frames planked with LVL on two levels of the tower.

CONSTRUCTIVE PROTECTION OF THE STRUCTURE
There are two different types of protection to guarantee the durability of timber: Chemical protection or constructive protection. Even without any chemical protection all timber structures can grow very old, as many buildings erected in the Middle Ages still prove today. Taking the spirit of the world-exposition - "Human-Being, Nature, Technique" - into account, it was one of the main goals to design a construction that would last a lifetime without any chemical protection.

The whole roof is covered by an impermeable membrane. Therefore structural considerations concerning the constructive protection of timber were intended for the design of many details. A distance of about five centimeters between the wood and the membrane ensures a permanent air conditioning of the structural members. Elements, which still get wet because of dewdrops, are able to dry out completely before decay can take place.

The most critical detail is the connection at the lower end of the columns, where the rain can directly get between timber and steel plate. The timber surface is covered with latex to close the fibers and to prevent the water from entering the timber by capillary forces.

Fig. 8: Connection of the lower end of the columns.

THE ASSEMBLY OF THE ROOF

The different elements of the roof were prefabricated at different places and transported as a whole to the construction site where the structure was erected. For easy mounting, the main joints between the elements were designed as steel-steel joints. The steel members also served as gauges for the prefabrication of the roof elements.

Fig. 9. Assembly of a ribbed shell.

The ribbed shells were produced in an exposition hall near the site in their entirety. Seven timber gauges were constructed to produce the 40 shells within three months. With an auxiliary steel frame one shell of 36 tons, covering a surface of 19 m x 19 m, was lifted out of the gauge and put on a truck to be transported to the construction site.

The mounting of an "umbrella" took place in several steps. First the columns and the timber frames of the tower were assembled and transported to the construction site. There they were put together and placed on the foundation. The welded central steel structure was produced in three parts and completed on site before they were lifted as a whole to the top of the tower. The cantilevers were completely prefabricated before they were delivered to the site. Because of their size they had to be transported by night on closed roads over a distance of about 700 km.

Once all four cantilevers were put in place, the ribbed shells with the temporary steel structure also could be lifted one after the other. To prevent excessive deformation, two of the cantilevers were temporarily supported. The temporary steel structures were taken away after the "umbrella" was completed and the four shells were tightened.

REFERENCES
1. NATTERER Ju., BURGER N., MÜLLER A., NATTERER Jo., Holzrippendächer in Brettstapelbauweis, Bautechnik, 77 (2000), H. 11, p.783-792.
2. ALEXANDROU C., GOULPIÉ P., HERTIG J-A. ,Generalisation of the aerodynamic admittance concept with application to the aeroelastic behaviour of bridges. 2 EACWE, Genova, Italy, 1997.
3. SANDOZ J-L, BENOIT Y., DEMAY L, Standing tree quality assessments using acousto-ultrasonic, International Symposium on Plant Health in Urban Horticulture BBA, 2000, pp. 172-179 (Heft 370). – IS.

Picture and drawing credits
Dipl. Ing Alan Müller Figure. 2, Figure 4
Dipl. Ing Kathrin Koetitz Figure 7, Figure 9
Dipl. Ing Johannes Natterer Figure. 1, Figure 8
IEZ Natterer GmbH Figure.3, Figure 5, Figure 6

Parametric design for the structural elements of timber rib shells

J. LEUPPI
ARUP, Los Angeles, USA

ABSTRACT

The presentation reveals a new method for determining the layout of the structural elements for timber rib shells, using parametric design.

The author has developed a computer program that generates lines of single curvature over any given surface. The goal was to provide a tool for simplifying the design process of ribbed timber shell structures built from a grid of planks running flush with the surface and, given the nature of a plank, curved in one direction only. Finding the correct geometrical layout of the structural elements is crucial for the construction of rib timber shells. In addition, it is beneficial for other methods of construction using curved surfaces. Currently, adequate computer tools are not available for the design of arbitrary shell structures, with the exception of membrane structures.

The proposed computer tool, combined with the parametric design option, promises to fill this void, providing the designer with a simple interactive tool that finds the optimal shape of structural elements for timber shell structures. The parametric design option is provided by CustomObjects, a Microstation sub program by Bentley. Automating this traditionally time-consuming development process will result in considerably reduced design costs.

INTRODUCTION

Contemporary examples in architecture show a growing preference for organic free shapes (e.g. Frank Gehry, Foster, Morphosis). These interesting and aesthetically appealing constructions demand more innovative thinking and present new challenges in the design and construction phases.

Unfortunately, the design team often views these challenges as handicaps that despite technical resolution would not be economical in terms of design effort or construction. Consequently, the original concept gives way to a standard solution. This commonly occurs for timber rib shell structures where the economical "standard" solution is selected over the original spatial shell structure. The difficulty in the execution of timber shell projects lies in designing the correct geometry of the ribs, which are the structural elements of the shell.

The design tool presented in this paper seeks to simplify the design of timber rib shell structures and produces simplified, consistent geometric solutions with single curved lines for what appears as a complex architectural layout. This single curved line approach considerably reduces design time, detailing, fabrication, transportation and installation on site.

TIMBER RIB SHELLS
Background to Timber Rib Shells
Timber rib shell structures are a relatively new construction method. A first prototype of this kind was erected in 1991 in Lausanne, Switzerland, called the 'Polydôme' (Hoeft, Ref 4).

Fig. 1. Example of a timber rib shell under construction: the 'Polydôme'. (Source: Hoeft, Ref 4).

Since then, numerous other roofs have been built in this manner, mainly in Germany, France and Switzerland. The most notable example of this type of structure is the Expodach for the world exposition in Hanover, by Herzog + Partner (Architect) and IEZ Natterer Gmbh (Structural Engineer), 2000 (Herzog, Ref 3).

Fig. 2. Roof structure of the Expodach by night. (Source: Herzog, Ref 3).

At this time, timber rib structures are becoming more attractive to designers because they are lightweight, aesthetically appealing, sustainable, long spanning (10 – 40m), and cost effective.

Fig. 3. Example of a barrel vault, Multimedia Pavilion, Gernsbach, Germany. (Source: www.aai-knapp.com).

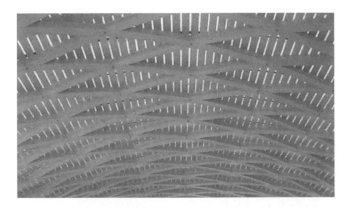

Fig. 4. Roof detail of a barrel vault shell. (Source: www.aai-knapp.com).

The main barrier to their wider use and acceptance is the lack of design tools. Presently the major difficulty lies in the planning of the shell structure. Designing a timber shell requires a significant level of expertise. In order to encourage and facilitate a wider use of them, the design costs need to be reduced. The proposed program tries to make a contribution in this area.

Construction of Timber Rib Shells
Timber rib shell structures form a complex and interesting topic.

This paper does not intend to explain timber rib shells in detail. However, in order to understand the significance of this design tool, the reader requires an understanding of the timber material used and the construction process employed for timber rib structures. This chapter outlines the relevant information.

Short, timber rib shell structures are three dimensional timber shells composed of wooden planks to form the ribs. Several interlocked layers are used to form each rib. The planks are screwed together; no glue is used. Each layer of planks is continuous in one direction. The gaps between the continuous layers are filled with filling boards, which contribute to the stiffness of the structures. The structure is assembled on site with planks that have been prefabricated, cut to the exact length and with predrilled holes for future connections.

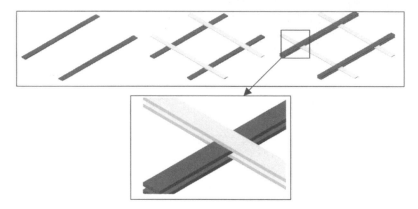

Fig. 5. Construction principal of stacked layers of planks (only the continuos layers are shown. The filling planks are not shown for clarity). (Source: Leuppi, Ref 6).

A scaffold on the site helps the construction and gives some fixed points for the future shell. Four to six layers of planks cross each other. In the intersection points the planks are connected with bolts.

Fig. 6. Example of a Timber Rib Shell under construction. (Source: www.arlesheim.ch).

Planning
In order to run a structural analysis, prefabricate the planks and set up the scaffold, the geometry of the rib structure has to be clearly defined.

Typically, the overall shape of the roof is defined by the architect or by other exterior conditions.

The ribs of the shell, which are the load bearing elements, need to lie flush with this given surface. The spacing and the general orientation of the ribs are defined by the engineer to suit stress distribution and avoid stress concentration. Typically the ribs are spaced 50cm – 100 cm, the spacing is narrower where stresses are higher and further apart where stresses are smaller.

Within these latter parameters, the exact geometry of the ribs is defined by the physical property of a timber plank. A plank can naturally be curved to a radius 'r' of about

r = d * 200   ('d' = thickness of the plank),

without the interior stresses being excessive. This means for a typical 3 * 16 cm plank, that it can be curved to r = 6 m in the weak axis of inertia, but only to r = 32 m in the strong axis of inertia. This property dictates the geometry in which the planks run over the surface.

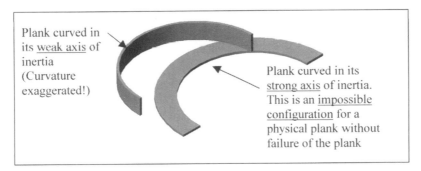

Fig. 7. Possible and impossible configuration of a plank. (Source: Leuppi, Ref 6).

Fig. 8. Illustration how planks naturally bend in their weak axis of inertia. (Source: Hoeft, Ref 4).

The plank has to follow a line for which bending normal to the surface is permitted; bending in the plane of the surface however is not possible.

As easy as it sounds, with the currently available drafting programs it is extremely time consuming, and sometimes not feasible, to define and visualize this line.

One solution to surround the problem is to compute these lines through mathematical equations. However, computing lines with single curvature over a given surface requires sophisticated mathematics and is only possible for relatively simple and mathematically defined surfaces. If the shape of the shell is modified during the progression of the project, the equations have to be adapted to the new conditions and the design and drafting process starts from scratch.

The difficulty of defining the geometry is the reason that mainly spherical and cylindrical roof structures have been built. The Expodach for the Main fair of the World Exposition (Herzog, Ref 3), which is a hyperbolic-paraboloid shape, is an exception. For this project a mathematician assisted the design team in finding the coordinates by mathematical computation.

Further information on timber rib shell structures can be found in texts by Natterer, Refs 7, 8, 9.

THE PLANK LINE
In this paper, the term "plank line" is used to define the line, which a physical plank would be occupying on a given surface, lying flush with the surface. On a spherical surface, this line coincides with the 'great circle' (refer to Figure 9). For other surfaces however, this is not as simple, for example with a cylinder (refer to Figure 10).

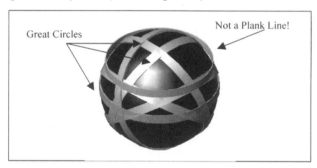

Fig. 9. Spherical surface with a series of great circles (plank lines) and a line, which isn't flush with the surface and therefore not a plank line. (Source: Leuppi, Ref 6).

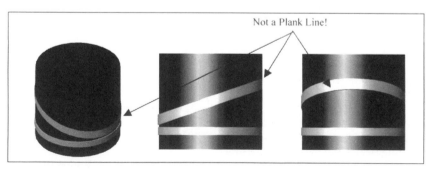

Fig. 10. Cylinder with an oblique line representing the intersection of the cylinder with a plane. This does not result in a plank line. (Source: Leuppi, Ref 6).

A plank cannot be bent in its strong axis of inertia, but it can, to a certain degree, be bent in its weak axis of inertia and it can also tolerate a certain level of twist (torsion). Knowing these latter statements, plank lines to the following three conditions can be created:

In a local coordinate system moving along the longitudinal axis of the plank, with 'X' in the longitudinal direction of the plank, 'Y' in the width of the plank, and 'Z' in the depth (thickness) of the plank (see Figure 11);

Condition 1: rotation around 'X' is permitted (torsion),

Condition 2: rotation around 'Y' is permitted (bending in the weak axis of inertia),

Condition 3: rotation around 'Z' is not permitted (bending in the strong axis of inertia).

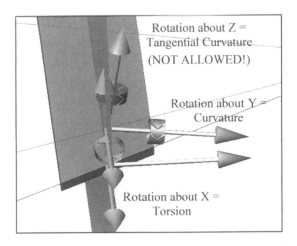

Fig. 11. Local Coordinate System of the plank line. (Source: Leuppi, Ref 6).

A fourth and last condition is due to the layers of planks intersecting each other in the opposite direction. In order for the assembly to be possible, the planks in each direction have to lie flat on the surface. Thus,

Condition 4: 'Y' has to be tangential to the surface in each instance.

Finite Element Approach
The four conditions are readily resolved using the finite element approach. By breaking down the surface into small elements, each element can be analysed by itself, knowing the conditions of the neighbouring elements. A plank line intersecting with one of the edges of a finite element can simply be projected to the other edge of the finite elements, by applying the previously mentioned four conditions. The method applied is purely geometrical by finding the intersection of planes cutting lines.

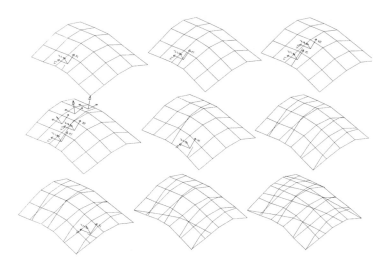

Fig. 12. Symbolic representation of the finite element approach. (Source: Leuppi, Ref 6).

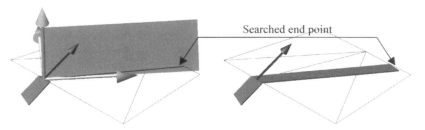

Fig. 13. Symbolic representation of the algorithm executed on a single finite element. (Source: Leuppi, Ref 6).

PARAMETRIC DESIGN

The algorithm described above has been translated into program code and integrated into CustomObjects, Ref 2, by Aish, Ref 1. CustomObjects, which is a subprogram of Microstation, is a 3D drafting tool with parametric design features.

Parametric Design allows designers to define 'intelligent' elements, which can modify themselves according to a set of rules defined by the designer. In other words, models are able to adapt themselves according to the results of an analysis. It is a relatively new area in the computer-aided design, with increasing applications in the architectural and construction communities.

In the case of timber rib shell structures, the algorithm for determining the plank line is hard coded into the objects defined in CustomObjects and consequently only requires input parameters (given by the designer).

In general terms, the "hard code" may contain a calculation as simple as to find the mid point between two given points or a more complex one such as the calculation of a truss between two points following geometrical and structural rules.

The input parameters for the plank line are as follows:

1) Two points (the origin and a second point which determines the direction of the line),

2) The surface on which the plank line lies

3) The width and the height of the plank.

4) Some other parameters, of less importance

The output is a plank, line or series of points (depending on the user selection), which is the optimal trace for a wooden plank element on the surface. In other words: a planar element, bent only in the weak axis but not in the strong axis of inertia, while lying flush with the surface.

But what happens if the shape of the surface changes?

Now the parametric approach becomes useful. In this case, since the plank line is associated with the surface (see input parameter 2 above), the surface is one of the "root objects" of the plank line that the plank "depends" on. By changing the shape of the surface, the plank lines associated with the surface will also be modified automatically.

In a normal design process, frequent changes occur as an optimal solution is realized. While a conventional drafting tool requires redrafting of everything that has been modified, the parametric design keeps the logic between the different objects.

The parametric nature and the programming of objects as outlined create a robust design tool. Further information on parametric design and CustomObjects in particular can be found in texts by Aish[1].

RESULTS
The author has developed a computer program called RIPS (Rib structure Interactive Plank System lines) (Leuppi, Ref 6 and Kensek, Ref 5) that will generate lines of single curvature over any given surface.

Figure 14 shows the results of RIPS. The program reads the dxf file for a specific surface drawn in AutoCAD, computes the plank line over the surface and creates the plank lines drawing as a different dxf file that can be visualized in AutoCAD again. The results show that the planks lines are accurate although it is difficult to manipulate the geometry. The parametric design features are missing in this program; modifying the plank lines or the surface is awkward and difficult.

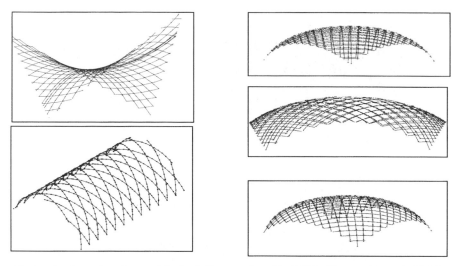

Fig. 14. A selection of surfaces with plank lines analysed in RIPS. (Source: Leuppi, Ref 6).

Figures 15 and 16 show the result of the plank line applied in the CustomObjects software, Ref 2. The line as well as the surface can interactively be moved and modified in its form; the plank lines automatically adapt to the new shape. The model shows a retrospective study of the Expodach (see Figure 2).

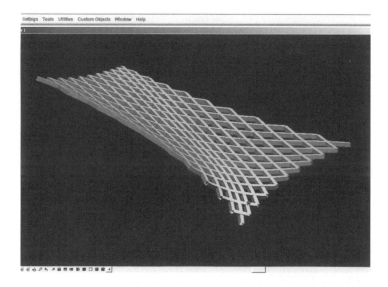

Fig. 15. Retrospective study of the Expodach, model in CustomObjects (axonometric view).

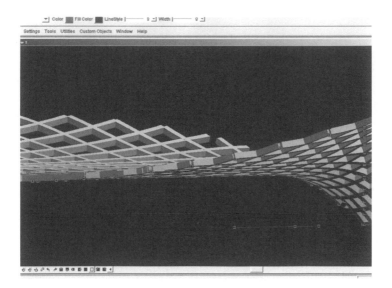

Fig. 16. Retrospective study of the Expodach, model in CustomObjects (elevation).

SUMMARY
Object based parametric modeling is the future of computer-aided design. It will allow us to define intelligent elements, which can modify themselves according to a set of rules defined by the designer. A perfect example for the application for parametric design is the computation of the ideal geometry for the structural elements of timber rib shells. The algorithm presented in this paper for calculating the ideal plank line over a given surface has been integrated in CustomObjects (a parametric design program by Bentley, subprogram of Microstation). The main advantages that interactive parametric modeling offers in comparison to conventional drafting tools are cost savings in planning, shop work and construction. As a consequence of the reduced design cost, timber shell structures may come into wider use.

REFERENCES
1. AISH R., Custom Objects: a model-oriented end-user programming environment, Workshop on Visual Languages for End User and Domain Specific Programming. University of Washington, September 2000.
2. CustomObjects, MicroStation, Bentley Systems Incorporated, (www.bently.com).
3. Herzog T., (ed) Expodach, Roof Structure at the World Exhibition Hanover 2000, Prestel, München London New York, 2000.
4. HOEFT M., KAELIN J., Ausstellungspavillon in Brettstapelbauweise, Holzbau, Fachbeilage zum Schweizer Baublatt, No.2, 1992, January 10.
5. KENSEK K., LEUPPI J., NOBLE D., Plank Lines of Ribbed Timber Shell Structures, ACADIA 2000, Catholic University, Washington, D.C., 2000.
6. LEUPPI J., Interactive C++ Program to Generate Plank System Lines For Timber Rib Shell Structures, a Thesis presented to the Faculty of the School of Architecture, University of Southern California, 2001.
7. NATTERER J., BURGER B., MÜLLER A., NATTERER J., Holzrippendächer in Brettstapelbauweise – Raumerlebnis durch filigrane Tragwerke, Bautechnik, No.77, 2000, pp783-792.

8. NATTERER J., Design and Construction of Timber Space Structures, International Symposium on Theory, Design and Realization of Shell and Spatial Structures, Nagoya, Japan, October 2001.
9. NATTERER J., HERZOG T., & VOLZ M., Construire en bois 2 (2nd ed.), Lausanne, Presses Polytechniques et Universitaires Romandes, 1998.

Analysis of stress and strain in wood beams

N. T. MASCIA and L. VANALLI
State University of Campinas, Brazil

ABSTRACT
The general elastic constitutive model formulated to describe the mechanical behavior of wood is the orthotropic model, due to the internal structure of wood, to have the axes of elastic symmetry longitudinal, tangential and radial. The study of the orthotropy implies in the knowing of the constitutive law that governs the elastic behavior of the material and in the determining the constitutive tensor, S_{ijkl}. For an application of this study to determine the distribution of stresses and strains in structural elements it was adopted Airy's stress function. This function involves the equations of equilibrium, the deformation-displacement relations and the constitutive equations, considering in this paper the plane state of stresses. The solution of this function in the form of polynomials leads to satisfactory results in the obtaining of the stress functions that are solutions of the plane problems in the elasticity. To carry out this work cantilever beams were used, made of *Ipê*- Brazilian wood species, under triangular and uniform distributed loads. From the obtained results, it was observed the concentration of normal and shear stress is different when comparing wood and isotropic material.

INTRODUCTION
On the whole, in order to solve a solid mechanics problem some conditions must be satisfied. These conditions are related to equations of equilibrium, strain-displacement relations and material constitutive laws. The first and second conditions do not depend on the characteristics of the material of which the solid is composed. Whereas the third, which relates stress to strain components at any point in the solid, is a function of the material. These laws may be simple or complex, depending on the material of the body. In fact, the behavior of the real material is not easy to comprehend. When trying to model mathematically that behavior, it is necessary to construct idealizations and perform simplifications, using a convincing theory and adequate experimental tests. The final result of modeling is to get an expression that can be used to predict a specific property, with an acceptable degree of reliability.

The most general elastic constitutive model formulated to describe the mechanical behavior of material is the anisotropic model. This kind of model implies that there is no material symmetry, and mechanical properties in certain directions are different. On the other hand, if there is material symmetry, the material can be denominated, for example, orthotropic or isotropic. In this context, the adequacy of a determined material for a certain elastic model is based on the existence of elastic symmetry axes. In these axes, denominated elastic principal axes, there is invariance of the constitutive relations under a group of transformations of coordinate axes.

In fact, the study of anisotropy implies in the knowing of the constitutive law that governs the elastic behavior of the material and consequently, determining the constitutive tensor, S_{ijkl}, and its components. In a completely elastic and anisotropic model this tensor has 81 unknown constants. By using adequate simplifications, this number can be reduced to 9 constants, which is denominated orthotopric model, or to 3 constants, the isotropic model. Among the construction materials, wood, because of its internal structure with axes of elastic symmetry longitudinal, tangential and radial, reveals an orthotropic pattern. Thus, there are 9 constants to be determined.

For a specific application of this theory, the distribution of stresses and strains in some structural elements, as for example, wood beams will be analyzed in this work. This way, looking for to determine functions of stress, which supply the solutions of the differential equations that shape some problems of elasticity, it will be applied the equations of equilibrium, the relations deformation-displacements and the constitutive equations, considering always the plane state of stresses. It is important to notice that the stresses and the strains could be very different of the isotropic solids. It also notices that the solutions in the form of polynomials lead to satisfactory results in the obtaining of the stress functions, which are solutions of the plane problems in the theory of the elasticity.

For a better inquiry of these solutions, theoretical and experimental comparisons, in terms of stresses, with structures of wood and isotropic material had been carried through. For this, it was used wood of the following Brazilian species, *Ipê*. From the obtained results, as for example, it was observed that in the free extremities of cantilever beams, under the triangular and uniform distributed loads, the concentration of normal and shear stress is different when comparing wood and isotropic material. This, certainly, supplies information that contributes for the application of wood in structures of the civil construction.

THEORETICAL ANALYSIS OF THE ELASTIC PROPERTIES OF ANISOTROPIC MATERIALS ELASTIC MODELS

According to Love (Ref 1), Chen and Saleeb (Ref 2) among others, the laws and equations that govern engineering problems are related to the stored energy in a solid. So, an elastic solid is capable of storing the energy developed by the external work and transforms it into potential elastic energy that is denoted as strain energy. During this process the body is deformed, but recovers its original shape and size.

In this condition, if no energy is dissipated during the process of deformation, under adiabatic and isothermal conditions, the derived equations from this supposition are termed elastic models of Green and the material that makes the body as hyperelastic material. Thus, a hyperelastic material is one that has a strain energy function, denoted by U_o.

The elastic material of Green is, in fact, a special case of the most general elastic material called elastic material of Cauchy, but considering the existence of the U_o, in order to maintain unaltered the laws of thermodynamics. These laws say that an elastic material produces no work in a closed loading cycle.

For an elastic body, the current state of stress depends only on the current state of strain. Mathematically, the constitutive laws can be written as:

$$\sigma_{ij} = F_{ij}(\varepsilon_{kl}) \tag{1}$$

in which: σ_{ij} is the stress tensor; ε_{kl} is the strain tensor, and F_{ij} is the response function. Notice that only small strains given by $\varepsilon_{ij} = (u_{i,j} + u_{j,i})/2$ are considered in the strain tensor. The term $u_{i,j}$ represents the partial derivative of displacement.

As has been emphasized in the literature, the elastic model described by (1) is both reversible and path independent since strains are uniquely determined from the current state of stress or vice versa.

We can set the response function as polynomial relations of n-degree, relating stress and strain, by:

$$\sigma_{ij} = \phi_0 \delta_{ij} + \phi_1 \varepsilon_{ij} + \phi_2 \varepsilon_{im} \varepsilon_{mj} + \phi_3 \varepsilon_{im} \varepsilon_{mn} \varepsilon_{nj} + \dots \tag{2}$$

where $\phi_0, \phi_1 \dots$ are elastic response parameters. (Desai (Ref 3)).

One can observe that the first term in (2) is related to the scalar state of stress or strain, the second term represents the first order model or linear model, the third term represents the second order or nonlinear model and so on.

Consider now an elastic solid in equilibrium, with conditions of respected compatibility, as we can see in Figure 1.

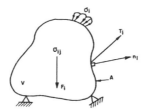

Fig. 1. Elastic solid in equilibrium.

The Principle of Virtual Work relates a series of equilibrium $F_i, T_i, \sigma_{ij}, u_i$ to a series of the virtual compatibility $\delta u_i, \delta \varepsilon_{ij}$ via the following equation:

$$\int_A T_i \delta u_i \, dA + \int_V F_i \delta u_i \, dV = \int_V \sigma_{ij} \delta \varepsilon_{ij} \, dV \tag{3}$$

where T_i is the surface force; F_i is the external body force; u_i is the displacement; A is the area; V is the volume and δ denotes variation.

The left side of the equation (3) represents the variation of external work δW, while the right side represents the variation of the strain energy delta δU. Using $U_o = \dfrac{U}{V}$ and since U_o is only a function the components of strains ε_{kl}, one can show that:

$$\sigma_{ij} = \frac{\partial U_o}{\partial \varepsilon_{ij}} \tag{4}$$

Thus the relationship among σ_{ij} , ε_{kl} and U_o can form the constitutive laws of the material. In general, these laws describe the behavior of the usual construction material.

Using the strain energy function and considering the Green elastic model of the first order, formulations of the constitutive laws for different classes of elastic materials can be established. So, consider a strain energy function given by:

$$U_0 = C_0\delta_{ij} + \alpha_{ij}\varepsilon_{ij} + \beta_{ijkl}\varepsilon_{ij}\varepsilon_{kl} \tag{5}$$

where C_0, δ_{ij}, β_{ijkl}, α_{ij} are constants.

In view of the strain energy formulation where the strain energy has a stationary value in relation to the strain tensor, it is possible to set $C_0 = 0$. It may also be taken into the account the fact that $\alpha_{ij} = 0$, since that the initial strain field corresponds to an initial stress free state. From equation (4) and (5), the stresses can be expressed by:

$$\sigma_{ij} = C_{ijkl}\varepsilon_{kl} \tag{6}$$

where $(\beta_{ijkl} + \beta_{klij})$ can be taken as C_{ijkl} .The C_{ijkl} is the tensor of material elastic constants.

Agreeing that $|C_{ijkl}| = 0$ the equation (12) can be expressed as:

$$\varepsilon_{ij} = S_{ijkl}\sigma_{kl} \tag{7}$$

where S_{ijkl} is the compliance tensor.

The C_{ijkl} has 81 constants to be determined and must be symmetrical due to Cauchy's second law of motion. In addition, since both σ_{ij} and ε_{kl} are symmetrical, the number of elastic constants is reduced to 21, with 18 independent ones. This implies that in an anisotropic material the principal stress directions may not coincide with the principal strain directions.

Plane Problems of the Theory of Elasticity of an Anisotropic Body
In accordance with Lekhnitskii (Ref 4), the state of stress in all the points of an elastic body is known if the stress components are known in three perpendicular planes between itself and the coordinate directions. The state of strain is determined by the components of strain, which depend on the three displacements in the coordinate directions. Consequently, to have the full determination of the states of stress and strain of an elastic body, which is submitted to the external forces, it is necessary to determine nine functions: six components of stress σ_{ij} (i, j = 1, 2, 3) and three components of displacements u_i (i = 1, 2, 3). To determine them nine independent equations are necessary, that are the three equations of equilibrium and the six equations expressing the generalized Hooke´s Law.

As a specific case, Lekhnitskii (Ref 4) considers a homogeneous anisotropic elastic plate that is in equilibrium and subjected to a plane state of stress. If we do not take into account the action of the body forces, the equations of equilibrium can be written:

$$\sigma_{i1,1} + \sigma_{i2,2} = 0 \quad (i = 1, 2) \tag{8}$$

Using the reduced to notation presented by Ting (Ref 5), the constitutive law (7) in a simplified form can be written:

$$\varepsilon_m = S_{mn}\sigma_n \quad (m, n = 1, 2, 6) \tag{9}$$

The equations of equilibrium are satisfied by the introduction of a function of stress F and considering that:

$$\sigma_{11} = F_{,22} ; \qquad \sigma_{22} = F_{,11} ; \qquad \sigma_{12} = F_{,12} \qquad (10)$$

where σ_{ij} is normal stress for $i=j$ and shear stress for $i \neq j$.

If we substitute the equation (10) in the constitutive equations (9) and this result in the compatibility equation (5):

$$\varepsilon_{1,22} + \varepsilon_{2,11} - \varepsilon_{6,12} = 0 \qquad (11)$$

it is obtained the differential equation:

$$S_{22}F_{,1111} - 2S_{26}F_{,1112} + (2S_{12} + S_{66})F_{,1122} - 2S_{16}F_{,1222} + S_{11}F_{,2222} = 0 \qquad (12)$$

Thus, Lekhnitskii (Ref 4) shows that a plane problem of the theory of the elasticity is reduced to the determination of a function of stress F, which satisfies the differential equation (12) and the boundary conditions of each problem in study.

For plane problems in the isotropic elasticity, Timoshenko and Goodier (Ref 6) had presented solutions using polynomial functions of diverse degrees to determine the stress functions.

Some Considerations about the Orthotropic Model Applied to Wood

The theory of elasticity applied to wood is based on the hypothesis that wood has three mutual planes of elastic symmetry according to its internal structure. Bodig and Jayne (Ref 7), in addition to this hypothesis, consider the material homogeneous. Therefore, the longitudinal- tangential surface is not a plane, but roughly cylindrical. The other two surfaces, the longitudinal-radial and radial-tangential are, truthfully, more straight. Thus, wood may be treated as a cylindrical orthotropic body.

Instead of adopting the cylindrical model, avoiding the mathematical complication and experimental difficult as well, we adopted the rectilinear orthotropic model only, that we can see it in the Figure 2 (Hearmonson, Ref 8).

In this way, arbitrating for wood the rectilinear ortrotropic model, with the three elastic principal axes denoted *L*, *R* and *T* the components of the compliance tensor are given by (7), replacing the indices *1,2* and *3* by *L*, *T* and *R*.

However, we notice that if a different coordinate system is considered, other components of S_{ijkl} will be non-zero and the constitutive laws will become more complicated to use.

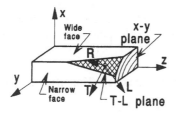

Fig. 2. Material axes and board axes for wood.

NUMERICAL EXAMPLES

With the purpose of studying the distributions of stresses on anisotropic cantilever beam, subject to firstly a triangular distributed load and secondly to a uniform distributed load, and of obtaining a stress function as well, the analytical method of Hashin (Ref 9) was applied.

Cantilever beam subject to a distributed triangular load

For the cantilever beam, considered anisotropic, of Figure 3:

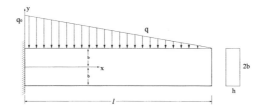

Fig. 3. Cantilever beam subject to a distributed triangular load.

The following conditions of contour are presented:

$$\phi(x_1,-b)=0 \tag{13}$$

$$\phi(x_1,-b)_{,2}=0 \tag{14}$$

$$\phi(x_1,b)=-\frac{q_0}{6}\left(l^2-3lx_1+3x_1^2-\frac{x_1^3}{l}\right) \tag{15}$$

$$\phi(x_1,b)_{,2}=0 \tag{16}$$

Equation (15) corresponds to the bending moment in a generic cross section of the cantilever beam of Figure 1. The greatest power of x_1 in the contour conditions (13-16) is $M=3$. Thus, $N=M+3=6$, and the stress function becomes:

$$\phi(x_1,x_2)=\sum_{m=0}^{m=3}\sum_{n=0}^{n=6}C_{mn}x^m x_2^n = \qquad m+n \le 6$$

$$= C_{00}+C_{01}x_2+C_{02}x_2^2+C_{03}x_2^3+C_{04}x_2^4+C_{05}x_2^5+C_{06}x_2^6+$$
$$+C_{10}x_1+C_{11}x_1x_2+C_{12}x_1x_2^2+C_{13}x_1x_2^3+C_{14}x_1x_2^4+C_{15}x_1x_2^5+$$
$$+C_{20}x_1^2+C_{21}x_1^2x_2^2+C_{22}x_1^2x_2^2+C_{23}x_1^2x_2^3+C_{24}x_1^2x_2^4+$$
$$+C_{30}x_1^3+C_{31}x_1^3+C_{32}x_1^3x_2^2+C_{33}x_1^3x_2^3 \tag{17}$$

The number of equations that are necessary for the determination of the coefficients, C_{mn}, is given by:

$$S=\frac{1}{2}(M+1)\cdot(M+8)=22 \tag{18}$$

With the 22 equations it is possible to solve a system to find the 22 unknown coefficients, C_{mn}, of the stress function (17), using a numerical technique.

Using the Airy's stress function it is possible to obtain the stresses in these beams and the following elastic constants:

$$S_{1111} = S_{11} = \frac{1}{E_x}; \; S_{1122} = S_{12} = \frac{-v_{xy}}{E_x}; \quad 2 \cdot S_{1212} = \frac{S_{66}}{2} = \frac{1}{2 \cdot G_{xy}}; \quad S_{1112} = \frac{S_{16}}{2} = \frac{\eta_{xy,x}}{2 \cdot E_x};$$

where: E_i = Young's Modulus, G_{ij} =Shear Modulus, v_{ij}=Poisson's Ratio and $\eta_{ij,i}$= the mutual influence coefficients of the first kind (Lekhnitskii[4]), we can write that:

$$\sigma_x = \frac{\partial^2 \phi(x,y)}{\partial y^2} = \frac{q\left[\left(3E_x y + 4G_{xy}\left(\frac{-12y \cdot \eta_{xy,x}^2 + 51 \cdot \eta_{xy,x} - 10x \cdot \eta_{xy,x} -}{-3 \cdot v_{xy} \cdot \eta_{xy,x} \cdot y}\right)\right)b^2\right]}{40b^3 G_{xy}} +$$

$$+ \frac{q\left[5y\left(2G_{xy}\left(\frac{3l^2 - 6xl - 6 \cdot \eta_{xy,x} \cdot yl + 3x^2 + 8 \cdot \eta_{xy,x}^2 \cdot y^2 +}{+ 2 \cdot v_{xy} \cdot y^2 + 12 \cdot \eta_{xy,x} \cdot x \cdot y}\right) - E_x y^2\right)\right]}{40b^3 G_{xy}}$$

(19)

$$\sigma_y = \frac{\partial^2 \phi(x,y)}{\partial x^2} = -\frac{q(2b-y)(b+y)^2}{4b^3}$$

(20)

$$\sigma_{xy} = \frac{\partial^2 \phi(x,y)}{\partial y \partial x} = \frac{q(3l - 3x - 4 \cdot \eta_{xy,x} \cdot y)(y^2 - b^2)}{4b^3}$$

(21)

For the comparison of the normal and shear stresses (taking anisotropic and isotropic beams into account) comparative diagrams were constructed, considering wood of the species: *Ipê* as example of anisotropic material of the beams (the elastic constants gotten in Mascia, Ref 10).

Table 1 . Elastic Constants of Brazilian wood species: *Ipê*

E_x (kN/m²)	G_{xy} (kN/m²)	v_{xy}	$\eta_{xy,x}$
18043900.00	620200.00	0.4345	1.3051

The sections of analysis in the beam are x = 0. It was also considered the beams with thickness h = 1 m and *l* = 1.50 m, q = 20 KN/m and b = 0.20 m. We can see some of these results in Figures 4 and 5.

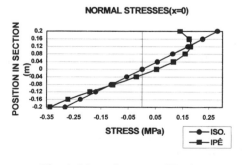

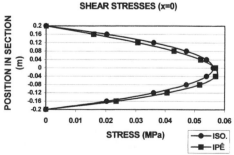

Fig. 4. Normal stress at X = 0.   Fig. 5. Shear stress at X = 0.

Cantilever beam subject to a Uniform Distributed Load
Consider, now, the following beam:

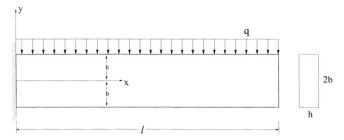

Fig. 6. Cantilever beam subject to a uniformly distributed load.

with the following conditions of contour:

$$\phi(x,-b) = 0 \tag{21}$$

$$\phi(x,-b)_{,2} = 0 \tag{22}$$

$$\phi(x,b) = \frac{q}{2}(l-x)^2 \tag{23}$$

$$\phi(x,b)_{,2} = 0 \tag{24}$$

Analogously to the Equation (15), the Equation (23) corresponds to the bending moment in a generic cross section of the cantilever beam of Figure 6. The greatest power of x_1 in the contour conditions (36 – 37) is $M = 2$. Thus, $N = M+3 = 5$, and the stress function becomes:

$$\phi(x,y) = \sum_{m=0}^{m=2}\sum_{n=0}^{n=5} C_{mn} x^m y^n = \qquad m+n \leq 5$$

$$= C_{00} + C_{01}y + C_{02}y^2 + C_{03}y^3 + C_{04}y^4 + C_{05}y^5 +$$
$$+ C_{10}x + C_{11}xy + C_{12}xy^2 + C_{13}xy^3 + C_{14}xy^4 + \tag{25}$$
$$+ C_{20}x^2 + C_{21}x^2y + C_{22}x^2y^2 + C_{23}x^2y^3$$

The number of equations that are necessary for the determination of the coefficients, C_{mn}, is given by:

$$S = \frac{1}{2}(M+1)\cdot(M+8) = 15 \tag{26}$$

With the 15 equations it is possible to solve a system to find the 22 unknown coefficients, C_{mn}, of the stress function.

$$\phi(x,y) = \frac{-q(b+y)^2\left(\begin{array}{l}10\alpha_{04}^2(2b-y)(l-x)^2 - 2\alpha_{13}^2(b-y)^2 y + \alpha_{04}(b-y)^2 \cdot \\ \cdot(5\alpha_{13}(l-2x)+2\alpha_{22}y)\end{array}\right)}{80\cdot\alpha_{04}^2\cdot b^3} \tag{27}$$

Using the Airy's stress function it is possible to obtain the stresses in these beams and the following elastic constants:

$$\alpha_{04} = S_{1111}; \quad \alpha_{13} = 4\cdot S_{1112}; \quad \alpha_{22} = S_{1122} + 2\cdot S_{1212}$$

we can write that:

$$\sigma_x = \frac{\partial^2 \phi(x,y)}{\partial y^2} = \frac{q\left[\left(3E_x y + 4G_{xy}\left(\begin{array}{c}-12y \cdot \eta_{xy,x}^2 + 5l \cdot \eta_{xy,x} - 10x \cdot \eta_{xy,x} - \\ -3 \cdot \nu_{xy} \cdot \eta_{xy,x} \cdot y\end{array}\right)\right)b^2\right]}{40b^3 G_{xy}} +$$

$$+ \frac{q\left[5y\left(2G_{xy}\left(\begin{array}{c}3l^2 - 6xl - 6 \cdot \eta_{xy,x} \cdot yl + 3x^2 + 8 \cdot \eta_{xy,x}^2 \cdot y^2 + \\ + 2 \cdot \nu_{xy} \cdot y^2 + 12 \cdot \eta_{xy,x} \cdot x \cdot y\end{array}\right) - E_x y^2\right)\right]}{40b^3 G_{xy}}$$

(28)

$$\sigma_y = \frac{\partial^2 \phi(x,y)}{\partial x^2} = -\frac{q(2b-y)(b+y)^2}{4b^3}$$

(29)

$$\sigma_{xy} = \frac{\partial^2 \phi(x,y)}{\partial y \partial x} = \frac{q(3l - 3x - 4 \cdot \eta_{xy,x} \cdot y)(y^2 - b^2)}{4b^3}$$

(30)

The sections of analysis in the beam are x = 0 and x = 1.50 m (free extremity of the beam). It was also considered the beams with thickness h = 1 m and l = 1.50 m, q = 20 KN/m and b = 0.20 m. Following we presented some of these results through Figure 7 and 8.

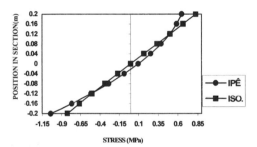
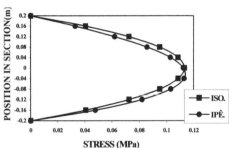

Fig. 7. Normal stress at X = 0.   Fig. 8. Shear stress at X = 0.

CONCLUSIONS
In this paper the general concepts of the orthotropic elastic model were described, particularly the rectilinear model, and the application of Airy's stress functions, in order to study the distributions of stresses in wood beams, considered as anisotropic material.

It is important to notice that the stresses in anisotropic beams are very different from those presented in isotropic beams. This is especially important when treating wood as an orthotropic material and considering heavy-timber construction that involves sophisticated engineering design.

Particularly, in the application of Hashin's method of polynomial solution it was evidenced that the stress polynomial functions lead to satisfactory results in terms of stresses in anisotropic beams (wood beams) under the distributed loads.

REFERENCES

1. LOVE, A. E. *A Treatise On The Theory Of Elasticity.* Dover Publications, New York,1944.
2. CHEN, E.F., SALEEB, A. *Constitutive Equations For Engineering Materials.* New York: John Wiley & Sons, Inc, New York, 1982.
3. DESAI, C. S., and Siriwardane, H. J. *Constitutive Laws for Engineering Materials -with Emphasis on Geologic Materials*, Prentice-Hall, New Jersey, 1984.
4. LEKHNITSKII, S.G. *Theory of Elasticy of an Anisotropic Body,* Mir Publishers, Moscow, 1981.
5. TING, T.C.T. *Anisotropic Elasticity. Theory and Applications,* Oxford University Press, New York, 1996.
6. TIMOSHENKO S. P., Goodier J. N. *Theory of Elasticity.* 3ª ed. 1970.
7. Bodig, J., and Jayne, B.A. *Mechanics of Wood and Wood Composites.* Van Nostrand, New York, 1982.
8. HEARMONSON, J.C., Stahl, D.C., Cramer, S.M., Shaler, S.M. "Transformation of Elastic Properties for Lumber with Cross Grain." *J. of Struct. Div.* 123(1), 1997, 1402-08.
9. HASHIN, Z. Plane Anisotropic Beams. *Journal Applied Mechanics*, Trans. of Asme, 1967, 257-262.
10. MASCIA, N.T. *Concerning Wood Anisotropy.* EESC-USP, Brazil, 1991. *In Portuguese.*

ACKNOWLEDGEMENT
The authors gratefully acknowledge CAPES and FAPESP, Brazilian Foundations for the financial supporting of this research.

On strength calculations of composite wooden bars

J. B. OBRĘBSKI
Warsaw University of Technology & University of Warmia and Mazury in Olsztyn, Poland

ABSTRACT
In contemporary civil engineering it can be pointed many structures, where the bars are built of some materials having different mechanical properties. The paper turns attention on possibility to apply new kind of strength calculations for the wood structures composed of the bars built of some different materials. The appearance of two or more materials in cross-section, different boundary conditions at bar ends and additional constraints on its length and/or variable bar strength on its length, needs to apply the more laborious method of calculation, which on the other hand should not be too complicated. There exist some numerical programs helping in this matter. The method is easy enough in application, to be used for usual engineering practice. Such method of calculation was elaborated for the discussed types of bars by J.B.Obrębski, and tested on many examples.

INTRODUCTION
There can be pointed out many examples of bars built of some materials having different mechanical properties. In the simplest case it can be a prismatic bar made of one piece of timber, but laminated outside, as it is shown in the Figure 1a. The other example can concern the bar composed of some layers or box girder, where can be applied different kinds of timber, as it is shown in the Figures 1b and 1c. There stronger kinds of timber can be combined with weaker kinds. As the next case we can point to the homogenous members – made of one block of timber but fastened in steel node as, e.g. it is shown in the Figure 1d and Figure 2. As the last example, it can be mentioned the timber beam is strengthened by banding steel, Figure 1e. Such as above and similar examples it can quoted many. In the case – shown in the Figures 1d and 2 - we have the bar with its rigidity variable on the bar length – strengthened parts at the ends.

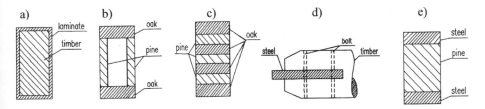

Fig. 1.

Such examples are quoted in many books and papers, by P.Huybers, Mielczarek (Ref 1), Werner & Zimmer (Ref 13) and by Stephane Du Chateau. See the exhibition in Delft (Ref 14), too.

In all above cases, on whole bar length its cross-sections are composed of two (or more) materials, or even there are the variable bar rigidity. These structural solutions point on the necessity to apply special, more exact manner of mechanical and strength calculations.

In the above cases we can say, that the bar cross-section is *composed* of some materials or that it is a **composite bar**. Structure built of such elements will be called as **composite structure**.

a)   b)   c)

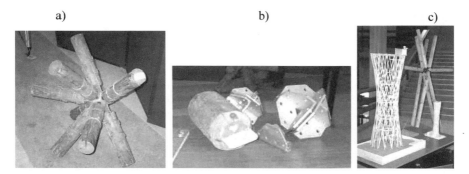

Fig. 2. Exhibition (Ref 14).

In the paper there are presented some specially interesting theoretical problems and results of calculations concerning determination of the behaviour of such bars: disposition of internal forces, displacements, stresses, critical, simple and combined loadings - compared with e.g. Euler's ones etc. All given results can be compared with traditional ones, and its percentage evaluation.

The paper presents some kind of proposal to apply new – more exact and more economical manner of mechanical and strength calculation for many contemporary civil-engineering objects – including first of all the wood structures.

ON THE APPLIED THEORY
The examples presented in the following sections are calculated by original theory elaborated for thin-walled bars enclosed in the book – Obrębski (Ref 2) and next for bars with any cross-section including compact, thick ones – Obrębski (Ref 3). Next theory was completed by many details and examples in papers of Obrębski (Refs 4 –12).

SIMPLE EXAMPLES OF STRESSES DISTRIBUTION
There is one general approach to calculation of normal stresses for the bar with **any** type of cross-section – including composite ones, as in the formula

$$\sigma_1 = \frac{E}{\overline{E}} \left[\frac{T_1}{\overline{A}} - \frac{M_3 \eta_2}{\overline{I_3}} + \frac{M_2 \eta_3}{\overline{I_2}} + \frac{B\hat{\omega}}{\overline{I_\omega}} \right] . \tag{1}$$

Here T_1, M_2, M_3, and B – are internal forces: longitudinal, bending moments and bimoment. Some particular cases of its application are given in the following sections. The problem of shearing stresses is a little more complicated and therefore, here is almost not discussed.

The internal force – bimoment B is calculated under the assumption, that for any kind of cross-section can be calculated its warping ω. If it is possible, the whole procedure known from Wlasov's theory for thin-walled bars can be applied. So, the bimoment can be calculated, too.

Normal stresses under action of longitudinal force

The normal stresses in the bar tensioned with force P, only, is calculated accordingly to formula (1) simplified here to shorter form as below $(2)_1$. Moreover, the elongation of the bar is calculated accordingly to $(2)_2$.

$$\sigma_i = \frac{PE_i}{\overline{E}\overline{A}} \quad \wedge \quad \Delta l = \frac{Pl}{\overline{E}\overline{A}}. \tag{2}$$

In both above formulae appear *reduced area* of cross-section, calculated as below

$$\overline{A} = \sum_{i-1}^{n} \frac{E_i}{\overline{E}} A_i = \sum_{i-1}^{n} \overline{A}_i \text{ , where } \quad \overline{A}_i = \frac{E_i}{\overline{E}} A_i \text{ .} \tag{3}$$

The symbols E_i and $\overline{E}$ mean Young's modulus and **assumed** general modulus for the whole bar. Calculated this way stresses for the tensioned bar with two materials (the first is twice stronger than the second one), are shown in the Fig 3 (Obrębski, Ref 3). There are visible jumps of stresses.

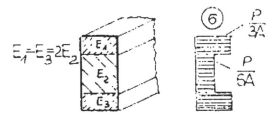

$$E_1 = \overline{E} = 2E_2$$

Fig. 3.

In the case of the column composed from three materials, as in Figure 4, calculated internal forces N_α have different values on diagrams of the Figure 4c and 4d, when the same materials are located in other sub-areas: 1, 2, 3 (Obrębski, Ref 3).

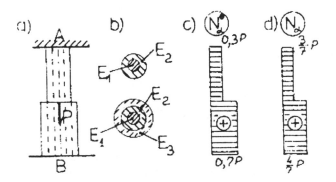

Fig. 4.

In this cases calculation of reduced area for particular cross-sections follow as below:

$$\bar{A}_1 = \sum_{i-1}^{n_1} \frac{E_i}{E} A_i = \sum_{i-1}^{n_1} \bar{A}_i, \quad \bar{A}_2 = \sum_{j-1}^{n_1} \frac{E_j}{E} A_j = \sum_{j-1}^{n_1} \bar{A}_j. \tag{4}$$

Normal stresses in bent bar
Accordingly to simplification of (1) – when bimoment is equal to zero, the formula on normal stresses gets the form

$$\sigma = \frac{E}{E} \left(\frac{T_1}{\bar{A}} - \frac{M_3 \eta_2}{\bar{I}_3} + \frac{M_2 \eta_3}{\bar{I}_2} \right) \tag{5}$$

where appears the *reduced moments of inertia*:

$$\bar{I}_2 = \int_A \eta_3^2 d\bar{A}, \quad \bar{I}_3 = \int_A \eta_2^2 d\bar{A}, \quad \text{where} \quad d\bar{A} = \frac{E}{E} dA, \tag{6}$$

By pure bending by moment M_2 of the beam composed of two materials (external are twice stronger) distribution of normal stresses is shown in the Figure 5c.

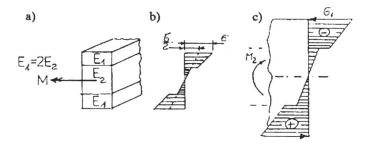

Fig. 5.

Example of normal stresses by combined loading
In the example of the Figure 6, of the beam uniformly loaded in two principal planes, with cross section combined of wood (10 x 10cm and steel (outside 5 x 10cm), by presence certain bimoment, the normal stresses are disposed as in the Figure 7 (see Obrębski, Ref 3).

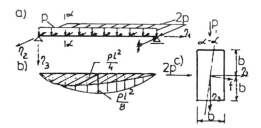

Fig. 6.

In this case, on effect of bimoment acting in cross-section warping stresses results in non-planar character of normal stresses. Jumps of stresses are as result of two materials cooperating in one cross-section.

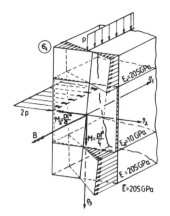

Fig. 7.

On more advanced calculation of stresses

As it was mentioned above, almost each torsioned cross-section obtain warping. In the case of rectangular cross-section, shearing stresses have character as in the Figure 8b. If we regard full cross-section as composed of some tubes located one inside the other, its warping calculated by means of theory of thin-walled bars is shown in the Fig 8a.

a)   b)

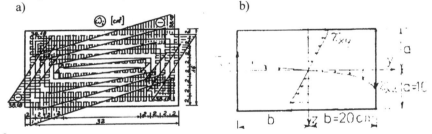

Fig. 8.

The shearing stresses, calculated by the same assumption, using theory of thin-walled bars, have the distribution shown in Fig 9. It is very similar! (Obrębski, Ref 3 and LSCE'95).

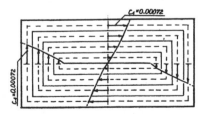

Fig. 9.

The other two examples of Figures 10 and11, show a steel channel section filled by concrete, but the same concern of the case with steel and wood. There are applied two different divisions of cross-section on tubes. In Figure 10a two materials cooperate without friction, and in Figure 10b both materials are form one monolithic bar. There in both cases distribution of shearing stresses will be different – as it shows dashed lines.

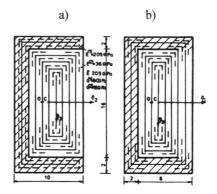

Fig. 10.

Similarly, three cross-sections with (almost) identical dimensions, have calculated different positions of *gravity centre*, *reduced gravity centre* O and *shearing centre* - A, Figure 11 (Obrębski, Ref 7).

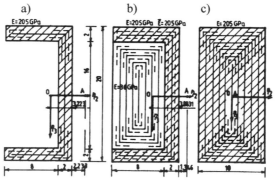

Fig. 11.

Calculated warping of thin-walled bar with stiffer corners, is shown in Figure 12, Obrębski, (Refs 8 and 9).

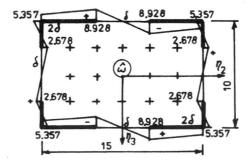

Fig. 12.

ON CALCULATION OF CRITICAL FORCES AND CRITICAL STRESSES

Generally, we can consider the bars with any cross-section and made of any materials, as described by the same equilibrium equations and all remaining relations. The only difference concerns details in calculation of geometrical characteristics of its cross-sections. So, theory derived very carefully in the book of Obrębski (Ref 2), is valid after detailed examination, for bars with compact – full cross-sections, too, (Obrębski, Ref 3). In the next sections are shown some examples of determination of critical forces and stresses, in the way used up till now for thin-walled bars, only.

Critical forces and critical stresses for single bar

The diagrams of critical stresses for Euler's problem, for identical columns, but made of steel a), of brass b), aluminium c) and wood – d) are different (Obrębski, Ref 3) as shown in Figure 13.

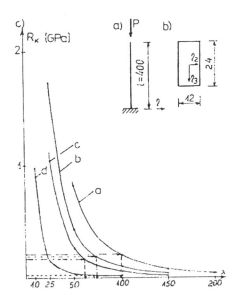

Fig. 13.

The above critical stresses were calculated according to formulae known for Euler's approach.

On the basis of almost identical formulae given below

$$R_{L} = \frac{\pi^2 E_i}{\lambda_j^2} \quad \wedge \quad \lambda_j = \frac{L}{i_j} = \frac{\mu l}{i_j} \quad \wedge \quad i_j^2 = \frac{\bar{I}_j}{A}.$$

were calculated critical stresses, Figure 14c, for steel (1) and wood (2) cooperating in one cross-section, as in the Fig 14b (Obrębski, Ref 3).

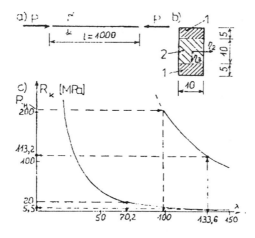

Fig. 14.

The other example of calculation of critical loading for beam loaded as in Figure 15 was presented by Obrøski, Ref 3, too.

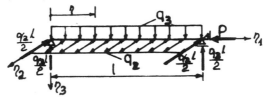

Fig. 15.

It is interesting, that two proposals were given by Obrøski, Ref 3, for correcting buckling coefficients for normal stresses. The first with four coefficients

$$\sigma_1 = \frac{E_1}{E}\left(m_{w1}\frac{T_1}{A} - m_{w3}\frac{M_3\eta_2}{I_3} + m_{w2}\frac{M_2\eta_3}{I_2} + m_{w4}\frac{B\omega}{I_\omega} \right) \rightarrow$$

$$\rightarrow \sigma_{red} = \sqrt{\sigma_1^2 + 3\tau^2} \leq aR ,$$

and the second, recommended, with one buckling coefficient, only.

$$\sigma_1 = \frac{E_1}{E}m\left(\frac{T_1}{A} - \frac{M_3\eta_2}{I_3} + \frac{M_2\eta_3}{I_2} + \frac{B\omega}{I_\omega} \right) \rightarrow$$

$$\rightarrow \sigma_{red} = \sqrt{\sigma_1^2 + 3\tau^2} \leq aR ,$$

Critical loading for composite beams and frames
In the paper of Obrøski, Refs 9 – 11, were given examples of calculation of the critical forces for composite bars made of reinforced concrete, Figures 15-17, having variable rigidity over its length. There some results were obtained for bars with different - variable rigidity.

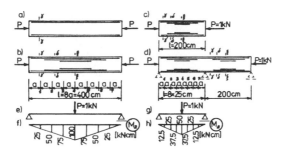

Fig. 16.

It is clear, that identical calculations can concern e.g. wooden beam strengthened by steel bands. There, bars are loaded eccentrically or bent, only.

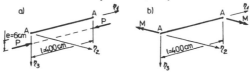

Fig. 17.

Now these numerical tests were not repeated for case of wood and steel.

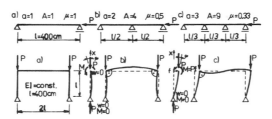

Fig. 18.

CONCLUSIONS

The paper turns attention on some possibilities of presented theory, only. Some given numerical examples of the more important phenomena, which can be described by such approach were presented. The other numerical examples, concerning the bars shown in the Figure 1, will be shown during the presentation. There will be presented some specially interesting results of calculations concerning determination of such bars behaviour: disposition of internal forces, displacements, stresses, critical simple and combined loadings - compared with e.g. Euler's ones etc. All given results will be compared with traditional ones, and its percentage evaluation.

REFERENCES

1.   MIELCZAREK Z. *Budownictwo drewniane* (In Polish – Wood engineering), Arkady, Warszawa 1994.
2.   OBRĘBSKI J.B., *Cienkościenne sprężyste pręty proste*, (In Polish - Thin-Walled Elastic Straight Bars). (lecture notes Warsaw University of Technology & University of Warmia and Mazury), Printed by Publishers of Warsaw University of Technology, Warsaw 1991, (pp.452) Second edition, Oficyna Wydawnicza Politechniki

Warszawskiej, Warszw 1999, pp.455, ISBN 83-7207-125-X.

3. OBRĘBSKI J. B., *Wytrzymałość materiałów*, (In Polish - Strength of Materials). (lecture notes Warsaw University of Technology & University of Warmia and Mazury), Micro-Publisher J.B.Obrębski Wydawnictwo Naukowe, Printed by AGAT,Warsaw 1997, pp.238.ISBN 83-908867-0-7.

4. OBRĘBSKI J. B., Uniform Criterion for Geometrical Unchangeability and for Instability of Structures. Proceedings of the International Conference on Stability of Structures. Zakopane, Poland, September 22-26, 1997.

5. OBRĘBSKI J. B., Mechanics and strength of composite space bar structures. Intern. IASS Congress on Spatial Structures in New Renovation Projects of Buildings and Constructions. Moscow, Russia, 22-26 June 1998 (general lecture).

6. OBREBSKI J. B., Some rules and observations on the composite bar structures mechanical analysis. Int. IASS 40th Anniversary Congress. Madrid, 20-24 September, 1999.

7. OBRĘBSKI J. B., Examples of determination of the non-conventional geometrical characteristics for composite bars with any cross-section. Local Seminar of IASS Polish Chapter on Lightweight Structures in Civil Engineering. Warszawa, 3 Dec.1999, pp. 64-77, ISBN-83-908867-3-1.

8. OBRĘBSKI J. B., Mechanical point of view on modelling of space structures made of composite bars. Int.IASS Symp. Istanbul, 29.05-2.06.2000, pp. 491-500.

9. OBRĘBSKI J. B., Non-linear character of the computations of composite bar structures. Proc. of Fourth Int. Colloquium on Computation of Shell & Spatial Structures, June 4-7,2000, Chania-Crete, Greece, CD-ROM 20 pages & abstr. vol. pp. 58-559.

10. OBRĘBSKI J. B., On the mechanics and strength analysis of composite structures. Structural Engineering, Mechanics and Computation, Cape Town, 2-4.04.2001, Edited by A.Zingoni, Elsevier Science Limited, Amsterdam-London-New York - Oxford - Paris- Shannon – Tokyo, 2001, pp.161-172.

11. OBRĘBSKI J. B., Examples of non-conventional analysis for composite bar structures. The 7TH International Conference on Modern Building Materials, Structures and Techniques. Wilno, Litwa, 16-18 May, 2001 pp. 289-290 & CD-ROM.

12. OBRĘBSKI J. B., Some new applications of the theory of thin-walled bars. 3-rd Intern. Conf. on thin-walled structures, Kraków 5-7.06.2001, pp. 321-328.

13. WERNER G., ZIMMER K., *Holzbau* t.1,2, Springer Verlag, Berlin, Heidelberg, New York 1996.

14. Exhibition during International IASS WG 15th Structural Morphology Conference Delft, 17-20.08.2000.

World's largest aluminium domes

F. CASTAÑO
Geometrica, Inc., Houston, TX, USA
D. HARDY
Alstec Ltd, Leicester, UK

BACKGROUND
In 1995 a joint venture between Taiwan Cement Corporation and CLP Power International Ltd of Hong Kong bid successfully to build, own and operate a 1320MW coal fired power plant at Ho-Ping on the east coast of Taiwan, and created Ho-Ping Power Company (HPC) to develop the project. HPC awarded separate contracts for Technical Services, Equipment Supply and Construction to three ALSTOM companies.

To prevent environmental pollution the project specification required that all storage and coal handling operations be conducted under cover. Two domes each to store 139,000 tons of coal would be built initially, with a possible third one to be built later. To achieve this capacity the domes would need to be approximately 145m in diameter. Given the corrosive marine environment of the site, aluminium was specified as the required structural material. Each dome was required to house the coal feed conveyor, the stacking and reclaiming equipment, and the coal transfer system to the exit conveyor.

With these requirements, the stage was set for the construction of the largest aluminum domes in the world.

SOLUTION SEARCH
In early 1999, ALSTEC Ltd (formerly ALSTOM Automation Ltd.) of Leicester, England was entrusted with the responsibility for the supply of the Coal, Ash & Limestone Handling Systems.

In recognition of the potential risks inherent in the enclosed storage of coal; including fire, explosion, accumulation of gases etc, a Hazard and Operability Study (HAZOP) of the proposed storage facility was required under the contract to be conducted in the early phase of the design.

One of the hazards identified in the study was the potential for dust accumulation on the members of the structure. This condition could create a potential for dust explosion if the dust became dislodged when an ignition source was also present. It was calculated that with a conventional I-beam structure, up to 150 tons of coal dust could accumulate on the domes over time. The HAZOP study concluded that the domes should be internally clad to eliminate this risk.

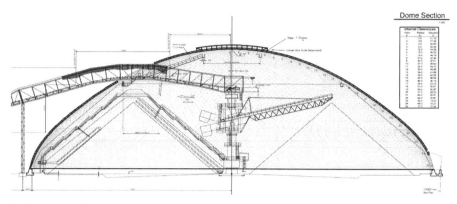

Fig. 3. Dome section.

The domes were designed in accordance with the provisions of the American Uniform Building Code and the Aluminum Association. Principal design loads were specified as: Live load – 59 kgf/m^2, Wind Speed – 212 km/h, Seismic zone 4 in the combinations given by the UBC.

The governing load combinations were those with wind, where a net force of more than 300 kg/m2 of uplift was considered on the upper region of the domes. Given the double innovation of the external structure and the parabolic profile, the design team requested a wind tunnel study from the Wind Science and Engineering Research Center at Texas Tech University in Lubbock, TX. This study showed that the wind loads would be somewhat lower than what had been assumed for design. Interestingly, the external structure had the effect of reducing substantially the dome's lift and somewhat increasing the drag, and the parabolic shape reduced both drag and lift as compared to a circular dome of the same height.

Dome structural members and joints were all of AA 6000 series aluminum alloys. Tubes ranged in diameter from 50mm to 100mm and most had 3mm wall thickness. Linear analysis indicated maximum deflections under live loads of less than 6cm and under wind loads of less than 30cm.

General buckling was checked using Dr. Douglas Wright's method, Ref.1. Because of the improved bending stiffness of the selected Vierendeel geometry, a factor of safety of 6 was achieved with respect to this phenomenon with a structural weight of only around 21 kgf/m^2 over the covered area.

MANUFACTURING
The dome was manufactured at Geometrica's facility in Monterrey, Mexico under strict adherence to a complete quality plan as part of Alstec's ISO 9000 procedures.

Geometrica bars are made from extruded 6061 aluminum tubes with a proprietary process and equipment. The bars are press-formed at each end controlling the resulting length and angles of pressing. Approximately 180,000 bars were manufactured for the Ho-Ping domes, each with its own unique identifying number. Bars were formed in the T4 temper and then heat treated to develop the higher strength of the T6 condition.

Connectors are slotted aluminum extrusions that match the tube end profile. They were cut and machined to the necessary shape for final assembly. All other components such as posts, purlins and hardware were also fabricated of aluminum and marked for ease of assembly.

All structure components were packaged in small crates of no more than 2 ton of weight each. Each crate contained only component immediately adjacent to each other in their final position on the assembled dome. Cladding panels were bundled in packets also weighing about 2 tons each and then the crates and packets were containerized and shipped to site. About 75 containers were required for each dome.

CONSTRUCTION
Construction of the plant was subcontracted by ALSTOM Projects Taiwan Ltd to CTCI Corporation who, in turn, subcontracted Triumstar, of Taipei, Taiwan for building the domes. Triumstar had much experience building Geometrica structures, including the prior world-record aluminum dome of 135m diameter at Taiwan Cement Corporations' cement plant also at Ho-Ping.

The domes were assembled using the perimeter-in method of construction. In this method, the dome is built ring by ring starting from the concrete foundation. Using safety equipment approved by local authorities, workers climb on the structure to build it. All bars are man-sized and light, so assembly can proceed without specialized equipment. The assembly sequence was designed to resist typhoon winds at any time during construction.

Cladding was installed using man-lifts from the inside after the dome was complete and properly anchored to the foundation. (Figure 4)

Fig. 4.

COMISSIONING

The first dome was commissioned in September 2001. The plant is the most modern in Taiwan and the domes are a fitting and visible reminder of this. (Figure 5 & 6).

Fig. 5.

Fig. 6.

REFERENCES

1. WRIGHT, D.T., Member forces and Buckling in Reticulated Shells, Proceedings ASCE Structural Division, 91, pp. 193-201, Feb. 1965.

Development of a new generation of single-layered spatial structures for fluid building designs

M. EEKHOUT
Faculty of Architecture, Delft University of Technology, the Netherlands
Director of Octatube Space Structures bv, Delft, the Netherlands

INTRODUCTION

The current architectural tendency towards 'fluid designs' produces a new type of buildings with a remarkable influence on thinking about spatial structures. An obstinate example of such a 'fluid' or 'Blob' building is the design of Kas Oosterhuis, Rotterdam, of a pavilion for the province Noord-Holland, the Netherlands, meant for the Floriade, which was realized in April 2002. In the design stage, the architect Kas Oosterhuis and structural designer Mick Eekhout developed the structural design, up to a point where the architectural and structural ideas and finances showed unbridgeable differences. This contribution is mainly concerned with the various design considerations, which strongly steered the development process. After all, the most radical decisions are made in the earliest design phases. The significance of design decisions in the final phases of design and engineering has much less impact.

Fig. 1. The NHP finished on site (© ONL).

THE EARLY DESIGN STAGES

The preliminary drawing of the architect looked like a collapsed Gouda cheese: a round building with rounded sides and a somewhat dented roof. The estimated dimensions in the floor plan were approximately 24 metres, the height of the edge was 7,5 metres and the centre height 5,5 metres. Although this was an obstinate draft with regard to designing, it was still very much related to the long years experiences of building dome structures because of its rotational symmetry. The negative curving in the roof would be a reason for a double-layered realization, while the rest of the dome would basically be single-layered and three directional.

The second version of the design of the architect showed a rounded triangular shape, much like a Brie cheese wedge. Other shape associations use the terms 'cobble' or 'potato', or more respectfully, a spacecraft. The dimensions in the floor plan were 27 x 20 metres and 5,4 metres in structural height and 6,3 metres as the total height. From the dialogue between Oosterhuis and Eekhout arose a structural concept for a single-layered space frame with universal joints, connected to bars into three directions. This concept would be capable of handling the irregularity of the geometry, but was also based upon a record of accomplishment. The cladding and the space frame would be parallel. The second design did not have any rotational symmetry anymore. It was fully arbitrary in its shape and so showed suddenly a serial size for the production of one piece for all components. By that, the industrialization factor between the two described models seemed backdated two centuries.

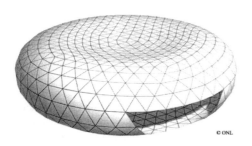 

Fig. 2, 3. Preliminary design 1 & 2 for the NHP.

The familiar dome models from the latest history of three-dimensional metal dome structures show the single-layered domes for smaller spans and double-layered domes for larger spans. Furthermore, the nineteenth century geometry is orthogonal, with radial ribs and horizontal rings, the type Schwedler. The twentieth century domes are all based upon various domes of Fuller are familiar: network models with horizontal rings and parallel lamella/delta girder models. The dome models that are composed of triangles, all have a similar material efficiency, which is much greater than that of the orthogonal models.

Usually, all these different dome models are half, or less, spherical, their height is less than half the diameter. Only rarely, domes are made ¾ spherical, i.e. radomes. In the eighties, Mick Eekhout did research into a 60 metres sphere in Rotterdam. The structural analysis and the soldered (1: 100) model proved that the greatest forces would occur in the bottom bars of the dome. Three-quarter domes suffer from weak knees. This phenomenon would clearly emerge in the Floriade design.

In the history of metal dome structures, the globular shape of the single-layered domes has always been fully synclastic. One of the few exceptions was the Multihalle of Frei Otto, Mannheim, Germany. There have been scientific analyses of the collapse of single-layered metal domes, where a small indentation in the synclastic surface, a local failure of bars and joints, had a catastrophic consequence for the stability of the entire dome and resulted in a total collapse. The Floriade design had a hollow, an indentation in the upper part of the design, to which the architect, from design considerations, was much attached. A solution to this intrinsic problem could be two-folded.
- The local extension of the single-layered system to a double-layered system;
- The removal of the indentation and make the entire surface synclastic again.

Overall, the above considerations illustrate that the architectural design was propelled by the possibilities of sculptural designing on the computer, as opposed to the acquired experiences and regularities of the design and building of domes in the last decades. Therefore, in the first discussions between the architect and the structural designer, these considerations came up extensively. For each experiment, the challenge to the designer is a motive of great importance. In this case, the challenge in structural sense was to make the improbable possible and feasible. Architect and structural designer soon agreed on the following basic principles:

- consider the object as a shell;
- make the shell rigid by the triangulation of bars and joints;
- make the shell rigid for loads perpendicular to the surface;
- to introduce either moment rigid connections;
- or to introduce shape rigid spatial angles at the connecting points.

The necessity of shell effects and triangulation was agreed upon in the first telephone conversation.

On the one hand, the dimensions of the triangles depended on long bars, rough angles and greater shape rigidity and on the other hand on the limit of the covering 'Hylite' aluminium panels of 1,5 mm. thickness. These panels for this project would be supplied free of charge by Corus, their main establishment being in IJmuiden, in the province Noord-Holland. The size of the trading plates was 3,0 x 1,5 m². The number of panels was limited. Each triangulation causes much waste (up to 50%), therefore, the engineering had to make optimal use of the material. The structural concern was: the rougher the connecting angles perpendicularly to the shell, the greater the external loads resistance. Small angles around the joints often cause failure. Therefore, a triangular coarse-mesh netting was preferred over a fine-mesh one.

The architect had in mind to project the model of an icosahedron on the envelope of the object, whereby the way of the subdivision of the primary axes of the icosahedron (five meridians) was characteristic of the dimension of the triangulation. Because of this optimization, four triangulation alternatives occurred, namely from a 5-piece to an 8-piece. With regard to the shell rigidity (per moment or per shape), it was decided to take the risk of applying a single-layered hinging space frame to these relatively small free spans, with maximal sized triangles. If the structural analysis would show that the shell would not be sufficiently rigid, possible additional actions would be taken, i.e. a moment rigid joint, instead of a ball hinge, or of internal cross bearers that were functionally useable in the floor plan to be tightened to, or five short frame rigidities, diagonally on the outer walls, or of bow strings reinforcements.

It was well-considered not to make a choice for the usual schedule of deep ribs perpendicular to the surface and in an orthogonal system, for reasons of the expected large consumption of material, the relatively banal simplicity of such a schedule and the fact that de development of an internally ribbed structure is quite a common thing in aircraft construction and the ship building industry. For a new way of designing in architecture, the structural designer desired to develop at least a new and unique way of a lightweight structure. The architect was interested in building the object, in whatever realization. The basic idea of the principal was to develop the design as 'economically feasible'.

Of course, in this phase the industrialization factor played a part too. The first design of the Gouda cheese was rotational symmetric. However, the architect already had made an arrangement of the icosahedron, which had internally exploded against the envelope of the object and so, the repetition factor was reduced. The result of these considerations is the pentagonal roof of the object. The walking about visitors of the Floriade cannot really see the roof. It was designed as a five-fold icosahedron roof: the five constituent icosahedron triangles were all equal, so that they provided a small serial profit with the production of both the skeleton and the skin of the object. The structure of the object has lost somewhat in consequence, in a scientific manner of speaking, but has won economically.

A subsequent consideration was the cladding. This would be made from 1,5 mm. thick panels. The initial considerations were:
   • the triangulation in the shape of the panels;
   • the cutting loss of the panels from rectangular trade plates;
   • the individualization of the panels and the edges;
   • the maximum size of the panels versus the thin material;
   • the relatively non-rigidity or flexibility of the panels;
   • the necessary water tightness for the entire skin of the object.

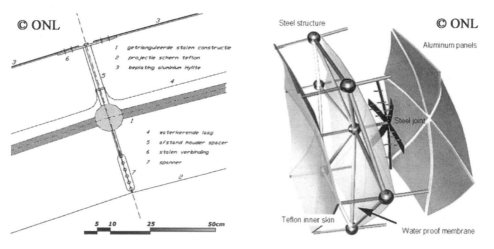

Fig. 4. Original detailing NHP.

The major concern was how to make the skin of the dome panels watertight. This was expected to be difficult, due to the high degree of individualization. The main task would be to make an envelope of fitting panels. As a second task, a waterproof membrane of PVC coated polyester fabric, suspended from the panels, would realize the water sealing.

From that concept, the thoughts of the architect and the structural designer led into two different directions. The architect thought that, if there is a watertight suspended membrane, then the seams between the panels do not have to be waterproof, or accurately connected. Moreover, in the space between the aluminium panels and the watertight skin, artificial lighting can be applied, so that the object would be line-shaped illuminated. The architect thought in terms of the object as an extravagant building with fanciful possibilities.

To the structural designer, the aim of the development of the panels was to make them fit as accurately as possible in the side connections, so that a sealed watertight could be applied. He considered the waterproof membrane as a second moisture barrier, not unusual for buildings.

It should be possible to put the progressing insights into practice for further assignments. After all, 'Blob' designs have a rising popularity and are worthy of a sound development to bridge the increasing gap between designing, engineering and production.

Meanwhile, the architect went on with the detailed design of the object. He found that, due to the largely visual character of the panels, the skin design could be semi-independent of the skeleton design. In terms derived from the car industry: the architect proposed to run the body independent of the chassis at the rear. This considerably complicated the individualizing of the panels. Initially, the panels would be parallel to the space frame. Now, suddenly spheres and hollows occurred in the cross-sections of the panels over the space frame, as a result of sculptural interventions in the shape model in the computer of the architect. The image of the American cars from the Sixties came up. This new wish of the architect, which gradually became a demand, would eventually result in an unbridgeable difference op opinion between the architect and the structural designer.

COMPUTER AIDED ENGINEERING
The architect Kas Oosterhuis is famous for his digital designs. He has worked in this field for over ten years and did not built over one design per year, but he published several colourful books. In the year 2000, the Delft University of Technology appointed him part-time professor for the duration of three years. Thanks to pioneers like himself, new 'Blob' designs are published all over the world. In general, architects explore the bounds of possibilities by means of their CAD expertise and increasingly advanced computer programs in which 3D designs can be made. However, these programs are not yet compatible enough to the usual engineering programs. For an example, the number of bars, joints and panels had to be derived from a generation list. The drawings were not sufficient. The DXF file, which transferred the computer file that was certified by the architect for the use of co-designers, showed many shortcomings. Only after many weeks of work, the geometric data became clearly readable on the, for engineering common AutoCAD 2000/14, but only for the computer operator. By the time the geometric data could be read, the structural data were still not suitable for a cost accounting. Because of the experimental nature of the draft design and the insecurity with regard to feasibility, the structural designer started work based on a commission for co-designing and pre-engineering.

The gap between the accelerating architect whose greatest focus was the design on the one hand, and the carefully operating structural designer whose greatest focus was on feasibility, became even greater. This gap cannot be bridged, because the computer programs the architect uses are not compatible with those of engineers and producers. Therefore, a classic dilemma emerges: to act alone (keep everything in one hand, basically how Octatube started some twenty years ago), or communicate in collaboration with appropriate means. Maybe, the architect could have generated the entire project by means of his own computer programs, so that the sum of the architectural design, the structural analysis and the component breakdown could have been developed by one hand, while the structural designer only gave advice and the producer made his cost account as accurate as possible.

In the course of the three months of cooperation (between the end of February 2002 and the end of May 2002), the choice for a parallel way of working was made: concurrent and sometimes collaborative. The architect worked out the design and tried his best at forming proposals for materializing. The structural designer and his engineers (Karel Vollers, Sieb Wichers and Freek Bos) tried, already in an early stage, to get acquainted with the essence of the work, the rigidity of the structure, the composition of the elements and components and the full water tightening, parallel to the work of the architect. They reached different conclusions with regard to their possible future responsibility and liability for structure and water tightness than the architect. The result was a continuous dialogue and discussion.

The total budget of the project was of a very decisive influence. The first estimate of Octatube varied for two alternative realizations from € 430,000 to € 630,000. Only one month later, when the commission was handed out, the actually available budget for the engineering, production and assembly of the structure, cladding, membrane, floor and two doors emerged: € 240,000. Calculated on a skin surface of 600 m², this meant a price of € 400/m², including two complex doors. This square metre price was hardly a realistic budget for an experimental project. The commission for co-design and pre-engineering was estimated at approximately 5% of the realistic budget, but was only accepted as an obligation regarding the means, not the obligation regarding the result!

The estimates were established as the sum of assessed individualized element component costs and the way of realizing the components. The great difference between the first estimate and the available budget proved that it was necessary to be modest in the degree of experimenting. Indeed, there is nothing wrong with a low budget, just as long as it is realistic. Nevertheless, in the current case, the estimate could only be verified after three months of design development, structural engineering and the development of the skin. Only by the time the layout of the space frame was established and computer analysis showed that the structure was sufficiently strong, rigid and stabile, the cost accounting of the space frame provided insight. Unit prices and the number of elements and components established this. Basically, there was no discussion on the economy of the space frame (56% of the budget), but all the more on the economy of the cladding.

STRUCTURAL ANALYSIS
The second half of April was used to make the structural analysis. The 'Blob' Graduation Building Technology student Freek Bos did this. Splines do not transfer correctly from Maya to AutoCAD. Therefore, much work was done by both parties to come to a proper computer communication. Via the pre-processing in FemGen, the structural analysis in DIANA and a post processing in FemView, insight was obtained in the behaviour of the frame shell. The indentation in the roof (with the possible local snap through as a result) was already removed in the network development.

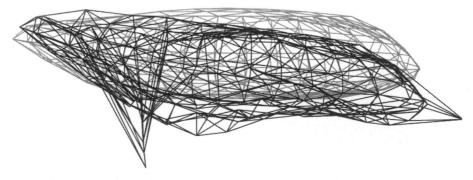

Fig. 5. Displacement (10x) of the original single layered space frame construction.

The introduced bars had the following diameters: 82,5 x 5,0 mm., 101,6 x 7,1 mm. and 203,0 x 8,0 mm., the heavy bars on the five main axes, the meridians.

Dead load, snow and wind loads were considered the main loads. As a result of the dead load and snow loads, the tail in the long diameter proved to flap up. Furthermore, the roof sagged by the deformation of the lower half of the bars: the knees. This was confirmed in the material model of the students. The remedies were:

- to enlarge the structural height of the shell in all cases with maximal 1000mm. from 5,4 to 6,4 metres;
- to make the bottom bars of the frame shell considerably heavier;
- to flange couple a number of bars on the five meridians, moment rigidly over the joints by means of welded flanges, by which tubular interconnected beams would occur with a diameter of 203,0 x 8,0 mm.
- to place five internal slender cross bearers on the meridians with a diameter of 101,6 x 7,1 mm., by which the entire shell would be parted into a flat roof and a round 'doughnut' wall;
- to introduce five internal meridional trusses to give the shell a great rigidity and which were acceptable to the architect;
- to introduce a number of shell rings to master the horizontal lateral thrusts, but through these, the two doors in the bottom edge would cut. This alternative was further neglected;
- to place five external outriggers at the sides of the meridians to strengthen the bottom;
  sidewall. The architect did not appreciate this, though it strongly looked like the landing gear of a space vehicle.

The conclusion from this phase of the structural analysis was that, by rising from 5,4 to 6,4 metres and the internal strengthening by means of welded beams with a diameter of 203 mm. on the meridians, a reasonably rigid frame structure was the result, with a maximum vertical replacement of 20 mm. in the middle over the shortest span of 20 metres, which means a 1/1000 of the span. In other words, with still some approach cycles ahead, the frame shell would not cause unexpected and insolvable problems. The total of the estimated dead load of the frame shell with strengthening would amount to 600 m² over the skin surface, approximately 6,000kg. (10 kg/m²).

CLADDING DEVELOPMENT

Problems seemed to concentrate at the side of the cladding. On the one hand, there was the strongly reduced budget of the client, and the architect who wanted a metal skin in its shape independent of the space frame on the other hand.

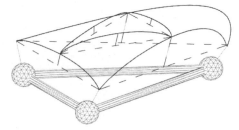

© ONL

Fig. 6. Digital model of the polygonal space frame in combination with the double curved metal skin.

Already from the first internal estimate, it could be concluded that there was too small a budget for an extended engineering of the skin, now that the skeleton was more or less roughly established. Though less desired by the architect, with the possibility of reaching a level of flat panels at the back of the mind, the quest for panels with a spatial curve began. After the architect had established the desired curves in the skin with regard to the centres of the joints and axes, by means of splines and nurb curves, it became clear that very complex 3D components had to be manufactured to fix the cladding to the required position in space, considering the space frame. Furthermore, a number of rather extreme cladding components had to be developed and manufactured. The required 3D nature of the cladding triangles was limited because the flat plates could only be folded or curved into one direction. Therefore, a triangle has a more or less flat centre and three points which can be curved upward or downward, independent of one another. The regularities of curving 2D panels into star-shaped 2,5D panels still have to be further exploited. Based upon these results, the definite offer was made with the bulged plating, a number of flat alternatives for panels and a stressed membrane as alternatives. Based on this outline of the costs, an agreement to go on could not be reached and the activities of Octatube ended there. Half a year later, the structural designer developed 3D aluminium panels and they were realized for the council pavilion in the Dutch Floriade, on which a different lecture for this conference [Ref lecture 200] will expand. The costs for that project proved to be even higher than those estimated above.

STUDENT PROTOTYPES
In May and June 2002, a group of ten third year Building Technology students has been occupied in the framework of an obligatory study part, named 'the prototype', with two different parts. Half of the students had to develop a regular cladding component, as mentioned above, based upon a box-shaped component of which the body would run independently of the chassis.

Fig. 7. Prototype made by students at the Faculty of Architecture TU Delft.

The other half was commissioned to design a workable door, fitting in the system, with a minimum width of 2,2 metres, opening turnable, swivelling, twistable or slewing. They derived their principles from the double hinging doors of civil airplanes. To obtain the necessary insight, a thread model was built first on a scale of 1 : 20, of 3 mm. soldered coppered welding wire. After all the virtual models on the computer, at last a material model existed. The students learned much in these eight weeks, but the results of their work were not satisfactory. Their work confirmed the assumption that, within the given preconditions of the required design of the cladding, the necessary engineering and production efforts and the available budgets, no satisfying compromise was possible.

Fig. 8. Scale model of the irregular space frame made by students.

FINAL REALISATION
After the leave-taking of the architect and the structural designer Eekhout/producer Octatube, the architect had to take another route. With the help of the consultancy D3BN, a load bearing structure was realized, based upon set steel strips of 20 to 30 mm. thick and 100 to 400 mm. high, in the familiar triangulated geometry. The strips functioned simultaneously as main support structure and as cladding fixation and were buckled to that purpose. The architect applied for a patent on this system. The cladding still had slightly bent flat panels 'Hylite' in a flat shape, fastened by means of steel braces to the steel strips, not being waterproof. The water tighting of the inner space came about by making the projection screen waterproof. By this, the project had indeed become a fancy and extravagant building. Yet, the building was realized. The dead load of the steel structure, manufactured and assembled on the building site by steel structure producer Henk Meijer, Serooskerke, the Netherlands, is over 100 tons of steel. The 'Hylite' skin weight is 3 tons. The eventual costs are not known. The intention is to dismount the pavilion when the Floriade exhibition has ended and to built it up again with a new watertight skin as a 'Blob' laboratory at the Faculty of Architecture of the Delft University of Technology.

Fig. 9. Assembly of the final steel structure (© ONL).

CONCLUSIONS

The lessons learned from the preliminary studies are fundamental and essential enough to put them before an international forum.

- The dramatic break in the development of systemized light-weight spatial structures, by unusual architectural 'Blob' structures.
- 'Blob' structures bring along a loss of systemizing and repetition in the material load bearing structure.
- The high degree of individualization in engineering and manufacturing of the individual space frame components requires further research.
- The shift of critical attention from spatial structures to spatial cladding calls for design energy.
- Computer Aided Design enables to architects to make 'Blob' designs and Computer Aided Engineering is vital for the establishment of the spatial complexity of the entire design and the individual establishment of the elements and components.
- In the near future, the digital bridge between Design and Engineering will determine a great part of the technical and financial feasibility of 'Blob' designs.

The structural making of the Eden Domes

K. KNEBEL, J. SANCHEZ-ALVAREZ, S. ZIMMERMANN
MERO GmbH & Co. KG, D-97084 Würzburg, Germany

INTRODUCTION
In the spring of 2001, on the south-western tip of England in Cornwall, the Eden Project was opened to the public. This project is, along with the dome and the ferries wheel in London, one of the largest British millennium projects.

In an outdoor area of 15 ha, and in two giant greenhouses, the modern Garden of Eden presents different climate zones of the world with their typical vegetation.

The steel structure of the two huge domes was developed from the MERO space frame system: pipes are bolted together by means of nodes. Due to very low tolerances and quick assembly, economical structures can be realised even for complex geometrical configurations.

Very light and transparent, but also durable air filled foil cushions were chosen for the cladding system. A cushion system of this size had never been built before.

The 125 million Euro project is a great success. Since the official opening in March 2001, thousands of visitors take pleasure in the gardens every day. The Garden of Eden has been called the eighth wonder of the world by the British press.

DESCRIPTION OF THE COMPLEX
The complex consists of several parts (**Fig. 1** and **Fig. 2**). Beside the outside area there are 4 main buildings. The entrance and **visitor centre** is located at the top of the clay pit. Here, several souvenir shops, restaurants and exhibitions are located. This building was completed and opened to the public about one year before the opening of the rest of the complex. About half a million people visited the Eden construction side during May 2000 and the official opening in spring of 2001 and watch as the complex grew.

The main building complex consists of three parts.

The biggest part is the Humid Tropic Biomes (HTB). Here plants from the subtopic part of the world like West Africa, Malaysia

Fig. 1. Overview of the complex.

or Oceanic are shown. The HTBiomes are comprised of 4 domes (ABCD) and the dome B is the biggest one of all. The diameter is almost 125 m and the free height inside is close to 55 m so that even big jungle trees have enough room to grow up. Since these plants need the most sunlight to grow, the location of these domes were located by the architects so that they catch the most sun power.

The other four Domes (EFGH) form the **Warm Temperature Biome** (WTB) with an internal length of about 150 m, width 56 m and a height up to 35 m. Here the typical plants of the dry and warm areas like south Africa, California or the Mediterranean area are located. The humidity and temperature is not as high as in the HTBiomes.

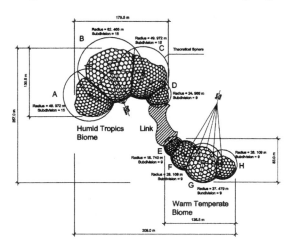

Fig. 2. Layout of the Biomes.

These two domes are connected by the **Link** building which is covered by a grass roof and is therefore almost invisible from the surface. The link building is the entrance to the Biomes. Beside sanitary facilities the visitor can have a rest in restaurants.

The central outside area is about 15 ha big. Outside many other plants from different areas of the world can grow due to the mild clime of Cornwall without shelter. Many other varying attractions and exhibitions make the Eden Project an attractive park for the visitors.

STRUCTURAL CONCEPT

The first layout made by the architects of Nicholas Grimshaw and Partners (NGP) together with the engineers of Anthony Hunt Associates (AHA) [1] was similar to the London Waterloo train station (**Fig. 3A**). Arches and purlins are the basic steel structure for the glazing elements. The disadvantage of this layout was the high steel weight and the small glass

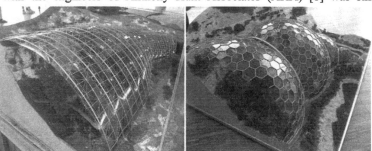

Fig. 3. First and second concept for the building.

elements which blocked too much sun light. Although this concept was difficult to fit to the varying natural surface of the clay pit.

After McAlpine was chosen as general contractor, it was also clear that this steel structure was too expensive to realise. Another concept had to be found. The architects and engineers of NGP and AHA then developed a single layered domes structure based on a hexagonal geometry (**Fig. 3B**). This layout had several benefits. It is easier to fit the structure to the ground surface and the size of the hexagon elements allows more sun light to enter. The visual appearance a hexagonal geometry is also close to many object found in nature. The next question was how to realise this structure on an economical basis.

In 1997 MERO joined the Eden Project. MERO has realised many complex structure in the past decades all over the world. After some preliminary studies it was found, that the single layered structure in this large dimensions could not be build economically and that the deformation were too large. Working closely with NGP and AHA, the geometry and structure was modified by MERO. The result of this optimisation was a double layered structure with the characteristic hexagonal top chord geometry (**Fig. 4**). As a cladding system of air filled

EFTE foil cushions was chosen. The very low weight of this material in contrast to glass allowed a further reduction of the necessary steel weight. This foil material allows much more UV light to pass into the Domes and also provides good heat insulation. This structural concept fulfilled all the required points. The weight and amount of the steel was minimised, the surface cladding was transparent for the sunlight, the entire interior area is free of columns, the optical appearance is attractive and the structure is

Fig. 4. The Hex-Tri-Hex Structure.

economical and fast to be build. In the spring of 1999 MERO began with the final technical design for this unique project.

GEOMETRY

The Eden domes are geodesic spherical networks. They are "spherical" because the elements of a network, normally the nodal points, lie on the surface of a sphere. These grids are called "geodesic" because they have the form, structure and the symmetry properties of the geodesic domes known through Buckminster Fuller, where not all the members follow true geodesic lines. Strictly speaking, geodesic lines are curves on any kind of surface and, from the innumerable lines that can connect two points on the surface, they represent the shortest distance between the two points. In the common geodesic domes, however, structural members are normally made straight, not curved, and only their end points, usually the centres of physical connectors, lie on the surface of a theoretical sphere.

Some other ways to project or map networks on spherical surfaces have been developed by Emde [2] in Germany, Fuller [3] in America and Pavlov [4] in Russia. The majority of spherical geodesic networks in building practice are derived from the platonic solids icosahedron and dodecahedron. An icosahedron is a regular polyhedron with twenty identical faces, which are regular triangles. A dodecahedron is a regular polyhedron with twelve identical faces, which are regular pentagons. Dodecahedron and icosahedron are duals from each other. If the midpoints of adjacent faces of a polyhedron are connected with lines, the resulting body is the dual of the first one and vice versa. It should be noted that the two polyhedrons placed as duals have a common centre and they can also be positioned concentrically within a circumscribing sphere. Thus, a geodesic network can be obtained by projecting or mapping in a prescribed way the tessellated faces of the polyhedron onto the surface of the sphere.

GENERATION OF THE GEOMETRIC MODEL

The structural network of a dome in the Eden Project consist of two concentric spherical networks with a prescribed radius difference or structural depth between them. Here, external and internal networks are interconnected with a set of lines called diagonals, thus giving rise to a double-layered spherical network with a three-dimensional carrying behaviour. The external grid is a hexagonal network, here referred to as "Hex-Net", whereas the internal grid consists of triangles and hexagons and is consequently called "Tri-Hex-Net".

The key steps to generate the Eden-geometry, which are dodeca-ico networks, are shown in **Fig. 5** and reference [3]. In order to generate a dodeca-ico (DI-) network, the two polyhedrons have to be placed as "duals" with respect to the centre of a sphere. The corners or vertices of the icosahedron in the resulting network can be recognised by their pentagonal symmetry and they correspond with the midpoints of the dodecahedron's faces. In **Fig. 5** a so called "characteristic" triangle (after Emde, [2]) is defined by the icosahedron point I2, the dodecahedron point D1 and the icosahedron's edge midpoint DI-1' projected on the surface of the sphere. This triangular surface region is the smallest symmetry part of the whole spherical network. In English speaking countries, this triangle is often known, after Fuller, as the "lowest-common-denominator-" or LCD-triangle. Through this approach, it is possible to subdivide the spherical surface in 120 minimal symmetry parts. The actual specifications of geometric and connectivity properties of the whole network can thus be reduced to this minimal triangle. In the Eden domes, the hexagons were obtained by

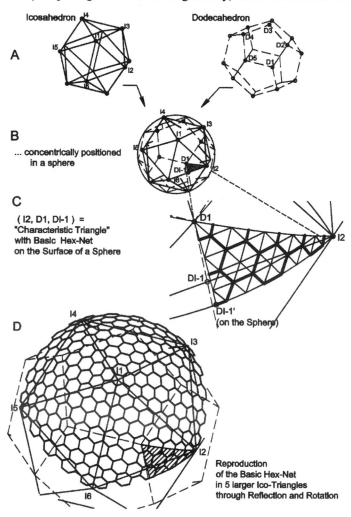

Fig. 5. The generation of the geometry.

omitting the appropriate elements of the minimal triangular net. The complete hexagonal network was subsequently generated by reflections and rotations on the surface of the sphere of the minimal network within the characteristic triangle.

The internal tri-hex net is obtained from the corresponding elements in the characteristic triangle. Here, the nodal points of the internal grid are derived from the external line

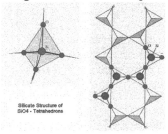

Fig. 6. Silicate molecular.

midpoints which have been projected concentrically onto the theoretical sphere carrying the internal grid. The tri-hex net is then generated by connecting the points corresponding to adjacent lines of the hexagonal grid. The inter-layer diagonals are obtained, in turn, by connecting every internal point with the corresponding endpoints of its external source line. The resulting spatial network strongly resembles the molecular organisation of certain minerals, like the silicates (SiO_4) as seen in **Fig. 6**. Among other properties, these natural crystalline formations present minimal energy paths with minimal material consumption. Similarly, the three-dimensional geometric arrangement of the Eden domes makes possible an economic structure with a visually attractive appearance.

PLANARITY OF THE HEXAGONS

The hexagons of geodesic domes, unless special measures are taken, are not normally plane. The selected cladding of foil pillows required the hexagons to be as planar as possible in order to facilitate the construction and assembly of the supporting edge frame and to prevent unplanned folding and wrinkles on the pillows. Based on the works of Emde [2], Fuller [3] and Pavlov [4], a special algorithm was developed to obtain hexagons as plane as possible within tolerable fabrication and installation deviations. "Tolerable" means for the Eden domes that a point can be 60 mm out of the average plane of the largest hexagon, which has edge lengths of up to 5.20 m. A further "flattening" of the hexagons would mean losses in the uniformity of the networks, which is measured as the ratio between the maximal and the minimal length of a grid. As an orientation, very homogeneous DI-geodesic networks have uniformity quotients of about 1.2, while the Eden networks vary around 1.26. Furthermore, alternative networks for the Eden domes with perfectly plane hexagons yield uniformity quotients of up to 2. In these extreme cases, that correspond to the finer subdivided networks, hexagons along the edges of the basis icosahedron tend to present larger distortions with the corresponding disadvantages for the structural system and a disturbing visual effect.

STATICAL CALCULATION

After the final design of the geometry, the calculation of the steel structure was done. The

geometry was transferred into a statical 3D computer model (**Fig. 7**). The calculation was carried out using the 3D analysis program RSTAB [5] based on second order theory. The top cord elements and the arches are beam elements, the bottom cord and diagonals are modelled using truss elements. The basic load cases and load combination are according to BS 5950. Due to the mild climate in Cornwall the basic snow load is only

Fig. 7. The static model of the HTBiomes.

0.3 kPa. The results of the wind tunnel test performed at BMF Fluid Mechanics LTD. in London showed, that due to the form of the clay pit, the wind is acting mostly as suction onto the Biome structure. In addition to the ordinary loads cases, some extraordinary load patterns had to be regarded. In the valleys of the cushions and especially in the valleys of the arches, drifting and sliding show can occur. Local snow loads up to 1.2 kPa were considered which resulted in some extra cables supporting the cushions in the area of the arches. A cushion that loses it's air pressure could result in a water filled area in the case of rain. Local load up to 250 kN per hexagon were considered for this case. The performed analysis proved the stability of the structure even when some members failed. The governing load case for the design was mostly the snow and snow drift load case. Changes in temperature are usually not critical for the stress design of dome structures since they can expand freely in radial direction. Also, support movements are causing no major forces. Therefore, the domes of Eden are build without any expansion joints.

THE BOWL NODE
The main design parameters for the connection in the top chord were :
- Rigid Connection for the three tubes with d=193 mm
- Hinged connection for the three diagonal members
- Fast and easy erection
- Minimum tolerances
- No side welding
- Possibility to fasten a rope for mounting the domes on the outside
- Architecturally pleasant

As a result of these requirements, a so called bowl node was chosen (**Fig. 8**). This type of node is an enhancement of the node type used by MERO when circular or rectangular hollow

Fig. 8. The top chord 'bowl' node.

tubes are joined together by bolts. The top of this connection is even with the pipes, so that the cladding can be put right on top of it. The bowl node is made out of cast iron (GGG40) and the weight is about 80 kg. The diameter of the 1100 nodes is about 400 mm and the wall thickness is 40 mm. Each node was cut and drilled by a computer aided machine which limited the tolerance to a minimum.

TOP CHORD BEAMS
The stress design of the top chord resulted in a tube diameter of 193.7 mm. In order to use the same connection to the node, all top chord pipes are the same diameter, but with different wall thickness according to forces and buckling length. Since all the necessary geometrical angles to form a dome are put in the node, the ends of the top chord beams are cut rectangular which allow fast and efficient manufacturing. At each end, an endplate is welded and at the top of the beam a erection hole cut (**Fig. 9**). High strength pre-stressed bolts (M27 and M36) were used for the connections of the beams to the bowl node. An additional bold M16 was

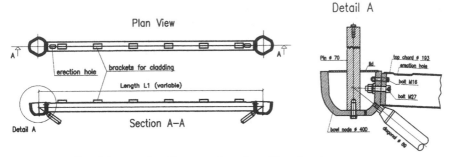

Fig. 9. Details of the top chord beam and section of bowl node.

used to fix the beam in it's right position and to transfer torsional moments. On top of the beams short brackets were welded which support the aluminium framing for the cushions.

BOTTOM CHORD AND DIAGONALS

The bottom chord tubes and diagonals are made of the classical and well known MERO space frame system. The performed stress and stability design lead to diameters between 76.1 and 168.3 mm. According to the BS Standard the buckling length was limited to 180. They are designed and manufactured according to the MERO technical approval [6]. The connections in the bottom layer are also classical MERO space frame nodes (**Fig. 10**) which allow a fast and easy erection of the structure.

Fig. 10. The bottom chord node.

THE ARCHES

Along the intersection of the domes a triangular truss girder is applied. The span is up to 100 m for the biggest one. The sections of the arches mesure 219.1 mm for the top beam, 159 mm for the two bottom beams and 101.6 mm for the diagonals. The top and bottom chord beams are bent. For manufacturing, the girders were welded in three pieces and after setting up the rest were welded on site. The arches rest hinged supported on heavy foundation concrete blocks. On top of the 219 mm tube a 10 mm steel plate was attached to form and fix the gutter. The connecting arches are flush with the dome top and bottom surface, and therefore, do

Fig. 11. View of the arch.

not interrupt the flow of the design

SUPPORTS

A challenging point was the design of the support system. Because the 800 m long foundation varies, each of the 187 support points is geometrically different. The supporting construction also consists of tubes with diameters of 193 mm which are welded together . The connecting top chord

Fig. 12. Support construction.

beams and diagonals are bolted together. The base plates are fixed to the foundation by anchor bolts M27 and M36 and the horizontal forces are transferred by shear blocks.

DOORS AND VENTS

To achieve the tropical climate inside the domes, a special ventilation system had to be used. The required openings were determined by Ove Arup & partners , London. On top of each of

Fig. 13. Vents and heating.

the 8 domes are vent openings. The 5 hexagons surrounding the top pentagon were divided into 3 triangles so that each dome has 30 openings operated by remote control. These windows are also covered by triangular air cushions. The substructure consists of rectangular hollow sections 140 x 70 mm. For the air inlet glass lamella windows are arranged around the edge of the domes (**Fig. 13**). Warm air can be blown inside the domes using heaters. Each dome also has some doors for maintenance and emergency exits only. The access for visitors is through the link building only. For maintenance the vents on top of each dome has a cat walk.

MANUFACTURING

Most of the steel structure was manufactured in the MERO workshop close to Wuerzburg in Germany. Only the arches and support point were fabricated elsewhere. The manufacturing of the MERO beams and nodes was done using a computer aided machine. The end plate and support brackets of the top chord beams were welded by hand. Each element and node has it's unique number which remained the same during the design, manufacturing and erection phase. For corrosion protection all steel elements are hot dipped galvanised. Due to their sizes the segments of the arches were galvanised by a company in France, which has one of the biggest galvanising tubs in Europe. The bowl nodes made out of cast iron GGG40, were also galvanised. With a general inspection every two years, the steel structure is designed to be maintenance free for 30 years.

CLADDING

The more than 800 hexagon elements are covered by air filled cushions. These cushions are made of transparent EFTE (Ethyltetrafluorethylene) foil. The basic material is between 50 μm and 200 μm thick with a width of 1.5 m. The foil material was cut and welded. The normal cushions are made up of three layers. The top and bottom layer form the cushion and carry the loads. An additional layer between them has the function of enhancing the

Fig. 14. Installation.

Fig. 15. Air hock up.

temperature insulation and also dividing up the airspace in case of leakage. In areas of high local wind suction the outer surface of the cushions was strengthened by using two layers of foil.

The cushions are attached on an aluminium frame to the top chord beams. Each cushion is also attached to an air supply system (Fig. 15). The pressure inside the cushion is about 300 Pa. The maximum height of the inflated cushion is about 10 to 15 % of the maximum span (Fig. 16)

Material EFTE has been used for more than 20 years. It is extremely light and transparent. The surface is also quite smooth so that the dirt on the outside is washed down by the rain.

Cushions in this project size had never been built. During the design stage, extensive studies and tests were performed by MERO, the consultant Ove Arup (London) and the foil subcontractor Foiltec in Bremen (Germany). Some of the tests were performed on a real 1 to 1 scaled model. The results of this studies lead to the important parameters for the design of the cushions with spans up to 11 m. In areas of high show load, like the arches, some additional cables were needed to support the cushions. After the design phase, the size of each of the 800 elements was calculated, cut, an manufactured. The gutter construction

Fig. 16. The installed cushion.

between the single domes is made out of insulated aluminium parts and is covered on the outside by foil. The rain water is saved and used for the plants inside the Biomes. The entire roof surface can be maintained by abseilers using ropes attached to steel pins which are attached to each bowl node of the structure (**Fig. 8**).

ERECTION

The erection of the steel structure began in November 1999. Extensive ground movement and the building of the 858 m long concrete foundation was done by the general contractor. The

foundation is 2 m wide and 1.5 m high. It rests on up to 12 m long concrete piles which were drilled into the ground. For the erection of the structure, a scaffolding was set up. This scaffolding has it's place in the Guinness book of records as the biggest and tallest free standing scaffolding of the world. Most of the hexagons are put together on the ground and then lifted into place by tower cranes and then bolted together (**Fig. 17**). The prefabricated pieces for the arches were about 13 m long.

Fig. 17. Erection of the HTBiomes.

They where also set up on the scaffolding and then welded together. After the erection of the steel structure was done, the scaffolding was removed. The installation of the foil cushions was done by abseilers (**Fig. 14**). The ground work inside the Biomes could be done parallel to the cladding work. The cladding was finished on time in September 2000 so that the Biomes could be heated and the planting could commerce during the winter. The Eden project opened the doors for public in time in March 2001.

For further information see www.edenproject.com and www.eden-project.co.uk

THE MAIN STRUCTURAL DATA

Total surface	39.540 m^2
Total steel weight	700 tons
Total length off all beams	36000 m
Steel weight per surface	less than 24 kg/m^2
Biggest hexagon area	80 m^2 at a span of 11 m
Biggest dome diameter (dome B)	125 m
Column free area	15590 m^2 WTB and 6540 m^2 for HTB

PARTICIPANTS

Client:	The Eden Project	(www.edenproject.com)
Architect	Nicholas Grimshaw & Partners Ltd, London	(www.ngrimshaw.co.uk)
General Engineer	Anthony Hunt Associates, Cirencester (www.anthonyhuntassociates.co.uk)	
Bauphysik	Ove Arup & Partners, London	(www.arup.com)
Wind channel test	BMT Fluid Mechanics Limited London	(www.bmtfm.com)
General Contractor	McAlpine JV	(www.sir-robert-mcalpine.com)
Steel & Cladding	MERO GmbH & Co.KG, Würzburg	(www.mero.de)
	subcontractor : Foiltec GmbH, Bremen	(www.foiltec.de)

REFERENCES

[1] ALAN C. JONES : Civil and Structure Design of the Eden Project,
   International Symposium on Widespan Enclosures at the University of Bath,
   26-28 April 2000.
[2] EMDE H.: Geometrie der Knoten-Stab-Tragwerke :
   Veröffentlichung des Strukturforschungszentrums Würzburg 1978.
[3] FULLER R.B. : In : The Dome Builder's Handbook :
   Edited by John Prenis; Running Press Philadelphia Pennsylvania ; 1973.
[4] PAVLOV G.N.: Determination of Parameters of Crystal Latticed Surfaces Composed
   of Hexagonal Plane Facets ; Int. Journal of Space Structures ;
   Vol.5 Nos. 3 & 4 1990 Multi-Science Publishing.
[5] RSTAB 5 : Ingenieursoftware Dlubal GmbH http://www.rstab.de.
[6] MERO technical Approval ; Z-14.4-10 erteilt vom DIBT Berlin.
[7] KLIMKE H., How Space Frames Are Connected, IASS Madrid 1999, pp B4.13 – B4.19.
[8] KLIMKE H.: Entwurfsoptimierung räumlicher Stabstukturen durch CAD-Einsatz,
   Bauingenieur 61 (1986) Seite 481 bis 489.
[9] LEHNERT S.: Das Eden Projekt ; Intelligente Architektur; Ausgabe Nov/Dez 2000.

From Olympiad to Commonwealth – The Evolution of the City of Manchester Stadium.

DARREN PAINE, Arup Manchester, UK.

INTRODUCTION.

This paper considers the evolution of a stadium design for Manchester over a period of around eight years from the original concept as an anchor for the City of Manchester's Olympic bid, via the desire to provide a home for the National Stadium, to the now complete Commonwealth Games Stadium, and ultimately as a home for Manchester City Football Club.

Each step along this path has resulted in defining and redefining performance requirements and public aspiration for the City of Manchester. Whilst placing the City on the world stage the newly created Sports facilities at the stadium site also act as a local focus for regeneration of the surrounding local areas. The development of the design also reflects the change in attitude to Stadium design as a whole. This world-class facility includes design innovations, which have added value to the scheme, but are all delivered within the confines of a limited budget and a public body as a Client.

One of the clearest icons for the City of Manchester Stadium is in the design of the roof. The scheme has seen many developments and changes in brief, but there has always been a desire to create a structure that is striking, dynamic and efficient.

The logistics of providing weather protection and cover to a large number of spectators is significant. The constraints on the design are generally outside the control of the Structural Engineer, as stadium roof design follows as much from the form of the stadium as from the function it serves. The days of columns sited within spectator viewing areas have long since past, and whilst it is every Engineers nightmare to be asked by the Architect to remove primary columns from the structure the stadium designer knows that some rules do not need to be written down or discussed. Many other factors also impact on the design solution, the key factors are:

- Spectator sight lines, everyone should have an uninterrupted view for play along the pitch and in the air.
- Weather protection, global warming has yet to provide the UK with the type of playing environment that doesn't require hats, gloves and coats to be worn on most occasions.
- Pitch growth, the spectators require cover but the grass would like to be out in the open. Without the pitch the spectators won't be there.
- Flood lighting, as stadia sizes increase the traditional four post floodlighting solution is becoming obsolete. The requirements of television producers for lighting levels on the pitch to be at a consistent maximum can make a significant difference to the income a club may receive.

With all this in mind what are the options?

Essentially large span structures for stadia fall into one of three structural types, these definitions are quite broad, but it is possible to break any structural form down into its most basic components in order for a category to apply:

1. Cantilever, probably the most dominant form of stadia roof solution. For example, Bari.
2. Goal post, generally consisting of a deep beam, arch, or truss, spanning the length of the pitch along the leading edge of the roof. Rafter elements then span between this and the rear of the terracing structure. For example, Hong Kong Stadium
3. Three-dimensional structures, this type of solution may be generated by the form of the bowl, or a desire for the form of the roof. For example, Birmingham National Stadium Scheme.

Bari, San Nicola Stadium: Cantilever truss and fabric roof from rear of bowl.

Hong Kong Stadium : Trussed Arch spanning the full length of the pitch. Individual rafters span from the rear of the bowl to the arch.

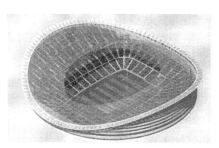

Birmingham National Stadium Scheme: A series of opposing cables tied to a compression ring around the rear of the bowl.

The more arena like forms of stadia with moving roofs generally fall into the category of the goalpost type, the majority of them consisting of beam/truss members spanning across the pitch to the rear of the terraces.

With these rational solution types as a guide the Stadium Designer can set about the task of the concept design.

THE OLYMPIC STADIUM SCHEME, FOR THE CITY OF MANCHESTER OLYMPIC BID 2000.

History has already shown that the 2000 Olympic Games, and the stadium that formed the central feature, was a massive success, both for the Games itself but also for the City of Sydney.

Manchester had long held an aspiration to raise its profile on the world stage. This process began with a failed bid for the Olympics in 1996. However, these decisions are made a long time in advance of the event date, and Manchester was undaunted in its desire. The Olympic 2000 bid brought leading designers to add a depth to the scheme that had not been present in the previous bid. Norman Foster was appointed as master-planning Architect and Arup, with Arup Associates, were appointed to design the stadium.

The stadium was to be a state of the art facility consisting of the following key brief elements:
- An uncompromising view of the action from all 65,000 seats
- No moving roof
- No moving pitch
- Moveable lower tier to bring people closer to the action at football events
- Efficient structural form
- Flexible approach to construction methodology

The iconic elements of the design consisted of:
- Masts to support the roof
- Circular ramps to access the concourse
- Saddle form overall geometry

Olympic Scheme Model.

From these elements a clear design philosophy was brought together. The structural diagram is logical and uncluttered.

The form of the roof was essentially a cantilever type system. Each mast supports a roof truss, which is supported near the base of the mast and reaches out to the leading edge of the roof. Each of these rafters forms a radial line on plan and reflects the near circular geometry of the outer edge of the stadium. Changing the inclination of each of the rafters forms the saddle shape geometry. Each mast is supported on an individual tower and access ramp. The primary rafters consist of open lattice trusses, which sit above the roof surface.

Secondary members spanning between the primary rafter trusses form the roof plate surface. These secondary members would be clad on each surface to create the single plate. As with most modern stadia the leading edge of the roof is clad in a transparent material to aid in natural light reaching the pitch.

Unfortunately for Manchester the Olympic scheme was never to get past the model and concept stage. The 2000 Olympics being won by Sydney and the equally dramatic Stadium Australia.

THE NATIONAL STADIUM SCHEME, 1998
The City of Manchester was not going to give up their dream quite so easily. When the idea of a replacement for a National Stadium was to be open to a competition the game was once again afoot....

The brief requirements for the National Stadium were similar in respects to the brief for the Olympic bid. The stadium had to be capable of hosting a possible Olympic Games in addition to being the home of National and International Football in England.

The brief now called for the following requirements:
- An increased capacity of 70,000 seats
- A fully closing roof
- Athletics and Football.

National Stadium Model.
©Andrew Putler

The same team responsible for the Olympic scheme developed the Manchester bid. Whilst the design certainly contained elements of the Olympic Scheme it had moved on in terms of the evolution of the design.

The retention of the masts and ramps was very much at the heart of the scheme. These were seen as something that gave Manchester's entry a certain iconic status, much as Wembley will always be remembered for the twin towers.

The main change in the design was the incorporation of the moving roof. The solution here was to opt for two roof support systems that interacted rather than one system that had to meet all criteria.

The primary roof plate was supported by a series of trussed rafters, which span from the masts to the leading edge of the roof. The leading edge of the roof consists of an arch, which spans from each end of the stadium. There are two parallel arches either side of the pitch. The moving elements of the roof use rails fixed alongside these rafters but are supported by an additional cable system, which spans the full width of the stadium, over the pitch. There are eight moving elements. Six elements are sat over each side roof of the stadium, and move together in pairs, joining along the centreline of the pitch. The last two segments are above the two end stands and move along the arches.

The intention is that the roof can provide a fully enclosed environment, or roof coverage for a football event, or roof coverage for an athletics event. This design also has the flexibility that it could be retrofitted as a later addition without significant enhancement of the primary structure.

But, as they say, history repeats itself, and Manchester lost out to the proposed redevelopment of Wembley.

THE STADIUM FOR THE COMMONWEALTH GAMES AND LEGACY STADIUM FOR MANCHESTER CITY FOOTBALL CLUB.

The desire to stage a world-class event had not diminished. The XVII Commonwealth Games in 2002 will be one of the major focal points to the Queens Golden Jubilee year. Once again Manchester had to put forward a bid to become the host city, the other contenders being London and Sheffield. Having gained significant experience in previous bids Manchester was chosen as the successful host city by the Commonwealth Games Council for England in 1998.

In developing any scheme it is important that each key stage is also reviewed against the requirements of the brief. The main criteria driving the roof design for the Manchester Stadium was the ability to hold football and athletics events within the same venue. In moving on to the final scheme and achieving a long-term viability for this stadium this requirement underwent serious review. The primary brief requirements could now be summarised as follows:
- 38,000 seats for the Commonwealth Games. (Subsequently increased to 41,000)
- 48,000 seats for the legacy football stadium.

The design starting point was obviously the evolution from the Olympic Stadium scheme. The concept would be similar, but downsized to the new stadium size. The masts and circular ramps would remain as the icon that would draw people to the stadium, and help them orientate themselves when circulating the perimeter of the building.

Whilst the brief had been interrogated to maximise the value of the proposed facility, the design had to be flexible enough to appear as a complete design at two distinct stages of its construction. The Commonwealth Games stadium was to consist of approximately 50% of the permanent structure. The remainder of the stadium would be built for the football configuration. This constraint would have significant implications when considering the uplift design case.

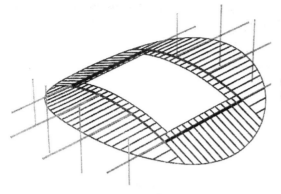

Original analysis model for the
Commonwealth Games Stadium.

The first scheme consisted of the mast and booms that had been proposed throughout each competition. The resistance to uplift in this scheme was provided by a cable, which ran parallel to pitch and connected to each boom. This straightforward system would adequately resist the uplift pressures and provide the flexibility of the phasing requirements. The second phase of the roof would infill above the ends of the stadium and the corners. The primary support for this system would be a truss spanning the full width of the pitch and supported by the end masts.

It can be seen from all of the schemes that the overall roof geometry had consisted of the saddle shape. From the outside the roof plate would mirror this desire with a very clean line. However, the underside looks inconsistent as the massing of each elements seems to bare little relation to the geometry.

It is at this point that the desire to continue can sometimes out way the desire to revisit the design. In satisfying the brief requirements the solution should also satisfy structural efficiency and logic. At the start the options for stadia roofs were defined as three categories. It is a well-recognised fact that the boundaries between these categories are a little blurred. The design solution for Manchester had to allow the flexibility of the two-stage construction. With this in mind the option of merging to roof systems together was explored.

The most straightforward solution for the individual stands is a cantilever type roof. This has been expressed in the form of the mast and cable system. A cable runs from the top of the mast to each rafter. As a series of components this can be likened to a crane like structure. Each rafter reflects the lower concrete bowl geometry by being placed on each individual gridline. Simplistically, it would be perfectly possible to install a minimum of two rafters and their tie cables and form a stable system.

Whilst the above is efficient at resolving vertical downward loads, it is totally inefficient at resisting uplift. The key to solving this was to provide a 3-dimensional cable structure anchored at the four corners of the stadium. This cable system is prestressed against the vertical support cables, and grounded using anchors to the underlying sandstone. The benefit to the phased construction is that this tie down cable system can be installed without any of the rafters as it is self-supporting.

The intended construction sequence is then to simply support each rafter from the cable catenary and support from the rear of the bowl. For the Commonwealth Games only those rafters that are required to cover stands are included. For the legacy football stadium the remaining rafters are installed to the existing cable net.

View of the East stand in Commonwealth games configuration, showing the erected roof and the cable catenary across the North Stand.

CONCLUSION

The design process is always about evolution. The important lesson to be learnt is that evolution may not be a series of linear decisions. The final design of the City of Manchester Stadium is unique, and an icon for the City, and is a clear response to the requirements of the brief. It can be compared to the previous bid schemes, and includes many of the original design intent features from the very first scheme. However, Arup's design took a bold change in direction to satisfy the structural logic, the architectural aesthetic and the client's aspiration.

©Hayes Davidson

Expandable structures formed by hinged plates

F JENSEN and S PELLEGRINO
Department of Engineering, University of Cambridge, UK

Abstract This paper presents a family of two-dimensional expandable structures formed by flat plates connected by cylindrical (scissor) joints. These structures are kinematically equivalent to previously known expandable bar structures. Special shapes of the plates are determined for which the plates do not overlap or interfere during the expansion of the structure, and for which the structure forms a gap-free disk in the closed configuration and an annulus in the open configuration.

INTRODUCTION

This paper is concerned with the geometric design of symmetric expandable structures consisting of rigid, flat plates connected by joints that allow only a relative rotation about one axis, i.e. cylindrical hinges or *scissor joints*. We are interested in assemblies of identical plates forming a complete ring, and which are able to move relative to one another between two extreme configurations. In the closed configuration these structures form a gap free disk and in the open configuration they form an annulus with, for example, a circular opening in the middle. These structures are visually pleasing and have potential applications in the design of retractable roofs.

A limited number of structures with properties broadly similar to those studied in this paper have been known for some time. Verheyen [1] made extensive studies of transformable structures based on pairs of overlapping elements connected by a cylindrical hinge in the middle, with neighbouring pairs connected by spherical joints. Wohlhart [2] and You [3] have considered expandable structures formed by rigid elements connected through hinged links. Due to the relative rotation between the elements in each pair in the former case, and between the main elements and the links in the latter case, these structures can vary in size while maintaining essentially the same geometric shape.

The approach that is presented in this paper starts from a two-dimensional expandable *bar structure* and replaces the bars with flat plates which are connected with scissor joints at exactly the same locations as the original bar structure. Thus, the kinematic behaviour of the bar structure is unchanged. Then, the shape and size of these plates are determined for which the largest motion of the plate structure is possible.

The layout of this paper is as follows. The next section briefly reviews existing solutions for expandable bar structures. A solution for *n*-fold symmetric structures consisting of multi-angulated bars is of particular interest, and the following section obtains expressions for the inner and outer radii of this type of structure in the open and closed configurations. Then, the section Cover Elements for the Bar Structure determines certain special shapes of flat plates that can be attached to the bar structure without affecting its range of motion. This solution is then used in the following section for the design of some practical structures. A brief discussion concludes the paper.

BACKGROUND: EXPANDABLE BAR STRUCTURES

Expandable bar structures based on the concept of scissor hinges have been known for a long time, initially only as two-dimensional, lazy-tong-type structures that extend linearly. More recently, curved structures that expand in three dimensions have been pioneered by the Spanish engineer Pinero [4] and further developed by Escrig [5] and Zeigler [6]. A major advantage of these structures compared to other expandable structures is the relative simplicity of their joints.

A considerable advance in the design of this kind of structures was made by Hoberman, with the invention of the simple *angulated element* [7]. In its simplest form, shown in Figure 1, this element consists of two identical angulated bars with central kinks of equal amplitude α ("angulated bars"), connected by a scissor hinge at the centre, i.e. at node E. This element has the special property that lines through the end nodes A, D and B, C subtend a constant angle when the angle between the two angulated bars is changed; in other words, an angulated element subtends a constant angle when it is opened and closed.

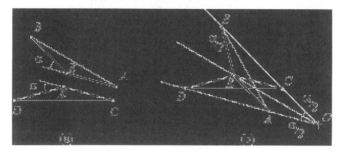

Fig. 1. Angulated element.

Because of this property, simple two-dimensional expandable structures can be obtained by forming one or more closed, concentric rings of identical angulated elements. For example, the expandable structure shown in Figure 2 could be made by joining with scissor hinges two such rings, each consisting of 12 elements. Three-dimensional expandable structures, such as the Hoberman sphere [8], are also formed from closed rings of identical angulated elements.

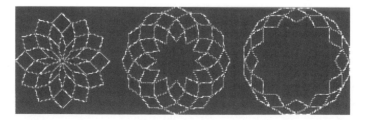

Fig. 2. Expandable structure consisting of multi-angulated bars.

You and Pellegrino [9,10] noted that consecutive angulated bars in Figure 2 maintain a constant angle equal to α when the structure is expanded, and thus can be replaced with a single *multi-angulated bar*. Thus, the structure of Figure 2 could also be made from a total of 24 bars, each having four segments with equal kink angles: 12 bars are arranged in a clockwise direction and 12 anti-clockwise. At each cross-over point, there is a scissor joint. Note that the whole structure is arranged according to a pattern of identical rhombuses, which are "sheared" when the structure expands. It is also possible to design much less symmetric expandable structures, e.g. with elliptical shape [10], but this is beyond the scope of the present paper.

A general expandable structure consisting of identical multi-angulated elements, such as that shown in Figure 2, is defined by the number of segments in each angulated element, k, and the number of angulated elements in each layer, n, plus the segment length. Thus, apart from a scaling factor, this structure is fully defined by the parameters n; k, see Figure 3 for an example. Note that, since the structure has n-fold symmetry

$$\alpha = \frac{2\pi}{n} \tag{1}$$

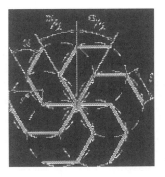

Fig. 3. One layer of *6;3* structure.

If the structure is allowed to rotate while it expands, i.e. the scissor hinges are not required to move on radial lines, the structure can be connected to n fixed points, which are the centres of the n circles of radius r^* defined by the multi-angulated elements that make up one layer, see Figure 3. Thus, the motion of one layer of the structure is a pure rotation of each of its elements about the centre of its corresponding circle. Using r^* to define the size of the structure, the segment length is

$$l = 2r^* \sin\frac{\alpha}{2} \tag{2}$$

Either rigid or flexible covering elements can be attached to the structures described above to create retractable roofs [8,11], but in previous studies these elements were allowed to overlap in some, or all configurations.

EXTREME CONFIGURATIONS OF BAR STRUCTURES

The total rotation undergone by a multi-angulated element during the motion of the structure from the fully-closed to the fully-open configuration will be called the *rotation angle*, β. This angle can be found from Figure 4, showing the closed position. Note that all of the angulated elements meet at the centre O, and their circles of motion also intersect at O. The open configuration is reached when neighbouring multi-angulated elements touch at the point of intersection of their respective circles of motion, P.

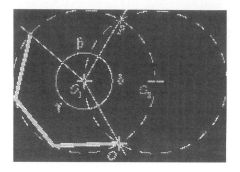

Fig. 4. Definition of angles for a multi-angulated element.

Because the angle subtended by a single bar on the circle of motion is equal to α, the *element angle* subtended by a multi-angulated element is

$$\gamma = k\alpha \tag{3}$$

The *limit angle* subtended by the intersection points of two neighbouring circles of motion, O and P, corresponds to the two limits for the motion of the multi-angulated element. It is found from:

$$\delta = \pi - \alpha \tag{4}$$

The rotation angle can then be calculated from:

$$\beta = 2\pi - \gamma - \delta = \pi + (1-k)\alpha \tag{5}$$

In practice, the bar structure will not be able to reach these extreme configurations, due to the physical size of elements and joints. Therefore, we introduce two *reduction angles*, γ_1 and γ_2, to denote the corrections needed respectively for the closed and open positions. The *reduced rotation angle* is thus

$$\beta^* = \pi + (1-k)\alpha - \gamma_1 - \gamma_2 \tag{6}$$

Now, consider the two extreme configurations of the bar structure. Denoting by r and R its inner and outer radii, respectively, and using the subscripts *min* and *max* for the open and closed configurations, respectively, the following expressions can be derived by analysing Figure 5:

$$r_{max} = 2r^* \sin\left[\frac{1}{2}\left(\pi + (1-k)\alpha - \gamma_2\right)\right] = 2r^* \cos\left[(k-1)\frac{\alpha}{2} + \frac{\gamma_2}{2}\right] \tag{7}$$

$$R_{max} = 2r^* \sin\left(\frac{\pi - \gamma_2}{2}\right) = 2r^* \cos\left(\frac{\gamma_2}{2}\right) \tag{8}$$

$$r_{min} = 2r^{*}\left(\frac{\gamma_{1}}{2}\right) \tag{9}$$

$$R_{min} = 2r^{*}\left(\frac{\upsilon + \gamma_{1}}{2}\right) \tag{10}$$

Joint Size Effects

Because the bar structure is composed of two distinct layers, the only possible interference is between elements in the same layer. Assuming the joints to be circular with radius r_{j} the two reduction angles are

$$\gamma_{1} = \gamma_{2} = 2\arcsin\left(\frac{r_{j}}{r^{*}}\frac{1}{2\sin(\alpha/2)}\right) \tag{11}$$

COVER ELEMENTS FOR BAR STRUCTURE

We begin by considering what limitations are imposed on the motion of only a small piece of the bar structure when covering plates are attached to it.

Consider a linkage consisting of two parallel bars $A_{i}A_{i+1}$ and $B_{i+1}B_{i+2}$ and a pair of parallel linking bars, as shown in Figure 6(a). Bar $B_{i+1}B_{i+2}$ is assumed to be fixed so no rigid body motions are allowed and this leaves *one* internal mechanism which allows the linking bars to rotate and bar $A_{i}A_{i+1}$ to translate. The top-left angle defines the rotation angle, β, as shown in Figure 6(a); this angle is positive clockwise.

Consider a rigid plate attached to bars $A_{i}A_{i+1}$ and $B_{i+1}B_{i+2}$. This rigid body eliminates the mechanism of the parallelogram. If a straight cut at an *inclination angle* θ is made in the plate, then the mechanism is restored; the line of the cut is called the *inclination line*. So, now we have two plates attached to the linkage, which are not allowed to overlap; depending on the inclination angle, β can either increase or decrease. In each case, the motion of the parallelogram has to stop when the gap between the two plates is closed again. The two limits on the rotation angle are denoted by β_{1} and β_{2}, and once they are known θ can be found from Figure 6(b) by considering the sum of the internal angles in ABC:

$$\theta = \frac{\pi - \beta_{1} - \beta_{2}}{2} \tag{12}$$

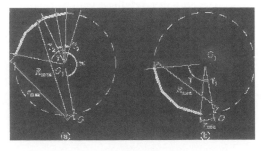

Fig. 5. Definition of maximum and minimum radii and reduction angles.

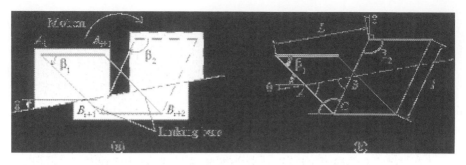

Fig. 6. Motion of simple linkage with two plates.

As no length variables are present in Equation 12, the position of the inclination line relative to the linkage does not affect the limits of the motion. Figure 6(b) shows that bar A_iA_{i+1} translates by a distance L parallel to the inclination line

$$L = \sqrt{l^2 + l^2 - 2l^2 \cos(\beta_2 - \beta_1)} = l\sqrt{2 - 2\cos(\beta_2 - \beta_1)} \tag{13}$$

Next, we consider two neighbouring multi-angulated elements, A_1-A_5 and B_1-B_5, which are part of an expandable bar structure with $k = 4$, as shown in Figure 7. Together with bars A_1B_2, A_2B_3, etc., these multi-angulated elements form three interconnected linkages, which are free to move within certain limits.

In the closed configuration, Figure 7(a), the bars A_1A_2 and B_1B_2 form an angle α, giving for the first linkage $\beta_1' = \alpha$. Since the kink angles in the multi-angulated element are equal to α, the third linkage has

$$\beta_1''' = \beta_1' + (k-2)\alpha = 3\alpha \tag{14}$$

The limit for the open position is found by noting that bars A_3A_4 and B_4B_5 of the third linkage become collinear and thus $\beta_2''' = \pi$, as shown in Figure 7(b). Therefore,

$$\beta_2' = \beta_2''' - (k-2)\alpha = \pi - 2\alpha \tag{15}$$

A rigid plate is then attached to these interconnected linkages and cut along a single straight line, as before. Now the inclination will be defined with respect to the line B_1B_2, hence the inclination angle is

$$\theta' = \frac{\pi - \beta_1' - \beta_2'}{2} + \alpha \tag{16}$$

Substituting $\beta_1' = \alpha$ and Equation 15 into Equation 16 we obtain

$$\theta' = (k-1)\frac{\alpha}{2} \tag{17}$$

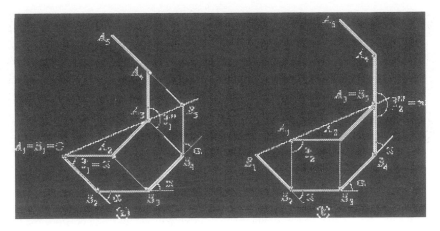

Fig. 7. Two extreme positions of neighbouring angulated elements.

Now, turning to the complete bar structure, we consider n inclination lines between the angulated elements. These lines define n identical wedge-shaped covering elements with wedge angle α. In the closed configuration these wedges meet at the origin O and form a gap-free surface; in the open configuration they form a gap-free annulus; and they never overlap in any intermediate configurations, as was first observed in Ref. [11].

The reduction angles are included in the definition of the inclination angle by modifying the limits for the linkages:

$$\beta_1^* = \beta_1' + \gamma_1 \tag{18}$$

$$\beta_2^* = \beta_2' - \gamma_2 \tag{19}$$

Substituting Equations 18 and 19 into Equation 16 the *reduced* inclination angle is obtained:

$$\theta^* = (k-1)\frac{\alpha}{2} - \frac{\gamma_1}{2} + \frac{\gamma_2}{2} \tag{20}$$

From this equation it can be seen that if the two reduction angles are equal, then the solution is a line parallel to that of the original solution in Equation 17.

Shape of Cover Elements

So far, only straight-edged cover elements have been considered, but in fact non-straight shapes of a periodic type are also possible. Consider the motion shown in Figure 6(a); for the upper and lower cover plates to fit together without any gaps or overlaps in both configurations, the boundary edges must also match in both configurations. Hence, non-straight features are allowed, provided that they repeat with period L, as shown in Figure 8. If the common boundary between the two plates is longer than L, then the same features of the lower plate must also be repeated in the upper plate. More generally, plates with common boundaries must be shaped such that all features have a periodic pattern.

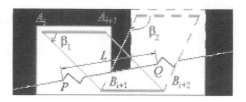

Fig. 8. Periodic pattern of non–straight boundary.

There are two important restrictions to the above periodicity rule, as obviously a boundary that deviates significantly from the original straight line would inhibit the motion of the linkage. First, when the plates are in contact the initial velocity of any point on the boundary, which is perpendicular to the linking bars, needs to form an angle with the inclination line greater than the slope of the boundary. This is to avoid that the plates jam when the motion is about to begin, Figure 9(a). Second, any deviations of the boundary shape from the original straight line need to lie within a region bounded by two circular arcs that pass through the extreme points P and Q of the repeating length of the boundary; the centres of these circles are defined by the intersections of lines through P and Q, and parallel to the linking bars in the open and closed configurations. This is to avoid interference between points on the boundary during the motion of the two plates, Figure 9(b).

The maximum distance, h, from the inclination line to the boundary of either plate can be shown to be given by:

$$h = l\left(1 - \sqrt{1 - \left(\frac{L}{2l}\right)^2}\right) = l\left(1 - \sqrt{\frac{1}{2} + \frac{1}{2}\cos\beta^*}\right) \tag{21}$$

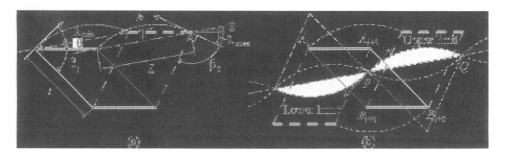

Fig. 9. (a) Direction of velocity vectors in the two extreme positions; (b) region defining possible boundary shapes.

DESIGN OF EXPANDABLE STRUCTURES

Having defined the limits within which the shape of the cover elements can be varied, it is possible to design many different expandable structures. Two different types of structures will be presented. The first is a bar structure covered by a layer of plate elements: as the motion of this structure is controlled by the underlying bar structure, here the cover elements can be fixed to the multi-angulated element in many different ways. The second is purely a plate structure, consisting of two layers of identical plates; this structure is designed such that it is kinematically equivalent to the previous bar structure.

Bar Structure Covered by Plates

For cover elements with straight edges the shape of the individual elements is that of a wedge bounded by the inclination lines, which in the closed configuration meet at the origin. Note that in this case, in general, a single cover element will not cover completely the angulated element beneath it.

For cover elements with non-straight edges it is possible to find particular designs that achieve various design aims. For example, it is possible to design a structure that forms a *circular opening*, in the open configuration. This requires the repeating part of the edge of the cover elements to be a circular arc, since in the open configuration the innermost repeating part forms the edge of the central opening, and the arc radius to be equal to the radius of the opening, $r_{opening}$. The smallest possible radius of the boundary arcs is l, hence the general condition to create a circular opening is

$$l \le r_{opening} \tag{22}$$

Consider a circle through the tips of the cover elements in the open configuration, which includes the effect of both reduction angles. Hence, effectively adding γ_1 to γ, the total reduction angle is defined as $\gamma_{tot} = \gamma_1 + \gamma_2$. The opening radius is then determined from Equation 7

$$r_{opening} = 2r^* \cos\left[(k-1)\frac{\alpha}{2} + \frac{\gamma_{tot}}{2}\right] \tag{23}$$

Substituting Equations 2 and 23 into Equation 22

$$2r^* \sin\left(\frac{\alpha}{2}\right) \le 2r^* \cos\left[(k-1)\frac{\alpha}{2} + \frac{\gamma_{tot}}{2}\right] \Rightarrow \gamma_{tot} \le \pi - k\alpha \tag{24}$$

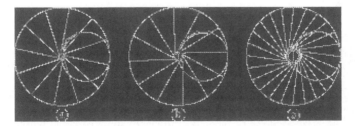

Fig. 10. Wedge shaped plate structures with (a) $k = 2$; (b) $k = 3$; (c) $k = 4$

Plate Structures

In a plate structure that does not rely on a separate supporting bar structure, it must be possible to fit all joints within the plates, which is often a significant challenge. To date, it has been impossible to construct structures with straight-edged plates and with $k > 5$; the solutions that have been found, with $k \le 4$, are shown in Figure 10. Note that for all these solutions the inclination lines do not meet at the centre, thus leaving a small hole at the centre of the structure in the closed configuration. It is possible to close this opening by modifying the basic wedge shape.

Using non-straight boundaries it is possible to design structures with a circular opening, as described above, and to minimize the gap, at the centre of the structure, that results from using plates with straight boundaries, see Figure 11.

Fig. 11. Model of expandable structure.

DISCUSSION AND CONCLUSION
An analytical approach for finding the shape of hinged plates that can execute large motions has been presented. It has been shown that a large variety of shapes are possible, which are all variants of a simple wedge shape. These expandable plate structures could be used in a wide variety of small and large scale applications where a continuous gap free surface is required that has the ability to execute a large shape change, such as toys, flow and sound control elements and roof structures.

REFERENCES
1. VERHEYEN H F, A Single Die-Cut Element for Transformable Structures, International Journal of Space Structures, Vol. 8, No. 1 & 2, 1993, pp 127-134.
2. WOHLHART K, Double-Chain Mechanisms, Proceedings of IUTAM-IASS Symposium on Deployable Structures: Theory and Applications (edited by S Pellegrino and S D Guest), Kluwer Academic Publishers, The Netherlands, 2000, pp. 457-466.
3. YOU Z, A new approach to design of retractable roofs, Proceedings of IUTAM-IASS Symposium on Deployable Structures: Theory and Applications (edited by S Pellegrino and S D Guest), Kluwer Academic Publishers, The Netherlands, 2000, pp. 477-483.
4. ESCRIG F, Arquitectura Transformable, Escuela Técnica Superior de Arquitectura de Sevilla, 1993
5. ESCRIG F and VALCARCEL J P, Geometry of Expandable Space Structures, International Journal of Space Structures, Vol. 8, No. 1 & 2, 1993, pp 71-84.
6. ZEIGLER T R, Collapsible self-supporting structures and panels and hub therefore, USA Patent no. 4,290,244, 1981.
7. HOBERMAN C, Reversibly expandable doubly-curved truss structure, USA Patent no. 4,942,700, 1990.
8. HOBERMAN C, Radial expansion/retraction truss structures, USA Patent no. 5,024,031, 1991.
9. PELLEGRINO S and YOU Z, Foldable ring structures, Space Structures 4, vol. 4 (edited by G A R Parke and C M Howard), Thomas Telford Publishing, London, 1993, pp. 783–792.
10. YOU Z and PELLEGRINO S, Foldable bar structures, International Journal of Solids and Structures, vol. 34, no. 15, 1997, pp. 1825–1847.
11. KASSABIAN P, YOU Z and PELLEGRINO S, Retractable roof structures, Proceedings Institution of Civil Engineers Structures and Buildings, vol. 134, Feb 1999, pp. 45-56.

The space frame roof for air India at Mumbai

K. K. KHANRA

SUMMARY
The paper describes the planning, design and construction of the largest Multi-layered space frame structure recently built at Mumbai for Air India for servicing two Boeing 747-400 Aircraft's at a time – one docked in 'tail-in' and the other in the 'nose-in' position. A unique feature of the hangar is that the docking systems and cranes are suspended from the space frame itself. The design involves a main pin-jointed space frame which is three layered, that is strengthened by the addition of two layers and made rigid to form a five-layered front girder with a clear span of 129.384 m. The front girder is designed to act as portal along with the two end columns of steel 25m high which are supported on concrete pedestals of 4.5 m height. Weighing only 65 kg/m², this long space & high altitude space frame hangar is believed to be the lightest of its class with suspended docking systems and craneage and built first class with suspended docking systems and craneage and built first time in the country. The entire hangar roof weighing about 1100 tons assembled on the ground and lifted in one operation using 12 nos. PSC centre hole jacks. The paper also briefly describes some of the special joints of space frame developed at site.

Although there is a growing demand to use space frame structure in the industry because of its aesthetic appeal, economy and faster construction, it has however been felt to develop extensive testing programme and research to arrive at appropriate factor of safety for node connectors and other aspects of space frame.

INTRODUCTION
A new 3 D tubular space frame hangar has recently been built at the old Airport Complex for Air India at Santacruz, Mumbai. The clear internal plan dimensions of this two bay hangar are 129.384 m x 88.663 m with a clear height of 21 m under the front girder and 23 m elsewhere. A 3500m² Annexe building with basement + ground + two storeys with provision for future extension by the addition of five more storeys is constructed at the rear of the hangar. The hangar is planned for the maintenance of two Boeing 747-400 airplanes – one aircraft docked in the 'tail-in' position and the other in the 'nose-in' position. The hangar is equipped in one bay – for the aircraft docked in the 'tail-in' position – with full suspended docking facility, undercarriage lifting platforms and fixed hydraulic systems. Two 10 ton and two 3 ton overhead cranes, compressed air, fire protection systems and other specialized ground support services for complete check-up on the airplanes are provided across the entire area covered by both the bays. The plan and cross-section of the hangar may be seen in Fig. 1 and Fig. 2.

DESIGN CONSIDERATIONS
The concept designs relating to planning and equipment services were prepared by M/s. Scott Wilson and Kirkpatrick, UK, giving due consideration to the constraints imposed by existing structures at the site including a maintenance hangar built in the late 70's on the East side and an engine workshop built in the early 30's on the West side which are approximately 125 m apart. To accommodate the new hangar, two bays each of 7.5m of the engine workshop, had to be demolished. Because the site is just 150 m away from the main runway, the overall height of the hangar had to be restricted to 30.7 m to avoid any interference with the transitional slope and the rays emitted by radar. The detailed engineering includes a study and investigation of various alternatives of roof structure. These alternatives included schemes based on portal frames, box main door girder spanning longitudinally in conjunction with a cantilever in front and transverse girders spanning between the door and rear girders. Alternative materials were also considered, including rolled sections, structural hollow sections, prestressed steel, reinforced concrete, prestressed concrete and combinations of these.

Each scheme was analysed and assessed in relation to efficiency in service, overall cost, maintenance, speed of erection and completion. The scheme finally selected involves the construction of the largest tubular space frame in the country with node connectors. The internal dimensions were adjusted to 129.384 m x 88.663 m x 23 m to fit the space frame grid of 4.222 metres. Because of its high inherent rigidity, this alternative enabled deflections to be kept within permissible limits.

STRUCTURAL ARRANGEMENT
The structure comprises of :

❖ A front girder of 129.384 m clear span and 8.444 m width supported at the ends on 25 m high steel columns to act as a portal. The steel columns, in turn, rest on R.C.C. columns 5.5 m high. The front girder is a five-layer rigid space frame with a structural depth of 8.683 m. Geometrically speaking, it is an extension of the three-layer pin-jointed space frame roof made rigid in this region. A top and bottom flange of heavy structural sections are added to make it five-layered.

❖ A rear girder of 129.384 m clear span and 12.667 m width, continuous over two intermediate R.C.C. columns, and resting on two R.C.C. columns at the ends. The rear girder is of the same depth as the main space frame roof. The only difference is that while the main space frame is pin-jointed, the rear girder is rigid.

❖ The triple layer main space frame with a structural depth of 4.871 m spanning between the front and rear girders which are 61.22 m apart.

❖ A rear space frame spanning between the rear girder and the columns of the annexe building in the rear.
❖ A space frame canopy of 4.222 m cantilevering from the front girder. A details of the structural arrangement may be seen in Fig. 2.

TOPOLOGY
The triple-layered main space frame has a diagonal over square over diagonal configuration. The top and bottom layers added to the triple layer to form the five-layered front girder are of square configuration.

LOADS
The loads for which the structure is designed are (a) Dead load comprising of self weight of the structure. Dead load of the roof cladding and purlins/stubs, Hangar services and cranes and docking systems, (b) Superimposed Loads consisting of Live loads as per Indian Standards, Live Load due to craneage and docking system and Wind Loads.

❖   Seismic Loads
Long span space frames of this type behave differently from traditional structures. The roof, as a whole, is rigid horizontally. However, in the vertical direction, the structure is flexible. Hence detailed response spectrum analysis was carried out using seismic coefficients applicable to Zone III of the seismic map appended to the Indian Standard IS : 875.

❖   Temperature Loads
The space frames were checked for a temperature variation ranging from 10°c minimum to 40°c maximum prescribed for the climatic conditions in and around Mumbai city.

The space frame is supported on POT 1 PFTE bearings on the columns common to it and the annexe building to ensure that only vertical loads are transmitted to them and no complex interaction problems arise.

STRUCTURAL ANALYSIS AND DESIGN
The space frame is supported on R.C.C. edge beams along the longitudinal sides which in turn rests on R.C.C. columns. The structural model comprising of 14721 members, 92 shell elements with 5527 nodes idealized as "Truss", "Beam" and "Plate/shell" elements, analysed using STA AS III structural analysis package and subsequently checked using "SAP 90" and "NISA" package.

Front Girder
The structural members used in the front girder are built-up closed sections comprising of channels and plates of Grade Fy 250 with an yield strength of 250 Mpa. The maximum forces in the bottom-most and top-most chords of the girder were 1500 tons tensile and 1600 tons compressive respectively. The maximum force developed in individual members meeting at a joint in the intermediate layers was 100 tons. To transmit this force and to facilitate connection with members, a special cylindrical node with fin plates radially welded to it was developed at site with the help of consultant as shown in photograph 1. This type of connector can joint up to a maximum of 12 members.

The front girder weigh 430 tons and the steel consumption is as low as 3.30 tons/m.

Rear Girder
The rear girder weight is 250 tons and the corresponding weight per metre is 1.91 tons. The cylindrical tubular joint with radial fins is used for making the connections.

Main Space Frame

A major part of the main space frame regarded as pin-jointed for purpose of analysis, is built of tubular sections with hollow tuball nodes of 135, 162.5 and 200 mm diameters. In the middle layer, solid nodes of 135 mm and 200 mm diameters are used. The bolts in solid nodes are tightened by cutting a slot in the tubes to provide access to a spanner. These slots are later closed with a cylindrical plate welded to the tube to restore its original tubular shape. To prevent any loosening of the bolt with the passage of time, a diaphragm plate is welded to the inner wall of the tubular member just above the bolt head. The tubes used have an yield strength of 220 Mpa and their sizes range from 60.30 x 3.65 mm, 328.90 x 6.30 mm and 273.0 x 9.50 mm. These large diameter tubes can carry a maximum force of 78 tons and 95 tons respectively. The tubes are connected to the hollow nodes by a single bolt and a contoured spring washer over a copper washer 0.50 mm thick is provided under the bolt head. The threaded ends of the connecting bolt of Grade 10.9 are secured to tapped solid plugs fitting to the ends of the tubes as shown in photograph 2. The effective length of compression members is assumed as 0.95 times of their actual length.

CONSTRUCTION

Fabrication and assembly of all components involves extensive welding. Hence, a very comprehensive three level quality management system was advised and implemented to ensure rigid compliance with various welding codes and practices. At every stage, the welds were visually inspected and dry penetration and ultrasonic tests were carried out. Radiographic test were also carried out at important locations to detect welding defects.

The fabrication and assembly of the front and rear girders were taken up first. Fabrication of tubular members were undertaken simultaneously at the site workshop. After the completion of the fabrication and assembly of the front and rear girders, the assembly of the space frame was carried out. The main space frame had both bolted and welded connections. The front girder was supported on steel props of varying heights with a built-in camber of 500 mm to offset dead load deflections. The space frame was assembled at a height of 2.5 m because of the depth difference between the space frame and the front girder. The rear girder had to be assembled in two stages because of the presence of the tow intermediate columns.

SPECIAL JOINTS

There is a highly stressed bank originating form the intermediate R.C.C. columns supporting the rear girder. The maximum force that spherical nodes can carry in tension is little less than 50 tons. In a narrow bank, starting from the interior girders supporting the rear girder, forces in members exceed 50 tons. In this region, the highly stressed member at a joint is run through and the other members at the joint are connected to it by means of gusset plates. This joint was developed by us and approved by a high level Expert Committee headed by Prof. Dr. Nooshin of the University of Surrey ser up to review all aspects of design and construction.

TESTS ON NODE CONNECTORS

Test procedures

Octabube N. K. had specified an uniaxial tension test as the acceptance test for tuball node connectors. Following the prescribed procedure, uniaxial tension tests up to failure were carried out on six samples of each diameter at the Indian Institute of Technology, Mumbai after construction of special testing frame is the laboratory of the Institute.

The objects of the test were :

❖   To arrive at breaking load
❖   To ascertain type of fracture (ductile or brittle)
❖   To study load-deformation behaviour. To this end, a total of nine strain gauges, three on the outside surface of the node where the failure is not expected, three on the outside surface where failure is expected and three on the outside surface of the bolts. The extension of the bolts was determined as the project of the bolt strain and its length. The extension of the entire rest assembly was also measured. The node extension is found by subtracting the bolt extension from the extension of the test assembly.

Observations

❖   The strain on the outside surface at both points of measurement were found to be compressive. This may at first glance, look surprising because the node is submitted to uniaxial tension. However, this behaviour is easily explained. The node under loading develops both direct and flexural stresses. The flexural stresses on the surface of the node are compressive. The flexural stresses being higher than the direct stresses, the net stress is compressive.

❖   An examination of the nodes after the test clearly showed that the fracture is brittle. This is to be expected because the load-deformation curve of the node is more or less linear right up to failure.

❖   The variation of the failure loads of the 135 mm diameter was less than those of the large 162 mm to 200 mm diameter nodes. This higher coefficient of variation may be attributable to the smaller degree of randomness to be expected in casting a smaller degree of randomness to be expected in casting a smaller diameter node.

❖   The failure patterns of all the nodes tested is the same. The failure surface passed the large opening for the node cap and the hole through which the space frame member is connected and the intermediate holes in between.

❖   Test result for spherical nodes carried out at IIT, Mumbai, may be seen in Table 1.

HOISTING AND ERECTION

The various options were considered for erection of Hangar Roof with an area of 10,000m² and weighing 1100 tons. Finally it was decided that entire roof assembled on the ground (expect members joining the support node) would be lifted in one piece to its final position by hydraulic jacks positioned on temporally towers.

Choice of lifting jacks

PSC Centre hole jacks were chosen because of its advantage that the jacks are comprehensive, completely automatic, remote control and monitoring of the jacks from the power pack locations, mechanically fail safe gripping mechanisms, automatic synchronized lifting speed, irrespective of individual jack loads, extremely precise adjustment of the lift for setting of weld gaps or the fitting of steel work connecting bolt etc.

Operation of lifting jack

Twelve jacking locations were selected for positioning jacks and the expected reactions at these points were computed by carrying out an erection analysis with the roof suspended from the jacks. The reactions add up to 1100 tons. It was therefore decided to have 12 nos. of L 180 PSC Heavy lift centre-hole jacks with a rated safe working load of 180 tons. Each power pack operates six jacks at 6m / hours. Hence, two power packs were engaged in the lifting operations.

PSC centre-hole jacks lift the roof in incremental steps equal to the stroke of the jack which in this instance is 500mm. The jade system is operated by self-contained electro-hydraulic jacks. The lifting strands are gripped. The jack is next retracted. The jacks have a fail-safe mechanism which ensures that in the event of any hydraulic failure, the load is automatically locked into the bottom anchor of the jacks. The far end of the lifting strands are secured by an anchor block of the lifting jacks. The entire hoisting operation was completed in 8 days. Photographs 3,4,5,6,7 & 8 show the hangar as it is being lifted.

Camber and Deflection
The front girder was precambered to 500 mm to offset dead load deflections. The computed deflection with the roof suspended from the jacks was 300 mm upwards. After the structure was connected to its end supports and all the jacks were released the measured downward deflection was 215 mm. Thus the front girder was still precambered to 285 mm. The final deflection of the structure measured again after completing the roof cladding and installing all services in the hangar roof were within the calculated deflection. The maximum theoretic deflection of hangar roof during erection and under service condition is 230 mm and 360 mm respectively. Deflections in three directions were accurately measured by using Total Station Survey equipment. Completed hangar is shown in photographs 9 &10.

LIMITATIONS
In view of the size and complexity of the hangar, an Expert committee was set up under the Chairmanship of Professor Nooshin of the University of Survey to review the design and construction of the project. One of the important issues addressed by the Committee, among others, was the appropriate factor of safety adopted in the design of hollow and solid spherical nodes. The node connectors in the hangar are in a triaxial state of stress. The Committee noted that there is no published literature of the appropriate factor of safety applicable to such a situation. Limited bi-axial tests carried out on hollow. Tuball spherical nodes by Octatube at Delft, however, had revealed that a bi-axial state of stresses with tension as long one axis and compression along the other may be more severe than an uni-axial state of stress.

Committee finally recommended that the chosen node for the force system should satisfy the following relationship.

$$2.5 \, (Tm+Cm) \quad Tu$$

where   Tm = Maximum tensile force,

   Cm = Ultimate load of the node in uniaxial tension.

The formula takes account of the fact that the condition of a node which is compressive force in other direction and is more severe than that of the same node under uniaxial tension. This criterion proposed by Professor Nooshin has been accepted for the design of spherical node connectors. It is however, recognized that more extensive testing programmes and research need to be carried out to arrive at more refined criteria to arrive at appropriate safety factors for node connectors. It is also felt that a code pertaining to the design and construction of space frame structure should be published in the near future for its wide end option.

It is also felt that an appropriate standard pertaining to the design and construction of space frame should be formulated in order to meet its growing demand in the construction industry. Test result for spherical nodes carried out at IIT, Mumbai.

Table 1. Hollow node diameter 200 mm.

S. No.	Sample No.	Date of testing	Failure Load ton	Type of Failure
1	1	July' 96	104.50	Shear failure in nuts
2	2	July' 96	98.60	Breaking of node
3	3	July' 96	108.80	Breaking of node
4	1	Jan' 97	94.50	
5	2	Jan' 97	116.30	
6	3	Jan' 97	109.60	
7	1	Dec' 97	122.10	Breaking of node
8	2	Dec' 97	117.60	Breaking of node
9	3	Dec' 97	123.60	Breaking of node

Hollow node diameter 162 mm

S. No.	Sample No.	Date of testing	Failure Load ton	Type of Failure
1	1	July' 96	55.00	Shear failure in nuts
2	2	July' 96	55.70	Breaking of node
3	3	July' 96	62.70	Breaking of node
4	1	Jan' 97	54.40	
5	2	Jan' 97	65.50	
6	3	Jan' 97	51.20	
7	1	Dec' 97	63.20	Breaking of node
8	2	Dec' 97	76.70	Breaking of node

Hollow node diameter 135 mm

S. No.	Sample No.	Date of testing	Failure Load ton	Type of Failure
1	1	Jan' 97	43.80	
2	2	Jan' 97	43.20	
3	3	Jan' 97	41.00	
4	1	May' 97	48.80	Bolt failure
5	2	May' 97	46.50	Bolt failure
6	3	May' 97	47.20	Bolt failure
7	1	Dec' 97	39.50	Breaking of node

Hollow node diameter 135 mm

S. No.	Sample No.	Date of testing	Failure Load ton	Type of Failure
1	1	Dec' 97	55.20	Tensile failure of bolt
2	2	Dec' 97	55.60	Tensile failure of bolt
3	3	Dec' 97	55.80	Tensile failure of bolt

Structural behavior of domes during construction

F. CASTAÑO
Geometrica, Inc., Houston, TX, USA

INTRODUCTION

Many Geometrica domes are built by the perimeter-in method, where construction starts at the support and continues ring by ring upwards and towards the center. A "hole" in the middle gets smaller and smaller until it is completely closed. A partially built dome behaves quite differently, structurally, than it does when complete. Recognizing this behavior is very important for successful construction.

Experience shows that it is easy to assemble domes during the initial stages, but after some progress it gets more difficult. After about half the dome is complete, the work front may show waves and it may appear that the dome is not fitting properly. Light towers with lifting cable pullers are then used to give the correct shape to the work front and continue. Assembly is most difficult when the dome is 75-90% complete, but, finally, the last few pieces before closing get easier to place.

This paper describes why all domes show this behavior during construction and how to control it for successful assembly.

STRUCTURAL LOADS

The following loads are acting on an unfinished dome, arguably in order of importance:

1. Edge forces
2. Self weight
3. Tower-applied lift
4. Other assembly loads
5. Manufacturing variations in member lengths
6. Temperature

The first arises out of the support conditions and may be due to a reaction to the other loads, or also due to self-straining because of an initial misalignment of the edge supports. The edge forces may be considered vertical and horizontal, but because the dome is similar to a membrane (in that it is much stiffer in its surface than out of it), we will also consider them as tangential or perpendicular to the dome's surface. The other forces are evident. We will concentrate on the first 3 only in this paper, but the others have similar effects.

To illustrate the effects of these loads on a dome, we will use the example in Figure 1. This dome has a Kiewitt geometry in a Vierendeel double layer with 6 segments and 20 divisions at each of these (thus the dome has 20 "rings"). All the bars were selected as 1-7/8" X 0.104". Nodes on the base ring (edge) are joined into a single layer.

Space Structures 5, Thomas Telford, London, 2002

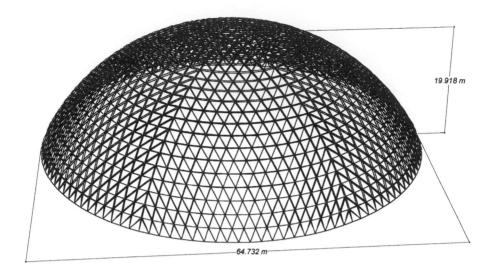

Fig. 1. Dome Model.

FORCES AND DISPLACEMENTS ON THE EDGE

In order to study edge effects, it is useful to remember that any shape of forces (or displacements) at the bottom ring may be approximated to any degree of accuracy as the sum of a Fourier series of periodic functions with selected coefficients. For load case n:

$$p_n = p_0 * \cos (n*\theta)$$

Where p_n is the force at the base located at an angle θ on the base. The cases where n = 0 and n = 1 are not very interesting. When n = 0, there is a constant force at the edge. If p radial, this is like temperature with a fixed support. If p is not radial, then this force is not self-equilibrating, but may balance a symmetric external load. When n = 1, the forces around the edge are not self-equilibrating.

For n = 2, things start getting interesting. In this case, the applied forces pull the dome along the x axis, and push it at 90° (along the y axis). For n = 3, there are three areas of pushing forces alternating with 3 areas of pulling forces. And so on for any n. Figure 2 shows two cases for radial loads. Of course, it is also possible to have similar shapes for loads vertical, tangential or normal to the dome's surface.

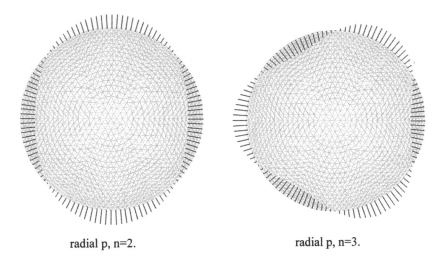

radial p, n=2.   radial p, n=3.

Fig. 2: Dome edge forces.

In this paper, we study enforced displacements rather than applied forces. To model enforced displacements, two additional bars were created in the computer model, stemming out from each of the base nodes, one radial and another vertical. Both bars have very high axial stiffness, and are supported at the end opposite the dome's base node. The radial bar was supported against all displacements except vertical, and the vertical bar only against vertical displacement.

High vertical and horizontal loads (of the required magnitude to extend 10 cm the stiff bars at $\theta = 0$) were applied at each of the base nodes. Different load sets were created for each of the terms of the Fourier series, and for each of the different directions (radial and vertical). Load combinations that combine vertical and radial displacements into tangential and normal to the membrane were also input.

THE FIRST FEW RINGS
In figure 3 the stress and deformation patterns for this dome are shown. The deformations are exaggerated 10 times and the bars with higher stresses are darker. This dome is quite flexible, and the enforced displacements cause little stress. The radial displacements stress only those bars that change in length due to the extension, mostly those at the bottom horizontal rings (there would be an out-of plane deformation that would produce almost no stress, if we allowed tangential displacements).

Note that the vertical deformation produces almost no stress, and the top ring shows only a small variation in elevation.

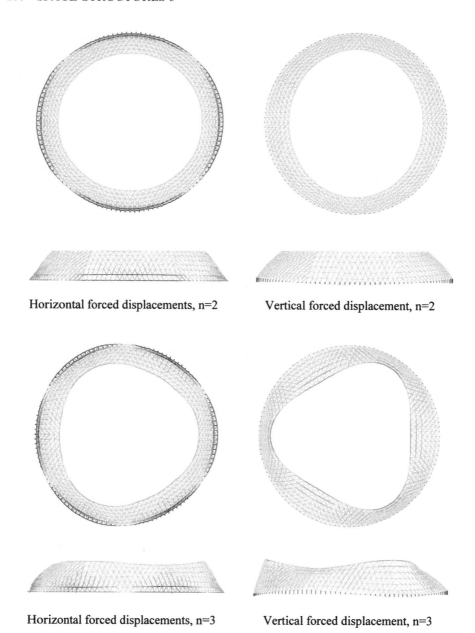

Horizontal forced displacements, n=2

Vertical forced displacement, n=2

Horizontal forced displacements, n=3

Vertical forced displacement, n=3

Fig. 3: Six ring dome.

The case where n = 3 shows a bit more stress at the work front, due to bending there, but still not very significant.

The reason the vertical displacement produces almost no stress is that the dome at this stage is almost developable. That is, its behavior is similar to that of a flat sheet under bending. The dome is free to deform to the new shape with only minor bending strains and almost no in-surface strain. Thus, in a practical application, this means that a dome can be assembled to this point very easily even if the base is not level: tubes fit quite well.

ASSEMBLY PROGRESS

For the next cases we will plot the tangential and the normal displacements rather than vertical and horizontal, and the deflection amplifier will be reduced to 4. The dark bars indicate more than 6 kip tension or compression. In the ten-ring dome, for every centimeter of maximum tangential displacement at the base, there is about 2.5 cm of horizontal and 1.5 cm of vertical displacement at the working front. For the fourteen- and sixteen-ring domes, the tangential displacements are multiplied by over 5 times at the working front, (both horizontally and vertically), and all members are more and more stressed. In all cases the normal displacements dissipate before getting to the front.

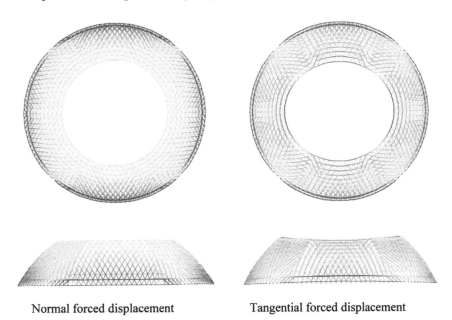

Normal forced displacement   Tangential forced displacement

Fig. 4: Ten ring dome.

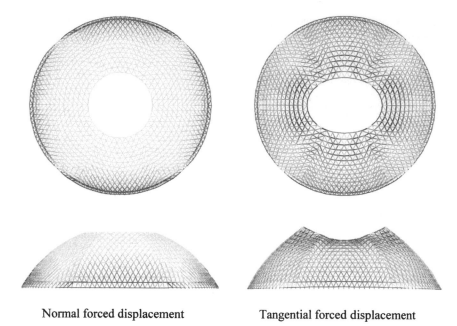

| Normal forced displacement | Tangential forced displacement |

Fig. 5: Fourteen-ring domes, n = 2.

Note that in the sectors that the dome is raised above the average level at the base, the front falls from its natural level, and where the base is lower, the front raises.

For n=3 or larger, the pattern is similar except that there is a maximum deformation reached earlier. Under tangential stress there are three "waves" at the front. In all cases, the tangential deformations at the base get multiplied several times into normal displacements when they get to the front, and the stresses due to these deformations do not dissipate as quickly as the perpendicular ones do.

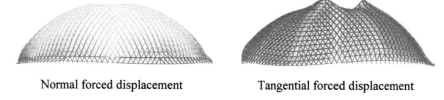

| Normal forced displacement | Tangential forced displacement |

Fig. 5: Sixteen-ring dome, n = 3.

If we were able to continue building, and the forced edge displacements were maintained, the dome would be stressed as shown in Fig. 6 when finished (for n = 2). By this time there is no front to "release" some of the tangential strains and the whole dome is highly stressed under this condition. Of course, in a real application, while the dome is not fixed at the base the edge displacements would be limited to those required to carry the dead load on the (uneven) support minimizing internal stresses.

| Normal forced displacement | Tangential forced displacement |

Fig. 6: Complete dome, n=2.

DEAD LOAD EFFECTS

Most domes under construction are supported on an uneven beam or foundation. If the dome were infinitely rigid, it would touch the beam only at the highest 3 points. But partially built domes are flexible enough that they follow the supports' profile.

As can be deduced from the above, the variation in shape of the beam that is perpendicular to the dome's surface would not be very important. The tangential variation, however, creates major problems when the dome progresses beyond a few rings.

In order to get a feel of the potential effect of an uneven support on the above dome, an analysis was run with a random variation in stiffness of the supports for the above model. The fake, stiff bars were removed, and force supports were added with 40, 10, 2.5, 1.25 and 0.31 kips/in spring constants. These were distributed in a near-random pattern around the perimeter. In a field case, there would be a strong correlation in elevations at adjacent nodes, but even without this strong correlation, the example illustrates the effect adequately. The results for a 16 ring dome are shown in Fig. 7.

The characteristic deformation under an uneven support is similar to that observed with forced displacements (although it had to be exaggerated more to show). The maximum support elevation difference under the random stiffness is only 2 cm at the base, but this gets multiplied to nearly 8 cm at the top, and the horizontal distance difference is changed by over 8 cm also. A few bars were removed from the front ring in order to see how the gaps would be affected by the front's deformation. It turns out that the missing bars in compression would appear to be 2 cm too long.

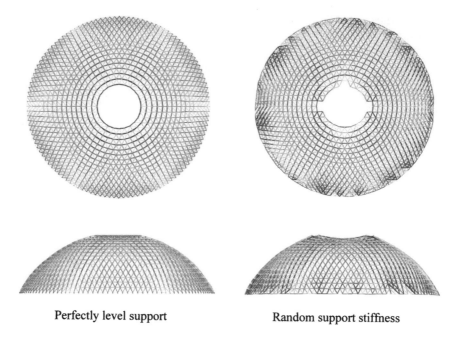

Perfectly level support   Random support stiffness

Fig. 7: 16-Ring dome under dead load.

TOWERS

Towers apply a load with a large component perpendicular to the surface, thus their effects are constrained to the small area near the tower. This is usually enough to relieve the deformations due to tangential edge load at the work front and to continue installing bars. Figure 8 illustrate the effect of towers, alone and also combined with the dead load and random edge support dome of Figure 7. Notice that the stresses at the front's spiders are reduced (except immediately next to the towers, where there is a local concentration). In this example, each of 4 towers applies a 0.75 metric ton vertical load.

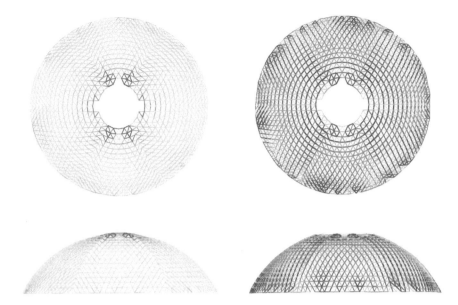

Towers only, random support stiffness   Dead + towers + random support
Fig. 8: 16-Ring dome with towers.

CONCLUSIONS

Small variations in support level or stiffness cause large deformations at the work front of partially built domes (Figure 9).

Fig. 9: 110 m dome during construction with wavy work front and towers.

Such variations do not affect ease of assembly in the early stages of construction, but make it difficult in the latter stages. Properly placed towers relieve front stresses locally to the point that assembly may proceed. Towers should be placed where the structure "hangs" at the front and, to avoid generating pre-stresses, the dome should not be fixed permanently to the base until it is complete.

These simple steps assure quick and trouble free construction of Geometrica domes by the perimeter-in method.

Structural design of Kyoto station atrium

Y KANEBAKO
Kanebako Structural Engineers, Tokyo, Japan
T KIMURA
Kimura Structural Engineers, Tokyo, Japan

OUTLINE OF BUILDING

The project of renewing the Kyoto station was proposed to commemorate the 1200-year anniversary of Kyoto being established as the ancient capital city of Japan. By means of an international design competition, the design by HIROSHI HARA had been selected for the new station complex. Including 16 floors, 3 basement levels and comprising of 230,000 square meters of floor area, this large station complex houses a train station, hotel, department store, and a theatre. One of the key design features of the building complex is the 'atrium', a concourse in the center of the building covered by a glass wall and roof, which was modeled according to the geography of Kyoto city.

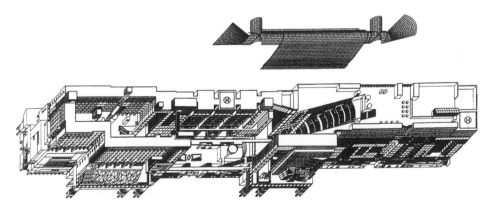

Fig. 1. Rendering of Kyoto station.

The basic structure of the building consists of three sections orientated horizontally and four sections placed vertically (See Figure 2). The horizontal sections are termed E-WING, M-WING and W-WING; E-WING consists of the hotel, convention center, M-WING the hotel and station facility, and W-Wing the department store and parking spaces.

In order to allow flexibility in the column layout of the building, a rigid transfer truss girder, the 'Structural Matrix' is placed between the levels of 10.5 meters and 15.5 meters. The Structural Matrix is a two way steel truss that is 4 meters in depth. Above the matrix is where the structure is separated into the wing sections, which is in turn tied together at the top portion to form an architectural gate. The upper and lower areas of the matrix consist of a steel moment frame with partial bracing, and the basement level is made of steel reinforced concrete.

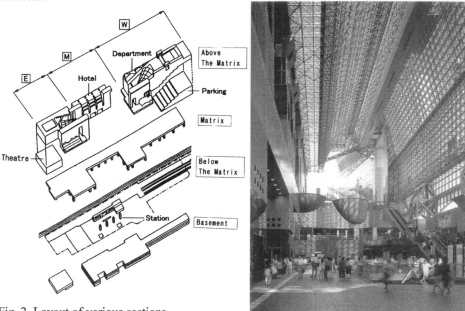

Fig. 2. Layout of various sections.

Fig. 3. Atrium interior

OUTLINE OF THE ATRIUM

The atrium roof and wall has a glass finish and its overall dimensions are 150 meters in length, 30 meters in width and 50 meters in height. The elemental form of the structure is a combination of a large vault, small vault, and a plane. In addition, intersectional vault forms are used in the either ends of the atrium. An observatory passageway termed "Skyway" is hung from the roof at the height of 45 meters above ground, to allow individuals to experience the dynamic view of the atrium from above. The structure consists of three parts which are separated by expansion joints: the center portion, and the half cone-shaped canopies at the east and west areas. This paper will mainly discuss the center part of the atrium structure.

The overall structure is illustrated in the plan and elevation, and section view in Figures 4 and 5, respectively. The support locations of the atrium are placed along several levels. The south perimeter of the atrium is continually supported at the 31 meters level of the M-WING, the north perimeter is supported at the 19 meters level by a rigid frame every 20 meters which forms the entrance gate to the concourse level. The edges of the east and west portions of the atrium is supported at the 45 meters level. The connection of the atrium structure to the building structure is firmly fixed, therefore the seismic movement is greatly influenced by the building and thermally induced stress plays a significant role in the structural design.

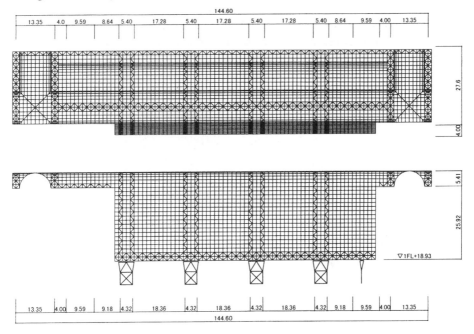

Fig. 4. Plan and elevation view.

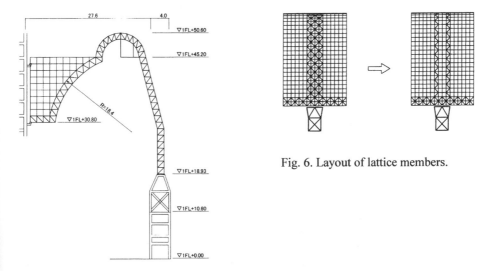

Fig. 5. Sectional view.

Fig. 6. Layout of lattice members.

The delicate grid structure of the atrium is comprised of thin steel elements forming a double-layered structure, with an approximate dimension of 1.44 meters in depth at a pitch of 1.44 meters. Each individual element is a box section with outer dimensions of 100mm x

100mm with varying thickness of 4.5, 6, 9, and 12mm.

The vertical load of the structure is supported by truss structures with lattice members spanning in the north-south direction. This arrangement allows the remaining east-west structure to be Vierendeel without lattice members, except local zones near the supports. In order to keep planar rigidity and strength, latticed bays are formed along the top length of the atrium. Based upon the structural characteristic of the grid--if the grid is made too stiff there is heavy stress build-up due to the deformation of the building, if made to flexible there is high stress buildup due to its own seismic loading, several lattice formations were investigated to find the optimal stiffness/stress balance of the latticed grid (See Figure 6).

The connections of the steel members are grouped into two types: directly welded and cast steel pieces. The welding of joints were designed according to the Architectural Institute of Japan, and later tested for performance. The cast steel pieces were utilized into joints with dense traffic of intersecting members and high stresses.

Fig. 7. View of joints.

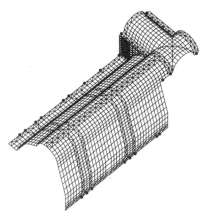

Fig. 8. Static analysis model.

DESIGN OF THE FRAME STRUCTURE

For the analysis of the atrium structure, a half model and an equivalent single layer model was utilized (See Figure 8). The supports to the building were modeled as pin connections and the supports at the north perimeter were modeled as elastic supports reflecting the stiffness of the substructure below.

The loading criterion of the atrium includes: dead and live loads of the Skyway, snow, temperature, seismic and wind loads on the atrium. The dead load of the roof and wall of the atrium is set as 160 kgf/m^2 and the dead load of the Skyway is set as 260 kg/m^2. The live load includes the people in the Skyway. The snow loads assumes a snowfall depth of 30 cm. Two seismic loading components were applied to the model: the inertial force of the atrium and the applied displacements of the building. The seismic coefficient for the inertial load is designated as 0.5 (seismic design is further discussed in the following section). There are three wind load cases according to the wind direction. The wind pressure is assumed to be approximately 240 kgf/m², calculated according to the former Japanese regulation codes (revised in 1999.5). As shown in Figure 9, the wind force coefficient is set from a range of 0.4 to 1.2.

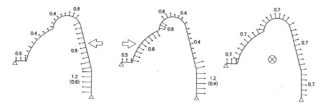

Fig. 9. Wind coefficients.

Temperature loading is set as ±30°C to consider the varying temperature from summer to winter. The ±30°C temperature loading is used in combination with dead and live loads. A temperature load of ±15 °C is used in combination with each wind load and earthquake load. The stress level of the members due the temperature loading was reduced by allowing its radial deformation in the north-south plane. However, since the movement of the members near the supports is restricted, the stress induced reaches a value of 50% of its allowable yield stress. The greatest vertical displacement of roof is 3-centimeters under dead load and live load.

SEISMIC DESIGN
Analysis Approach
Although all the stress analysis was done with the static model, the effectiveness of the design load and the earthquake resistance safety were verified by conducting dynamic analyses. The shear coefficient of the seismic load of 0.5 that is used in the static analyses of the structure is a value obtained by the results of the dynamic analyses.

Natural Period and Vibration Mode
A three-dimensional model shown in Figure 10 was used for the modal analysis of the building. Each floor of each wing was modeled as a single mass, which is in turn connected by vertical and horizontal members. As another model, each wings are separated is used. The result is shown in Table 1.

Table 1. Natural period of building (sec).

Mode	Total model	Wing Independent Model		
		W-Wing	M-Wing	E-Wing
1	1.460 (Y: 1st)	1.470 (Y: 1st)	1.524	1.295 (X: 1st)
2	1.401 (X: 1st)	1.440 (X: 1st)	1.510 (Y: 1st)	1.160
3	1.369 (X: 1st)	1.264	1.410 (Y: 1st)	1.129 (X: 1st)

X: east-west direction   Y: north-south direction

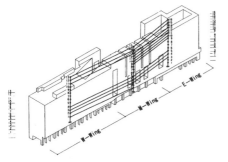

Fig. 10. Dynamic analysis model.

For the analysis of the atrium, two types of models were utilized. The "Independent Model" is the atrium only model and the "Connected Model" consists of the atrium and building that are modified into masses and stiffness. In both models, the atrium was modeled intensive about the east-west direction.

The natural periods and vibration modes of Independent Model are shown Figure 12. The first natural period is 0.44 second, which is significantly shorter than the natural first period of building. Therefore it was assumed that the vibration of the building would not amplify the vibration of the atrium structure. The first mode is the horizontal movement in the north-south direction, the second is the horizontal east-west direction, and the third is a vertical movement. The natural periods and vibration modes of Connected Model are shown Figure 14. The first and second modes are similar as the first and second mode of building.

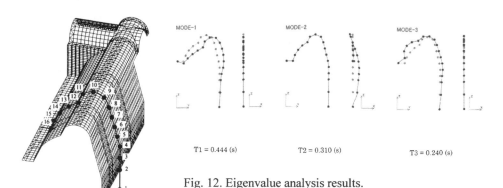

Fig. 12. Eigenvalue analysis results.

Fig. 11. Independent model.

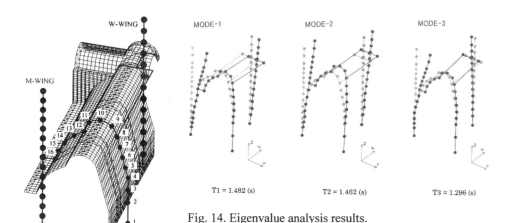

Fig. 14. Eigenvalue analysis results.

Fig. 13. Connected model.

Earthquake Response Analysis

A linear seismic response analysis for the horizontal wave inputs was conducted using the Connected Model. The seismic waves used for the analysis were EL CENTRO 1940 NS (204.3gal), TAFT 1952 EW (198.2gal), and HACHINOHE 1968 NS (132.0gal). These magnitudes are equivalent of serviceability level (Level 1). The assumed damping coefficient of 2% about the first vibration mode is proportional to its stiffness. The results of analyses are shown Figure 15 and 16. The maximum response acceleration about north-south direction of structure is under 320gal, and under 350gal in the east-west direction. The inertial earthquake shear coefficient of 0.5 was obtained from these results.

The Independent model was used for seismic response analysis for a vertical wave input. For this analysis, the seismic waves of EL CENTRO 1940 UD (102.2gal), TAFT 1952 UD (99.4gal), and HACHINOHE 1968 UD (66.0gal) have been used. The vertical magnitude of ground acceleration was assumed to be half of horizontal magnitude. Similar to the horizontal analysis, the damping was taken as 2% for the first mode. The result of analyses is shown Figure 17. The maximum response acceleration of 420gal was marked at the interface zone from vault e to the plane area.

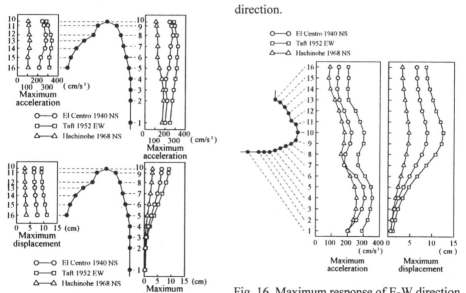

Fig. 16. Maximum response of E-W direction.

Fig. 15. Maximum response of N-S.

Seismic Analysis by Applied Displacements of Building

Another seismic force is applied to the atrium model by means of the deformation of the support points that are attached to the building. Horizontal displacement values taken from the analysis results of the building only model was applied to the supports of the atrium structure. The horizontal displacement value was found to be 0.4% of the height at each level.

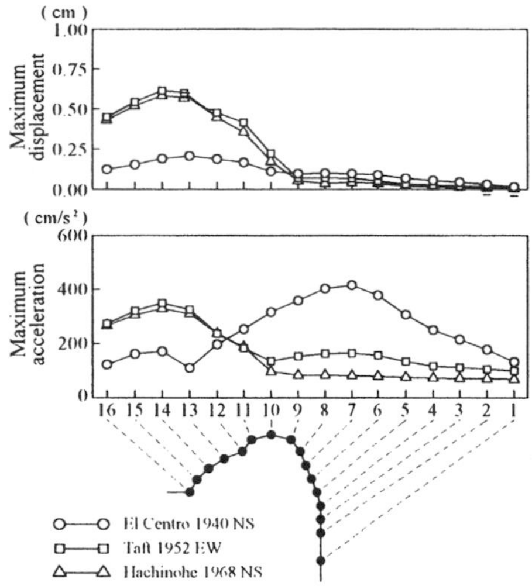

Fig. 18. Construction.

Fig. 17.   Response of vertical direction.

The Investigation for Earthquake Load of Safety Limit Level
In order to verify the seismic safety of the atrium structure, a loading of twice the serviceability level magnitude was estimated. The analysis considered the inertial force of the atrium and the force induced by the deformation of the building. It was observed that for the inertial force loading, the stress level of all the members were under the yield stress level. The loading due to the deformation of the building resulted in several members exceeding the yield stress level. To verify the structure would be able to support itself under dead and live loads with the members yielded, a static analysis was conducted with a model minus the yielded members. The results confirmed the safety of the atrium structure under severe seismic events.

CONSTRUCTION PROCESS
The atrium structure was divided seven blocks for construction. Both the end parts were constructed using temporary fixed supports. The center five parts had same section, so they were constructed by the moveable temporary stage method. The stage has the same plan size as a single part of block, approximately 20 meters by 30 meters. Two rails set along both ends of the 31 meters level supported this stage. The temporary supports were placed on the stage, and the structural members were assembled from the support. After assembling one block, they were jacked-down to allow the stage to be moved to the next block. The utilization of this method reduced the number of supports, and allowed work to be carried out simultaneously underneath the stages.

REFERENCES
1. KIMURA T, KANEBAKO Y, et al. Structural Design of Kyoto Terminal Building, GBRC, vol. 18, no 4, 1993, pp3-14.

Load carrying capacity of a single layer latticed structure with metallic sheets

H. TANAKA
Technical Research Institute, Takenaka Corporation, Japan
Y. TANIGUCHI
Graduate School of Engineering, Osaka City University, Japan
T. SAKA
Graduate School of Engineering, Osaka City University, Japan

INTRODUCTION

Space frames as a latticed structure are widely used to cover large areas without the need for intermediate supports and usually designed under the assumption that the bending stresses of members are negligible. In order to achieve this assumption, a separate sub framing system has to be utilized for a roofing system. In this system members of a sub frame are subjected to laterally distributed loads and transmit those loads to joints of a main latticed frame. In the future if the possibility of a large space structural system may be pursued, a new structural system must be needed, barring unexpected developments of materials. One of possible structural systems that have been developed is the so-called hybrid system containing the composite action with a space frame structure and a cladding, for examples, timber plates (Ref.1), steel panels or membrane (Ref.2, 3). In this structural system, latticed members of a main frame are directly subjected to laterally distributed loads and have to be designed for a combination of axial and bending stresses. Therefore latticed members are larger than those in a separate sub framing system.

In order to estimate the nonlinear behavior of the structural system subjected to a distributed laterally load, slope-deflection equations for uniformly or sinusoidally distributed laterally loaded members under axial force have been presented by Ref.4, 5. Tangent stiffness equations for uniformly or sinusoidally distributed laterally loaded members under axial force have been presented by Ref.6, in which a numerical study has carried out using these equations to investigate the load-deformation relationship and the elastic buckling load of a parallel chord latticed beam with a laterally distributed load, and buckling behavior of a latticed beam with upper chord members subjected to a distributed load bas been theoretically and experimentally clarified. Furthermore, tangent stiffness equations for three evenly-spaced laterally loaded members under axial force have been presented by Ref.7, and load-carrying capacity of double-layer latticed grids and cylindrical latticed roof shells with three evenly-spaced laterally loads bas been theoretically made clear.

In this paper, the hybrid system that combines a latticed structure with metallic sheets is treated. The load bearing capacity and the buckling behavior are theoretically investigated and compared with the results of a separate sub framing system. A latticed dome and a cylindrical roof shell are treated as a single layer latticed structure.

Space Structures 5, Thomas Telford, London, 2002

FEM MODEL

A commercial finite element computer program, MSC/NASTRAN for Windows Version 4.0 is used for FEM analysis. A latticed frame is modeled with beam elements and a sheet is modeled with a quadrilateral plate element. In order to investigate the load-bearing capacity and buckling behavior, two kinds of analyses are carried out, one is elastic buckling analysis considered geometric nonlinearity due to change of joint coordinates, and another is elastic-plastic analysis considered geometric nonlinearly and material nonlinearity due to yielding. In elastic-plastic analysis, the stress-strain relationship is assumed to be a bi-linear type shown in Figure 1 and the von Mises criterion is used for the yield criteria. The modified Newton Raphson method is used for the nonlinear iteration.

A distributed load is replaced by a series of concentrated loads shown in Figure 2 and the value of them is determined in proportion to the area shared among nodes.

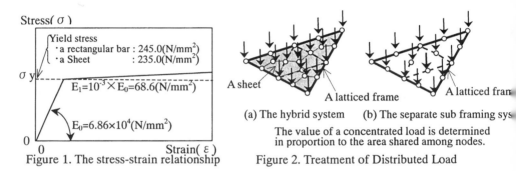

Figure 1. The stress-strain relationship

Yield stress
· a rectangular bar : 245.0(N/mm^2)
· a Sheet : 235.0(N/mm^2)

$E_1 = 10^{-3} \times E_0 = 68.6$(N/mm^2)

$E_0 = 6.86 \times 10^4$(N/mm^2)

(a) The hybrid system   (b) The separate sub framing system

The value of a concentrated load is determined in proportion to the area shared among nodes.

Figure 2. Treatment of Distributed Load

NUMERICAL ANALYSIS

Geometry and loading condition

Single layer latticed domes and single layer cylindrical roof shells shown in Figure 3 are treated. Single layer latticed domes have a regular hexagonal plan and they are pin supported at boundary nodes. In single layer cylindrical roof shells, two boundary conditions are treated. One is that boundary nodes shown by △ and ▲ in Figure 3 are pin supported and another is that boundary nodes shown by △ in Figure 3 are pin supported. A list of models is shown in Table 1. In Figure 3 the solid lines denote primary members of a latticed frame and the dotted lines denote secondary members. Model DHi, CHiT or CHiO (i = 2, 4 or 6) represents the hybrid system composed of a rigidly jointed latticed frame and sheets connected with a latticed frame rigidly at numerical nodes. The third letter "i" represents the thickness of a sheet. In model CHiT or CHiO, the last letter "T" or "O" represents the boundary condition. If the last letter equal to "T", boundary nodes shown by △ and ▲ in Figure 3 are pin supported and if the last letter equal to "O", boundary nodes shown by △ in Figure 3 are pin supported. Model DF or CF represents the separate sub framing system. In each member of a latticed frame three intermediate nodes are created in order to evaluate the flexural deformation of a member, and in each sheet seven intermediate nodes for one structural unit are created in order to evaluate that of a sheet. A latticed frame in the hybrid system is as same as that in the separate sub framing system. The mechanical properties of members and sheets are shown in Table 2.

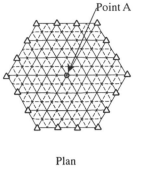

Point A

Point B Point A

Plan

Plan

— : A primary member
--- : A secondary member

6.0deg. « 6.0deg. H=734
R=15,000
L=9,270

8,158

H=734

L=9,270

Elevation and section

Elevation

△▲ : Supporting nodes
O   : Nodes whose deformation are shown in load-deformation relationship

(a) Single layer latticed dome (model DF)   (b) Single layer cylindrical roof shell (model CF)

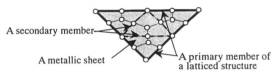

A secondary member

A metallic sheet

A primary member of
a latticed structure

O : Numerical nodes

(c) One structural unit of the hybrid system, model DHi, CHiT or CHiO

Figure 3. Single layer latticed structures

Table 1. List of models

A configuration of a structure	Boundary condition	A structural system	A thickness of a sheet	Model
A single layer latticed dome	Boundary nodes shown by △ in Figure 3(a) are pin supported.	The hybrid system	t=6	DH6
			t=4	DH4
			t=2	DH2
		A separate sub framing system	-	DF
A single layer cylindrical roof shell	Boundary nodes shown by △ and ▲ in Figure 3(b) are pin supported.	The hybrid system	t=6	CH6T
			t=4	CH4T
			t=2	CH2T
		A separate sub framing system	-	CFT
	Boundary nodes shown by △ in Figure 3(b) are pin supported.	The hybrid system	t=6	CH6O
			t=4	CH4O
			t=2	CH2O
		A separate sub framing system	-	CFO

Table 2. The mechanical properties of members and sheets

A rectangular bar B×D(mm)	Section area A (mm²)	Moment inertia I (mm⁴)
A primary member 20×100	1,999	1.66×10^6
A secondary member 16× 75	1,198	5.62×10^5
Modulus of elasticity E		$6.86 \times 10^4 (N/mm^2)$
Yield stress σ y	a)A rectangular bar :	$245.0(N/mm^2)$
	b)A sheet	$: 235.0(N/mm^2)$

The load-bearing capacity due to yielding and buckling behavior for single layer latticed dome
The load-deformation relationship is shown in Figure 3(a), and the relationship between load divided by the weight of a latticed frame and sheets and deformation divided by the span is shown in Figure 3(b). In Figure 3(a) the ordinate represents the total vertical load and in Figure 3(b) the ordinate represents that divided by the weight of a latticed frame and sheets. In Figure 3 the abscissa represents the deformation of the point A in the vertical direction or that divided by the span. The load-bearing capacity due to yielding and the buckling load are shown in Table 3. The buckling modes are shown in Figure 4.

In Figure 3(a), the initial stiffness of model DHi is higher than that of model DF and this tendency is remarkable in proportion to the thickness of a sheet. The maximum load of model DHi is bigger than that of model DF. In model DH6 or DH4, the elastic buckling load is bigger than the load-bearing capacity due to yielding. The load-bearing capacities due to yielding of these models are about between 85 and 86% of the elastic buckling loads. However in model DH2 or DF, the elastic buckling load is smaller than the load-bearing capacity due to yielding. The elastic buckling loads of these models are about between 95 and 98% of the load-bearing capacities due to yielding. As compared model DHi with model DF, the maximum load of model DH6 is about 234% of that of model DF, that of model DH4 is about 221% of that of model DF, and that of model DH2 is about 182% of that of model DF. In Figure 3(b), the initial stiffness of model DHi is higher than that of model DF, and the maximum load divided by the weight of a latticed frame and sheets of model DHi is bigger than that of model DF, too. The maximum load divided by the weight of a latticed frame and sheets of model DH6 is about 137% of that of model DF, that of model DH4 is about 156% of that of model DF and that of model DH2 is about 155% of that of model DF. Therefore, model DH4 is the most effective of all models. In Figure 4, the buckling mode of model DF is member buckling of primary members near the six corners and secondary members near the apex in shell plane. This is because the axial force of primary members connecting the six corners and the apex is bigger than that of other members in model DF. The buckling mode of model DH2 is local buckling of a sheet and that of model DH4 or DH6 is overall buckling. The axial force of model DHi is more uniformly distributed than that of model DF and in model DHi member buckling is restrained by the composite effect.

Tab.3 The load-bearing capacity due to yielding and the elastic buckling load for single layer latticed dome

Model	DH6	DH4	DH2	DF
The weight of a latticed frame and sheets : Wa	20.6kN	17.6kN	14.5kN	12.1kN
Load-bearing capacity due to yielding : Py	2,818kN	2,663kN	2,373kN	1,237kN
Elastic buckling load : Pe(kN)	3,320kN	3,200kN	2,200kN	1,206kN
Maximum load : Pmax(=Minimum [Py, Pe])	2,818kN (234%)	2,663kN (221%)	2,200kN (182%)	1,206kN (100%)
Py / Wa	136.8	156.0	163.5	102.4
Pe / Wa	161.2	182.3	155.0	99.8
Pmax / Wa(=Minimum [Py / Wa, Pe / Wa])	136.8 (137%)	156.0 (156%)	155.0 (155%)	99.8 (100%)

The values in a round bracket note percentages

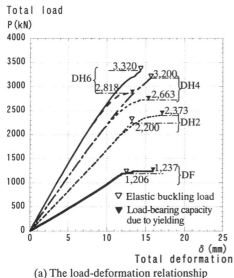

Total load
P (kN)

(a) The load-deformation relationship

Total load / weight of a latticed frame and sheets
P/W

(b) The load / weight of a latticed frame and sheets
-deformation / span relationship

Figure 3. The load-deformation relationship of the point A in Figure 3 for single layer latticed domes

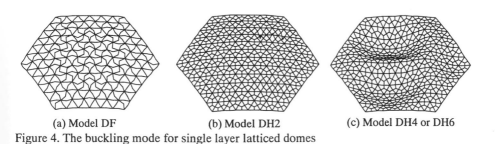

(a) Model DF   (b) Model DH2   (c) Model DH4 or DH6

Figure 4. The buckling mode for single layer latticed domes

The load-bearing capacity due to yielding and buckling behavior for single layer cylindrical roof shell

The load-deformation relationships of the shells whose boundary nodes shown by △ and ▲ in Figure 3(b) are pin supported are shown in Figure 5(a) and Figure 6(a), and that of the shells whose boundary shown by △ in Figure 3(b) are pin supported is shown in Figure 8. The relationships between load divided by the weight of a latticed frame and sheets and deformation divided by the span of the shells whose boundary nodes shown by △ and ▲ in Figure 3(b) are pin supported are in Figure 5(b) and 6(b). In Figure 5(a), 6(a) or 8, the ordinate represents the total vertical load, and the abscissa represents the deformation in the vertical direction. In Figure 5 or 8, the deformation of the point A in Figure 3 is shown, and in Figure 6 that of the point B is shown. In Figure 5(b) or 6(b), the ordinate represents the total vertical load divided by the weight of a latticed frame and sheets, and the abscissa represents the deformation in the vertical direction divided by the span. The load-bearing capacity due to yielding and the buckling load are shown in Table 4 and 5. The buckling modes are shown in Figure 7 and 9.

In Figure 5(a) and 6(a), the initial stiffness of model CHiT is higher than that of model CFT, and this tendency is remarkable in proportion to the thickness of a sheet. The maximum load of model CHiT is bigger than that of model CFT. In model CHiT, the elastic buckling load is bigger than the load-bearing capacity due to yielding and the load-bearing capacities due to yielding of these models are about between 97 and 98% of the elastic buckling loads. However in model CFT, the elastic buckling load is smaller than the load-bearing capacity due to yielding and the elastic buckling load is about between 99% of the load-bearing capacity due to yielding. As compared model CHiT with model CFT, the maximum load of model CH6T is about 123% of that of model CFT, that of model CH4T is about 117% of that of model CFT, and that of model CH2T is about 109% of that of model CFT. In Figure 5(b) and 6(b), the initial stiffness of model CHiT is higher than that of model CFT. However, the maximum load divided by the weight of a latticed frame and sheets of model CHiT is smaller than that of model CFT. The maximum load divided by the weight of a latticed frame and sheets of model CH6T is about 75% of that of model CFT, that of model CH4T is about 83% of that of model CFT, and that of model CH2T is about 92% of that of model CFT. In Figure 7, the buckling mode of model CFT is member buckling in shell plane, and that of model CHiT is overall buckling.

In Figure 8, the initial stiffness of model CHiO is higher than that of model CFO, and this tendency is remarkable in proportion to the thickness of a sheet. However, the maximum load of model CHiO is almost as same as that of model CFO. In Figure 9, buckling modes of the shells whose boundary nodes shown by △ in Figure 3(b) are pin supported are overall buckling in all models.

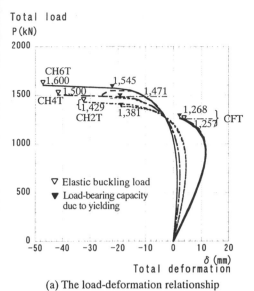

(a) The load-deformation relationship

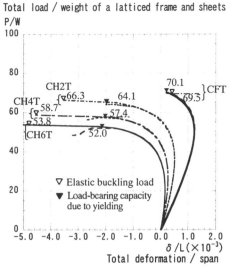

(b) The load / weight of a latticed frame and sheets-
deformation / span relationship

Figure 5. The load-deformation of relationship of the point A for single layer cylindrical roof
shells whose boundary shown by △ and ▲ in Figure 3(b) are pin supported

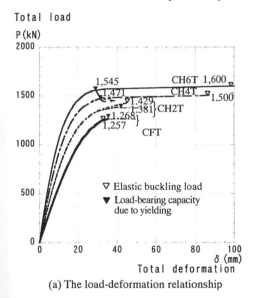

(a) The load-deformation relationship

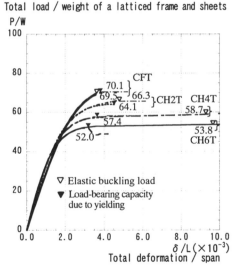

(b) The load / weight of a latticed frame and sheets-
deformation / span relationship

Figure 6. The load-deformation of relationship of point B for single layer cylindrical roof
shells whose boundary nodes shown by △ and ▲ in Figure 3(b) are pin supported

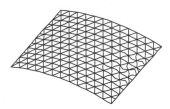
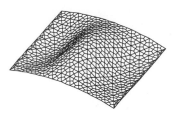

(a) Model CFT   (b) Model CHT

Figure 7. The buckling mode for single layer cylindrical roof shells
whose boundary nodes shown by △ and ▲ in Figure 3(b) are pin supported

Table 4 The load-carrying capacity and the elastic buckling load for single layer cylindrical roof
shell whose boundary nodes shown by △ and ▲ in Figure 3(b) are pin supported

Model	CH6T	CH4T	CH2T	CFT
The weight of a latticed frame and sheets : Wa	26.7kN	22.6kN	18.6kN	14.5kN
Load-bearing capacity due to yielding : Py	1,545kN	1,471kN	1,381kN	1,268kN
Elastic buckling load : Pe(kN)	1,600kN	1,500kN	1,429kN	1,257kN
Maximum load : Pmax(=Minimum [Py, Pe])	1,545kN (123%)	1,471kN (117%)	1,381kN (109%)	1,257kN (100
Py / Wa	52.0	57.4	64.1	70.1
Pe / Wa	53.8	58.7	66.3	69.5
Pmax / Wa(=Minimum [Py / Wa, Pe / Wa])	52.0 (75%)	57.4 (83%)	64.1 (92%)	69.5 (100%)

The values in a round bracket note percentages

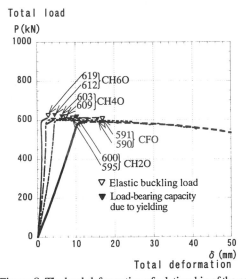

Total load
P(kN)

∇ Elastic buckling load

▼ Load-bearing capacity
due to yielding

δ (mm)
Total deformation

Figure 8. The load-deformation of relationship of the point A
for single layer cylindrical roof shells
whose boundary nodes shown by △
in Figure 3(b) are pin supported

Figure 9. The buckling mode for single laye
cylindrical roof shells whose bou
nodes shown by △ in Figure 3(
pin supported (model CH6O)

Table 5. The load-carrying capacity and the elastic buckling load for single layer cylindrical roof shell whose boundary nodes shown by △ in Figure 3(b) are pin supported

Model	CH6O	CH4O	CH2O	CFO
The weight of a latticed frame and sheets : Wa	26.7kN	22.6kN	18.6kN	14.5kN
Load-bearing capacity due to yielding : Py	619kN	603kN	600kN	590kN
Elastic buckling load : Pe(kN)	612kN	609kN	595kN	591kN
Maximum load : Pmax(=Minimum [Py, Pe])	612kN (104%)	603kN (102%)	595kN (101%)	590kN (100%)
Py / Wa	23.2	26.7	32.3	40.7
Pe / Wa	22.9	26.9	32,0	40.8
Pmax / Wa(=Minimum [Py / Wa, Pe / Wa])	22.9 (56%)	26.7 (66%)	32.0 (79%)	40.7 (100%)

The values in a round bracket note percentages

CONSIDERATION

In comparing the hybrid system composed of a latticed frame and sheets with a separate sub framing system, the initial stiffness of the hybrid system is generally higher than that of a separate sub framing frame system, and composite effect for the initial stiffness is proportional to the thickness of a sheet. The maximum load of the hybrid system is bigger than that of a separate sub framing system, and the composite effect for the maximum load of the single layer latticed dome is bigger than that of the single layer cylindrical roof shell, and especially that of the single layer cylindrical roof shell whose boundary nodes shown by △ in Figure 3(b) are pin supported is the smallest. This is because for the single layer cylindrical roof shell the bending stress of the structure is superior to the axial stress, and for the single layer latticed dome the axial stress is superior to the bending stress. Furthermore, for the single layer latticed dome, the maximum load divided by the weight of a latticed frame and sheets of the hybrid system is bigger than that of the separate sub framing system, but the composite effect of that isn't proportional to the thickness of a sheet. The maximum load divided by the weight of a latticed frame and sheets of model DH4 is the biggest of all models. For the single layer cylindrical roof shell, the maximum load divided by the weight of a latticed frame and sheets of the hybrid system is smaller than that of a separate sub framing system. Therefore, for the single layer latticed dome, model DH4 is the most effective of all models, but for the single layer cylindrical latticed shell, a separate sub framing system is more effective than the hybrid system.

In comparing the load-bearing capacity due to yielding with the elastic buckling load, for the single layer latticed dome if the thickness of a sheet is more than 4 millimeter, the elastic buckling load of model DHi is bigger than the load-bearing capacity due to yielding. For the single layer cylindrical roof shell, the elastic buckling load is almost as same as the load-bearing capacity due to yielding.

SUMMARY AND CONCLUSIONS

In this paper, the load-bearing capacity and the buckling behavior for the hybrid system that combines a latticed structure with metallic sheets are theoretically made clear, and results of the hybrid system are compared with a separate sub framing system. In comparing the hybrid system that combines a latticed frame and sheets with a separate sub framing system, the initial stiffness of the hybrid system is generally higher than that of a separate sub framing system. The maximum load of the hybrid system is bigger than that of a separate sub framing system. The composite effect for the maximum load of the single layer latticed dome is bigger than that of the single layer cylindrical roof shell. For the single layer latticed dome, the maximum load divided by the weight of a latticed frame and sheets of the hybrid system is bigger than that of a separate sub framing system. However, for the single layer cylindrical roof shell, the maximum load divided by the weight of a latticed frame and sheets of the hybrid system is smaller than that of a separate sub framing system. Therefore, the single layer latticed dome is more effective structure for the hybrid system than the single layer cylindrical roof shell.

REFERENCES

1. ZHAO H. L., CAO X. M., QIAN M. Q., YANG Q., ZHAO J. MA and C.Q., Theoretical Analysis of the Sheet Space Structure System and Engineering Practice, Proceedings of the Fourth International Conference on Space Structures, Space Structures 4, Vol.2, 1993, pp.1726-1734.

2. TANIGUCHI Y., SAKA T. and MAEHATA T., Restraint Effect of Membrane on the Member Buckling for Spatial Structures, Proceedings of the Sixth Asian Pacific Conference on Shell and Spatial Structure, October, 2000, pp.89-95.

3. TANIGUCHI Y., SAKA T. and MAEHATA T., Effect of Membrane Arrangement on Member Buckling of Spatial Structures, Proceedings of IASS Symposium, Nagoya, 2001, TP099.

4. BOGNAR, L STRAUBER, B.G., Load-Shortening Relationships for Bars, Journal of Structural Engineering, ASCE, Vol.115, No.7, 1989, pp.1711-1725.

5. MCCONNEL R.E., Force Deformation Equations for Initially Curved Laterally Loaded Beam Columns, Journal of Structural Mechanics, ASCE, Vol.118, No.7, 1992, pp.1287-1302

6. TANIGUCHI T., SAKA T. and TANAKA H., Tangent Stiffness Equations for Laterally Distributed Loaded Members, Journal of Engineering Mechanics, ASCE, Vol.125, No.5, 1999, pp.537-544.

7. EL-SHEIKH A., Behavior of Space Frame Roof Structures with Direct Member Loading, Journal of the International Association for Shell and Spatial Structures, Vol.41, No.133, 2000, pp.75-90.

Modeling the semi-rigid behaviour of the MERO jointing system

J. VASEGHI AMIRI
Assistant Professor of Mazandaran University, Iran
M. R. DAVODI
Lecturer of Mazandaran University, Iran

ABSTRACT

The semi-rigidity of the connections has an important effect on the behaviour of space structures. A number of researchers have developed some computer programs that include the effects of the semi-rigidity of the connections in the structural analysis. The aim of this paper is to propose a simple method based on the use of the ANSYS software suite that is easy to use in practical work. The MERO system has been considered in the present work. A small beam element together with some nonlinear springs elements are used to represent the behaviour of the connections. The stiffness of the rotational springs represents the semi-rigidity of the connections. A large value of the stiffness relates to a rigid condition and a small value relates the pin condition. A number of examples for both static and dynamic cases are included. The numerical results obtained by the proposed method are in a reasonable agreement with experimental and analytical result.

INTRODUCTION

The MERO jointing system is a multidirectional system allowing up to eighteen tubular members to be connected together through a spherical node at various angles. The jointing system is suitable for space structures of various geometry including those with curved and irregular shapes. Consider Figure 1a where four tubular elements are connected together by means of a MERO jointing system. A view of the connection is shown in Figure 1b and its section is shown in Figure 1c.

As may be seen from Figure 1 the MERO jointing system consists of a forged steel ball with a number of threaded borings, a forged conical end piece, a high tensile threaded bolt, and a sleeve with a window and a dowel pin. The ball is located at the intersection of the longitudinal axes of members. The conical end piece is welded to the end of the tube element. A high tensile threaded bolt passes through the conical end piece and is screwed into the ball by means of a sleeve. A dowel pin is used to constrain the bolt to the sleeve in order to allow the turning of the bolt. A window on the sleeve permits the inspection of the penetration of the bolt into the ball.

BEHAVIOUR OF THE MERO JOINTING SYSTEM

A MERO connector in a space structure can be subjected to tensile, compressive and bending effects. When an element that connects to the joint is in tension then the tension is transmitted to the ball through the bolt and the sleeve is inactive in this case. In contrast, if the element is in compression then the force is transmitted to the ball by direct bearing through the conical end piece, sleeve and the ball. In this case, the sleeve is under compression and the bolt is inactive. In the pure bending state, because of the bending deformation, a portion of the sleeve will be separated from the ball and the conical end piece. Thus, this part of the sleeve does not contribute to the transmission of any force. Therefore, the bolt and a part of sleeve that is in contact with other components of the connection transmit the bending moment.

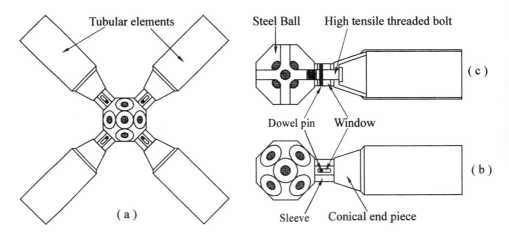

Fig. 1. The MERO jointing system.

The above three cases of load transfer are simplified models. In practice, the loads will be either a combination of a tensile force together with some bending or a combination of a compressive force together with some bending. The interaction of the axial forces and bending moment in the connection is a complex phenomenon and there are hardly any publications for the description of the behaviour of the connection in this case. The axial forces have important effects on the moment-rotation relationship of the connector. In general, a compressive axial force increases the bending rigidity of the connection and a tensile force decreases this rigidity Refs 1,3. However, in order to analyse a space structure with consideration of the semi-rigidity of the connections, the moment-rotation relationships of the connections must be found.

STRUCTURAL ANALYSIS INCLUDING THE EFFECTS OF SEMI-RIGIDITY OF CONNECTIONS

In general, there are two approaches for the analysis of space structures with consideration of semi-rigidity of the joints:
- Development of a computer program that includes the effects of the semi-rigidity of the connections or
- Using an available commercial package for structural analysis.

In the first approach, a general member stiffness matrix has to be developed that includes the semi-rigidity effects. This stiffness matrix is incorporated into a computer program for structural analysis. In the second approach, which is adopted in the present work, by using the facilities of a structural analysis package the behaviour of the connection is appropriately modelled.

As shown in Figure 2a, a typical component of the MERO system consists of a steel tubular element together with the end connectors. This component is connected to the rest of the structure through the end balls. In what follows the modelling of the component part in Figure2 is discussed. The model will be used for the analysis based on the following considerations.

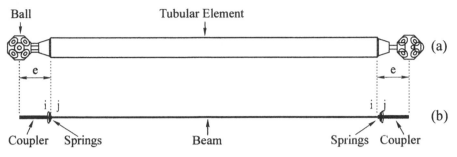

Fig. 2. (a)A typical component. (b) An analytical model.

It is assumed that the tubular elements have a uniform cross-section and material. The element is modelled using a single beam element. Henceforth, in this paper, the term beam element is used to refer to a general spatial beam element involving tension, compression, shear force, torsion and bending moments. This beam element has six degrees of freedoms at each end node consisting of translations in the x, y and z axes and rotations about x, y and z axes. One approach for considering the semi-rigidity of a connection is based on the use of springs to represent the flexibility of the connection. In the present work, a connection is modelled using a set of six springs and a "coupler" at each end of the tubular element, as shown in Figure2. The coupler is a small beam element that connects the springs to the other elements of the structure. The couplers of the elements of the structure are connected together rigidly.

The springs at each end consist of three axial springs and three rotational springs. The axial springs relate to the axial and shear displacements and the rotational springs relate to the bending and torsional rotations. It is assumed that the two ends i and j of each of the six springs have the same coordinates. The axial springs are used to relate the axial and shear displacements of the tubular element and the coupler. The rotational springs represent the semi-rigidity of the connections. The stiffnesses of the springs are calculated from force-displacement or moment-rotation relationships of the connection. These relationships can be obtained either by using an analytical approach (finite element method) or by using an experimental approach. The most commonly used approach is to fit a curve on the experimental data. In this paper, it is assumed that force-displacement and moment-rotation relationships of the connectors are known for different degrees of freedom and have been obtained analytically or experimentally.

ANALYSIS PROCESS USING ANSYS COMPUTER PACKAGE
The ANSYS software suite is a powerful and easy-to-use simulation software package that is used for structural analysis. This package has a large number of library elements that can be used to model a structure with semi-rigid connections. The PIPE20 and COMBIN39 are two of these library elements that are employed in the present work. The PIPE20 is a plastic straight pipe element, which has six degrees of freedom at each end. This element is used to model the tubular elements as well as the end couplers. In order to obtain accurate analytical results by using the proposed model, suitable cross-sectional properties for couplers should be

estimated. Numerical analyses by the authors on a number of available examples have shown that when the ratio of the second moment of area of the coupler and tubular element is above 5000, the analytical results tend to the experimental data. The required high rigidity of the couplers is due to the fact that the flexibility of the connections is represented by the springs and the coupler elements simply act as a rigid links.

The COMBIN39 is another ANSYS library element with non-linear generalised force-displacement capability. The element behaves as either an axial spring or a rotational spring. Three separate COMBIN39 elements represent the axial springs and relate to the axial and shear displacements. Since the nodal displacements at the connection of the tubular elements and the couplers have to be the same, the stiffness of the COMBIN39 elements is taken as infinite. Alternatively, these three translational degrees of freedom between the end nodes of a tubular element and a coupler can be related using a "coupling" command. In the ANSYS program, all coupled nodes are forced to assume the same displacement values in the specified nodal coordinate directions. Three other COMBIN39 elements represent rotational springs and relate to bending and torsional rotations. These elements represent the semi-rigidity of a connection. The stiffness of these elements is calculated from the moment-rotation relationship of the connector. This relationship for the MERO jointing system is non-linear.

VERIFICATION OF THE PROPOSED ANALYTICAL MODEL
The verification of the proposed analytical model for the semi-rigid analysis has been carried out based on a fully rigid upper limit, a pin jointed lower limit and comparisons with the available experimental and analytical results. In the proposed model when the stiffness of the COMBIN39 rotational elements is large as compared with EI/L of the tubular elements the behaviour of the structure is nearly the same as a rigidly connected structure. Also, using the "coupling" command in ANSYS, the nodal rotational degrees of freedom between tubular elements and the couplers can be related together and the structure demonstrates a rigid jointed response. On the other hand, if the stiffness of the spring elements is small in comparison with EI/L of the tubular element, due to the short length of the couplers, the behaviour of the structure will be close to a pin connected structure.

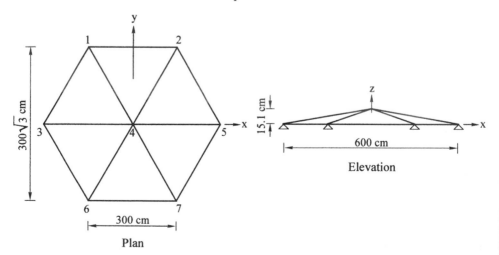

Fig. 3. Plan and elevation of dome.

The authors have examined a number of numerical examples for both static and dynamic cases. These show that when the ratio of the stiffness of the springs K and the flexural stiffness EI/L of the tubular element is above 1000 the structure behaves as a rigidly jointed structure. Also, when this ratio is less than 0.01 the structure exhibits a pin jointed response. In order to compare the efficiency, accuracy and capability of the proposed model with other numerical and experimental results, the following examples are considered.

Example for Static Analysis
A single layer lattice dome has been studied experimentally and analytically by Shibata et al Ref 2. The plan and elevation of this dome together with its dimensions is shown in Figure 3. The dome has 12 steel members connected by means of a MERO-type jointing system. The members are tubular with an outside diameter of 139.8 mm and a thickness of 4 mm. The dome has seven joints and it is supported at all the boundary nodes.

The nodes 1, 2, 6 and 7 are supported only in the Z direction. Node 5 is constrained in the Y and Z directions and node 3 is constrained in the X, Y and Z directions. The only unconstrained node is node number 4, at the crown. An incremental concentrated vertical load is applied to the node number 4. The exact properties of the tubular elements and the jointing system are given in Ref 2. The moment-rotation curve of the connections has been obtained by the experimental data, as shown in Figure 4.

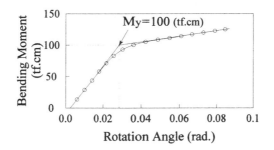

Fig. 4. Moment-rotation curves obtained by Shibata *et al.*

Figure 5 shows the load-deflection relationships for node number 4 with different connection types. In this figure the circles and squares show the results of analysis for rigidly and pinned jointed domes. These results are obtained using conventional beam and truss elements in ANSYS (PIPE20 and LINK8, respectively). The solid lines in the Figure 5 show the results of analyses using the method proposed in this paper. The figure indicates that the results obtained by the proposed method are in good agreement with the results based on the conventional method. The triangles in the figure show the test results obtained by Shibata et al. The analytical results obtained by using the proposed method are shown by solid line. It can be seen that the structural analysis using the proposed method could predict the semi-rigid behaviour of the structure very well.

Example for Dynamic Analysis
In order to illustrate the validity of the proposed model, a dynamic analysis of a simple lattice dome is carried out. The dome is composed of 12 steel elements. The beams are connected by

means of a MERO-type jointing system. The members are tubular with a diameter of 60.5 mm and a thickness of 1.6 mm.

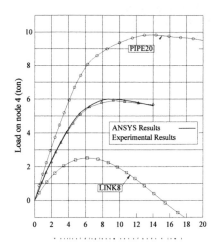

Fig. 5. Vertical displacement [node 4].

The dome is supported at the boundary nodes is the same manner as explained for the example of Figure 3. A plan and elevation of the dome together with the loading history applied at the central node are shown in Figure 6. The exact mechanical properties of the tubular elements and different parts of jointing system together with the applied load are given in Ref 1. The moment-rotation relationship (M-θ relationship) of the connections is given as

$$M = 2218 \times \left(1 + e^{-2}\right) \times \left(e^{-\frac{1}{\sqrt{}}\left(10^{1.6} \times \theta + 0.5\right)} - e^{-2} \right)$$

where, e is the base of natural logarithm, M is in kN.mm and θ is in radian. The specific gravity and yield stress of the material is 7850 hg/m^3 and 2400 kg/m^2, respectively. The analytical results from the proposed method show a good agreement with the analytical results in Ref 1.

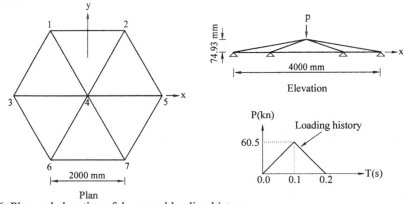

Fig. 6. Plan and elevation of dome and loading history.

Also, time-displacement response of the dome is shown in Figure7. In this figure the circles and triangles show the results of analyses for pinned and rigidly jointed domes, respectively, using conventional truss and beam elements. The solid lines show the results of the analyses with the proposed method when the rotational stiffness of the connection is very small or infinite, respectively, for lower and upper limits. The results indicate that the proposed method satisfies the upper and lower limits satisfactorily.

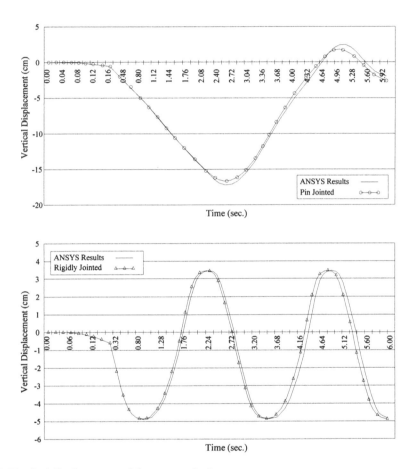

Fig. 7. Vertical displacement of dome at node 4.

CONCLUSION

An analytical method that can take account of the effects of semi-rigidity of connections in structural analysis has been proposed. The method is based on the use of an available commercial package that is easy to use in practical work. A small beam element together with some nonlinear springs elements are used to represent the behaviour of the connections. The stiffnesses of the springs may be chosen such that the connections behave as fully rigid joints or as pin joints. The comparison of the analytical results with the experimental results and the data obtained by other researchers verifies the efficiency and accuracy of the proposed method.

REFERENCES
1- CHENAGHLOU M R, Semi-Rigidity of Connections in Space Structures, Ph.D. Thesis, University of Surrey, UK, October 1997.
2- SHIBATA R, KATO S and YAMADA S, Experimental Study on the Ultimate Strength of Single-Layer Reticular Domes, Proceeding of the Fourth International Conference of Space Structures, University of Surrey, UK, 1993.
3- YAMADA M, SATO Y, OBATA A, MOGAMI K and YAMAGUTI I, Three Dimensional Stress and Deformation Analysis of Joint Parts Used in Truss System, Sixth Asian Pacific Conference on Shell and Spatial Structures, Seoul, Korea, October 2000.
4- ANSYS Manual, ver5.4, October 1997.

Designed for 1000 Years - the Maitreya Buddha project

D. HOOPER Mott MacDonald Ltd. London UK

INTRODUCTION
The Maitreya Buddha Project presents some interesting challenges as we seek to create a world monument designed to last for many future generations. This paper describes some of the implications of designing a structure to last 1000 years with particular regards to the derivation of statistically determined loads and durability of materials.

DESIGN BRIEF
It is intended to construct a Statue of the Maitreya Buddha in Northeast India. This will not only be the largest statue in the world but also a truly magnificent monument that will provide spiritual enlightenment to inspire future generations for at least the next 1000 years. The project is essentially a spiritual development intended to inspire and educate those who visit. People from all walks of life, countries, religions and race will be welcome and should feel a spiritual inspiration emphasising the universality of loving kindness and compassion.

Maitreya Buddha	Ushiku Statue	Statue of Liberty	Lantau Buddha
India	Japan	USA	Hong Kong

Fig. 1. Comparison with other world monuments.

The entire complex will contain religious Buddhist art which itself will further inspire and educate. The tone of the complex is to be quintessentially uplifting and inspiring and must appeal to Buddhists of all traditions as well as people of other religions and those with no religion. The Maitreya project is also intended to become a major international tourist destination, developed to the highest international standards.

DESIGN CRITERIA
Dead & Imposed Loading
Dead and imposed loading throughout the Statue and Throne was derived from the more onerous of British or Indian Standards. Close collaboration with the Art procurement was particularly important given that the Maitreya and Shakayamuni Buddha statues alone, located within the Thone, weigh up to 46.5tonnes and 17.3tonnes respectively.

Wind Loading
The analysis for wind loading was based on the National Building Code of India, 1983. This code bears a close similarity to British Standards Institution CP3: Chapter V: Part 2:1972 "Code of Basic data for the design of buildings Chapter V. Loading, Part 2. Wind Loads" but is amended to reflect conditions in India. The basic wind speed at the Maitreya site was 39m/s. This increased to 47m/s within two zones bordering to the north, east and south.

Climate change, in the form of global warming, is now a recognized phenomenon. The primary effects of this are an increase in sea level and an increase in the frequency of climatic extremes. Unfortunately, global circulation models have only produced climate change scenarios for about 50 to 200 years into the future. Thus any attempt to specify how the climate will change over the design life of the Maitreya Buddha is essentially educated guesswork. To determine the effects of climatic change on wind data requires a rational approach. The dominant wind system that affects India is the tropical cyclone, with cyclones moving from east to west across the Indian Ocean before hitting the Indian coast. As they meet the coast, their strength decreases markedly and this is reflected in the wind speed map of the Indian code, which shows high wind speeds in thin bands along the coast. If one of the effects of climate change is to raise sea level, then it is possible that this could effectively bring the coastline tens of kilometres inland. At present the site of the Maitreya Buddha is in a region of reference wind speed of 39m/s, but 100km to the east there is a region of reference wind speed of 47m/s. Assuming the wind zones remain of the same thickness, and that the other effect of global warming is to increase the frequency and intensity of wind storms, the assumption that the Maitreya Buddha should be in a reference wind speed zone of 47m/s does not seem unreasonable.

Figure 2 illustrates the design wind pressures varying with height for basic wind speeds of 39 m/s and 47 m/s and 50 year and 1000 years design life. At a height of 100 metres the respective design wind pressures are   1.29 kN/m^2 and   3.49 kN/m^2, equating to design speeds of 46.4 and 76.2 m/s.   The design wind pressure for the statue and throne is therefore significantly higher than for the would be used for a normal building and of the increase roughly 46 % is due to the climatic change allowance and 54 % is due to the long life span (note that these percentages vary at different heights etc)

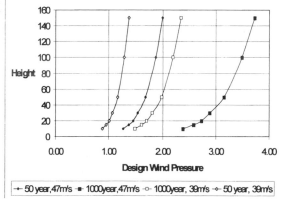

Fig. 2. Design wind pressure profile.

The design wind speed selected for this project has a return period of 1000 years, the probability that this wind speed will be exceeded during the working life of the structure is 63.2% i.e. more likely than not. However, in design for the ultimate limit state it is normal to apply load factors (γ_f) which result in an overall probability of the wind load being exceeded during the design life that is very different to that of the wind speed being exceeded. The lifetime risk of the load being exceeded (P_N') for different values of the safety factor is shown in Table 1.

Table 1. Effect of safety factor on risk.

γ_f	1.0	1.1	1.2	1.3	1.4	1.5	1.6
P_N'	0.632	0.439	0.290	0.187	0.120	0.077	0.050

The approach taken for the Scheme Design of the Maitreya Buddha is that the design wind speed will be such that its risk of it being exceeded during the design life is 0.632. In order to ensure that the lifetime risk of the load being exceeded is reduced to 5% at ultimate limit state design loads, to be consistent with the level of risk adopted for seismic design at ultimate limit state, the load factors used in design have been increased.

British Standards are consistent in their approach to load factors for wind. When wind loading is combined with dead loading only, $\gamma_{f\,W} = 1.4$ and $\gamma_{f\,DL} = 1.4$. The reduced probability of full dead, imposed and wind loading occurring simultaneously is reflected in reduced partial safety factors when this load combination is considered; these partial factors are: $\gamma_{f\,W} = 1.2$, $\gamma_{f\,DL} = 1.2$ and $\gamma_{f\,IL} = 1.2$. In order to achieve lifetime risk of the load being exceeded of 5% the following increases to partial safety factor for wind loading will be adopted:

Load combination - dead + wind: $\gamma_{f\,W} = 1.6$
Load combination - dead + imposed + wind: $\gamma_{f\,W} = 1.4$ $= (1.2/1.4) \times 1.6)$

Seismic Loading
The magnitude of the design earthquake is directly related to the return period considered. The longer the return period, the greater will be the magnitude of the design event. The return period is defined as the period of time during which an earthquake of a particular magnitude will have a 63.2% chance of being exceeded once. The magnitude of the design earthquake for the Maitreya Buddha is selected on the basis of an appropriate earthquake return period commensurate with the design life of 1000 years and a probability of being exceeded commensurate with the importance of the Statue.

For structures with normal usage and design life, the seismic design codes of practice give pre-determined values of hazard. Seismic design codes generally assume a design life of 50 years and set the design load such that the probability of these loads being exceeded during the design life is 10%. To assess an acceptable level of risk to be adopted for the seismic design of the Maitreya Buddha, it is useful to draw comparisons with current design practices for other typical structures such as buildings, bridges and nuclear installations.

Table 2 presents preliminary values for Peak Ground Acceleration and respective return periods, derived from a site specific seismic hazard assessment for the Maitreya project. This facilitates a direct comparison between the seismic design parameters that would be applicable to buildings, bridges and nuclear installations built at the site with those proposed for the Maitreya Buddha.

Table 2. Comparison of return periods for typical structures.

Design Life	__	__	__	__	__	Peak Ground Acceleration (pga) at Maitreya site (g)										
	1	5	10	25	50	100	120	475	1,000	2,500	9,500	10,000	19,500	20,000	49,500	99,500
						0.047		0.083	0.107	0.182	0.195	0.229	0.285		0.410	0.470
	Probability of exceedance															
1	0.632	0.181	0.095	0.039	0.020	0.010	0.008	0.002	0.001	0.000	0.000	0.000	0.000	0.000	0.000	0.000
5	0.993	0.632	0.393	0.181	0.095	0.049	0.041	0.010	0.005	0.002	0.001	0.000	0.000	0.000	0.000	0.000
10	1.000	0.865	0.632	0.330	0.181	0.095	0.080	0.021	0.010	0.004	0.001	0.001	0.001	0.000	0.000	0.000
25	1.000	0.993	0.918	0.632	0.393	0.221	0.188	0.051	0.025	0.010	0.003	0.002	0.001	0.001	0.001	0.000
50	1.000	1.000	0.993	0.865	0.632	0.393	0.341	0.100	0.049	0.020	0.005	0.005	0.003	0.002	0.001	0.001
100	1.000	1.000	1.000	0.982	0.865	0.632	0.565	0.190	0.095	0.039	0.010	0.010	0.005	0.005	0.002	0.001
120	1.000	1.000	1.000	0.992	0.909	0.699	0.632	0.223	0.113	0.047	0.013	0.012	0.006	0.006	0.002	0.001
475	1.000	1.000	1.000	1.000	1.000	0.991	0.981	0.632	0.378	0.173	0.049	0.046	0.024	0.023	0.010	0.005
1,000	1.000	1.000	1.000	1.000	1.000	1.000	1.000	0.878	0.632	0.330	0.187	0.095	0.050	0.049	0.020	0.010
2,500	1.000	1.000	1.000	1.000	1.000	1.000	1.000	0.995	0.918	0.632	0.231	0.221	0.120	0.118	0.049	0.025
9,500	1.000	1.000	1.000	1.000	1.000	1.000	1.000	1.000	1.000	0.976	0.632	0.613	0.386	0.378	0.175	0.091
10,000	1.000	1.000	1.000	1.000	1.000	1.000	1.000	1.000	1.000	0.982	0.651	0.632	0.401	0.393	0.183	0.096
19,500	1.000	1.000	1.000	1.000	1.000	1.000	1.000	1.000	1.000	1.000	0.872	0.858	0.632	0.623	0.326	0.178
20,000	1.000	1.000	1.000	1.000	1.000	1.000	1.000	1.000	1.000	1.000	0.878	0.865	0.641	0.632	0.332	0.182
49,500	1.000	1.000	1.000	1.000	1.000	1.000	1.000	1.000	1.000	1.000	0.995	0.993	0.921	0.916	0.632	0.392
99,500	1.000	1.000	1.000	1.000	1.000	1.000	1.000	1.000	1.000	1.000	1.000	1.000	0.994	0.993	0.866	0.632

Annotations:
- Wind speed UK, India and others
- Bridges in Europe, Buildings to EC8 & UBC
- Major Bridges in USA with a 100 year design life
- This value would comply with many building codes at 10% risk for the design life
- Nuclear installations in the UK and USA
- Proposed design case for Maitreya Project "with light and repairable damage"

For the design of buildings and bridges in the UK and Europe, it is common to accept a 10% risk of the design forces being exceeded during the design life (EC8). This gives a return period earthquake of 475 years. Buildings designed using these parameters will not collapse during the design earthquake, thus safeguarding life, but may suffer damage and be unserviceable after the earthquake. For more important buildings such as hospitals, it is common to limit damage through the use a lower probability that the design loads will be exceeded. This may be taken into account either directly if the data is available, or implicitly within the codes of practice through the use of Importance Factors. These values vary between codes, but generally range between 0.8 and 1.5. For the design of structures such as bridges, which have a design life of 120 years (BS 5400), it would seem sensible to adopt an appropriate return period. However, in the UK and Europe, Importance Factors are again used to take this into account, although in the US it is noted that for important bridges, an earthquake with a return period of 2500 years should be used (AASHTO-1998). For a design life of 100 years, this results in design loading with a 3.9% risk of being exceeded. This lower probability reflects the more severe consequences of a bridge becoming unserviceable, resulting in disruption to infrastructure at the time when emergency services and subsequent rebuilding work rely on it.

For the nuclear industry in the UK, it is stipulated (NII, 2000) that for natural hazards, the uncertainty of data may prevent reasonable prediction of events for frequencies less than once in 10,000 years. Hence nuclear plants should be designed to a design-basis earthquake that has a frequency of being exceeded once every 10,000 years. It is also stated (NII, 2000) that there should be no disproportionate increase in risk from an appropriate range of events that are more severe than the design-basis earthquake. An operating basis earthquake is also defined as the event (below the design-basis earthquake) the repeated occurrence of which would not impair the plant, system or structure. The principles do not spell out though how the latter two conditions may be met, their relationship to the design-basis earthquake or their return periods. In the USA, the common practice is to design for the 10,000year return period earthquake (US NRC, 1975, 1990 amongst others). This also seems to be adopted by other guidance notes (ASCE, 1998), although not explicitly stated. It is therefore concluded that an earthquake with a return period of 10,000 years is the current standard for seismic design of nuclear plant, whilst less onerous loading scenarios may be checked to ensure the safe operation of the plant and its components. It is noted though that although the demand is lower for the operating basis earthquake, the limit state for verification is tighter than in the case of the design-basis earthquake. A 10,000year return period results in a 1% probability of being exceeded in a design life of 100 years. The Statue and Throne will be designed for a Peak Ground Acceleration of 0.285g. This corresponds to an earthquake with a return period of 19,500 years, with a 5% probability of being exceeded in a design life of 1000 years.

Thermal Loading
Annual mean global solar radiation on the horizontal surface is around 200W/m² and is relatively constant throughout the year regardless the season. The bronze skin to the Buddha is expected to reach temperatures as high as 60°C in January and 80 °C in August. The skin and its supporting structure must therefore be designed to accommodate the full range of temperatures over the course of 1000 years, with a maximum differential temperature approaching 50°C.

MATERIALS
Reinforced Concrete
The fact that Roman structures in which masonry was bonded by mortar, such as the Collosseum in Rome and the Pont du Gard bridge near Nîmes, and concrete structures such as the Pantheon have survived to this day is a testament to the potential durability of concrete. However, most of today's structural concrete is reinforced and it is the steel in concrete that is its Achilles' heel. Although, initially, the inherent alkalinity of concrete protects steel against corrosion, in the presence of sufficient moisture, oxygen and chloride and/or carbon dioxide it can start to rust. Eventually the expansive rust can crack and spall off the overlying concrete cover. The corrosion of reinforcement is the most common cause of premature deterioration of concrete structures throughout the world and is the subject of intense research and investigation. It is not possible to guarantee modern reinforced concrete and materials for a service life of 1,000 years. However, correctly designed and placed unreinforced, mass concrete which comprises good quality materials will achieve a service life greater than that for reinforced concrete and may last for 1,000 years.

Pathways for the degradation of concrete are well understood for most known environmental conditions and constituent materials. Potential problems with workmanship are also well understood. Deterioration due to environmental conditions may be either chemical or physical, but the rate of degradation can be controlled through the correct selection of materials and careful design. The external environment will affect durability both during the production of the concrete and through subsequent exposure conditions. For concrete below ground, chemical composition of the ground and ground water together with the level and mobility of the ground water will affect the exposure conditions. Concrete within the Statue and Throne will be subject to changes in temperature, relative humidity and most importantly carbon dioxide. Table 3 overleaf presents a matrix of the options available to produce durable concrete, with relative durability rating: A – Best: E – Worst.

Table 3. Comparison of relative durability ratings and costs for various durability-enhancing materials and measures.

Concrete	Cement/Blends				Admixtures		Reinforcement						Additional Protection				Durability Rating (2)	Comparative Costs (3)
	Portland Cement	Pulverized-Fuel Ash (pfa)	Ground granulated balst furnace slag (ggbs)	Silica Fume (sf)	Corrosion Inhibitors	Self-Compacting Concrete	Conventional	Epoxy Coated	Galvanised	Stainless Steel	Non-Ferrous	Fibres	Controlled Permeability Forms	Coatings	Surface Treatments	Cathodic Protection		
Mass Concrete																		
Control	x																F	1.0
PC/Pfa 1	x	x															E	0.9
PC/Pfa 2	x	x				x											D	1.5
PC/Pfa 3	x	x				x						x					C	1.8
PC/ggbs 1	x		x														E	0.9
PC/ggbs 2	x		x			x											D	1.5
PC/ggbs 3	x		x			x						x					C	1.8
PC/ggbs/sf 1	x		x	x													D	0.9
PC/ggbs/sf 2	x		x	x		x											B	1.5
PC/ggbs/sf 3	x		x	x		x						x					A	1.8
Reinforced Concrete (Conventional High Yield Steel)																		
Control	x						x										F	1.0
PC/ggbs 1	x		x		x		x										E	1.1
PC/ggbs 2	x		x		x		x					x					E	1.2
PC/ggbs 3	x		x		x	x	x									x	C	2.0
PC/ggbs/sf 1	x		x	x	x		x										D	1.2
PC/ggbs/sf 2	x		x	x	x		x					x					D	1.3
PC/ggbs/sf 3	x		x	x	x	x	x									x	C	2.5
Reinforced Concrete (Epoxy Coated)																		
PC/ggbs 1	x		x		x			x									D	1.5
PC/ggbs 2	x		x		x			x				x					D	1.6
PC/ggbs 3	x		x		x	x		x								x	C	2.4
PC/ggbs/sf 1	x		x	x	x			x									D	1.6
PC/ggbs/sf 2	x		x	x	x			x				x					C	1.9
PC/ggbs/sf 3	x		x	x	x	x		x								x	B	2.9
Reinforced Concrete (Galvanised Steel)																		
PC/ggbs 1	x		x		x				x								D	1.6
PC/ggbs 2	x		x		x				x			x					D	1.7
PC/ggbs 3	x		x		x	x			x							x	C	2.5
PC/ggbs/sf 1	x		x	x	x				x								D	1.7
PC/ggbs/sf 2	x		x	x	x				x			x					C	1.8
PC/ggbs/sf 3	x		x	x	x	x			x							x	B	3.0
Reinforced Concrete (Stainless Steel)																		
PC/ggbs 1	x		x		x					x							C	2.2
PC/ggbs 2	x		x		x					x		x					C	2.4
PC/ggbs 3	x		x		x	x				x						x	B	4.0
PC/ggbs/sf 1	x		x	x	x					x							C	2.4
PC/ggbs/sf 2	x		x	x	x					x		x					B	2.6
PC/ggbs/sf 3	x		x	x	x	x				x						x	A	5.0

The most durable concrete that can be achieved would comprise of cement blended from portland cement and ground granulated blast furnace slag with the addition of silica flume, non-ferrous fibres and admixtures for self-compacting concrete. The inclusion of embedded steel will require a maintenance and replacement strategy to be developed to achieve the required design life due to the corrosion of the steelwork. Life to first maintenance can be significantly improved with the use of corrosion resistant reinforcement, particularly stainless steel, together with cathodic protection. However, the cost of such measures is 5 times higher than conventional concrete reinforced with conventional steel, or 2.6 times higher with the omission of cathodic protection. It should be noted that cathodic protection is not required to be active until the protection from the concrete cover is lost and, with attention to detailing, may be retrofitted at a future date as part of the maintenance strategy.

At present, all concrete within the Statue and Throne is to be reinforced with stainless steel, although a lifetime cost analysis supports the potential benefits of using conventional reinforcement to all elements that can be replaced with relative ease.

Structural Steel
The invention of the 'Crucible Process' in around 1740 made possible the manufacture of large quantities of steel with fairly consistent properties. By the mid-1800's the 'Bessemer Process' was developed, providing further advances in mass production and thus availability, of steel. Although a widely used and versatile construction material with a long history of use, steel has changed significantly over the period since its introduction. Although there are some metallic artefacts that have lasted far in excess of the design life of the Maitreya Buddha, such as the 1600 year old Iron Pillar of Delhi in India, the composition and properties of modern steels are very different to those used in antiquity. It must therefore be considered that, in specifying grades and types of steel for use in the Maitreya Buddha, direct analogies with the long-term performance of steel elements in existing structures are of little relevance.

Steel corrosion in most environmental conditions is well understood. The susceptibility of elements of the structure to uniform, pitting, crevice, galvanic and stress corrosion, cracking and corrosion fatigue will be dependent upon the environmental conditions, type of steel and form of construction adopted. The heating and cooling cycles within the Statue will maintain a humid internal climate and promote the formation of condensation on the inside of the skin. Work carried out to date suggests it is unlikely that condensation will form directly on any structure greater than 600mm from the inside of the skin, but all steelwork may be subject to run off and dripping condensation. These factors will create ideal conditions for corrosion within the Statue and makes the selection of steel and protection against corrosion very important.

Comparisons of the relative durability of materials and potential protection techniques, in general terms, are given in Table 4. In this table each available combination of material and protection option has been assigned a relative 'Durability Rating' based upon the letters A-H, with A likely to exhibit the least inherent durability, and H indicating the highest inherent durability. Ratings have been assigned on the basis that the internal environment of the Maitreya Statue is anticipated to be hot and humid, with frequent wetting of the surfaces by condensation.

Table 4. Relative durability of construction material.

Material	Durability Enhancement Method /Durability Rating (A-H)					Comments
	None	High performance coating	Metal spraying and coating	Galvanising	Galvanising and coating	
Carbon steel	A	B	C	D	E	Basic structural steelwork material. Low inherent corrosion resistance, requiring additional protection in almost all but the mildest exposure environments.
Stainless steel	D/E	F	n/a	n/a	n/a	High corrosion resistance austenitic grade assumed (316L). Stainless steels do not have sufficient strengths to be substituted for structural elements.
Duplex steel	F/G	H	n/a	n/a	n/a	Good quality, high corrosion resistant grade assumed. Duplex steels have sufficient strengths to be substituted for structural elements.

Notes:
Durability Ratings are relative, indicated by the letters A-H, with A indicating the lowest intrinsic durability within the anticipated Maitreya Statue environment, and H the highest.
Certain factors, such as localised environmental effects and design details, may increase the corrosive attack of particular options. Certain corrosion mechanisms can initiate preferential attack of particular alloys.
Corrosion inhibitors may have an additional restricted protective effected for the internal surfaces of closed structures, such as blind box section members. They have not been included here.
Quality of joining, particularly in welding and bolting design, can have significant local deleterious effect on durability; it is not considered here.

Two options have been considered for the skeleton framework supporting the skin. Mild steel protected with a high performance paint system will be the most economical in terms of capital cost. It is possible that an initial coating of either sprayed metal or galvanising could be applied, although the extra protection would be relatively insignificant compared to the overall life of the structure and these coatings create problems with welding and maintenance. Metal sprays may also be site applied to avoid problems with welding and may also be used to protect bolted connections. However, the application of such a spray is costly and is quite an advanced technology and as such may prove difficult to acquire in India. To avoid hidden surfaces that cannot be inspected, all sections in mild steel will be hot rolled universal column (H section) or beam sections (I section) of sections fabricated from welded plate. Particular attention to the detailing of joints will be required to avoid crevice corrosion and drainage holes will need to be strategically placed to avoid water traps. All sharp edges will need to be removed to ensure adequate thickness of coatings at external corners of the section and all areas will need to be provided with adequate access for inspection, maintenance and replacement. As an alternative, the use of duplex stainless steel offers many advantages. Circular hollow sections, which for the skeleton are the most efficient shape, could be fabricated from plate by cold rolling into half round sections and seam welding together. For durability, sections would be designed with a sacrificial thickness similar to the bronze skin and the use of protective coatings could therefore be avoided, thus negating concerns over hidden surfaces. A framework constructed entirely of fully welded circular hollow sections avoids water traps and negates the risk of crevice corrosion and architecturally, provides an elegant structural solution. This solution would require a higher initial capital investment, but whole life cost and risk analysis may have a significant influence on the choice of material.

Steelwork to the spine will be protected from dripping condensation and is therefore within a more protected environment. For the spine, mild steel protected with metal spraying and a high performance paint system is considered to be adequate. The shrines are designed as insulated self-contained pods with adequate fire resistance to prevent the break out of fire. Elements comprising the spine will be visible for inspection from within the Statue void, however cladding around steelwork within the shrines will need to be demountable for maintenance purposes. Concrete encasement to these elements has been avoided to minimise the mass in the structure and reduce seismic loading.

For the A-frames and box trusses, mild steel encased in concrete will provide adequate fire protection and corrosion resistance and has many other structural benefits, such as increasing the stiffness of the A frame structure to displacement under lateral loading. Steelwork to the tabletop box truss structure at top of Throne level will be similarly encased in concrete to reduce the need for maintenance. Additional protective coatings to the steel offer no real advantage when considering life to first maintenance.

All steelwork structure will require varying levels of inspection and maintenance with provision for replacement, in order to achieve a design life of 1000 years. Where access for maintenance is impracticable duplex stainless steel will have to be adopted.

Bronze Skin
From the client's perspective, the skin to the Maitreya Buddha is the most important component of the building. It is irreplaceable and therefore the choice of a suitable material that will last 1000 years was paramount. Initial investigations reviewed all options for the selection of materials and the associated processes involved in manufacture and construction. The use of stone for example, was discounted due to its poor strength to weight ration, given the scale of the structure and the selection of material soon settled on the use of bronze, a traditional material used for the construction of statues. This was chosen due to its appearance; good casting and welding properties; weight; strength and durability.

There were many bronze alloy to choose from, each with their own relative merits. Tin alloys and gunmetals for example, have severe limitations with weldability. Mechanically sound welds are very difficult to achieve and both strength and elongation of welds is known to be considerably lower than the base material. In specifications of the UK Ministry of Defence, repair welding on gunmetals is only allowed "for cosmetic repair". Although many of the traditional bronzes are known as "easily castable", this generally refers to filling the mould and replicating the details in the mould. On a technical level however, these alloys lack reproducibility during casting due to factors beyond control and in general are prone to internal defects particularly due to gas pick-up from both the atmosphere and from the mould itself. Silicon Bronze (nominally 4% Si) appeared a viable option, as it is said to have good casting and in particular superior welding properties. Relative disadvantages are inferior corrosion resistance, considerably lower strength, and the fact that only very limited experience exists with this alloy. Nickel Aluminium Bronze (type AB2 as per BS1400) was finally selected for its excellent corrosion properties, good weldability and the alloy is quite well known in the foundry industry for technical applications. Casting properties are not especially good due to a strong tendency to shrinkage and oxide formation, but the alloy is known to produce consistent quality once a good method is developed.

A number of studies on the chemical and metallurgical effects of corrosion on copper and copper alloys were reviewed. Copper and copper alloys are by nature relatively inert to corrosion; in many environments they behave similar to stainless steels and nickel alloys. However, some corrosion does occur, leaving a layer of corrosion products on the surface (patina). The effect of the patina on further corrosion is a complicated process as chemical reactions take place between the metal, the various corrosion products and the environment. For example, patinas of certain compositions can form a sealing layer, which protects the underlying metal. Other patinas can be porous and actually enhance corrosion.

The aluminium bronze alloys are relatively new (some 100 years), and the specific alloy composition of AB2 has only been widely used in the last few decades. Although well understood when used in marine environments, little research has been carried out on corrosion rates under normal atmospheric inland conditions. In general, one could say that the alloy tends to behave equal or better than pure copper in most environments. It is not certain however, if effects like de-alloying (especially of aluminium) can have a significant effect on the long-term corrosion. Therefore, the results of the limited research that has been carried out should be approached with caution. In the literature, overall long-term corrosion rates of about 1 μm per year or less are indicated, leading to a mere 1 mm in 1,000 years, but this appears quite optimistic. The limited information on pitting corrosion indicates that localised corrosion is usually limited to 2 mm or less, and no local accelerated corrosion takes place after the formation of the initial pitting. For design purposes, a corrosion rate of 2.5mm over the design life of 1000 years has been adopted.

With particularly aggressive conditions adjacent to the skin due to condensation and to avoid bimetallic corrosion, the connectors between the skin and supporting frame will be fabricated from aluminium bronze plate. A suitable material will need to be selected to electrically isolate the connectors from the supporting frame to prevent bimetallic corrosion. Sections fabricated from aluminium bronze will also be used to brace sections of the skin that are inaccessible for maintenance, for example, within the folds of the cloak.

Maintenance & Replacement Strategy
Regular inspection maintenance and repair will be essential to achieve a design life of 1000 years. To ensure that the design is fail safe in the event of lapses in planned maintenance, the structure is provided with sufficient access and redundancy to facilitate replacement of individual structural members. The frequency of inspection and maintenance and methods of repair will vary depending upon the materials and form of construction used and exposure conditions. The strategy for maintenance and repair has been developed throughout the scheme design stage and will continue to develop through detailed design.

REFERENCES
1. Mott MacDonald Limited 'Scheme Design Report, Structural Engineering' May, 2001.

Skytech SYSTEM 2000

G. A. GAMANIS
Skytech Space Frames, Greece

SUMMARY
This paper addresses the use in buildings of Skytech SYSTEM 2000, a modular space frame system, composed of spherical nodes, structural members and connection mechanisms. It specifies the design/development and fabrication requirements of tubular, statically loaded steel space frames. Starting point is an introduction of specific examples of applications. In these examples, materials and manufacturing processes for nodes, as well as structural members and connection mechanisms are specified. Following, is an outline of the methods for design and analysis, together with a procedure for the determination of the length of engagement of bolt threads. Finally, a corrosion protection system is presented.

INTRODUCTION
Skytech SYSTEM 2000 is an outgrowth of earlier forms of space frame systems. Evolution of the system over the past 10 years has been spectacular. Result is that this system has prominent status in architectural designs today.

This cannot only be attributed to its aesthetic potential, its impressive simplicity and its unlimited possibility for three-dimensional expansion with standardized elements at a minimum of space obstruction, but also to economic considerations.

Development of the system combined with: a) efficient selection of structural materials; b) use of appropriate manufacturing processes; c) use of growing computer power in designs, analysis and manufacturing with low computer-related costs; resulted in reduction of system prices, thus increasing its competitiveness.

This system is appropriate for use in design/development and fabrication of tubular statically loaded steel space frames[1]. It is especially adaptable to buildings requiring special aesthetic structure forms such as sports arenas, exhibition halls, shopping centers, schools, museums, libraries, etc.

Figure 1 shows the universal Skytech SYSTEM 2000 combined with the newest *Skytech Structural Glass System.*

Note 1.   Per *Guidelines for the Design of Double-Layer Grids,* Task Committee on Double-Layer Grids, Cuoco, D.A., ed., ASCE, New York (1997):
"It is common practice to apply the 'space frame' designation to structures that would more accurately be categorized as 'space trusses', i.e., structures that are pin-connected at the joints, or nodes."

Fig. 1. Shopping center, Kavala, Greece.

DESCRIPTION OF THE SYSTEM

Skytech SYSTEM 2000 is a modular space frame system, which comprises of nodes, structural members and connection mechanisms. The nodes are solid steel spheres bearing radial threaded holes. The number of threaded holes and their orientation and size, are determined by geometry of the structure and node space positioning.

Fig. 2. SYSTEM 2000.

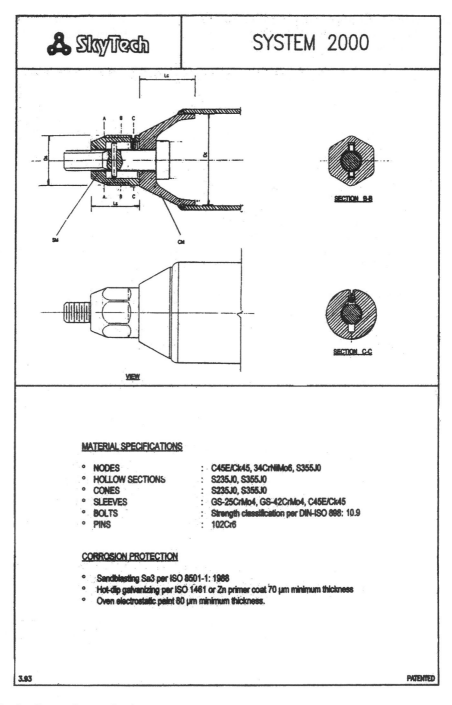

Fig. 3. Connection mechanism.

Each structural member consists of a circular hollow section and two hollow cone ends welded to the hollow section. After welding the welds are ground smooth using automatic grinding machines. The welds are then invisible after painting.

Each connection mechanism (Figure 3) consists of a bolt with a pin and a spanner sleeve equipped with two opposite internal grooves wherein the pin ends move. By rotating the sleeve a torque is transmitted to the bolt through the pin screwing the bolt into the node. When the bolt is fully screwed, the sleeve contacts the node and the structural member. Compression from the structural member to the node is transmitted through the sleeve; tension is transmitted through the bolt.

A key feature of this design is a system of enhanced visual appearance. There are no visible welds, bolts or pins.

STRUCTURAL COMPONENTS

The standard structural components of Skytech SYSTEM 2000 are listed in Table 1. For each node, Table 1 gives the maximum size of bolt and the size of the corresponding circular hollow section (CHS) that can be connected to the node.

Table 1. Standard structural components of Skytech SYSTEM 2000.

Node Diameter [mm]	Bolt External Thread	Pin Diameter [mm]	Sleeve	Cone	CHS D×T [mm]	
60	M12×1.75	3	SM12	CM12	33.7×2.5	S235J0
75	M16×2	4	SM16	CM16	60.3×2.5	
90	M20×2.5	5	SM20	CM20	76.1×3	
105	M24×3	6	SM24	CM24	88.9×3	
120	M27×3	6	SM27	CM27	101.6×3	
135	M30×3.5	6	SM30	CM30	114.3×3	
150	M36×4	8	SM36	CM36	139.7×4	S355J0
180	M42×4.5	8	SM42	CM42	168.3×4	
210	M48×5	8	SM48	CM48	168.3×6	
240	M56×5.5	10	SM56	CM56	219.1×6	
270	M64×6	10	SM64	CM64	219.1×8	
300	M72×6	10	SM72	CM72	273×8	

Table 2 shows the relationship between node diameter and node internal thread size. The corresponding design ultimate resistance, for standard manufactured external threads of grade 10.9, is also given in Table 2.

Table 2. Relationship between node diameter and node internal thread size.

Node Internal Thread	M12	M16	M20	M24	M27	M30	M36	M42	M48	M56	M64	M72
Node Diameter [mm]	Design ultimate resistance[1] [kN]											
	53.8	96.4	151	218	292	374	520	750	990	1360	1810	2360
60												
75												
90												
105												
120												
135												
150												
180												
210												
240												
270												
300												

[1] for standard manufactured external threads having strength grade 10.9

MATERIALS
Nodes
Nodes are manufactured using:
   a) heat-treatable steel C45E/Ck 45 (Material number 1.1191) according to EN 10083-1 for nodes subject to standard stresses;
   b) heat-treatable steel 34CrNiMo6 (Material number 1.6582) according to EN 10083-1 for nodes under heavy loads, over 210 mm $\varnothing$ up to 300 mm $\varnothing$;
   c) structural steel S355J0 (Material number 1.0553) according to EN 10025 if weldable nodes are specified.

Structural members
Circular hollow sections
Normally, use is made of longitudinally welded cold-formed circular hollow sections according to EN 10219-2. If however, it is specified by the client, hot-finished circular hollow sections according to EN 10210-2 are used.

For the aforementioned hollow sections, the most commonly used steel grades are:
   a) structural steel S235J0 (Material number 1.0114) according to EN 10025 for members subject to low stresses;
   b) structural steel S355J0 (Material number 1.0553) according to EN 10025 for members of intermediate and high stresses.

Other grades of structural steel according to EN 10025 may be used if it is required by considerations and approved by client's engineer.

Cones
Cone ends are manufactured by:
  a) structural steel S235J0 (Material number 1.0114) according to EN 10025 for components subject to low stresses;
  b) structural steel S355J0 (Material number 1.0553) according to EN 10025 for components of intermediate and high stresses.

Connection mechanisms
Bolts
Bolts are manufactured using:
  a) heat-treatable steel 42CrMo4 (Material number 1.7225) according to EN 10083-1, strength grade 10.9, which conforms with Eurocode 3 (ENV 1993-1-1);
  b) heat-treatable steel 34CrNiMo6 (Material number 1.6582) according to EN 10083-1, strength grade 10.9, which conforms with Eurocode 3 (ENV 1993-1-1), for bolts over M64.

Pins
Pins are manufactured using cold work tool steel 102Cr6 (Material number 1.2067) according to EN 10132-4.

Spanner sleeves
Spanner sleeves are manufactured using:
  a) heat-treatable steel C45E/Ck 45 (Material number 1.1191) or 25CrMo4 (Material number 1.7218), according to EN 10083-1 for components subject to low stresses;
  b) heat-treatable steel 42CrMo4 (Material number 1.7225) according to EN 10083-1 for components of intermediate and high stresses.

MANUFACTURING PROCESSES
Nodes
Nodes are hot-forged from hot rolled material, normalized, machined and marked. Normalizing of forged nodes is undertaken in order to attain isotropy and improved machinability. Machining and marking of forged nodes is performed automatically on multi-tasking machine tools, which incorporate the latest technology and operate as an integral part of the CAD-CAM system. In some cases, the nodes are machined from hot rolled material using the above mentioned multi-tasking machine tools. Each node is assigned specific part number.

Note 2. According to *ASM Handbook*:
normalizing •Heating a ferrous alloy to a suitable temperature above the transformation range and then cooling in air to a temperature substantially below the transformation range.
isotropy •The same values of mechanical properties in all directions.

Structural members
Circular hollow sections are cut to the exact length according to shop drawings. The cone ends are hot-forged from hot rolled material, normalized and machined. In some cases, the cone ends are machined from hot rolled material on CNC machine tools. Welding of circular hollow sections with cones is undertaken using the MIG-metal inert gas fusion welding process. The electrode feed is the weld metal, which is continuously supplied. Welding using the MIG-process is performed by means of any of the following methods:
  a) automatic welding on CNC machines;
  b) semi-automatic welding on adjustable speed rotating tables with the direct action of the welder.

Both methods give a size tolerance of ±0.5 mm for structural members of normal size. After welding the welds are ground in automatic grinding machines. This results in invisible welds after painting and enhances the visual appearance of the space frame.

Welding of circular hollow sections with cones conforms with Eurocode 3: Part1.1 (ENV1993-1-1) and all ENs, prENs or ISO Standards (if available) listed in Part1.1 (see Annex B to ENV 1993-1-1, Reference standards).

Each structural member, with the corresponding connection mechanisms attached, is marked with its assigned part number.

Connection mechanisms
Bolts
Bolts are manufactured according to European standards by experienced manufacturers with specialized know how and quality assurance systems certified according to ISO 9001.

Pins
Pins are machined from cold work tool steel in CNC machine tools equipped with cut-feeder system.

Spanner sleeves
Spanner sleeves are manufactured using any of the following processes:
  a) multi stage hot forging (near-net-shape) on machines of the latest generation, normalizing and minor machining;
  b) casting using the shell (Croning's) process, normalizing and machining;
  c) machining from hot rolled material to finished part on multi-tasking machine tools equipped with cut-feeder system.

DESIGN AND ANALYSIS
System components
General
Design, analysis and optimisation of SYSTEM 2000 components is carried out using ANSYS, a finite element analysis program. In order to calibrate the methods of analysis and better assess the load bearing capacity of structural components, design and analysis is assisted by testing. Design of SYSTEM 2000 components conforms with Eurocode 3: Part 1.1 (ENV1993-1-1) and all ENs, prENs or ISO Standards (if available) listed in Part1.1 (see Annex B to ENV 1993-1-1, Reference standards).

Length of engagement of bolt threads
The length of engagement of mating threads, that is the screwed length of the bolt into the node, is calculated by the following formula:

$$L = L_e + u + c + t \tag{1}$$

where L = length of engagement;
  L_e = the required length of engagement to prevent stripping of either the external or internal thread according to *Formulas for Stress Areas and Length of Engagement of Screw Threads* given in *Machinery's Handbook, 24th Edition, Industrial Press Inc., New York*;

u = length of incomplete external thread;
c = chamfer of the threaded hole;
t = bolt nominal length tolerance (absolute value).

Notes:
3   Incomplete thread u is given by bolt manufacturer. In any case, u ≤ 2 P, where P = pitch of thread.
4   Chamfer c ≤ P/2, where P = pitch of thread, according to shop drawings.
5   Nominal length tolerance t according to bolt manufacturer's drawings.
6   The length of engagement of mating threads L determined using the above formula is sufficient
to carry the full load necessary to break the bolt without the threads stripping.

Table 3 gives the design tension resistance $F_{t.Rd}$ of bolts according to Eurocode 3 and the
length of engagement L of bolt threads obtained by the use of Formula 1.

Table 3. Design tension resistance of bolts and length of engagement of bolt threads.

d [mm]	P [mm]	d_{pin} [mm]	A [mm²]	A_S [mm²]	A_{d3} [mm²]	A_{pin} [mm²]	$F_{t.Rd}$ [kN]	L_e [mm]	L [mm]
12	1.75	3	113.1	84.3	76.2	74.7	53.8	10.4	15.3
16	2	4	201.1	156.7	144.1	133.9	96.4	14.2	19.7
20	2.5	5	314.2	244.8	225.2	210.2	151.3	17.4	24.1
24	3	6	452.4	352.5	324.3	303.6	218.6	20.7	28.7
27	3	6	572.6	459.4	427.1	405.2	291.7	23.9	31.9
30	3.5	6	706.9	560.6	519.0	520.9	373.7	26.0	35.2
36	4	8	1017.9	816.7	759.3	722.7	520.3	31.3	41.8
42	4.5	8	1385.4	1120.9	1045.1	1041.0	749.6	36.4	48.2
48	5	8	1809.6	1473.1	1376.6	1416.0	991.1	41.6	54.6
56	5.5	10	2463.0	2030.0	1905.2	1891.8	1362.1	48.8	63.1
64	6	10	3217.0	2676.0	2519.5	2564.2	1814.1	56.1	71.6
72	6	10	4071.5	3459.7	3281.5	3337.1	2362.7	64.4	79.9

Symbols and their explanation: d = Nominal size of thread, P = Pitch of thread, d_{pin} = Pin diameter,
A = Nominal bolt area, A_S = Bolt stress area, A_{d3} = Area at nominal minor diameter, A_{pin} = Bolt net
area at pin hole, $F_{t.Rd}$ = Design tension resistance of the bolt to EC3 (ENV 1993-1-1), L_e = Required
length of engagement to Machinery's Handbook 24th Ed., $L = L_e + u + c + t$, where u = 2P, c = P/2
and t = ±0.5 mm.
Threads: Metric M profile threads (ISO 68) of tolerance class 6H/6g (ISO 965/1).
Materials: Bolts are 42CrMo4 (Material number 1.7225) to EN 10083-1 having strength grade 10.9,
nodes are C45E/Ck45 (Material number 1.1191) to EN 10083-1. The nominal value of ultimate
tensile strength f_{ub} of bolts (which is adopted as characteristic value in calculations) is 1000 MPa.
The nominal value of ultimate tensile strength f_u of nodes is 620 MPa.

Modular space frame

Design, analysis and optimization of modular space frame conforms with Eurocode 3: Part1.1
(ENV 1993-1-1) and its supporting standards, in conjunction with the National Application
Document (NAD) valid in the country where the structure is built.

Design, analysis and optimization of modular space frame is performed using a "FIRST
ORDER ELASTIC GLOBAL ANALYSIS AND DESIGN OF PIN-JOINTED SPACE FRA-
MES" program, which is interfaced with the CAD-CAM system.

If more refined analysis is justified, then the design that results from the above mentioned method is further analysed using ANSYS.

SURFACE PROTECTION OF STEEL

There are many corrosion protection systems available for the surface protection of Skytech SYSTEM 2000 components and it is up to the client's engineer to indicate the process to be applied in each case. In general, SKYTECH modular space frames are galvanized and additionally protected with a powder or wet coating. Circular hollow sections are hot-dip galvanized according to DIN 50976 with a zinc coating of 75 µm. Nodes, sleeves, bolts and pins are galvanized in a galvanic bath according to DIN 50961 with a 25 µm zinc coating. After electrolytic galvanization, bolts are baked at a temperature of 180-220 °C for 10-12 hours in order to prevent hydrogen-induced brittle fracture. Circular hollow sections receive an additional electrostatically applied polyester powder coating of 75 µm followed by sintering at a temperature of 200 °C. Nodes and sleeves receive an additional wet paint coating of total thickness of 60 µm applied in three layers; a primer suitable for galvanized surfaces and two layers of two component acrylic resin.

QUALITY ASSURANCE

SKYTECH modular space frames are fabricated by an industrial projects company with quality assurance systems certified according to ISO 9002.

Acknowledgements: The author would like to thank Dr A G PRAKOURAS for his review of the manuscript.

Behaviour of tubular space truss connections with stamped end bars

A. S. C. SOUZA
The School of Engineering, University of Lins, Brazil
R. M. GONÇALVES, and M. MALITE
Department of Structural Engineering, University of São Paulo at São Carlos, São Carlos-São Paulo, Brazil

ABSTRACT

This paper presents a theoretical and experimental analysis of the behaviour of connections of tubular bar space trusses. Results of four space trusses prototype with plan dimensions of 7.5m x 15.0m using three types of joint system most commonly used in Brazil are presented. *System 1* comprises a superposition of bars with stamped ends fastened together by a single bolt. In *system 2*, the stamped ends of the bars are connected to nodes made of welded steel plates while in *system 3*, the connection between bars and node is through a transition plate welded to the ends of the tube.

The main objective of the present analysis are: to determine the failure modes of these connections, establish limits for their use and propose modifications in geometry to improve their performance. Numerical analyses using the ANSYS finite element code allowed the formulation of representative mechanical models of the physical behaviour of the connection and whole structure, making it possible to extrapolate the analyses beyond the models tested at laboratory. The aforementioned types of space trusses have been widely used in Brazil. However, structural damages have been found to occur due to the inadequate use of these joint systems mainly in large span structures. Test results show that the failure of the connection is the main factor responsible for observed structural failures at load levels below the load carrying capacity of the bars.

INTRODUCTION

In recent years, there has been a wide increase in the use of space trusses in Brazil. This can be attributed mainly to the increase publishing of research studies together with the development of computational aids, which have eased and made analyses more precise. Space trusses have a high stiffness with a reduced self-weight, which permits a complete prefabrication of these structures, hence easing their production and erection. The structural and aesthetic characteristics of space trusses could permit architects to design for large areas, thus attending to demands of space while the structure constituting the architectural solution.

The joint system between bars, or simply the node, has been the main difficulty encountered in the development of space trusses. This is due mainly to factors inherent in the constructive and structural aspects in joint system as well as related economic factors, the latter representing more than 25 percent of the total cost of the finished structure (Ref. 1).

A variety of connection systems for space trusses have been developed in several countries (Ref.2) in an attempt to combine structural efficiency and easy fabrication and assembly with reduced costs. Many of these systems have resulted in international patents, such as the German MERO system.

As patented joint systems usually involve high costs. Simpler and less costly alternative solutions (Fig. 1), also called low technology joints, have been used and investigated (Refs. 3 to 6). Similar Brazilian joints, have been widely used and studied.

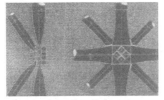  

Waco – De MARTINO (1992) Octatube – GERRITS(1994)   COOD(1984)
Fig. 1 – Joint systems used in space trusses.

The use of patented joints such as the MERO system in Brazil is very restricted due to the relatively high cost compared to more simplified systems. The joint assemblies are comprised of a superposition of flattened end of circular cross section bars fastened together by a single bolt – *system 1* (Fig. 2a). Alternatively, the joint assembly may be composed of welded plates to which the bars are connected by bolts - *system 2* (Fig. 2b), in which case the bars could have their ends flattened (Fig. 2b) or the end plates could be welded in the bars ends - *system 3* (Fig. 2c).

a) *system 1*   b) *system 2*   c) *system 3*
Fig. 2 – Joint systems most common in Brazil.

The joint system 1 has a low cost of fabrication and high erectibility since it requires no special devices for its construction. Not withstanding, eccentricities and the stiffness variations of the bar ends result in stress concentrations in the neighbourhood of these end regions with the consequent slipping of the bars. This failure mode represents the ultimate limit-state of the whole structure.

In joint systems 2 and 3, problems of eccentricities and slipping of the bars are eliminated. In fact, this results in a joint with satisfactory stiffness and a good structural behaviour. It is however necessary, in this case, to consider the bar ends cross section variation responsible for the reduction of the normal compressive strength of the bars when calculated without taking the inertia variation into account (a common design practice in Brazil). Besides, fabrication imperfections of the joint may lead to significant reduction in the load carrying capacity of the structure.

Brazilian researches on the behaviour of space trusses are limited and still recent (Refs. 7 to 13), having been stimulated only by the necessity to investigate the partial and total structural failures in accidents that have occurred during the last years (Refs. 14 to 16). These studies started with compression tests on isolated flattened end bars and bar end plates (Refs. 7 to 9). Later research was carried out on space truss (Ref. 14) and then on prototype with spans of 7.5m x 7.5m (Refs. 11 and 12). The main objectives of these research studies were: to characterise the behaviour of space truss joint systems commonly used in Brazil, evaluate and propose changes in the analytic models adopted as well as in the detailing of such joints.

TEST PROGRAM
Description of the tested structures
Three types of space trusses with the square on offset square configuration and pyramidal modules of length 2.5 x 2.5m, height of 1.5m and plan dimensions of 7.5m x 15m and supported on four vertices by tubular columns fixed to the base were tested.

Results of four of the nine prototype space trusses differentiated from one another by the joint type and the cross section of the support diagonal element are presented. Circular cross section tubes of diameters ϕ 76 x 2.0 and ϕ 60 x 2.0 were used for the chord and diagonal elements respectively as shown in Table 1.

Table 1– Summary of tested prototype space trusses.

MODEL	JOINT	SUPPORT DIAGONAL	Observation
Truss 1	*System 1*	ϕ 60x2,0	
Truss 2	*System 1*	ϕ 88x2,65	
Truss 3	*System 1*	ϕ 88x2,65	*In system 2,* only the top joints of the support diagonals
Truss4	*System 3 – extreme with the end plate*	ϕ 88x2,65	

The dimensions adopted for the tested models were motivated by the necessity to reproduce the behaviour of structures normally used in Brazil, associated to the necessity to verify the influence of bending on the joint behaviour.

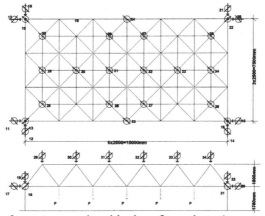

Fig. 3. – Dimensions of prototypes and positioning of transducer (measurements in mm).

Test equipments

The load was applied on ten nodes of the lower tie bars by means of a 300kN nominal capacity Enerpac hydraulic jack. Displacements transducers with a sensitivity of 0.05mm and gage range of 50mm and 100mm were used to measure the displacements. The bar strains were measured at the mid and extreme sections by means of Kyowa strain gages with a base reading of 2mm. Figs. 3 and 4 show the instrumentation set-up.

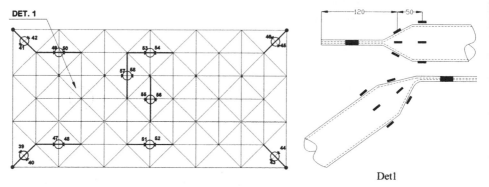

Fig. 4. – Instrumentation set-up for reading of deformations.

The photos of figure 5 show the general aspects of the structures and detail instrumentation set-up.

Fig. 5. – General view of test set-up and details of the instrumentation of a node region.

MATERIALS

The steel tubes used in the present study were cold formed, circular cross section, ASTM A570 profile with a nominal yield strength of 290 MPa and Young's modulus of 205000MPa. The joints were made of ASTM A36 steel. The bolts were ASTM A325 with diameters of ϕ16mm and ϕ19mm.

THEORETICAL ANALYSIS

Traditionally, designers use either the ideal truss model or the space frame model under linear elastic analyses to estimate the displacements and forces in space trusses. In the present paper this model will be denominated *TM1*. Table 2 shows predicted values of loading for the *TM1* theoretical model compared to experimental values of ultimate force. In this case the predicted failure mode is the buckling of the compression elements.

The authors, using shell and volume finite elements (Refs. 11 to 13) modelled the local behaviour of the studied joint systems. Figure 6 shows a comparison of the final configuration of a joint according to the proposed theoretical model and that obtained experimentally.

Fig. 6. – Tri-dimensional model of *system 1*.

In references 11 to 13, it is shown that the analysis of the global behaviour of trusses with such joint systems as those studied in the present paper should consider some characteristics inherent to the joint system such as eccentricities, cross section variation of the bar ends and non-linearity material. The consideration of these factors would lead to more realistic results, which represent a better approximation to the real structural behaviour. Numeric analyses using a finite element method code, ANSYS, considering these variables was performed. This theoretical model will herein be denominated *TM2*.

To simulate the effect of cross sectional variation in the finite elements, the bars were divided into three segments: segment 1 corresponding to the circular section, segment 3 representing the flattened bar end and segment 2, a linear interpolation of segments 1 and 3 (Fig. 7).

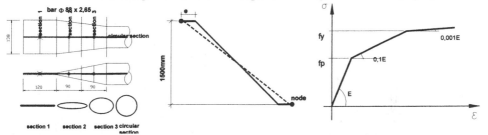

Fig. 7. – Variation of bar cross sections, eccentricity and constitutive model.

The finite element, Beam 24, with six degrees of freedom at each node, which permits the use of cross sections of any format was adopted. Figure 7 shows the constitutive model adopted for this analysis. A residual stress of 70MPa was adopted according to the Brazilian code. Test results were also compared to the ideal truss model under linear elastic conditions.

RESULTS AND DISCUSSION

Table 2 shows the observed test results of the maximum vertical load compared to predicted theoretical values for model *TM 1* (the ideal truss) and *TM 2* (truss considering eccentricity, bar cross section variation and non-linear analysis).

Table 2 – Theoretical and experimental values of ultimate load and displacements.

	$F_{exp.}$ (kN)	$F_{theoretical}$ (kN)		D_{exp} (cm)	$D_{theoretical}$(cm)		$F_{exp}/F_{theoretical}$		$D_{exp}/D_{theoretical}$	
		TM1	TM2		TM1	TM2	TM1	TM2	TM1	TM2
Truss1	93	124,0	76,7	4,68	1,84	3,2	0.75	1,21	2,54	1,46
Truss2	71	151,4	123,8	5,20	1,40	3,46	0.47	0.57	3,7	1,50
Truss3	106	151,4	175	5,01	2,1	4,3	0.70	0.61	2,38	1,16
Truss4	142,0	151,4	-	3,65	2,81	-	0.94	-	1,30	-

TM1 – ideal linear theoretical model
TM2 – theoretical model considering eccentricity and bar cross section variation in non-linear analysis
D – maximum vertical displacements   F – total applied load.

The main difference between *truss 1* and *truss 2* models stems from the fact that in the latter the diameter of the support diagonal element is larger. However, contrary to what is expected, its strength was lower than that of the former. This is because larger bar diameters indicate larger lengths of the flattened bar end, lower joint stiffness and consequently larger strains in the joint region. Results show that the average strains of the flattened sections of the bars are of the order of ten times those observed at the mid-section.

In model *truss 3*, the presence of joint *system 2* at the four vertices nodes was sufficient to increase the load carrying capacity of the structure. In this case failure occurred in the plastic range by yielding of the end plate. This was due essentially to imperfections inherent in the fabrication process and erection of the prototype.

Better results were however obtained for *system 3*. The difference between the theoretically predicted and experimentally observed ultimate force in this case was of the order of 6%. Failure of the structure was characterise by buckling of top chord, which is in conformity with results obtained from the theoretical analysis of the ideal truss model under linear elastic conditions.

In all tested structures except *truss 4*, failure occurred due to failure in the joint system. *System 1* shows a stiffness reduction in node and is thus sensitive to the effects of eccentricities, which are the main factors responsible for the excessive strains at the flattened bar. These strains cause the rotation of the node, slipping of bars, increase of vertical displacements and consequently the failure of the structure at low levels of loading. The photographs of figure 8 show the failure modes observed.

Truss 1 Failure of corner node (support diagonal).

Truss 2 Failure of corner node (support diagonal)

Truss 3 collapse of joint system 2 node (support diagonal).

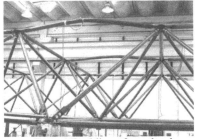
Truss 4 Buckling of top chord.

Fig. 8. – Observed failure modes.

Figure 9 presents the average vertical displacements along the larger span for truss 1. It can be observed that the structure maintains its symmetry even right up to the last stages of loading. The deflection/span ratio was observed to be of the order of 1/300 for truss 1, truss 2 and truss

3 and 1/400 for truss 4. It is worth remembering here that the Brazilian codes limit this ratio to 1/360.

The Figure 10 shows results of the experimentally obtained maximum vertical displacements for all tested trusses compared to predicted theoretical values for *TM1* (the ideal truss). It can be observed that the trusses with the joint *system 1* show a pronounced non-linear behaviour, the results being significantly different from those obtained in Table 2 for the ideal truss model.

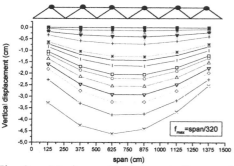 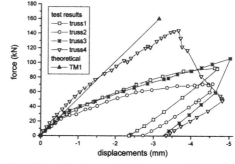

Fig. 9. – Maximum displacements for tested trusses.

Fig. 10. – Maximum displacements for tested trusses.

The proposed theoretical analysis (*TM2*) for trusses with joint *system 1* showed good correlation with experimental results (Fig. 10). However, the performance index of the bars varied between 45 % to 89%, reducing with the increase in bar diameter. A close analysis of the displacements of the structure under service loading with *TM2* considering a linear elastic analysis (Fig. 11) shows a difference of 14% between theoretical and experimental results for truss 1 and 30% for truss 3.

Figure 12 shows a comparison between experimental and theoretical results obtained for *TM2* considering material non-linearity and adopting a residual stress of 70MPa at the flattened ends of the bars for truss 1. From the figure, it can be observed that the theoretical and test results are in good agreement despite the theoretically predicted ultimate load being lower than that experimentally observed. For all the tested structures, bar strains also showed very good agreement between experimental and theoretical values.

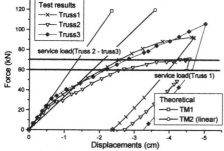 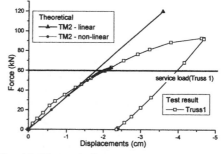

Fig. 11. Theoretical and experimental displacements.

Fig. 12. Theoretical and experimental displacements.

CONCLUSIONS
It was observed experimentally the joint *system 1* shows a non-linear load-displacement behaviour resulting in large stresses and strains in the structure. The collapse is characterised by the joint failure. Slipping of the bars around the node region was also observed together with high stress concentration causing a sudden yielding of the flattened region of the bar.

The proposed numeric model for trusses with joint *system 1*, which permits the consideration of eccentricities at the node *(TM2)*, variation of bar cross section and non-linear effects was found to be satisfactory in predicting the behaviour of the structure.

It was also observed that in the truss with joint *system 2*, the failure of the joint led to the collapse of the structure for load levels below predicted values. Moreover, for this system, failure due to buckling of the compressed bars as predicted. However, imperfections in the fabrication of the joint could be the factor responsible for the modified failure mode and reduced strength capacity of the structure. Additional tests are currently being carried out to evaluate the effects of node imperfections on the behaviour and collapse mode of these structures.

The behaviour of trusses with *system 3* was in agreement with predictions. This means that the load-displacement behaviour can be satisfactorily obtained from a linear elastic analysis of the ideal truss model or space frame.

Further studies are currently being carried out in the School of Engineering at São Carlos, which will permit drawing definite conclusions and developing more realistic theoretical models capable of predicting the behaviour of these joint systems.

ACKNOWLEDGEMENT:
The authors gratefully acknowledgement to FAPESP (São Paulo State Research Foundation) for the financial support necessary in carrying out this research.

REFERENCES
1 Iffiland, J., Preliminary planning of steel roof space trusses. *Journal of the Structural Division*, v.108, n.11, p.2578-2589, Nov, 1982.
2 Makowski, Z.S., Review of development of various types of double-layer grids, *In: MAKOWSKI, Z.S., ed. Analysis, design and construction of Double-layer grids*, Applied Science., p.1-55,1981.
3 De Martino, A., Relazione generale: progettaziopne, lavorazione e montaggio. **Costruzioni Metalliche**, n.1, p.14-54, 1992
4 Gerrits, J.M. (1984). **Space structures in the Netherlands since 1975.** *In: INTERNATIONAL CONFERENCE ON SPACE STRUCTURES*, 3., Guildford, UK, Sept. 1984, **Proceedings.** London/New York, Elsevier Applied Science. p.28-32, 1984.
5 Cood, E.T., Low technology space frames. *In: INTERNATIONAL CONFERENCE ON SPACE STRUCTURES, 3.*, Guildford, UK, Sept. 1984, **Proceedings.** London/New York, Elsevier Applied Science. p.955-960,1984.
6 El-sheikh, A.I., Experimental study of behaviour of new space truss system. *Journal of Structural Engineering*, v.122, n.8, p.845-853, Aug, 1996.
7 Gonçalves, R.M. ; Fakury, R.H. ; Magalhães, J.R.M., Performance of tubular steel sections subjected to compression: theoretical and experimental analysis. *In: INTERNATIONAL COLOQUIUM ON STRUCTURAL STABILITY, 5,. Stability problems in designing, construction and rehabilitation of metal structures: Proceedings.* p.439-449, Rio de Janeiro, August 5-7, 1996.

8. Malite, M; Sáles, J.J.; Gonçalves, R.M. and Takeya, T. *Experimental analysis of the steel compression tubular members with stamped ends* (in Portuguese), Technical Report, University of São Paulo at São Carlos campus – Brazil, 1996.

9. Malite, M.; Gonçalves, R.M. and Sáles, J.J. Tubular section bars with flattened (stamped) ends subjected to compression – a theoretical and experimental analysis. *Proceedings of the SSRC Annual Technical Session and Meeting*, Toronto – Canada, 1997.

10. Malite, M. et al. Space Structures in Brazil. *Proceedings of the 2th World Conference on Steel in Construction*, San Sebastian - Spain. Oxford, Elsevier Science, 1998.

11. Souza, A.C. *Contribution to the study of steel space structures* (in Portuguese). MsC Thesis, University of São Paulo at São Carlos campus – Brazil, 1998.

12. Maiola, C.H. *Theoretical and experimental analysis of steel space structures composed of members with stamping ends* (in Portuguese). MSc Thesis, University of São Paulo at São Carlos campus – Brazil, 1999.

13. Vendrame, A.M. *Contribution to the study of braced domes with steel tubular members.* (in Portuguese). MSc Thesis, University of São Paulo at São Carlos campus – Brazil, 1999.

14 Fakury, R.H. ; et al., Investigation of the causes of the collapse of a large span structure. *In: FOUTH INTERNATIONAL CONFERENCE ON STEEL AND ALUMINIUM STRUCTURES*, , **Proceedings.** Elsevier Applied Science. p.617-624, Finland, jun, 1999.

15 Batista, R.C.; Batista, E.M., Experimental determination of the mechanisms of collapse of a typical joint of metallic space structures. In: JOURNEYS SUDAMERICANAS OF ESTRUCTURAL INGENIERIA, 28., São Carlos, Brazil, 01-05 September 1997. **Proceedings** v.3, p.665-674.

16 Batista, R.C.; PFEIL, M.S.; Carvalho, E.M.L., Qualification through reinforcement of the metallic structure of a great spherical dome. In: JOURNEYS SUDAMERICANAS OF ESTRUCTURAL INGENIERIA, 28., São Carlos, Brazil, 01-05 September 1997. **Proceedings.** v.3, p.1127-1137.

Recent Developments in CUBIC Structures

L A KUBIK, Kubik Enterprises Ltd, UK

SYNOPSIS

This paper outlines recent developments in CUBIC Structures, with particular emphasis on the new CUBIGRID Space Frame. This is an exciting development of the original CUBIC Space Frame double layer grid concept. It uses spaced apart bracing members to extend the range of applications, whilst still offering the full depth service openings for which CUBIC Structures are already known. The versatility and economy of the resulting structure are outlined, together with a number of factors that can influence design and cost.

INTRODUCTION

The first two members of the CUBIC Structures family, the CUBIC Space Frame and CUBIC Composite Floor, have been described in previous papers (Refs 1-4). They offer a number of key benefits that are able to add value to buildings during design, construction and, equally importantly, use. For example, they offer specifiers the opportunity to create economical but highly versatile large span buildings, with considerable freedom to develop and adjust the architectural and service details independently of the main structural frame. They also offer developers and clients the opportunity to have buildings with few internal columns, in which alterations to services and internal layout can be easily accommodated with minimal constraints imposed by the building structure.

Since the first application of the CUBIC Space Frame, development work has continued to improve the products and ways in which they can be applied, addressing the needs of ever more demanding clients both inside and outside the construction industry. This has now led to the introduction of two exciting new members of the CUBIC Structures family that will be described below, with particular emphasis on the CUBIGRID Space Frame.

THE CUBIC SPACE FRAME AND CUBIC COMPOSITE FLOOR

The CUBIC Space Frame is a modular roof or floor frame without diagonal bracing members (Refs 1, 2). This means that the full structure depth between top and bottom chords is unobstructed (Fig 1), providing excellent access for service installation and maintenance, or for accommodation. Using three basic prefabricated modules, it is possible to assemble large structures using only nuts and bolts.

The CUBIC Space Frame has already been applied in a wide variety of buildings including the large-span shallow-depth roof over the diamond shaped Stansted airport hanger; flexible building forms with few internal columns for fast-build superstores and factories; versatile food factories with plant located and maintained within the roof void; and theatres and leisure centres capable of accommodating large loads suspended at various points over the roof span (Fig 2).

Figure 1. Full depth void available with CUBIC Space Frame

Figure 2. Versatile CUBIC Space Frame roof

The CUBIC Composite Floor is a development of the CUBIC Space Frame intended primarily for larger span floor construction (Ref 3). The system consists of a CUBIC Space Frame in which the top chords are embedded in the structural concrete topping slab (Fig 3). This leads to improved economy due to composite action without the need for separate shear connectors.

Figure 3. CUBIC Composite Floor

CONTINUING DEVELOPMENT

Ever since the first CUBIC Space Frame was first constructed over 20 years ago, opportunities for developing and improving the offerings have been continually explored, whilst retaining the emphasis on providing versatile structures with large full-depth service openings. Initially this led to the development of the CUBIC Composite Floor in recognition of the considerable benefit to be gained by using the concrete to resist compressive loads.

Now continued, and indeed increasing, demand from clients and specifiers for more versatile buildings capable of future adaptation, as well as the ever-present requirement for better value for money, has resulted in some exciting new opportunities associated with the introduction of two new products – the CUBIGIRDER and the CUBIGRID Space Frame (Figs 4 & 5). These new products are suitable for both roof and floor structures. They retain the full depth service openings of the CUBIC Space Frame but considerably extend the range of applications for which CUBIC Structures can now be cost-effectively adopted.

Figure 4. Typical CUBIGIRDER with full depth service void

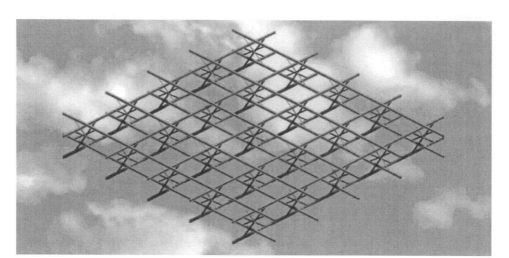

Figure 5. The CUBIGRID Space Frame

Since the CUBIGIRDER is a planar girder rather than a space frame it will not be considered further in this paper. However it can be seen from Figs 4 and 5 that the CUBIGRID Space Frame can crudely be regarded as an assembly of CUBIGIRDER frames.

THE CUBIGRID SPACE FRAME

The CUBIGRID Space Frame is a development of the original CUBIC Space Frame double layer grid concept. It combines good access through the structure for service installation with excellent economy due to design efficiency and simplified construction details, suited to modern manufacturing methods in the steel construction industry.

It utilises pairs of intersecting bracing members that are bolted between upper and lower grids of chords in place of the posts in the CUBIC Space Frame (Fig 5). Each pair of bracing members is spaced apart from the adjacent bracing members, thereby creating a full depth void that can be used for service runs and maintenance access. The grid size and depth can be varied to suit project requirements. Also, the section types used for members can be chosen to suit applied forces and any special aesthetic requirements.

The inclination of the bracing members and the location of the point of intersection between bracing members can be adjusted to suit available spare capacities within the chosen upper and lower grid of chords, as well as the required clear space for service installation or access. Fig 6 shows an extreme case with the point of intersection coincident with the lower chord.

Figure 6. CUBIGRID with bracing members intersecting at lower chord level

The grid size, structure depth and member strengths can be chosen to suit project requirements. Since the structure can be delivered to site in pieces, even depths greater than the 4m previously used for the CUBIC Space Frame do not present a problem.

An enlarged detail of the structure shown in Fig 4, with the preferred configuration using structural hollow section bracing and continuous I-section chords, can be seen in Fig 7. Note the use of continuous chords. Alternative sections can be used for bracing members and chords if required, including circular hollow section or angle sections for the bracing and circular or structural hollow sections for the chords, although the latter will complicate connection details.

Figure 7. Enlarged detail of CUBIGRID showing chord and bracing intersections

Whilst the previous comments have been related to steel construction, the construction method of the CUBIGRID Space Frame with bolted connections between chords and bracing members means that it is also eminently suited to construction in other materials. This can encompass, for example, construction in cold-formed steel and aluminium, as well as combinations of materials such as timber or concrete members for chords with steel bracings.

The basic configuration enables square, rectangular and skew grids to be easily catered for, as can cambered, sloping and tapered structures. Small cambers can be introduced simply during manufacture by simply adjusting the position of the connections between chords and bracing members, although larger cambers may require special curving of the chords.

THE BENEFITS OF CONTINUOUS CHORDS
The use of continuous chords reduces the number of connections that have to be made either during manufacture or on site, the benefits of which have previously been recognised (Ref 5). There are also further benefits in making the connections between bracing members and chords rather than by splicing the chords – the bracing to chord connections typically carry

REFERENCES

1. KUBIK M L & KUBIK L A, An Introduction to the CUBIC Space Frame, Int. J. Space Structs, 1991, Vol. 6, No. 1, pp 41-46.

2. KUBIK L A, The first 10 years of the CUBIC Space Frame, Space Structures 4, Thomas Telford, London, 1993, pp 1383-1391.

3. THOMPSON K W & KUBIK L A, Development and testing of the CUBIC Composite Floor, Space Structures 4, Thomas Telford, London, pp 1510-1517.

4. SHANMUGANATHAN S & KUBIK L A, Optimization of CUBIC Space Frame Structures, Space Structures 4, Thomas Telford, London, pp 1766-1773.

5. CODD E T, Low Technology Space Frames, Proceedings of the Third International Conference on Space Structures, Elsevier, 1984, pp 955-960.

6. BRITISH STANDARDS INSTITUTION, BS 5950-1:2000: Structural use of steelwork in building - Part 1: Code of practice for design – Rolled and welded sections, BSI, 2000.

7. DOOLEY, J F, The Torsional Deformation of Columns of Monosymmetric I-section, with restrained axis of twist, under doubly eccentric load, Int. J. Mech. Sci., Vol. 9, 1967, pp 585-604.

8. HORNE M R & AJMANI J L, "Stability of Columns supported laterally by side-rails, Int. J. Mech. Sci., Vol. 11, 1969, pp 159-174.

A novel modular constructional element

M. SAIDANI, M. W. L. ROBERTS, and R. HULSE
Civil Engineering Group, School of Science and the Environment, Coventry University,
Priory Street, Coventry CV1 5FB, England, U.K.

ABSTRACT
The paper reviews a novel modular constructional element for the building and construction industries. The modular construction comprises hexagonal elements assembled together to form a constructional assembly of honeycomb form secured by fixings which clamp the sides of adjacent elements. Concrete, or other appropriate material, is disposed within the elements providing a composite type of action. The novel type of modular construction has potential applications in the building, construction, and highways sectors. It could be used as a permanent or temporary structure for roofs and walls, it may also be used for concrete pavements, tunneling, and retaining wall structures.

KEYWORDS
Modular construction, hexagonal elements, assembly, construction.

INTRODUCTION
With increased pressure on the construction industry to reduce costs of building and managing structures, extensive research into this area has been undertaken in the past. One such kind of project was concerned primarily with ways of optimising construction procedures through standardisation methods aimed at making more efficient and effective use of resources from the feasibility study to the execution stage Refs 1-3. Another approach was based on attempts to produce what has become more commonly known as modular construction, through the development of fast and cheap ways of building structures Refs 4-5. A cost-effective structure (or building) means speed of executing of all phases at the lowest possible price without compromising the safety of the structure to be built. The use of modular systems in large projects and, when produced in large quantities, was often found to provide the answer Ref 4. Obviously, the efficiency of any of these methods relies on an optimal use of resources taking into account any constraints associated with the execution of a particular construction project. The key to such methods is the use of what is more commonly known in the construction sector as "resource sharing" Ref 6, Ref 7, and Ref 8.

The current project is concerned with the study of the behaviour of a new modular constructional element. Individual hexagonal elements or "cups" are assembled together to form different structural components such as, slabs, walls, columns or even pavements. As is explained in the following sections, the element has some advantages over traditional type of construction in terms of speed and cost reduction.

THE INDIVIDUAL MODULE AND ELEMENT ASSEMBLY

The constructional individual element comprises a hexagonal laminar base, each side of the hexagonal base being provided with integral upstanding walls with gaps in between (see Figure 1) Ref 9.

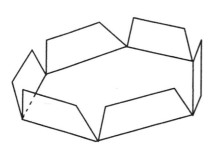

Fig. 1. Individual hexagonal module (or "cup").

The presence of the gaps will allow fluid concrete (or any other liquid material) to flow from one element to another when individual modules ("cups") are assembled together to form a structure. It will also allow reinforcing bars to be laid across in case the concrete is to be reinforced to provide extra strength (Figure 2).

The assembly of individual elements may be planar (slabs, walls, etc.) or curved to form a domed structures, in this case tie-bars or setters may be used to allow for the slight curvature. Individual elements may be assembled together using rivets or screws/bolts or threaded pipe and nuts. The "cups" may also be provided with additional holes in the bases for passage of liquid material. Thus if two or more such elements are stacked vertically a column or a wall structure could be formed. The assembling the elements vertically or horizontally may be used to produce a wall.

The assemblage of elements using ties could produce curved surfaces that may be used as formwork to curved structures (such as domes or tunnels) or be used to actually build such structures (figure 3).

Fig. 2. Assemblage of hexagonal "cup" elements to form planar structural elements.

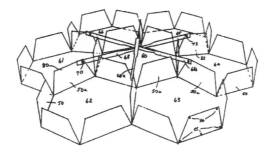

Fig. 3. Assemblage of hexagonal "cup" elements linked with ties to form dome-like structural elements.

POTENTIAL
The patented system (Ref 9) presented in this paper (see Figure 1) may potentially be used in a number of civil engineering applications using variants of the same component geometry, such as:

- Floors (elements filled with reinforced concrete);
- Roofs;
- Temporary support structures for roofs and floors;
- Walls;
- Concrete pavement for aircraft runways;
- Tunnels.

The new modular system provides a cheap and efficient way of erecting structural elements but also to maintain them in the longer term. The advantage of the system, in addition to its speed of erection and low cost, is that, in a structural situation, only individual modules ("cups") need to be repaired in case of local deterioration if these are filled with concrete for example. Cracks propagation is limited to individual modules and repair is possible. This is particularly important in concrete highways where the cost of repair is very high.

ONGOING RESEARCH ON THE SYSTEM
A number of static tests are being undertaken at the Civil Engineering laboratories of Coventry University. Figure 4 shows a typical slab measuring 1.75mx2.17mx76mm that was tested to failure. The hexagonal elements, fabricated from 1.5mm thick steel sheets, were assembled together using rivets and filled with reinforced concrete (diameter of reinforcement bars = 6mm). The slab is simply supported along its shorter edges and was loaded centrally up to failure.

Preliminary results on one such slab show that the system performs well compared with traditional slabs. The slab failed at 21kN with a maximum deflection of 14 cm. In addition the failure of the specimen was ductile and not catastrophic. The main cracking in the concrete had followed the lines of the joints between the modules. It was also apparent that some of the rivets joining the modules together had failed as a result of the increased load and deformations.

Areas of improvement are being investigated as the testing programme progresses with the aim of achieving higher capacity for the new system. This includes providing better jointing system between the individual modules and reducing the weight of the assembled structure so that deflections could be reduced. Other improvements regarding the system geometry and reinforcement size and mesh are under review.

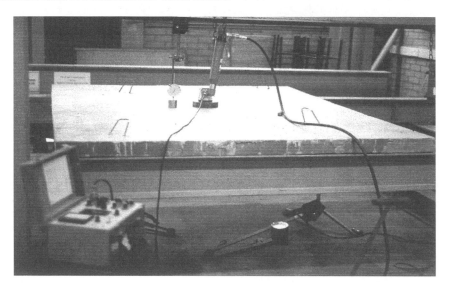

Fig. 4. Slab made of hexagonal elements being tested.

The research programme is still in its early stages and there are already good indications that the system has potential to compete with traditional methods of construction.

The planned programme of testing is summarized in Table 1.

Table 1. Planned testing programme.

Test Number	Test Designation	Comment
HE1	Hexagonal assemblage without concrete	Single point load and
HE2	Hexagonal assemblage with lightweight aggregate concrete	uniformly distributed load. Assemblage is made into a
HE3	Hexagonal assemblage with ordinary concrete	rectangular slab (as in Figure 4)

As may be seen from Table 1, the experimental programme will include testing slabs made using the hexagonal "cups". Two kind of concrete will be considered: lightweight aggregate concrete and ordinary concrete. The aim of combining these tests is to produce a cheap and strong lightweight structure. Different reinforcement meshes will also be investigated for the best design. In the longer term, another series of tests envisaged will be to test the element resistance to lateral loading (for example vertical walls subjected to wind loading). Comparison will be made with traditional types of elements in terms of behaviour at service loads and ultimate strength. The economy of the system will be investigated. It is also envisaged to conduct dynamic testing.

CONCLUSIONS AND THE FUTURE

The system has great potential in construction with possible applications for a variety of structures, slabs, roofs, floors, walls, tunnels, pavements, etc. It may also be used as a permanent or temporary formwork. It is anticipated that the system will represent a speedy and low cost construction process in addition to its cheap maintenance costs. Potentially the system could compete with traditional forms of construction, especially in large projects. The authors are already engaged in discussions with parties from the construction industry interested in the system.

Current and future work at Coventry University will be focussing on the structural performance of the system through an elaborate experimental and numerical modelling research programme. In the short term, this will consist of static loading for different structural components with varying parameters (type and thickness of material used for the "cups", size of reinforcing bars and mesh, type of concrete used (or any other material used for filling), etc.). In the longer term in dynamic testing to study the response of the system to vibrations and earthquake loading will be carried out.

ACKNOWLEDGEMENTS

This project is funded by the EPSRC Industrial CASE Award Scheme. The Authors are very grateful to EPSRC for their financial support, and to Specialist Welding Ltd, Mr Frank Cooke, for the generous financial and technical support provided. The tests are conducted in the Structures laboratories at Coventry University, England.

REFERENCES

1. STEEHUIS M, GRESNIGT A M, and WEYNAND K, Strategies for Economic Design of Unbraced Steel Frames, Proceedings of the 2nd World Conference on Steel in Construction, Paper 69 on CD-ROM, 1998.
2. BRISCOE G, The Economics of the Construction Industry, Batsford & CIOB, 1988, p.61.
3. BOTTOM D, GANN D, GROAK S, and MEIKLE J, Innovation in Japanese Prefabricated Housebuilding Industries, London, CIRIA Special Publication 139, 1996.
4. POLLALIS S N, Country Club Village, Hawaii: A Case Study on Daewoo's Multi-room Modular Construction System, Harvard Graduate School of Design, Cambridge, Massachusetts, December 1997.
5. ATKINSON J, PAYLING B, LUCAS B., and GRAY K, Affordable Housing Adaptations Using Modular Construction - A Case Study In Applied Project Management. CIB World Building Congress on "Performance in product and practice", Wellington, New Zealand, Vol.1, 2001, pp.535-545.
6. SOU-SEN L and SHAO-TING H, Optimal Repetitive Scheduling Model with Shareable Resource Constraint, ASCE Journal of Construction Engineering and Management, 127(4), July/August 2001, pp.270-280.
7. ASHLEY D B, Simulation of repetitive-unit construction, ASCE Journal of Construction Engineering and Management, 106(2), 1980, pp.185–194.
8. BARTUSCH M, MOHRING R H, and RADERMACHER F J, Scheduling project networks with resource constraints and time windows. Ann. of Operations Res., Basel, Switzerland, 16, 1988, pp.201–240.
9. ROBERTS M W L, Hexagonal Constructional Element, UK Patent GB 2335211B, The Patent Office, Application No 9825350.3, 1998.

Fatigue Properties of KT-Space Truss System with Threaded Spherical Nodes

S TSUJIOKA, Professor, Fukui University of Technology, Fukui, Japan,
K IMAI, Professor, Osaka University, Osaka, Japan,
K WAKIYAMA, Professor, Osaka Sangyo University, Osaka, Japan,
T FURUKAWA, Associate Professor, Osaka University, Osaka, Japan, and
R KINOSHITA, Manager, Kawatetsu Civil Co. LTD., Kobe, Japan

INTRODUCTION

Generally, in any system truss, the tubular member is connected to the threaded spherical node with the single-bolted joint assembly. The system truss is an economical method for elegant structural form to cover large space. Its main applications are the stadium roof structure, the large span gymnasium and the high-rise building roof etc. In case of the complex curved-roof structures and the open-air structures, it is necessary to consider the fatigue fracture because of the stress fluctuation caused by strong winds. Since the thread root of the joining bolt is considered as a sharp mechanical notch, the catastrophic collapse of the system truss results from the fatigue failure of the bolt. Especially, the size effect of the bolt in the fatigue is of great importance. The high quality control and the careful heat treatments are demanded for the manufacturing of the large diameter bolt applied for the large space structure. Therefore, in order to evaluate the fatigue characteristics of the system truss joint, the clarification of the fatigue behavior of this joining bolt under the repeated loading is indispensable.

This paper reports on the fatigue tests of the full-scale system truss joints (Refs.1-3) in which use M20-M48 bolts produced in the thread machining before heat treatment. Furthermore, to improve the fatigue strength, it reports on the fatigue tests of the joints using M30-M42 bolts produced in four manufacturing types.

SYSTEM TRUSS JOINT

The composition of typical joint is shown in Figure 1. The joining bolt is the high-strength bolt, and has the threads at both ends and the hexagonal boss for the torque transmission in the

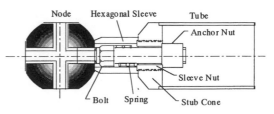

Figure 1. KT-System Truss Joint

middle of the shank. As a result, high clamping force, which significantly improves the fatigue characteristics of the joint, can be effectively introduced. For large diameter bolts, the fine pitch thread is used and the mechanical property class of the bolt is strength grade 9.5.

FATIGUE TESTS OF MACHINED THREAD BOLT

Test details

Ninety-one practical system truss joints consisting of the threaded spherical node, M20-M48 high-strength bolts, the hexagonal hollow sleeves and the thick plates as shown in Figure 2 are tested under the fluctuating type loading (sinusoidal wave of 3-5Hz). The mechanical properties of the materials are listed in Table 1. The test variables are the bolt size, the stress ratio R (algebraic ratio of minimum stress to maximum stress) and the initial clamping force of joining bolt. The 50tf servo-controlled fatigue machine was used for testing (see Figure 2). The system truss joints on the stress ratio 0.7 or 0.8 correspond to the clamped joints. The high-strength bolts are manufactured by the thread machining.

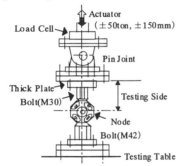

Figure 2. Test setup

Table1. Mechanical properties

Material, Grade		Tensile Strength (MPa)	Yield Stress (MPa)	Elong. %	Hardness HRC
Bolt M12-M36	SCM435	≧1000	≧900	≧15	29~33
M42-M56		≧950	≧850	≧15	29~33
Anchor Nut	SCM435	≧900	≧800	≧15	29~36
Node	SCM435	≧900	≧800	≧15	27~36
Sleeve Nut	SCM435	≧700	≧500	≧17	25~30
Hexagonal Sleeve	S45C	≧700	≧500	≧17	25~30
Stub Cone	SS400	≧410	≧240	≧23	-

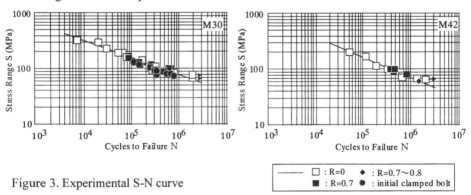

Figure 3. Experimental S-N curve

Test results

The resulting test data provide the relationship between the applied stress range S of the bolt and the number of cycles to failure N. Typical relationship is referred to as the log-log S-N diagram shown in Figure 3. The mean S-N curve is plotted as a straight line. The stress range S (algebraic difference between maximum stress and minimum stress) used to plot the test data is determined by the effective area of the bolt.

Although the data plotted in Figure 3 shows considerable scatter, it indicated the stress ratio does not significantly influence the fatigue behavior. The fatigue fracture in the system truss joint occurred in the bolt thread part at either the node surface side or the anchor nut side. It was obtained that the failure was relatively observed in the anchor nut side thread in M30 and M36 bolt, and in many case the failure of the node side thread was observed in M20, M42, and M48 bolt. However, there is no difference in the S-N curve by the failure position since the test data are on the same curve as mentioned above.

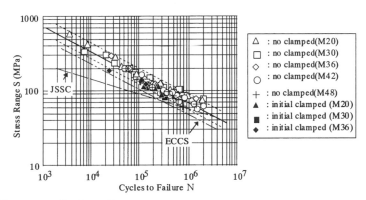

Figure 4. Summary of test results

Fatigue design curve
All available test data on the system truss joints fabricated from M20-M48bolts are plotted in Figure 4. The test data are affected in the variability in the mechanical properties, the heat treatment, the size, the initial clamping force, the stress ratio and other variables. For a given stress range, a slightly lower life is observed for specimens introduced the initial clamping force in bolt as compared with no-clamping ones.

For design purposes, the test data are compared with several fatigue design curves. The JSSC fatigue design curve (Class K5) (Ref.4) is greatly different from the actual test data. However, it is apparent that the ECCS fatigue design curve (Category 36) (Ref.5) provides a reasonable lower bound to the test data.

To develop conservative recommendations for the design of the system truss joints subjected to the repeated type loading, the mean S-N curve and the 95% confidence limits to all the test data are also shown in Figure 4. Moreover, the fatigue limit is made the maximum stress range where the repetition of 2 million cycles is endured. After all, from this lower bound, the proposed fatigue design curve is given by Equation (1).

$$\log S = -0.3450 \times \log N + 3.803 \quad (S > 53.6) \tag{1}$$

FATIGUE TEST OF THREAD ROLLING BOLT

Test details

The specimens are the same as the above-mentioned system truss joints and consist of three series, that is series I-III. The test variables are the stress ratio R (0, 0.7 and 0.8), the manufacturing process types and the size of the heat-treated joining bolt. The initial clamping force does not introduced in the bolt. The joining bolts are M30x3 and M42x3 using Grade SCM435 steel. The bolt threads are produced in four manufacturing process types, namely the thread rolling before heat treatment (RH), the thread rolling after heat treatment (HR), the thread machining before heat treatment (MH) and the thread machining after heat treatment (HM). Each specimen is subjected to sinusoidal wave of frequency 2-6Hz, by using the 50tf servo-controlled fatigue machine

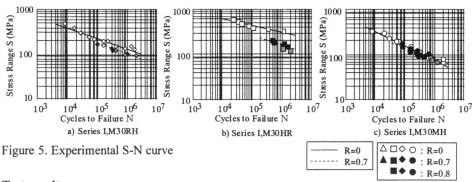

a) Series I,M30RH   b) Series I,M30HR   c) Series I,M30MH

Figure 5. Experimental S-N curve

Test results

The fatigue test results are shown in Figure 5 in the form of the log-log S-N diagram on the stress range. The mean S-N curves are also shown in this figure to straight line.

The fatigue strength is greatly influenced by the manufacturing process and the stress ratio, and cannot be approximated in one S-N curve. That is, the test data of the heat-treated bolt are shown as almost one straight line and, compared with the machined thread bolt, the fatigue strength of the bolt of the thread rolling before heat treatment is equal to or greater. Moreover, on the bolt of the thread rolling after heat treatment, the fatigue strength is greatly improved and, in addition, the fatigue strength of the stress ratio 0.7, 0.8 provides the lower bound, and there is a tendency to which the fatigue limit decreases.

Nextly, on the fatigue limit, the influence of the residual stress and the mean stress on the thread rolling bolt is examined. The fatigue limit diagram (relationship between fatigue limit and mean stress) is shown in Figure 6. The test data of the fatigue limit in each stress ratio are also indicated in Figure 6. The solid line represents the predicted curve obtained by applying the Yoshimoto's hypothesis (Refs.6-7) concerning the fatigue limit of the thread. This curve on the bolt of the thread rolling after heat treatment passes a test data of the stress ratio 0. The predicted curve assumed that the fatigue limit does not depend on the mean stress is shown in the dotted line. One point chain line shows the modified Goodman relationship. It is apparent that the test data of the machined thread bolt are almost constant and the mean stress has a significant effect on the fatigue limit of the thread rolling bolt. While, the modified Goodman relationship is almost approximate in the fatigue limit of the bolt of the thread rolling before heat treatment and the machined thread, and overestimates the fatigue limit of the bolt of the thread rolling after heat treatment at the high stress ratio.

Therefore, The fatigue limit of the bolt of the thread rolling before heat treatment and the machined thread, and the bolt of the thread rolling after heat treatment can be estimated by the Yoshimoto' hypothesis considered the residual stress, the Yamamoto's method (Ref.7) doesn't depend on the mean stress, respectively.

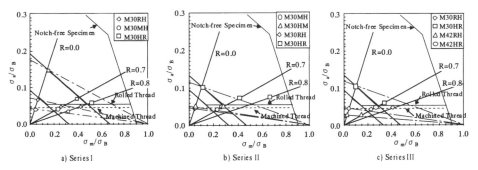

Figure 6. Fatigue limit diagram

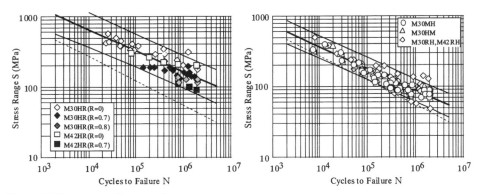

Figure 7. Summary of test results

Fatigue design curve

For the fatigue design, the S-N curve and the fatigue limit of each stress ratio are necessary. Since the stress ratio is not always constant under the repeated loading, providing for the fatigue limit depending on the stress ratio is a dangerous side. The fatigue limit is almost constant in the high mean stress. Therefore, the fatigue design S-N curve is assumed not to receive the influence to the size and the stress ratio of the bolt. It divides the joining bolts into the bolt of the thread rolling after heat treatment and other manufacturing process bolts. The mean S-N curve and the 95% confidence limits are shown in Figure 7. The fatigue strength of the bolt of the thread rolling after heat treatment appears to be greater than that of other manufacturing process bolts. However, the slop of the mean S-N curves is almost identical.

On the other hand, in Figure 7, the fatigue design curve shown by Equation (1) is also illustrated in the dotted line. The fatigue design curve almost corresponds to the lower bound of test data of bolt of the machined thread and the thread rolling before heat treatment. Equation (2) can be led as more practical fatigue design curve from the lower bound formula, and the fatigue strength is improved by using bolt of the thread rolling after heat treatment, and rises as a result by 38.5%.

(Bolt of thread rolling after heat treatment) $\log S = -0.2927 \times \log N + 3.762$ (S>89.4)   (2-a)
(Other manufacturing process bolts) $\log S = -0.3080 \times \log N + 3.631$ (S>48.8)   (2-b)

CONCLUSION
In this paper, the tests are conducted to evaluate the fatigue characteristics of the system truss joints under constant-amplitude repeated loading. In the fatigue tests of the system truss joints using M20-M48 machined thread bolts, the following matters are obtained.
1) The size and the stress ratio of the bolt do not significantly influence the fatigue behavior and the S-N curve.
2) The ECCS fatigue design curve (Category 36) provides a reasonable lower bound to the test data.
And, a fatigue design S-N curve is proposed from the lower 95% confidence limit for the fatigue test data.

Moreover, the fatigue tests of system truss joints consisting of M30-M42 joining bolts are described to improve of the fatigue strength of the system truss joints., the following matters are obtained.
3) The fatigue strength is greatly influenced by the manufacturing process and the stress ratio, and cannot be approximated in one S-N curve.
4) On the bolt of the thread rolling after heat treatment, the fatigue strength is greatly improved and rises as a result by 38.5%.
And, proposed fatigue design S-N curves concerning the improvement of fatigue strength in the system truss joints are presented.

REFERENCE
1. IMAI K et al., Proposing A New Joint System (KT-System) Of Space Frame With Threaded Spherical Nodes And It's Fatigue Characteristics, Madrid, 1989.

2. IMAI K et al., The KT Space Truss System, FORTH INTERNATIONAL CONFERENCE ON SPACE STRUCTURES, vol. 2, GUILDFORD, 1993, pp.1374-1382.

3. IMAI K et al., Development of the KT-Wood Space Truss System, IABSE Conference, Lahti, Finland, August, 2001

4. JSSC, Recommendation for Fatigue Design of Steel Structures, GIHOUDO, 1993. (in Japanese)

5. ECCS-Technical Committee 6 – Fatigue, Recommendation for Fatigue Design of Steel Structures, FIRST EDITION, 1985.

6. Edited by YOSHIMOTO I, A Point of Design of Bolted Joints, JSA, 1992, pp.136-150. (in Japanese)

7. YAMAMOTO A, Theory and Design of Bolted Joints, YOKENDO LTD., 1995, Cap.5. (in Japanese

Evaluation Procedure for Strength of Reticulated Shells

I MUTOH, Dept. of Architecture and Civil Engineering, Gifu National College of Technology, JAPAN
S KATO, Dept. of Architecture and Civil Engineering, Toyohashi University of Technology, JAPAN

INRODUCTION

Fundamental characteristics of reticulated shells have been investigated against buckling collapse in view of both analytical and experimental researches. The structure is one of representative for the instability problem in both elastic and elastic-plastic behaviors. The results obtained have been compared with the behavior of related continuum shells, Refs1-3.

There is several problem among results in consideration of imperfection sensitivity: some results shows less sensitive rather than boundary effects and/or load distributions, however, of intrinsic contributions from construction/manufacture accuracy which was called, *an initial geometric imperfection* in mechanically designed reticulated shells. The members and joints in such structure has been designed and realized in a way of systematic fabrication without so-called material and/or structural imperfections, like residual stress and welding strain.

Thus the investigation into a way for evaluating the capacity and strength which may be easier to evaluate themselves in every confidence which was accounted for metal member framing taller buildings and plane trusses. Thus the investigation into a way for evaluating the capacity and strength which may be easier to evaluate themselves in every confidence which was accounted for metal member framing taller buildings and plane trusses. You can go for that if the way for evaluation of strength in such structures, since the procedure for evaluation might be accustomed to assess the overall load carrying capacity for designed frames with appropriate sections, Ref 4.

For reticulated shells of large span, how to evaluate appropriate sections of members in shells, insofar as engineers can be understood the process for design and construction

STRENGTH ESTIMATION (REVIEW)

Buckling strength of overall reticulated shell has been studied in view of shell-like non-linear elastic bifurcation with a typical eigen-mode imperfection in nodal initial displacement. On the other hand some has been investigated an effect of member initial crookedness considering additionally elastic member buckling as well as large nodal displacement modes.

Basic shell (metal shell) stress formula which was normalized by yield stress f_y, has been related to the shell slenderness R/t, the radius to thickness ratio, and derived in a way of experimental lower bound curves, Ref 4. Stability specification design provisions for metal shells are reflected the initial geometric imperfections and material non-linearity with residual stress in estimation of load capacity. If the reticulated shells will be replaced through equivalent membrane stress, then the relation between limit stress σ_u and external pressure load p_u can be attained by a characteristic slenderness parameter R/t. On the other hand, the nature similar to frame structures may be dominant, member buckling and frame buckling as a whole must be recognized and the strength of both member section and overall frame have to be evaluated analytically by means of a second-order non-linear numerical computation.

The elastic strength and load carrying capacity depend on both the member section properties and mesh density for a specific configuration with boundary supports under uniform vertical loads. The "spongy" nature as stated by Wright (Ref 1), equivalent buckling half-wavelength becomes larger than length of constituent member, the behavior is close to the replacement continuum shell. The section strength may be calculated by means of conversion from membrane stress to axial force. On the other hand, buckling half-wavelength becomes rather short, then flexural buckling of each member is dominant. However more dense mesh with shorter members can be controlled as the behavior like continuum shells. Actual reticulated shells have been designed for compound behavior: shell-like and compressed flexural member element in reticulated shells.

The problem from a viewpoint of homogenization scheme tends to be like the method for optimal configuration under a specific condition for continuum shell: how to make voids over shell surface considering elastic stability criteria, Ref 5. Then the outcome is seemed so that "spongy-nature" of reticulated shell can be attained through micro-to-macro member arrangement with many voids. However the structure constructed behaves like shells or framework, these consideration is difficult. At beginning ground-structure behaves like shells but the last configuration will behave like frames. Anyway, overall strength/load carrying capacity must be evaluated.

The load carrying capacity (elastic buckling load) was arranged as a way for buckling criteria which has been derived from replacement of reticulated shells into an equivalent continuum, with the aid of characteristic slenderness, called shell-likeness S by Yamada, Ref 6 and Lindh's α (characteristic parameter which was defined as the openness and slenderness of lattice), Ref 3. Both the slenderness measure was known as one of elastic buckling half-wave length, Kato has proposed also the same measure in order to characterize the elastic behavior of lattice dome and cylindrical roof. The auxiliary factor ξ is defined in Ref 7, as follows. This factor means also one of effective buckling length for equivalent continuum shells derived from Wright's replacement scheme for reticulated dome with an equilateral triangulated mesh. The normalized linear buckling load η with $\eta = P_{cr}^{lin}/EA\theta_0^3$ is also introduced as below:

$$\xi = 12\sqrt{2}/(\theta_0, \lambda_0), \text{ where } \xi_{cr} = 48/\pi^2, \text{ critical factor for Euler's flexural buckling.}$$

Here another relationships between equivalent buckling half-wavelength and member length can be derived as $l_{cr}/l_0 = \sqrt{\pi * \pi / 48\xi}$ with $\eta_E = \xi^2 \pi^2/48$, the normalized Euler load with buckling length of member length. And here $l_{cr} \propto 3.5\sqrt{Rte}$ is assumed for spherical shells with t_e of an equivalent shell thickness.

How do you determine and evaluate member section sizes and overall strength of reticulated shells? There are two way for estimation and/or prediction: (1) semi-empirical formula for load capacity based on "classical" linear theory and member strength adopted from flexural buckling formula, or strength interaction theorem and (2) numerical analysis, second-order elastic plastic analysis on computer.

RESISTANCE AND STRENGTH EVALUATION

After such exact non-linear numerical computations, the results for load capacity and member stress or force distributions must be evaluated. Thus reference measure for evaluation of the outcome is necessary to be proposed. As well-known scheme the interaction between critical load and/or critical stress has been discussed. For example, the interaction equation as shown below, is most popular and appeared in many design codes and specifications.

$$\sigma_u / (\alpha \, \sigma_{cr}) + (\sigma_u / f_y)^2 = 1 \qquad (1)$$

$$\sigma_u = f_y / (a\lambda^4 + b\lambda^2 + 1)^{1/2} = f_y \times \chi(\lambda) \qquad (2)$$

here in Eq(1) α and Eq(2) a and b mean a specific factor. Also limit stress σ_u is related to f_y with a function of characteristic slenderness $\chi(\lambda)$. In case for continuum shells the plastic reduction function $\varsigma(\lambda_s)$ like below is often proposed as a function of shell slenderness λ_s.

$$\varsigma(\lambda_s) = \lambda_s^2 (0.25\,\lambda_s^2 + 1)^{1/2} - 0.5\,\lambda_s^2 \qquad (3)$$

Then prediction of ultimate strength for section thickness, σ_{CR} can be given as follow.

$$\sigma_{CR} = \varsigma(\lambda_s) \times \alpha / \lambda^2 \qquad (4)$$

On the other hand, load carrying capacity is related to both the linear buckling P_{cr}^{lin} for overall structure and elastic plastic maximum load $P_{cr0}^{el}, P_{cr0}^{pl}$, which is detected on an equilibrium path as a peak appeared first. The former, a specific coefficient as shown below Eq(5) is derived as a measure for fundamental behavior of structures. Here

$$c(\theta_0, \lambda_0) = P_{cr0}^{pl} / P_{cr}^{lin} \qquad (5\text{-}1)$$

$$c_{el}(\theta_0, \lambda_0) = P_{cr0}^{el} / P_{cr}^{lin} \qquad (5\text{-}2)$$

The ultimate load Pu is predicted through carrying out once linear buckling analysis, as follow.

$$Pu = \rho_{pl} \times c(\theta_0, \lambda_0) \times P_{cr}^{lin} \qquad (6)$$

Here the reduction ("knock-down" factor) is introduced and is given as:

$$\rho_{pl} = P_{cr}^{pl} / P_{cr0}^{pl} \qquad (7\text{-}1)$$

$$\alpha = \rho_{pl} \times c(\theta_0, \lambda_0) \qquad (7\text{-}2)$$

Eq(7-2) means well-known "knock-down" factor from linear buckling load, as shown in Eq(1), for example.

Then the member force Nu is predicted as below, insofar as the relation between load and force may be linear and attained even to a limit state.

$$Nu = \chi(\lambda) \times N_y = \rho_{pl} \times c(\theta_0, \lambda_0) \times N_{cr}^{lin} \qquad (8)$$

Finally, prediction of ultimate load Pu can be obtained with relation to Nu in reticulated shells, by carrying out once linear stress analysis which gives P_{d0}/N_{d0}.

$$P_u = Nu/N_{d0} \times P_{d0} = \chi(\lambda) \times N_y \times P_{d0}/N_{d0} \qquad (9)$$

$$\chi(\lambda) = \rho_{pl} \times c(\theta_0, \lambda_0) / \Lambda^2 \qquad (10)$$

Next we investigate Eqs (9) and (10) and propose the evaluation procedure.

EVALUATION PROCEDURE-PROPOSAL

The reduction factor ρ_{pl} against elastic plastic buckling load for perfect geometry is assumed as a function of imperfection parameter $\varepsilon_g = w_0/t_e$ with maximum imperfection amplitude w_0. Then a "knock-down" factor as mentioned by IASS recommendation normalized by linear buckling load α_{IASS} can be approximated as below, but denoted by ρ.

$$\rho = \exp(-1.6\sqrt{\varepsilon}\, g) = \alpha_{IASS} \qquad (11)$$

Also, reduction ρ_{pl} to plastic buckling load like curves given by Hutchinson is denoted as:

$$\rho_{pl} = \exp(-0.8\sqrt{\varepsilon}\, g), \quad \text{for Hutchinson} \qquad (12)$$

In this paper the notation as below is proposed:

$$\rho^*_{pl} = 1 - [1 - \exp(-1.6\sqrt{\varepsilon}_g)] / \sqrt{\xi} \qquad (13)$$

Here again the auxiliary factor ξ is used. As defined above ρ^*_{pl} is calculated against linear buckling load, the effect of elastic buckling criteria may be reflected. Notice to a combination of parameters θ_0 and λ_0 which characterize the degree of "sponginess" as well as elastic behaviors.

Thus the evaluation equations for member strength are derived as follows. We consider the notation adopted from design curve for shell buckling in ECCS.

$$\sigma_{CR} = 1 - \Lambda^2/[\rho^*_{pl} \times c(\theta_0, \lambda_0)\sqrt{\xi}\,], \quad \Lambda \leq 1 \quad (14\text{-}1)$$

$$\sigma_{CR} = \rho^*_{pl} \times c(\theta_0, \lambda_0)/(\Lambda^2\sqrt{\xi}), \qquad \Lambda > 1 \quad (14\text{-}2)$$

From discussion before, the fundamental behavior of reticulated shells may be elastic and each member flexural buckling without large nodal displacements, the another reference equation given below, is appropriate. The effect of Euler strut behavior can be judged by the auxiliary factor, ξ less than about 4.0.

$$\sigma_{CR} = \rho^*_{pl} \times c(\theta_0, \lambda_0)/\Lambda^2 \qquad (15)$$

Next some comparison between evaluation method and numerical calculations for a specific reticulated dome will be provided.

Reticulated dome of about 200 meters span composed member of length 10 meters with mechanical ball joints with pin-supports under uniform vertical loading was computed by means of second-order plastic hinge analysis. The initial geometric imperfections are assumed as initial nodal displacements which are set over circular area with diameter of an equivalent shell's buckling half-wavelength l_{cr} with imperfection parameter ranged from 0 to 1.0. Also the auxiliary factor ranges from 4 to 16 with (θ_0, λ_0) ;1deg and 2deg, 60 and 120, respectively. Fundamental characteristics are summarized as in Tab 1 and depicted as shown in Fig 1. The evaluation for "knock-down" factor based on plastic effects is listed as in Tab 2. Tab 3 shows an example of evaluation equations.

Table 1

parameter	marks	$c(\theta_0, \lambda_0)$	ξ	$1/\sqrt{\xi}$	$\sqrt{\xi}$	Λ
1deg 60	◆	0.64	16.2	0.25	4	0.98
1deg 120	□	0.67	8.1	0.35	2.83	1.45
2degs 60	△	0.51	8.1	0.35	2.83	0.72
2degs120	×	0.98	4	0.5	2	1.2

Table 2

ρ_{pl}^{*}		marks	$\sqrt{\xi}$	$1/\sqrt{\xi}$	Λ	$\varepsilon_g=0$	0.2	0.4	0.5	0.8	1
1deg	60	◆	4	0.25	0.98	1	0.9	0.87	0.83	0.81	0.8
1deg	120	□	4.83	0.35	1.45	1	0.86	0.82	0.76	0.73	0.72
2degs	60	△	2.83	0.35	0.72	1	0.86	0.82	0.76	0.73	0.72
2degs	120	×	2	0.5	1.2	1	0.8	0.74	0.66	0.62	0.6
$1-\exp(-1.6\sqrt{\varepsilon_g}) = 1 - \alpha_{IASS}$						0	0.51	0.64	0.68	0.76	0.8

Table 3

strength	marks	Λ	A	B
1deg 60	◆	0.98	$1-1/\rho_{pl}^{*}*0.15$	
1deg120	□	1.45	$\rho_{pl}^{*}*0.11$	$\rho_{pl}^{*}*0.32$
2degs60	△	0.72	$1-1/\rho_{pl}^{*}*0.09$	
2degs120	×	1.2	$\rho_{pl}^{*}*0.34$	$\rho_{pl}^{*}*0.68$
Note: Slenderness bounds $\sqrt{2}$			Eqn 14	Eqn 15

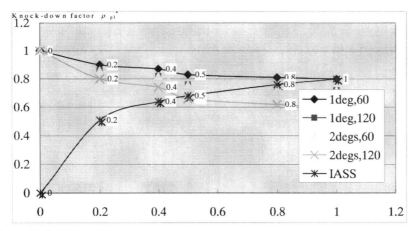

Imperfection parameter
Figure 1

The effect of c(θ_0, λ_0) is illustrated as a function of generalized slenderness parameter Λ as shown in Fig 2. From the figure, the plot apart from regression curve is in case of $\xi < 4.0$.

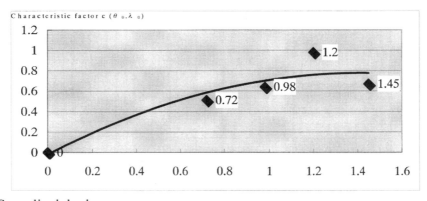

Generalized slenderness
Figure 2

Finally, the normalized member strength as a function of Λ is illustrated as shown in Fig 3 with insertion of numerical results for ε_g=0.2 and 0.5. Here again switching to ordinary Euler curve for the plots in case of (θ_0, λ_0) =(2degs with 120), Eq(15) shows as reasonable. Here the marks plotted are different from the marks listed in Tabs 1-3. However please refer to the value of each slenderness parameter Λ. The generalized slenderness reflects an effect elastic behavior of perfect geometry, thus the degree of equivalence in-between shells and frames is considered on the horizontal axes in Figs 2 and 3. The difficulty of implication of fundamental structural behavior on such arrangement in figures is still discussed further. One of the others will be investigated by means of modified slenderness to Λ considering non-linear nodal displacement for shell-like imperfection sensitivity directly. The mean of reduced slenderness of shell's characteristic slenderness adopted in ECCS buckling curve. However the knock-down factor has to be explicitly described with nominal parameters.

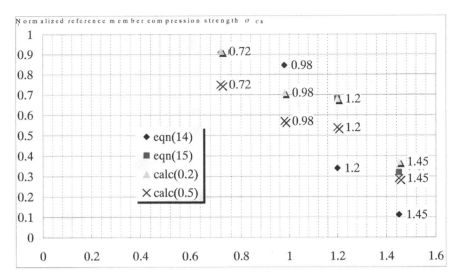

Normalized reference member compression strength σ_{cr}

Generalized slenderness
Figure 3

However the others except the mark ◆ may be appropriate in view of evaluation equation, Eq(14) even results for imperfections.

The discussion about the utility of interaction formula is still of necessary to study into practical engineers' view, instead of analytical consideration for exact non-linear numerical computations. As known, *e.g.* in Ref 8, a kind of Dunkerley type theorem and formula is based on the reciprocal theorem for the least critical load parameter of an elastic structure subjected to a complex load system. Also Strigl's investigation into generalizing Dunkerley formula found that original Dunkerley formula can result in approximation both the safe and unsafe cases. On the other hand, Rankine formula which relates to elastic-plastic structures does not belong to the Dunkerley type formula. Originally, for a compressed elastic-plastic bar, the failure load relates to plastic collapse load with geometric parameters. This paper induces the evaluation problem to more convenient and user-friendly procedure but the structure will be analyzed through an exact non-linear numerical program on computer with perfect geometry.

CONCLUSIONS
The paper proposes the evaluation procedure for design of reticulated shells. The strength of both the member section and overall shell has to be assessed even in case of complex configuration. However the engineer may often compare the outcome with previous results, then the previous results must be understood in a way for application further with some evaluation reference measure for proportioning members and predicting load carrying capacity of the structure. The method is very accustomed to structural engineers who have been worked to carry out basic shell elements and/or fundamental configuration of reticulated lattices over curved surface. The proposal has to be modified and arranged with further data in order to confirm the evaluation formula to actual outcomes. Notice given is the large reticulated shells which should be constructed and fabricated in pure quality and inspected exactly. Then the proposed evaluation procedure will be workable to structural engineers at least to make the design parameters be qualified.

REFERENCES
1. WRIGHT D T, Membrane Forces and Buckling in Reticulated Shells, Journal of the Structural Division, ASCE, vol.91, Feb.1965, pp.173-201
2. FORMAN S E & HUTCHINSON J W, Buckling of Reticulated Shell Structures, International Journal of Solids Structures, vol.6, 1970, pp.904-932
3. LIND N C, Local Instability Analysis of Triangulated Dome Frameworks, The Structural Engineer, vol.47, No.8, Aug. 1969, pp.317-324
4. BEEDLE L S, Stability of Metal Structures A World View 2nd Edition, Structural Stability Research Council, 1991, 921pages.
5. BLETZINGER K –U, MAUTE K & RAMM E, Structural Concepts by Optimization, Proceedings Conceptual Design of Structures, Stuttgart, vol.1, Oct. 1996, pp.169-177
6. YAMADA M, An approximation on the Buckling Analysis of Orthogonally Stiffened and Framed Spherical Shells, Shell and Spatial Structures Engineering, IASS Symposium, Rio de Janeiro, Pentech Press, 1983, pp.177-193
7. MUTOH I & KATO S, Comparison of Buckling Loads between Single-Layer Lattice Domes and Spherical Shells, Space Structures 4, vol.1, Thomas Telford, London, 1993, pp.176-185
8. TARNAI T, Summation Theorems concerning Critical Loads of Bifurcation, Structural Stability in Engineering Practice, ed. L.Kollar, E & FN SPON, 1999, pp.23-58

Appendix : **NOTATIONS**
fy :=Yield Stress
P : =Load (at each nodes)
N : =Internal Force (axial)
Nu: =Ultimate Force (axial)
Ny: =Yield Force
c(#,#): =Characteristic Coefficient as a Function of θ_0 and λ_0

α = Knock-down Factor

ξ = Auxiliary Factor as a function of θ_0 and λ_0

λ = Slenderness Parameter,(member slenderness)

ς = Plastic Reduction Factor

σ = Stress, $_{CR}$ (critical), $_u$(ultimate), ^{pl}(plastic), and $_{cr0}$(critical for perfect shape)

$\chi(\#)$ = Normalized Strength Function of Slenderness

ρ = Load Reduction Factor, $_{el}$ (elastic load), $^*_{pl}$(prediction for evaluation)

$\theta0$ = Half-Subtended Angle of Members in Unit at Apex

εg = Imperfection Parameter, normalized to equivalent shell thickness

Λ = Generalized Slenderness defined by the ratio of Ny to $N_{cr}{}^{lin}$ calculated by eigen-value analysis

Multiobjective Shape Optimization of Shells Considering Roundness and Elastic Stiffness

M OHSAKI, Dept. of Archi. and Architectural Systems, Kyoto University, Japan,
T OGAWA, Dept. of Archi. and Building Engng., Tokyo Inst. of Tech., Japan, and
R TATEISHI, Dept. of Archi. and Building Engng., Tokyo Inst. of Tech., Japan

SUMMARY

A method is presented for generating round surfaces by directly specifying the center of curvatures. The surface is defined by a single parametric surface, and discontinuities in tangent vectors and curvatures are allowed. The minimization problem of the distance of the center of curvature from the specified point is converted into a maximization problem of the reciprocal objective value. The parameter region where integration is to be carried out is restricted in view of the sign of the curvature. This way, shells with and without ribs can be generated within the same problem formulation and from the same initial shape. Optimal shapes are also found under constraints on the compliance against static loads, which is a measure of elastic stiffness. A set of Pareto optimal solutions between roundness and compliance is generated by using the constraint method.

INTRODUCTION

In the conventional approaches of generating smooth or fair curves and surfaces, the square norms of curvature and variation of curvature are minimized, respectively, to obtain the minimum energy curve and the minimum variation curve [1, 2]. Subramainan and Suchithran [3], presented a method for adjusting the knot vector of a B-spline curve based on the derivative of curvature in the process of designing ship hulls. For surfaces, the principal curvatures, mean curvature, Gaussian curvature, and their derivatives can be used for formulating the fairness metrics [4–7].

Several advanced formulations have recently been presented for designing smooth curves and surfaces in view of roundness, rolling and flattening [8, 9], which have been applied to ship hull design [10]. Ohsaki and Hayashi [11] presented a modified version of the roundness metric by Rando and Rourier for optimizing ribbed shells. It has been pointed out, however, that those fairness metrics do not always conform to the human impressions [11, 12].

In the process of designing complex surfaces, the surfaces are divided into several regions, each of which is defined by a parametric surface such as Bézier patch and B-spline patch. In this case, constraints should be given for continuity and intersection between the adjacent regions [5, 13]. Boundary conditions should also be given to formulate a constrained optimization problem [14, 15]. The formulations for continuity in curvatures

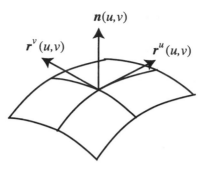

Figure 1. Unit tangent and normal vectors of a surface.

of surfaces, however, are very complicated, and it is inconvenient that the constraints should be modified and optimization problem itself should be reformulated depending on the types of the desired surfaces; e.g., sometimes discontinuity is allowed between the tangent vectors in the adjacent regions.

Parametric surfaces such as Bézier surfaces and B-spline surfaces are also very useful for structural shape optimization. Ramm [16] optimized a shell defined by the Bézier surface considering stress deviation or fundamental frequency. The method has been extended to the problem with buckling constraints [17] and optimization of membrane fabric structures [18]. Ohsaki *et al.* [19] presented a trade-off design method between smoothness and elastic stiffness of an arch-type truss. Ohsaki and Hayashi [11] presented a method for generating round ribbed shells. In their method, however, the number of ribs should be defined in advance, and the shell should be modeled by different number of Bézier surfaces depending on the number of ribs.

In this paper, a method is presented for generating round surfaces allowing discontinuities in tangent vectors and curvatures. The distance of the center of curvature from the specified point is used for formulating the objective function which is a continuous function of the design variables through convex and concave shapes. Optimal shapes are also found under constraints on compliance that is regarded as a mechanical performance measure. A set of Pareto optimal solutions are obtained by using the constraint approach to generate a trade-off design between roundness and mechanical performance. It is also shown that a ribbed shell can easily be generated by specifying the center of curvature of the surface or the isoparametric curves.

SHAPE OPTIMIZATION OF SURFACES

Consider a surface defined by parameters u and v as $\mathbf{x}(u, v)$. The unit normal vector of the surface is denoted by $\mathbf{n}(u, v)$. A line in u- or v-direction in the parameter space corresponds to a curve in the physical space which is called an *isoparametric curve*. Let $\mathbf{r}^u(u, v)$ and $\mathbf{r}^v(u, v)$ denote the unit tangent vectors of the isoparametric curves in u- and v-directions, respectively. $\mathbf{n}(u, v) = \mathbf{r}^u(u, v) \times \mathbf{r}^v(u, v)$ is defined as shown in Figure 1.

Let $\kappa_i(u, v)$ $(i = 1, 2)$ denote the two principal curvatures at a point on a surface. The centers of curvatures can be defined for $\kappa_i(u,v) \neq 0$ as

$$\mathbf{p}_i(u,v) = \mathbf{x}(u,v) + \frac{1}{\kappa_i(u,v)}\mathbf{n}(u,v), \quad (i=1,2) \tag{1}$$

The definition (1), however, can not be conveniently used for specifying the desired shape, because two centers of curvatures exist at a point on the surface. If the Gaussian curvature $K(u,v) = \kappa_1(u,v)\kappa_2(u,v)$ is used, the center of curvature can be uniquely determined as follows for $K(u,v) > 0$ [4]:

$$\mathbf{c}(u,v) = \mathbf{x}(u,v) + \frac{1}{\sqrt{K(u,v)}}\mathbf{n}(u,v) \tag{2}$$

Note that the surface of Figure 1 has two negative principal curvatures and $K(u,v) > 0$.

The surface $\mathbf{x}(u,v)$ is defined by the tensor product Bézier surface as [20]

$$\mathbf{x}(u,v) = \sum_{i=0}^{n}\sum_{j=0}^{m}\mathbf{R}_{i,j}^{*}B_i^n(u)B_j^m(v) \tag{3}$$

where $\mathbf{R}_{i,j}^{*} = (R_{i,j}^{*x}, R_{i,j}^{*y}, R_{i,j}^{*z})$ $(i=0,\ldots,n; j=0,\ldots,m)$ are the control points. It is convenient to define $\mathbf{R}_{i,j}^{*}$ by an affine transformation

$$\mathbf{R}_{i,j}^{*} = \mathbf{TR}_{i,j} \tag{4}$$

for investigating optimal shapes on various quadrilateral regions. $\mathbf{R}_{i,j}$ is defined in the unit cell and is the design variables for shape optimization. The linear transformation matrix $\mathbf{T}$ is fixed during optimization. The components of $\mathbf{R}_{i,j}$ are combined to a vector $\mathbf{R}$, and all the properties of the surface are functions of $\mathbf{R}$.

Let $\mathbf{c}_0$ denote the specified center of curvature. The determinant of the first fundamental matrix of the surface is denoted by $g(u,v;\mathbf{R})$. The parameters u and $v \in [0,1]$ are divided into regions with uniform interval by Δu and Δv, respectively, and the parameter values at the centers of the regions are denoted by u_i and v_j. The square of distance between $\mathbf{c}(u_i,v_j;\mathbf{R})$ and $\mathbf{c}_0$ is defined as

$$d(u_i,v_j;\mathbf{R}) = ||\mathbf{c}(u_i,v_j;\mathbf{R}) - \mathbf{c}_0||^2 \tag{5}$$

The objective function to be minimized may be given as

$$D(\mathbf{R}) = \sum_i\sum_j d(u_i,v_j;\mathbf{R})\sqrt{g(u_i,v_j;\mathbf{R})}\Delta u\Delta v \tag{6}$$

An optimal shape as illustrated in Figure 2 can be obtained by minimizing $D(\mathbf{R})$ from the initial shape as shown in Figure 3. Details of the mathematical formulations will be shown in the examples. The optimal ribbed surface as shown in Figure 4, however, has to be reached from the initial shape Figure 3 through an intermediate shape as illustrated in Figure 5 of which the center of curvature around the center $(u,v) = (0.5, 0.5)$ is far from $\mathbf{c}_0$ and even in the opposite side of the surface. Therefore, the optimal solution of Figure 4 cannot be obtained by minimizing $D(\mathbf{R})$ from the initial surface of Figure 3.

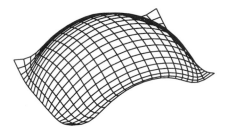

Figure 2. Optimal solution for $\mathbf{c}_0 =$ $(35, 25, 0)$.

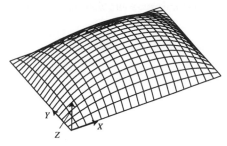

Figure 3. Initial shape.

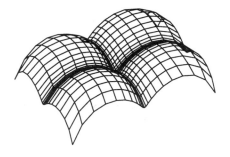

Figure 4. Optimal solution for $\mathbf{c}_0 =$ $(17.5, 12.5, 0)$.

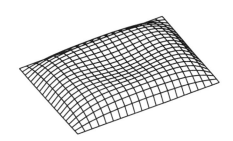

Figure 5. Intermediate shape.

Let U denote the region where both of the principal curvatures are negative. Note that $\mathbf{n}(u, v; \mathbf{R})$ defined in Figure 1 is directed to the upper side of the surface irrespective of the sign of the principal curvatures. It is possible that the summation of $d(u_i, v_j; \mathbf{R})$ is to be carried out only over the region U. In this case, however, there is no region for integration if $\kappa_1(u, v; \mathbf{R}) > 0$ and $\kappa_2(u, v; \mathbf{R}) > 0$ are simultaneously satisfied throughout the region. Therefore, a surface with positive principal curvatures in anywhere in the region may be obtained because objective function to be minimized is nonnegative and such a solution has vanishing objective value that leads to an obvious and meaningless optimal solution. The problem may alternatively be formulated to maximize

$$D^*(\mathbf{R}) = \sum_{(u_i, v_j) \in U} \frac{1}{d(u_i, v_j; \mathbf{R})} \sqrt{g(u_i, v_j; \mathbf{R})} \Delta u \Delta v \qquad (7)$$

In the process of maximizing $D^*(\mathbf{R})$, the design variables are to be modified so that $d(u_i, v_j; \mathbf{R})$ is reduced and the both of the principal curvatures are positive in wider region. Therefore, as shown in the following examples, the optimal surface in Figure 4 is successfully obtained from the initial shape of Figure 3. Note that the objective function diverges if $d(u_i, v_i; \mathbf{R}) = 0$ is satisfied at a point. In order to prevent the divergence, $1/d(u_i, v_i; \mathbf{R})$ is replaced by $\bar{d}$ if $1/d(u_i, v_i; \mathbf{R}) > \bar{d}$, where $\bar{d}$ is a small positive number. Finally the objective function to be maximized is given as

$$\hat{D}^*(\mathbf{R}) = \sum_{(u_i, v_j) \in U} \min\left\{\frac{1}{d(u_i, v_j; \mathbf{R})}, \bar{d}\right\} \sqrt{G(u_i, v_j; \mathbf{R})} \Delta u \Delta v \qquad (8)$$

and the optimization problem is formulated as

$$\text{P1: Maximize} \quad \hat{D}^*(\mathbf{R}) \tag{9}$$
$$\text{subject to:} \quad \mathbf{R}^L \leq \mathbf{R} \leq \mathbf{R}^U \tag{10}$$
$$\mathbf{H}(\mathbf{R}) \leq \mathbf{0} \tag{11}$$

where $\mathbf{R}^U$ and $\mathbf{R}^L$ are upper and lower bounds of $\mathbf{R}$, and $\mathbf{H}(\mathbf{R}) \leq \mathbf{0}$ denote the geometrical constraints given if necessary.

A round shape is generated by solving P1. The mechanically optimal shape, however, is quite different from a round shape. Therefore, we next consider the trade-off between roundness and mechanical property defined by compliance (external work) against static loads. A standard finite element method is used for finding the responses against static loads. Let $\mathbf{F}$ and $\mathbf{U}$ denote, respectively, the vectors of nodal loads and nodal displacements obtained by solving

$$\mathbf{K}\mathbf{U} = \mathbf{F} \tag{12}$$

where $\mathbf{K}$ is the linear elastic stiffness matrix, and dependence of all the variables on $\mathbf{R}$ is assumed. The compliance W is defined by

$$W = \mathbf{F}^\top \mathbf{U} \tag{13}$$

The specified structural volume is denoted by $\bar{V}$. The thickness t is then given by $t = \bar{V}/A$ where A is the total area of the surface. The optimization problem is formulated as

$$\text{P2: Minimize} \quad W(\mathbf{R}) \tag{14}$$
$$\text{subject to:} \quad \mathbf{R}^L \leq \mathbf{R} \leq \mathbf{R}^U \tag{15}$$
$$\mathbf{H}(\mathbf{R}) \leq \mathbf{0} \tag{16}$$

Since the roundness and the elastic stiffness defined by using the compliance are conceived as conflicting performance measures, a multiobjective optimization problem can be formulated for optimizing the two objectives [21]. There are many approaches to obtaining a set of Pareto optimal solutions or to selecting the most preferred solution among the set of Pareto optimal solutions. In this paper, the constraint method is used [21]. Let $\bar{W}$ denote the specified upper bound for W, and consider the following problem:

$$\text{P3: Maximize} \quad \hat{D}^*(\mathbf{R}) \tag{17}$$
$$\text{subject to:} \quad W(\mathbf{R}) \leq \bar{W} \tag{18}$$
$$\mathbf{R}^L \leq \mathbf{R} \leq \mathbf{R}^U \tag{19}$$
$$\mathbf{H}(\mathbf{R}) \leq \mathbf{0} \tag{20}$$

Note that $\bar{W}$ is given in view of the values of W of the optimal solutions of P1 and P2. A set of Pareto optimal solutions can be generated by solving P3 for various values of $\bar{W}$.

The center of curvature can alternatively be defined by the isoparametric curves for generating a ribbed shell. Let $\kappa^u(u; v)$ denote the curvature of the isoparametric curve in u-direction, where the argument $(u; v)$ indicates that the parameter is u, but the curve is defined for each specified value of v. The unit normal vector is denoted by $\mathbf{n}^u(u;v)$. Note that $\kappa^u(u;v)$ of a curve in three dimensional space always has nonnegative value. The center of curvature of the isoparametric curve is given for the region $\kappa^u(u;v) \neq 0$ as

$$\mathbf{c}^u(u;v) = \mathbf{x}(u,v) + \frac{1}{\kappa^u(u;v)}\mathbf{n}^u(u;v) \tag{21}$$

$\mathbf{c}^v(u;v)$ can be defined similarly.

EXAMPLES

Optimal shapes have been found from the initial shape as shown in Figure 3. The (X, Y, Z)-coordinates are defined as shown in Figure 3. In the following, the unit of length is m, which is omitted for brevity. The transformation matrix $\mathbf{T}$ is diagonal as

$$\mathbf{T} = \begin{bmatrix} 70 & 0 & 0 \\ 0 & 50 & 0 \\ 0 & 0 & 20 \end{bmatrix} \tag{22}$$

Therefore, the plan of the surface is a rectangle, and the span lengths in X- and Y-directions are 70 and 50, respectively. The Bézier surface of 5×5 degrees is used; i.e. $n = m = 5$ in (3). We only consider the surfaces that are symmetric with respect to the planes defined by $X = 35$ and $Y = 25$. In this case, the surface is defined by nine control points $\mathbf{R}_{i,j}$ ($i = 0, 1, 2; j = 0, 1, 2$). The initial shape of Figure 3 is given as

$$\begin{aligned}
\mathbf{R}_{0,0} &= (0,0,0), & \mathbf{R}_{0,1} &= (0,0.2,0), & \mathbf{R}_{0,2} &= (0,0.4,0), \\
\mathbf{R}_{1,0} &= (0.2,0,0), & \mathbf{R}_{1,1} &= (0.2,0.2,0.4), & \mathbf{R}_{1,2} &= (0.2,0.2,0.4), \\
\mathbf{R}_{2,0} &= (0.4,0,0), & \mathbf{R}_{2,1} &= (0.4,0.2,1), & \mathbf{R}_{2,2} &= (0.4,0.4,1)
\end{aligned} \tag{23}$$

The three components of $\mathbf{R}_{0,0}$ are fixed during the optimization process. The control points along the boundary can move only in the vertical planes in which the boundary curves are located. The upper and lower bounds for $\mathbf{R}_{i,j}$ are 1 and 0, respectively, for X, Y-coordinates, and 3 and -1 for Z-coordinates. The parameters are divided by $\Delta u = \Delta v = 0.0125$. The upper bound $\bar{d}$ in (9) is 1.0. Geometrical constraints are given so that the X- and Y-components of the tangent vectors of the isoparametric curves in u- and v-directions, respectively, have nonnegative values at the points on the planes of symmetry defined by $u = 0.5$ and $v = 0.5$. These constraints are written explicitly as

$$R^x_{0,j} + 3R^x_{1,j} + 2R^x_{2,j} \leq 3S^x, \quad (j = 0, 1, 2) \tag{24}$$
$$R^x_{i,0} + 3R^x_{i,1} + 2R^x_{i,2} \leq 3S^y, \quad (i = 0, 1, 2) \tag{25}$$

where S^x and S^y are the span lengths in X- and Y-directions, respectively.

The optimal solution for $\mathbf{c}_0 = (35, 25, 0)$ is as shown in Figure 2. Note that there exist concave regions at four corners. Roundness in almost all the domain, however, has been increased by sacrificing smoothness at the corners of the optimal shape. An optimal shape of Figure 4 has been found for $\mathbf{c}_0 = (17.5, 12.5, 0)$; i.e., a ribbed shell with discontinuity in the tangent vector can be found from a convex initial shape. Therefore, optimal shape with various curvature distributions can be generated by solving P1.

Consider next a problem of minimizing the compliance under static loads. The curved shell is assumed to be sufficiently thin so that only membrane stresses should be considered. The standard nine-degree-of-freedom triangular element with uniform stresses and

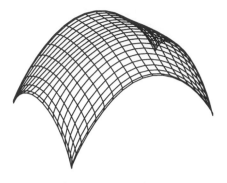

Figure 6. Optimal shape for minimizing compliance ($W = 2.0093 \times 10^4$ kNm).

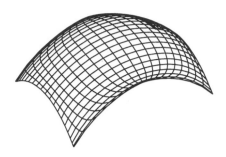

Figure 7. Optimal shape under compliance constraint ($W = 3.0 \times 10^4$ kNm).

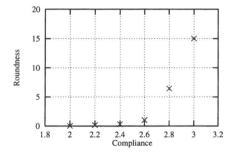

Figure 8. Relation between compliance W and roundness $\hat{D}$ of the optimal solution of P3.

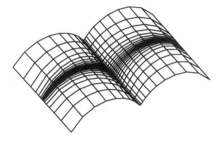

Figure 9. Optimal shape for specified center of curvature of isoparametric curve.

strains is used [22]. The parameter space (u, v) is divided into 20×20 regions with same interval. The shell is subjected to distributed load 100.0 N in negative Z-direction per unit area of the surface. All the displacement components including rotations are fixed along the boundary. The material is steel where the elastic modulus is 200.0 GPa. The specified structural volume is 10.0 m³.

The optimal solution is as shown in Figure 6, where the compliance is 2.0093×10^4 kNm. It is seen from Figure 7 that the optimal shape is a kind of cylindrical shell with parabolic cross-section in the direction of the shorter span. The value of W for the shell in Figure 4 is 6.7084×10^4 kNm. The trade-off solution between roundness and elastic stiffness by solving P3 for $c_0 = (35, 25, 0)$ and $\bar{W} = 3.0 \times 10^4$ kNm is as shown in Figure 7 which is between the shapes in Figures 4 and 6.

Figure 8 shows the relation between W and $\hat{D}$ of the optimal solution of P3 for various values of $\bar{W}$. Note that W is to be minimized and $\hat{D}$ is to be maximized. It is seen from Figure 8 that a set of Pareto optimal solutions has been generated by solving P3.

Finally, a ribbed shell is obtained by using the center of curvature of the isoparametric curve given by (21). The Z-coordinates of $\mathbf{R}_{0,1}$ and $\mathbf{R}_{0,2}$ are fixed at 0. The inverse of the square of the distance between $\mathbf{c}^u(u; v)$ and the line defined by $X = 17.5, Z = 0$ for the region $u \in [0, 0.5]$ and $v \in [0, 1]$ has been minimized to obtain the optimal shape in Figure 9, where $\bar{d} = 1.0$ has also been used, and the integration has been carried out only for the region with $\kappa^u(u; v) < 0$. It is observed from Figure 9 that a ribbed shell can be generated by specifying the center of curvature of the isoparametric curve without any modification of modeling method of the surface.

CONCLUSIONS

A unified approach has been presented for generating round shells with and without ribs. The conclusions drawn from this study are summarized as follows:

1. Round surfaces with different numbers of ribs can be generated by specifying the center of curvature without any modification of problem formulation or modeling method.

2. The optimal shape for minimizing compliance under constraint on structural volume has been shown to be a nearly singly curved shell that consists of a cylinder with quadric cross section.

3. The constraint approach can be successfully used for obtaining a trade-off design between round and mechanically efficient shapes.

4. A ribbed cylindrical shell can be generated by specifying the distribution of the center of curvature of the isoparametric curves.

REFERENCES

1. MEIER M and NOVACKI H, Interpolating Curves with Gradual Changes in Curvature, Comput. Aided Geom. Des., vol 4, 1987, pp 297-305

2. MORETON H P and SÉQUIN H S, Functional Optimization for Fair Surface Design, Computer Graphics, vol 26(2), 1992, pp 167-176.

3. SUBRAMAINAN V A and SUCHITHRAN P R, Interactive Curve Fairing and Bi-Quintic Surface Generation for Ship Design, Int. Shipbuild. Progr., vol 46(446), 1989, pp 189-208.

4. HANGEN H, HAHMANN S and SCHREIBER T, Visualization and Computation of Curvature Behaviour of Freeform Curves and Surfaces, Comput. Aided Des., vol 27(7), 1995, pp 545-552.

5. BARNHILL R E (ed.), Geometry Processing for Design and Manufacturing, SIAM, 1994.

6. SARRAGA R F, Recent Methods for Surface Shape Optimization, Comput. Aided Geom. Des., vol 15, 1998, pp 417-436.

7. GREINER G, Variational Design and Fairing of Spline Surfaces, Computer Graphics Forum, vol 13(3), 1994, pp 143-154.

8. RANDO T and J.ROULIER J, Designing Faired Parametric Surfaces, Comput. Aided Des., vol 23, 1991, pp 492-497.

9. ROULIER J and RANDO T, Measures of Fairness for Curves and Surfaces, In: SPADIS N S (ed.), Designing Fair Curves and Surfaces, 1994, pp 75-122, SIAM.

10. NOWACKI H and REESE D, Design and Fairing of Ship Surfaces, In: BARNHILL R E and BOHEM W (eds.), Surfaces in CAGD, 1983, pp 121-134, North-Holland.

11. OHSAKI M and M HAYASHI, Fairness Metrics for Shape Optimization of Ribbed Shells, J. Int. Assoc. Shells and Spatial Struct., vol 41(1), 2000, pp 31-39.

12. GEROSTATHIS T P, KORAS G D and KALKIS P D, Numerical Experimentation with the Roulier-Rando Fairness Metrics, Mathematical Engineering in Industry, vol 7(2), 1999, pp 195-210.

13. DU W-H and SCHMITT F J M, On the g^1 Continuity of Piecewise Bézier Surfaces: A Review with New Results, Comput. Aided Des., vol 22(9), 1990, pp 556-573.

14. NOWACKI H, LIU D and LÜ X, Fairing Bézier Crves with Cnstraints, Comput. Aided Geom. Des., vol 7, 1990, pp 43-55.

15. WELCH W and WITKIN A, Variational Surface Modeling, Computer Graphics, vol 26(2), 1992, pp 157-166.

16. RAMM E, Shape Finding Methods of Shells, Bulletin of Int. Assoc. for Shell and Spatial Struct, vol 33, 1992, pp 89-98.

17. RAMM E, BLETZINGER K-U and REITINGER R, Shape Optimization of Shell Structures, Bulletin of Int. Assoc. for Shell and Spatial Struct, vol 34(2), 1993, pp 103–121.

18. BLETZINGER K-U, Structural Optimization and Form Finding of Lightweight Structures, Proc. 3rd World Congress of Structural and Multidisciplinary Optimization (WCSMO3), 1999.

19. OHSAKI M, NAKAMURA T and ISSHIKI Y, Shape-Size Optimization of Plane Trusses with Designer's Preference, J. Struct. Engng., ASCE, vol 124(11), 1998, pp 1323-1330.

20. FARIN G, Curves and Surfaces for Computer Aided Geometric Design, Academic Press, 1992.

21. COHON J L, Multiobjective Programming and Planning, vol 140 of Mathematics in Science and Engineering, Academic Press, 1978.

22. ZIENKIEWICZ O C and TAYLOR R L, The Finite Element Method, McGraw-Hill, 1989.

Elastic buckling behaviour of latticed shell roofs composed of the ("triangle and hexagon)-and-hexagon" space trussed units

T. KONISHI, T. SAKA, Y. TANIGUCHI
Graduate School of Engineering, Osaka City University, JAPAN
H. KONISHI
Tomoe Corporation, JAPAN

INTRODUCTION

Latticed shell roofs composed of the "(Triangle and Hexagon)-and-Hexagon" space trussed units [Refs 1-2] attract visually and designedly. The "(Triangle and Hexagon)-and-Hexagon" denotes both lattice patterns of the (Triangle and Hexagon)-on-Hexagon and the Hexagon-on-(Triangle and Hexagon). The double-layer grids of these lattice patterns for the pin-jointed case are internal instability, which have well-known unstable deformation modes due to the relative rotations of triangular pyramidal units as showed in Figure 1. In designing the latticed shell roofs composed of these units, it is important to investigate the influence of the unstable modes on the stability and the stiffness for them [Refs 3-4].

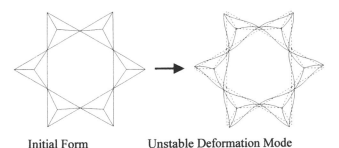

Initial Form   Unstable Deformation Mode
Fig. 1. Unstable deformation mode.

In this paper, the cylindrical and spherical latticed shell roofs composed of the "(Triangle and Hexagon)-and-Hexagon" space trussed units are analyzed under uniform vertical loads at the all joints of the upper layer, in which the joints are treated as rigid-joints. The elastic buckling loads and corresponding buckling modes, the load-deflection relationships and member's sectional forces are investigated by the elastic buckling analysis of space frames. The equivalent stress resultants and stress couples acting on the latticed shell elements as a continuum shell are estimated by the member's sectional forces and are considered as for the structural property of common continuum shells. Further, in order to increase the stiffness and elastic buckling loads of the latticed shell roofs, hybrid space structures composed of the space trussed units and thin plates are treated. The advantages of these structural performances are showed.

CYLINDRICAL LATTICED SHELL ROOFS
Analytical Models

The forms of cylindrical latticed shell roof are showed in Figure 2. TH-H Type is composed of the "(Triangle and Hexagon)-on-Hexagon" space trussed units, and H-TH Type is done of the "Hexagon-on-(Triangle and Hexagon)" units. The length of span of the two shell roofs is 76.5m. These roofs consist of the tubular members with the same section, of which mechanical and section properties are showed in Table 1. The loading condition is the uniform vertical loading at all joints in the upper surface as showed in Figure 3. The perimeter nodes of the domes are supported at the pins or rollers. The pin-support is denoted as "-pin" and the roller support is done as "-roller".

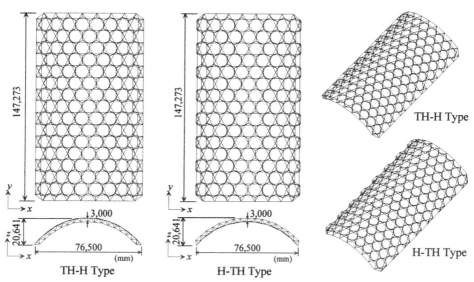

Fig. 2. Forms of cylindrical latticed shell roofs.

Table 1. Mechanical and section properties of constituent members.

Steel Pipe	
Sectional Dimension (mm)	Diameter 190.7 x 7.0
Cross Sectional Area (cm^2)	40.4
Geometrical Moment of Inertia (cm^4)	1710
Young's Modulus (kN/cm^2)	2.058 x 10^4
Poisson's Ratio	0.3
Weight of Unit Length (kN/m)	31.7

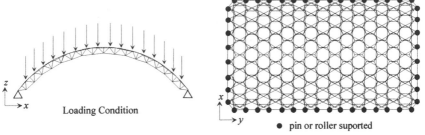

Fig. 3. Loading and support conditions.

Numerical Results

As numerical results of four cylindrical latticed shell roofs, axial forces in members, load-deflection relationships and elastic buckling loads and buckling modes are showed as follows.

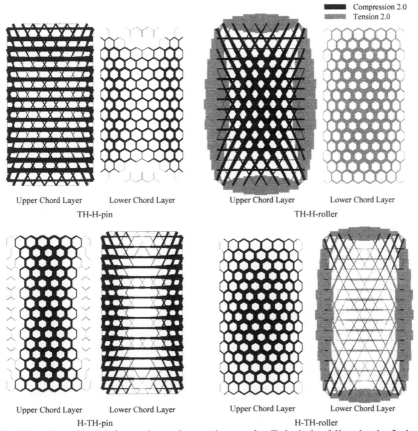

Fig. 4. The ratio of axial forces in each member to the Euler's buckling load of pin ended members.

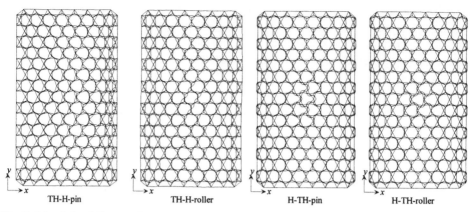

TH-H-pin   TH-H-roller   H-TH-pin   H-TH-roller

Fig. 5. Elastic buckling modes.

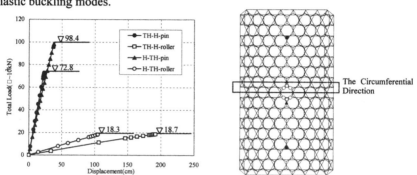

Fig. 6. Load-deflection relationship of the maximum vertical displacement nodal point.

TH-H-pin   TH-H-roller   H-TH-pin   H-TH-roller
(Total Load = 100kN)

Fig. 7. Equivalent resultant normal force (N_θ) in the circumferential direction.

TH-H-pin   TH-H-roller   H-TH-pin   H-TH-roller
Figure 8. Equivalent bending moment (M_θ) in the circumferential direction. (Total Load = 100kN)

Table 2. Total elastic buckling load and initial stiffness.

Model	Total Elastic Buckling Load (x10³kN)	Initial Stiffness (x10²kN/cm)
TH-H-pin	72.86	32.16
TH-H-roller	18.73 (25.7%)	1.06
H-TH-pin	98.44	28.54
H-TH-roller	18.25 (18.5%)	1.80

SPHERICAL LATTICED SHELL ROOFS

Analytical Models

The forms of spherical latticed shell roof are showed in Figure 9. TH-H Type is composed of the "(Triangle and Hexagon)-on-Hexagon" space trussed units, and H-TH Type is done of the "Hexagon-on-(Triangle and Hexagon)" units. The length of span of the two shell roofs is 76.5m. The size of each analytical model is showed in Table 3. These roofs consist of the tubular members with the same section, of which mechanical and section properties are showed in Table 1. The loading condition is the uniform vertical loading at all joints in the upper surface as showed in Figure 10. The perimeter nodes of the domes are supported at the pins or rollers.

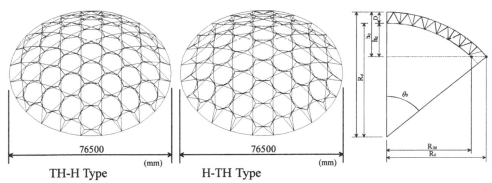

TH-H Type   H-TH Type

Fig. 9. Forms of spherical latticed shell roofs.

Table 3. The size of each analytical model.

Model	R_0(m)	R_{0d}(m)	R(m)	R_d(m)	h_0(m)	h_d(m)	D(m)	θ_0(deg)	h_0/R_0
TH-H	38.25	35.96	42.23	60.71	19.87	18.35	1.51	47.1	0.52
H-TH	38.25	35.96	42.23	60.71	19.87	18.35	1.51	47.1	0.52

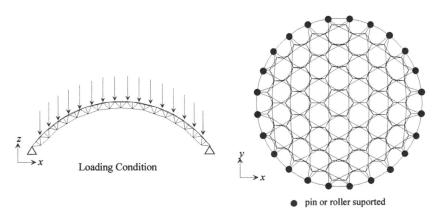

Loading Condition

pin or roller suported

Fig. 10. Loading and supported conditions.

Numerical Results

As numerical results of each spherical latticed shell roofs, axial forces in members, load-deflection relationships and elastic buckling loads and buckling modes are showed as follows.

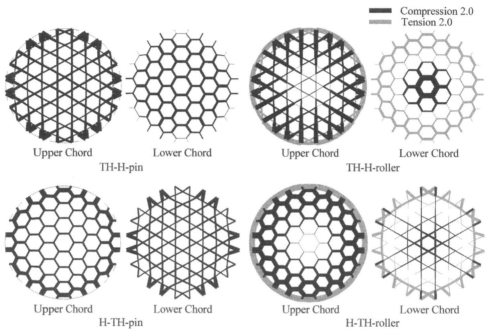

Fig. 11. The ratio of axial forces in each member to the Euler's buckling load of a pin ended member.

Table 4. Total elastic buckling load and initial stiffness of the nodal points.

Model	Total Elastic Buckling Load ($\times 10^3$kN)	Initial Stiffness ($\times 10^2$kN/cm)			
		Node 1	Node 2	Node 3	Node 4
TH-H-pin	42.19	46.00	56.37	71.08	131.73
TH-H-roller	24.97 (59.2%)	7.83 (17.0%)	5.69 (10.1%)	5.81 (8.2%)	24.93 (18.9%)
H-TH-pin	42.83	67.19	69.37	69.31	95.95
H-TH-roller	26.55 (62.0%)	5.59 (8.3%)	4.76 (6.9%)	5.09 (7.3%)	14.15 (14.7%)

( )□The ratio of the elastic buckling load or initial stiffness of pin supported condition to that of roller supported condition.

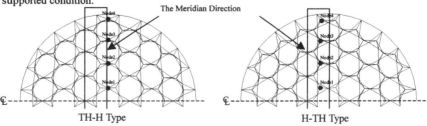

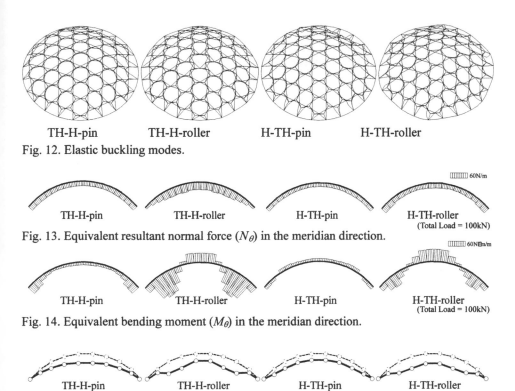

TH-H-pin   TH-H-roller   H-TH-pin   H-TH-roller

Fig. 12. Elastic buckling modes.

TH-H-pin   TH-H-roller   H-TH-pin   H-TH-roller
                                                          (Total Load = 100kN)

60N/m

Fig. 13. Equivalent resultant normal force (N_θ) in the meridian direction.

TH-H-pin   TH-H-roller   H-TH-pin   H-TH-roller
                                                          (Total Load = 100kN)

60NBn/m

Fig. 14. Equivalent bending moment (M_θ) in the meridian direction.

TH-H-pin   TH-H-roller   H-TH-pin   H-TH-roller

Fig. 15. Vertical nodal displacement in the meridian direction.

CONSIDERATION

Elastic Buckling Behavior of Cylindrical Latticed Shell Roofs

a) TH-H Type

In Figures 7 & 8, the large compressive forces are appeared in the circumferential direction on the pin-supported case. On the roller-supported case, the compressive forces in the members are showed in the central part on the upper surfaces and the large tensile forces in the members are showed in the perimeter members on the lower surfaces.

The buckling modes on the pin- and roller-supported cases include the mode corresponding to the unstable higher-order deformation. The ratio of the total buckling load on the roller-supported case to the pin-supported case is 25.7%.

b) H-TH Type

On the pin-supported case, the large compressive forces in the members are showed in the central part on the upper surfaces and near edge part on the lower surfaces. On the roller-supported case, the large compressive forces in the members are showed in the central part on the upper surfaces and the large tensile forces in the members are showed in the perimeter members on the lower surfaces. The total buckling modes on the pin- and roller-supported cases are as well as TH-H Type. The ratio of the buckling load on roller-supported to on the pin-supported is 18.5%.

c) Comparison of Equivalent Resultant Normal Force (N_θ) and Bending Moment (M_θ) of TH-H Type and H-TH Type

The equivalent resultant normal force of each type is almost uniform except the ones of TH-H-roller Type. On the pin-supported case, the equivalent bending moment of TH-H Type is different from the one of H-TH Type. On the roller-supported case, the equivalent bending moment of TH-H Type is as well as the one of H-TH Type.

Elastic Buckling Behavior of Spherical Latticed Shell Roofs

a) TH-H Type

In Figures 13 & 14, the large compressive forces in the members are appeared in the near by the boundary upper surface, on the pin-supported case. On the roller-supported case, the compressive forces in the members are showed in the near by the boundary on the upper surface and the central part in the lower surface, and the large tensile forces in the members are showed in the perimeter members in the lower surfaces.

The buckling mode of the pin-supported case is member buckling of members on the line near by the boundary. The buckling mode of the roller-supported case includes the mode corresponding to the unstable higher-order deformation in the central part. The ratio of the buckling load on the roller-supported case to the pin-supported case is 59.2%. The initial stiffness of the nodal points on the roller-supported case is from 8.2% to 18.9% of the pin-supported case.

b) H-TH Type

On the pin-supported case, the large compressive member forces are showed in the near by the boundary on the upper and lower surfaces. On the roller-supported case, the large compressive member forces are showed in the near by the boundary on the upper surface and the large tensile forces in the members are showed in the perimeter members on the lower surfaces.

The buckling mode on the pin-supported case is as well as TH-H Type. The one of the roller-supported case includes the mode corresponding to the unstable higher-order deformation on the all surface. The ratio of the buckling load on roller-supported to one on the pin-supported is 55.0%. And the ratios of the initial stiffness of the nodal points on the roller-supported case to the pin-supported case are from 6.9% to 14.7%.

d) Comparison of Equivalent Resultant Normal Force (N_θ) and Bending Moment (M_θ) of TH-H Type and H-TH Type

The equivalent resultant normal force (N_θ) of each type is almost uniform except the ones of TH-H-roller Type. On the pin-supported case, the equivalent bending moment of TH-H Type is different from the one of H-TH Type. On the roller-supported case, the equivalent bending moment of TH-H Type is as well as the one of TH-H Type.

Elastic Buckling Behavior of Hybrid Space Structure

In order to increase the stiffness and elastic buckling loads of the H-TH type spherical latticed shell roofs, hybrid space structures composed of the space trussed units and thin plates as showed in Figure 16 are treated. This thin plate is the polycarbonate plastic plate as showed in Table 5.

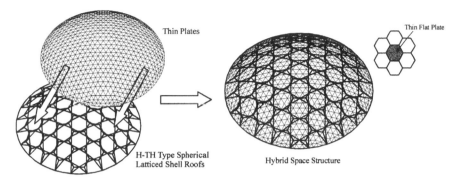

Fig. 16. Hybrid space structure.

Table 5. Mechanical and section properties of the thin plate.

Polycarbonate Plastic Plate	
Thickness (mm)	15.0
Compressive Strength (kN/cm^2)	8.751
Tensile Strength (kN/cm^2)	6.517
Young's Modulus (kN/cm^2)	236.7
Poisonn's Ratio	0.38
Rative Dencity	1.2

As numerical results of each hybrid spherical latticed shell roofs, axial forces in members, principal stresses in thin plate, load-deflection relationships and elastic buckling loads and buckling modes are showed as follows.

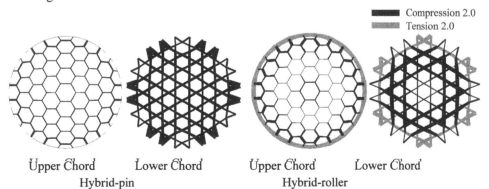

Fig. 17. The ratio of axial forces in each member to the Euler's buckling load of a pined ended member.

Table 1. Geometric & material data for dome structure.

DEFINITION	SYMBOL	VALUE	UNIT
Geometry			
Dome Diameter	Δ	160	m
Dome Height	H	86	m
Sectional Area	A	0.3	m2
Moment of Inertia	I	0.24	m4
Steel			
Elasticity Modulus	Ec	2.1e5	Mpa
Tangent Modulus	$E\tau$	0.2	Mpa
Poisson Ratio	v	0.3	-
Density	g	20000	kg/m3
Steel Yield Stress	$\Phi\psi$	320	Mpa
Spectral Modulus			
Damping Parameter	a	3.10	-
Damping Parameter	b	0.51	-
Damping Parameter	c	0.54	-
Damping Parameter	Fr	4.0e4	Mpa

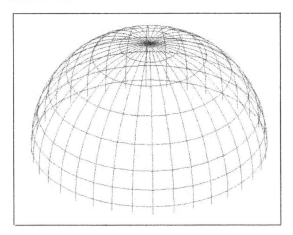

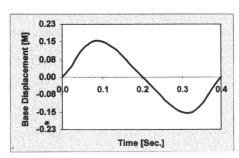

Fig. 1. Finite element model & its time history load for space dome structure.

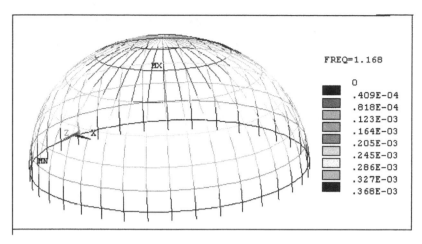

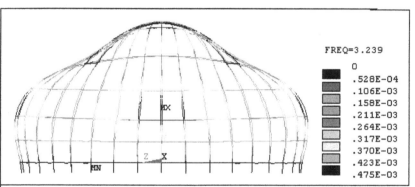

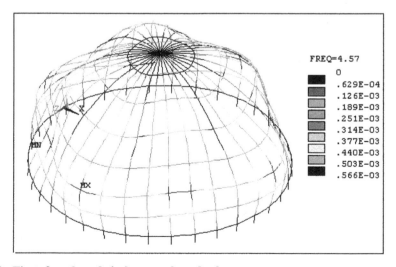

Fig. 2. First, fourth and sixth natural mode shapes.

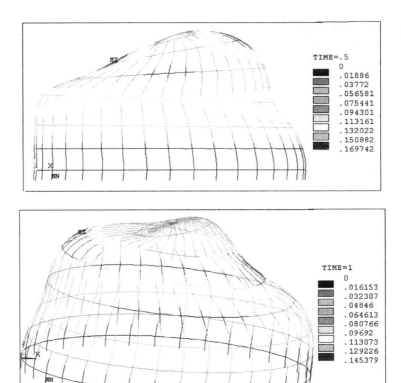

Fig. 3. Displacement contours for t=0.5 and t= 1 sec.

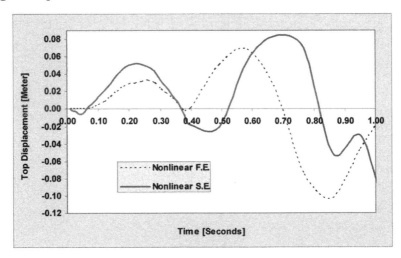

Fig. 4. Comparison of time history displacements for two models at top node.

CONCLUDING REMARKS
It can be concluded that the frequency domain equations of motion for damped three dimensional space structures can be constructed and solved in a straightforward manner when mechanical characteristic of the material are portrayed using the fractional derivative model. The fractional derivative non-linear model allows for the continuous transition from the viscous state to the solid state. The theoretical basis of the calibrated Timoshenko spectral element method of analysis has briefly been presented in the first part of the paper. The non-linear damping solution has also been shown, and the consistency of method has been investigated. As far as stability and efficiency are concerned, it can be assessed that even though the assemblage of global matrices has to be repeated for all frequency components in comparison with the single assemblage procedure for the conventional finite element method, the spectral element approach completely outperforms the conventional method in the global nonlinear dynamic analysis of space structures.

ACKNOWLEDGMENTS
The financial support by the PaymaBargh Power Engineering Contractor Company is gratefully acknowledged.

REFERENCES
1.   HORR, A. M., (1995), Energy Absorption in Structural Frames, Ph.D. Dissertation, University of Wollongong, Australia.
2.   HORR, A. M. and Schmidt, L. C. (1996), Dynamic Response of a Damped Large Space Structure: A New Fractional-Spectral Approach, Int. J. of Space Struct., Vol. 10, No. 2, p. 134-141.
3.   HORR, A. M. and Schmidt, L. C., (1996), Modeling of Non-Linear Damping Characteristic of a Viscoelastic Structural Damper, Int. J. Engrg. Struct., Vol. 18, No. 2, p. 154-161.
4.   HORR, A. M. and Schmidt, L. C., (1995), Closed Form Solution for the Timoshenko Theory Using A Computer Based Mathematical Package, Int. J. Computers & Struct., Vol. 55, No. 3, p. 405-412.
5   HORR, A. M. and Schmidt, L. C., (1997), Complex Fractional-Spectral Method for Space Curved Struts: Theory and Application, Int. J. Space Struct., Vol. 12, No. 2, p. 59-67.

Structural analysis of a braced dome and a torus grid by means of non-coplanar trihedra

Z. S. HORTOBÁGYI, J. SZABÓ and T. TARNAI
Budapest University of Technology and Economics, Hungary

ABSTRACT

For analysis of space grids composed of straight bars and frictionless spherical joints, several methods have been developed. In our method presented here, bars of a space grid are grouped in a way that 3 non-coplanar bars are associated to each node. The configuration determined by these 3 bars is called a trihedron. Trihedra are selected to form a statically determinate structure associated to the original statically indeterminate one. Remaining bars not incorporated in the system of trihedra are redundant bars. The advantage of the triherdon composition is present at tracing the change of state of the structure subjected to increasing load. If bars in trihedra buckle, they get out the trihedron system, and they should be replaced by redundant bars. This operation modifies the basic statically determinate structure, and the inverse of its equilibrium matrix can be obtained by simple modification of the previous one, that speeds up the process. This technique has been used for space grids above a rectangular base. In this paper we will extend it to braced domes and torus grids.

INTRODUCTION

The aim of this paper is the static equilibrium analysis of single-layer and double-layer space grids under continuously increased one-parameter load of arbitrary arrangement. The space grids are composed of straight bars and frictionless spherical joints. The material of the structure is linearly elastic. It is supposed that force in tensional bars can be increased up to their ultimate strength where fracture occurs. Force in compressed bars can be increased up to their critical value where the bars buckle. The force in the buckled bars, with good approximation, is considered constant. This fact is taken into account in the way that a buckled bar is removed from the structure but it is replaced at its ends with its buckling force. This force is considered as external load which is kept constant until the distance between the end points of the buckled bar reaches again the length of the bar prior to buckling.

From bars of the grid, we compose trihedra. A trihedron is defined by three non-coplanar bars having one end point in common, and this common point is considered as starting point of the bars in the trihedron. That means that each node is the starting point of three bars, but destination point of more other bars, which can be bars of neighbouring trihedra or can be redundant bars. In Fig. 1(a), bars in a trihedron at each node of a triangular grid are shown by arrows pointing outwards from the node. The structure composed of trihedra is a statically determinate structure associated to the original statically indeterminate one. Therefore, this statically determinate structure defines – in mathematical terms – a directed graph.

Space Structures 5, Thomas Telford, London, 2002

We have used the trihedron principle in Refs 1, 2 before, but only for space grids above a rectangular ground plan. In this paper, we will extend this technique for single- and double-layer grids fitted to multiply connected surfaces, such as braced domes with skylight opening and torus grids, where it is easy to develop a system of trihedra due to the cyclic property of the structure.

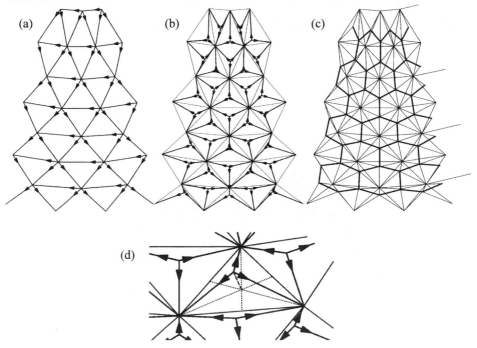

Fig. 1. Part of the bar network of a braced dome: (a) triangular network of the external layer with the trihedra; (b) trihedra of the connecting bars are added to (a); (c) the internal layer is added to (b), bars in the internal layer are emphasized by thick lines. (d) Construction of one node in the internal layer, and the respective trihedron.

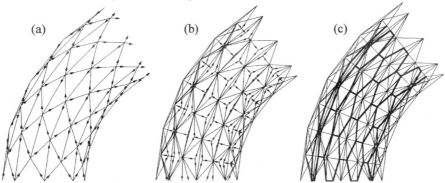

Fig. 2. Part of the bar network of a torus grid: (a) triangular network of the external layer with the trihedra; (b) trihedra of the connecting bars are added to (a); (c) the internal layer is added to (b), bars in the internal layer are emphasized by thick lines.

AUTOMATIC GENERATION OF TRIHEDRA

First, the external layer of the double-layer grid in question is generated. The external layer is a triangular net that is not obtained by projection of a regular triangular lattice given in the ground plan, but it is generated directly on the surface. Here, due to the cyclic character of the network, the trihedron system is easily defined and programmed. In the obtained system, to each of the internal joints (not laying on the foundation) of the external layer there corresponds a trihedron such that each bar of the external layer appears only in one trihedron. In the case of a complete torus grid, the static and kinematic determinacy of the structure determined by trihedra is guarantied if there are trihedron bars among the bars supporting the structure. This requirement can cause local disturbance in the regular cyclic arrangement of trihedra. In the case of a braced dome with skylight opening, if it has both rotational symmetry and a plane of symmetry, it is required that the number of vertices of the regular polygon determined by the foundation joints should be an odd number. Otherwise the structure determined by the trihedra will be both statically and kinematically indeterminate (Refs 3, 4). Figs 1(a) and 2(a) show a part of the external layer of a braced dome and a torus grid, respectively, together with the trihedra associated to the respective nodes.

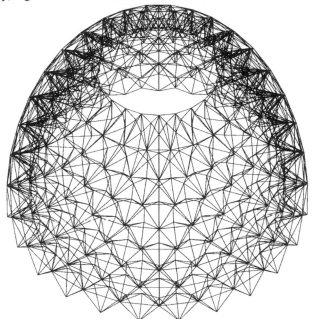

Fig. 3. Bar network of a double-layer braced dome with skylight opening.

Each node of the internal layer corresponds to a triangle in the external layer: it is positioned on the line passing through the centre of gravity of the triangle, perpendicular to the plane of the triangle, at a constant distance from the plane of the triangle. Then, each of the nodes of the internal layer is connected to the vertices of the respective triangle by three bars forming a trihedron (Figs 1(b) and 2(b)). In this way, bars of trihedra associated to the nodes of the internal layer all are bars connecting the internal layer to the external one. Bars in the internal layer do not belong to any trihedra, that is, they are redundant bars. Bars in the internal layer form hexagons as shown by thick lines in Figs 1(c) and 2(c) for a part of a braced dome and a torus grid. The network of the complete grid structure is presented in Fig. 3 for the dome and in Fig. 4 for the torus.

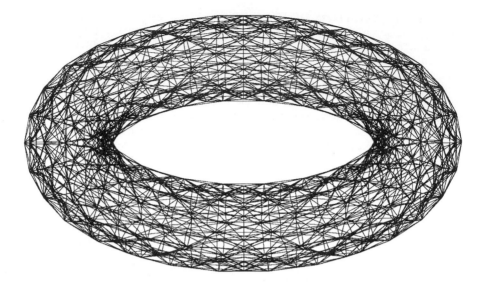

Fig. 4. Bar network of a double-layer complete torus grid.

INVESTIGATION OF EQUILIBRIUM

Matrix expressions of the force method of an elastic bar structure can be derived from the partitioned form of the state equation (Ref. 5)

$$
\begin{bmatrix} & \mathbf{G}_1 & \mathbf{G}_2 \\ \mathbf{G}_1^T & \mathbf{F}_1 & \\ \mathbf{G}_2^T & & \mathbf{F}_2 \end{bmatrix} \bullet \begin{bmatrix} \mathbf{v} \\ \mathbf{s}_1 \\ \mathbf{s}_2 \end{bmatrix} + \begin{bmatrix} \mathbf{q} \\ \mathbf{t}_1 \\ \mathbf{t}_2 \end{bmatrix} = \mathbf{0}
$$

(1)

where the following notation is used:

$\mathbf{F}_1$, $\mathbf{F}_2$ flexibility matrices of trihedron bars and redundant bars,

$\mathbf{G}_1$, $\det(\mathbf{G}_1) \neq 0$ equilibrium matrix of the statically determinate structure defined by the unit vectors of bars in trihedra,

$\mathbf{G}_2$ equilibrium matrix composed from the unit vectors of redundant bars,

$\mathbf{q}$ vector of loads on nodes,

$\mathbf{s}_1$, $\mathbf{s}_2$ vectors of forces in trihedron bars and redundant bars,

$\mathbf{t}_1$, $\mathbf{t}_2$ vectors of kinematic loads of trihedron bars and redundant bars,

$\mathbf{v}$ vector of displacements of nodes.

Let us introduce the additional notation

$\mathbf{Q}_1 = \mathbf{G}_1^{-1}$,

$\mathbf{Q}_2 = \mathbf{G}_1^{-1} \bullet \mathbf{G}_2 = \mathbf{Q}_1 \bullet \mathbf{G}_2$.

The first equation in (1) is the equilibrium equation $\mathbf{G}_1 \bullet \mathbf{s}_1 + \mathbf{G}_2 \bullet \mathbf{s}_2 + \mathbf{q} = \mathbf{0}$, from which $\mathbf{s}_1$ can be expressed in the form

$$
\mathbf{s}_1 = -\mathbf{Q}_2 \bullet \mathbf{s}_2 - \mathbf{Q}_1 \bullet \mathbf{q}.
$$

(2)

The second equation in (1) is the compatibility equation for the trihedron bars: $G_1^T \bullet v + F_1 \bullet s_1 + t_1 = 0$, from which, after replacing s_1 by (2), v can be expressed as

$$v = Q_1^T \bullet F_1 \bullet Q_2 \bullet s_2 + Q_1^T \bullet F_1 \bullet Q_1 \bullet q - Q_1^T \bullet t_1. \tag{3}$$

The third equation in (1) is the compatibility equation for the redundant bars: $G_2^T \bullet v + F_2 \bullet s_2 + t_2 = 0$. By introducing (3) to it, we have

$$\left(Q_2^T \bullet F_1 \bullet Q_2 + F_2 \right) \bullet s_2 + Q_2^T \bullet F_1 \bullet Q_1 \bullet q - Q_2^T \bullet t_1 + t_2 = 0, \tag{4}$$

that in a short form is

$$A \bullet s_2 + a_0 = 0, \tag{5}$$

where A is the matrix of the flexibility coefficients and a_0 is the vector of load coefficients. By solving equation (5), that is (4), forces s_2 in redundant bars are obtained. Substitution of s_2 into (2) yields forces s_1 in bars of trihedra.

CHANGE OF STATE

In producing the coefficient matrix A in the force method, matrix Q_1 – that is the inverse of the equilibrium matrix G_1 composed of the unit vectors of the bars in trihedra – plays an important role. During the loading process, both trihedron bars and redundant bars are continuously rearranged. If in a trihedron, a bar ceases to exist by breaking or buckling, then the incomplete trihedron should be supplemented so, that the missing bar is replaced with the nearest redundant bar directly or with the help of a chain of rearranged trihedra leading to the nearest redundant bar (Ref. 2). In this way, direction of some of the bars in trihedra can change. As a result of the rearrangement, matrix G_1 changes, and consequently, matrix Q_1 also continuously changes as well. In the different load steps, very often we have a different statically determinate structure associated to the original structure, and together with it a different matrix G_1 as well. However, thanks to our method, it is not necessary to invert directly the new G_1 at each step, but the required inverse matrix Q_1 can be produced by modifying the inverse obtained in the previous step, without determining the new matrix G_1 itself. This procedure needs very little computation time even for a large system.

A detailed account of handling of the change of state can be found in Ref. 6. Here only the most important relationship based on a result of Rózsa (Ref. 7) is presented to show the advantage of the trihedron built up in the modification of matrix Q_1. Let matrix G_1 be modified by a matrix given in the form of a product $F \bullet H$. Then the inverse of the modified matrix is

$$(G_1 + F \bullet H)^{-1} = Q_1 - Q_1 \bullet F \bullet D^{-1} \bullet H \bullet Q_1$$

where

$$Q_1 = G_1^{-1}; \quad D = E + H \bullet Q_1 \bullet F; \quad \det(G_1) \neq 0; \quad \det(D) \neq 0.$$

Number of rows of H and number of columns of F is identical to the number of trihedra involved with the rearrangement.

CONCLUSIONS

It is easy to generate automatically the network of a space grid fitted to an elliptic surface of revolution and to a torus if the equation of the surface is known. Forming trihedra can be done easily due to the cyclic symmetry of the structure. Generation of the second layer and arrangement of bars connecting the two layers to each other and forming their trihedra is also automated. On the basis of experiences we can say that forming trihedra is not a problem for either single-layer or double-layer space grids.

It is easy to trace the change of state of the grid under an increasing one-parameter load, thanks to the effectiveness of the procedure that comes from the trihedron principle and the technique producing the inverse of the modified equilibrium matrix of the statically determinate part of the structure. This procedure is efficient even in the case of a structure composed of more thousand bars. The load can be increased until the number of buckled and fractured bars is greater than the number of redundant bars, when the structure becomes kinematically indeterminate, that is, it works as a one- or more-degree-of-freedom mechanism. Analysis of such a mechanism, however, is another chapter of investigation of change of state (Ref. 8).

ACKNOWLEDGEMENTS

The research reported here was supported by OTKA Grant Nos. T031931 and T029640 awarded by the Hungarian Scientific Research Funds. Partial support by the János Arany Foundation is also gratefully acknowledged.

REFERENCES

1.   SZABÓ J and TARNAI T, General theory and numerical analysis of single- and double-layer pin jointed space grids, Space Structures 4 (Ed.: Parke G A R and Howard C M), T Telford, London, 1993, vol. 1, pp. 715-722.
2.   SZABÓ J and TARNAI T, Analysis of critical and post-critical states of pin-jointed double-layer space grids fitted to surfaces of double curvature, Proc. of the Int. Colloquium on Computation of Shell and Spatial Structures (Ed.: Z B Yang), Taipei, 1997, pp 121-126. [Reprinted: Journal of IASS, vol. 39, 1998, pp. 91-95.]
3.   FÖPPL A, Vorlesungen über technische Mechanik, II. Band, Graphische Statik, Oldenbourg, München, 1942, pp. 257-266.
4.   TARNAI T, Simultaneous static and kinematic indeterminacy of space trusses with cyclic symmetry, International Journal of Solids and Structures, vol. 16, 1980, pp. 347-35
5.   SZABÓ J and ROLLER B, Anwendung der Matrizenrechnung auf Stabwerke, Akadémiai Kiadó, Budapest, 1978.
6.   HORTOBÁGYI ZS, Analysis of the critical state of double-layer space grids fitted to a flat surface, Proc. IASS Int. Conf., LSCE, Warsaw, 1998, pp. 198-201.
7.   RÓZSA P, Linear Algebra and its Applications (in Hungarian), Tankönyvkiadó, Budapest, 1991, pp. 294-295.
8.   HORTOBÁGYI ZS, Numerical analysis of inextensional kinematically indeterminate assemblies, Periodica Polytechnica, Civ. Engng, vol. 44, 2000, pp. 43-55.

Polyhedral Space Structures and the Periodic Table of the Polyhedral Universe

MICHAEL BURT, Faculty of Architecture and Town Planning, TECHNION, I.I.T. Israel.

POLYHEDRAL SPACE STRUCTURES-INTRODUCTION

Space Structures are those which owe their structural performance mostly to the mode of their material distribution in space, rather than to the amount of the invested material or its kind. Structural performance, its balance and equilibrium, level of it's potential energy, and it's stability and rigidity, is predominantly govered by morphological-topological constraints and the way space and matter throughout the structure are manipulated.

Morphologically, all space structures and their force patterns in 3-D space are related to the domain of polyhedral forms and phenomenology. Understanding their nature and perceiving their governing order and the resulting hierarchical classification is an intellectual task of immense importance and is at the core of this presentation.

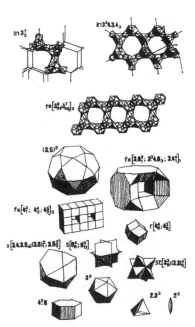

Fig.1 Types of polyhedra within the polyhedral universe–finite, toroidal and infinite.

Polyhedral space structures, namely those which could be characterized as finite, or (the so called) infinite, lattice (bar and joint) structures, or plate (plane or curved panel) structures are of special interest and importance, because of their role and abundance in our built environment. (Refs. 1-4).

The number of such polyhedral structures, differing in some of their structural-energetical and topological-geometrical parameters, amounts to infinity, turning their hierarchical and typological classification into a major professional concern.(Fig. 1).

Throughout history many made important contributions to their morphology understanding. To mention just few:

-Rene Descarts (17-th century) stated that: "The total angular deficit, or the sum of all the angular deficits (∂), taken over all the vertices of convex polyhedron, equals 4π for all regular polyhedra".

(1) $\partial v = (2\pi - \Sigma\alpha)v = 4\pi$ ($\Sigma\alpha$ as the sum of angles in a vertex).

Leonhard Euler (18-th century), in a most celebrated theorem, stated that:
The number   V-E+F=K (V,E,F stand for

Vertices, Edges and Faces, respectively, with K- the Euler characteristic of the manifold) is the same for all permissible maps on the manifold. In general, for the 3-D space, this theorem could be presented as:

(2) $V-E+F = 2(1-g)$;   with $2(1-g)$ as the Euler characteristic-K (g-stands for genus of the manifold).

The Descartes formula could be expanded into the following:

(3) $\partial v = v(2\pi - \Sigma\alpha) = 2\pi K = 4\pi(1-g)$, where ∂ - is the angular disparity (positive or negative) value; $\Sigma\alpha$. is the average sum of angles in a vertex. This formula is valid for every 3-D polyhedron (non-periodic as well, and for all g-domains), when defined as follows:

"A polyhedron-P is a connected, unbounded, 2-D manifold, formed by a set of simply connected polygonal regions of order n, for $n \geq 0$, arranged so that each edge of each region is matched with exactly one other edge of the same, or another region (and vertices are matched only as required by the matching of edges. It implies that one and the same, or two, and no more than two distinct polygonal regions (faces) meet at each edge " (paraphrasing Stewart). (Ref. 3).

The Descartes formula establishes a consequential relation between V and $\Sigma\alpha_{av.}$ for any given-g. From it and the a.m. definition it follows that:

(4) $Pav = \dfrac{2\pi . val_{av}}{\pi . val._{av} \Sigma\alpha_{av}}$, where Pav. Is the average polygonal face (the polygon's number of edges-vertices) and val_{av} Is the average valency over the vertices – (the average number of edges meeting in a vertex).

THE PERIODIC TABLE OF THE POLYHEDRAL UNIVERSE

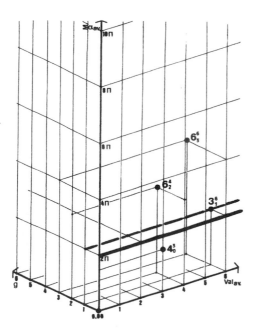

Fig.2 Periodic Table of the Polyhedral Universe.

'The Periodic Table of the Polyhedral Universe' is a relational-referential system of all possible polyhedra in 3-D space. It deals with all polyhedral envelopes which comply with the celebrated Euler's formula of $V - E + F = 2(1-g)$, with Vertices, Edges, Faces and genus of the polyhedral envelope, respectively, and represents the formula's visual animation,(Ref-2) The Table is constructed on the basis of the polyhedral primary parameters, considered, by the author, to be the Val.av. (average valency); $\Sigma\alpha av$ (the average sum of angles of the polygons around a vertex) and g (genus of the polyhedral envelope). When taken as coordinates of a Cartesian 3-D space, these primary parameters provide for a 3-D environment, in which every conceivable 3-D polyhedron, whether topologically spherical, toroidal or hyperbolical (sponge-like), has a unique point representation, (Fig-2).In case of a non-periodic polyhedron these parameters should be deduced from it's macro-characteristics, in terms of average values.

It transpires that while the g-domains appear as evenly interspaced parallel planes, $\Sigma\alpha$ and Val. coordinates evolve into much more intricate patterning which is at the core of the prevailing order of the 'Periodic Table' and which accounts for its periodic   nature.

The principal, topologically different polyhedral phenotypes are discernible through their distinct location characteristics, with all the toroidal 2- manifolds, assuming the single-line locus of: g=1; $\Sigma\alpha av. = 2\pi$,   which sharply subdivides between the topologically 'spherical-finite' polyhedra of the g=0 domain and the topologically 'hyperbolic' polyhedra of the g ≥ 2 domains. All conceivable structural-topological properties, shared by a group of polyhedra and expressible in terms of the a.m. parameters, disclose characteristic location patterns, and every discernible pattern of polyhedral location points represents a distinct shared property, (Fig-3).

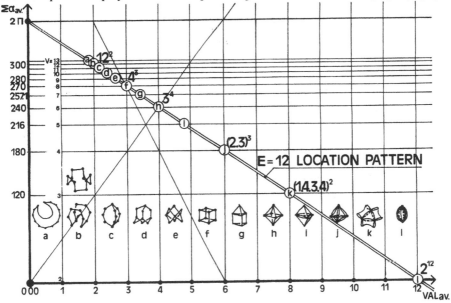

Fig.3 E=12 location pattern within the g=o domain.

From the nature of polyhedral 2 – manifolds it is evident that: (5) $Pav.F=2E=Val_{av.} V$
This set of equations, when joined with the Euler's and Descartes' formulas, gives rise to the following:

(6) $V = \dfrac{4\pi(1-g)}{2\pi - \Sigma\alpha_{av}}$

(7) $E = \dfrac{2\pi.Val._{av}(1-g)}{2\pi - \Sigma\alpha_{av}}$.

(8) $F = \dfrac{2(\pi.Val._{av} - \Sigma\alpha._{av.})(1-g)}{2\pi - \Sigma\alpha._{av.}}$

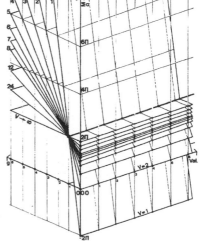

Fig.4 Quadratic, doubly-curved surface as a location pattern of all polyhedra with E=12 edges only .

It should be noted that these equations, when applied to 'Periodic Infinite Polyhedra', should refer to their translation unit.

Each of the equations leads to a specific universal location pattern, meaning: embracing all the polyhedral universe, throughout all the g-domains. Location patterns of polyhedra sharing the same number of vertices take the form of a set of planes sharing the (g=1; $\Sigma\alpha = 2\pi$) line, (Fig-4). Location patterns of E and F are in the form of quadratic (hypar) surfaces. Location pattern of E, sharing the (Val.=0; $\Sigma\alpha=2\pi$) and (g=1; $\Sigma\alpha=2\pi$) lines and location pattern of F, sharing the (Val=2; $\Sigma\alpha=2\pi$) and (g=1; $\Sigma\alpha = 2\pi$) lines.

It is of great interest and importance to try and to discover, describe and comprehend these location patterns, especially those which relate to various significant structural families and categories. By doing so, we can gain significant insights into their structural-topological nature, their (theoretical) range of existence and their bounding topological constraints.

LOCATION CHARACTARISTIC OF STRUCTURAL PROPERTIES

Morphologically speaking, structural properties of lattice 'bar and joint' structures, their stability, rigidity and stiffness, are mostly affected by bar & joint density and curvature, when considered as 3–D surface structures (lattice or membrane – plate like).

In a physical polyhedral configuration, bar and joint density is associated with $Val._{av.}$ and $P_{av.}$ and is directly affected by the absolute values of angular deficiency $-\partial$. It is gradually increasing when moving in the direction of the product vector of the increasing ∂ and $Val_{av.}$ (Fig-5).

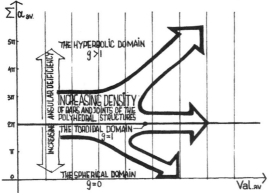

Fig.5 The development of density of polyhedral lattice structures.

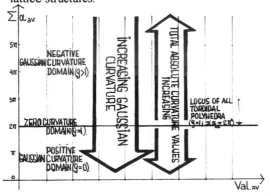

Fig.6 The development of polyhedral envelope curvature-absolute and Gaussian, within the periodic table's space.

Gauss was the one to originate a method of representing the curvature at a point of a surface by a single number (K), which depends on the principal curvatures at this point (k,& k_2), and equals their product: $K = k_1 \cdot k_2$. The Gaussian curvature has the highly important property of remaining invariant if the surface is subjected to an arbitrary bending.

It could be easily perceived that the topological-structural behavior of polyhedra, as revealed

through the 'Periodic Table of the Polyhedral universe', is consistent with the gradual development of the Gaussian curvature over the polyhedra within the'Table's entire space, from the positive curvature of the g =o domain, to the zero curvature of the g =1 domain and into the region of negative curvature values, of

$g \geq 2$ domains. Since resistance to surface buckling depends on the absolute curvature of the polyhedral envelope, it is assumed that the farther the location point of a polyhedron from the $\Sigma \alpha = 2\pi$ zone, the greater its stiffness and surface buckling resistance. (Fig-6).

STABILITY OF 3-D POLYHEDRAL PLATE AND LATTICE STRUCTURES AND THE GRAND DIVIDES

Polyhedra, as physical structures, may appear as pure lattice or plate structures or their hybrids. Pure lattice structures consist of (ideal) swivel joints (nodes, vertices) and bending-resistant bars-edges, with no faces-plates, and external loads (including supports) act along the edges only.

Pure plate structures consist of plates (faces), joined together by hinged (shear-resistant) edges and can transfer external load forces through the rigid plates only.

Polyhedral structures can be stable or non-stable and that due to the number of force transmitting elements and their mode of interrelation". (Ref. 5;6)."By stability it is meant to imply a pure lattice or plate structure's condition in which all relative motion between the nodes or the plates is completely constrained. In 'just stable'structures, with no redundancy of elements whatsoever (global or local), it is impossible to remove any of the edges (bars or shear-lines, lines of support) without causing total or local collapse.

Stability of finite, topologically spherical polyhedral structures (in the $g=0$ domain) is generally associated with triangulation, in the case of lattice structures, and 3-way edge valency in the case of plate structures, (Ref. 7). For polyhedra of $g \geq 1$,stability is dictated by g, $\Sigma \alpha$ and Val. Values. Non-stability or hyper stability of polyhedral structures is associated with edge disparity- Ed. (deficiency or redundancy of edges), which represents the difference between the actual number of edges of a given polyhedral structure (E), and the required number of edges for its stability (Est.). When Ed. = E – Est. = 0, it means that the structure is ' just stable' = statically determinant.   Ed.equations are as follows:

$$(9) \quad Ed.FIN.L. = \frac{2\,(\pi\,Val._{av} - 3\,\Sigma\alpha_{av}) + 2\pi\,g(6 - Val._{av})}{2\pi - \Sigma\alpha_{av.}} \qquad \text{for finite lattice structures.}$$

$$(10) \quad Ed.FIN.P. = \frac{(6\Sigma\alpha_{av.} - 4\pi\,Val_{av.})(1 - g)}{2\pi - \Sigma\alpha_{av.}} + 6 \qquad \text{for finite plate structures.}$$

$$(11) \quad Ed.INF.L. = \frac{2\pi\,(1 - g)(Val_{av.} - 6)}{2\pi - \Sigma\alpha_{av.}}$$

$$(12) \quad Ed.INF.P. = \frac{(6\,\Sigma\alpha_{av.} - 4\pi\,Val_{av.})(1 - g)}{2\pi - \Sigma\alpha_{av.}}, \qquad \text{for infinite lattice and plate structures.}$$

The location patterns for any of the Ed. Values of all polyhedral lattice and plate structures, whether finite or infinite, are in the form of quadratic, hypar surfaces, crossing through all the g-domains. They are combined into universal, periodic multi-surface arrays, crossing through all the g-domains in a very typical fan like pattern, and converging in each of the g-domains on points: the Ed.L ($Val = 6; \Sigma\alpha = 2\pi$)   for the lattice structures and the Ed.P ($Val = 3; \Sigma\alpha = 2\pi$) for the plate structures. Ture Wester has devised a geometrical solution of 3 valency tiling on a hyperbolic surface, with plane (non regular and non uniform) hexagonal tiles.

All the location points of polyhedral lattice and plate structures, with Ed.=0, form into four continuous distinct patterns, in the form of two quadratic (hypar) surfaces for finite polyhedra, and two plane surfaces for infinite polyhedra. These surface-loci define the Grand Divides of the polyhedral universe, dividing the Periodic Table's space between the non-stable (with negative Ed. values) and the potentially hyper-stable structures, having redundancy and positive Ed. values.(Fig.-7).

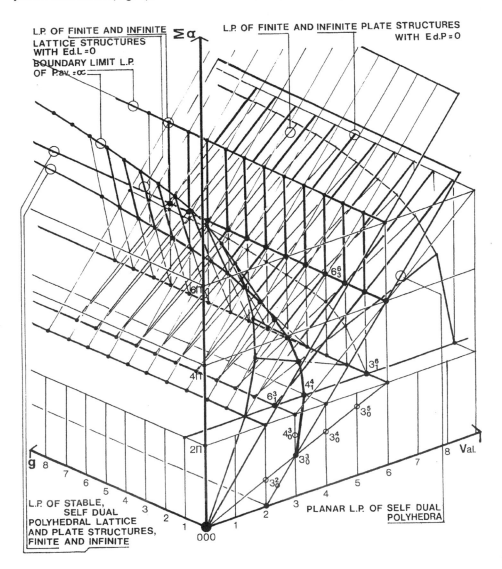

Fig.7 The grand divides of the polyhedral universe; location patterns of all 'Just stable' lattice and plate structures.

It should be noted that physical polyhedral plate and lattice structures are just a subset of all conceivable polyhedral geometries in 3-D space. Geometrically speaking,3-D polyhedra, complying with the Euler's formula, may assume the forms of what might be called 'pseudo polyhedra', or what the author calls – ' Floral Polyhedra': with digons, dihedrons, curved edges and faces, in numerous cases with V<4; Pav.<3; Val.av.<3 and even zero or negative $\Sigma\alpha_{av.}$ values, for polyhedra with two or even one vertex only (in the g-0 domain).

Their structural-statical nature is ambiguous although the 'Periodic Table' displays no discrimination whatsoever; casts on them the light of legitimacy and lets them to participate in the grand ordering game.

DUALITY OF LATTICE AND PLATE STRUCTURES

The duality of polyhedral lattice and plate structures was postulated and developed by Ture Wester (and in parallel by Whitely) back in 1984 (Ref.7). Morphologically, the essence of this duality is in the following:

1. If a specific polyhedral plate structure is stable, than it's polyhedral dual, as a lattice structure, will be stable as well, and vice versa. It means that if Pav.of the first equals –3, than Val.av. of the latter will equal-3 also.

2. The Ed. Value of a polyhedral lattice structure is the same for its dual, acting as a Polyhedra plate structure, and vice versa.

The lattice-plate duality receives a very clear and convincing illustration through the ordered structure of the 'Periodic Table' (Ref.6), and, as it seems, the Self-Dual Polyhedra (S.D.P) location pattern plays in it a central role.

All the mutually dual polyhedral pairs, stable, unstable or hyperstable, finite or infinite, in all g - domains   and   on all V- levels, demonstrate a 'reflection – like symmetry' about   this plane.(Fig-8).

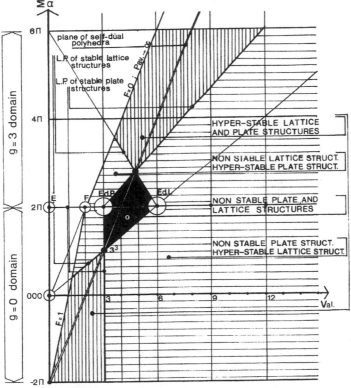

Fig.8   Regional distribution of structural properties of finite polyhedra in the g=o and the g-3 domains.

REPRESENTATION OF STRUCTURE TYPES, WITH POLYHEDRAL CHARACTERISTICS
Some natural and artificial structure types, with an affinity to specific groups and types of polyhedra, may be represented within the 'Periodic Table' and through their mere location characteristics and location patterns, bring significant general and some particular insights into the nature of their topological-structural-energetical behavior. To some extent the Periodic Table' may also facilitate their morphological classification and hierarchical order.
Various physical structure types, natural and artificial, when morphologically generalized into structural polyhedral categories, emerge to possess distinct location characteristics.

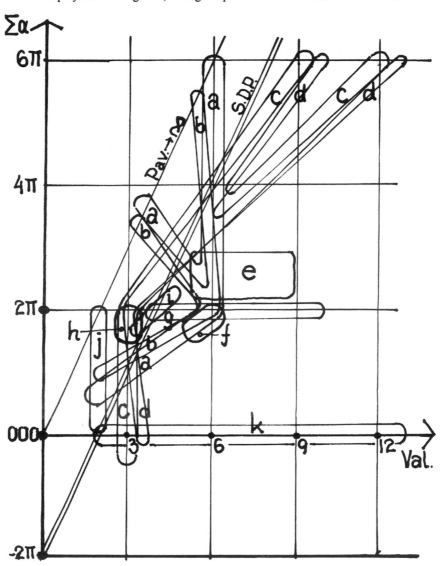

Fif.9   Artificial and natural polyhedral structures and their location within the
Periodic table of the polyhedral universe.

Of special note are:
- (a) Pure 'just stable' lattice (finite and infinite) structures.
- (b) Polyhedral lattice mechanisms.
- (c) Pure 'just stable' plate (finite and infinite) structures.
- (d) Polyhedral plate mechanisms.
- (e) I.P.L (Infinite Polyhedral Lattice)space structures.
- (f) Single layer geodesic dome lattice structures.
- (g) Spherical radiolarian structures.
- (h) Buckminster Fullerens molecular carbon structures.
- (i) Quadrangulated grid shell structures.
- (j) Pneumatic dihedral (natural and artificial) structures.
- (k) Pneumatic polyhedral – bi-polar (natural and artificial structures.

Some of these categorically defined structural phenotypes are confined to small areas and distinct single line location patterns within the Table's space, but some refer to a region or occupy a space blob, spreading over more than a single g-domain (Fig-9).

IN CONCLUSION

'The Periodic Table of the Polyhedral Universe' is a way of point- representing and interrelating all possible 3-D polyhedra which comply with the celebrated Euler's theorem and formula: $V - E + F = K$.

Constructed on the basis of polyhedral 'primary parameters', considered to be: $\Sigma\alpha.av$ Val.av and g, it relates all polyhedral properties of a topological nature and presents them as distinct location patterns, some of which are limited in extent and some extending over all the g-domains.

Of special interest are the polyhedral phenotypes and their topological-structural-energetical nature, the related properties of density, curvature and the resulting stability, rigidity and stiffness, as represented by their location pattern characteristics within the Table's domain. While delving into these location patterns, many significant deep insights are gained into the polyhedral nature and behavior as physical structures.

REFERENCES

1. M.Burt (1966). 'Spatial Arrangement and Polyhedra with Curved Surfaces and their Architectural Application'. Technion Publication, Israel

2. M.Burt (1955), 'The Periodic Table of the Polyhedral Universe'. Technion Publication, Israel.

3. R.Stewart (1970). 'Adventures Among the Toroids', Published by the author, Michigan.

4. A.Wachman,M.Burt,M.Kleinman. (1874). 'Infinite Polyhedra'. Technion Publication, Israel.

5. T.Wester,M.Burt (1997). 'The Basic Structural Content of the Periodic Table of the Polyhedral Universe'.Proceedings, IASS conf. San Francisco.

6. M.Burt,T.Wester. (1997), 'The Periodic of the Polyhedral Universe and the Plate-Lattice Duality'. Proceedings of IASS Conference, Singapore.

7. T.Wester, (1984). 'Structural Order Space, the Plate-Lattice Dualism'. Danish Royal Academy Publication

A Limited and Biased View of Historical Insights for Tessellating a Sphere

JOSEPH D. CLINTON, Clinton International Design Consultants / Educational Assistance Resource Center

INTRODUCTION

Richard Buckminster Fuller, in the preface to his book *"No More Secondhand God and Other Writings"*, wrote:

"My philosophy requires of me that I convert not only my own experiences but whatever I can learn of other men's experiences into statements of evolution trending....that I at least attempt to solve the problems by inanimate invention of comprehensive anticipatory design science rather than by yielding to the easier behavior of problem discovery and the exhortation of others to solve these problems." Ref 1

The mind is a collective, made up of those in the past who passed on knowledge to those in the present, who will add to it and pass it on to those in the future. The focus of this paper is to follow a number of random contributions from great mines that have helped form the collective mind in the evolving quest for understanding the principles applied to geodesic tessellations of the sphere.

THALES of MILETUS (636?-546? BC)

We begin our trip with Thales of Miletus around 636?-546? BC who is thought to have used the method of gnomonic projection as a means of charting the heavens.

The method was developed by the projection of points on the surface of a sphere onto a tangent plane using projection lines radiating from the center of the celestial sphere. See figure 1, Ref 2.

Figure 1. Gnomonic projection

ABRAHAM SHARP (1717)

Abraham Sharp was a maker of astronomical instruments for the Royal Naval dockyard in Chatham and was associated with Flansteed at the Greenwich Observatory. In 1699 he calculated PI to 72 decimal places and, in 1717 he published a book *Geometry Improve'd*. Ref 3 His Plate II, illustrated in figure 2, from the book, shows how to cut polyhedral solids from cubes of wood. Note: the icosahedral symmetry.

If we look at them with eye of Fuller's geometry in mind we can see several frequencies of the Class I and Class II geodesic subdivision. If the diamonds are converted to triangles and

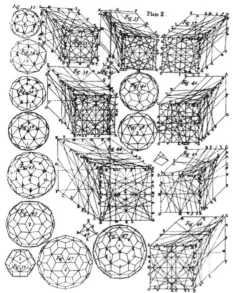

their vertices are projected to the surface of the same sphere others also have the appearance of Fuller's geodesic forms. At this point a comparison has not been done to see if they are in fact the same as Fuller's original geometry. But they do share the same topological characteristics.

AIDA YASUAKI (late 1700's – early 1800's)

This early sketch of a polyhedron approximating a sphere composed of hexagons and pentagons appeared in "*A Mathematical Collection of Polyhedra (Sanpo-Kir-iko-Shu)*."

A Japanese mathematician Aida Yasuaki drew it in the late 1700's or early 1800's. Ref 4

Figure 2. Sharp's polyhedra models

The illustration, figure 3, is a truncated pentakis dodecahedron. The pentakis dodecahedron is the dual of the truncated icosahedron. The implication is that Yasuaki was following the series of truncations of the familiar polyhedra. In the eyes of modern Fuller geodesic geometry, we have a 2ν Class II geodesic icosahedron hexagonal tessellation.

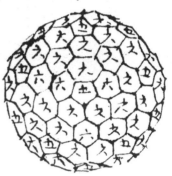

Figure 3. Yasuaki's polyhedron

SCHWARZ (1873)

A spherical triangle, which is 1/6th of a spherical icosahedral face or 1/120th of a sphere, is one of 44 Schwarz triangles, figure 4. In Fuller's icosahedral geodesic forms he refers to this triangle as the "*Basic Disequilibrium 120 LCD*". Other Schwarz triangles are used that are appropriate to other polyhedra to generate geodesic spheres, such as those based on the tetrahedron, octahedron etc. Ref 5, 6

Figure 4. Schwarz triangle

IMPERIAL PALACE CHINA (1885)

The Imperial Palace, or Forbidden City, in Beijing China was constructed during the Ming and Qing Dynasties (1368-1911). The lion guard, shown in figure 5, at the Nurturing the Heart Gate is similar to one at China's Summer Palace, near Peking, built in about 1885. If one looks closely, under one paw, you can see what some believe is a model of the celestial sphere in the form of a geodesic subdivision of the icosahedron. Ref 7, 8

Figure 5. The celestial sphere

WALTER BAUERSFELD (1922)

In 1922, the Carl Zeiss optical works of Germany unveiled the first planetarium projector, figure 6, and working with the firm of Dyckerhoff and Widmann the first thin-shell concrete structure. Both were under the direction of Walter Bauersfeld. These remarkable achievements used the same geometrical subdivision of the icosahedron as the basis of their mechanical and structural system. Ref 9, 10

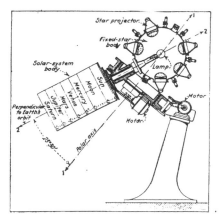

Figure 6 Zeiss star projector

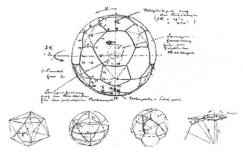

Figure 7. Bauersfeld's geometrical solution

In the words of Walter Bauersfeld, "If one starts with the familiar regular solid whose surface consists of 20 equilateral triangles, and makes a straight cut across each of the 12 vertices which this solid possesses, then 20 hexagons and 12 pentagons are formed on the surface. With the cuts in the right places, it is easy to ensure that the circles circumscribing the pentagons and hexagons are all equal. If one then imagines the edges of this solid projected out from the center onto a spherical surface with the same center, then the division of the sphere as described is formed." See fig 7. Ref 11-12

The method of projection onto the surface of the sphere describes the classic Gnomonic Projection used to create maps of the stars.

Since it was desirable to have all the lenses for the projector with the same cone of projection, a regular truncated icosahedron would not be the best choice. Thus, an irregular form was chosen having congruent circles circumscribing the hexagons and pentagons. This way a uniform configuration would emerge for the location of each required lens. This is a brilliant solution to a complex optical problem.

If we follow the logic in the design of the projector, then the choice for the geometry of the. lightweight reinforcing bars for the first thin-shell concrete structure followed naturally.

By subdividing the hexagon and pentagon facets into smaller elements and projecting them to the surface of the sphere by Gnomonic projection, a lightweight lattice dome provides enough stiffness to support the concrete while being applied. It also, provides the necessary reinforcement for the shell after curing. In contemporary terms the result is clearly a Class I Method 1 triangulated geodesic geometry. The topology was the same as those that would appear later. But, as we will see, it is not the same geometry that Fuller would introduce to us in another 25 years. The geometry was dropped for the thin shell concrete structures by Dyckerhoff and Widman early on in favor of reinforcement meshes that did not demand the same accuracy.

GOLDBERG (1934-1937)

In 1937 Michael Goldberg introduced "a class of multi-symmetric polyhedra" consisting of twelve pentagons, eight quadrilaterals or four triangles and all additional faces being hexagons. Thus he introduced the fact *"trihedral polyhedra which posses the same number of hexagonal faces in addition to 12 (8, or 4) regularly and symmetrically disposed pentagons* (quadrilaterals, or triangles) *can be topologically different."* "Topologically, the arrangement of the hexagons in a triangular patch is the same as in a 30° sector of a regular honeycomb arrangement of hexagons…Using a, b as the inclined coordinates (60° between axes) of the vertex of a patch, the square of the distance from the center of the patch to the vertex is equal to $a^2 + ab + b^2$…the total number of faces bounding the *'polyhedron'* is $10(a^2 + ab + b^2)+2;"$ *for the icosahedral system, $4(a^2 + ab + b^2)+2$ for the octahedral system, and $2(a^2 + ab + b^2)+2$ for the tetrahedral system.* See figure 8. Ref 13, 14

Often today's classifications of geodesic spheres are based on Goldsberg's notation.

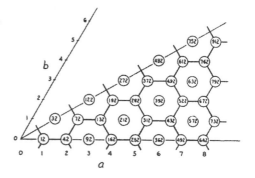

Figure 8. Goldbergh's polyhedra

GINZBURG (1936)

Dr. Pavlov of the Civil Engineering Institute of Gorky, USSR points out in an article by Gainsburg published in 1936, that "Gainsburg proposed a geometry of regular crystal forms : that of tetrahedron, octahedron, hexahedron, dodecahedron, and icosahedron for the purpose of giving uniform distribution of facets on the surface of a latticed dome". REF 15, 16

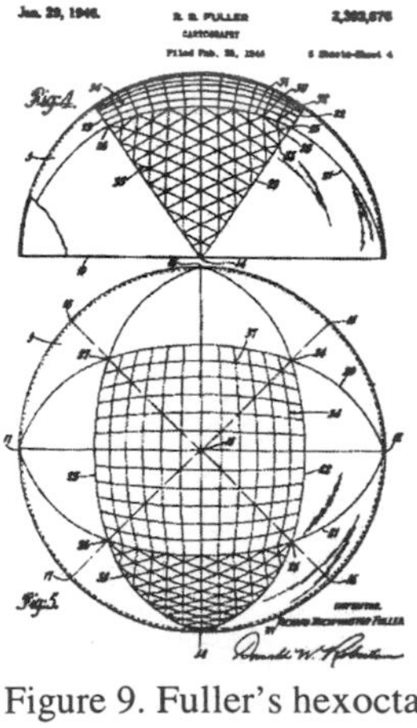

FULLER (1943-1948)

Fuller published his Dymaxion (Hexoctahedron) Map in Life magazine in March of 1943. His patent, figure 9, describes a two-way great circle grid for the square faces and a three-way great circle grid for the triangular faces. Due to using graphical

Figure 9. Fuller's hexoctahedral map

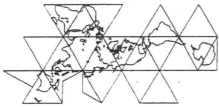

Figure 10. Fuller/Gray icosahedral map

cartography techniques in creating his map Fuller failed to recognize the three way grid did not have coincident intersections at all crossings. An icosahedral version of the map was drawn by Shoji Sadao in 1952, still using the graphical cartography technique. However, by that time Fuller realized the windows were appearing. A correct mathematical solution of the icosahedral projection was published in 1980 when Grip and Kitrick used computer cartographic methods for its production. Ref 17-19 Robert Gray published the actual algorithms for the Fuller projection in 1994. Figure 10 illustrates the Fuller/Gray map.

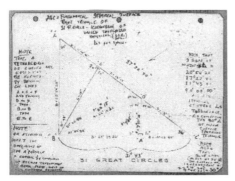

Figure 11. Fuller's great circle models

Fuller's studies of the thirty-one great circles, figure 11, first appeared in 1947. These studies, emerging from the development of his map projection and his studies on Energetic-Synergetic geometry, would become the topological basis for his subdivision of the icosahedron into smaller facets that would describe his geodesic domes, figure 12. Ref 20, 21

The $^1/_{120}$th Schwarz right spherical triangle or, LCD as Fuller called it, of the icosahedron is broken into 4 additional right spherical triangles. See figure 13.

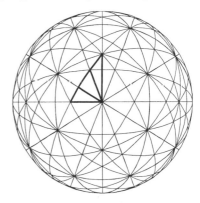

Figure 12. Fuller's LCD

Figure 13. Fuller's 31 great circle calculations

While teaching at Black Mountain College in 1948, Fuller continued to develop the geodesic principles of further subdivision of the Schwarz triangles into smaller and smaller units of right spherical triangles, figure 14. These triangles gave him the coordinates for constructing the three-way triangular grid we are all familiar with today. Ref 22, 23

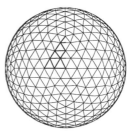

Figure 14. Fuller's 1st Geodesic grid

TUPOLEY (1948-1949)

At around the same time, in Russia, the researcher, M.S. Tupolev, professor of the Moscow Architectural Institute was experimenting with a similar grid surface division of the sphere. His inspiration came from earlier work of the mathematician A. M. Ginzburg. According to G.N. Pavlov in his article appearing in Tibor Tarnai's book <u>Structural Grid Structures: Geometric Essays On Geodesic Domes</u>, Ginzburg was awarded a patent on his Crystal domes in 1948-1949. Ref 15 I have been unable to examine the patent; therefore, I will not make any comments about its similarities to Bauersfeld's or Fuller's work.

FULLER/STUART/ RICHTER/WAINWRIGHT (1950-1955)

The original 1948 three-way grid calculated by Fuller, had an irregular pattern. It was modified in 1950 by Duncan Stuart while working in the Raleigh office of Geodesics Inc., to a more uniform geometry, figure 15. This method became known as the "Regular" triangulated grid, figure 17, and was generated by constructing perpendiculars to the edges of the triangle from regular subdivisions along its edges. It is the one described in Fuller's 1954 Patent. Duncan Stuart also discovered an even more efficient grid in 1951. The "Triacon" grid, figure 19, solved the problem of windows in the "Regular" grid. It was based on the rhombic triacontahedron instead of the icosahedron.

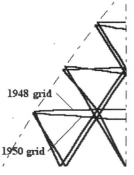

1948 grid

1950 grid

Figure 15. Fuller grids

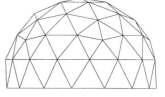

Figure 16. Alternate truncatable

During this same time period one of Fuller's students, Don Richter, developed the "Alternate" grid, figure 18, and Bill Wainwright of the Cambridge office of Geometrics, Inc. developed the "Alternate Truncatable" grid, figure 16.

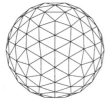

Figure 17. Regular

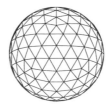

Figure 18. Alternate

Figure 19. Tricon

The term "geodesic" was applied to all of the geometrical methods developed by Fuller and his associates. All of the geometrical methods were based on his LCD and calculated using spherical trigonometry as compared to the gnomonic projection method used by Bauersfeld. Prior to the late 1950's the geometry was hand calculated. However during the late 1950's proprietary computer programs began to emerge. Notably Fuller's associate, Bernard Taylor of Synergetics, Inc., wrote a program to calculate the geometry for the dome frame of the American Society of Metals project. Others involved in those early proprietary programs were: Duncan Stuart, T.C. Howard, Chizko-Kojima, Joe Steinborn and R. Lewontin. Ref 24-28

CLINTON (1964-1971)

The research effort under a NASA contract at Southern Illinois University placed the mathematics of the geodesic geometry into the public realm. Perhaps the most influential distribution of the information was due to the premature release through the publication of Dome Books I and II. Clinton introduced the notion of classes to the topological characteristics of the tessellation of the sphere. The 'regular',' tricon' and 'modified tricon' of Fuller's structures fell in the Class II group. The 'alternate' and 'alternate truncatable' fell in the Class I group. The Class III group is composed of skewed forms of polyhedra. Ref 29

The NASA research also investigated 13 geometrical methods for describing the three-way grid of the geodesic polyhedra. Six of the seven methods along with three additional methods published are shown in figure 20 as an overlay of a Class I spherical triangle based on the icosahedron. This was the beginning of a departure from using the icosahedron and the rhombic tricontahedron as the basic polyhedra from which to derive the geodesic polyhedra. The spherical triangle may be the face of a polyhedron or any triangle on the surface of the sphere. Ref 30-31

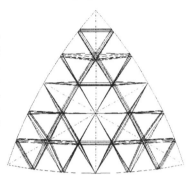

Figure 20. Geometry comparisons

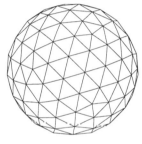

Figure 21. $\{3,5\}_{1,3}$

COXETER (1972)

In his paper *"Virus Macromolecules and Geodesic Domes"*, the mathematician H.S.M. Coxeter proposed Michael Goldberg's notation of $T = b^2 + bc + c^2$ as applied to the class of $\{p,q\}b,c$ polyhedra, figure 21, could be used to describe the topology of Casper and Klugs Virus's and Fuller's domes. This notation helps further classify the skew forms as proposed by Clinton. Ref 32, 33

E.MAKAI & T. TARNAI (1970's)

Through the early to mid 1970's, E. Makai, Jr. and T. Tarnai published several papers on the morphology of spherical grids and circle packing on the surface of a sphere. They proposed an optimum grid to be one with a ratio of the longest edge to the shortest edge of $\eta = 1.1756$. Two such geometries have been found: one by Clinton and one by Kitrick. Ref 15, 16

KENNER (1976)

There was a lot of activity and interest in geodesic structures during the 1970's. Geodesic domes were finding their place as an architectural from. And, they were being established as a symbol for a better future. Hugh Kenner's book provided a simple solution for calculating the necessary geometry for a variety of methods. He presented it in a form that could be used by anyone owning an inexpensive hand calculator. Ref 34

PAVLOV (1970's)

A little known but very important work through the mid to late 1970's, is G. N. Pavlov's studies of crystal domes first proposed by Ginzburg in 1936 and later expanded upon in the late 1940's to early 1950's. An excellent description of Pavlov's work is given in his article *"Compositional Form-shaping of Crystal Domes and Shells"*. Ref 15, 16

NOOSHIN, et al. (1978-)

A mathematical system which is ideally suited for configuration processing called 'formex algebra' was first introduced in the mid 1970's. As a means of using 'formex' as a convenient medium for configuration processing to help solve complex problems an interactive programming language "Forman" began development in the early 1980's and continues through the present. Ref 35, 36

WENNINGER (1979)

Magnus J. Wenninger closed out the 1970's by completing the classification of the geodesic forms started by Clinton and improved upon by Coxeter. Ref 32, 37

HUYBERS/WESTER/ LALVANI/KITRICK/MIYAZAKI (1980's – 1990's)

The 1980's - 1990's ushered in an even more accelerated interest in the geodesic research .
Many contributions to the study of 'geodesics' were made by several people. Dr. Pieter Huybers, as an example, expanded the geodesic forms to include forms generalized by the equations of the tri-axial ellipsoids. His computer program continued to be expanded and improved upon through the 1990's. Ref 38

Ture Wester introduced the notion of plate-lattice dualism. In lattice structures, geometrical stability is maintained in a triangulated grid. In plate structures, geometrical stability is maintained in a three-way vertex connection, thus, the duality. Ref 39

Haresh Lalvani defined a "unified morphological method for generalizing subdivisions of periodic surfaces." "The surfaces are characterized by symmetry, frequency, and the topology of subdivision." Ref 40

Chris Kitrick provided us with a unified approach to generating the geometry of the Class I, II, & III geodesic forms. He did this by defining a set of mathematical algorithms for 10 different geometrical methods of generating geodesic triangulated grids. Several of these had not been previously found in the literature. Ref 41

Koji Miyazaki introduced us to the Hypergeodesic Polytopes, as a four-dimensional analogue of the geodesic forms in 3-space. Ref 42

THE NEW MILLENNIUM

As we move into the new millennium I would expect to see new contributions into the geometry of expanding geodesic forms such as the work by Chuck Hoberman. The work being done by Charalambos Gantes, Félix Escrig, José Sanchez, Juan Pérez Valcarcel, S. Pellegrino, to name a few, should also advance the understanding of geodesic forms. We should also see contributions into the use of the principles of tensegrity in the ways Emmerich, Fuller and Snelson initially saw them. New ways are emerging from the research being done by R. Motro, R. Grip, M. R. Barnes, R. Connelly, A. Hanaor, Donald E. Ingber. Ref 43 The"International Conference on Discrete Global Grids" met in March 2000 to discuss the need to change the traditional latitudes and longitudes to a new international standard. A good candidate for the new standard is a system composed of hexagons; this will improve on the flow of digital data. Ref 44 A geodesic icosahedron tessellated into hexagons consisting of equal length edges may be a solution. Ref 45 From low frequency spherical transducers of the 70's that can be plunged to depths of 1000ft in the ocean...to super balls for the golfer in the 90's...to visual display systems with images better than the resolving power of the eye in 2000....We can expect to continually see advancements in technical innovations utilizing the contributions of exploring minds in the quest to understand the principles of geodesics in the new millennium. The natural sciences have made many contributions throughout the history of geodesics forms. We should expect to see many more coming from the developments in virology, chemistry, microbiology, such as the work of Dr Ingber..., and, the investigations into the Buckminsterfullerenes, Boron ..., and, others we have no knowledge of yet. With the Internet playing such an important role in our daily lives, we can expect a wider distribution of information on geodesics than ever before. Programs such as Tekstar's, newly released TekCad, coupled with its design modeling kit, should provide wide access to geodesic design tools for the Architect, Engineer, Scientist, Educators. The packaging of Rick Bono's program 'Dome' with the "Linux" operating system will go a long way to introducing a wide group of people to the subject of geodesics. The continual expanding of information available on geodesics from all subject areas will no doubt accelerate interest and contributions to the subject.

REFERENCES

1. FULLER B, No More Second Hand God and Other Writings, Southern Illinois University Press, Carbondale Illinois, 1963, ppv-xii.
2. SNYDER J, Map Projections – A Working Manual, US Geological Survey Professional Paper 1395, US Government Printing Office, Washington DC, 1987, pp164-169.
3. SHARP A, Geometry Improv'd: 1. By a large and Accurate Table of Segments of Circles 2. A Concise Treatise of Polyhedra or Solid Bodies of Many Bases, London, 1717, pp65-96. (Thanks are due the US National Observatory for their allowing access to this rare book)
4. MIYAZAKI K, A History of Polyhedra in Japan, Kobe University, Kobe, 1979.
5. SCHWARZ H, Zur Theorie der Hypergeometrischen Reite, J. reine angew. Math. Vol75, 1873, pp292-335.
6. FULLER B, Synergetics, Macmillan Publishing Company, New York, 1975, pp480-487.
7. The Forbidden City, http://users.aol.com/alisonnyt/tour/china-fc2-6.jpg.html
8. KAHN L, Shelter, Shelter Publications, Bolinas, 1973, pp111.
9. DYCKERHOFF & WIDMAN, 1979, private correspondence.
10. BAUERSFELD W, Projection Planetarium and Shell Construction: James Clayton Lecture, The Institute of mechanical Engineers, 10 May 1957.
11. LICHTENSTEIN C, 1996, private discussion.
12. KRAUSSE J, The Miracle of Jena, Arch+, vol116, Mär4z, 1993, pp50-59.
13. GOLDBERG M, The Isoperimetric Problem for Polyheda, Tôhoku Mathematical Journal, vol40, 1934, pp226-236.
14. GOLDBERG M, A Class of Multi-Symmetric Polyheda, Tôhoku Mathematical Journal, vol43, 1937, pp104-108.
15. TARNAI T ed, Structural Grid Structures: Geometric Essays on Geodesic Domes, Hungarian Institute for Building Science, Budapest, 1987.
16. TARNAI T private discussion, 1993.
17. FULLER B, Cartography, US Patent Number 2,393,676, US Patent Office, Washington DC, January 29, 1946.
18. SADAO S, taped interview 19 November 1978.

19. GRAY R, Fullers Dymaxion™ Map, Cartography and Geographic Information Systems, vol21, no4, 1994, pp243-216.
20. FULLER B, taped interviews 4, 12 October and 7 January 1979.
21. FULLER B, photographs of original documents in R. B. Fuller archive, 1978.
22. RICHTER D, taped interview 14 November 1978.
23. RITCHER D, photographs of original documents in Don Richter's archive, 1978.
24. FULLER B, his archive materials.
25. STUART D, private collection of Duncan Stuart's work during the period from 1952 through 1955.
26. STUART D, On the Orderly Subdivision of Spheres, Student Publication of the School of Design, North Carolina State University, Raleigh, vol 5, No1, 1954, pp23-33.
27. RICHTER D, private collection of Don Richter's work during the period from 1948 through 1955.
28. ZUNG T, Buckminster Fuller, St. Martin's Press, New York, 2001, pp22-42.
29. KAHN L, Domebook 2, Pacific Domes, Bolinas, 1971, pp106-113.
30. CLINTON J, Advanced Structural Geometry Studies: Part I-Polyhedral Subdivision Concepts for Structural Applications, NASA Contractor Report NASA CR-1734, NASA, Washington DC, 1971.
31. CLINTON J, private collection of unpublished documents and NASA research notes on the math behind geodesic forms.
32. COXETER HSM, Virus Macromolecules and Geodesic Domes, A Spectrum of Mathematics, ed J C Butcher, Auckland University Press, 1972, pp98-108.
33. CASPER DLD, KLUG A, Physical Principles in the Construction of Regular Viruses, Cold Spring Harbor Symposia on Quantitative Biology, volXVII, 1962.
34. KENNER H, Geodesic Math and How to Use IT, University of California Press, Berkeley, 1976.
35. NOOSHIN H, Algebraic Representation and Processing of Structural Configurations, International Journal of Computers and Structures, vol5, 1975, pp119-130.
36. NOOSHIN H, DISNEY P, YAMAMOTO C, Forman, Multi-Science Publishing Company, Brentwood, 1993.
37. WENNINGER M, Spherical Models, Chambridge University Press, Cambridge, 1979, pp120-124.
38. HUBERS P, Polyhedral Shapes Visualized With CAD/CAM, IASS-Congress, Madrid, 1989, pp81-91.
39. WESTER T, Structural Order in Space: The Plate-Lattice Dualisum, Royal Academy of Arts, School of Architecture, Copenhagen, 1984.
40. LALVANI H, Continuous Transformations of Subdivided Periodic Surfaces, International Space Structures, vol5, nos3&4, 1990,pp255-279.
41. KITRICK C, A Unified Approach to Class I, II & III Geodesic Domes, International Space Structures, vol5, Nos3&4, 1990, pp223-246.
42. MIYAZAKI K, primary Hypergeodesic Polytopes, international Space Structures, vol5, nos3&4, 1990, pp309-323.
43. Collection of unpublished works on the development and history of geodesic geometry.
44. International Conference on Discrete Globe Grids, US National Center for Geographical Information and Analysis, Santa Barbara, 2000.
45. CLINTON J, A Group of Spherical Tessellations Having Edges of Equal Length, manuscript submitted for publication in the Proceedings of the Fifth International Conference on Space Structures, University of Surrey, Surrey, 2002.

The Concept of Pellevation for Shaping of Structural Forms

I S HOFMANN, Ing.-Büro Braun & Partner GbR, Pforzheim, Germany

INTRODUCTION

In this paper, a new concept is introduced which serves as a useful tool to create many different types of structural shapes in a quick and efficient way. The basic idea can be imagined as follows: A geometric object is used as a die to 'deform' any part of a given configuration (shell or lattice structure) by pushing the die upwards or downwards. This process may be repeated on the 'deformed' configuration using the same or another die. This process is referred to as pellevation. The new concept may be used in many different areas, however, it is of particular relevance to the designers of grid domes and shell structures. The initial idea of the concept has first been introduced in Ref 1.

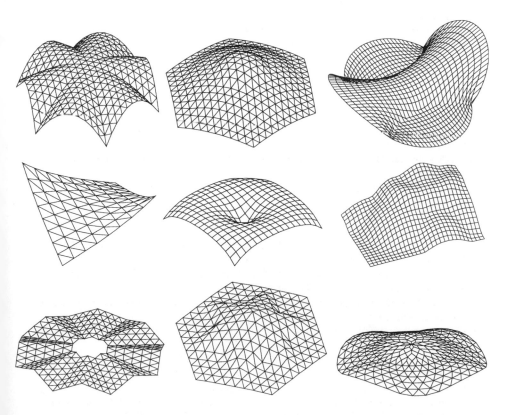

Figure 1. Examples involving pellevation

Space Structures 5, Thomas Telford, London, 2002

For design purposes, several families of pellevations have been created which supply the designer with a number of pre-defined geometric objects as dies. Additionally, two types of pellevation have been set up which enable the designer to create his or her own die by different means. Figure 1 shows some examples that involve the application of pellevation.

In this paper, the main features of pellevation are briefly outlined and illustrated by some examples. A complete guide including all facilities, derivations of mathematical formulations, discussion of details and numerous applications is given in Ref 2.

CAP PELLEVATIONS

The first family of pellevations is called 'cap pellevations'. Within this family, there are several cap pellevations the first of which is referred to as 'circular cap pellevation with circular base'. This cap pellevation can be imagined to describe the properties of a part of a sphere.

Take, for example, the planar grid shown in Figure 2(a) lying in the xy-plane. The top part of the sphere which is given in Figure 2(b) is used as a die. Such a die is called a 'pellevant' and its shape and position are defined by a number of parameters, namely, the coordinates of its centre in the xy-plane, the span of its base and its height at the centre as shown in the figure.

If this pellevant is used to push up the central region of the configuration of Figure 2(a), the result will be as shown in Figure 2(c). This is an example of the process of pellevation. Here, it can be seen that outside the boundaries of the pellevant, the configuration remains unchanged while within the scope of the pellevant, the nodes of the configuration experience a 'push-up'. That is, the z-coordinates of the nodes are changed according to the shape of the pellevant.

Typically, a cap pellevant is defined by
- a 'pellevation code' which characterises its basic shape in section and plan
- the position of its centre in the xy-plane with respect to a Cartesian coordinate system
- the size of the boundaries in plan
- the height of the pellevant at the centre

The function used to obtain a circular cap pellevation with circular base, is of the general form

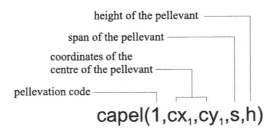

height of the pellevant ⎯⎯⎯⎯
span of the pellevant ⎯⎯⎯⎯
coordinates of the
centre of the pellevant ⎯⎯⎯
pellevation code ⎯⎯⎯⎯⎯

$$\text{capel}(1, cx_1, cy_1, s, h)$$

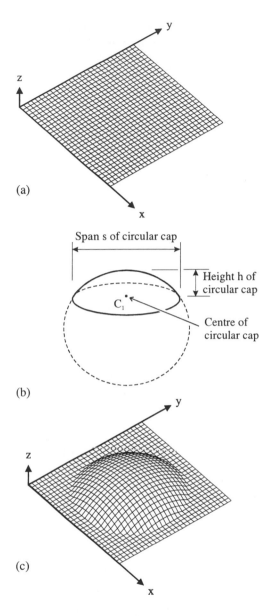

Figure 2. Typical cap pellevation process
(a) initial configuration, (b) pellevant, (c) pellevated configuration

As can be seen, the function requires a list of parameters, the first of which is called the 'pellevation code'. The circular cap pellevant with circular base has been allocated pellevation code 1 (+1 for upwards movement and −1 for downwards movement of the pellevant). Further parameters are as indicated in Figure 2(b): The coordinates of the centre of the circular cap in plan C_1 (cx_1, cy_1), the span s in plan and the height h at the centre.

The complete formex formulation of the example in Figure 2 is as follows:

```
E = rinid(31,30,1,1) | [0,0,0; 0,1,0] # rinid(30,31,1,1) | [0,0,0; 1,0,0];
F = capel(1,15,15,24,5) | E;
use &, vm(2), vt(1), vh(-3,-5,4,0,0,0,0,0,1);
clear; draw F;
```

Here, formex E denotes the initial planar 30 by 30 configuration. Formex F represents the configuration after the pellevation where the parameters used are as explained above. The next line contains some use-items which set the view options, and the last line produces the actual picture shown in Figure 2(c). If the reader is not familiar with Formian, Ref 3 may be consulted where these basic concepts are explained in detail.

Altogether, eight cap pellevation functions, all referred to as capel function, have been defined using pellevants with two different plan views in combination with four different cross-sections, namely, these are the

- circular cap pellevant
- elliptic cap pellevant
- parabolic cap pellevant
- wedge cap pellevant.

Each one of these can be used with either a circular or an elliptic base. Table A (at the end of the paper) shows the general characteristics and required parameters of these cap pellevants. In this context, point C_1 (cx_1, cy_1) always represents the centre of the base, points C_2 (cx_2, cy_2) and C_3 (cx_3, cy_3) mark the ends of major and minor axis of the elliptic base and h_e is a parameter unique to all pellevants with an elliptic cross-section. It represents the length of the vertical semi-axis of the ellipse in section and, therefore, $h_e \geq h$ should apply in all cases.

Figure 3 shows some further examples which involve the use of the capel function, where, alongside the picture, the corresponding formex formulation can be seen.

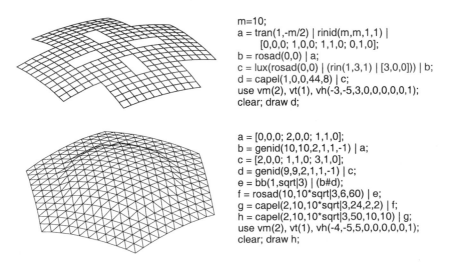

```
m=10;
a = tran(1,-m/2) | rinid(m,m,1,1) |
    [0,0,0; 1,0,0; 1,1,0; 0,1,0];
b = rosad(0,0) | a;
c = lux(rosad(0,0) | (rin(1,3,1) | [3,0,0])) | b;
d = capel(1,0,0,44,8) | c;
use vm(2), vt(1), vh(-3,-5,3,0,0,0,0,0,1);
clear; draw d;
```

```
a = [0,0,0; 2,0,0; 1,1,0];
b = genid(10,10,2,1,1,-1) | a;
c = [2,0,0; 1,1,0; 3,1,0];
d = genid(9,9,2,1,1,-1) | c;
e = bb(1,sqrt|3) | (b#d);
f = rosad(10,10*sqrt|3,6,60) | e;
g = capel(2,10,10*sqrt|3,24,2,2) | f;
h = capel(2,10,10*sqrt|3,50,10,10) | g;
use vm(2), vt(1), vh(-4,-5,5,0,0,0,0,0,1);
clear; draw h;
```

Figure 3. Examples involving cap pellevations (cont'd)

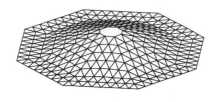

```
h1 = [0,0,0; 2,0,0; 1,1,0];
h3 = genid(8,7,2,1,1,-1) | h1;
h3 = [2,0,0; 1,1,0; 3,1,0];
h4 = genid(7,7,2,1,1,-1) | h3;
h5 = bb(1,tan|67.5)|(h2#h4);
h6 = rosad(8,8*tan|67.5,8,45) | h5;
h7 = capel(4,8,8*tan|67.5,30,5) | h6;
use vm(2), vt(1), vh(-4,-5,3,0,0,0,0,0,1);
clear; draw h7;
```

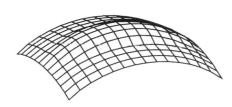

```
a = [0,0,0; 1,0,0; 1,1,0; 0,1,0];
b = rinid(14,12,1,1) | a;
c = capel(6,7,6,7,-7,-2,6,4,8) | b;
c1 = capel(6,7,6,7,3,0,6,0.7,3) | c;
use vm(2), vt(1), vh(-2,-5,2,0,0,0,0,0,1);
clear; draw c1;
```

Figure 3. Examples involving cap pellevations

BARREL PELLEVATIONS

The second family of pellevations is referred to as barrel pellevations. Altogether, there are eight barrel pellevations. Similar to the case of the cap pellevations, the barrel pellevations can be divided into two groups according to the shape of the base of the pellevants. As can be seen in Table A, the two types of bases are referred to as 'rectangular' and 'trapezoidal' with either circular, elliptic, parabolic or wedge cross-section.

Barrel pellevations serve to create many interesting shapes by multiple application of the barrel pellevant, Figure 4. However, they are also of major use in combination with the cap pellevations, Figure 5.

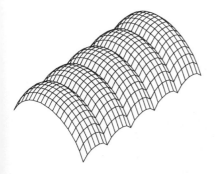

```
a = rinid(6,18,1,1) | [0,0,0; 0,1,0; 1,1,0; 1,0,0];
b = bapel(2,3,0,3,18,6,2,3) | a;
c = rin(1,5,6) | b;
d = bapel(2,0,9,6,9,20,14,14) | c;
use vm(2), vt(1), vh(-5,-5,8,0,0,0,0,0,1);
clear; draw d;
```

Figure 4. Example involving multiple use of barrel pellevations

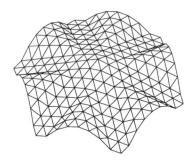

```
a1 = {[0,0,0; 2,0,0; 1,2,0]};
a2 = genid(8,8,2,2,1,-1) | a1;
b1 = {[2,0,0; 1,2,0; 3,2,0]};
b2 = genid(7,8,2,2,1,-1) | b1;
c = a2#b2; d = bapel(7,8,10,8,0,1,12,0.5,4)|c;
e = bb(1,sqrt|3/2) | d;
f = rosad(8,sqrt|3/2*16,6,60) | e;
g = capel(4,8,sqrt|3/2*16,32,8) | f;
use vm(2), vt(1), vh(-5,-5,9,0,0,0,0,0,1);
clear; draw g;
```

Figure 5. Example involving combination of barrel and cap pellevations

Some interesting shapes may be obtained using 'inclined pellevation'. To obtain 'inclined pellevation', the initial configuration is rotated about the x- or y-axis prior to the application of a pellevation function. The configuration is afterwards partially or fully rotated back into its original position. Figure 6 shows such an example and a possible practical application.

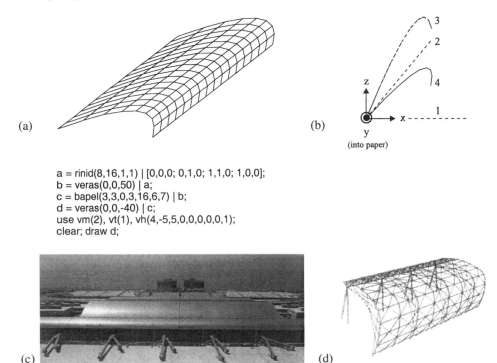

(a)

(b)

```
a = rinid(8,16,1,1) | [0,0,0; 0,1,0; 1,1,0; 1,0,0];
b = veras(0,0,50) | a;
c = bapel(3,3,0,3,16,6,7) | b;
d = veras(0,0,-40) | c;
use vm(2), vt(1), vh(4,-5,5,0,0,0,0,0,1);
clear; draw d;
```

(c)

(d)

Figure 6. Example involving inclined pellevation
(a) pellevated configuration and formex formulation, (b) 'process' in elevation,
(c) + (d) Kansai International Airport Passenger Terminal Building – picture and technical detail of roof, Ref 4

HALO AND SADDLE PELLEVATIONS

Further pellevations which are based on pre-defined pellevants are the halo and saddle pellevations. In the first case, the base of the pellevant is either a circular or an elliptic ring. In the latter case, the base is formed by parabolic or hyperbolic curves. As before, these bases are available in combination with either circular, elliptic, parabolic or wedge cross-section, Table B.

Figure 7 gives one example of each of these pellevation types together with their formex formulations.

(a)

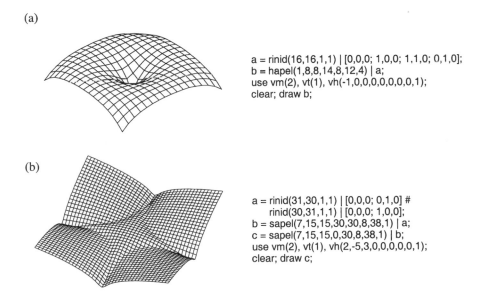

```
a = rinid(16,16,1,1) | [0,0,0; 1,0,0; 1,1,0; 0,1,0];
b = hapel(1,8,8,14,8,12,4) | a;
use vm(2), vt(1), vh(-1,0,0,0,0,0,0,0,1);
clear; draw b;
```

(b)

```
a = rinid(31,30,1,1) | [0,0,0; 0,1,0] #
       rinid(30,31,1,1) | [0,0,0; 1,0,0];
b = sapel(7,15,15,30,30,8,38,1) | a;
c = sapel(7,15,15,0,30,8,38,1) | b;
use vm(2), vt(1), vh(2,-5,3,0,0,0,0,0,1);
clear; draw c;
```

Figure 7. Examples involving
(a) halo pellevation (in combination with cap pellevation)
(b) parabolic saddle pellevation with hyperbolic base

RAFT AND MANDATE PELLEVATIONS

Raft and mandate pellevations are different from the previously described types because they do not incorporate any pre-defined types of pellevant. Here, it is up to the designer to create his or her own die which will then serve as a pellevant.

In the case of the raft pellevation, the surface of the pellevant is defined by a number of points. These points may, e.g., be the output of an experiment, Figure 8.

The general form of the raft pellevation function is given by:

   F = rapel (1,raft) | E

In this context, formex *raft* represents the group of points.

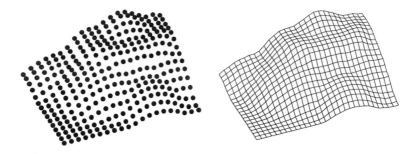

Figure 8. Pellevant (number of points) and pellevated configuration (initially flat)

In the case of the mandate pellevation, the surface of the pellevant is described by a mathematical function. The example shown in Figure 9 involves two sine-functions, where the first one is parallel to the x-axis and the second is parallel to the y-axis.

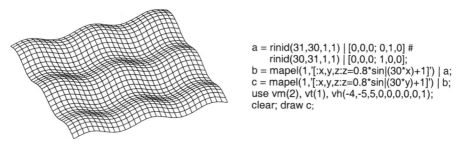

```
a = rinid(31,30,1,1) | [0,0,0; 0,1,0] #
    rinid(30,31,1,1) | [0,0,0; 1,0,0];
b = mapel(1,'[:x,y,z:z=0.8*sin|(30*x)+1]') | a;
c = mapel(1,'[:x,y,z:z=0.8*sin|(30*y)+1]') | b;
use vm(2), vt(1), vh(-4,-5,5,0,0,0,0,0,1);
clear; draw c;
```

Figure 9. Pellevated configuration and formex formulation for a mandate pellevation

REFERNECES

1. NOOSHIN H, A Technique for Surface Generation, Proceedings of the International Symposium on Conceptual Design of Structures in Stuttgart, October 1996, pp 331-338.

2. HOFMANN I, The Concept of Pellevation for Shaping of Structural Forms, PhD Thesis, University of Surrey, 1999.

3. NOOSHIN H & DISNEY P, Formex Configuration Processing, Parts I, II and III, International Journal of Space Structures, Vol 15, No 1, 2000, Vol 16, No 1, 2001 and Vol 17, No1, 2002.

4. Kansai International Airport Passenger Terminal Building, The Japan Architect No.15, Autumn 1994-3.

Table A. Characteristics of Pellevant and General Form of Function for Cap and Barrel Pellevation

Section of Pellevant	Base of Pellevant	General Form of Function for CAP PELLEVATION	Section of Pellevant	Base of Pellevant	General Form of Function for BARREL PELLEVATION
circular	circular	$capel(\pm1,cx_1,cy_1,s,h)$	circular	rectangular	$bapel(\pm1,cx_1,cy_1,cx_2,cy_2,s,h)$
elliptic	circular	$capel(\pm2,cx_1,cy_1,s,h,h_e)$	elliptic	rectangular	$bapel(\pm2,cx_1,cy_1,cx_2,cy_2,s,h,h_e)$
parabolic	circular	$capel(\pm3,cx_1,cy_1,s,h)$	parabolic	rectangular	$bapel(\pm3,cx_1,cy_1,cx_2,cy_2,s,h)$
wedge	circular	$capel(\pm4,cx_1,cy_1,s,h)$	wedge	rectangular	$bapel(\pm4,cx_1,cy_1,cx_2,cy_2,s,h)$
circular	elliptic	$capel(\pm5,cx_1,cy_1,cx_2,cy_2,cx_3,cy_3,h)$	circular	trapezoidal	$bapel(\pm5,cx_1,cy_1,cx_2,cy_2,s_1,s_2,h_1,h_2,h_{e_1})$
elliptic	elliptic	$capel(\pm6,cx_1,cy_1,cx_2,cy_2,cx_3,cy_3,h,h_e)$	elliptic	trapezoidal	$bapel(\pm6,cx_1,cy_1,cx_2,cy_2,s_1,s_2,h_1,h_2)$
parabolic	elliptic	$capel(\pm7,cx_1,cy_1,cx_2,cy_2,cx_3,cy_3,h)$	parabolic	trapezoidal	$bapel(\pm7,cx_1,cy_1,cx_2,cy_2,s_1,s_2,h_1,h_2)$
wedge	elliptic	$capel(\pm8,cx_1,cy_1,cx_2,cy_2,cx_3,cy_3,h)$	wedge	trapezoidal	$bapel(\pm8,cx_1,cy_1,cx_2,cy_2,s_1,s_2,h_1,h_2)$

Table B. Characteristics of Pellevant and General Form of Function for Halo and Saddle Pellevation

Section of Pellevant	Base of Pellevant	General Form of Function for HALO PELLEVATION	Section of Pellevant	Base of Pellevant	General Form of Function for SADDLE PELLEVATION
circular	circular	$hapel(\pm1,cx_1,cy_1,cx_2,cy_2,s,h)$	circular	parabolic	$sapel(\pm1,cx_1,cy_1,cx_2,cy_2,s_1,s_2,h_1)$
elliptic	circular	$hapel(\pm2,cx_1,cy_1,cx_2,cy_2,s,h,h_e)$	elliptic	parabolic	$sapel(\pm2,cx_1,cy_1,cx_2,cy,s_1,s_2,h_1,h_e)$
parabolic	circular	$hapel(\pm3,cx_1,cy_1,cx_2,cy_2,s,h)$	parabolic	parabolic	$sapel(\pm3,cx_1,cy_1,cx_2,cy_2,s_1,s_2,h_1)$
wedge	circular	$hapel(\pm4,cx_1,cy_1,cx_2,cy_2,s,h)$	wedge	parabolic	$sapel(\pm4,cx_1,cy_1,cx_2,cy_2,s_1,s_2,h_1)$
circular	elliptic	$hapel(\pm5,cx_1,cy_1,cx_2,cy_2,cx_3,cy_3,s,h)$	circular	hyperbolic	$sapel(\pm5,cx_1,cy_1,cx_2,cy_2,s_1,s_2,h_1)$
elliptic	elliptic	$hapel(\pm6,cx_1,cy_1,cx_2,cy_2,cx_3,cy_3,s,h,h_e)$	elliptic	hyperbolic	$sapel(\pm6,cx_1,cy_1,cx_2,cy_2,s_1,s_2,h_1,h_{e1})$
parabolic	elliptic	$hapel(\pm7,cx_1,cy_1,cx_2,cy_2,cx_3,cy_3,s,h)$	parabolic	hyperbolic	$sapel(\pm7,cx_1,cy_1,cx_2,cy_2,s_1,s_2,h_1)$
wedge	elliptic	$hapel(\pm8,cx_1,cy_1,cx_2,cy_2,cx_3,cy_3,s,h)$	wedge	hyperbolic	$sapel(\pm8,cx_1,cy_1,cx_2,cy_2,s_1,s_2,h_1)$

Formian for art and mathematics

T. ROBBIN
Independent Artist

Implementing formex algebra, the computer program Formian was developed by H. Nooshin, P. Disney, and their collaborators at the Space Structures Research Centre as a tool to design and study space frame structures. The computer program is so open and flexible, however, that it has applications beyond architecture, and is wonderfully useful for art and mathematics.

Thirty years ago when I formulated my aesthetic goals of depicting an abstract space in painting that would embody our contemporary experience, I settled on the idea of using patterns to represent space. The problem for the artist who wants to depict space is that space is what is not there; using telephone poles and railroad tracks to make space leaves one with a picture of telephone poles and railroad tracks instead of a picture of space itself. The Islamic tradition of pattern making, which revels in the intricate symmetries and implicit structures of space, is ideally suited to my artistic purpose: it is possible to see patterns as markers for space alone. No doubt I was drawn to this tradition having grown up in Iran, as did Professor Nooshin; Formian can be seen to be the continuation with modern technology of an ancient tradition.

Formian began as a solution to the practical problem of loading data about architectural space frames into the then new engineering analysis programs. Though these programs could quickly compute the axial and bending forces on nodes and rods, and tell whether or not a hypothetical structure was stable, hours were required to prepare stacks of key punch cards to load the programs. Formian was to automate this loading process. Nooshin, Disney and their collaborators soon realized, however, that an algebra of patterns and a program to implement that algebra would go far beyond the requirement to simply reproduce existing simply-conceived space frames. Instead, such a program could be an engine for the design of novel space frames, could generate complexly curved lattices, could emulate shell and membrane structures, and could be a tool for the study of patterns for their own sake. Over the years Formian has evolved into a modern, Windows-based program augmented by the Ph.D. research of many engineers at the Space Structures Research Centre, Ref 1.

In 1952, A. Einstein instructed us to imagine contemporary space, the space of Special Relativity, by visualizing a large box, inside of which there are many small boxes. Each of these small boxes is moving in a different direction and at a different speed, and as a consequence each has a different internal structure - a different metric (and a different clock), Ref 2. Now we are to imagine each of the small boxes expanding to fill up the larger box. Many spaces in the same space at the same time - this is the geometry of four spacial dimensions projected into three-dimensional space.

To my surprise, Formian can do four-dimensional geometry. As I demonstrated in the Journal of Hyperspace, off-the-shelf Formian can have nodes of with an arbitrary number of coordinates, can connect these nodes into higher-dimensional figures, and can rotate these figures in higher dimensional space, Ref 3. This remarkable capability of Formian is a result of its open computer-architecture; one need not use preset arrays to define lattices or even be consistent with the arrays once they are established. Rather, one may expand and contract not only the length of arrays but also their dimensions by using standard Formian functions, named *pan* and *dep*.

Formian writers pride themselves on writing short code so that just a few condensed instructions generate the most complicated patterns with thousands of members. Such code, however, is hard to read, and the examples that follow have been spelled out to allow us to see the functioning of the program, see Fig 1.

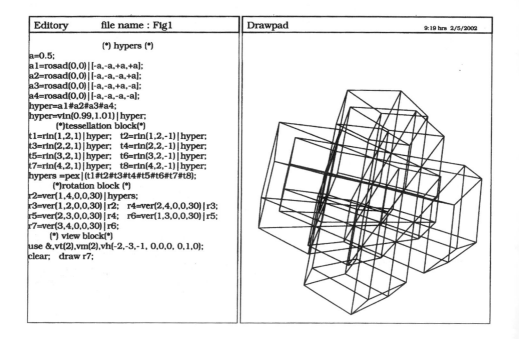

Editory	file name : Fig1	Drawpad	9:19 hrs 2/5/2002

```
                    (*) hypers (*)
a=0.5;
a1=rosad(0,0)|[-a,-a,+a,+a];
a2=rosad(0,0)|[-a,-a,-a,+a];
a3=rosad(0,0)|[-a,-a,+a,-a];
a4=rosad(0,0)|[-a,-a,-a,-a];
hyper=a1#a2#a3#a4;
hyper=vin(0.99,1.01)|hyper;
            (*)tessellation block(*)
t1=rin(1,2,1)|hyper;   t2=rin(1,2,-1)|hyper;
t3=rin(2,2,1)|hyper;   t4=rin(2,2,-1)|hyper;
t5=rin(3,2,1)|hyper;   t6=rin(3,2,-1)|hyper;
t7=rin(4,2,1)|hyper;   t8=rin(4,2,-1)|hyper;
hypers =pex|(t1#t2#t3#t4#t5#t6#t7#t8);
            (*)rotation block (*)
r2=ver(1,4,0,0,30)|hypers;
r3=ver(1,2,0,0,30)|r2;   r4=ver(2,4,0,0,30)|r3;
r5=ver(2,3,0,0,30)|r4;   r6=ver(1,3,0,0,30)|r5;
r7=ver(3,4,0,0,30)|r6;
            (*) view block(*)
use &,vt(2),vm(2),vh(-2,-3,-1, 0,0,0, 0,1,0);
clear;   draw r7;
```

Fig. 1.   A pattern of 9 close packed hypercubes constructed with Formian.

After defining a unit distance, the four *rosad* functions generate the vertices of four squares on the origin at plus and minus unit distances in the third dimension and plus and minus unit distances in the fourth dimension. (I am told that the original purpose of having an arbitrary number of dimensions to a point was to allow for markers to distinguish groups within the totality of points, for example to identify different materials for a set of three-dimensional nodes.) The next two <u>hyper</u> statements group these vertices together and join all of them that can be connected with lines of one unit length; this makes one complete hypercube: the four dimensional analogue of the cube composed of eight cubic cells. The instructions in the tessellation block move this central hypercube forward and backward one unit length in each of the four directions previously defined. Duplicate points are removed by the *pex* function. The rotation block rotates the tessellation of hypercubes about the origin. The *ver* function requires that the plane of rotation be specified first. Since there are six combinations of four dimensions when taken two at a time, there are six possible rotations in four dimensions and thus six *ver* statements. All these rotations are about the origin, and all are 30 degrees. The last instructions establish the view of the object and draws a plot of it. In this example the figure is drawn in isometric projection from the fourth to the third dimension; then that figure is drawn in perspective. The result of this program is a tessellation of nine hypercubes: eight around a central hypercube, where each outer cubic cell of the central hypercube is a cell of just one other hypercube. (The three-dimensional analogue would a stack of seven cubes, where the central cube's faces are also faces of just one other cube).

One more example of a four dimensional tessellation is given using the 24-cell, see Fig 2. Closest to the semi-regular cuboctahedra, this regular four dimensional figure is without an exact analogue in three dimensions; its octahedral cells are obtained by joining the centres of the faces of hypercube, Ref 4. It is not quite the dual of the hypercube, however, as it is not obtained from the centers of the cells of the hypercubes, as is the (tetrahedral)16-cell that is considered to be the four dimensional octahedron.

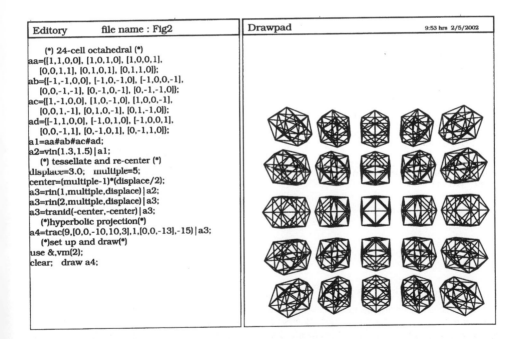

Editory	file name : Fig2	Drawpad	9:53 hrs 2/5/2002

```
    (*) 24-cell octahedral (*)
aa={[1,1,0,0], [1,0,1,0], [1,0,0,1],
    [0,0,1,1], [0,1,0,1], [0,1,1,0]};
ab={[-1,-1,0,0], [-1,0,-1,0], [-1,0,0,-1],
    [0,0,-1,-1], [0,-1,0,-1], [0,-1,-1,0]};
ac={[1,-1,0,0], [1,0,-1,0], [1,0,0,-1],
    [0,0,1,-1], [0,1,0,-1], [0,1,-1,0]};
ad={[-1,1,0,0], [-1,0,1,0], [-1,0,0,1],
    [0,0,-1,1], [0,-1,0,1], [0,-1,1,0]};
a1=aa#ab#ac#ad;
a2=vin(1.3,1.5)|a1;
    (*) tessellate and re-center (*)
displace=3.0;  multiple=5;
center=(multiple-1)*(displace/2);
a3=rin(1,multiple,displace)|a2;
a3=rin(2,multiple,displace)|a3;
a3=tranid(-center,-center)|a3;
    (*)hyperbolic projection(*)
a4=trac(9,[0,0,-10,10,3],1,[0,0,-13],-15)|a3;
    (*)set up and draw(*)
use &,vm(2);
clear;  draw a4;
```

Fig. 2.   An array of 24-cells projected to a hyperbolic surface by Formian.

In this program, the 24-cell is first constructed by long lists of its four-dimensional coordinates (more elegant but more confusing code could have been written here), and these coordinates are connected by the *vin* function. Five copies are made three unit distances apart; the 24-cell can be close packed (in a different orientation), but it is hard to resolve when seen this way. Copies of this row are stacked. Then the tessellation is re-centred on the origin by the *tranid* function. Using the powerful tractation routine written by O. Champion, the set of figures is projected onto a hyperbolic surface from a central point.

Four-dimensional polytopes, the four-dimensional analogues of the Platonic solids, are becoming less and less arcane every day. In one completely unexpected example, the physicist P. K. Aravind, Ref 5, has found them in the matrices that sum up the quantum interactions of particles. In our own discipline, K. Miyazaki, Ref 6, and H. Lalvani, Ref 7, have each shown that four-dimensional figures can lead to novel lattices for space frame structures, and I have built a Quasicrystal structure (that can be seen as a projected six-dimensional lattice). Having a program that can easily manipulate four-dimensional structures will be a boon to many diverse disciplines. Moreover, Formian's export capability allows the artist or mathematician to assemble multiple curved Formian lattices of many dimensions, to maintain these as vector structures, and to import them into other programs for display, analysis, or further manipulation, for example into Adobe's *Illustrator*, as I have done for the artist's drawing that accompanies this article.

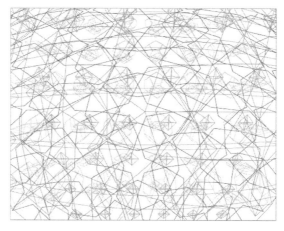

Tony Robbin #21, September 1, 2001

Fig. 3.   An artist's drawing interweaving curved lattices, each made with Formian.

In Drawing #21 in Fig 3, six patterns made with Formian are superimposed: A two-dimensional Persian pattern has been curved onto a dome form. A tessellation of rhombic dodecahedra has been projected onto a hyperbolically curved surface. A field of octahedra and also a field of dodecahedra have been projected to two separate perspective planes. A tessellation of octahedra joined at their points (sometimes called a J-truss) has been placed on a nosecone surface. And finally there is a geodesic pattern, as from a Fuller dome. All these patterns were first generated by Formian as flat patterns, then curved by standard Formian functions. They were exported as *AutoCAD* files, preserving their nature as vector drawings.

Illustrator imports them onto separate layers, where each can be scaled, rotated, colored, filled and in general manipulated separately. Finally the combined drawing can be sent to printers or projectors to scale up for paintings.

After a visit to Scott Carter at the Mathematics Department of The University of Southern Alabama in Mobile in 2000, I began to think more about these drawing and paintings as four-dimensional knot diagrams: my flowing hyperplanes, which have not only thickness but an internal structure as well, are like the sheets of Carter's topology that flow through a space of four dimensions. As the topology teaches us, the hyperplanes braid in ways that are impossible in three dimensions but are the natural consequences of projecting higher-dimensional structures into lower-dimensional spaces, Ref 8.

A new tool can be considered a great tool when it is used for purposes that its creators could never imagine. This is the case with Formian. As the patterns that define space are more and more seen to be multidimensional, Formian will become more and more ubiquitous.

Footnotes:
1)   Formian can be downloaded from the Space Centre's website:
      http://www.surrey.ac.uk/CivEng/research/ssrc/formian.htm
      My four dimensional programs can be downloaded from:
      http://TonyRobbin.home.att.net
2)   EINSTEIN A, Relativity, Crown Publishers, New York, 1961, p. 138.
3)   ROBBIN T, Hypercubic Tessellations Using Formian, HyperSpace, Volume 6, Number 3, 1997, p. 41-6, see also U. S. Patent 5,603,188 Feb. 18. 1997, and A Quasicrystal for Denmark's COAST, Space Structures 4, Surrey Proceedings, vol 2,1993, p 1980.
4)   Buckminster Fuller thought that the cuboctahedron was a special figure because it is the only regular or semi-regular figure where the edge length is equal to the distance of the vertices to the origin. The 24-cell preserves this special feature: the 4d distance from the origin to the vertices is equal to the 4d edge length.
5)   ARAVIND P. K, Physics Letters A 262 (1999) p.282-286.
6)   MIYAZAKI K, An Adventure in Multidimensional Space, Wiley & Sons, New York, 1986.
7)   LALVANI H, Hyper-Geodesic Structures: Excerpts from a Visual Catalog. IASS Atlanta Proceedings, 1994, p 1053.
8)   CARTER S and SAITO M, Knotted Surfaces and Their Diagrams, American Mathematical Society, Providence, 1998.

Frames of nested polyhedra

P. HUYBERS
Delft University of Technology, The Netherlands

ABSTRACT
The envelope of a polyhedron can be formed by the rotation of polygons. This rotation takes place along circular routes around the X- and Y-axis of the co-ordinate system, but not before the polygonal face has been translated over a certain distance along the Z-axis. In some cases an initial rotation around the Z-axis is also necessary. This procedure has been described by the author on a few previous occasions. He developed the computer programme CORDIN (formerly called CORELLI). Thus formed polyhedra can in their turn be put together in spatial arrangements and for that purpose a number of options are available. In stead of flat polygonal faces, objects of any kind can be shifted in space or rotated. If this is done in the same way as the faces in a polyhedron, a second generation of polyhedra is formed, which takes the place of the polygonal faces of the first. This can be done repeatedly. Thus, different levels of polyhedra are obtained that are nested. If arbitrary figures or polyhedron related figures, such as prisms, spheres or ellipsoids, are rotated in space, interesting configurations can be obtained. This can be used for the formation and for the realistic presentation of spatial structures, where polyhedron related elements are used on different levels: macroform or overall shape, microform or internal structure, space frames, nodes and struts. Some examples are shown in this paper.

INTRODUCTION
In the context of this paper we consider the polyhedra with regular faces as sufficiently well-known. Many publications are available on this subject. The author himself reported often on the geometry and the formation by computer of the regular and semi-regular polyhedra [Ref. 1], their duals [Ref. 5], star-polyhedra [Ref. 3] and prism based forms [Ref. 8]. He calculated the geometric properties of most of these forms and their digital data were collected and are now available for further elaboration. It is the main aim of this paper: to see what can be done with these basic forms, if they are multiplied, modified and combined in space. This is done with the help of the above mentioned computer programme.
The directly available tools in this programme for the modification of any element are:
– Initial rotation around X-, Y- and Z-axis
– Initial translation along X-, Y- and Z-axis
– Additional rotation of the whole around X-, Y- and Z-axis
– Additional translation of the whole along X-, Y- and Z-axis
– Magnification factor in 3 directions
– Triangular or circular compression
This is called the zero-functionality. The input can either be: number of sides of a regular polygon, number of sides of a regular polygram (star-polygon), an arbitrary figure which is formed by the input of co-ordinates and its connecting planes, and finally a readily made figure derived from storage.

MOVEMENT, FORMATION AND REPRODUCTION OF ELEMENTS IN SPACE
In the following the main possible manipulations to move or to copy elements in space are discussed.

Translation and rotation
A) Translation, shifting along straight lines without copying
B) Linear repetition, moving along straight line leaving a copy at the place of departure
C) Rotation, moving along circular lines without copying
D) Circular repetition, moving along circular line leaving a copy at the place of departure

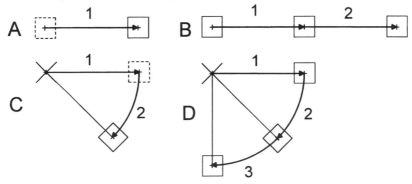

Fig. 1. Principle of translation and rotation with and without copying

Matrices
Matrix production, building a cubical set with given numbers at given distances in three directions.

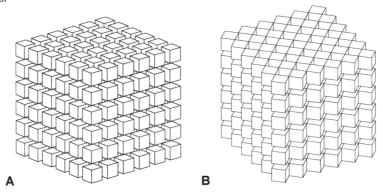

Fig. 2. A) Matrix of cubes, frequency 6 x 6 x 6, distances 1.41. B) Matrix of cubes previously rotated 45° around Y-axis, frequency 6 x 6 x 6, all distances 1.41.

Helixes
Helicoidal progression, moving along circular lines while being lifted at a given distance from the circular plane and leaving a copy at the place of departure.

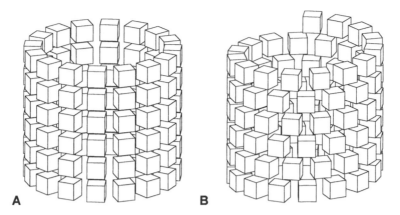

Fig. 3. A) Cylinder of 96 cubes: first a circle is formed and put in a matrix of only two dimensions, distance 1.41, or: a vertical column of 6 cubes rotated over angle 22.5°. B) Helix of 96 cubes wit rotation angle = $(360 + 360/32) : 16 = 23.203°$, translation per step 2.41: 16 = 0.088

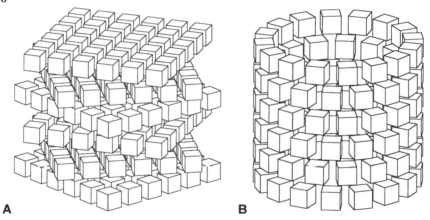

Fig. 4. A) Helix of 6x36 cubes with rotation angle 45° and translation per step of 1.41. B) Helix of 16 cube circle, rotation angle of 11.25° and vertical distance 1.41.

Polyhedral rotation
The formation of polyhedra is done with given sets of rotation angles around X- and Y-axis for each of the polygon types occurring in the different polyhedra gathered in seven cases, leaving in these positions copies of all faces of the regular and semi-regular polyhedra, duals, prisms, antiprisms and star-polyhedra. The final formation of polyhedra is a product of reproduction: one polygonal face of a certain kind is placed in the centre of the XY-plane, shifted along the Z-axis, rotated around the X- and Y-axis respectively. This is done as many time as this specific polygon occurs in the polyhedron and copied at that place. The rotation angles for all polygons of one kind are kept in store and such a combination of angles and number of elements is called here: rotation case. 7 rotation cases are sufficient for the formation of all existing polyhedra [Ref. 1], the 5 regular or Platonic polyhedra and the 13 (or actually 15, as two of them occur in a left-handed and a right-handed version) semi-regular or

Archimedean polyhedra.

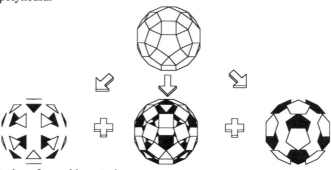

Fig. 5. A polyhedron formed by rotation.

All uniform polyhedra can be formed using any of the following sets of rotation angles belonging to the faces of the Tetrahedron, the Cube, the Octahedron, the Dodecahedron, the Icosahedron, the Truncated Cuboctahedron or the Truncated Icosidodecahedron. The input can either be: number of polygon or polygram sides, data of dual faces and of star-faces. Also must be given the distance of this face and an eventual initial rotation around the Z-axis.

CLOSE PACKINGS OF POLYHEDRA

Polyhedra can be packed in various combinations to fill space entirely. The Truncated Octahedron is self-filling (Fig. 6). If it is combined with the Truncated Cuboctahedron, cubes in three different positions are necessary to make a complete space filling (Figs. 7-9)

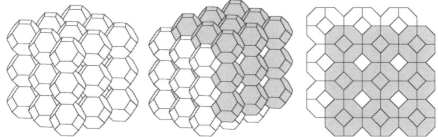

Fig. 6. Close packing of Truncated Octahedra, placed in two matrices 3 x 3 x 3, at distances of 2.83, the second matrix shifted along 1.415 in three directions.

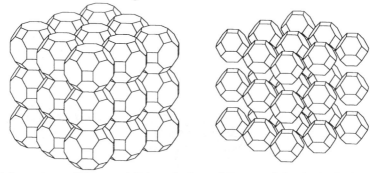

Fig.7. 3x3x3 matrices of Truncated Cuboctahedra and Truncated Octahedra both at distances of 3.83 and the second shifted along the axes: X = Y = -Z = 1.915

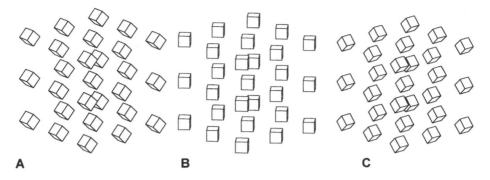

Fig. 8. 3x3x3 matrices of cubes at 3.83 distance with initial X-, Y- or Z- rotation = 45° and shifted: A) X = 3.83, Y = -Z = 1.915, B) X = -Z = 1.915, Y = 0, C) X = Y = 1.915, Z = 0.

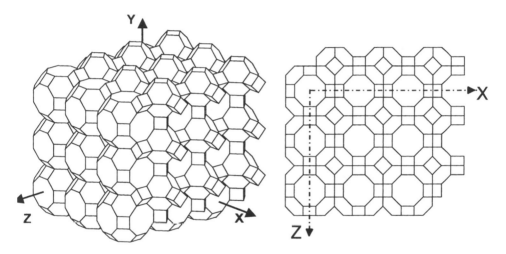

Fig. 9. Combination of the 5 matrices in Figs. 7 and 8 to a fully closed packing.

PRISMS AND ANTIPRISMS

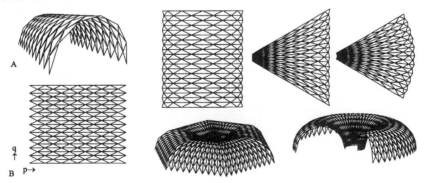

Fig. 10. An antiprismatic cylinder in triangular and circular compression.

The input of the side numbers of either a polygon or a polygram is sufficient for the automatic production of prisms and antiprisms. Cylindrical antiprismatic forms are generated by the input of values of characteristic properties and element numbers in transverse and length direction.

DUAL POLYHEDRA

For polyhedra the input normally is a regular polygon, but it is also possible to rotate an object of any kind in a similar way. This is for instance done with the duals of the polyhedra [Ref. 2]. Their faces are flat, but in many cases they have a form that is not regular.

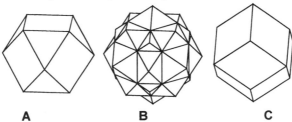

A   **B**   **C**

Fig. 11. A typical polyhedron A (the 'quasi-regular' Cuboctahedron), its dual C (the Rhombic Dodecahedron) and the compound B of both.

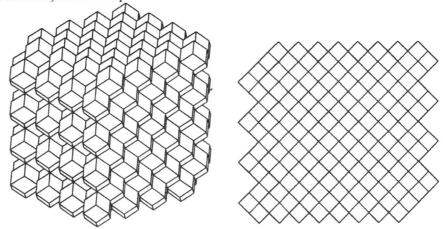

Fig. 12. Four 4x4x4 matrices forming a so-called honeycomb of 256 Rhombic Dodecahedra.

REALISTIC VISUALISATION OF SPACE FRAMES

The computer program CORDIN has specifically been developed for the generation of polyhedron based forms. Therefore space frames are initially obtained in the form of polyhedral packings, the faces of which are formed by flat solid plates. A conversion routine has been developed, converting these planar forms into wire frames. The struts then can be shaped such that they obtain the appearance of real structural elements, eventually with additional nodes (fig. 13). Some of these manipulations can be combined and the results can be stored to be used again for further application. The above principles have to a great extent been applied for the visual presentation of a tower building system, that has been constructed built of roundwood poles in 1999. The basic form has an octagonal basis and it consists of three antiprisms stacked on top of each other. It has its top floor at a height about 9m (Fig. 14).

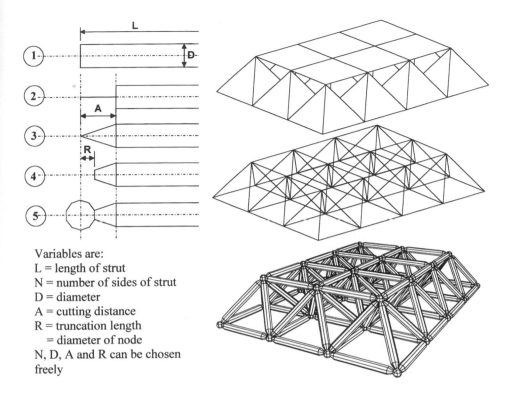

Variables are:
L = length of strut
N = number of sides of strut
D = diameter
A = cutting distance
R = truncation length
  = diameter of node
N, D, A and R can be chosen
freely

Fig. 13. Space deck converted by a fully automatic routine into a realistic space frame.

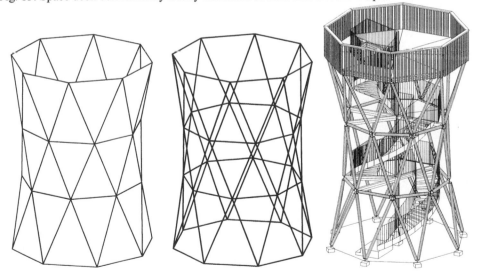

Fig. 14. Sketch of an antiprismatic tower structure of roundwood, built in May 1999 in Kootwijk, The Netherlands, comprising many of the techniques described above.

POLYHEDRAL PATTERNS AND SPHERE SUBDIVISIONS

Before the rotation takes place, any pattern can be projected upon the polyhedron face. This leads to interesting, and sometimes very practical solutions [Ref. 9]. These patterns can also be so-to-say 'exploded' so that all their points get the same distance from the centre and that the envelope thus becomes spherical. Most of the sphere subdivision methods are realized by a further subdivision of the triangles in the Octahedron or of the Icosahedron. These subdivided triangles, which have generally also a triangular pattern on them, are subsequently rotated and projected upon the surrounding spherical envelope. Similar procedures can be followed for the semi-regular polyhedra, where the polygons with more than three sides must previously be covered with a suitable pattern. The spheres themselves can be rotated similarly as before.

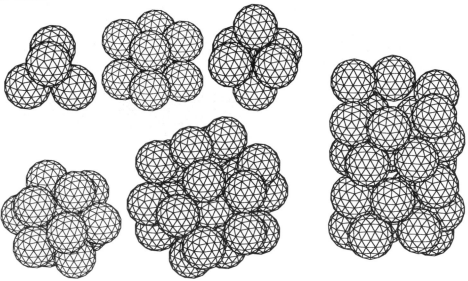

Fig. 15. Spheres in the corners of the regular solids.   Fig. 16. Spiral packing of spheres.

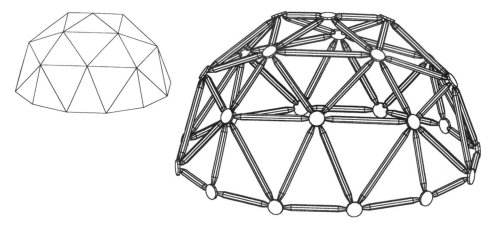

Fig. 17. Dome structure with circular nodes.

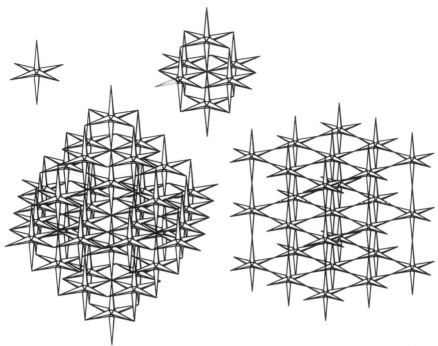

Fig. 18. Spheres with a very low exponent of the two main elliptical generators – in this case 0.3 in stead of 2 – give a 'spikey' figure. This is rotated here, one and two times according the Octahedron. It is also put in a 3x3x3 cubic matrix.

STELLATED POLYHEDRA

Stellated polyhedra are formed by the extension of the polyhedral faces in space until they intersect again. This is so when the dihedral angle between these faces is convergent. The Tetrahedron and the Cube do not have stellated forms, as the planes through their faces will never intersect. The Octahedron is the first to have one stellation: the Stella Octangula. In the Dodecahedron the stellation procedure can be continued three times and in the Icosahedron as many as 58 times [Ref. 3]. All these versions must be formed by the rotation of parts of planes touching the original polyhedron.

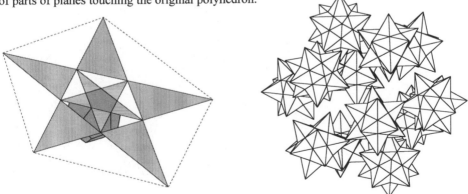

Fig. 19. The stellation of a Dodecahedron a dodecahedral cluster of its first stellation.

CONCLUSIONS
All pictures in this paper have been made with the programme CORDIN, being developed by the author in co-operation with G. van der Ende. A computer programme like CORDIN may offer ways for the visual presentation of spatial forms without the need to build them in reality. It is based on the generation of facial objects, but these can be transformed into skeletal frames. Once these frames and their nodes having been formatted in a proper way, they can be materialised in the form of a real and realistic configuration of struts and nodes.

REFERENCES
1. HUYBERS P, The formation of polyhedra by the rotation of polygons. 4th Int. Conference on Space Structures, Guildford, 6-10 September, 1993, p. 1097-1108.
2. HUYBERS, P., Form generation of polyhedric building shapes, Int. Journal of Space Structures, Special issue on Morphology and Architecture, Vol. 11, Nos 1&2, November, 1996, p.173-181.
3. HUYBERS P and VAN DER ENDE G, Star-polyhedra, International Symposium of IASS on 'Shell and Spatial Structures: Design, Construction, Performance & Economics', Singapore, 10-14 November, 1997, Vol. 1, p. 325-334.
4. HUYBERS P, The visualisation of spatial structures with computer techniques, Fourth International Conference on Computational Structures Technology, 18-21 August 1998, Edinburgh, Scotland. p.B1-B14.
5. HUYBERS P, The formation of the reciprocal polyhedra by rotation, LSA'98 Conference on 'Lightweight Structures in Architecture, Engineering and Construction', 5-9 November, 1998, Sydney, Australië, p. 1019-1028.
6. HUYBERS P, The materialisation of space frames, IASS Congress on 'Shells and spatial structures: from the recent past to the next millennium', Madrid, 20-24 September, 1999, p. F33-40.
7. HUYBERS P, Spheres and spheroids, IASS/BLS Conference, 29/5-2/6/2000, Istanbul, p. 547-556
8. HUYBERS P, Prism based structural forms, Engineering Structures, No. 23, 2001, p. 12-21.
9. HUYBERS P and VAN DER ENDE G, Polyhedral Patterns, Book of extended abstracts, IASS 2001 Int. Symposium on Theory, Design and Realization of Shell and Spatial Structures, October 9-13, 2001, Nagoya, Japan, p. 290-291, Paper TP129.

Topological optimisation of double layer grids using a genetic algorithm

H. E. FARSANGI
Space Structures Research Centre, University of Surrey, UK

ABSTRACT
The ground structure method is one of the approaches that have been used in topology optimisation of trusses. This approach usually leads to the creation of a structure that may not have been reduced in an acceptable way. That is, the resulting structure may not be suitable from the viewpoint of construction or architecture. As an alternative approach, the densest possible configuration in a particular family of a commonly used structure is taken as the base structure. In this paper, the suggested approach is used in relation to the 'double layer grids' which are one of the most commonly used families of space structures. In practice, it is usual, that some of the elements of a dense double layer grid may be eliminated for reducing the weight of the structure or meeting some architectural requirements. Usually, this process of element removal is based on experience and engineering judgment. To establish a more reliable criterion for choosing the elements to be removed, an approach is developed to find the 'degree of importance' of the elements of a double layer grid in carrying the loads. In this method, a fitness function is defined and a genetic approach is used for minimising the fitness function. Using the degree of importance of the elements, a designer can decide more reliably about the most suitable patterns for reducing a grid.

INTRODUCTION
Genetic algorithm may be used in a wide variety of structural optimisation problems. In particular, it may be employed effectively in structural topology optimisation [1-3]. Some approaches have been used for this purpose one of which is the 'ground structure approach'. In this approach, a structure containing all possible elements connecting all pair of existing nodes is considered as the ground structure. Then, using the genetic algorithm some of the elements are removed to find the topology associated with minimum weight or cost for the structure. This approach may lead to a structure that may not be suitable from the construction and architectural viewpoint. In this work, an approached is developed by which the suitable reduced grids that may be derived from a double layer grid are found. This double layer grid which is used for generating the other reduced grids is referred to as the 'base grid'.

The fundamental objective of this study is to investigate the possibility of assigning a degree of 'importance' to each element of the base grid. It is desirable that the importance of each element be expressible quantitatively, such that two different elements of the base grid may be compared in terms of their corresponding 'importance'. A question may arise at this point, namely, what could be the 'sense' of importance for an element of the grid in the present context?

In general, the configuration of a double layer grid is chosen based on architectural requirements and is designed for transferring the loads to the supports through its elements. The importance of an element may then be regarded as the degree of its contribution in carrying the load. That is, the elements that carry more forces may be regarded as more important.

However, carrying the load should be associated with a reasonable behaviour for the structure. That is, the stresses and displacements should not exceed the allowable values. To have a grid with a desirable behaviour, the grid is designed and the cross-sectional areas are allocated to the elements of the grid. The cross-sectional areas may be allocated through an optimisation process. In this sense, an element with greater cross-sectional area may be regarded as a more important element. It is seen that assigning a degree of importance to an element of the base grid may be carried out with different viewpoints. However, in this work the attention is focussed on the investigation of the importance of the elements in carrying the loads.

ILLUSTRATIVE EXAMPLE
Consider the double layer grid with a square on square pattern given in Fig 1a. In this figure the top layer elements are indicated with thick lines while the bottom and the web elements are shown with thin lines. This grid which is considered as a base grid, is supported at four bottom layer nodes indicated by little circles as shown. Support S1 is constrained in the X, Y and Z directions and support S2 is constrained in the Y and Z directions. The other supports are constrained only in Z direction. This grid consists of 72 elements all of which are of the same length L. For all the members of the grid being of the same length, the depth of the base grid should be equal to $\sqrt{2} L/2$. The elements of the grid are numbered as shown in Fig 1a. The base grid of Fig 1a is a statically indeterminate structure with 4 degrees of indeterminacy. The grid is subjected to vertical concentrated loads of magnitude of P at all the top layer nodes.

GENETIC ALGORITHM IN TOPOLOGY OPTIMISATION
The principles of the genetic algorithm have been described in many texts in particular by Goldberg[4] and Davis[5]. The genetic algorithm starts by creating an initial population of members which are generated randomly. This population evolves using the 'pairing', 'crossover' and 'mutation' operations. The procedure for implementing the genetic algorithm in relation to the evaluation of reduced grids is explained step by step in the sequel.

Initiation
The genetic process starts with an initial population of members each of which is a chromosome. A chromosome represents a reduced grid which is derived from the base grid of Fig 1a. In the reduced grid the support arrangements and the load conditions are the same as for the base grid. Fig 1b indicates a reduced grid that has been generated randomly. As shown in this figure, the reduced grid has four missing elements that are shown by dotted lines. Each chromosome consists of a number of '1' and '0' digits. The number of these digits is equal to the number of the elements of the base grid. For the case under consideration, the number of the digits is equal to 72. A digit 1 in a chromosome indicates the presence of an element of the base grid in the reduced grid and a digit 0 indicates the absence of an element of the base grid in the reduced grid. The digits 1 and 0 are the genes of the chromosome. The chromosome representing the reduced grid of Fig 1b is shown in Fig 2. The leftmost gene corresponds to the element number 1 of the base grid. Also, the rightmost gene corresponds to the element number 72. The other genes in the 2nd to 71st positions correspond to the elements 2 to 71. Comparing the grids of Figs 1a and 1b, it is seen that the elements removed from the base grid

are elements 8, 36, 59 and 63. The genes corresponding to these elements are 0. Therefore, the genes in positions 8, 36, 59 and 63 in the chromosome of Fig 2 are 0 and the rest of the genes are 1.

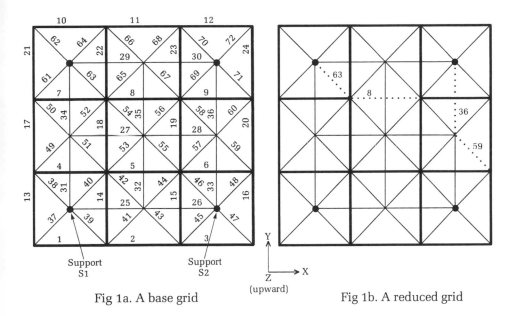

Fig 1a. A base grid

Fig 1b. A reduced grid

Fig 1. A base grid and a reduced grid

To generate a chromosome, each gene is created randomly. To this end, a random number between 0 and 1 is generated for each gene. If this random number is smaller than a value, called 'element presence factor', then the digit 1 is considered for the gene. Otherwise, the gene will be taken as 0. The element presence factor is denoted by C. Thus, the elements remain in the reduced grid with a probability equal to C and are eliminated with a probability equal to (1-C). The choice of C is discussed below.

1111111011111111111111111111111111111111110111111111111111111111111111011101111111111

Fig. 2. A chromosome.

Studies carried out in the present work show that if C is too low, then too many elements are removed and the resulting reduced grid has a high probability of being statically unstable. On the other hand, the evolution of the population depends on the existence of some chromosomes representing statically stable grids in the initial population and therefore it is not desirable for the value of C to be too low. Experience shows that a suitable percentage for the members of the initial population representing stable grids is about 10%. For the case of the base grid of Fig 1a, in one occasion it was found that a suitable value for the element presence factor C was 0.92. In this occasion, nine members of the initial population of one hundred members were the chromosomes which represented stable grids.

Fitness Function

The genetic process proceeds by selecting members of the initial population for mating. However, before the mating process is performed, the members of the population are 'evaluated'. The evaluation of members is carried out based on the 'fitness values'. The fitness value assigned to a member is used to measure how 'good' that member is in a particular context. For the problem under consideration, a fitness function is defined as explained below.

Consider two different reduced grids that are obtained by random selection from the base grid of Fig 1a. Suppose that these grids are statically stable, and let the grids be subjected to linear analysis with support conditions and loads the same as the base grid. These two reduced grids are compared in terms of their internal forces. The grid with smaller summation of absolute values of internal forces is chosen as the 'better' grid. The fitness function is given as

$$\text{Fitness} = \frac{s}{S} \qquad\qquad 1$$

Where, s and S denote the summations of the absolute values of the internal forces of the reduced grid and that of the base grid, respectively. In what follows the phrase 'summation of absolute values of internal forces' is abbreviated to 'summation of internal forces'. Table 1 lists the fitness values of some of the members of the initial population with 100 members, generated in one occasion during the running of the genetic process. These fitness values are related to the members representing stable reduced grids. The fitness value for a member representing an unstable grid will be discussed later.

Returning to the discussion concerning the procedure of the genetic algorithm, the process works based on the selection of the members of the population for mating. This selection is based on the fitness values of the members. With Eqn 1 as the fitness function, a member with a smaller fitness value will have more chance of being selected.

Table 1. Initial population.

Member number	Number of elements of reduced grid	Fitness value
21	71	1.00
36	68	2.83
43	68	2.33
48	70	1.67
53	68	1.00
58	70	1.00
91	70	1.00
94	68	5.08
98	69	1.00

Here, it should be noted that Eqn 1 can not be used for calculating the fitness of a member of the population whose corresponding grid is not statically stable. A member that represents a statically unstable grid may be dealt with in different ways. Normally one may discard such a member and only accept the members representing statically stable grids. However, the policy adopted in this work is that a member of the population that represents a statistically unstable grid is given a low probability of selection for mating, rather than being discarded. To elaborate, a suitably large fitness value is given to such a member so that it has a small chance of being selected for mating. This fitness value is larger than the fitness values of all members representing statistically stable grids.

Pairing

In this work, the tournament pairing is employed. That is, two chromosomes are randomly chosen from the population and they are compared in terms of their fitness values. The chromosome with the smaller fitness value will be chosen as a parent chromosome. Then, the same procedure is repeated to choose the other parent chromosome.

Mating

Mating process is performed with two parent chromosomes selected by tournament pairing. The mating process consists of two steps, namely 'crossover' and 'mutation'.

Crossover

The two selected parent chromosomes (selected through the tournament pairing) are operated upon with some of their genes being exchanged by subjecting them to the process of crossover. Here, a type of crossover, called, 'uniform crossover' is used. To perform the uniform crossover a string of 1 and 0 digits, called a 'mask', is generated randomly. The number of the digits of this string is equal to the number of the digits of a chromosome in the population. For generation of each digit of the mask a random number between 0 and 1 is generated. If this random number is greater than 0.5, the corresponding digit of the mask will be equal to 1, otherwise, it will be equal to 0. For the example under consideration, the mask consists of 72 digits. To perform the crossover, those genes of the parents whose corresponding digits in the mask is 0, are exchanged and those that correspond to a digit 1 in the mask are left alone. An example of two parent chromosomes together with a mask is shown in Fig 3. Also shown in this figure, are the resulting two child chromosomes. The exchanged genes in the example of Fig 3 are shown in bold. Each of the child chromosomes represents a new reduced grid.

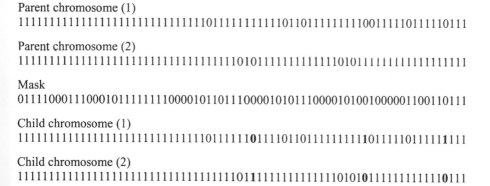

Parent chromosome (1)
111111111111111111111111111111101111111111110110111111111110011111011111101111

Parent chromosome (2)
111111111111111111111111111111111111110101111111111111110101111111111111111111

Mask
011110001110001011111111000010110111000010101110000101001000001100110111

Child chromosome (1)
111111111111111111111111111111101111110111101101101111111110111110111111111111

Child chromosome (2)
111111111111111111111111111111111110111111111111101010111111111110111

Fig. 3. Uniform crossover.

Mutation

Mutation takes place occasionally and changes some of the genes of the child chromosomes randomly. The frequency of the mutation is specified by the 'mutation rate' and each gene of the child chromosomes may be subjected to the mutation. To perform the mutation, for each gene a random number between 0 and 1 is generated. If this random number is less than the mutation rate, then the corresponding gene is mutated. That is, if the selected gene is 1, then it will be changed to 0 and if it is 0 it will be changed to 1. For instance, with a value of 0.005

for the mutation rate, at one occasion the child chromosome 2 of Fig 3 was mutated by the gene at the 40th position from the left being changed from 1 to 0, as shown below.

11111111111111111111111111111111111110111**0**1111111111010101111111111110111

The mutated gene is shown in bold. This new chromosome represents a new reduced grid.

Replacement
The two child chromosomes created by the mating process, replace the worst two chromosomes in the current population. This operation is carried out even if the child chromosomes are not better than the two worst chromosomes. The reason for this is that, it is possible for the child chromosomes to have some good genes that may improve the performance of the genetic process.

The process of pairing, mating and replacement are repeated until the number of the generated child chromosomes is equal to the number of the members of the initial population. That is, the cycle is continued until one generation is completed. For instance, in the current example, the cycle will be repeated 50 times to generate 100 new member of the population.

Termination
When one generation is completed, the best solution of that generation, that is, the chromosome with the least value of fitness is found. Experience shows that normally the fitness value of the best solution becomes smaller gradually. Here, a criterion is needed to terminate the process. In this work, the following equation is used as the convergence criterion.

$$\frac{b_{i-1} - b_i}{b_i} \leq \varepsilon \qquad\qquad 2$$

In the above equation, b_{i-1} and b_i are the fitness values for the best members of the previous and the current generations, respectively, and ε is a small value, typically, 0.001. If this equation is true for a number of successive generations, typically, 10, then the process is terminated. However, in some cases, the genetic algorithm does not converge and it should be terminated with another criterion. The maximum number of generations is chosen as this criterion which is used for termination. Then, if Eqn 2 is not satisfied after the maximum number of generations, the process is terminated.

Portancy Map
Actually, one of the goals of the procedure developed here is to establish a method by which the importance of an element of a double layer grid may be expressed numerically. This method is explained in the sequel.

Suppose that the genetic process is carried out once. The result is a reduced grid with the least value of the summation of the internal forces. For instance, Fig 4b shows a reduced grid that has been obtained for an application of the genetic process in which the grid of Fig 4a has been considered as the base grid. In this reduced grid the eliminated elements are indicated by dashed lines. The force distribution corresponding to this reduced grid is shown in Fig 5a. Now, let the genetic process be carried out again. Comparing the force distribution for the resulting optimum reduced grid with the force distribution obtained for the first running of the genetic process, it is seen that they are similar but not identical (Figs 5a and 5b). The reason for the differences is that the genetic algorithm works in a stochastic manner. Therefore, by executing the process using different initial randomly generated populations the results are probabilistically different and the analytical results are bound to be different.

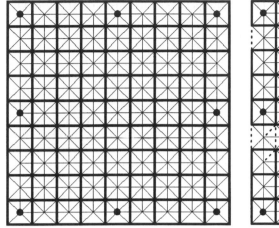

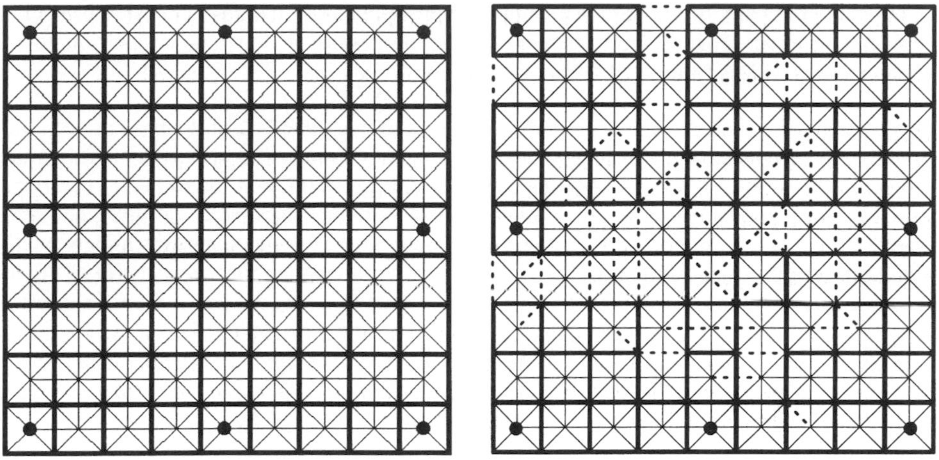

Fig 4a. Base grid   Fig 4b. Optimum reduced grid

Fig 4. Base grid and an optimum reduced grid

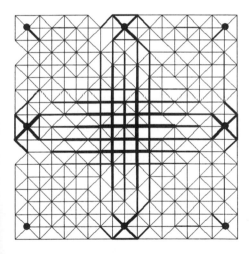 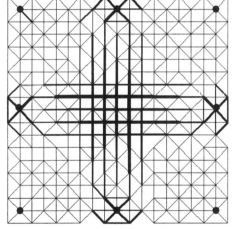

Fig 5a. Force distribution in   Fig 5b. Force distribution in
the reduced grid of Fig 4b.   another optimum reduced grid

Fig 5. Force distributions in optimum resuced grids

Suppose that the genetic process is executed several times, say 'r'. Then, there are r reduced grids each of which is an optimum solution obtained by the genetic algorithm. Now, for each element of the base grid, the absolute values of the internal forces of that element in these r reduced grids are added together. For the cases in which this particular element is absent from a reduced grid, the value of zero is considered for its internal force. Then, the value corresponding to each element of the base grid is divided by the maximum of the values for all the elements of the base grid. The result of the division will be a number between 0 and 1 for each element. Each of these values is referred to as the 'portancy' of an element. To obtain a general appreciation of the portancy values of the elements of the base grid, the results are graphically shown. That is, each element of the base grid is drawn in proportion to its portancy. The result is called a 'portancy map'.

The portancy map indicates the degree of importance of each element of the base grid. In this context, two aspects of importance are considered for an element. One aspect is the role of an element in the statically stability. Another aspect is the contribution of the element in carrying the loads, while the fitness function is minimised. A portancy map which is produced based on several optimised grids indicates the relative importance of the elements from view points of both of the above considerations. For instance, the portancy map for the base grid of Fig 4a that have been obtained based on 94 times repetition of the genetic process is given in Fig 6.

Fig 6. A portancy map

A portancy map may be imagined as a map representing an average of the results obtained by repeating the optimisation process. A question may arise at this point, namely, how many times the process should be repeated in order to give rise to a reliable portancy map. In this relation, one may carry out a statistical investigation. To elaborate, the force distribution corresponding to each resulting reduced grid may be considered as a sample that has been obtained in a stochastic procedure. Then, the required number for repetition of the process may be found based on a statistical analysis such that a 'good' average is obtained for these samples. However, in this work a simple procedure is proposed for this purpose which is based on the symmetry expected for the portancy map. To elaborate, referring to the force distributions given in Figs 5a and 5b, it is seen that these figures do not show a high degree of symmetry. On the other hand, as explained before, the portancy map is obtained from a number of such force distributions each of which relates to an optimum reduced grid. If the portancy map is made based on a few numbers of the optimum reduced grids, then it may have a low degree of symmetry. However, experience shows that the symmetry is improved while the number of the repetitions of the process is increased, although, complete symmetry may not be achieved. In this work, the anticipated symmetry for the portancy values of the elements is used as a criterion to find the number of the times that the genetic process should be repeated. To elaborate, each resulting reduced grid which is the best solution given by the genetic algorithm, is obtained in a random manner. The symmetry of the portancy map is an assurance that the genetic algorithm has explored the whole of the domain of the search space randomly and it has no bias toward some particular parts of the search space. Then, the symmetry may be considered as a suitable criterion for the determination of the number of the times necessary for repeating the genetic process. Since the complete symmetry cannot be achieved, then the symmetry is checked for the elements with more contribution in carrying the loads. That is, the number of the repetitions of the genetic process is controlled such that the symmetry is satisfied for the elements with the highest portancies.

The portancy map provides an overall view of the element portancies. So, it cannot be used to find the accurate portancy of a particular element. However, the accurate values of the portancies of the elements of the base grid are saved in a file and may be used when the accurate values are of interest. Referring to the portancy map of Fig 6, the elements with high portancies are the elements that are expected to have large internal forces. For example, the web elements near the supports and the top and bottom layer elements in the middle region of the grid are usually the elements with large internal forces. These elements are seen to be represented by thick lines in the portancy map of Fig 6. For a particular boundary condition and load arrangement, the portancy of an element of the base grid is a specification that describes how important that element is at its position in the grid. However, the portancy of a particular element may change when the boundary conditions and the external loads are changed.

One of the main benefits of the portancy map is that a designer may obtain an overall view of the contribution of the elements in carrying the loads. This overall view is useful in designing an efficient grid. However, in relation to finding the optimum topology, the practical reduced grids derived from a particular base grid are compared using the portancy values. Suppose that the grids shown in Fig 7 are both acceptable from architectural viewpoint. Each of these

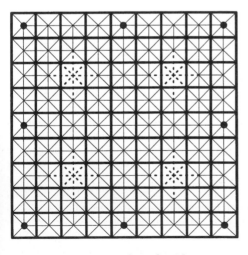

Fig 7a. A reduced grid 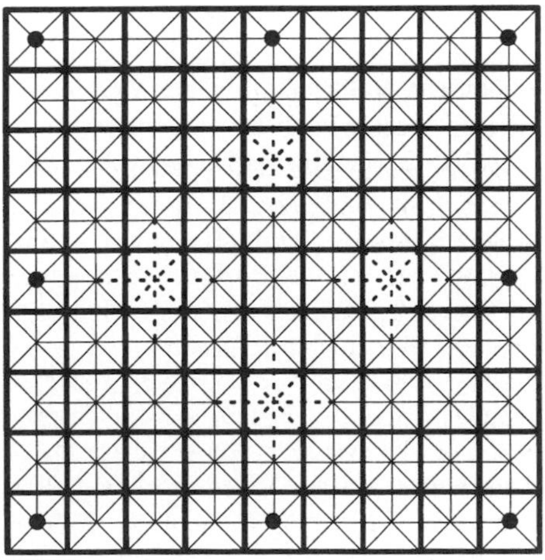   Fig 7b. Another reduced grid

Fig 7. Two possible reduced grids

grids are obtained by removing 32 elements from the base grid of Fig 4a. The removed elements are indicated by dashed lines. Using the portancy values obtained for the portancy map of Fig 6, the summation of the portancy values of the removed elements for the grids of Figs 7a and 7b will be equal to 1.5 and 8.2, respectively. This implies that the grid of Fig 7b has been obtained by removing the elements with higher portancies. That is, more important elements have been eliminated from the base grid for generating the grid Fig 7b. Now, using the genetic algorithm, the minimum weight of the structure for each of these grids is calculated. To this end, each grid is supposed to have dimensions 36m×36m in plan and a depth of 1.8 m. The grids are subjected to vertical concentrated loads of magnitude of 32KN at all the top layer nodes. The elements of the grids are taken to have hollow tabular cross-sections with a thickness equal to 1/15 of the outside diameters of the tubes. The outside diameters of the tubes that vary in the range 60 to 150 mm are considered as independent variables of the problem. Running the genetic process it is seen that the grid of Fig 7a has a less minimum weight. So, it is concluded that by comparing different topologies derived from a particular base grid in terms of their corresponding portancy values of the removed elements, the most suitable topology may be chosen such that it is associated with the least weight.

CONCLUSION

In this paper a method is presented for topology optimisation of the double layer grids. In this method a procedure is developed for assigning a numerical value to each element of the double layer grid. Using this numerical values, the reduced grids derived from a particular double layer grid may be compared with each other and the most suitable topology is found.

REFERENCES

1. RAJEEVE S and KRISHNAMOORTHY C S, Genetic Algorithms-Based Methodologies for Design Optimisation of Trusses, Journal of Structural Engineering, vol 123, No.3, 1997, pp 350-358.
2. GRIERSON D E and PAK W H, Optimal Sizing, Geometrical and Topological Design Using a Genetic Algorithm, Structural Optimisation, 6, 1993, pp 151-159.
3. HAJELA P and LEE E, Genetic Algorithm in Truss Topological Optimisation, International Journal of Solids and Structures, vol 32, No 22, 1995, pp 3341-3357.
4. GOLDBERG D E, Genetic Algorithm in Search, Optimisation and Machine Learning, Addison-Wesley Publishing Company, Inc. New York, 1989.
5. DAVIS L, Handbook of Genetic Algorithms, Van Nostrand Reinhold, New York, 1991.

RBF and BP neural networks used for the design of domes

A. KAVEH
Iran University of Science and Technology, Narmak, Tehran, Iran
M. RAEISSI DEHKORDI
Iran University of Science and Technology, Tehran, Iran

ABSTRACT
In this paper, efficient neural networks are trained for predicting the deflection of domes using the Backpropagation and Radial Base Functions networks. Radial basis functions network is used for simultaneous prediction of deflection and weight of the structure. Configuration processing is performed using FORMIAN, analysis and design are carried out employing SAP2000. Programs are developed for distribution of applied forces on the nodal points and connecting different programs to each other.

INTRODUCTION
Space structures are often used for covering large spans. Sport stadiums, assembly halls, exhibition centers, swimming pools, shopping arcades and industrial buildings are typical examples of structures where large unobstructed areas are essential and where minimum interference from internal support is required. Space structures are often classified as grids, domes and barrel vaults.

Configuration processing and data generation for space structures can be simplified using the concepts from Formex algebra [1]. For preprocessing for efficient solutions, theory of graph can be employed, Ref [2]. An optimal analysis can be carried out using usual matrix displacement method [3]. For large-scale problems the analysis is usually time consuming due to a large number of equations being involved. Therefore approximate methods are needed in the process of nonlinear analysis, design and optimization.

Neural networks provide a powerful tool for approximate analysis and design of space structures. Such networks are trained using backpropagation [4-5] and counterpropagation networks, Refs [6-9].

In the present paper efficient neural networks are trained for design of domes using *NeuralWorks* software, Ref [10]. Single layer domes with varying the spans with an increment of 2.5 meters between 42.5 and 65 meters are considered. Backpropagation (BP) and Radial Basis Functions (RBF) networks are employed for training these nets. Configuration processing is performed using *FORMIAN*, analysis and design are carried out employing *SAP2000*. Additional programs are developed for the correct distribution of applied forces on the nodal points and linking different programs to each other.

STRUCTURAL MODELS

The model considered in this paper is a single layer dome with elements joined together rigidly, Figure 1. The spans are varied between 42.5m and 65m with steps of 2.5 meters. The opening angle of the domes varies between 45 and 67.5 degrees. A typical ribbed dome used in this study is shown in Figure 1(a). Simple supports are considered for the exterior ring of the dome. The number of distances between the ribs is considered as 20 and for rings this number is taken as 8.

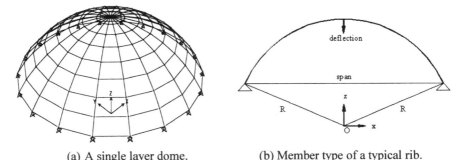

(a) A single layer dome.   (b) Member type of a typical rib.

Fig. 1. A single layer ribbed dome and members of a typical rib.

In order to simplify the design, and due to symmetry, 8 types of cross sections are considered for the members of a typical rib. The span and deflection at the apex of the dome are illustrated in Figure 1(b). A total number of 100 structures are generated and analyzed from which 80 structures are used for training and 20 pairs are employed for testing the networks.

CONFIGURATION PROCESSING, ANALYSIS AND DESIGN

The structural models are generated using *Formian2* developed by Nooshin et al [11]. The analysis is performed by *SAP2000* developed by Wilson [12], and a program is written for design of the structures, which iteratively improves the design. The following data is used:
1. The sum of dead load and live load equal to 2.5kN/m^2 is applied to the nodes.
2. All the nodes are considered as rigid.
3. Elements are selected from the tube sections available in European profile list.
4. Linear analysis is performed.
5. Design is carried out according to AISC-89 code.

BP and RBF NEURAL NETWORKS

Recently neural networks are extensively employed in structural mechanics, Refs [4-9]. In the following, some basic concepts of neural networks is summarized, however, the interested reader may refer to textbooks on this subject [13-15].

Basic concepts

Artificial neural networks (ANNs) try to simulate the functions of the brain through mathematical and computational models. ANNs are structured from a set of artificial neurons or processing units, which are arranged on a set of layers. A processing element of an ANN is a simple device that approximates the function of a biological neuron. The topology of ANNs may be in the form of feed-forward or recurrent. Learning may be supervised or unsupervised depending on the topology. Feed-forward neural networks are usually trained using supervised training procedures.

In a processing element, the input signals are translated to an output signal by an appropriate transfer function. Weights, representing the synapses, are given to input signals according to their importance. These weights w_{st} are the strength of connections between processing elements. After a processing element receives all of its input signals, through its input path, it computes the net sum of input according to:

$$net_{pt} = \sum_{s} w_{st} o_{ps},$$ (1)

where w_{st} is the connecting weight from the source "s" layer to the target "t" layer and o_{ps} is the output produced from pattern p as the result of the input $o_{p(s-1)}$.

The net sum is applied to an activation function. The most commonly used activation function is the sigmoid function defined as

$$o_{pt} = \frac{1}{1 + e^{-\frac{(net_{pt} + bias)}{\theta}}}$$ (2)

where bias serves as a fictitious weight of a unit value which always fires ($o_{pt} = 1$). This function is bounded within the range (0,1) and has a continuous first derivative. The output of a unit is obtained by passing the value of the net-sum through the activation function as

$$o_{pt} = f_t(net_{pt}).$$ (3)

Backpropagation algorithm

The backpropagation learning algorithm was discovered by Parker [16], and it was popularized and developed into a workable process by Rumelhart et al [15]. Numerous applications have been found for backpropagation since 1986.

The method of training adopted in this paper is the backpropagation. This approach is one of the most successfully and widely used algorithm among artificial neural networks. In backpropagation, learning is carried out when a set of training patterns is propagated through a network consisting of an input layer, one or more hidden layers and an output layer. Each layer has its corresponding units and weight connections. The topology of a BP network is illustrated in Figure 2.

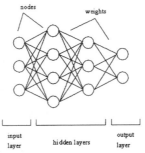

Fig. 2. The topology of a BP network.

At the output layer, the computed outputs are subtracted from the desired (target) output and the squared sum of the output difference is a measure of the error. This measure indicates the level at which the network has learned the input-output data, and may also be used to determine the gradient of the learning procedure.

The sum of the squares of the error for a single pattern at the output units is measured by:

$$E_p = \frac{1}{N_{output}} \sum_{outputs}^{network} (t_p - o_p)^2, \qquad (4)$$

where E_p is the measure of the error of pattern p, t_p is the target output, o_p is the observed output at the output nodes, and N_{output} is the number of the output units.

After all patterns have been processed by the network, the total sum of the error, is given by:

$$E = \frac{1}{PN_{output}} \sum_P \sum_{outputs}^{network} (t_p - o_p)^2, \qquad (5)$$

where P is the total number of patterns in the training set.

The delta signal for the output layer is defined by:

$$\delta_{pt} = (t_{pt} - o_{pt}) f_t'(net_{pt}), \qquad (6)$$

and for the hidden layer:

$$\delta_{pt} = f_t'(net_{pt}) \sum_k \delta_{pk} W_{kt}. \qquad (7)$$

The derivative of the activation function that is used in Eq. (3) is given by:

$$f_t' = \frac{\partial o_{pt}}{\partial net_{pt}} = o_{pt}(1 - o_{pt}), \qquad (8)$$

where k represents an upper layer unit (the output layer is the uppermost and the input layer is the lowermost layer).

The learning rule associated with the backpropagation method is known as *generalized delta rule* [17]. According to this rule, for changing the weights in the network, for a given pattern p, the following relation may be used:

$$\Delta W_{st}(n) = \eta \delta_{pt} o_{ps} + \alpha W_{st}(n-1), \qquad (9)$$

where n is the input-output presentation number and η is the *learning rate* and α is the *momentum*.

The momentum, α, is a fraction which is multiplied by the previous weight change and added to the current weight change; α should be positive and less than unity. The two terms are not independent, and should be chosen and modified as a pair. In general, dynamic adjustment of both parameters can accelerate convergence, however, a large value for either η or α, may lead to instability in the training process [13].

Once the change to the weights are found, the new weights are computed as:

$$w_{st}(n+1) = w_{st}(n) + \Delta w_{st}(n). \tag{10}$$

The errors are usually high at the beginning of the learning process. This necessitates bigger changes in weights in the early stages of the learning process. Using the backpropagation procedure, the network calculates delta signals for the output layer and hidden layers, using Eqs (6) and (7), respectively. These deltas are employed to compute the changes for all the weight values according to Eq. (9) and the subsequent weight values using Eq. (10).

Radial basis functions networks
The backpropagation algorithm can be viewed as an application of optimization method known in statistics as *stochastic approximation*. While in RBF a different approach is adopted, namely a neural network is designed for *curve fitting* (approximation) problem in higher dimensional space. According to this viewpoint learning is equivalent to finding a surface in a multidimensional space that provides a best fit to the training data, and generalization corresponds to the use of multidimensional surface to interpolate in multidimensional space.

Broomhead and Lowe [17] were the first to exploit the use of radial basis functions in the design of neural networks. Other major contributions include papers by Moody and Darken [18], and Poggio and Giroso [19].

The idea of Radial Basis Functions networks derives from the theory of function approximation. The main features of these networks are as follows:

1. They are two-layer feed-forward networks.
2. The hidden nodes implement a set of radial basis functions (e.g. Gaussian functions).
3. The network training is divided into two stages: first the weights from the input to hidden layer are determined, and then the weights from the hidden to output layer.
4. The training/learning is very fast.
5. The networks are very good at interpolation.

The exact interpolation of a set of N data points in a multi-dimensional space requires every D dimensional input vector $\mathbf{x}^p = \left\{ x_i^p : i = 1,..., D \right\}$ to be mapped onto the corresponding target output t^p. The goal is to find a function $f(\mathbf{x})$ such that

$$f(\mathbf{x}^p) = t^p \text{ for every } p = 1,..., N \tag{11}$$

The radial basis function introduces a set of N basis functions, one for each data point, which takes the form $\phi(\|\mathbf{x} - \mathbf{x}^p\|)$ where $\phi(.)$ is some non-linear function. Thus the pth such function depends on the distance $\left\| \mathbf{x} - \mathbf{x}^p \right\|$, usually taken to be Euclidean, between $\mathbf{x}$ and $\mathbf{x}^p$. The output of the mapping is then taken to be a linear combination of the basis function, i.e.

$$f(\mathbf{x}) = \sum_{p=1}^{N} w_p \phi \left(\left\| \mathbf{x} - \mathbf{x}^p \right\| \right) \qquad (12)$$

The idea is to find the weights w_p such that the function goes through the data points.

The equation that defines the weights comes from combining the above equations:

$$f(\mathbf{x}^q) = \sum_{p=1}^{N} w_p \phi \left(\left\| \mathbf{x}^q - \mathbf{x}^p \right\| \right) = t^q \qquad (13)$$

This can be written in a matrix form as $\mathbf{w} = \mathbf{\Phi}^{-1} \mathbf{t}$, where $\mathbf{w} = \{ w_p \}$, $\mathbf{t} = \{ t^p \}$ and $\mathbf{\Phi} = \left\{ \mathbf{\Phi}_{pq} = \phi \left(\left\| \mathbf{x}^q - \mathbf{x}^p \right\| \right) \right\}$. Then provided the inverse of $\mathbf{\Phi}$ exists, one can use any standard matrix inversion technique to give

$$\mathbf{w} = \mathbf{\Phi}^{-1} \mathbf{t} \qquad (14)$$

Consider a set of N data points in a multi-dimensional space with D dimensional inputs $\mathbf{x}^p = \{ \mathbf{x}_i^p : i = 1,..., D \}$ and corresponding K dimensional target outputs $\mathbf{t}^p = \{ t_k^p : k = 1,..., K \}$. The output data can generally be generated by some underlying functions $g_j(\mathbf{x})$ plus random noise. The goal is to approximate the $g_j(\mathbf{x})$ with functions $y_k(\mathbf{x})$ of the form

$$y_k(\mathbf{x}) = \sum_{j=0}^{M} w_{kj} \phi_j(\mathbf{x}) \qquad (15)$$

There are good computational reasons to use Gaussian basis functions

$$\phi_j(\mathbf{x}) = \exp \left(-\frac{\left\| \mathbf{x} - \mu_j \right\|}{2\sigma_j^2} \right) \qquad (16)$$

in which we have basis centers $\{ \mu_j \}$ and widths $\{ \sigma_j \}$ that are generally obtained by unsupervised methods. The output weights $\{ w_{kj} \}$ can then be found analytically by solving a set of linear equations. This makes the training very quick. RBF mapping can be illustrated as a neural network as shown in Figure 3.

The hidden layer to output connections are as in a standard feed-forward network, with the sum of the weighted hidden unit activation giving the output unit activations. The hidden unit activations are given by the basis functions $\phi_j(\mathbf{x}, \mu_j, \sigma_j)$, which depend on the weights $\{ \mu_{ij}, \sigma_j \}$ and input activations $\{ x_i \}$ in a non-standard manner.

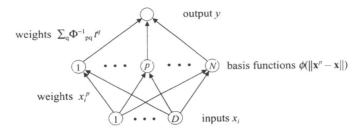

weights $\Sigma_q \Phi^{-1}_{pq} t^q$

basis functions $\phi(\|\mathbf{x}^p - \mathbf{x}\|)$

weights x_i^p

output y

inputs x_i

Fig. 3. The RBF network architecture.

Comparison of RBF networks with BP networks

There are several similarities and differences between RBF and BP:

Similarities:
1. Both are non-linear feed-forward networks.
2. Both are universal approximators.
3. Both are used in similar application areas.

Difference:
1. An RBF network has a single hidden layer, whereas BP can have any number of hidden layers.
2. RBF networks are usually fully connected, whereas it is common for BP to be only partially connected.
3. In BP the computation nodes in different layers share a common neuronal model, though not necessarily the same activation function. In RBF networks the hidden nodes (basis functions) operate very differently, and have a very different purpose, to the output nodes.
4. In RBF networks, the argument of each hidden unit activation function is the distance between the input and the weight (RBF centers), whereas in BP it is the inner product of the input and the weight.
5. BP nets are generally trained with a single global supervised algorithm, whereas RBF networks are usually trained one layer at a time with the first layer unsupervised.
6. BP construct global approximations to non-linear input-output mapping with distributed hidden representations, whereas RBF networks tend to use localized non-linearities at the hidden layer to construct local approximations.

Although, for approximating non-linear input-output mappings, the RBF networks can be trained much faster, BP may require a smaller number of parameters.

TRAINING AND TESTING OF THE NETWORKS

Software for training

In this paper *NeuralWorks* is employed, Ref [10]. In the process of training, both sigmoid and tangent hyperbolic functions are employed. *NeuralWorks* is capable of representing useful information in the process of training. In *Neuralworks* an additional hidden layer is considered between the first hidden layer (prototype) and the output layer. This layer makes it possible to weaken the mappings. In this software the following facilities are available:

Learning rules for output and the second hidden layer are Delta-Rule, Norm-Cum-Delta, Ext. DBD, Quickprop, Maxprop, and Delta-Bar Delta. The functions for translation consists of Linear, Tangh, Sigmoid, DNNA, and Sine.

Input Data

The input data for the trained NNs are the span and the height of the dome. Pairs for training and testing are selected randomly from 100 domes being analyzed and designed. Table 1 contains typical information on the pairs for training.

Table 1: Typical data for training

Span (m)	Height (m)	Deflection (m)	Opt. Weight(kg)
42.5	8.80	0.00738	10391.06
45.0	15.03	0.00719	12128.78
52.5	10.87	0.01307	18104.55
65.0	21.72	0.02011	30668.43
60.0	18.20	0.02587	24201.22
47.5	15.14	0.00831	12958.53
55.0	14.31	0.01286	18966.36
60.0	14.79	0.01677	23889.07
62.5	12.94	0.02102	29259.07
45.0	12.35	0.00730	11756.92

Neural network for predicting deflection
RBF network

In this net the number of input layer is 2 and output consists of one node. The results of training 12 different topologies are illustrated in Table 2. The network defr902 with 90 nodes in the prototype layer performed the best among the considered topologies. Using tangh as the translation function, the correlation factor approaches to unity and the net is trained in a short period with a high convergence rate. For all these nets 100000 cycles of trainings are performed the desirable correlation factor were obtained after 20000 training cycles. This coefficient for defr902 net was obtained as 0.9994 after 20000 cycles and approached to unity after 100000 cycles.

Table 2: Training by RBF using DBD learning rule.

Net name	Pair numbers		Layer no.	Pes in each layer				Corol. Factor		Trans. Funct.
	Train	test		Input	Proto.	Hid.	output			
defr500	80	20	2	2	50	0	1	0.9990	0.9972	tangh
defr500s	80	20	2	2	50	0	1	0.9970	0.9935	sigmoid
defr502	80	20	3	2	50	2	1	0.9990	0.9979	tangh
defr502r	80	20	3	2	50	2	1	0.9966	0.9926	sigmoid
defr602	80	20	3	2	60	2	1	0.9987	0.9968	tangh
defr602s	80	20	3	2	60	2	1	0.9970	0.9915	sigmoid
defr702	80	20	3	2	70	2	1	0.9999	0.9981	tangh
defr702s	80	20	3	2	70	2	1	0.9980	0.9909	sigmoid
defr802	80	20	3	2	80	2	1	0.9999	0.9983	tangh
defr802s	80	20	3	2	80	2	1	0.9981	0.9927	sigmoid
defr902	**80**	**20**	**3**	**2**	**90**	**2**	**1**	**1.0000**	**0.9975**	**tangh**
defr902s	80	20	3	2	90	2	1	0.9986	0.9935	sigmoid

• Using tangh function increase the speed of convergence and leads to better correlation factors compared to sigmoid function.
• Increasing the nodes in the prototype layer increases the learning rate.
• Increasing the nodes in the second hidden layer results in a better learning rate.

It should be noted that the range of mapping for the input and output for tangh functions were considered as (-1,+1) and (-0.8,+0.8), respectively. The ranges for input and output mappings are taken as (-1,+1) and (+0.2 +0.8) for the sigmoid function, respectively.

BP Network

Using backpropagation algorithm different nets are designed as shown in Table 3. For all these nets 100000 cycles of trainings are performed and in the second stage 200000 cycles are employed. Comparison of the results with those of RBF shows the superiority of RBF.

Table 3: Training by BP using DBD learning rule (100000 cycles for the first 4 rows and 200000 for the last 4 rows).

Net name	Pair numbers		Layer	Pes in each layer				Corol. Factor		Trans.
	Train	test	no.	Input	Proto.	Hid.	output			Funct.
Defr40t	80	20	2	2	4	0	1	0.9982	0.9977	tangh
Defr40s	80	20	2	2	4	0	1	0.9981	0.9984	sigmoid
Defr42t	80	20	3	2	4	2	1	0.9978	0.9911	tangh
Defr42s	80	20	3	2	4	2	1	0.9952	0.9967	sigmoid
Defr402t	80	20	3	2	4	0	1	0.9983	0.9980	tangh
Defr402s	80	20	3	2	4	0	1	0.9996	0.9996	sigmoid
Defr422t	80	20	3	2	4	2	1	0.9962	0.9962	tangh
Defr422s	80	20	3	2	4	2	1	0.9962	0.9974	sigmoid

Simultaneous prediction of deflection and optimal weight of the structure

In this net the output layer consists of two nodes. 12 different topologies are trained using RBF as shown in Table 4. For all the nets 100000 cycles of trainings are employed, and previous ranges are used for input and output mappings.

Table 4: Training by RBF using DBD learning rule with 80 pairs for training and 20 for testing.

Net name	No. of layers	Pes in each layer				Corellation factor				Trans funct.
		input	Proto	hidden	output	training		testing		
						Out. 1	Out 2	Out. 1	Out. 2	
DW500	2	2	50	0	2	0.9993	0.9991	0. 9980	0.9976	tangh
DW500s	2	2	50	0	2	0.9975	0.9966	0.9944	0.9963	sigmoid
DW502	3	2	50	2	2	0.9994	0.9990	0.9981	0.9989	tangh
DW502s	3	2	50	2	2	0.9971	0.9959	0.9953	0.9951	sigmoid
DW504	3	2	50	4	2	0.9993	0.9992	09980	0.9980	tangh
DW504s	3	2	50	4	2	0.9975	0.9968	0.9949	0.9952	sigmoid
DW604	3	2	60	4	2	0.9998	0.9997	0.9975	0.9985	tangh
DW604s	3	2	60	4	2	0.9978	0.9970	0.9935	0.9947	sigmoid
DW804	3	2	80	4	2	0.9999	0.9999	0.9968	0.9983	tangh
DW804s	3	2	80	4	2	0.9977	0.9977	0.9903	0.9923	sigmoid
DW1004	**3**	**2**	**100**	**4**	**2**	**0.9999**	**0.9999**	**0.9931**	**0.9974**	**tangh**
DW1004s	3	2	100	4	2	0.9983	0.9978	0..9896	0.9914	sigmoid

The net DW1004 with 100 nodes in the prototype layer and 4 nodes in the second hidden layer had the best performance leading to correlation factor of 0.999. After 20000 cycles of training, this net led to correlation factor of 0.993 and 0.990 for deflection and weight, respectively.

CONCLUDING REMARKS

Efficient neural nets are trained and tested for predicting the deflection of ribbed dome using data RBF and BP neural networks. RBF is used for simultaneous prediction of deflection and eight of the domes. The bounds considered for span length and structural height can easily be extended to cover a wider range. Similar nets can be trained for other types of space structures.

REFERENCES

1. NOOSHIN H, *Configuration Processing in Structural Engineering*, Elsevier Applied Science Publisher, London, 1984.

2. KAVEH A, *Structural Mechanics; Graph and Matrix Methods*, RSP, UK, 1995.

3. KAVEH A, *Optimal Structural Analysis*, Research Studies Press, UK, 1997.

4. JENKINS WM, Neural network-based approximations for structural analysis, *Developments in Neural Networks and Evolutionary Computing for Civil and Structural Engineering*, Edit. BHV Topping, Civil-Comp Press, Edinburgh, 1995.

5. WASZCZYSZYN Z, Some recent and current problems of neurocomputing in Civil and structural engineering, Edit. BHV Topping, Civil-Comp Press, UK, 1996, pp. 43-58.

6. KAVEH A AND SERVATI H, Design of double layer grids using artificial neural networks, Proc. Civil-Comp 99, Oxford, 1999.

7. HAJELA P AND BERKE L, Neurobiological computational models in structural analysis and design, *Computers and Structures*, vol 41, 1991, pp 657-667.

8. KAVEH A AND IRANMANESH A, Comparative study of backpropagation and improved counterpropagation neural nets in structural analysis and optimization, *International Journal of Space Structures*, vol 13, 1998, pp 177-185.

9. IRANMANESH A AND KAVEH A., Structural optimization by gradient base neural networks, *Int. J. Numer. Meths Engng.*, vol 46, 1999, pp 297-311.

10. NeuralWare, *Using NeuralWorks*, NeuralWare, USA, 1993.

11. NOOSHIN H, DISNEY P. AND YAMAMOTO C. *Formian*, Multi-Science Publishing Company, UK, 1993.

12. WILSON E. *Structural Analysis Program, SAP90*, Berkeley, California, 1990.

13. HAYKIN S, *Neural Networks; A Comprehensive Foundation*, Prentice Hall, New Jersey, 1994.

14. BULLINARIA JA, Introduction to Neural Computation, Lecture Notes, http://www.cs.bham.ac.uk/jxb/~inc.html, 2001.

15. RUMELHART DE, HINTON, GE AND WILLIAMS RJ, Learning representations by backpropagation errors, *Nature (London)*, vol 323, 1986, pp 533-536.

16. PARKER DB, Learning logic, TR-47, Center for Computational Research in Economics and Management Science, MIT, Cambridge, M.A. 1985.

17. BROOMHEAD DS AND LOWE D, Multivariable functional interpolation and adaptive networks, *Complex Systems*, vol 2, 1988, pp 321-355.

18. MOODY JE AND DARKEN CJ, Fast learning in networks of locally-tuned processing units, *Neural Computation*, vol 1, 1989, pp 281-294.

19. POGGIO T AND GIROSI F, Networks for approximation and learning, *Proceedings of the IEEE*, vol 78, 1990, pp1481-1497.

Comparative study of neural networks for design and analysis of double-layer grids

J. KEYVANI
Ministry of Science, Research and Technology, Tehran, Iran
M. A. BARKHORDARI
Department of Civil Engineering, Iran University of Science and Technology, Tehran, Iran

ABSTRACT

In this paper, application of artificial neural networks in predicting deflection and weight of double layer grids is discussed. In addition, the analysis of this structure using neural networks is carried out. Two algorithms are employed and the results are compared and discussed. The results show that this technique can be applied for preliminary design and analysis of double layer grids.

INTRODUCTION

Increasing construction of large span buildings such as stadiums, auditoriums, temples and hangars has required more studies for analysis and design of double layer grids covering such structures. Although the exact analysis of space structures is possible by using different softwares, the optimization process needs considerable time and effort. So, developing approximate methods using artificial neural networks to find a preliminary design is very useful.

In the beginning of twentieth century, the scientists studied the mechanism of information processing of human nerves, and then others tried to simulate biologic neural networks in solving problems [1]. In the 40th decade, W. McCulloch and Walter Pitts introduced the artificial neural networks technique. The technique was developed by Donald Hebb and others and resulted in introduction of Perceptron network by Frank Rosenblatt [2]. In the same time, Bernard Widrow developed Adaptive Linear Element (Adeline) network [3]. Although these methods provided a good start but they were able to classify only linear data separation [4]. In 1980s, two new trends caused important development in neural networks techniques:
a. Random mechanism used for recurrent networks capable of saving information, introduced by John Hopfield.
b. Error Back propagation Algorithm developed by David Rummelhurt and James McClelland [5-6].

In this study, two main algorithms namely Back-propagation (BP) and Radial-Basis Function (RBF) are employed and trained to predict weight and deflection of double layer grids and also for analysis of these types of structures. Furthermore, the outputs of networks are compared with exact solution and discussed.

NEURAL NETWORKS IN STRUCTURAL MECHANICS

In the last decade, a wide range of research has been carried out and many papers published in using neural netwerks for analysis and design of structures. Hajela and Berke applied the neural networks in analysis of structural Mechanics [7-8]. Jenkins used the neural networks method as an approximation approach for stuctural analysis [9]. Adeli and Park applied the Counterpropagation Nets (CPN) in structural engineering [10]. Kaveh and Iranmanesh further improved the CPN algorithm for structural optimization [11]. Kaveh and Servati applied BP network for the design of double layer grids using a grouping method [12].

A neural network system consists a number of processing cells that are arranged to form a group called layer. Each network consists at least two layers (input and output). Additional layers known as hidden layers may be presented. One of the most important characteristics of neural networks is learning. Through learning procedure, the weight vectors of the networks are adjusted. In general, the network is exited by input signals and the weight vectors improve the data to produce desirable output vectors. By repeating this process, when an input vector is given, the network will make a decision to give an output vector based on previous experience. The accuracy and convergence of the system depends on many factors such as initial weight vector, activation function, and learning factor.

PREDICTING THE DEFLECTION OF DOUBLE LAYER GRIDS

In this study, among different configurations of double layer grids, due to its common use, square on square type of double layer grids is studied
For the prediction of maximum deflection, 168 double layer grids were analysed using SAP 90. Following data are employed:
Span (L) = 32, 40, 48, 50, 56 (m)
Depth of structure (H) = $\dfrac{L}{12}, \dfrac{L}{16}, \dfrac{L}{20}$
Number of bays (B) = 7, 9, 11, 13
Support condition (S): As shown in Figure 1

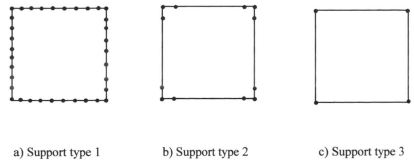

a) Support type 1   b) Support type 2   c) Support type 3

Fig. 1. Different types of supports.

Eighteen of the samples were randomly selected for testing and 150 for training. The deflections obtained from analysis were used to train BP and RBF networks.

The architecture of a BP network is not completely constrained by the problem to be solved. The number of inputs to the network and the number of neurons in the output layer is problem dependent, but the number of hidden layers and the sizes of layers are up to designer. In current problem the number of inputs is 4 and the size of output layer is 1. In order to obtain the most efficient architecture for BP network, a program is developed in MATLAB. The network having 6 cells in the first hidden layer and 3 cells in the second is obtained as the best architecture. Optimal architecture of the networks is one that has the least error. Training of BP and RBF networks is done for obtaining maximum deflections of the samples.

Table 1 shows the results of BP and RBF networks for 18 testing samples and compares them with the exact solution. Due to this table, the results show that the error of networks outputs is often low. The average error of BP outputs is 1.51% and of RBF is 5.11%.

Table 1. Maximum deflectoin obtained by different methods.

No.	L (m)	B	H (m)	S	D (Exact)	Output Of BP	Error of BP (%)	Output of RBF	Error of RBF (%)
1	56	11	4.667	2	116.0	115.6	0.34	114.8	1.03
2	40	13	3.333	1	57.0	57.6	1.05	59.4	4.21
3	32	11	2.667	3	82.0	76.7	6.46	69.1	15.73
4	56	9	2.8	1	136.0	132.3	2.72	131.8	3.08
5	48	7	2.4	3	210.0	211.5	0.71	203.3	3.19
6	48	11	3	1	87.0	88.4	1.59	86.4	0.69
7	40	9	2	3	180.0	180.3	0.16	176.8	1.77
8	40	7	2.5	2	77.0	77.8	1.03	80.1	4.02
9	32	9	1.6	2	98.0	95.6	2.44	96.7	1.32
10	32	7	2.0	1	46.0	47.2	2.61	42.0	8.69
11	50	11	2.5	1	141.0	137.6	2.41	126.8	10.07
12	56	13	4.667	3	171.0	173.1	1.22	175.0	2.34
13	50	15	4.167	2	142.0	143.2	0.84	127.2	10.42
14	50	11	3.125	3	217.0	217.4	0.27	197.3	9.07
15	48	13	4.0	2	108.0	108.3	0.27	109.7	1.57
16	50	15	2.083	3	342.0	345.3	0.96	354.5	3.65
17	32	13	1.6	3	143.0	143.1	0.07	153.0	6.99
18	40	7	3.333	1	44.0	43.1	2.04	42.2	4.09

L=Span length in meter, B=Number of bays, H=Height in meter,
S=Type of support, D= Deflection in millimetre

It should be noted that the error for some models is rather high. Maximum error obtained by BP is 6.46% for model No. 3 and maximum error obtained by RBF is 15.73% for the same case.

PREDICTING THE WEIGHT OF DOUBLE LAYER GRIDS
The same 168 models of double layer grids are also studied for prediction of weight. Again, 150 models are selected for training and 18 for testing. The most efficient architecture for BP is the one having 6 cells in the first and 4 cells in the second hidden layer.

Table 2 shows the results obtained by Back–propagation and RBF networks used for the weights of structures and compared with the exact solution. These results show that the error of the networks is often low but the error of few samples of RBF outputs is rather high. The average error of BP outputs is 1.61% and of RBF is 2.40%. This is notable that maximum error by BP is 4.59% for the model No. 18, but the maximum error by RBF is 11.82% for the same case.

Table 2. Weight of the structures in kN obtained by different methods.

No.	L	B	H	S	W (Exact)	Output of BP	Error of BP (%)	Output Of RBF	Error of RBF (%)
1	56	11	4.667	2	892.10	907.59	1.70	892.20	0.01
2	40	13	3.333	1	370.20	372.45	0.61	347.75	6.06
3	32	11	2.667	3	280.80	288.03	2.57	297.74	6.03
4	56	9	2.8	1	635.30	642.10	1.07	635.51	0.03
5	48	7	2.4	3	823.50	817.17	0.76	821.25	0.27
6	48	11	3	1	465.60	452.41	2.83	436.84	6.17
7	40	9	2	3	555.60	546.78	1.58	553.84	0.32
8	40	7	2.5	2	267.80	271.67	1.45	264.64	1.17
9	32	9	1.6	2	192.70	193.54	0.44	195.31	1.36
10	32	7	2.0	1	144.60	140.12	3.09	147.48	1.99
11	50	11	2.5	1	528.10	523.71	0.83	522.17	1.12
12	56	13	4.667	3	1325.60	1302.83	1.71	1327.11	0.11
13	50	15	4.167	2	878.40	901.72	2.65	872.08	0.72
14	50	11	3.125	3	994.30	986.53	0.78	997.65	0.33
15	48	13	4.0	2	692.40	708.01	2.25	688.97	0.49
16	50	15	2.083	3	1437.90	1438.38	0.03	1425.23	0.88
17	32	13	1.6	3	347.70	344.12	1.03	332.60	4.34
18	40	7	3.333	1	219.20	229.26	4.59	245.11	11.82

L=Span length in meter, B=Number of bays, H=Height in meter, S=Type of support, W= Weight in kN

ANALYSIS OF DOUBLE LAYER GRIDS
In order to analyse a double layer grid and obtain axial forces in all members, 21 samples having different span and height subjected to 2.5 kN/m^2 uniform loads is studied. Due to

symmetry in configuration, Bays and supports, 1/4 of the structures have been analysed. 21
samples are considered with following characteristics:
- Span Length (L): 32, 36, 40, 44, 48, 52 and 56 (m)
- Depth of structure (H) = $\dfrac{L}{12}, \dfrac{L}{16}, \dfrac{L}{20}$
- Number of bays: 9
- Type of support: Type 2 as shown in figure 1
 Among the 21 structures, the samples number 5 and 10 were randomly chosen for testing
and other 19 for training.
Structure number 5 having   L= 36m and H=2.25m
Structure number 10 having   L= 44m and H=3.667m
Each square on square double layer grid with 9 bays consists of 648 members. Based on
symmetry, 171 members need to be analised.

For training process, the results of the analysis of the samples are used.The analysis carried
out by using SAP90. The AISC Code including buckling limitations is considered in all
analysis procedures. Training and testing process is done by employing the BP and RBF
algorithms as follows:

Training and Testing by BP
The forces of members obtained by analysis is submitted to the sytem to train BP network.
The architecture of a network is defined by the number and the size of layers. As mentioned
previously, the number of inputs to the network and the number of neurons in the output layer
is problem dependent, but the number of hidden layers and the sizes of layers are up to
designer. In current problem the number of inputs is 2 and the size of output layer is 171.

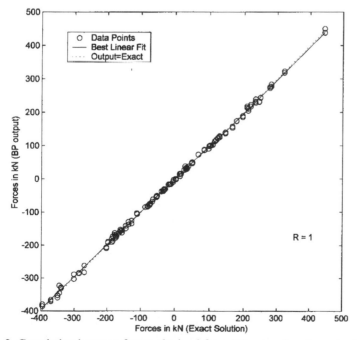

Fig. 2. Correlation between forces obtained from BP network and exact solution.

To obtain the most efficient architecture of network, a program is developed in MATLAB. The most efficient architecture of networks is one which minimizes the root mean square (RMS) of outputs errors. Performing the program, the following results obtained:
The number of neurons in the first hidden layer: 8
The number of neurons in the second hidden layer: 10

The results of testing process show that the root mean square of error of all members for sample number 5 is 74.78 kN and of sample number 10 is 67.44 kN. Due to having 171 member in a structure, the mean error of a member is 0.43 kN for sample number 5 and 0.39 kN for sample number 10. The correlation coefficient between BP outputs and exact solution obtained for samples number 10 is illustrated in figure 2.

Training and Testing by RBF
Process of training and testing of network is repeated by using RBF algorithm. The results show that the root mean square of error of all members for sample number 5 is 103.72 kN and for sample number 10 is 104.38 kN. The mean error of a member for both test samples is 0.60 kN. The correlation coefficient between RBF outputs and exact solutions obtained for sample number 10 is illustrated in figure 3. The corrolation coefficient is 0.999 for this case.

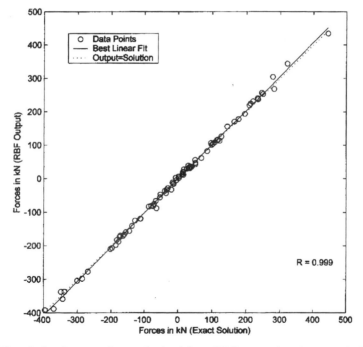

Fig. 3. Correlation between forces obtained from RBF network and exact solution.

Comparing the exact solution and outputs of BP and RBF, some members of structure are selected and the forces of them obtained by BP, RBF and exact solution for 2 test samples (5 and 10) are shown in table 3.

Table 3. Forces obtained from BP and RBF nets and exact solution for the members of samples 5 and 10 in kN.

No. of Member	Sample No. 5			Sample No. 10		
	BP	RBF	Exact	BP	RBF	Exact
5	-244.936	-248.316	-253.600	-285.767	-297.936	-285.700
15	33.174	39.983	32.600	49.777	47.833	50.500
25	88.827	99.794	93.400	125.693	125.796	127.000
35	-150.412	-151.375	-155.500	-185.827	-182.240	-183.500
45	35.008	38.638	32.600	50.644	43.373	50.500
55	3.371	1.571	-2.400	-2.682	-4.162	-4.500
65	14.438	13.457	14.600	15.340	14.870	13.600
75	30.549	31.993	34.400	38.110	36.114	39.000
85	27.685	29.241	28.800	33.304	33.469	32.300
95	-123.842	-126.725	-127.600	-128.387	-139.799	-136.900
105	-112.067	-107.964	-109.900	-133.723	-124.004	-128.200
115	-50.317	-51.103	-55.000	-54.517	-55.746	-53.900
125	-185.851	-197.123	-187.100	-205.993	-209.708	-202.500
135	39.187	36.178	40.700	31.469	36.618	35.800
145	163.293	169.825	170.300	174.242	178.336	176.900
155	342.174	351.500	339.700	438.181	434.894	443.800
165	108.707	114.311	113.400	114.139	115.273	113.900

CONCLUSIONS

Observing the results obtained and comparing the outputs of BP and RBF networks with exact solution, the following points are concluded:

- The neural network technique can be used as an aproximate method for analysis, and design of double layer grids
- Although the Back Propagation technique and Radial–Basis Function, both give acceptable results, the BP is more accurate and the RBF is trained faster.
- Study of the solutions show that while the average error is low, in some cases, error is rather high especially in RBF outputs. Thus, we can conclude that Neural Network technique can be used for preliminary design of space structures, not for final design.

REFERENCES
1. DAYHOFF J E, *Neural Network Architectures (An Introduction)*, VNR, New York, 1989.
2. ROSENBLATT F, *Principles of Neurodynamics: Perceptrons and the theory of Brain Mechanics*, Spartan Press, Washington, 1961.
3. WIDROW B AND HOFF M E JR, Adaptive switching circuits, *IRE Western Electric show and Convention Record*, 4, 1960, pp 96-104.
4. HASSOUN MOHAMMAD H, *Fundamentals of Artificial Neural Networks*, Massachusetts Institute of Technology, 1995.
5. RUMELHART D E, MCCLELL and, J L, and the PDP research group, *Parallel Distribute Exploration in the Microstructure of cognition*, vol.1, MIT Press, Cambridge Mass 1968.
6. RUMELHART D E HINTON, G E, and WILLIAMS, R J, Learning internal represetation propagation, in *Parrallel Destributed Processing: Explorations in the Microstructure of cognition*, D. E. Rumelhar, J. L. McClelland, and PDP Research Group eds. MIT Press, Cambridge, 1986.
7. HAJELA P and BERKE L, Neural networks in Structural analysis and design: an overview, *Computing Systems in Engineering*, 3, 1992, pp 525-538.
8. BERKE L, and HAJELA P, Applications of artificial neural nets in structural mechanics", *J. Structural Optimization*, 4, 1992, pp 90-98.
9. JENKINS W M, Neural network-based approximation for structural analysis, In: Topping B.H.V., editor, *Developments in Neural Networks and Evolutionary Computing for Civil and Structural Engineering*, Edinburgh: Civil-Copm Press, 1995.
10. ADELI H and PARK H S, Counterpropagation neural networks in structural engineering, *J. Structural Engineering*, 121, 1995, pp 1205-1211.
11. KAVEH A and IRANMANESH A, Comparative study of backpropagation and improved counter propagation neural nets in structural anslysis and optimization, *Int. J. Space Structures*,13, 1998, pp 177-185.
12. KAVEH A and SERVATI H, Design of double layer grids using back propagation neural networks, *Computers and Structures*, 79, 2001, pp 1561-1568.

Optimum design of space structures by combining genetic algorithms and simulated annealing using a response surface approximation

E. SALAJEGHEH, K. LOTFI,
Department of Civil Engineering, University of Kerman, Kerman, Iran

ABSTRACT

Optimum design of structures is achieved by a modified genetic algorithm. Some features of the simulated annealing are used to control various parameters of the genetic algorithm. The new evolutionary algorithm is employed to design space structures. The method improves the computing efficiency of the large-scale optimization problems and enhances the global convergence of the design process. To reduce the computational work, some approximation concepts are employed. Response surface method is used to approximate the functions under consideration. A double layer grid is designed for optimal weight and the results are compared with conventional genetic algorithm.

INTRODUCTION

Optimum design of structures is to select the design variables systematically such that the weight of the structure is minimized while all the design constraints are satisfied. The design variables are normally considered as the member cross-sectional areas, which are chosen from a set of available values (discrete variables). The design constraints are bounds on member stresses and joint displacements. The optimum design problem is formulated as a mathematical nonlinear programming problem and the solution can be found by various numerical optimization methods. In this work, genetic algorithm (GA), is used to evaluate the optimal solution. The GA method has the capability of finding the global optimal solution. It is a reliable method for optimum design of structures and the method does not require the derivatives of the functions under consideration. The standard GA has been employed for optimum design of structures [1-3].

The probabilistic nature of the standard GA makes the convergence of the method slow. This is due to the fact that the control probabilities for some of the GA operations such as crossover and mutation are chosen constant during the optimization process. In this study, an improvement in the operations of the algorithm is obtained if these controllers are adapted during the process. The adaptation is obtained by employing some features of the method of simulated annealing (SA). Some adaptive rules are outlined based on the SA method, which results in smooth convergence of the GA method. Simulated annealing is also a derivative-free method for optimization and has been used for both continuous and discrete optimization problems [4-5]. Some kind of modification of GA by using SA has been introduced by researches in the field of Electrical Engineering [6-7].

Another aspect of the GA technique is that the computational cost of the process is high. For some problems such as space structures with large number of degrees of freedom, the structural analysis is time consuming. This makes the optimal design process very inefficient. To overcome this difficulty, the functions under investigation are approximated by using a response surface (RS) approach [8]. The RS is a global function approximation and in this study a second order polynomial function is used. By introducing such approximation, the analysis of the structure is not necessary during the optimization process. Thus an efficient approach is presented for optimum design of space structures.

In this paper, first the basis of the standard GA and SA methods are explained and then some features of the SA are combined with GA in the optimization process. Approximation concepts with a second order RS will be discussed and some numerical examples for space structures will be presented. The convergence of the proposed method is compared with standard GA.

DESIGN PROBLEM FORMULATION
The most popular optimization problem in structural design is to minimize the weight. The structure is subject to constraints imposed on the member stress and joint displacement. This is mathematically shown as

$$\text{Find X to minimize F(X)}$$
$$\text{Subject to } g_j(X) \le 0, j=1,\dots,m. \tag{1}$$

In this formulation, $X^T=\{x_1, x_2, \dots, x_n\}$ is the vector of design variables with n variables. In this study X is considered as the cross-sectional areas of the elements. The objective function, F(X) is normally taken as the structural weight. The m design constraints imposed on the design problem are shown as inequalities of the form $g_j(X) \le 0$. To solve the above-mentioned constrained optimization problem by the GA method, first the problem must be converted into an unconstrained optimization problem. There are various methods and a simple method is achieved through exterior penalty function method as

$$\varphi(X) = F(X) + r_p \sum_{j=1}^{m} \{\max[0, g_j(X)]\}^2 \tag{2}$$

The scalar r_p is a multiplier and by changing this multiplier and minimizing $\phi(X)$, the minimum of $\phi(X)$, approaches minimum of F(X). Genetic algorithm is based on maximization of a positive unconstrained function. This can be achieved as

$$\Phi(X) = C - \varphi(X) \tag{3}$$

where $\Phi(X)$ is the fitness function and C is a positive constant and its value must be greater than the largest value of $\phi(X)$ in a generation to ensure the fitness function to be positive. The multiplier r_p is increased in each generation. This can be achieved by different approaches and the following formula is proposed in Ref. [9].

$$r_p = r_1[1 + 0.2(p-1)] \tag{4}$$

where r_1 is a given initial value at first generation and p is the generation number. The penalty value increases gradually until it reaches $4r_1$ and then remains constant for the remaining process.

GENETIC ALGORITHM

Genetic algorithm is a derivative-free stochastic optimization method based mainly on the concepts of natural selection and evolutionary process. The method was first proposed in 1975 [10] and was extended in Ref. [11]. The main feature of GA is that it can be used for both continuous and discrete optimization problems. In addition, because of stochastic nature of the method and using a population of design points in each generation usually gives rise to the global optimum.

Genetic algorithms encode each point in the design space into a binary bit string called a chromosome and to each point a fitness function such as Eq. (3) is associated. Instead of a single point, GA usually creates a set of points as a population, which is then evolved repeatedly toward a better solution. In each generation, the GA produces a new population using genetic operators such as crossover and mutation. Design points with higher fitness values are more likely to survive and to participate in crossover operations. After a number of generations, design points with better fitness values are obtained. Major components of GA include encoding schemes, fitness evaluations, parent selection, crossover operators and mutation operators. These are briefly explained.

Encoding schemes transform points in design space into bit string representations. By using binary coding, the value of each variable is encoded as a gene composed of a number of binary bits. To each point a fitness function is allocated. This function includes the effect of the objective function and the constraints imposed on the problem. A simple formulation of the fitness function is presented by Eq. (3). After creating a population, the fitness values of all members are evaluated. After fitness evaluation, a new population is created from the current generation. The selection operation determines which parents participate in producing offspring for the next generation and it is analogous to survival of the fittest in natural selection. Usually members are selected for mating with a selection probability proportional to their fitness values. The most common way to implement this is to set the selection probability equal to $\Phi_i / \sum_{k=1}^{k-q} \Phi_k$, where q is the population size.

Crossover operators are used to generate new chromosomes that we hope will retain good features from the previous generation. Crossover is usually applied to selected pairs of parents with a probability equal to a given crossover rate. One-point crossover is the most basic crossover operator, where a crossover point on the genetic code is selected at random and two parent chromosomes are interchanged at this point. In two-point crossover, two crossover points are selected and the part of the chromosome string between these two points is then swapped to generate two children. One can define n-point crossover in a similar manner. Some researchers use uniform crossover, which is based on a randomly created binary string, called a mask [12]. Parent strings are exchanged if the corresponding bit of the mask is zero, otherwise, the exchange is not performed.

Mutation operation is employed for the possibility that nonexisting features from both parent strings may be created and passed to their children. By this operator each bit of a string is changed from 0 to 1 or vice versa based on mutation probability. The mutation probability is usually kept low so good chromosomes obtained from crossover are not lost.

The main steps in the standard GA can be summarized as follows:

Step 1: Initialize a population with randomly generated members and evaluate the fitness value of each individual.

Step 2:

(a) Select two members from the population with probabilities proportional to their fitness values.

(b) Apply crossover with a probability equal to the crossover rate.

(c) Apply mutation with a probability equal to the mutation rate.

(d) Repeat (a) to (d) until enough members are generated to form the next generation.

Step 3: Repeat steps 2 and 3 until a stopping criterion is met.

SIMULATED ANNEALING

Simulated annealing (SA) is another derivative-free optimization method that can be used for both continuous and discrete problems [13]. The principle behind SA is analogous to what happens when metals are cooled at a controlled rate. The slowly falling temperature allows the atoms in the molten metal to line themselves up and form a regular crystalline structure that has high density and low energy. But if the temperature goes down too quickly, the atoms do not have time to orient themselves into a regular structure and the result is a more amorphous material with high energy. In SA, the value of an objective function that we want to minimize is analogous to the energy in a thermodynamic system. At a given temperature, T, the algorithm perturbs the position of an atom randomly and evaluates the resulting change in the energy of the system, ΔE. if the new energy state is lower than the initial state, then the new configuration of the atoms is accepted. Otherwise, the new state that increases the energy, may be accepted or rejected based on an acceptance function, $P(\Delta E,T)$. There are several acceptance functions and the most frequently used function is the Boltzmann probability distribution

$$P(\Delta E,T) = \frac{1}{1 + \exp(\Delta E / cT)} \tag{5}$$

where c is a system-dependent constant. The criterion to accept or reject the new state when $\Delta E \geq 0$, is that a number in the interval (0, 1) is randomly selected and compared with $P(\Delta E,T)$. if the number is less than $P(\Delta E,T)$, then the perturbed state is accepted, otherwise it is rejected. The basic steps in SA are as follows:

Step 1: Choose a start point X and a high starting temperature T.

Step 2: Evaluate the objective function, $E=\phi(X)$.

Step 3: Select ΔX, randomly and set the new point $X_{new}=X+\Delta X$.

Step 4: Calculate the new value of the objective function, $E_{new}=\phi(X_{new})$.

Step 5: Set X to X_{new} and E to E_{new} with probability determined by the acceptance function, $P(\Delta E,T)$, where $\Delta E= E_{new} -E$.

Step 6: Reduce the temperature T, according to the annealing schedule by multiplying T by α, where α is a constant between 0 and 1.

Step 7: Repeat steps 3 to 7 until a stopping criterion is met.

The starting temperature and the method of reducing temperature are important part of the method. This is usually application specific and requires some experience. In an optimization problem, T is just a control parameter that has no physical concepts and it only regulates the convergence of the process.

COMBINING GENETIC ALGORITHM AND SIMULATED ANNEALING

In this paper, the optimization is based on the genetic algorithm but the probability of the genetic algorithm's operations (crossover and mutation) is controlled with Boltzmann probability distribution criterion (Eq. 5). In this method, the crossover and mutation operations are accepted, if the offspring's objective function values are less than their parents or the criterion specified by Eq. (5) is satisfied.

When number of generation increases the control parameter T (imaginary temperature) is reduced according to cooling progress and the chance of acceptance of the offspring with large value of objective function is reduced. Figure 1 shows the process of crossover operation. In this algorithm, η is a random number, F_{obj} is the objective function and P_M is the probability of acceptance. A similar process can be considered for the mutation operation. In fact, by employing the features of the simulated annealing, we create an adaptive probability rate, which is different in each generation of the genetic algorithm. The numerical results show that by employing this type of adaptive control probabilities for crossover and mutation, a smooth convergence is achieved in the process of GA. The combination of the two methods is referred to as GA-SA method.

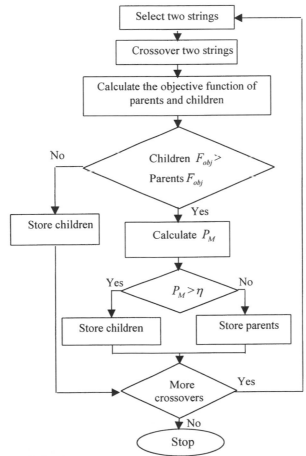

Figure 1. Process of adaptive crossover

APPROXIMATION METHOD

All the optimization methods that do not need functional derivatives information are in general, slower than derivative-based optimization techniques. Instead, they rely exclusively on repeated evaluations of the objective function and the design constraints. In particular, in GA, a population of points is involved in each generation. For the design problems with great number of degrees of freedom, the computational work is high, thereby, using GA for optimization, makes the process inefficient. In practical optimization methods, it is necessary to approximate the functions that are computationally expensive to evaluate [14]. The method of approximation mainly depends on the optimization approach. If the optimization method is based on line search, then some local approximation such as Taylor series expansion of the functions under consideration is adequate. However, for methods such as GA that relay on points spread all over the design space, a global approximation should be employed. Response surface methodology (RS) is one of the approaches for global function approximation and is especially suitable when the sensitivity information is not available [15]. In this method an approximate function is created by using the function values at a number of points. The method of RS is suitable for GA, as after completion of some generations, the function values are available. In this work, a second-order polynomial approximation is used. For a given function y(X), the following approximation is employed.

$$y(X) = b_0 + \sum_{i=1}^{n} b_i x_i + \sum_{i=1}^{n}\sum_{j=1}^{n} b_{ij} x_i x_j \qquad (6)$$

where x_i and x_j denote design variables and n is the number of independent variables. The unknown parameters b_0, b_i and b_{ij} are estimated regression coefficients. Equation (6) can be written in matrix form as

$$Y = \mathbf{X} B \qquad (7)$$

where $\mathbf{X}$ is the data matrix, B is the unknown parameter vector with dimension of $(n+1)(n+2)/2$. In order to estimate B, a set of data for $\mathbf{X}$ and Y should be available. By substitution of these data into Eq. (7) and finding the sum of the squires of the residuals, we can write

$$\varepsilon = \sum (y_i - X_i B)^2 = (Y - \mathbf{X}B)^T (Y - \mathbf{X}B) \qquad (8)$$

where $Y^T = (y_1, y_2, \ldots)$ and $\mathbf{X}^T = (X_1, X_2, \ldots)$.

to minimize the error ε, by the idea of least squire method, we can differentiate Eq. (8) with respect to B and equate the derivatives to zero, the unknown vector B is estimated as

$$B = (\mathbf{X}^T \mathbf{X})^{-1} \mathbf{X}^T Y \qquad (9)$$

In the above formulation, bold, capital and small letters are used to represent matrix, vector and a component of a vector, respectively.

The required set of data to evaluate B is 3^n. It can be observed that as the number of design variables increases, the number of trial data would also increase. Some attempts have been made to decrease the number of data points and the results of some of the methods are summarized in Ref. [16].

The approximate functions can be constructed for member forces, stresses, displacements, constraints, fitness function or any other criterion that is difficult to evaluate. In the present work, the fitness function is estimated after completion of a number of generations in GA. The number of required generations depends on the number of data points. Of course, one can construct the RS for the functions under investigation prior to optimization. In this case, a set of suitable data must be provided. The combined method of GA-SA and RS is referred to as GA-SA-RS method.

NUMERICAL EXAMPLE

A double-layer grid of the type shown in Fig. 2 is chosen with dimensions 16×12 m for top-layer and 12×8 m for bottom-layer. The height of the structure is 0.5 m and is simply supported at the corner joints 5, 7, 26 and 28 of the bottom-layer. The loading is assumed as 150 kg/m^2 on the top-layer and it is transmitted to the joints acting as concentrated vertical loads only. The problem is designed with stress constraints and the allowable stress is assumed as ±1260 kg/cm^2. The material properties are given as Young's modulus, E=2.1×10^6 kg/cm^2 and weight density, ρ=0.008 kg/cm^3. The set of available discrete values considered for the cross-sectional areas of the members are as follows:

$$\{4.6, 6.5, 8.3, 10.8, 15.5, 20.5, 24.4, 28.1\} \ \ (cm^2)$$

The members are grouped into 19 different types as shown in Table 1. The optimal results for the case GA-SA are presented in Table 1 with a total weight of 1890 kg. The convergence history of the problem for the cases GA, GA-SA and GA-SA-RS are shown in Figs. 2, 3 and 4, respectively. The iteration histories of the problem show that GA-SA converges uniformly with lower optimal weight. In the case GA-SA-RS, the final weight is 1944 kg, however, the number of required generations is about 30 with 10 generations less than GA-SA.

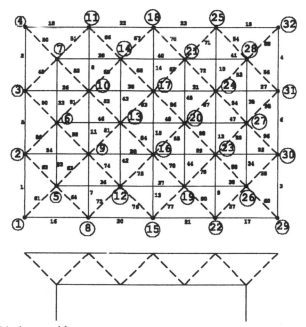

Figure 2. Double-layer grid

Table 1. Member grouping and Results of optimization (cm²)

Group No.	Member No.	Area	Group No.	Member No.	Area
1	36-41	6.5	11	28-31	10.8
2	32-35	4.6	12	7-10	4.6
3	42-45	4.6	13	13-14	4.6
4	46-47	4.6	14	15	4.6
5	48	4.6	15	11-12	6.5
6	16-19	4.6	16	49-64	15.5
7	20-23	8.3	17	65-80	4.6
8	1-4	4.6	18	81-88	4.6
9	5-6	4.6	19	89-96	4.6
10	24-27	4.6			

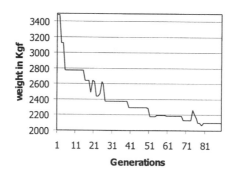

Figure 2. Convergence history (GA)

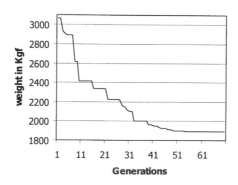

Figure 3. Convergence history (GA-SA)

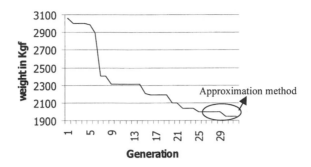

Figure 4. Convergence history (GA-SA-RS)

CONCLUSIONS

From the numerical results, the following points can be concluded:

(a) If the probabilities of the crossover and mutation operators in standard GA are controlled by SA features, then a smooth convergence would be achieved.
(b) The number of required generations in GA-SA is less than standard GA.
(c) The final objective function in GA-SA is less than that of GA.
(d) Combining both GA and GA-SA with approximation concepts reduce the overall optimization cost.
(e) RS methodology is an effective global approximation and works well with GA.

REFERENCES

1. LIN C Y and HAJELA P, Genetic Algorithms in Optimization Problems with Discrete and Integer Design Variables, Engineering Optimization, Vol. 19, No. 4, 1992, pp. 309-327.

2. RAJEEV S and KRISHNAMOORTHY C S, Discrete Optimization of Structures using Genetic Algorithm, Journal of Structural Engineering, ASCE, Vol. 118, No. 5, 1992, pp. 1233-1250.

3. ADELI H and CHENG N T, Integrated Genetic Algorithms for Optimization of Space Structures, Journal of Aerospace Engineering, Vol. 6, No. 4, 1993, pp. 315-328.

4. KIRKPATRICK S, GELATT C D Jr and VECCHI M P, Optimization by Simulated Annealing, Science, Vol. 220, 1983, pp. 671-680.

5. BALLING R J and MAY S A, Large-Scale Discrete Structural Optimization: Simulated Annealing, Branch and Bound, and Other Techniques, Proceedings of the AIAA/ASME/ASCE/AHS/ASC 32nd Structures, Structural Dynamics and Materials Conference, Long Beach, CA, 1990.

6. GHOUDJEHBAKLOU H and DANAI B, A New Algorithm for Optimum Voltage and Reactive Power Control for Minimizing Transmission Lines Losses, International Journal of Engineering, I R I, Vol. 14, No. 2, 2001, pp. 91-97.

7. LI W and LI Q, Parallel Genetic Optimization Design of Antenna Structure with Simulated Annealing Mechanism, Proceedings of the Forth World Congress of Structural and Multidisciplinary optimization (WCSMO-4), Dalian, China, 2001, (in CD ROM).

8. BOX G E P and DRAPER N R, Empirical Model -Building and Response Surface, Wiley, New York, 1987.

9. CHEN T Y and CHEN C J, Improvements of Simple Genetic Algorithms in Structural Design, International Journal for Numerical Methods in Engineering, Vol. 40, 1997, pp. 1323-1334.

10. Holland J H, Adaptation in Natural and Artificial Systems, University of Michigan Press, Ann Arbor, Michigan, 1975.

11. GOLDBERG D E, Genetic Algorithms in Search, Optimization and Machine Learning, Addison-Wesley Publishing Co., Inc., Reading, Massachusetts, 1989.

12. SYSWERDA G, Uniform Crossover in Genetic Algorithms, Proceeding of the 3rd International Conference on Genetic Algorithm, 1989, pp. 2-9.

13. LAARHOVEN P J M and AARTS E, Simulated Annealing: Theory and Applications, D. Reidel Publishing, Dordrecht, The Netherlands, 1987.

14. SALAJEGHEH E and SALAJEGHEH J, Optimum Design of Structures with Discrete Variables using Higher Order Approximation, Computer Methods in Applied Mechanics and Engineering, Vol. 191, No. 13-14, 2002, pp. 1395-1419.

15. KHURI A I and CORNELL J A, Response Surfaces: Designs and Analyses, Marced Dekker Inc., New York, 1987.

16. UNAL R, LEPSCH R A, ENGLAND, W and STANLEY D O, Approximation Model Building and Multidisciplinary Design Optimization using Response Surface Methods, Proceedings of the 6th AIAA/NASA/ISSMO Symposium on Multidisciplinary Analysis and Optimization, Part 1, Bellevue, WA, 1996, pp. 592-598.

Strategy to establish a reasoning engine to predict the buckling behaviour of single layer reticular domes

H. TAKASHIMA
Department of Architecture, College of Engineering, Kanto Gakuin University, JAPAN

INTRODUCTION

How to utilize the accumulated results being yielded from numerical simulations or experimental tests, on a practical design, should be one of noticeable headings to find an effective solution. It would be so avail for all designers when the method to extract easily the digitized certain knowledge was established.

The present author has carried out elastic-plastic numerical simulations for single layered reticular domes and a lot of results have also been issued (for example Ref 2). To utilize or lump together the results, using an artificial life algorithm so-called "soft computing" which includes a neural computing and a genetic algorithm, should be attracted because of these human-like behaviours. And the neural network is one of compact and powerful tools that can involve knowledge bases relative easily. The noticeable advantage of the neural network as the reasoning engine would be that the network can interpolate within many given parameters that imply the dome geometries, material properties and flexibility of the connections et al.

Then the applicability to establish the reasoning engine that can predict the buckling behaviour of reticular domes, based on soft computing including both the neural computing and genetic algorithm will be analyzed and discussed in the present paper. The report dealing with how to create reasoning engine by using the neural network and its reasoning ability, has also been found in former studies Refs 3 and 4. In addition to them, the primitive expert system that has adopted the neural network as the reasoning engine has been introduced in Ref 5. In that paper, a graphical interface between the reasoning engine and users has also been developed by Hyper Text Mark-up Language and Perl script in the system. As the recent study, Ref 6 has also described the results of an application of the soft computing to the optimization for structural designs.

However, a few important problems to be solved have come out through the investigations (Refs 3-5).

(1) The network behaviours are sensitive with changing initial parameters at drawing up the neural network.
(2) The ability of extrapolation out of the learning area would be uncertain.
(3) An effective method to update the network is needed to learn other new data.

In Refs 3 and 4, the most adaptable neural network was searched by manually with changing a lot of parameters in the neural network. When we supposed that the method with a high degree of efficiency to determine the parameters of the neural network and to find the most effective network, the above problem (3) would be closed up. The present paper will introduce and investigate the considerable solutions based on the genetic algorithm.

PROPOSED EXPERT SYSTEM

The present expert system can indicate the ultimate strength and the buckling deformation pattern of a steel single layered reticular dome. The parameters of the dome are shown in Table 1 and also illustrated in Fig.2. The framework of this system is shown in Fig.1. The expert system is roughly divided into two parts, an user-interface coded by Hyper Text Mark-up Language and Perl script, and the reasoning engine based on the neural network. The ability of the system has already been performed in Ref 5.

Figs 2-4 show screen shots of the system. Fig.2 introduces a guideline at preparing input parameters needed from the system. The screen at filling the parameters into blanks is indicated in Fig.3. The final response including the given condition and predictions, is shown in Fig.4. The system can only deal with the 3-way grids single layered reticular dome that has a sphere surface and all the constitutive members have the same performances.

Table 1. Dome geometries.

Dome Parameters	Values
Number of members on diagonal line	2,4,6
Slope angle of the member on the apex	2, 2.5, 3, 3.5, 4 deg.
Member slenderness ratio	60, 96.20
Boundary condition	Simply support, Roller Support
Joints	Rigid, Semi-rigid, Pin
Yield stress of the member	2.4 tf/cm^2
Yield stress of the connector	2.4, 2.7 tf/cm^2

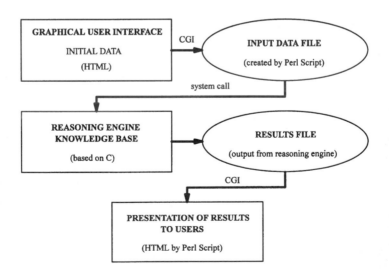

Fig. 1. Framework of proposed expert system.

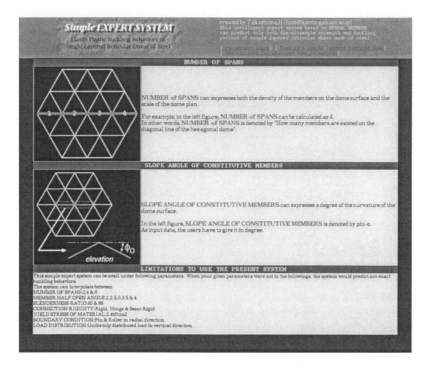

Fig. 2. Descriptions for input parameters.

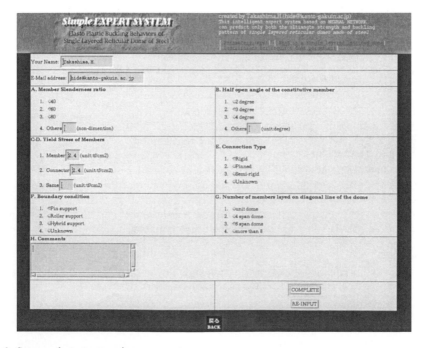

Fig. 3. Screen shot at preparing parameters.

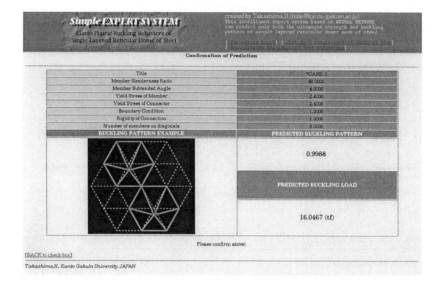

Fig. 4. Presentation of predicted results.

The system has several restrictions above, but, the proposing tool in which the soft computing algorithms are adopted as the reasoning procedure, could give some suggestions for designers through bringing them in contact with the past designed samples stored as a digital knowledge.

Through investigating the expert system, two distinctive problems to be solved have been observed. They are (2) and (3) found in the previous chapter. The problem (2) would be so difficult to be solved immediately. In (3), there are necessities to propose the strategy that can assure the ability to reconstruct the connectivity strengths and geometry. Then in the next chapter, the one solution for the second one will be introduced.

STRATEGY TO FIND ADAPTABLE NEURAL NETWORK
In this chapter, a simple method how to search for parameters and geometry of the neural network is described. Several proposals for such a problem are found in Ref 7 *et al*.

In order to find the appropriate configure of the neural network, several parameters have to be decided experimentally and arbitrarily. And the optimal value of these parameters has to be searched in the wide variable ranges. When there is a necessity to experiment huge number of cases in order to get a relative optimal one, it is quite natural that one supposes applying the genetic algorithm (GA) as an effective method on such problems.

Assumptions
The flow chart of the proposed strategy is shown in Fig.5. In this procedure, following parameters utilized in the neural network are selected as targets to be optimized by GA.

(1) ε : incremental value to update the connectivity strengths
(2) ε_b : incremental value to update the thresholds
(3) μ_0 : slope of the sigmoid function

(4) Δm : accelerated coefficient in the iterative procedure
(5) N_1 : number of nodes in the 1st associative layer
(6) N_2 : number of nodes in the 2nd associative layer
(7) N_3 : number of nodes in the 3rd associative layer

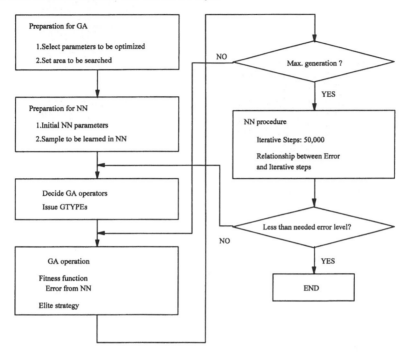

Fig. 5. Proposed strategy based on GA.

Essentially, the number of associative layers has to be variable and be determined as an appreciate value. However, the consideration will force an enormous computing time on searching the geometry of the network. Then the number of associative layers is fixed as 3 according to the results obtained from the former study (Ref 1).

As the basic condition, the ranges of the above variables are assumed as follows.

$$0.1 < \varepsilon < 0.5$$
$$0.1 < \varepsilon_b < 0.5$$
$$0.1 < \mu_0 < 0.5 \tag{1}$$
$$0.01 < \Delta m < 0.2$$
$$1 < N_1, N_2, N_3 < 10$$

These ranges are divided by 4 bits chromosome. The operators in GA are fundamentally assumed as Table 2.

"Uniform crossover" is adopted as the crossover operation. In the GA reproduction, the error magnitude issued from the neural network through IS-th iterative step, is utilized in the fitness function. The fitness value is evaluated by

Fitness $= 1 / ($ ERROR at IS-th step $)$ \hspace{2cm} (2)

where with decreasing the error magnitude, the fitness gets greater.

Table 2. GA operators.

Operator	Default value
Maximum number of generations	50
Number of chromosomes	10
Crossover ratio	80%
Mutation ratio	10%
Elite survive ratio	10%
Chromosome length	4 bits for each parameter

INVESTIGATIONS OF PROPOSED STRATEGY
In the Ref 1, the author has shown some tendencies of the proposed procedure with parametric calculations changing the number of generations, number of bits of chromosome for each parameter, initial value for random seeds and mutation ratio. Through the tendencies, avail parameters can be chosen to carry out the proposed GA procedure. In the present chapter, further investigations around the GA parameters and discussion on the performances of the selected network will be stated.

Effects of number of iteration steps in fitness function on issued error level
To confirm the effectiveness of the number of iterative steps, IS in Equation (2), further calculations are carried out in addition to the analyses in Ref 1. Fig. 6 shows the relationship between the error magnitude issued from the neural networks and IS. IS are changed from 100 to 3,000. The error magnitude is issued from the most applicable network selected through the present strategy and is observed at the 50,000th iterative step in the back-propagation procedure. Increasing the number of iterative steps, the error magnitude becomes smaller. However, the relational curve is not proportional with increasing the number of iterative steps. When the same values of the plots are found, in those cases, the same geometry and parameters for the network are obtained from the GA procedure. From the present result, the number of iterative steps in the fitness function should be selected as about 1,000 or 1,500.

Comparison between the selected network and former one
Fig. 7 shows the predicted ultimate strengths of the reticular dome. The dome periphery is restricted by simply support, the number of the member along the ridge is 6 and the slope angle of the member at the apex varies from 2 to 4 degrees by 0.5 degrees. In this graph, squares with a solid line denote given results to be learned in the neural network. They are correct answers. Circles are prediction plots obtained from the former network that has been drawn up in Ref 3. The network is called as one "manually" obtained from a lot of parametric case studies. The expression "Manual" on the plots in this paper means that.

Both the GA prediction and manual one are quite similar to the given strengths. As one of the reason, it can be pointed out that the ultimate strengths are relatively in direct proportion to the slope angle of the members. The GA prediction seems to be the same as the manual one. However, there are slight differences between them through severely observing Fig.7. The prediction by the network obtained from the GA procedure exhibits inconsiderable small values in comparison with the correct answers. Meanwhile, the prediction by the manual procedure shows little bit larger values than the correct ones. That is caused by differences in the network geometries and parameters. The differences are denoted in Table 3. The error magnitude issued from the network obtained from the GA procedure is smaller than the manual one. As the results, the GA procedure also brings a slight improvement to the ability of the prediction by the network.

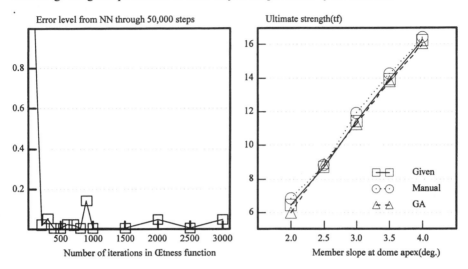

Fig. 6. Error level & number of iterative steps.   Fig. 7. Predicted & given ultimate strengths.

Table 3. Network parameters.

Parameter	Manual	GA
ε	0.300	0.180
ε_b	0.300	0.367
μ_0	0.450	0.473
Δm	0.100	0.124
N_1	8	8
N_2	6	8
N_3	4	5

Prediction for variable boundary conditions (1)
The predictive ability of the reasoning engine for other parameters of the reticular dome will be investigated in this section. As the objective parameter, the boundary condition is picked up. This parameter varies from 0 to 1 and the results to be learned are only two cases for 0 denoting a roller support and 1 expressing a simply support. For instance, when a half of the periphery of the dome is restricted as a pin support, the value of the restriction parameter becomes 0.5. Treated examples in this investigation are illustrated in Fig.8.

Fig.9 shows a comparison between the predicted plots and the ultimate strengths obtained from elastic-plastic simulations. Two squares in the middle of the graph express the ultimate strengths obtained from the numerical simulations for the boundary conditions in Fig.8. These are almost on the line that connects the given data linearly. The given data is also plotted as square marks.

In this figure, Lmax denotes a denominator that divides the ultimate strength to reform them to normalized values to be learned in the neural network. The initial value for Lmax is 20. But, through observing the plots in Fig.9, the ability is not enough to predict the ultimate strengths for variable boundary conditions. In order to improve the predicting curve, the normalizing factor Lmax is chosen because the curve at Lmax=20 is similar to the upper part of the sigmoid function. For each Lmax=18,20,25 and 30, the GA reproduction where IS in Equation (2) is assumed as 1,000, is carried out. Contrary to the expectation, the curvatures between the plots with Lmax=25 and 30 tend to more severe than the cases for Lmax=18 or 20. Consequently, such adjusting way to improve the predictive ability is judged to be not so effective in the present case.

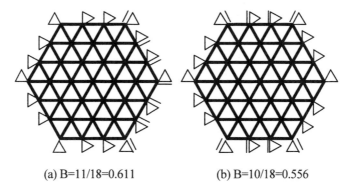

(a) B=11/18=0.611   (b) B=10/18=0.556

Fig. 8. Variable boundary condition with partial roller supports.

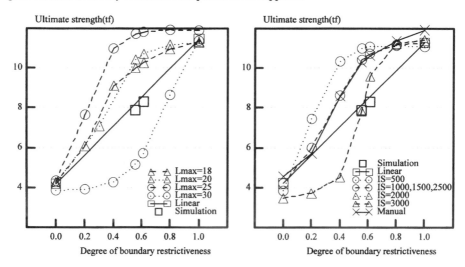

Fig. 9. Prediction changing normalizing   Fig. 10. Prediction changing fitness function
under variable boundary conditions.   under variable boundary conditions.

Prediction for variable boundary conditions (2)
In this section, the other parameter is chosen to improve the predictive ability of the reasoning engine for the above-mentioned boundary conditions.

In Fig. 10, IS is found as a new parameter and is same as the one in Equation (2). The iterative step IS is varied as 500, 1000, 1500, 2000, 2500 and 3000. For the cases at IS=1000, 1500 and 2500, the same neural network is survived through the GA procedure, as shown in Fig.6. They are also same as the network obtained for the case of Lmax=20. The error magnitude yielded by the network is the smallest in all present trial cases. Meanwhile, the curves crossing the linear line is obtained for the cases of IS=2,000 and 3,000. The ability to interpolate the given data looks like better than the other cases. But, the ability on the reproduction of the given data is spoiled. It should be stated that the larger IS does not always give an improvement and there are necessities to give results to be leaned for the intermediate values of the boundary condition.

Through the present investigations, in order to search the most adaptable neural network, further analyses are needed for various network parameters and GA operators.

CONCLUSIONS

The present study aims at establishing the expert system that can predict and reproduce the buckling behaviour of single layered reticular domes. The reasoning engine in the expert system is produced based on algorithms concerned with an artificial life. In the present paper, the method with a genetic algorithm to obtain the adequate geometry and parameters of the neural network has been introduced. Through the investigations, it is confirmed that the method would be avail to construct the adaptable neural network and some problems to be solved are still existed. Several remarks obtained in this paper are listed.

1. In order to search for the adaptable geometry and parameters of neural network being the reasoning engine in the expert system, a simple procedure based on the genetic algorithm is proposed.
2. The performance of the proposed simple fitness function gets greater with increasing adopted iterative steps where the fitness is evaluated. However, it has to be noticed that over the appropriate value, the performance is not always better than small number.
3. Through the estimative calculations for the interpolative ability, the reasoning engine can quite adequately predict the ultimate strengths under changing the slope angle of the member at the dome apex.
4. For the various boundary conditions of the dome, the ultimate strengths could not be estimated exactly by the present system. The improvement for such prediction is tried under changing the selected parameters. But, the tried parameters are insufficient to interpolate the given data.

There are still a lot of parametric case studies to be tried. It is expected that through the studies, the proposed method would be sophisticated and then the reasoning engine would also behave more exactly.

REFERENCES

1. Takashima,H., PROBLEMS TO ESTABLISH AN EXPERT SYSTEM FOR PREDICTING BUCKLING BEHAVIORS OF STEEL SINGLE LAYERED RETICULAR DOMES, Proceedings of IASS 2001 International Symposium on Theory, Design and Realization of Shell and Spatial Structures, CD-ROM, Oct., 2001.
2. Takashima,H. & Kato,S., Numerical simulation of elastic-plastic buckling behaviour of a reticular dome, SPACE STRUCTURES 4, Vol.2, Thomas Telford, Sept.,1993, pp.1314-1322.
3. Takashima,H. & Kato,S., CONCEPTS TO CREATE DATA-BASE/REASONING SYSTEM FOR SPACE DOME BUCKLING BASED ON NEURAL NETWORK, Spatial Structures: Heritage, Present and Future, IASS International Symposium 1995 - Milano Italia, vol.1 pp.119-126, Jun.1995.

4. Takashima,H., How to Create Reasoning Engine to Predict Elastic-Plastic Behaviors of Space Domes based on Neural Network, Proceedings of the International Colloquium on Computation of Shell & Spatial Structures, Taipei, Nov., 1997, pp.79-84.
5. Takashima,H., Proposals to Create an Expert System to Predict Elastic-Plastic Behaviors of Single Layered Lattice Domes. Lightweight Structures in Architecture, Engineering and Construction IASS/IEAust/LSAA International Congress, Sydney, Vol.1, Oct., 1998, pp.473-480
6. Jenkins,W.M., On the application of natural algorithms to structural design optimization, Engineering Structure, Vol.19, No.4, 1997, pp.302-308.
7. Kitano,H., Designing Neural Networks Using Genetic Algorithms with Graph Generation System, Complex Systems 4, 1990, pp.461-476.

Design, detailing and realization of recent glass walls with tension trusses and point supported glass in the United States

I. COLLINS, C. H. STUTZKI
Mero Structures, Menomonee Falls, USA

INTRODUCTION

Recently something new appeared in the American cities. Architectural eye catchers in exposed locations. It is not a traditional aluminum curtain wall system. It is not a glass wall with the traditional door and window hardware. It is not even new. Some critiques call it "Designer Glass Walls". They are a type of structure evolved from roots back in Old Europe, Paris and London.

Designer Glass Walls are a fusion of formerly separated industries and professions: they were developed by a team of architects, engineers, designers and manufacturers. The new type of glass wall represents the combination of all of these different knowledge realms: artistic design, architecture, manufacturing secrets and engineering (common sense with a pinch of mathematical methods).

CURRENT SITUATION

The Designer Glass Walls are just about to enter the architectural market of the United States after being developed and built for 20 years in Europe, spreading out from Paris and London all over the continent.

James Carpenter, a New York based architect, whose glass designs have been recognized internationally, notes still stronger appreciation for these types of buildings in Europe: "I see a real pride of ownership and an appreciation of the natural qualities of glass as a thing of beauty. This applies to architecture at all levels and not just the high-end residences or the monumental public spaces" [1].

One possible reason for the reluctant use of Designer Glass Walls is certainly a glass industry in the United States moving very carefully into new areas. Don McCann, manager of Architectural Design at Viracon, the leading manufacturer of insulated glass in the US warns: "Manufacturers and fabricators want to come up with good products to open up creative uses, but we focus on two things: cost effectiveness and liability issues. We have to make sure that a product has been tested to do safely what we've said it will do- and that there is a market for the product. This makes a one-of-a-kind experimentation more problematic" [2].

Hundreds of beautiful glass buildings in Europe would have remained unbuilt if European companies would not have taken the economic- not the safety- risk, and gone through painstaking test procedures and large test programms. Most major labs in Europe have now a glass testing facility, and codes for point supported glass are well on its way.

FUNCTIONS

Working with glass had entered a new level when Pilkington invented the float glass process about thirty years ago. Together with the tempering process of glass a material was created which is a fundamentally different material than the glass known for thousands of years. Glass has matured into a building material and we all are just at the beginning to understand and use glass in this sense.

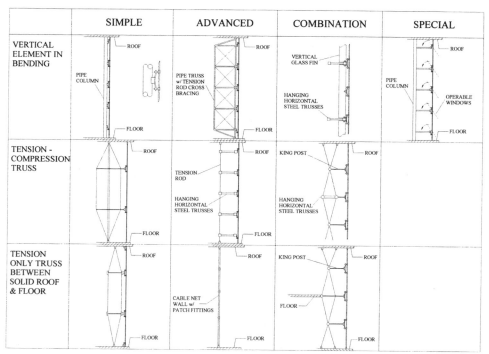

This paper wants to take a snapshot at the state of the art in the US, not in order to summarize the end of a long lasting development rather than to point out the beginning of a new branch of evolution in building envelopes.

How glass walls work is simple: they provide a vertical plane of glass over an opening. There are a lot of ways how to do that, and here we are talking about glass walls continuing the tradition of medieval glass windows in gothic cathedrals. The paper tries to find a systematic approach for the different ways of structural design. One way of doing this is a matrix showing three different concepts, each in a simple, advanced, and combined version. Also special applications have its place in this scheme. (Fig. 1)

All these examples refer to actual structures that have been built in the United States within the past three years or are currently under construction. They comply with all codes and standards, get all approvals and passed all required tests, in particular seismic tests for the structures built in California.

There are many more possibilities, academic ones, tensegrity structures as invented by Buckminster Fuller, still waiting to leave the labs and art studios. But even in the range of our current experience, as to structural performance and economic value, the architectural possibilities are immense. The budgets for these combinations of tempered glass, silicone, and high strength tension rods are not as experimental any more as they were for the first point supported glass wall fifteen years ago in Europe. (The costs are between three to four times of a conventional mullion system curtain wall)

MATERIALS AND MANUFACTURING
Glass
The glass for the designer glass walls is nothing special, nothing exotic. Based on ANSI standards [4], [5], [6], the most commonly used glass is based on float glass, as made by Guardian, Asahi, Pilkington or Saint Gobain, just to name the fabulous four. One architectural tendency is obvious: the design intent is to have a maximum of transparency. This leads to very large glass panels. The size in return requires thick glass which appears slightly greenish because of its thickness. Therefore many architects suggest low iron glass (white glass). That always raises a serious question about additional costs. How much more is the owner willing to spend? However it is worth considering it. Once you saw a curtain wall with 'white' low iron glass you know why it is used more and more in architectural glazing.

Point supported glass always need to be tempered in order to get the required strength. Either the so called heat treatment (a mild tempering) or the full tempering process is required. The full tempering has the advantage of producing very strong glass, but with the disadvantage of slowly growing nickel-sulphite crystals which can cause spontaneous breakage. The so called heat soaking process tries to detect the nickel-sulphite crystals prior to erection. [7]

The decision between monolithic and laminated glass is mostly a safety issue. Curtain walls do not require laminated glass, but very high curtain walls of lobbies have glass panels that are high up in the air. Therefore a similar safety approach as for overhead glazing should be considered using laminated glass. Laminated glass is described in industry standards world wide and is well suitable for point supported curtain walls [8]. Since there is not yet a standard for point supported glass the thickness of each glass panel needs to be determined with the help of an engineered stress analysis for each particular application.

Materials for Cold Formed and Machined Stainless Steel
Typical stainless steel materials are X2 Cr5Ni 18 10 (304L) and X2CrNiMo 17 13 3 (316L) with less than 0.03 % Cu. Instead of the yield strength the 0.2% plastic elongation is used which is between 26,000 and 27,000 psi. This level can be enhanced considerably through cold working of the material, especially for tension rods [9].

The most elegant types of tension rods currently made are the ones made by Tripyramid Inc., Massachusetts, USA. Coming from the equipment supplier for high performance sailing yachts Tripyramid tension rods were first used for architectural purposes. That was in Ieoh Pei's Louvre pyramid in Paris. Since then it became more and more popular to use stainless steel tension rods with their extreme strength/ diameter ratio. The tensile strength of these materials is 140,000 psi and 200,000 psi respectively [10]. The structural calculation should be based on the standard [11].

The glass fasteners (rotules) are made from 316L stainless steel except the threaded rod. This threaded rod is highly stressed by bending which is critical with the threaded surface. Therefore the material is a special alloy with a yield strength of 70,000 psi and a high ductility.

The glazing arms and other structural parts are typically made from bent plates using 304L stainless steel with a minimum yield strength of 33,000 psi.

Cast glazing arms are typically manufactured using the lost wax form process. This requires a clay mould for each produced glazing arm, made from a wax positive which is then melted out of the clay mould in order to get the cavity for each casting.

Finishes of Stainless Steel Parts

Tension rods are brushed carefully in order to give it satin like brushed finish. The end fittings and the rotules have a machined finish. The grinding and machining can be controlled in such a way that the fine grooves from the grinding tools create a shining surface similar to a brushed surface [12].

The surface of the glass arms has to match a fine brushed finish. A mirror polish finish can be achieved but is normally not desirable. It is important to have similar finishes for all stainless steel parts- tension rods, fittings, compression struts and rotules.

Cast glazing arms are normally sandblasted and electro-polished in order to get a very smooth surface, free of irregularities and burrs. Electro-polishing creates a shiny and very bright but slightly grainy surface. Further surface treatment with careful glass bead blasting creates a very smooth and slightly dull surface.

Silicone

The major producers of Silicone Sealant in the US are General Electric and Dow Corning. For some types there might be a compatibility problem between the silicone and the interlayer of laminated glass. Therefore it is common practice to use a combination of extruded silicone profiles which cover the interlayer, and liquid silicone that cures and is applied in the field. The advantage of this combined joint seal is also that it can be installed more economically than a joint filled completely with liquid silicone. The combination of extrusion and field applied liquid silicone fulfills also the rule of the two barriers against water for curtain walls.

COMPONENTS
Tension Rods and Fittings

In order to minimize the diameter of tension rods a special detail is used at the ends. The termination is not done with a threaded end but with a cold formed head sitting like a wedge in a cone shaped saddle of the fitting. This way the full area of the tension rod can be used to determine the strength. Together with the elegant jaw fitting these tension rods are an important element of Designer Glass Walls. An alternative is the bolt- like end fitting which allows an economic and small termination but opens a variety of connections to structural parts like to a Mero space frame node.

Monolithic and Laminated Glass Panels

Typically point supported glass panels are supported at 4 points. Many calculations showed that supporting a panel with six points reduces the span but very high stresses occur in the mid span rotules penalizing what you gain by reducing the span. In order to have a statically determinate support in the plane of each glass panel the glass is hung from its two top points, with one fixed and one horizontally sliding support and two bi-directional sliding supports at the bottom, taking only the wind load. This type of support is important to avoid any residual or temperature stresses between fixed supports. If these rules of supporting a panel are not followed the glass is most likely to crack within a short time. The edge treatment is important: polished edges of the glass allow a better permanent adhesion of the silicone than ground edges.

Insulated Glass Panels

The manufacturing of point supported glass panels is still not very common. Only a couple of companies in Europe and Japan are able to manufacture and warrant this product.

Several performance criteria control the design of an insulated glass panel:
- structural safety of each glass plate as to the specified wind load
- strict deflection limits of the edges in order not to overstress the seal at the perimeter. The edge could be torn apart under shear stress due to deflections.
- Permanent air tightness not only at the perimeter seal but also at the ring seals around the rotules.
- No condensation at the rotule by heat transfer through the metal parts.
- Rotational capability – typically 10 degrees in all directions in order to avoid local stresses when the glass deflects (plate deflections cause rotations at the support points).

These requirements can be met only with very careful design and manufacturing. That explains why there are only very few manufacturers who fabricate these delicate devices and give a warranty of ten or twelve years.

Following the architectural demand for ultra transparent curtain walls the German company OKALUX [13] developed an insulated glass panel with glass spacers, using glass clear structural silicone to glue the glass panels and the glass spacer bars together. The impression when looking at a glass façade using this glass together with stainless steel components can be called stunning.

Glazing Arms

A wide variety of custom designed glazing arms is found in Designer Glass Walls being built so far in the US. Besides some simple standard forms, many custom made glazing arms have been made and will be designed in the future. Some have several swivel parts in order to prevent glass damage in case of earthquake. The design rules follow the regular codes for structural steel since there is no unusual material or mechanism involved. Here is the place where creative design starts- in the combination of glazing arms, connections of tension rods, termination of glass fins and the transition to the heavier structural parts. Casting, machining, and welding of stainless steel can produce show case pieces- together with light weight structural design- which enhances the value of a building enormously.

Glass Fasteners

Point fasteners are settled into holes in the glass panels; the weight of the glass is supported by a bearing ring of a material softer than steel, in direct contact with the glass. In the housing of the fastener is a ball-and-socket-joint to allow for three dimensional rotation. The treaded rod of the glass fastener is mounted on the glazing arm.

EXAMPLES

From the many existing buildings here are some representative examples:

Glass Museum in Corning, New York
This project consists of five glass walls, 19 feet high, with a total length of 225 feet. The walls are mostly sloped in different angles. Plus an entrance area with point supported glass in overhead position. The stainless steel structure is similar to sail boat masts with rigging, designed by the New York based Architects Smith-Miller+Hawkinson with the project architect Ingalill Wahlroos. The glass is ¾" thick clear monolithic fully tempered. The steel masts every 10 feet and the glass are hanging from the roof, with sliding supports at the bottom. The glass arms are custom designed made from cast stainless steel. The masts, tension rods and fittings were detailed and manufactured by Tripyramid, Inc. This project is a good example of a successful integration of tension rods, glazing arms, and truss parts, machined in one piece (Fig. 2).

Fig. 1. Glass Museum in Corning, New York.

Fig. 2a, 2b. Glass Museum in Corning, New York.

Corporate Headquarter of Boeing, Seattle, Washington

A curved glass wall, approximately 200 feet long and 45 feet high is the accent for the entrance lobby of the Boeing headquarter. Designed by the architects Loschky Marquardt & Nesholm, this project is a combination of horizontal tubular trusses and vertical tension trusses with king posts. The insulated point supported glass panels have a screen print and are 7.5 feet high and 4 feet wide. (Fig. 3) Three curved tubular trusses are the support structure to withstand the wind forces; vertical tension trusses with king posts have the two functions of wind support at the mid points between the tubular trusses, and hanging the whole structure- tubular trusses and glass- from the roof. The cast stainless steel glazing arms are custom designed, the stainless steel tension rods are by Tripyramid, Inc.

Fig. 3: Lobby of the Boeing Headquarter, Seattle.

Fourth District Police Department, New York, New York
This project in lower Manhattan consists of a vertical glass wall 15 feet wide and 24 feet high. The architect is the New York Housing Authority with the project architect Bogdan Pestka. A canopy is suspended with tension rods from steel fins penetrating the glass wall at the glass joints. The tension trusses are prestressed horizontally and vertically between strong beams at top, bottom, and the two sides. All tension rods are of ¾" diameter (by Tripyramid, Inc.), an adjustable connection between compression post and glazing arms allowed an easy installation (Fig. 4).

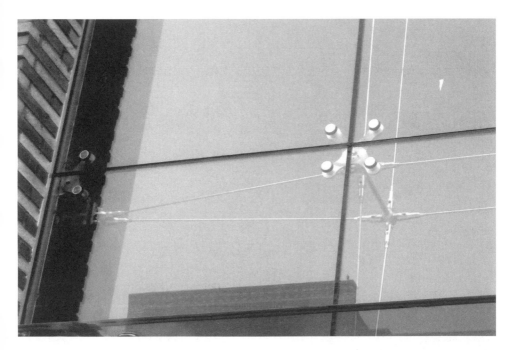

Fig. 4. Manhattan Glass structure, New York.

Seattle Madison Financial Center
This innovative project is still in the design phase. The architect is the New York based glass artist James Carpenter. This project will be the first large prestressed cable net in the United States. The technology was developed about 15 years ago by the German engineers Joerg Schlaich and Hans Schober. The structure consists of a net of prestressed cables eliminating all beam type members. The Seattle project is based on this idea, adding decorative glass fins at the inside and at the outside:

Fig. 5. Computer rendering of the glass and cable node.

Design for a point supported glass wall in Illinois
The curtain wall is a typical example for the first category in the chart (Fig.1), simple approach. A wall about 25 feet high, with glass panels 5' x 10' large, laminated glass. In addition there is a small roof of 8 feet width. The structure is a row of pipe columns, with glazing arms attached to them.

Fig. 6. Computer rendering for the glass wall of a project in Illinois.

REFERENCES

[1]   Glass Magazine, July 1998, p.58.
[2]   Glass Magazine, July 1998, p.60.
[3]   Powell, More: Nicholas Grimshaw & Partners. Berlin 1999.
[4]   TM C 1036-91, Standard Specification for Flat Glass.
[5]   ASTM C 1948- 92, Standard Specification for Heat-Treated Flat Glass.
[6]   Engineering Standards Manual, by: The Glass Association of North America.
[7]   Heat Soaking Program, in: VIRACON Tech Talk, February 1988 and April, 1985.
[8]   ASTM C 1172- 91, Standard Specification for Laminated Architectural Flat Glass.
[9]   Design Manual for Structural Stainless Steel; Nickel Development Institute, 1994.
[10]   Tripyramid Structures, Inc., Technical Data Sheets, Massachusetts 1999.
[11]   ASCE-8-90, Specification for the Design of Cold-Formed Stainless Steel Structural Members.
[12]   Metal Finishes Manual, The National Association of Architectural Metal Manufacturers, 1988.
[13]   Okalux Kapillarglass GmbH, Marktheidenfeld- Altfeld, Germany.

Dynamic loading experiment performed on a wooden single layer two-way grid cylindrical shell roof

T. FURUKAWA, Graduate School of Eng., Osaka Univ., Osaka, Japan
K.IMAI, Graduate School of Eng., Osaka Univ., Osaka, Japan
M. FUJIMOTO, Graduate School of Eng., Osaka City Univ., Osaka, Japan
N. KOMEDANI, Graduate School of Eng., Osaka Univ., Osaka, Japan
R. INOUE, Graduate School of Eng., Osaka Univ., Osaka, Japan
K. OKAMOTO, Graduate School of Eng., Osaka Univ., Osaka, Japan and
Y. FUJITA, Fujita Architect Office, Osaka, Japan

INTRODUCTION

Wind-induced vibration of wide-spanned structures, especially of single layer space frames, is one source of concern among structural engineers and designers of such structure. Frequency response characteristics and damping capacity are most important properties that affect such phenomenon. However, very few studies and measured data of vibration test have ever been reported (Ref. 1). Moreover most of the reports are on steel structure.

Recently, the authors have been developing an innovative space frame system using small/mid-sized round timber as space frame members. The detail of the system and static behaviors have been reported in a previous paper (Ref. 2). The aim of this paper is to evaluate frequency response characteristics and damping capacity of a wooden space frame system. In order to evaluate these properties, a full sized wooden single layer two-way grid cylindrical shell roof model was manufactured. Both the free vibration test and the dynamic loading experiment using vertical dynamic actuators were carried out. Furthermore, microtremor of the experimental model are also measured. By comparing the modal shapes solved from the eigenvalue analysis with the time-history motion, which were acquired by the dynamic loading experiment, actual modal frequencies of the experimental model are identified. Based on the results of these studies, free vibration test, and microtremor measurement, the frequency response function and the modal damping factors of the experimental model are estimated. Moreover, it also reports system identification analysis, which shows good agreement with the observed records of the experiment using linear time invariant multi-degree of freedom model.

STRUCTURAL OVERVIEW OF EXPERIMENTAL MODEL

The experimental model structure subjected to the vibration tests and the microtremor measurement was wooden single layer two-way grid cylindrical sell roof with half open angle 45°, 4m×4m. Photo1 shows the experimental apparatus. As shown in Figure1, the experimental model structure was supported by rigid reaction steel frame on six outermost nodes. This model consisted of wooden members and metal nodes of KT wood truss system (Ref. 3). Details of this system are shown in Figure2. The members are made of cypress with a diameter of six centimeters. The nodes of φ90/85 are used as support joints, while the nodes of φ75/68 are used as other joints.

Photo 1. Experimental model

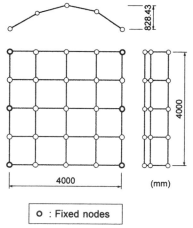

o : Fixed nodes

Fig. 1. Structural system.

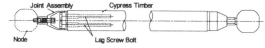

Fig. 2. KT wood truss system.

TEST METHOD

The free vibration test, the steady-state excitation test and the microtremor measurement on this experimental model were carried out.

Five servo velocity meters and nine laser gap sensors were installed on the underside of the shell roof. Figure3 shows the location of the measuring instruments. The velocity meters and the laser gap sensors measured the motion of the experimental model in the vertical direction only. On every test, output signal from measuring instruments was acquired in a laptop computer directly via A/D converter unit.

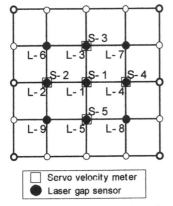

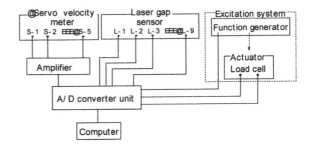

Fig. 3. Measured sensor location.   Fig. 4. Measuring system.

Free vibration test

In the free vibration test, free vibrations were generated by pulling the central node of the experimental model with a wire and suddenly cutting it. Vertical velocity and displacement at each measuring point were measured for the period of 40 seconds. Sampling frequency was 500Hz. The test was carried out three times under same condition, taking into consideration the effect of experimental noise. Each model impulse response was extracted by using appropriate band-pass filter acquired from the time history response of the free vibration tests. The damping factors for each mode were estimated by performing exponential approximation of the envelop of the waveforms. The damped natural frequencies were determined as the average period of one cycle motion of extracted model impulse response.

Steady-state excitation test

In the steady-state excitation test, steady-state vibrations were generated by two vertical dynamic actuators (maximum excitation force 30kgf per one unit) on the central node of the experimental model. Vertical velocity and displacement were measured for the period of 20 seconds. Sampling frequency was 500Hz. In this test, for the information of excitation force both the output signal from the load cells, which are installed in the actuators, and the output signal from the function generator were measured. Excitation frequency was varied from 1Hz to 35Hz at regular interval of 0.2Hz. The peak of response amplitude was estimated roughly by processing the obtained data. After that, small-step excitations were performed at frequencies in the vicinity of the estimated response amplitude peak. Resonance curve obtained from the vibration amplitude at different frequencies were used to calculate the damping factors.

Photo 2. Dynamic actuators.

Microtremor measurement

In the microtremor measurement, microtremor of the experimental model was measured for the period of 800 seconds. Sampling frequency was 500Hz.

The damping factor was estimated by fitting transmissibility (the ratio of the response acceleration amplitudes to the input force amplitudes) to the Fourier spectra acquired from time derivatives of the time history velocity response of the microtremor measurement. In this calculation, the input acceleration was assumed to be white noise, which have constant amplitude level at every frequency.

RESULT

Figure 5 shows the Fourier spectra obtained from the free vibration tests as one sample. There are clear peaks at a number of frequencies, in which five peaks were selected and the vibration characteristics were estimated at the peaks.

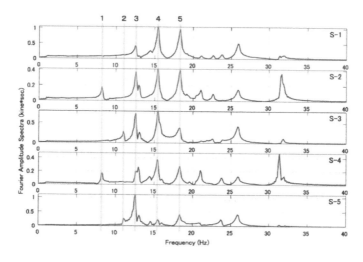

Fig. 5. Fourier spectra (free vibration).

2nd mode (8.44Hz)

4th mode (11.03Hz)

7th mode (13.17Hz)

9th mode (15.48Hz)

13th mode (17.17Hz)

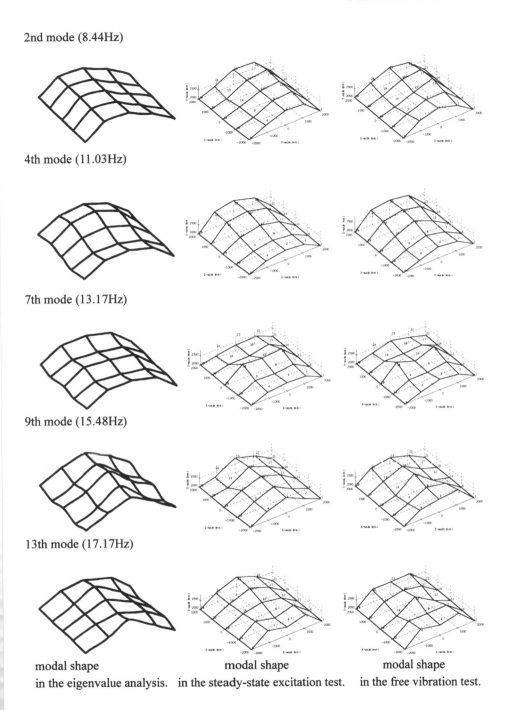

modal shape
in the eigenvalue analysis.

modal shape
in the steady-state excitation test.

modal shape
in the free vibration test.

Fig. 6. Modal shapes at each mode.

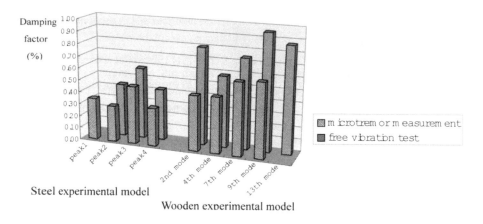

Fig. 8. Damping factor on wooden and steel experimental model.

SYSTEM IDENTIFICATION ANALYSIS WITH THE EXPERIMENTAL MODEL USING SUBSPACE IDENTIFICATION METHOD

System identification using subspace identification method was carried out in order to investigate the applicability of the method. Subspace identification method estimates state-space model directly from input-output data. The code used for the identification was System Identification Toolbox of MATLAB (Ref. 5). For informative and accurate identification using this method, input signal should contain many component waves, which excite every resonance frequencies of the identifying object. Therefore, the data for the identification were virtually synthesized from 26 different cases of input-output data of steady-state excitation test by superposing each time history data successively. From the viewpoint of the high efficiency and rapid estimation, the sampling rate of the data for the identification was reduced at 50Hz.

Figure 9 shows the comparison the identified frequency response function with the experimental one. It is clear that they show good agreement, within the frequency range under 20Hz. Therefore, it seems reasonable to suppose that the experimental model could be identified with subspace identification method.

Moreover, the modal damping factors were estimated by simulation with the identified model. The free vibration responses of the identified model after stopping harmonic inputs, which have each modal frequency, were simulated. In order to estimate the modal damping factors, their positive peaks fitted to the logarithmic decrement curves. Table4 shows the estimated damping factors, and Figure10 shows the bar diagrams of them. For comparison, this figure shows the same values estimated with the other approaches, too. The modal damping factors were estimated at about 1% appropriately, and their tendency, which is that the estimated value about 4th mode is small and those about 2nd mode and 13th mode are large, was similar to that of the free vibration test and steady-state excitation test.

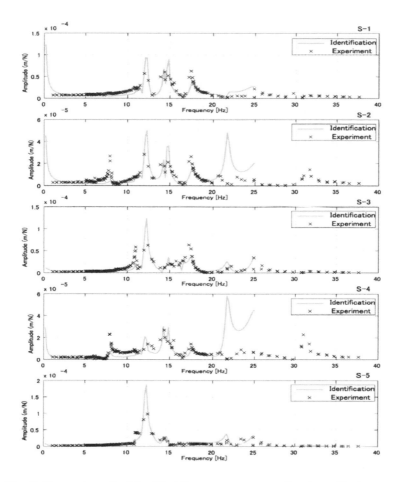

Fig. 9. Comparison the identified frequency response function with the experimental one.

Table 4. Damping factors and damped natural frequency obtained from subspace identification method.

Mode order	Subspace identification method	
	Damped natural frequency(Hz)	Damping factor (%)
2nd	7.94	1.42
4th	10.94	0.43
7th	12.20	0.91
9th	-	-
13th	17.54	1.55
Average	-	1.08

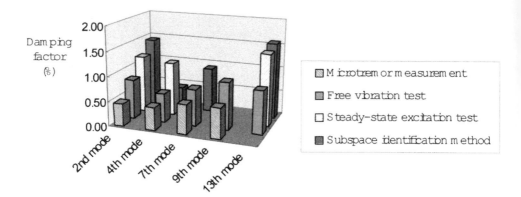

Fig. 10. Estimated damping factors using subspace identification method and experimental data

CONCLUSION
In order to evaluate the dynamic characteristics, especially damping capacity, of wooden space frame system, vibration testing and microtremor measurement of wooden single layer two-way grid cylindrical shell roof model were carried out. As the result, following results were obtained.

1) The estimated damping factors varies depending on excitation level
2) It cannot be seen such tendencies that estimated modal damping factor increase in proportion to the order of the mode.
3) The damping capacity of the wooden space frame is superior to that of conventional circular hollow steel tube space frame.

Moreover, it also reported system identification analysis, which shows good agreement with the observed records of the experiment.

REFERENCES
1. TATEMICHI I., HATATO T., ANMA Y., and FUJIWARA S. Vibration tests on a full-size suspen-dome structure, International Journal of Space Structures, Vol. 12, Nos. 3&4, 1997, pp 217-224.
2. FUJIMOTO M. IMAI K. FURUKAWA T. KUSUNOKI M. KINOSHITA R., Buckling Experiment of Single Layer Two-way Grid Cylindrical Shell Roof under Centrally Concentrated Loading, 5th international conference on space structures 2002, (Submitted)
3. IMAI K., FUJITA Y., WAKIYAMA K., TSUJIOKA S., FUJIMOTO M., WATANABE H., INADA M., and MORITA T. Further Development of the KT-Wood Space Truss System, IASS Symposium 2001, Nagoya, 2001, pp 350-351.
4. ARCHITECTURAL INSTITUTE OF JAPAN. Standard for Structural Design of Timber Structure, 1996, pp 17.
5. LENNART LJUNG. System Identification Toolbox -For Use with MATLAB, The Math Works. Inc., pp 4-118 – 4-120.

Glass space structures

M. OVEREND
Whitby Bird & Partners, London, UK

G.A.R. PARKE
University of Surrey, Guildford, Surrey, UK

ABSTRACT
Recent combinations of glass and cable trusses have resulted in exciting minimalist glass structures. However, to date there is still no agreement on a definite mathematical model that accurately predicts the failure of glass, hence resulting in disjointed and sometimes conflicting design data for this brittle *unforgiving* material. This induces engineers to adopt large factors of safety and devise inefficient glass structures. A design methodology is therefore advanced to determine the strength of annealed and tempered glass panels. This paper provides an overview of this method and gives examples of its application.

INTRODUCTION
Glass has fascinated people ever since its discovery. It appeared as a fulfilment of the apparently impossible requirements of transparency and durability. Of equal importance is that it is made from the melting and cooling of silica - one of the Earth's most abundant minerals. Despite these outstanding properties, its brittleness and relatively low tensile strength have made it an unattractive load bearing material (Figure 1).

The molecular structure of glass is made up of oxides of various elements - mainly silicon, sodium, potassium, calcium, magnesium and aluminium - which at high temperatures form a viscous magma. Cooling of this magma is carried out at such a rate that prevents crystal growth. As the glass cools its viscosity increases to such a high value that it effectively becomes a solid, and the randomly oriented molecules are *frozen* into the structure. Unlike metals, the random molecular structure of glass lacks crystallinity or long range order and has no slip planes or dislocations to allow yield before fracture, consequently glass exhibits brittle fracture at a theoretical value of 21,000 N/mm^2.

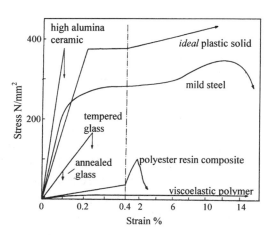

Fig. 1. Stress-strain curves for existing and ideal engineering materials.

However, the ultimate tensile strength of glass given by the European Code is 45 N/mm^2 [Ref. 1] and weathered window glass was reported to fail at stress levels of around 25 N/mm^2 [Ref. 2]. Furthermore, the strength of glass subjected to long-term loading may be as low as 8N/mm^2 [Refs. 3,4].

This large inconsistency between the theoretical and actual strength of glass was explained by A. A. Griffith in 1920 [Ref. 5]. Griffith argued that fracture did not start from a pristine surface, but from pre-existing flaws (now named Griffith flaws) on that surface. These flaws are also present on other materials such as steel, however, since glass cannot flow by plastic deformation, a sharp notch stays sharp, and the stress at the tip is limited only by the atomic bond strength. These flaws have been found to be atomically sharp [Ref. 6] and consequently experience a stress magnification such that a relatively low applied stress could produce bond rupture at the crack tip. Furthermore, when surface flaws are exposed to tensile stresses in the presence of water vapour, flaws corrode more rapidly in depth than in width, leading to higher stress concentrations and consequently failure at progressively lower loads. This phenomenon is commonly referred to as static fatigue or stress corrosion.

The most common way of reducing the deleterious effect of the flaws is by tempering the glass. In this process, the glass is heated and then rapidly quenched, thus introducing a parabolic stress gradient within the thickness of the glass whereby the outside surface is in compression (Figure 2). Any externally applied force must first neutralise the surface compressive stress before any surface tensile stress can be set up. Tempered glass with residual stresses of up to 120 N/mm^2 is commercially

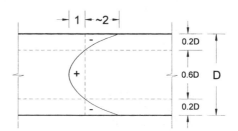

Fig. 2. Parabolic stress distribution through thickness, D, of tempered glass.

available, however the presence of other surfaces such as at plate edges, corners and holes may distort the parabolic stress distribution by up to ±15% [Ref. 7].

Tempered glass is therefore a relatively strong material and provides the opportunity for its use as part of the primary load bearing structure. For example: two sheets of 12mm thick tempered glass held at 600mm apart vertically provide the same moment of resistance as a 533x210x109kg/m universal beam supported at 1000mm centres.

EXISTING & POSSIBLE STRUCTURAL GLASS SYSTEMS
The majority of contemporary patent glazing adopted up to the 1970's employed a metal framework of mullions, which was visible both internally and externally.

The term *structural glazing* was first adopted in the United States in the early 1970's. This refers to the removal of the external cover bead and the use of elastomeric silicone sealants to bond the glass to the metal sub-frame (Figure 3) and may be more appropriately called structural silicone glazing.

Aesthetic requirements for a more transparent facades and the improved strength brought about by the tempering process for glass were the main factors behind the more recent form of structural glass. This is achieved by detaching the metal support structure from the line of the glass or by replacing it with a glass support structure (Figure 4). The glass is connected to the support structure by means of countersunk bolts and elegant castings.

The bolts are commonly located toward the corners of glazing panels and sometimes additionally at intermediate points along the edges. These bolted connections range from the earlier stud and patch-plate fittings to the more elegant countersunk connections

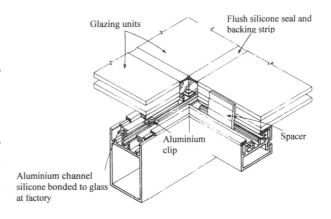

Fig. 3. Four sided structural silicone glazing.

(Figure 5). In addition to transmitting the lateral wind loads to the support structure, the glass is taking on a more structural role by carrying its own weight and that of other panels beneath or above it, depending on whether the assembly is suspended or ground-based. However, the global lateral strength of these systems is still provided by the support structure.

Fig. 4 Structural glass façade in Kuala Lumpur.

Fig. 5 Structural glass connection detail in Kuala Lumpur.

However, recent research suggests that it may be possible to use glass such that it contributes to the load bearing strength of a structure [Refs. 8, 9, 10]. An attractive way of using glass structurally is to combine it with a space structure such as double layer grid in which the top layer members are entirely replaced by structural glass panels. Such a structure may be designed to exhibit elasto-plastic failure by ensuring that the steel members yield before the glass breaks. The success and efficiency of this approach invariably depends on the efficiency of the glass/steel connections and on the ability to determine the strength of the glass panels.

THE STRENGTH OF GLASS

Traditional glass design procedures rely on the empirical representations of glass strength and rules of thumb to determine the stress distribution caused by the applied loads. These empirical rules have stood the test of time because glass was predominately used in short span window infill applications. However, with the development of the curtain wall, glass has evolved into a more important structural component of the building envelope. These developments led the glass design community to propose the first analytically derived failure prediction models for glass [Ref. 11]. From the various numerical and physical tests carried out [Refs. 11, 12, 13, 14, 15], it may be concluded that the strength of annealed glass depends on the following parameters:

(i)   load duration;
(ii)   surface area of glass exposed to the tensile stress;
(iii)   environmental conditions, especially humidity;
(iv)   magnitude and distribution of load-induced surface tensile stresses in glass;
(v)   ratio of major and minor principal tensile stresses on the surface of the glass.

More recently, a number of crack growth models have emerged as the most accurate representation of glass failure and strength [Refs. 16, 17, 18]. These models, derived from the application of linear elastic fracture mechanics, have been developed in response to the growing structural role of glass. The models provide a more accurate representation of glass failure, but are generally restricted to laterally loaded rectangular plates simply supported along their edges or do not account for all factors known to affect glass strength. Furthermore, the crack growth models are inherently unattractive for manual computation.

A design methodology based on the principles set out by the draft European Code [Ref. 1] and on previous research by the authors [Ref. 19] has been developed. This method quantifies the surface tensile strength of glass, σ_f, by adopting a proposed General Crack Growth Model (GCGM) developed by the authors [Ref. 20]. This is compared to the equivalent applied uniform stress, σ_p, which is computed by a specially written computer algorithm. Structural adequacy is ensured by:

$$\sigma_f \geq \sigma_p$$ Eq. 1

Effective Strength

A spread of strength values is always obtained when a batch of nominally identical test pieces of a glass is broken in a carefully controlled way. If enough test pieces are broken, it is found that the strength values lie on some shape of a distribution curve.

This variability is best represented by a 2-parameter Weibull distribution [Refs. 21, 22]. The two parameter Weibull distribution provides a description of the strength of glass in which a distribution function containing two interdependent parameters is used to predict the probability of failure P_f of a specimen, for a given load duration:

$$P_f = 1 - \exp\left(- kA\sigma_s^{\ m}\right)$$ Eq. 2

Equation 2 gives the probability of failure for a given instantaneous stress σ_s, applied uniformly over an area A, in terms of the two surface strength parameters: m, which is a measure of the variability of flaws within the specimen and k, which indicates their absolute size and density. The surface strength parameters are material constants and are determined by experiment.

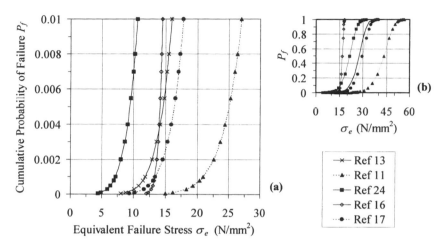

Fig. 6. Probability distribution function for a 60-second equivalent stress σ_e, uniformly applied over a 1m² annealed glass. (a) Part (0<P_f<0.01) and (b) Full (0<P_f<1).

Since tempered glass is used in the majority of structural glass applications the basic probability of failure expressed in equation 2 may be extended to tempered glass:

$$P_f = 1 - \exp\left[-kA(\sigma_s - \sigma_r)^m\right]$$

Eq. 3

which acknowledges that failure may only occur when the surface precompressions, σ_r, induced by the tempering process have been overcome.

For a given probability of failure and by conservatively adopting a 100% humidity level, the strength of glass, σ_f, for any load duration and surface area may be obtained from:

$$\sigma_f = k_{mod}\sigma_s + \sigma_r/\gamma_v$$

Eq. 4

Where σ_s is the instantaneous strength of annealed glass obtained from the probability distribution function (Equation 2); k_{mod} is the stress corrosion modification factor ($0.346 \leq k_{mod} \leq 1.0$); σ_r is the surface precompression induced by the tempering process (generally, $90N/mm^2 \leq \sigma_r \leq 120N/mm^2$); and γ_v is the safety factor to account for variations in the surface precompressions induced by the tempering process (typically, $\gamma_v \approx 1.5$).

The contribution of the annealed glass strength, $k_{mod}\sigma_s$, is an interaction between load duration, stressed area and surface tensile stress and may also be expressed by means of a strength envelope for a given probability of failure as shown in Figure 7. A top view of this strength envelope results in a design chart used to obtain $k_{mod}\sigma_s$ (Figure 8).

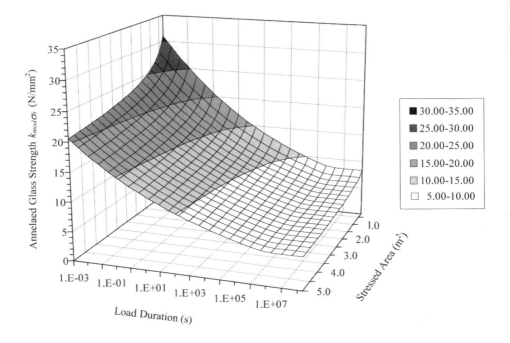

Fig. 7. Failure envelope for annealed glass for $P_f = 1/1000$.

Equivalent Applied Stress

A glass plate may be assumed to be uniformly subjected to the maximum tensile stress encountered anywhere on its surface ($\sigma_p = \sigma_{max}$). However, this is considered to be too conservative. A more accurate representation of the applied stresses is that the risk of failure of a glass plate is related to the summation of the contributions of all the stresses present on the glass surface. The summation of all the stresses may be transformed to an equivalent uniformly applied stress by:

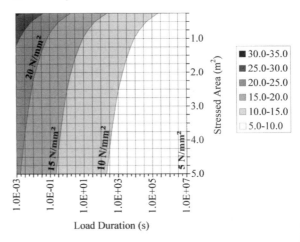

Fig. 8. Annealed glass design chart for $P_f = 1/1000$.

$$\sigma_p = \left[\frac{1}{A} \int_{area} \left(c_b \sigma_{max} \right)^m dA \right]^{1/m} \qquad \text{Eq 5}$$

where A is the total surface area, dA is the area of the subdivision, σ_{max} is the major principal tensile stress, m is the surface strength parameter ($m = 7.3$) and c_b is the biaxial stress modification factor that may conservatively be taken as unity.

As with other recent crack growth models, the accuracy of this method relies on density of subdivisions, dA, and subsequent summation of their contributions to the failure of the glass panel. This calculation is unattractive for manual computation. A Visual Basic algorithm, *Glasstress*, has therefore been developed to perform these calculations automatically from the results file of Finite Element (FE) analysis software. Input to the computer consists of the co-ordinates of the surfaces to be appraised. *Glasstress* conveniently calculates the surface area, dA, and average principal tensile stress, σ_{max}, for each element of the FE model. The equivalent uniformly applied stress, σ_p, for the whole surface is also calculated automatically from Equation 5 and elements that are subjected to a compressive stress are eliminated from this summation. In addition, the algorithm also creates a spreadsheet containing a listing of these calculations and a summary of the entire surface. *Glasstress* may be used with a number of commonly used elements ranging from 3-noded triangular elements to 20-noded brick elements.

This value obtained for the equivalent uniformly applied stress, σ_p, may be compared to the strength of glass, σ_f, obtained from Equation 4 to ensure structural adequacy as set out in Equation 1.

VERIFICATION OF THE FAILURE MODEL
The ability of the existing failure models and the proposed GCGM to portray the strength of glass was investigated for the following stress distributions:

(i)   concentrated stresses generated by a ring-on-ring arrangement.
(ii)   more uniform stresses in laterally loaded simply supported rectangular glass plates.

A series of ring-on-ring tests were undertaken by the authors on specimens measuring 300x300x6mm. The test was performed by placing the glass plate on a circular steel reaction ring and applying on its opposite surface a load transmitted through a steel loading ring until failure occurs. The purpose of this test is to achieve a uniform tensile stress field within the loading ring that is independent of edge effects (Figure 9).

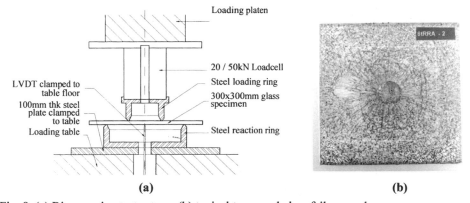

Fig. 9. (a) Ring-on-ring test set-up; (b) typical tempered glass failure mode.

Forty-nine ring-on-ring tests were successfully performed. These were composed of 30 annealed glass specimens and 19 tempered glass specimens (BS 6206 class A) and tested by means of a 51mm diameter steel loading ring and three steel reaction rings with 65mm, 127mm and 200mm diameters (Figure 9).

The proposed GCGM showed good agreement with the annealed and tempered glass ring-on-ring test data (Figure 10). The variability in strength of the tempered glass specimens was larger than expected and may be attributed to the variations in surface pre-compressions in the tempering process.

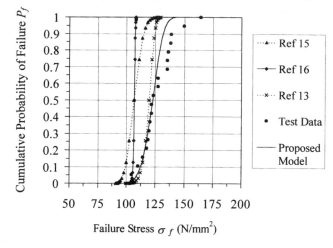

Fig. 10. Test results & failure models for the ring-on-ring tempered glass tests.

Laterally Loaded Plate Investigations

Two sets of published annealed glass failure data [Refs. 14, 23] and one set of published tempered glass failure data [Ref. 15] where used to test the validity of the GCGM and the *Glasstress* algorithm.

This was achieved by comparing the predictions obtained from FE analysis and the proposed glass design method with the independent test data.

The predicted relationship between the 60-second equivalent loads, P_{60}, and the probability of failure P_f are in excellent agreement with the annealed and tempered glass test results (Figure 11). This is particularly the case at the low probabilities of failure generally used in practice.

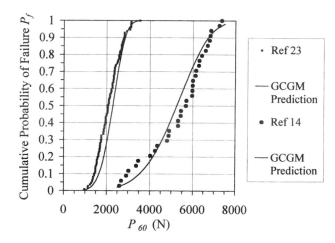

Fig. 11. Test results & GCGM predictions for laterally loaded annealed glass.

OPTIMISATION OF THE GLASS / SPACE STRUCTURE CONNECTION

Most of today's structural glass connections are direct descendants of structural steelwork bolted connections. Arguably, it is unwise to adopt this direct technology transfer approach as the resulting bolt holes and macroscopic flaws brought about by the drilling process are deliberate weak spots in the glass. There is a general lack of published research on the behaviour of bolted connections in glass, however the little research available confirms the view that the load bearing capacity of bolted glass assemblies is limited by the high stresses around the bolt holes [Refs. 8, 24, 25].

Parametric Optimisation

A parametric study of the various factors affecting the strength of bolted connections was performed by using FE analysis and the proposed glass design method. This study was also adopted to propose an alternative adhesive connection.

The ultimate tensile strength of a connection, P_t, may be expressed generically by:

$$P_t = k_{shape} \times k_{edge/end} \times k_{ecc} \times P_{nom} \qquad\qquad \text{Eq. 6}$$

where k_{shape}, $k_{edge/end}$ and k_{ecc} are modification factors which account for variations in shape of the bolt or adhesive patch-plate, edge/end distances for both corner and edge locations and eccentricity of load respectively. While P_{nom} is the load bearing capacity of a bolted or adhesive connection with a fixed set of parameters and fixed mechanical properties of the liner and the adhesive.

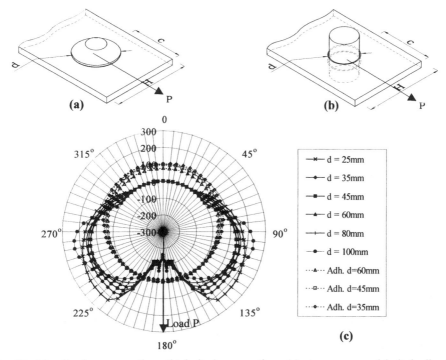

Fig. 12. (a) adhesive connection; (b) bolted connection; (c) stresses around bolt / adhesive perimeter for c/H = 100/150.

The comparative performance of the bolted and adhesive connections is best observed by overlaying the major principal stresses obtained from the FE analyses of the bolted connections onto those obtained for the adhesive connections. The resulting *radar* graph (Figure 12) represents the principal surface stresses around the hole or adhesive perimeter, whereas the linear graph represents the surface stresses with increasing distance from the centre of the connection (Figure 13).

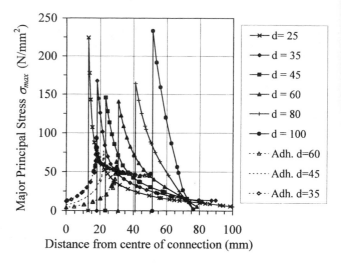

Fig. 13. Maximum surface stresses with increasing distance from centre of connection for $c/H = 100/150$.

From the analysis of the results it is evident that the maximum principal stresses induced by adhesive connections is in the order of 31% to 58% of the stress induced by the bolted connections. Furthermore, from the peak stress results it is possible to determine optimum geometrical arrangement that minimises surface tensile stresses.

From the FE analyses and corresponding results, it was also possible to conclude that:

(i)   For the bolted connections investigated, the tight fit through-bolt with a nylon liner results in the lowest major principal tensile stress. The optimum hole diameter depends on the end and edge distances adopted. For a $c/H = 100/150$ the optimum hole diameter is approximately 60mm.

(ii)   For the adhesive connections investigated, a tapered-edge circular coin with a low modulus of elasticity adhesive produces the lowest major principal tensile stress. However, the strength of the connection depends on the yield strength of the adhesive. It may be therefore more advantageous to opt for a stiffer adhesive with a higher yield strength.

(iii)   All double-lap adhesive connections investigated produced substantial lower tensile stresses in the glass when compared to the corresponding bolted connections.

Strength Predictions

Failure load predictions of the proposed test specimens were computed by performing Finite Element analysis and the applying *Glasstress* to determine the effective uniformly applied stress, σ_p, at load increments of 5kN. This applied stress was then converted to a cumulative probability of failure by using the Weibull's relationship. (Figure 15).

Experimental Investigations

The bolted test specimens (Figure 14) consisted of 6mm thick fully tempered glass (σ_r = 120N/mm^2) with 58mm diameter through bolts and a brass or PTFE liner. Failure occurred without warning in the 25kN to 32.3kN. The results show excellent agreement with the predicted failure loads (Figure 15).

A series of steel-to-glass single-lap adhesive tests were performed to identify a suitable adhesive. The best performing adhesive was a two-part toughened acrylic as it combined the high strength associated with single part acrylics and an amount of plastic deformation prior to failure generally displayed by modified epoxies.

The double-lap adhesive specimens failed at high loads in the order of 3 times the failure load of the bolted specimens. The number of double-lap adhesive tests performed is insufficient to prove the accuracy of the strength predictions, however, the results plotted in Figure 15 are in good agreement with the *Glasstress* predictions.

Fig. 14. Double shear bolted test set-up.

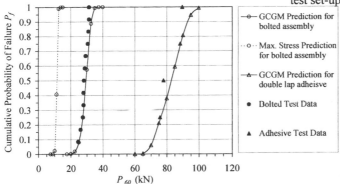

Fig. 15. Strength predictions and test results for bolted and adhesive assemblies.

CONCLUSION & ON-GOING RESEARCH

The proposed glass design method was successfully used to determine the failure strength of glass panels. The method was also used to determine the behaviour and failure load of existing bolted connections in glass, to propose an alternative adhesive connection and optimise bolted and adhesive connections in glass. *Glasstress*, the computer algorithm for determining the applied effective stress within the proposed method, was also successful in reducing the computation time required for the rigorous design of glass elements. Furthermore, the application of the proposed glass design method allows the accurate prediction of glass failure thus facilitating the appraisal of the contribution of glass to the load bearing strength of a space structure.

On-going research to supplement and verify the proposed glass design method includes a parametric analysis of structural glass plates within a space structure that are subjected to a combination of in-plane and lateral loads, additional investigations on single-lap and double-lap adhesive connections and load bearing cruciform glass-adhesive columns.

REFERENCES

1. CEN/TC129/WG8 *'Glass in Building – Design of Glass Panes – Part 1: General Basis of Design'*, Draft European Standard prepared by technical committee CEN/TC129, European Committee for Standardisation, Brussels, 1997.
2. BUTTON D. & PYE B. (eds.), *'Glass in Building'*, Butterworth Architecture, UK, 1993.
3. PILKINGTON GLASS CONSULTANTS, *'Design of Glass'*, Recommendations for the design of glass, Pilkington UK Ltd., 1997.
4. INSTITUTION OF STRUCTURAL ENGINEERS, *'Structural Use of Glass in Buildings'*, The Institution of Structural Engineers, London, 2000.
5. GRIFFITH A. A. 'The Phenomena of rupture and flow of solids', *Theoretical Transactions of the Royal Society of London*, 221: 163-179, 1920.
6. LAWN B.R., HOCKEY B.J. & WIEDERHORN S.M. 'Atomically sharp cracks in brittle solids – an electron microscopy study using indentation flaws', *J. Mater. Sci.*, 15, 207, 1980.
7. LAUFS W. *Private correspondence*. Mauro Overend with Wilfried Lanfs RWTH, Aachen, 1998.
8. OVEREND M. *'The Structural Use of Glass'*, MSc dissertation, University of Surrey, UK, 1996.
9. PYE A. *'The Structural Performance of Glass-Adhesive T-Beams'*, PhD Thesis, University of Bath, UK, 1999.
10. VEER F.A. & PASTUNINK J.R. 'Developing A Transparent Tubular Laminated Column' In: *Proceedings of Glass Processing Days*, Tampere, Finland, 1999.
11. BEASON W.L. AND MORGAN J.R. 'Glass Failure Prediction Model', *ASCE J. Struct. Eng.*, 110(2): 197-212, 1984.
12. CHARLES R. J. 'Static Fatigue of Glass 1&2', *J. App. Phys.*, Vol. 29, No. 11, 1549-1560, 1958.
13. BROWN W. G. *'A Practicable Formulation for the Strength of Glass and its Special Application to Large Plates'*, National Research Council of Canada, Publication no. NRC 14372, Ottawa, Nov 1974.
14. DALGLIESH W. A. & TAYLOR D.A., 'The strength and testing of window glass', *Canadian Journal of Civil Engineering*, 17, 752-762, 1990.
15. NORVILLE H. S., BOVE P.M. & SHERIDAN D.L. *'The strength of new thermally tempered window glass lites'*, GRTL, Texas Tech University, TX, 1991.
16. SEDLACEK G., BLANK K. AND GUSGEN J. 'Glass in structural engineering', *The Structural Engineer*, 73(2), Jan, 17-22, 1995.
17. FISCHER-CRIPPS A.C. & COLLINS R.E. 'Architectural glazings: Design standards and failure models', *Building and Environment*, 30, 1, 29-40, 1994.
18. PORTER M. I. & HOULSBY G.T. 'Development of Crack Size Limit State Design Methods for Edge-Abraded Glass Members'. *The Structural Engineer*, 79(8), 29-35, 2001.
19. OVEREND M., BUHAGIAR D. & PARKE G.A.R 'Failure prediction-What is the true strength of glass?' In: *Proceedings of Glass in Buildings*, Bath, 1999b.
20. OVEREND M. *'The Appraisal of Structural Glass Assemblies'*, PhD Thesis, University of Surrey, UK, 2002.
21. BEHR R.A., KARSON M.J. & MINOR J.E. 'Reliability Analysis of Window Glass Failure Pressure Data', *Structural Safety*, 11: 43-58, 1991.
22. WEIBULL W. 'A Statistical Distribution Function of Wide Applicability', *J. Applied Mech.*, 18, 293-297, 1951.
23. BEASON W.L. *'A failure prediction model for window glass'*, NTIS Accession no. PB81-148421, Institute for Disaster Research, Texas Tech University, 1980.
24. RAMM E. AND BURMEISTER A. 'Glass as Structure' *Proceedings of Int. IASS Symposium on Design Construction Performance and Economics*, Singapore 10-14 November, 1997.
25. BALDACCHINO D. *'Adhesive-Bolted Connections in Glass'*, Undergraduate Dissertation, Dept. Of Building Engineering, University of Malta, 1999.

Making double curved forms with the use of a 3D fabric

D. C. PRONK
Delft University of Technology, Faculty of Architecture
S. L. VELDMAN
Delft University of Technology, Faculty of Aerospace Engineering
R. HOUTMAN
Delft University of Technology, Faculty of Civil engineering and Geosciences.

INTRODUCTION

Inflatable structures have seen applications in area such as rescue equipment, civil-, maritime- and aerospace structures. Applications are derived from unique advantages that inflatable structures offer. Inflatable structures provide: low structural weight, low storage volume, ease of deployment, and low manufacturing cost (Ref. 9). For space applications these advantages result in significant mission cost reductions in some cases up to six times. (Ref 10). The requirements to space applications are very stringent because of high cost involved. Therefore research encompasses area to ensure a successful mission. Typical space research area membrane surface accuracy and deployment modelling.

Inflatable structures in civil engineering can be divided into two categories: air-supported structures and air-inflated structures. An air-supported structure can be regarded as a single membrane barrier structure i.e. only a single membrane separates the atmosphere from the internal atmosphere. The membrane is held in place by and internal overpressure. From the nineteen-sixties onwards these structures have been used for sport stadiums, air-houses, and exhibition halls. The following figure shows an example of an air-supported structure. Fig 1-2

Fig. 1 and 2 The Pontiac silverdome.

The Pontiac silverdome has got a 10 acre Teflon-coated Fiberglas roof. It was built in 23 months time at a cost of $ 55.7 million. The first game was played in August 1975.

Air-inflated structures are built up from an inflatable tubular support structure. The pressure in the interior is the same as the atmospheric pressure so there is no need for special air-locks.

Typical applications for structures like these are emergency shelters. The objective of this paper is to demonstrate new possibilities of inflatable technology in civil engineering. The inflatable technology is used as a building method. A sandwich blob structure can be created using an inflatable mould or a double wall can be used that is inflated with foam. This paper focuses on the latter building technology.

BLOB ARCHITECTURE

The similarity between form active structures (Ref. 2) like tent structures and pneumatics on one hand and blobs on the other hand is so substantial that it is obvious that we should try to make a blob with the same techniques. In 1994 writes K. Michael Hays (Ref. 3) that in reaction to fragmentation and contradiction there is a new movement in architecture which propagates a combination not only in form but also between media like film, video, computers, graphics mathematics and biology. He recognises that architecture under the development of increasingly complexity of information and communication has chanced into information and media. This development has lead to a kind of smoothness, which is called blob architecture. Fig 3-4

Fig. 3 and 4 by Michael Bittermann.　　　　　　　　　　　Fig 5. by Eckert Eckert Architekten AG.

The characteristics of a blobs are; smoothness, irregularity and having a double curved skin. Frei Otto already demonstrated the possibilities of influencing the form of pneumatic constructions by stretching nets and cables over it (Ref. 8). Another example of manipulating a tensile-form is the combination of cloth and a pneumatic structure in a blobby design. (Ref. 4) For instance in the floating theatre at the Expo 1970 in Osaka, designed by Yutaka Murata. One of the latest examples of transforming the form of a pneumatic construction is the tensile-structure for the Swiss pavilion (Fig. 5) The edges of the structure are transformed by using bending stiff elements.

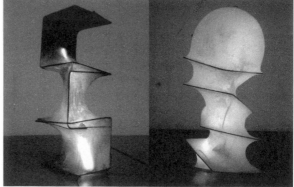

Figs. 6, 7 and 8 by Arno Pronk.

At the Technical University of Delft we have formed a group who wants to take the challenge of finding a way to make blobs. This group exists out of Professors researchers and students. In our research we made a model from balloons and a wire-frame in a panty. In this way it is possible to make all kind of forms. (Fig 6-8) We did some experiments with the possibilities of this technology. After modelling the shape we rigidized the form with glue.

With some students we had a module dealing with the problem of realising a blob like structure. One of the designs (Fig 9) influenced the form of the pneu by making as much as possible strings at the surface of the pneumatic structure. The strings where grouped and bound to an inner frame. The principal of this pneumatic structure works like a parachute.
If at the opposite of the outside of structure is put an inner structure. And if the strings of both sides are bound together there becomes a sandwich.

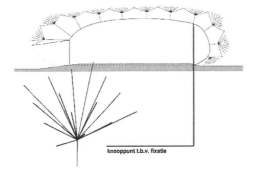

Fig. 9. by Paddy Sienwerts.

3D FABRICS
In order to inflate a dual wall structure, both walls need to be fixated. A 3D fabric provides a suitable way of fixating both facings. Vertical drop treads ensure that the distance between each facing remains constant.

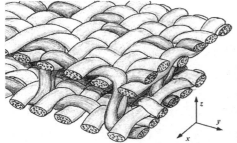 

Fig. 10. 3D fabric layout (Holtsmark, 2001) Figure 11 Goodyears Inflatoplane.

The picture above shows a layout of a parabeam 3D fabric. (Ref. 5) The distance between each facing can be varied to suit the application. When both facings are coated, an airtight space is obtained which can be inflated by either air or foam. Goodyear already produced an inflatable aeroplane using rubberised airmat in the mid nineteen-fifties. It is an example of the high-performance applications of a 3D fabric. (Ref.1)

Inflation of a 3D fabric can be done by a variety of media such as air, water or foam. A sandwich construction can be created when foam is used as an inflation medium. Research to use of foam inflation methods has been applied in space industry. (Ref. 7) In aerospace-engineering foam inflation is a way of rigidizing the structure to prevent the need for inflation gas for longer periods of time. A variety of foams can be applied but main requirement to the foam is that it has a long reaction time to allow for controlled inflation of the product. The inflation strategy should be such that the entire product is filled with foam.

Another group of students looked at the 3D fabrics. This material is used in lightweight-structures since a long time (Ref. 6). Pneumatic use of 3d fabrics is mostly used as a plane structure. We looked at the possibilities of making single and double curved forms by cutting out small pieces at one site of the fabric one side is becoming smaller as the other one and the result of that is a curved form. After fixing the cutting parts together we got the following result. (Fig.12-13) By cutting out a small part in the middle we also cut through the strings in that part. As a result of that the surface of the structure gives a bubble at the other side of the cut-out.

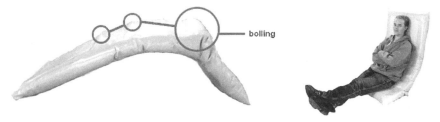

Figs.   12-13 by Alex de Ruiter, Jasper Felsch and Haiko Cornelissen.

A DOUBLE CURVED 3D FABRIC
After this experiment we studied the possibility of making a double curved model. The 3d model of the blob is generated with the software package EASY (Fig 14). The boundary of the blob is at one horizontal plane, which creates a total flat area when only prestress in the membrane is applied. To influence the shape of the membrane in a form-active way, the surface is loaded with normal pressure during formfinding. The pressure is not equally distributed on the surface, which creates the desired organic shape (Fig 14).

Fig. 14. by Freek Bos.   Fig. 15. by Rogier Houtman.

The material that is to be used has a double layer structure, so the model used to create cutting patterns should have a double layer structure too. The distance between the two layers has a fixed length of 100 mm. To model the inner layer, the outer layer is scaled in such a way that the inner layer approximately is 100mm away from the outer layer. Because the shape is not a sphere, the distance of each point to the centre is not the same and therefore the position of the innerlayer is only approximated.

The cutting patterns are determined by means of geodesic lines. Because geodesic lines can have many positions on spherical-like shapes, only half of the blob shape is used to create the patterns. Afterwards they are mirrored to obtain the full cloth. Picture 15 shows the result of the 3d cutting patterns.

The 3d patterns are transformed into 2d strips by means of the Cut&Grow procedure of EASY. The result of this is a set of individual strips which in normal situations are welded together to form the surface envelope. In this case, while using the 3d fabric, a different approach is taken. The fabric is not actually cut into the different patterns. The strips are shifted as much as possible towards each other, which gives a pattern layout according to fig 17. The remaining gaps between the strips are folded away and covered by a straight strip of fabric. By doing this, the fibres in the fabric are not cut trough and consequently at the other side of the fabric no bubbles occur like the ones in picture 12. Picture 18 shows the result of the experiment. The inside of the blob looks as we expected. The outside is less smooth. A reason for this can be the fact that the scaling of the outer fabric was not completely appropriate and consequently the outer fabric is too large.

Fig. 16. Making the cutting pattern.

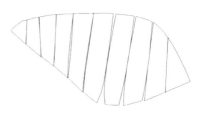

Fig. 17.   by Arno Pronk.

Fig. 18. Inflating the model.

ACKNOWLEDGEMENTS
The authors wish to thank Firma Buitink for their contribution in carrying out this experiment.

REFERENCES
1 Beukers A, O V Molder, C.A.J.R. Vermeeren; Inflatable structures in space engineering. Journal of the int. association for shell and spatial structures, 2000, no 3, p. 177-190.
2 Engel H. Structure Systems (1997) Stuttgart.
3 Hays K. M. I'm a victim of this song / good spirit come over me. 1994.
4 Herzog T. Pneumatische Construktionen, Bauten aus Membranen und luft 1976.
5 Holtsmark, A. The conceptual design of a continuous suction laminar flow control system. [M.Sc. Report] Faculty of Aerospace Engineering, 2001.
6 Huybers P. See Through Structuring page 16 –21 July 1972 Delft.
7 Jenkins, C. H. M. Gossamer spacecraft: Membrane and inflatable structures technology for space applications. Reston: American Institute of Aeronautics and Astronautics. 2001.
8 Pronk A. Veldman S. Making Blobs with aircushions Proceedings of the international symposium on Lightweight structures in civil engineering Warsaw 2002.
9 Veldman, S. L., Mölder, O.V., Lightweight architecture. Architectural Design. John Wiley & sons, London UK. Jan-Feb. 2002.
10 Veldman, S. L., Vermeeren, C. A .J. R., "Inflatable structures in aerospace engineering – An overview" Proceedings of the European conference on spacecraft structures, materials and mechanical testing, Noordwijk, the Netherlands 29 November – 1 December 2000, (ESA SP-468, March 2001), 2001.

The stainless steel structure of a sport stadium in Quart

LUIS SANCHEZ-CUENCA. Dr Architect.Universitat de Girona. SPAIN

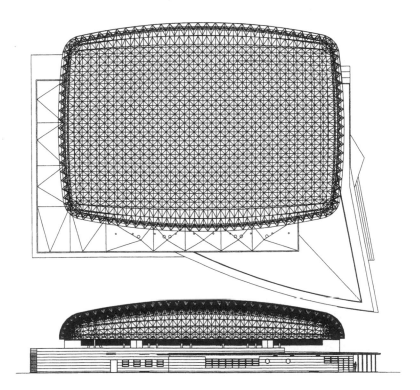

1. THE BUILDING AND ITS ENVIRONMENT

Quart is a small village of 2500 inhabitants, close to the town of Girona in the Northeast of Spain.

The Sports Hall is located in the middle of a forest park. It is a multipurpose pavilion that contains 4 changing rooms for sport participants, other 2 for referees, and rows of seats for 250 spectators. It also contains a gymnasium and different service spaces as store rooms, offices, a bar, and toilets. The closed built surface is about 2800 m2.

The sports arena and the rows of seats have been covered by a geodesic dome specially designed for this building. It will be described later in detail. In Girona the fire regulations are specially strict, in such a way that they made unfeasible to build an interior spatial truss structure. The solution has been to build the structure outside and hang the roof from it, so that this roof would act as a barrier to protect the structure against the fire.

This roof is composed (from the outside to the inside) of a flexible sheet of FPA, an insulation of 4 cms of rockwool, and a standard galvanised nerved plate supported on beams made of omega shaped galvanised steel 3 mm thick. Below it there is a rockwool false roof 85 mm thick. This set was successfully proved against the fire in a homologated official laboratory.

The structure is a double layer spatial grid. Since this structure was to be built outside it was made of stainless steel of quality AISI 304. All the elements of the structure such as bars, screws, nuts, etc., were made of this inox steel. These bars were preconformated with a hole in each border.

The rest of the building is a standard construction with reinforced concrete in pillars and slabs, and walls made of local brick.

2. GEOMETRY OF THE STRUCTURE

Geodesic domes are , in general, spherical surfaces. As a consequence, their basis are circular or regular polygons such as squares, pentagons, etc, able to be inscribed in a circle. In our case the covered space of the sport hall is a multipurpose arena which shape is not circular, neither a square nor other regular polygon. It is a rectangle, so we wanted to design a geodesic dome on a rectangular basis. At the same time the usual geodesic domes, being spherical, are rather high. Their height usually corresponds to the radius of the sphere. In our case, we also wanted to make a *low profile* geodesic dome.

To obtain a geodesic dome with these two conditions (a rectangular basis and a low profile) we used a technique that consists on what can be called to make a *carpanel dome*. In fact it consists on making in 3D the same operation we do in 2D in the *carpanel arc*. For making a *carpanel arc* we reduce the height of an arc by curving the borders of the arc with other arcs of smaller radius. In the Quart dome we have reduced the height of a spherical dome by curving their lateral parts, which correspond to the sides of a rectangular basis. Then, the two characteristic cross sections of the dome are two *carpanel arcs*. The final result is a *low profile geodesic dome on a rectangular basis*.

It could be interesting to mention that it is possible to use this technique with any polygonal base, of any number of sides, whether they are identical or not. That is to say that any regular or non-regular polygon can be used as the basis for a similar dome to the one we are describing. Moreover, it will be possible to adapt its final height to our needs.

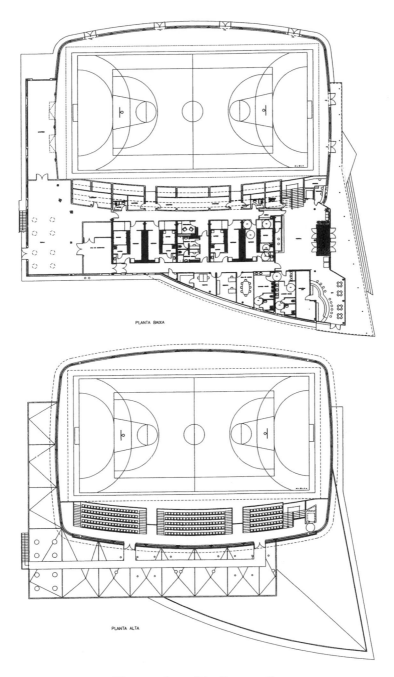

PLANTA BAIXA

PLANTA ALTA

The two plans of the Sport stadium.

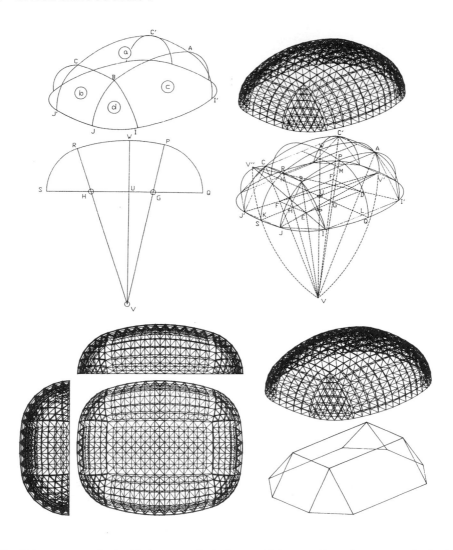

The full geometry of a double layer geodesic dome with low profile and a rectangular basis.

The dome is composed of a central piece, 4 lateral pieces and 4 corners. The central piece and the corners are spherical. The lateral ones are compound surfaces, but made with circle arcs in one of the two grid directions. In all of these pieces the structural grid lines are geodesic lines. That is to say, they are lines of minimal length. This is why they are the optimal lines for distributing the external forces. These geodesic lines have been obtained by projecting the straight lines of a poliedric grid on circle arcs of a sphere.In the central piece

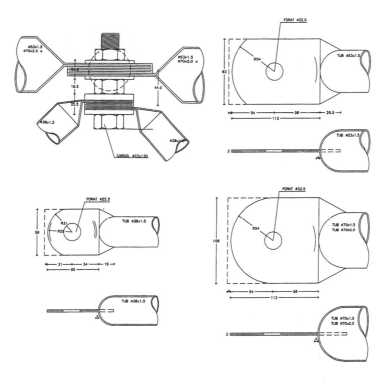

Constructive details of the bars of the stainless steel structure.

and in the corners, all of them spherical, the projection centres are fixed in the centres of those spheres. In the lateral pieces the projection centres are mobile and tour at the same time as those lateral pieces rotate. As a result, the obtained grid is a continuous double curvature surface being, at the same time, absolutely geodesic. The curve radii vary between 100m and 3.20m in the internal layer. And between 101.20m and 4.40m in the external layer. The constant distance between the two layers is 1.20m.

3. THE STRUCTURE AND ITS CONSTRUCTION

The material for the structure is stainless steel. Fine, but how could we make an stainless steel spatial structure at a reasonable price?. The answer is that we have used an extremely simple joint. In fact the bars are jointed together by one single screw in each node. So the bars are, due to its inox material, rather expensive, but the cost of the node is minimal, and the conjunct has been made at a cost of about 120 Euros/m2.

The bars are inox tubes of two basic diameters. One of 53 mm for the two layers. The other of 35 mm for the diagonals. There is a third diameter of 70 mm for the bars that are super stressed in any of the two layers. The bars were preconformed in factory with a 22.5 diameter hole in each border. The screws were of 22 mm diameter. Some special pieces were designed in order to assure the correct behaviour of the joints.

Since the grid is bi-directional, the nodes are composed by 4 layer bars and 4 diagonal bars. Between them there is a nut. The axis of the 8 bars practically converge in a geometrical point.

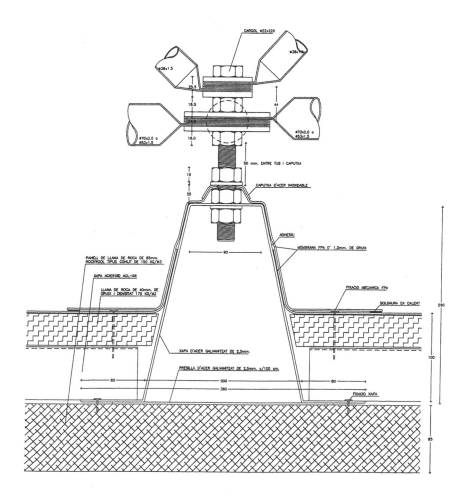

Detail of the connection between the roof and the structure.

Some aspects of the construction of the structure on the ground.

The geodesic dome in its final position.

Due to the fact that the length of every bar was fixed by the two holes in their borders, the construction of the structure had to begin by the highest node, going down toward the lower nodes. The assembly was made on the ground by only three workers, and without any special equipment. The full structure was built in two identical half parts. Once erected, the two half parts were *sewn* to complete the dome.

4. SUMMARY OF THE MAIN CHARACTERISTICS
Full geodesic dome on a rectangular basis.
Double layer 1.20 m thick.
Fully made with AISI 304 inox steel.
Diameters of bars: 35x1.5 mm, 53x1.5 mm, 70x2 mm.
Joints with 22 mm diameter screws.

Developed surface	2000 m2
Covered surface	1550 m2
Main transversal spans	48 m and 34.5 m
Number of bars	10456
Number of different length bars	625
Medium length bar	1450 mm
Number of joints	2565
Total weight of the structure	30000 kgs

The construction team
AGT was the general contractor
TROCOMPSA prepared the inox bars
CASANOVAS made the structure
ACIEROID made the roof

The technical team
Luis Sánchez-Cuenca, Dr. Architect
Miquel Llorens, Arch. structural consultant
Eduard Bonmatí, Technical Architect
Francesc Rodriguez, Technical Architect

Guildford, august 2002.

Effect of support flexibility in circular suspended cable roofs

J. K. JAIN
Department of Applied Mechanics, Maulana Azad College of Technology, Bhopal, India

SYNOPSIS
In the present paper, a formulation for analytical study of cable roof system is presented wherein the flexibility of supports bounding the roof is accounted for. The analytical work is based on Finite Element method and accounts for non-linearity for of cable assembly using Newton Raphsons method. The supporting structure is however, treated as being linear. Using this approach, a generalised computer program has been developed for analysis of a bicycle wheel type cable roof. A parametric study by changing stiffness of beam and load intensities has been carried out on a experimental model of 3-meter diameter and the effect of flexibility of supporting structure on maximum deflection, prestress in cable roof and bending moment in the ring beam has been studied in particular. Finally the developed algorithm has been used to verify the maximum deflection of Auditorium Roof, Utica, N.Y., U.S.A..

INTRODUCTION
Tension structures using high strength steel cables as structural elements are being increasingly employed to provide large unobstructed spaces required for sports halls, exhibition halls, aircraft hangers etc. These structures, not only are aesthetically pleasing, but are structurally sound and economical. Cable roofs are minimal structures using only a small proportion of total material used in the structure, a major proportion being utilized in the support and anchor system. The economics of a cable roof project is affected to a large extent by the design of roof supports and anchor systems. Practical evidence suggests that the supporting structures designed are usually heavy and rigid. In such cases the behavior of the cable roof under load will hardly be affected by the nature of the supports and this explains the comparative lack of activity in the study of the problem. In the interest of economy of material, however, lighter supporting systems need to be considered thus requiring an understanding of the nature and extent of interaction between the roof and the support.

DEVELOPMENT IN ANALYSIS
The development in the field of theoretical analysis of tension structures started only in about 1960. The advent of high speed electronic computers during the past two decades has enabled engineers to carry out a considerable amount of research in the filed of design and analysis of these structures. Cable structures are very light and flexible. As such they can undergo appreciable deformations when subjected to external loading. In order to define the position of equilibrium, the method of analysis should, therefore, cater for the possible change in geometry due to load and the material nonlinearity introduced due to nonlinear stress-strain behavior of material of construction used.

It is evident from existing theoretical work that most of the development in the analysis of suspension cable systems has been done using displacement approach, because of its simplicity in application. Very little work is reported using the flexibility (or force) approach. During the last decade the finite element method has been used in solving such systems. Most analytical studies of cable roof systems have assumed the supporting structure of the roof to be usually heavy and rigid. In the interest of economy of material, however, lighter supporting systems need to be considered thus requiring an understanding of the nature and extent of interaction between the roof and support. Evaluation of the effect of support flexibility has therefore engaged the attention of researchers during the last decade, or so. In the discrete approach it is possible to idealize the supporting structure by beam elements which in turn will interact with the cable elements. In this case it will be necessary to carry out a combined analysis of cable assemblies and beam elements (which are comparatively stiff).

FINITE ELEMENT METHOD
The finite element method is essentially a generalization of standard structural arfalysis procedure, which permits the calculation of stresses and deflections in elastic continuum or otherwise. The use of finite element method in solving linear problems is quite simple but the linear theory will yield satisfactory results for a very restricted class of problems, and there are some problem areas for which linear theory is totally inadequate. Nonlinear problems, for which finite elements analysis proves a valuable technique arise in many realms of engineering. This nonlinear behavior may be due to large displacements (geometric nonlinearity) or due to inelastic material behavior (material nonlinearity).

FORMULATION OF ANALYSIS

Stiffness Matrix for the Cable-Truss Member
For an axial element let ε^o be the initial strain at some intermediate tension P and ε^a be the additional strain as further deformation takes place. The total strain ε is given by:

$$\varepsilon = \varepsilon^o + \varepsilon^a \tag{1}$$

Let u, v and w be the displacement along the three coordinate directions. The strain displacement equation can be written as:

$$\varepsilon^a = du/dx + \tfrac{1}{2}(dv/dx)^2 + \tfrac{1}{2}(dw/dx)^2 + \tfrac{1}{2}(du/dx)^2 \tag{2}$$

Since only small strains are involved, the term $(du/dx)^2$ is dropped as compared to (du/dx) but since rotations are of interest the other terms are retained and the definition of strain is modified to:

$$\varepsilon^a = du/dx + \tfrac{1}{2}(dv/dx)^2 + \tfrac{1}{2}(dw/dx)^2 \tag{3}$$

The total strain energy increment U may then be written as:

$$U = \int_v \left(\int_{\varepsilon^o}^{\varepsilon^o+\varepsilon^a} \sigma\, d\varepsilon \right) dv \tag{4}$$

Where dv is the volume of the element. For linearly elastic material the integral of equation (4) can be integrated with respect to ε after introducing Hooke's law, $\sigma = E\,\varepsilon$. Further, substituting for ε^a from equation (3) and retaining only terms upto second order, the strain energy increment becomes:

$$U \;=\; \tfrac{1}{2} \int_0^L AE\,(du/dx)^2\,dx \;+\; \int_0^L AE\,\varepsilon^0\,\tfrac{1}{2}\,(dv/dx)^2\,dx$$

$$+\; \int_0^L AE\,\varepsilon^0\,\tfrac{1}{2}\,(dw/dx)^2\,dx \;+\; \int_0^L AE\,\varepsilon^a\,(dv/dx)\,dx \tag{5}$$

The first three terms of the right hand side will only contribute to the total stiffness $[K]$. The first integral yields the conventional small deflection theory stiffness matrix $[K^{(0)}]$. The second and the third integrals depend upon the initial stress and yield a second matrix called the geometric stiffness matrix $[K^{(1)}]$. The total stiffness matrix is given by:

$$[K] \;=\; [K^{(0)}] \;+\; [K^{(1)}]. \tag{6}$$

The matrices $[K^{(0)}]$ and $[K^{(1)}]$ are derived for a cable element in general and are modified in the global form using direction cosines. These matrices are updated in iterative procedure for obtaining optimum value of cable forces.

Stiffness matrix for a beam, beam-column element
The stiffness matrices for beam and beam-column elements are standard ones and are well documented in the literature. These matrices can be easily incorporated in the computer programs giving due regard to sign conventions employed.

Assembly of equations
The overall stiffness matrix is assembled according to sequence of nodes in the structure by adding the individual stiffness to the locations, in conformity with requirement of one-to-one correspondence between the nodes of the elements and those of the assemblage. In any structure a coefficient on the leading diagonal of the overall stiffness matrix represents the direct stiffness of a node i.e. the load required at the node to produce unit displacement of that node, all other nodes being fixed. The contributions to individual elements of the overall matrix are always such as to maintain symmetry, so that in practice it is needed only to consider the assembly of the leading diagonal coefficients and those above them the upper triangle as it is usually called.

In case of cable structures the overall assembly of stiffness matrix of the combined structure will be having some distinct features. Since the structure consists of cable elements having only the three translations, and beam elements having three translations and three rotations at each node, it is necessary to ensure that stiffness contributions are transferred at appropriate places while assembling the stiffness of cable-beam joints.

SOLUTION OF EQUATIONS
It is well known that for the solution of any nonlinear problem, the incremental or iterative method can be used. In the incremental method the total applied load is divided into number of increments, and a linear analysis is performed for each increment of load until the total load is applied. Each time the structure is solved, the current displacements are added to the

current coordinate points and a new geometry obtained for next increment of load. It is assumed that this series of linear solutions approximates the non-linear response. In the iterative method, loads are applied in one step and the displacements are improved iteratively. The initial set of displacement is usually assumed on the basis of the unstressed configuration of the structure. After every iteration and equilibrium of the structure is checked and the next iteration is carried out for the corrective loads, the procedure is continued until the corrective load is small percentage of the originally applied load.

Optimization of cable force

The iteratative procedure used for analysis is similar to the Newton-Raphson method used for solving nonlinear equations. The solution usually starts by assuming a set of displacement $[r^0]$. The stiffness equation for the operation would be:

$$[r] \quad = \quad [r^0] \quad + \quad [K^0]^{-1} \quad [p\text{-} p^0] \tag{7}$$

Where $[r^0] \& [r]$ are initial guess and final nodal displacements respectively

$[K^0]$ = The tangent stiffness matrix of the structure as already discussed

$[P^0]$ & $[P]$ are incremental load and total load applied respectively

The analysis is initiated by the use of initial stiffness, using new set of displacements to be used for the second iteration. The new stiffness matrix is generated and member forces are calculated. The applied loads and member forces are assumed at each joint and the result is the unbalanced load. If the unbalanced loads are large compared to applied load, then another iteration is performed. The procedure continues till the unbalanced load becomes small.

Computer Program

A mixed finite element computer program has been developed for determination of forces and moments in the ring beam elements and optimized cable forces in various cable elements.

MODEL FOR EXPERIMENTAL INVESTIGATIONS

A 3 meter diameter skeleton of circular roof structure consisting of eight radially spaced doubly convex trusses was selected for the experimental verification of analytical results evaluated for consideration of effects due to flexibility of the support. A cage type tension ring is provided for anchoring the cables at the centre of the roofs. The model reassembles a bicycle wheel. A general view of the model is shown in Fig-1.

Figure 1 : General view of the experimental model

This experimental structure is not a scaled model of any particular prototype. Cable tensions. beam thrust and moments are evaluated from the measurements made under symmetrical and unsymmetrical load intensities of 300, 400 and 500 N/m^2.

Design of model
The model design and fabrication is divided into three parts – i) the trusses, ii) the outer ring beam and iii) the central tension ring. The design has been done for the maximum load intensity of 500 N/m^2.

Trusses
Eight doubly convex radial trusses are provided in the roof model. The sag for the load carrying lower cables is fixed as span/20 = 15cm, whereas the sag for the upper cables is fixed as 10cm. Each truss is provided with 6 equally spaced struts. The pretensions in the upper and lower cables are determined as per normal methods considering various aspects of design and were calculated as 500 N in the lower cable and 750 N in the upper cable. A cold drawn high tensile steel wire of 2.05mm diameter is provided for lower and upper cables. It was also observed that the breaking strength was much higher than the required breaking strength of wires for expected load.

Six hangers are provided at a distance of 34.5cm c/c. The hangers are designed as compression members for maximum load caused due to pretension and dead load. 9mm diameter aluminium rods are used as hangers which gave allowable compressive strength much more than required.
The complete truss system after application of the maximum load was also tested for the maximum deflection due to symmetrical and unsymmetrical loads. It was observed that the deflections were within the permissible limits.

Ring beam
The outer ring beam has to be designed for radial loads of intensity 1250 N due to lower and upper cable, in addition to the tensions caused due to super imposed loads. A pretensioning sequence was also observed to determine the maximum thrust and moments due to pretensoining of the trusses. The circular ring beam was designed for maximum thrust of 3500 N and for a bending moment of 60000 N-cm arising through these calculations. A mild steel Indian Standard channel section ISLC 80 x 40 @ 70.5 N/m was used for the ring beam.

Central tension ring
A 24cm diameter cage type tension ring consists of two circular rings separated by spacers is used as central tension ring. The maximum values of axial tensions and bending moment for the two rings are evaluated in the same way as for the outer compression ring. A mild steel flat section of 31.75mm x 9.5mm is used for two rings. The tension ring is also checked against extreme combinations of tension and moment.

Fabrication of model
The various components of model are fabricated as per the design requirements. The significant points in the fabrication are:
1) The high tensile wire used for cables is provided with 8 BA threads on both sides for about 5cm. Square nuts made from high carbon steel are used for anchoring purposes.
2) The hangers are made with a provision of screw at the top and loading hook at the bottom. One important feature of the hangers is that holes for passing the cables are

drilled to the particular slope of cable at that point, which gave proper shape to upper and lower cables.

3) For supporting the 3m diameter ring beam 16 supporting columns made from mild steel channel section are used.

4) For pre-tensioning cables, pre-tensioning barrels as shown in Fig-2. are used. 32 such pre-tensioning barrels are used for pre-tensioning the cables at sixteen locations.

5) To have free lateral movement of the ring beam, the beam is supported on rollers. However to avoid accidental slipping of ring beam from rollers, push back type lateral supports have been provided as shown in Fig-3. at four equidistant points.

6) The central ring is like a cage. The 20cm diameter, 2 rings are fabricated from mild steel flats. Sixteen spacers are provided, which keep the two cables passing through upper and lower rings. The general view of the central tension ring is shown in Fig-4.

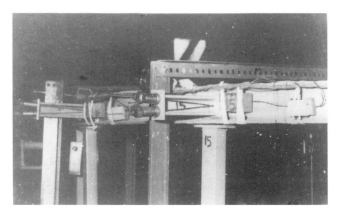

Figure 2 : A view of pre-tensioning barrels

Figure 3 : A view of lateral supports for ring beam

Figure 4 : A view of central tension ring

Instrumentation

A total 64 strain gauges are mounted on the model, 32 on the cables and remaining 32 on the ring beam for evaluation of cable tensions and ring beam thrust and moments. The strain gauges are connected to digital strain indicator through multiway junction box. 68 dial gauges are also used for the measurement of vertical deflection of cables and radial displacement of ring beams.

NUMERICAL STUDY

Two numerical examples have been studied – on a model suspended cable roof and another of an auditorium roof built at Utica, New York. Parametric study have been carried for the model roof under symmetrical loading due to dead or live load as well as unsymmetrical loading which could occur due to wind or snow.

Model Structure

The effect of support flexibility has been studied considering the range of stiffenesses of the supporting ring beam in terms of change in cable tension and deflection – and thrust and bending moment in beam for symmetrical and unsymmetrical load intensities of 300, 400 and 500 N/m2. The behaviour of the model has also been studies with constant beam stiffness but for different load conditions. The idealized structure consists of cable elements and beam type elements is shown in Fig-5.

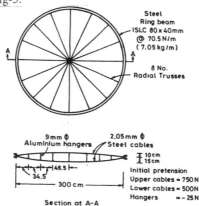

Figure 5 : Details of theoretical model adopted in the analysis

To study the effect of support flexibility on the behaviour of the structure. the beam stiffness has been varied taking stiffness of the model as reference value. Six values of beam stiffness have been selected and each case has been analyzed for 3 intensities of symmetrical and unsymmetrical loading. The values of maximum cable tensions for some typical members for various beam stiffnesses for theoretical model and experimental model are shown in Table-1 whereas a comparison of theoretical and experimental results for maximum deflection, beam thrust and beam moment for a symmetrical load intensity of 400 N/m^2 are shown in Table-2.

Table-1: Maximum cable force (N) in for various beam stiffness for symmetrical load

S. No	Load N/m^2	Tens -ions	A/Aref =0.1	0.2	0.5	1.0 (Model) Theore- tical	Experi- mental	2.0	5.0	∞
1.	300	H_u	512.6	565.8	602.2	615.3	610.0	685.0	626.1	628.7
	300	H_l	721.0	755.7	779.6	788.2	810.0	792.6	795.3	797.2
2	400	H_u	485.1	538.8	574.4	588.6	590.0	595.4	599.5	602.4
	400	H_l	826.7	861.1	884.8	893.3	910.0	897.7	900.3	902.2
3	500	H_u	460.6	514.9	552.0	565.3	570.0	572.2	576.3	579.7
	500	H_l	933.2	967.4	991.0	999.5	1010.0	1004.0	1006.0	1008.8

Table-2: Comparison of theoretical and experimental results of cable tensions, deflections beam thrust and moments for symmetrical/unsymmetrical load of 400 N/m^2

	Load on full span		Load on half span	
	Theoretical	Experimental	Theoretical	Experimental
Upper cable Force (N)	588.6	590.0	694.2	705.0
Lower cable	893.3	910.0	786.4	800.0
Max. deflection (cm)	0.788	0.757	1.56	1.45
Beam thrust (N)	3746.8	3990.0	3709.0	3990.0
Beam moment (N-cm)	-	-	3340.0	3591.0

Auditorium roof, Utica, New York
For verification of the combined analysis developed, a numerical study of Utica auditorium roof of 73.2 m (240 ft.) diameter of bicycle wheel type constructed at Utica, New York, as shown in Fig-6 is considered with following details.

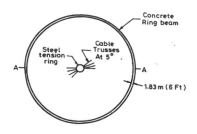

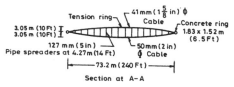

Section at A-A

Figure-6 : Details of Utica auditorium roof

Trusses

Spacings	=	5^0 c/c
Lower Cables	=	50mm ϕ, with initial pre-tension as 612.3 KN
Upper cables	=	41 mm ϕ with initial pre-tension as 739.8 KN
Struts	=	127 mm(5 in) diameter steel pipes at 4.27 m (14 ft) apart.
E for cables	=	16860 KN/cm² (24x10 psi)

Ring beam

Size	=	1.83 x 1.52 m (6 x 5 ft)

Loads

Dead Load	=	683.5 N/m² (14 psi)
Live load	=	1464.5 N/m² (30 psi)

The actual roof consisted of 72 trusses with each truss having 14 hangers. To reduce the computational work, each 3 trusses and each 2 hangers have been clubbed together in the present study.

Results

	Symmetrical load (2148 N/m²)	Unsymmetrical Load (2148 N/m²)
Max. Upper cable tension	507.07 KN	589.17 KN
Max. Lower cable tension	961.99 KN	879.90 KN
Max. Hanger compression	10.98 KN	14.56 KN
Central cable deflection	9.87 cm*	-0.72 cm
Max. Cable deflection	20.52 cm	27.47 cm

* Available result of central cable deflection = 10.0 cm.

CONCLUSION

The effect of support flexibility on performance of roof structure is marked for highly flexible supporting beams and tapers off rapidly as the stiffness of the support increases. It is seen that for the experimental model the inclusion of support flexibility influences the theoretical results for the cable assembly under superimposed loads to the extent of less than 6 percent. If the beam stiffness is doubled, the influence is approximately halved; and if the beam is made 10 times more flexible the effect borders around 15 percent. For the theoretical analysis of auditorium roof truss at Utica, NY, the analytical results agree closely with the available values. Also the inclusion of support flexibility on this structure has an effect varying between 5 to 7 percent.

The study suggest that the bicycle wheel type roofs as designed normally have supports which are rather rigid and the inclusion of support flexibility has therefore, only a small effect.

ACKNOWLEDGMENT

The paper is based on the findings of Ph.D. thesis of the author and therefore, the author wishes to extend his deep sense of gratitude to his guides Dr. Premkrishna, Dr. P.N. Godbole, and Dr. P.C. Jain of deptt. of Civil Engg., University of Roorkee, India.

REFERENCES

1. JAIN J K, 'Secondary Effects in Cable Roofs' Ph.D. Thesis, Civil Engg. Deptt., University of Roorkee, Roorkee, 1982.

2. JAIN J K, Godbole P N, Jain P C and Premkrishna, 'Finite Element Modelling of Cable Structure', Second International Conference on Computer Aided Analysis and Design in Civil Engineering, University of Roorkee, Roorkee,Feb. 1985.

Second-Order Cone Programming for Shape Analysis and Form Finding of Cable Networks

Y. KANNO, Dept. of Architecture and Architectural Systems, Kyoto University, Japan
M. OHSAKI, Dept. of Architecture and Architectural Systems, Kyoto University, Japan

SYNOPSIS

A method is presented for equilibrium shape analysis and form finding of cable networks considering geometrical nonlinearities as well as stress-unilateral behavior. The Second-Order Cone Programming (SOCP) problem which has the same solution as that of the minimization problem of total potential energy is solved to obtain the equilibrium configuration. Since no assumption of stress state is needed in the proposed method, no process of trial-and-error is required. Equilibrium shape or member initial lengths are obtained by solving an SOCP problem. Numerical examples are shown by using the well-developed software based on the primal-dual interior-point method.

INTRODUCTION

Various papers have been published concerning equilibrium shape analysis of finite dimensional cable networks or continuum cable models considering geometrical nonlinearity [1, 2]. Argyris and Scharpf [1] proposed the incremental method based on the tangent stiffness. Few studies, however, have explicitly dealt with the fact that cables are not capable of transmitting compression forces, which is referred to as the stress-unilateral behavior. Panagiotopoulos [3] formulated variational inequality of cable networks. The complementary energy principle for a cable member with distributed loading has been presented [4]. Atai and Steigmann [5] proposed a relaxed formulation for minimizing the total potential energy of cable networks.

In the existing methods based on the tangent stiffness, the stress-unilateral behavior is modeled by absence of stiffness of slackening members. Therefore, most of numerical procedures ever addressed except Cannarozzi [4] require an assumption whether each member will be in tensile or slackening state at the next step of the incremental computation. Since the assumed stress state may conflict with the results of obtained increments, the trial-and-error process which is similar to that of elastoplastic analysis should be carried out. Such assumption and trial-and-error process, however, sometimes cause the divergence of solution.

In this paper, a method is proposed for equilibrium shape analysis of cable networks with frictional joints based on Second-Order Cone Programming (SOCP) [6]. An SOCP formulation for form finding of cable networks is also presented. Recently, methods and theories of nonlinear optimization have received extensive interests to apply to structural engineering [2, 7]. Especially, SOCP is known to have various fields of application [8] and efficient polynomial-time algorithms referred to as the primal-dual interior-point methods [6].

Space Structures 5, Thomas Telford, London, 2002

The problem considered in this paper is non-linear because of the stress-unilateral behavior as well as large deformation. The frictional joints are assumed to obey Coulomb's friction law [7]. The difficulties of the considered problem possibly arise from the determination of stress states of cables, the singularity of the tangent stiffness matrix, and the non-differentiability of friction energy. To overcome these difficulties, an SOCP problem is formulated so as to have the same optimal solution as that of the minimization problem of total potential energy, and the SOCP problem is solved by the primal-dual interior-point method.

The cable networks are usually more flexible than other structures such as rigid frame structures and trusses. Moreover, the dependence of configuration on stress distribution prevents us from designing an arbitrary configuration of cable network. Therefore, the efficient form finding method is required to be developed which realizes the specified stress distribution at the equilibrium state. The force density method was proposed for the case such that the ratio of axial force to member length is specified for each member [9]. However, it may be desired in a practical point of view that the axial forces should be directly specified. Lewis and Gosling [10] proposed the dynamic relaxation method for form finding of cable networks and membrane structures. For the design of membrane structures, it is well known that the equilibrium shape with uniform stress distribution can be found as the stable minimal surface spanning a given closed boundary [10], and such a surface can be obtained by solving an SOCP problem [8]. Since a cable network can be regarded as an appropriately discretised fabric, it may be naturally conjectured that the form of cable network with uniform stress distribution at the self-equilibrium state can be obtained by minimizing the total member lengths. However, in the case of non-uniform stress distribution, it remains unclear if an appropriate minimal surface problem can be solved to obtain a solution of form finding problem of fabric or cable network [10].

In this paper, an SOCP problem is proposed for form finding of cable networks with specified axial forces. It shall be rigorously shown that the solution of the presented SOCP problem exactly satisfies the equilibrium equations considering finite rotations of members. The initial lengths of cables as well as the equilibrium configuration are obtained by solving the SOCP problem.

EQUILIBRIUM SHAPE ANALYSIS BASED ON SOCP

Consider a cable network in three dimensional space. A physically continuous cable connecting supports and pin-joints is simply referred to as *cable*, and assumed not to be capable of transmitting compression force. Each cable is divided into several *members* by *frictional joints*, which can move along a cable with tangential friction. Throughout the paper, we make assumptions of linear elastic material and small strain.

Let n^d denote the number of degrees of freedom of displacements. $x \in \Re^{n^d}$ and $f \in \Re^{n^d}$ denote, respectively, the vectors of coordinates of internal nodes that are not supported and the corresponding external dead loads. Let $x^0 \in \Re^{n^d}$ and $\bar{l}_i^0$ denote the vectors of coordinates of internal nodes and the unstressed length of ith member, respectively, at the initial state. Note that the initial state of the cable network is identified by x^0 and $\bar{l}^0 = (\bar{l}_i^0) \in \Re^{n^m}$, which does not satisfy the equilibrium conditions in general. In this section, we shall find x at the equilibrium state for the specified nodal coordinates of the supports, f, x^0, and $\bar{l}^0$.

Let F_h $(h = 1, \cdots, n^f)$ denote the tangential traction due to friction acting between the hth frictional joint and the corresponding cable, where n^f denotes the number of frictional joints. Suppose that the upper bound of the friction is given as $\bar{F}_h > 0$. z_h denotes the relative displacement of the hth frictional joint against the cable. We assume Coulomb's classical friction law [7]; i.e., F_h and z_h are assumed to satisfy

$$F_h = \bar{F}_h \implies z_h \le 0, \tag{1a}$$

$$|F_h| \le \bar{F}_h \implies z_h = 0, \tag{1b}$$

$$F_h = -\bar{F}_h \implies z_h \ge 0. \tag{1c}$$

The external work W_h^F done by friction F_h is written as

$$W_h^F = -\bar{F}_h |z_h| \quad (h = 1, \cdots, n^f). \tag{2}$$

Let n^m denote the number of members. l_i and $\bar{l}_i$, respectively, denote the lengths of the ith member at the deformed and the initial states; i.e.,

$$l_i = \|\boldsymbol{B}_i \boldsymbol{x} - \boldsymbol{b}_i^0\|, \quad \bar{l}_i = \|\boldsymbol{B}_i \boldsymbol{x}^0 - \boldsymbol{b}_i^0\| \quad (i = 1, \cdots, n^m). \tag{3}$$

Here, $\boldsymbol{B}_i \in \Re^{3 \times n^d}$ and $\boldsymbol{b}_i^0 \in \Re^3$ $(i = 1, \cdots, n^m)$ are constant matrix and vector, respectively. Note that (3) gives the geometrically exact relation between l_i (or $\bar{l}_i$) and $\boldsymbol{x}$. The relation between $\bar{l}_i$ and l_i is given by

$$l_i = \bar{l}_i + \boldsymbol{d}_i^\top \boldsymbol{z}, \tag{4}$$

where $\boldsymbol{d}_i = (d_{ih}) \in \Re^{n^f}$ $(i = 1, \cdots, n^m)$ is a constant vector defined such that d_{ih} is equal to either of $\{-1, 1\}$ if the hth joint can move along the ith member, otherwise $d_{ih} = 0$. Let l_i^0 denote the unstressed length of the ith member at the deformed state. By using the assumption of small strain and (4), we can see that

$$l_i^0 \simeq \bar{l}_i^0 + \boldsymbol{d}_i^\top \boldsymbol{z}, \tag{5}$$

is satisfied approximately. The elongation c_i of the ith member is written as

$$c_i = l_i - l_i^0, \tag{6}$$

where l_i is given by (3). Then the strain energy w_i is written in terms of c_i and l_i^0 as

$$w_i(c_i) = \begin{cases} \dfrac{1}{2} \dfrac{EA_i}{l_i^0} c_i^2 & (0 \le c_i), \\ 0 & (c_i < 0), \end{cases} \tag{7}$$

where E is the elastic modulus, and A_i is the given cross-sectional area of the ith member.

From (2), (3), (5), (6), and (7), the minimization problem of total potential energy can be written as

$$(\Pi_\mathrm{F}): \quad \min \sum_{i=1}^{n^m} w_i - \boldsymbol{f}^\top \boldsymbol{x} + \sum_{h=1}^{n^f} \bar{F}_h |z_h|$$

$$\left. \begin{aligned} \text{s.t.} \quad & w_i = \begin{cases} \dfrac{1}{2} \dfrac{EA_i}{l_i^0} c_i^2 & (c_i \ge 0), \\ 0 & (c_i < 0), \end{cases} \\ & c_i = \|\boldsymbol{B}_i \boldsymbol{x} + \boldsymbol{b}_i^0\| - l_i^0, \\ & l_i^0 = \bar{l}_i^0 + \sum_{h=1}^{n^f} d_{ih} z_h, \ l_i^0 \ge 0 \quad (i = 1, \cdots, n^m). \end{aligned} \right\} \tag{8}$$

Consider the following SOCP problem:

$$(\mathrm{P_F}): \quad \min \quad \sum_{i=1}^{n^m} w_i - \boldsymbol{f}^\top \boldsymbol{x} + \bar{\boldsymbol{F}}^\top \boldsymbol{\zeta}$$

$$\text{s.t.} \quad w_i l_i^0 \geq \frac{EA_i}{2} y_i^2, \ y_i + l_i^0 \geq \|\boldsymbol{B}_i \boldsymbol{x} + \boldsymbol{b}_i^0\|,$$

$$l_i^0 = \bar{l}_i^0 + \sum_{h=1}^{n^f} d_{ih} z_h, \ l_i^0 \geq 0 \quad (i = 1, \cdots, n^m),$$

$$\zeta_h \geq |z_h| \quad (h = 1, 2, \cdots, n^f),$$

$$\tag{9}$$

where the independent variables are $\boldsymbol{w} = (w_i) \in \Re^{n^m}$, $\boldsymbol{x} \in \Re^{n^d}$, $\boldsymbol{l}^0 = (l_i^0) \in \Re^{n^m}$, $\boldsymbol{y} = (y_i) \in \Re^{n^m}$, $\boldsymbol{z} = (z_h) \in \Re^{n^f}$, and $\boldsymbol{\zeta} = (\zeta_h) \in \Re^{n^f}$. Let $\boldsymbol{c}^{\mathrm{II}} = (c_i^{\mathrm{II}}) \in \Re^{n^m}$. In the similar manner to Lemma 4.2 in Kanno *et al.* [11], we can show that $(\widehat{\boldsymbol{w}}, \widehat{\boldsymbol{y}}, \widehat{\boldsymbol{l}}^0, \widehat{\boldsymbol{x}}, \widehat{\boldsymbol{z}}, \widehat{\boldsymbol{\zeta}})$ is an optimal solution of $(\mathrm{P_F})$ if and only if $(\boldsymbol{c}^{\mathrm{II}}, \widehat{\boldsymbol{l}}^0, \widehat{\boldsymbol{x}}, \widehat{\boldsymbol{z}})$ is an optimal solution of (Π_F) satisfying

$$\widehat{y}_i = \begin{cases} c_i^{\mathrm{II}} & (c_i^{\mathrm{II}} \geq 0), \\ 0 & (c_i^{\mathrm{II}} < 0). \end{cases} \tag{10}$$

Therefore, the equilibrium configuration can be obtained by solving $(\mathrm{P_F})$ instead of (Π_F). In this paper, we solve $(\mathrm{P_F})$ by using the primal-dual interior-point method to obtain the equilibrium configuration.

In order to investigate the property of the solution of $(\mathrm{P_F})$, the remaining part of this section is devoted to show the optimality conditions of $(\mathrm{P_F})$, which can be obtained in the similar manner to the case of eigenvalue optimization problem discussed by the authors [12]. Suppose that an optimal solution of $(\mathrm{P_F})$ satisfies $\widehat{l}_i^0 > 0$ $(i = 1, \cdots, n^m)$. By using the generalized Karush–Kuhn–Tucker conditions [13] and the convexity of $(\mathrm{P_F})$, we can see that $(\widehat{\boldsymbol{w}}, \widehat{\boldsymbol{y}}, \widehat{\boldsymbol{l}}^0, \widehat{\boldsymbol{x}}, \widehat{\boldsymbol{z}}, \widehat{\boldsymbol{\zeta}})$ is a global optimal solution of $(\mathrm{P_F})$ if and only if there exist $\widehat{k}_1, \cdots, \widehat{k}_{n^m} \in \Re$, $\widehat{q}_1, \cdots, \widehat{q}_{n^c} \in \Re$, and $\widehat{\boldsymbol{v}}_1, \cdots, \widehat{\boldsymbol{v}}_{n^m} \in \Re^3$ satisfying

$$\widehat{w}_i = \frac{1}{2} \widehat{k}_i \widehat{y}_i^2, \ \widehat{k}_i = \frac{EA_i}{\widehat{l}_i^0}, \ \widehat{q}_i = \widehat{k}_i \widehat{y}_i \quad (i = 1, \cdots, n^m), \tag{11}$$

$$\widehat{l}_i^0 - \bar{l}_i^0 - \boldsymbol{d}_i^\top \widehat{\boldsymbol{z}} = 0, \ \widehat{l}_i^0 > 0 \quad (i = 1, \cdots, n^m), \tag{12}$$

$$\widehat{y}_i + \widehat{l}_i^0 \geq \|\boldsymbol{B}_i \widehat{\boldsymbol{x}} - \boldsymbol{b}_i^0\|, \ \widehat{q}_i = \|\widehat{\boldsymbol{v}}_i\| \quad (i = 1, \cdots, n^m), \tag{13}$$

$$\widehat{q}_i(\widehat{y}_i + \widehat{l}_i^0) + \widehat{\boldsymbol{v}}_i^\top(\boldsymbol{B}_i \widehat{\boldsymbol{x}} - \boldsymbol{b}_i^0) = 0 \quad (i = 1, \cdots, n^m), \tag{14}$$

$$\sum_{i=1}^{n^m} \boldsymbol{B}_i^\top \widehat{\boldsymbol{v}}_i + \boldsymbol{f} = \boldsymbol{0}, \tag{15}$$

$$\left| \sum_{i=1}^{n^m} d_{ih} \widehat{q}_i \left(1 + \frac{1}{2} \frac{\widehat{y}_i}{\widehat{l}_i^0} \right) \right| = \bar{F}_h, \ \widehat{\zeta}_h = |\widehat{z}_h| \quad (h = 1, \cdots, n^f). \tag{16}$$

It is interesting to claim that $\widehat{q}_i$ and $\widehat{\boldsymbol{v}}_i$, respectively, coincide with the axial force and the internal force vector of the ith member. $\widehat{k}_i$ corresponds to the extensional stiffness of the ith member at the deformed state. We can see that the conditions (11) and (15) are constitutive laws and equilibrium equations, respectively.

Since $\widehat{y}_i$ is related to c_i^{II} by (10), $0 \leq \widehat{y}_i/\widehat{l}_i^0 \ll 1$ is obtained from the assumption of small strain. It is also easy to see that $\widehat{q}_\alpha > \widehat{q}_\beta$ if and only if $\widehat{y}_\alpha/\widehat{l}_\alpha^0 > \widehat{y}_\beta/\widehat{l}_\beta^0$. Hence, the first equality in (16) is reduced to

$$\left| \sum_{i=1}^{n^m} d_{ih} \widehat{q}_i \right| \leq \bar{F}_h, \quad \left| \sum_{i=1}^{n^m} d_{ih} \widehat{q}_i \right| \simeq \bar{F}_h. \tag{17}$$

Notice here that $\sum_{i=1}^{n^m} d_{ih} \widehat{q}_i$ corresponds to the residual nodal force; i.e., the difference of axial forces of two members on which the hth frictional joint can move. Therefore, (17) implies that the solution to (P$_F$) satisfies the friction law (1), and the magnitude of obtained friction at each joint is close to the upper bound $\bar{F}_h$.

FORM FINDING PROBLEM BASED ON SOCP

Consider the form finding of pin-jointed cable network with the specified axial force $\bar{q}_i > 0$ of the ith member at the self-equilibrium state; i.e., $\boldsymbol{f} = \boldsymbol{0}$. In this section, our purpose is to obtain the initial length l_i^0 of each cable and the equilibrium configuration $\boldsymbol{x}$. Letting $\widehat{\varepsilon} > 0$ denote the strain of the ith member, we obtain

$$\bar{q}_i = EA_i \widehat{\varepsilon} \quad (i = 1, \cdots, n^m).$$

It follows that the strain energy of the ith member is written as $w_i = EA_i \widehat{\varepsilon}^2 l_i^0 / 2$. Consider the following SOCP problem, which minimizes the total potential energy of the cable network:

$$(\mathrm{D}): \quad \min \ \sum_{i=1}^{n^m} \frac{1}{2} EA_i \widehat{\varepsilon}^2 l_i^0 \\ \mathrm{s.t.} \quad (1+\widehat{\varepsilon}) l_i^0 \geq \| \boldsymbol{B}_i \boldsymbol{x} - \boldsymbol{b}_i^0 \| \quad (i = 1, \cdots, n^m), \left.\right\} \tag{18}$$

where variables are $\boldsymbol{l}^0 = (l_i^0) \in \Re^{n^m}$ and $\boldsymbol{x} \in \Re^{n^d}$. The following theorem is our main result in this section:

Theorem 1. *Let $(\widetilde{\boldsymbol{l}}^0, \widetilde{\boldsymbol{x}})$ denote an optimal solution of* (D). *Consider the pin-jointed cable network with the initial unstressed length $\widetilde{l}_i^0$ of ith member, where $\boldsymbol{B}_i$ and $\boldsymbol{b}_i^0$ remain unchanged. Then, $\widetilde{\boldsymbol{x}}$ and $\widehat{\varepsilon}$, respectively, coincide with the nodal coordinates and the strain of the ith member of the cable network at the equilibrium state.*

Proof. First, we show the optimality conditions of (D) by using the generalized KKT conditions for non-differentiable convex optimization problems (See, e.g., Rockafellar [13, Theorem 31.3]). We can see that $(\widetilde{\boldsymbol{l}}^0, \widetilde{\boldsymbol{x}})$ is a global optimal solution of (D) if and only if there exist the Lagrange multipliers $\widetilde{q}_1, \cdots, \widetilde{q}_{n^m} \in \Re$ and $\widetilde{\boldsymbol{v}}_1, \cdots, \widetilde{\boldsymbol{v}}_{n^m} \in \Re^3$ satisfying

$$EA_i \widehat{\varepsilon} = \widetilde{q}_i \quad (i = 1, \cdots, n^m), \tag{19}$$

$$\sum_{i=1}^{n^m} \boldsymbol{B}_i^\top \widetilde{\boldsymbol{v}}_i + \boldsymbol{f} = \boldsymbol{0}, \tag{20}$$

$$\widetilde{q}_i \geq \| \widetilde{\boldsymbol{v}}_i \|, \quad (1+\widehat{\varepsilon}) \widetilde{l}_i^0 \geq \| \boldsymbol{B}_i \widetilde{\boldsymbol{x}} - \boldsymbol{b}_i^0 \| \quad (i = 1, \cdots, n^m), \tag{21}$$

$$\widetilde{q}_i (1+\widehat{\varepsilon}) \widetilde{l}_i^0 + \widetilde{\boldsymbol{v}}_i^\top (\boldsymbol{B}_i \widetilde{\boldsymbol{x}} - \boldsymbol{b}_i^0) = 0 \quad (i = 1, \cdots, n^m). \tag{22}$$

Consider the cable network where the initial unstressed length of the ith member is given as $\tilde{l}_i^0$. From Lemma 3.1 [11], we can see that $c_i^{\mathrm{I\!I}}$ and $\boldsymbol{x}^{\mathrm{I\!I}}$ are cable elongation and nodal coordinates at equilibrium, respectively, if and only if $(\hat{\boldsymbol{y}}, \boldsymbol{x}^{\mathrm{I\!I}})$ satisfying (10) is an optimal solution of the following problem:

$$
\left.
\begin{array}{ll}
\min & \displaystyle\sum_{i=1}^{n^m} \frac{1}{2}\tilde{k}_i y_i^2 - \boldsymbol{f}^\top \boldsymbol{x} \\
\text{s.t.} & y_i + \tilde{l}_i^0 \geq \|\boldsymbol{B}_i \boldsymbol{x} - \boldsymbol{b}_i^0\| \quad (i = 1, \cdots, n^m).
\end{array}
\right\}
\tag{23}
$$

Here, $\tilde{k}_i$ and y_i can be written as

$$
\tilde{k}_i = \frac{EA_i}{\tilde{l}_i^0}, \quad y_i = \varepsilon_i \tilde{l}_i^0 \quad (i = 1, \cdots, n^m).
\tag{24}
$$

By substituting (24) into (23), we have

$$
(\widetilde{\mathrm{P}}): \quad
\left.
\begin{array}{ll}
\min & \displaystyle\sum_{i=1}^{n^m} \frac{1}{2} EA_i \varepsilon_i^2 \tilde{l}_i^0 - \boldsymbol{f}^\top \boldsymbol{x} \\
\text{s.t.} & (1 + \varepsilon_i)\tilde{l}_i^0 \geq \|\boldsymbol{B}_i \boldsymbol{x} - \boldsymbol{b}_i^0\| \quad (i = 1, \cdots, n^m),
\end{array}
\right\}
\tag{25}
$$

where variables are $\boldsymbol{\varepsilon} = (\varepsilon_i) \in \Re^{n^m}$ and $\boldsymbol{x} \in \Re^{n^d}$. Let $(\tilde{\boldsymbol{\varepsilon}}, \tilde{\boldsymbol{x}})$ denote an optimal solution of $(\widetilde{\mathrm{P}})$. The equivalence of $(\widetilde{\mathrm{P}})$ and (23) guarantees that $\tilde{\varepsilon}_i$ and $\tilde{\boldsymbol{x}}$, respectively, coincide with the strain of the ith member and the nodal coordinates at equilibrium. It can be shown, from the KKT conditions [13], that the optimality conditions of $(\widetilde{\mathrm{P}})$ are obtained as (19)–(22) by replacing $\hat{\varepsilon}$ with $\tilde{\varepsilon}_i$. It follows that $(\tilde{\boldsymbol{l}}^0, \tilde{\boldsymbol{x}})$ is an optimal solution of (D) if and only if $(\hat{\boldsymbol{\varepsilon}}, \tilde{\boldsymbol{x}})$ is an optimal solution of $(\widetilde{\mathrm{P}})$, which completes the proof. □

Theorem 1 implies that, by solving (D), the initial length of each member can be obtained such that the axial force attains the specified value $\bar{q}_i$ at the self-equilibrium state. Since (D) is an SOCP problem, we can solve (D) efficiently by using the primal-dual interior-point method.

Remark 2. Suppose $\bar{q}_i = \bar{q}$ $(i = 1, \cdots, n^m)$; i.e., $A_i = \bar{A}$ $(i = 1, \cdots, n^m)$. Then, after simple manipulation, (D) can be reduced to the following problem:

$$
(\mathrm{D}_0): \quad
\left.
\begin{array}{ll}
\min & \displaystyle\sum_{i=1}^{n^m} l_i \\
\text{s.t.} & l_i \geq \|\boldsymbol{B}_i \boldsymbol{x} - \boldsymbol{b}_i^0\| \quad (i = 1, \cdots, n^m),
\end{array}
\right\}
\tag{26}
$$

where l_i corresponds to the member length at the deformed state. The problem (D_0) implies that minimizing the sum of member lengths leads to the design of cable network where the axial forces of all members become the same value at the self-equilibrium state.

EXAMPLES

Consider two cable networks as shown in Figs. 1 (a) and (b), which are referred to as initial configurations of Models (I) and (II), respectively. The elastic modulus is $E = 205.8$ GPa and cross-sectional area is $A_i = 10$ cm^2 for each member. Each cable network projected

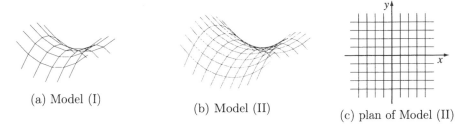

(a) Model (I)

(b) Model (II)

(c) plan of Model (II)

Figure 1: Initial configuration.

to the horizontal plane makes a grid with 1 m × 1 m squares. The (x, y)-axes are defined along the grid, and the origin is set at the center of each cable network as shown in Fig. 1. Then the z-coordinate of the nodes are given as $z = (x^2 - y^2)/\alpha$, where $\alpha = 6$ and 10 (m), respectively, for Models (I) and (II). The two ends of all the cables are supported. The initial unstressed length of each member is 99% of the member length at the initial configuration.

Equilibrium shapes are computed by solving (P_F) for Models (I) and (II) with frictional and/or frictionless joints. (D) is also solved for form finding of Model (II) with pin joints. Self-equilibrium shapes with $f = 0$ will be found in all the examples. NUOPT Ver. 4 [14] is used, which is an implementation of the primal-dual interior-point method for nonlinear programming.

Cable networks with frictional joints.
Suppose that all the internal nodes of Model (I) are frictional joints. The initial configuration is as shown in Fig. 2 (a), which does not satisfy equilibrium conditions. Let all the interior nodes have the same maximum static friction. Two cases of $\bar{F}_h = 98.0$ kN and 9.8 kN $(h = 1, \cdots, n^f)$ are considered in the friction conditions (1), and the obtained configurations are as shown in Figs. 3 (a) and 4 (a), respectively. For comparison purpose, the equilibrium configuration with frictionless conditions $\bar{F}_h = 0$ is also obtained as shown in Fig. 5 (a). By comparing these configurations, it can be seen that the smaller value of $\bar{F}_h$ leads to more difference between the equilibrium and initial configurations. The residual nodal forces at all the nodes of the solutions Fig. 3 (a) and Fig. 4 (a) satisfy (17), which guarantees that these solutions are really at equilibrium.

The stress state of each case is as shown in Figs. 2 (b)–5 (b), where the width of each member is proportional to its axial force. The maximum and minimum values of axial forces at the initial configuration are 4601.7 kN and 3570.0 kN, respectively. At the equilibrium configuration with $\bar{F}_h = 9.8$ kN, much smaller values are observed; i.e., 1509.4 kN for maximum and 590.3 kN for minimum. In the case of $\bar{F}_h = 98.0$ kN, the maximum and minimum values are 2469.5 kN and 1153.5 kN, respectively, which are between those of the previous two cases. In this way, the nodal locations and the axial forces of the solutions to (P_F) for various values of $\bar{F}_h$ have been shown to be between those of the initial state and the frictionless solution.

Frictional-frictionless joints.
Consider a cable network Model (II). Suppose that all the internal joints are frictionless in xz-plane and frictional in yz-plane; i.e., the upper bounds of frictions are $\bar{F}_h = 9.8 \times$

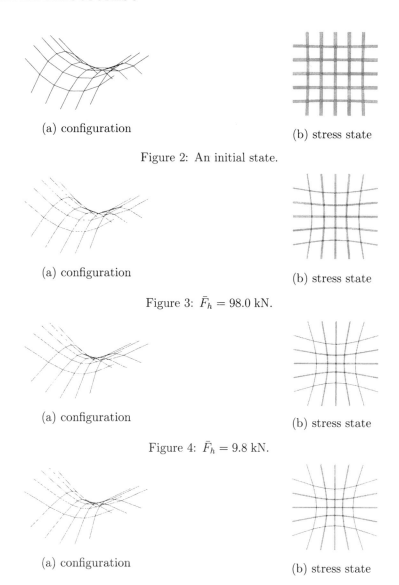

(a) configuration

(b) stress state

Figure 2: An initial state.

(a) configuration

(b) stress state

Figure 3: $\bar{F}_h = 98.0$ kN.

(a) configuration

(b) stress state

Figure 4: $\bar{F}_h = 9.8$ kN.

(a) configuration

(b) stress state

Figure 5: Frictionless joints.

10^{-3} kN for cables in xz-plane and 9.8×10^3 kN for those in yz-plane. The obtained equilibrium shape is as shown in Fig. 6. For comparison purpose, the frictionless cable network with $\bar{F}_h = 0$ has also been solved. The equilibrium configuration is obtained as shown in Fig. 7.

It can be observed that the configuration shown in Fig. 6 is between those of the initial configuration and the perfectly frictionless case. The cable networks actually constructed may be close to the configuration shown in Fig. 6 rather than that in Fig. 7, which agrees with the fact that the nodes are partially relaxed in the actual construction process.

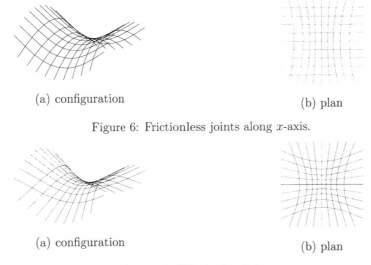

(a) configuration

(b) plan

Figure 6: Frictionless joints along x-axis.

(a) configuration

(b) plan

Figure 7: Frictionless joints.

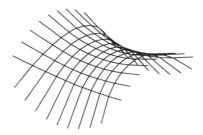

Figure 8: Cable network with pin joints.

Form finding of cable network with pin joints

For form-finding problem, we specify the topology and the coordinates of the supports similar to those of Model (II). The axial force at the equilibrium state is specified as $\bar{q}_i = 24.70$ kN for a cable in xz-plane, and $\bar{q}_i = 10.29$ kN for a cable in yz-plane; i.e., the strain of each cable is specified as $\hat{\varepsilon} = 1.0 \times 10^{-4}$, where the member cross-sectional area is $A_i = 12.0$ cm^2 for a cable in xz-plane, and $A_i = 5.0$ cm^2 for a cable in yz-plane. The initial unstressed length of each member as well as the equilibrium configuration is obtained by solving (D). The obtained configuration is as shown in Fig. 8. For the purpose of verification, the equilibrium shape analysis has been carried out for the cable network with computed initial lengths of cables. At the equilibrium state, the axial force of each cable agrees with the specified value within 5 digits, and the norm of the residual force at each node is less than 10^{-6} of the mean axial force. Therefore, the obtained model satisfies the equilibrium conditions within high accuracy, which confirms the result of Theorem 1.

CONCLUSIONS

An SOCP formulation which gives the same optimizer as that of the problem of minimum total potential energy for cable networks has been presented, and the equilibrium configuration has been obtained as a solution of the SOCP by using a primal-dual interior-point method. A form finding problem has also been formulated as SOCP, which provides the solution satisfying the equilibrium conditions exactly. In the proposed formulation, the geometrical nonlinearity as well as the cable slackening behavior is considered. Since no assumption on stress state is included, the proposed algorithm does not involve any processes of trial-and-error.

In addition to these advantages, SOCP can be solved effectively by using well-developed softwares, therefore our task is only to input the geometrical and material properties of cable networks and no effort is required to develop any analysis softwares.

REFERENCES

1. ARGYRIS, J.H. and SCHARPF, D.W., Large deflection analysis of prestressed networks, *ASCE,* Vol. 98(ST3), 1972, 633–654.

2. COYETTE, J.P. and GUISSET, P., Cable network analysis by a nonlinear programming technique, *Engng. Struct.,* Vol. 10, 1988, 41–46.

3. PANAGIOTOPOULOS, P.D., A variational inequality approach to the inelastic stress-unilateral analysis of cable-structures, *Comput. Struct.,* Vol. 6, 1976, 133–139.

4. CANNAROZZI, M., A minimum principle for tractions in the elastostatics of cable networks, *Int. J. Solids Structure,* Vol. 23, 1987, 551–568.

5. ATAI, A.A. and STEIGMANN, D.J., On the nonlinear mechanics of discrete networks, *Arch. Appl. Mech.,* Vol. 67, 1997, 303–339.

6. MONTEIRO, R.D.C. and TSUCHIYA, T., Polynomial convergence of primal-dual algorithms for the second-order cone program based on the MZ-family of directions, *Math. Programming,* Vol. 88, 2000, 61–83.

7. MISTAKIDIS, E.S. and STAVROULAKIS, G.E., *Nonconvex Optimization in Mechanics,* Kluwer Academic Publishers, Dordrecht, 1998.

8. VANDERBEI, R.J. and YURTTAN, H., *Using LOQO to solve second-order cone programming problems,* SOR-98-9, Statistics and Operations Research, Princeton University, 1998.

9. SCHEK, H.-J., The force density method for form finding and computation of general networks, *Comp. Meth. Appl. Mech. Engng.,* Vol. 3, 1974, 115–134.

10. LEWIS, W.J. and GOSLING, P.D., Stable minimal surfaces in form-finding of lightweight tension structures, *Int. J. Space Structures,* Vol. 8, 1993, 149–166.

11. KANNO, Y., OHSAKI, M. and ITO, J., Large-deformation and friction analysis of nonlinear elastic cable networks by second-order cone programming, *Int. J. Numer. Meth. Engng.,* 2002, to appear.

12. KANNO, Y. and OHSAKI, M., Necessary and sufficient conditions for global optimality of eigenvalue optimization problems, *Struct. Multidisc. Optim.,* Vol. 22, 2001, 248–252.

13. ROCKAFELLAR, R.T., *Convex Analysis,* Princeton University Press, 1970.

14. NUOPT User's Manual Ver. 4, Mathematical Systems Inc., 1998.

Analysis of Various Types of Tensegrity Steel/Glass Roof Structures

L ZHANG, J WARDENIER, Faculty of Civil Engineering and Geosciences, Delft University of Technology, 2600 GA Delft, The Netherlands, and
A C J M EEKHOUT, Faculty of Architecture, Delft University of Technology, 2600 GA Delft, The Netherlands

BACKGROUND/OBJECTIVE

Tensegrity structures, according to their inventor Richard Buchminster Fuller, are structures where the tensile components are continuous, while the compression components are not. In the last decade, a number of tensegrity structures have been made, composed of parallel compression studs and tensile rods, covered by glass panels.

These structures have often been designed in fixed frames or within fixed boundaries. The current research was carried out within the framework of the "Zappi" program and was concentrated on the possibility of substituting steel compression studs by glass members and optimising the geometry of a roof module.

Architecturally, a glass covered tensegrity roof structure could be very attractive [Refs 1-3].

DESIGN RESISTANCE OF MEMBERS

The Modelling of compression members and modelling of tension members can be found in [Ref 1]. Because the compression bars of the structures described in this paper are glass tubes, failures of these members are considered for one extreme case only, i.e., with brittle type strut buckling.

Table 1: Design resistance of a member

Design resistance		Stainless steel member	Glass member
Yield stress f_y (N/mm^2)		200	100 (ultimate)
Buckling resistance	$N_{b,Rd} = \chi \beta_A A f_y / \gamma_{M1}$	$\beta_A = 1$ $\gamma_{M1} = 1.1$	$\beta_A = 1$ $\gamma_{M1} = 4/1.4 = 2.85$
	$\chi = \dfrac{1}{\Phi + \left[\Phi^2 - \overline{\lambda}^2\right]^{0.5}}$ $\Phi = 0.5[1 + a(\overline{\lambda} - 0.2) + \overline{\lambda}^2]$ $\overline{\lambda} = (\lambda/\lambda_1)\sqrt{\beta_A}$ $\lambda = l/i \qquad i = \sqrt{I/A}$ $\lambda_1 = 93.9\varepsilon \qquad \varepsilon = \sqrt{235/f_y}$	$a = 0.21$ (Bucking curve **a**)[*]	$a = 0.49$ (Bucking curve **c**)

*hot formed tubes

The buckling resistance of compression members according to Eurocode 3 [Ref 5], and the yield stresses of the steel and the ultimate stress of the glass members considered are indicated in table 1.

For glass the γ_{M1} value has not been defined in codes and standards. Up to now, a total safety factor of 4 is usual in working stress design. Therefore here this value is divided by an average load factor 1.4 to obtain the γ_{M1} value.

NUMERICAL ANALYSIS FOR TENSEGRITY STRUCTURE
Geometry at initial state
Four 10m span tensegrity roof structures shown in Fig. 1 with a circular and a square layout respectively with rigid boundary conditions are analyzed for comparison. The surface is paraboloid. The maximum central rise and central sag are taken 0.5m and 1.0m respectively. It is assumed that the upper and lower chords can not resist compression. In other words, all of them are assumed to be cable elements which can only resist tension. According to this, the horizontal component of the forces in the upper rods is taken 18.4kN, and the corresponding component in the lower rods is taken 9.2kN. These values are determined on the basis of no member failing before reaching any design load given in this paper.

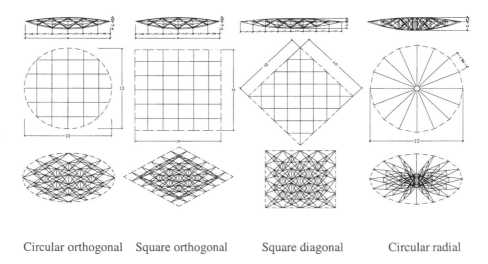

Circular orthogonal   Square orthogonal   Square diagonal   Circular radial

Fig.1. Four geometries considered in this paper

Loading

Various design loads are given in table 2 according to [Ref 6].
Four load cases are considered as indicated in Fig.2.

Table2: Various design loads according to [Ref 6]

Load type	Load value (kN/m²)	Shape factor for compression	Shape factor for suction	Load factor
dead load	0.7			1.2
wind load	0.75	0.4	-1.3	1.5
snow load	0.7	0.8		1.5

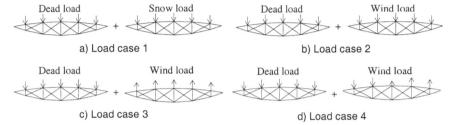

a) Load case 1   b) Load case 2

c) Load case 3   d) Load case 4

Fig.2. Four load cases

Dimensions and material properties
The dimensions and material properties of the elements are given in table 3.

Table3: Dimensions and material properties of elements in structures

	Steel member		Glass member	
	Square layout circular orthogonal layout	circular radial layout	Square layout Circular orthogonal Layout	circular radial layout
Upper chord	Φ16 mm, A=201.1 mm²			
Lower chord	Φ18 mm, A=254.5 mm²			
Compression bar			Φ40*3.2, A=369.8 mm²	Φ30*2.8, A=239.1 mm²
Diagonal Rod	Φ25 mm, A=490.9 mm²	Φ 20 mm, A=314.2 mm²		
Ring element		Φ 25 mm, A=490.9 mm²		
Elastic modulus (N/mm²)	2.0×10⁵		7.0×10⁴	
Yield stress (N/mm²)	200		100	

Computation results
The computation is based on the theory provided in [Ref 1]. Brittle type strut buckling (B) analysis is applied for structural analysis of the glass elements. A plot of the maximum vertical displacement versus the applied load and the corresponding collapse patterns are shown in Figs. 3 to 6. For every load case, the load reserve factor means the loading divided by the design load of the load case considered.

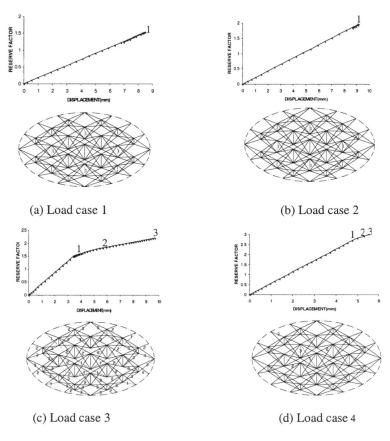

Fig.3 Load-displacement curve and collapse mechanisms for brittle type strut bucklin(B)
-analysis of circular orthogonal tensegrity structure; 1, 2, 3… failure sequence-

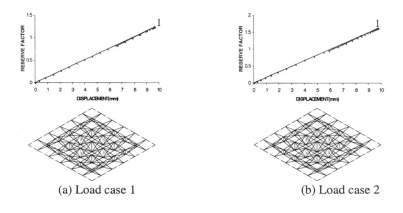

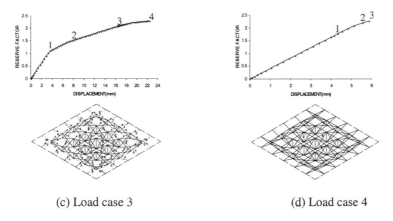

(c) Load case 3   (d) Load case 4

Fig.4 Load-displacement curve and collapse mechanisms for brittle type strut buckling (B)
-analysis of square orthogonal tensegrity structure; 1, 2, 3… failure sequence-

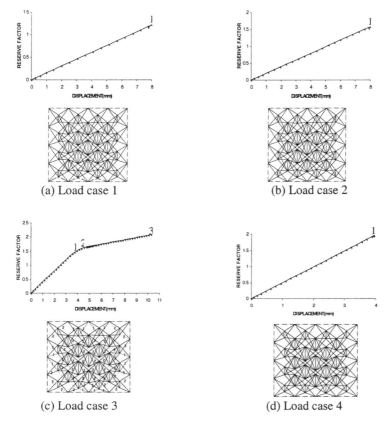

(a) Load case 1   (b) Load case 2

(c) Load case 3   (d) Load case 4

Fig.5 Load-displacement curve and collapse mechanisms for brittle type strut buckling (B)
-analysis of square diagonal tensegrity structure; 1, 2, 3… failure sequence-

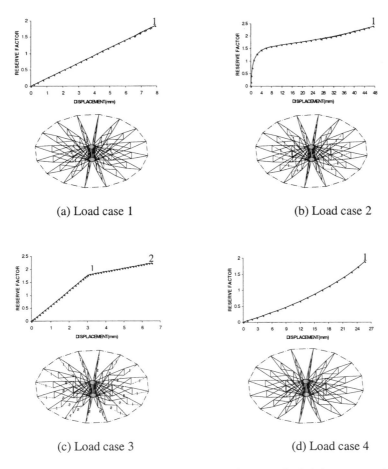

(a) Load case 1   (b) Load case 2

(c) Load case 3   (d) Load case 4

Fig.6 Load-displacement curve and collapse mechanisms for brittle type strut buckling (B) -analysis in circular radial tensegrity structure; 1, 2, 3... failure sequence-

Discussion of the results

The reserve factors for the first member failure and for the whole structure failure and the weight of each structural configuration are given in table 4.

In the brittle type strut buckling (B) analysis, if the vertical compression bars fail first, the collapse mechanism of the whole structure is determined by failure of these members. The values of the reserve factor for the first member failure and for the whole structure failure are the same. If the first failed members are not the vertical compression bars but other members, even though weakened by the failure of individual members, a tensegrity system is able to carry additional loading until the whole structure fails.

Under wind load in suction (load case 3), the collapse mechanisms of the whole structure are caused by the failure of the bottom chord members

From table 4 it can be seen that the behaviour of tensegrity structure with a geometry in circular radial layout and with a square diagonal layout behave better than those with a square orthogonal layout.

Because of its lower buckling resistance, the sizes for glass tubes are larger than those for steel tubes in reference [1], but the changing tendency of structural behaviour with increasing load is similar. For the glass structures, the B analysis should be used for the glass members.

Table 4: Limit reserve factor and total weight in each kind of structural form

Geometry and material	Limit reserve factor (a) for the first member failure (b) for whole structure failure								Weight (kg)	Weight (kg/m^2)
	Load case 1		Load case 2		Load case 3		Load case 4			
	(a)	(b)	(a)	(b)	(a)	(b)	(a)	(b)		
Circular orthogonal layout	1.49	1.49	1.97	1.97	1.48	2.20	2.70	3.03	8.269	0.105
Circular radial layout	1.84	1.84	2.38	2.38	1.80	2.23	1.89	1.89	6.405	0.082
Square orthogonal layout	1.24	1.24	1.62	1.62	1.01	2.30	1.69	2.27	9.941	0.099
Square diagonal layout	1.22	1.22	1.57	1.57	1.40	2.07	1.92	1.92	10.155	0.102

SUMMARY AND CONCLUSIONS

Based on the procedure for following the progressive failure of tensegrity structures presented in [Ref 1], failures of glass members in the paper is considered for one extreme case, i.e., with brittle type strut buckling. Four types of tensegrity steel/glass roof structures have been analyzed under symmetrical and asymmetrical loads. The following conclusions can be drawn.

1. In the brittle type strut buckling (B) analysis, under symmetrical and asymmetrical loads with exception under wind load in suction, the collapse mechanism of the whole structure is governed by failure of certain vertical compression members.
2. Under wind load in suction, the collapse mechanisms of the whole structure is caused by failure of the bottom chord members
3. The behaviour of tensegrity structure with a geometry in a circular radial layout and with a square diagonal layout are better than those of a square orthogonal layout.
4. The structural behaviour of tensegrity roof structures with glass compression members behaves similar to that of the structures with steel compression members presented in [Ref 1]. B analysis should be used for glass structural analysis.

REFERENCES

1. ZHANG L, WARDENIER J, EEKHOUT A C J M, Analysis of various types of tensegrity steel roof structures, to be presented at the Fifth International Conference on Space Structures, University of Surrey, 19-21 August 2002.
2. EEKHOUT A C J M, Architecture in Space Structures, Publisher 010, Rotterdam, 1989.
3. EEKHOUT A C J M, Product Development in Glass Structures, Publisher 010 Rotterdam, 1990.
4. EEKHOUT A C J M, et al. The Glass Envelope, Published by Delft University of Technology, Faculty of Architecture, Delft, 1992.
5. CEN:ENV 1993-1-1: Eurocode 3: Design of steel structures, Part 1.1 General rules and rules for buildings. Comité Européen de Normalisation, 1992.
6. NEN 6702: Loadings and Deformations. Netherlands Standardization Institute (*in Dutch*), 1990.

A simple procedure for the analysis of cable network structures

V. F. ARCARO
College of Civil Engineering, UNICAMP, Brazil

ABSTRACT
This text presents a mathematical modeling of a cable finite element. It includes a total Lagrangian description using the Engineering strain definition and assumes an elastic material (linear or nonlinear). A procedure to analyze a cable network in the presence of conservative forces and small deformations is summarized. Mathematical programming makes the use of stiffness matrix pointless. The web page http://www.arcaro.org/tension/ publicizes the computer code and examples related to this text.

NOTATION
The following applies unless otherwise specified or made clear by the context. A Greek letter expresses a scalar. A vector is always a column matrix and a lower case letter expresses it. An upper case letter expresses a matrix.

FINITE ELEMENT DEFINITION
The geometry of a one-dimensional cable element is shown in Figure 1. The nodes are labeled 1 and 2. The strain is assumed constant along the element and the material homogeneous and isotropic.

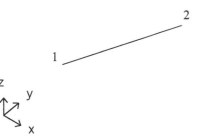

Fig. 1. The geometry of a one-dimensional cable element.

Deformed length
In Figure 2, the λu vector, where u is a unity vector, represents the cable element in a configuration with zero nodal displacements. It is easy to understand that λ represents the distance between the nodes of the element in this configuration. However, as will be explained latter, this distance will not always represent the undeformed length of the element. The vector l represents the element in its deformed configuration. The vectors p and q represent the nodal displacements vectors.

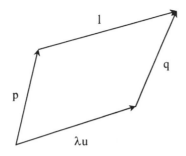

Fig. 2.

The deformed length can be found as follows:

$$\lambda u + q - l - p = 0$$

$$l = \lambda u + q - p$$

$$z = \frac{q - p}{\lambda} \Rightarrow l = \lambda(u + z)$$

$$\delta = 2u^T z + z^T z \Rightarrow \|l\| = \lambda\sqrt{1 + \delta}$$

Imposing a constant cut
Consider an element with undeformed length less than the *initial distance* of its nodes, where initial distance is defined as the distance of the two nodes with zero nodal displacements. This element can be pictured with a *cut* in its undeformed length. The result is that this element will show tension in any rigid body motion that preserves the initial distance of its nodes.

The Initial configuration of a cable network is defined as the configuration of zero nodal displacements for all its nodes. Applying imaginary cuts to selected elements of a cable network in its initial configuration is an easy way to apply tension to this cable network. Notice that if no cuts are present, the initial configuration is also the undeformed configuration.

It is worth mentioning that effects due to temperature change also can be treated through an imaginary cut in the undeformed length of the element.

Strain
Considering μ as the value of the cut in the undeformed length of an element, the *cut length* of this element can be written as:

$$\lambda_\mu = \lambda - \mu$$

After applying a cut, consider a change Δt in temperature. The coefficient of thermal expansion is denoted by α_t. This sequence leads to the *strain-free length* as:

$$\lambda_0 = \lambda(1-\rho)(1+\varepsilon_t)$$

Where,

$$\rho = \frac{\mu}{\lambda} \, , \, \varepsilon_t = \alpha_t \Delta t$$

The strain can be written as:

$$\varepsilon = \frac{\|1\| - \lambda_0}{\lambda_0}$$

$$\varepsilon = \frac{\sqrt{1+\delta} - (1-\rho)(1+\varepsilon_t)}{(1-\rho)(1+\varepsilon_t)}$$

In order to avoid severe cancellation, the strain expression should be evaluated as:

$$\varepsilon = \frac{\dfrac{\delta}{\sqrt{1+\delta}+1} + \rho + \rho\varepsilon_t - \varepsilon_t}{(1-\rho)(1+\varepsilon_t)}$$

Potential strain energy

Considering σ as the conjugate stress to the engineering strain ε and α as the undeformed area of the element, the potential strain energy and its gradient can be written as:

$$\phi = \alpha\lambda_0 \int_0^\varepsilon \sigma(\xi)d\xi$$

$$\frac{\partial\phi}{\partial p_i} = \alpha\lambda_0\sigma(\varepsilon)\frac{\partial\varepsilon}{\partial p_i} = -\frac{\alpha\sigma(\varepsilon)(u_i + z_i)}{\sqrt{1+\delta}}$$

$$\frac{\partial\phi}{\partial q_i} = \alpha\lambda_0\sigma(\varepsilon)\frac{\partial\varepsilon}{\partial q_i} = +\frac{\alpha\sigma(\varepsilon)(u_i + z_i)}{\sqrt{1+\delta}}$$

Geometric interpretation

Considering Figure 2, a unit vector v parallel to the element in its deformed configuration can be written as:

$$v = \frac{1}{\|1\|} = \frac{u+z}{\sqrt{1+\delta}}$$

Using vector v, the gradient of the potential strain energy can be written as:

$$\frac{\partial\phi}{\partial p_i} = -\alpha\sigma(\varepsilon)v_i$$

$$\frac{\partial\phi}{\partial q_i} = +\alpha\sigma(\varepsilon)v_i$$

Figure 3 shows the geometric interpretation of the gradient of the potential strain energy as forces acting on nodes of the element. These forces are known as internal forces.

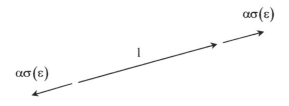

Fig. 3.

Imposing a constant tension

Consider the following scalar function ϕ and its gradient:

$$\phi = \alpha\sigma_0\lambda\sqrt{1+\delta}$$

$$\frac{\partial\phi}{\partial p_i} = -\alpha\sigma_0 v_i$$

$$\frac{\partial\phi}{\partial q_i} = +\alpha\sigma_0 v_i$$

The gradient can be interpreted as internal forces, with constant modulus, acting on nodes of an element. The scalar function can be interpreted as the corresponding potential strain energy.

An element with imposed constant tension can be defined by choosing positive value for the constant stress σ_0. The result is that this element will show constant tension in any displacement of its nodes. Applying constant tension to selected elements of a cable network in its initial configuration is another easy way to apply tension to this cable network.

Equivalence between constant cut and constant tension
A constant cut value is equivalent to a constant tension value in the sense that they both produce the same internal forces. To find the equivalence between them, consider a cable network at a known configuration.

To find the constant cut value equivalent to the constant tension value, first find the strain ε according:

$$\sigma(\varepsilon) = \sigma_0$$

Then, find the cut value μ according:

$$\mu = \frac{\lambda}{(1+\varepsilon_t)(1+\varepsilon)}\left(\varepsilon + \varepsilon_t + \varepsilon_t\varepsilon - \frac{\delta}{1+\sqrt{1+\delta}}\right)$$

To find the constant tension value equivalent to the constant cut value, first find the strain ε according:

$$\varepsilon = \frac{\dfrac{\delta}{\sqrt{1+\delta}+1} + \rho + \rho\varepsilon_t - \varepsilon_t}{(1-\rho)(1+\varepsilon_t)}$$

Then, find the tension σ_0 according:

$$\sigma_0 = \sigma(\varepsilon)$$

Constitutive relationship
A linear stress strain relationship is assumed according to the following expression:

$$\sigma = E\varepsilon$$

Where E is the Young's modulus. The potential strain energy can be written as:

$$\phi = \alpha\lambda_0 \int_0^\varepsilon \sigma(\xi)\,d\xi = \frac{1}{2}\alpha\lambda_0 E\varepsilon^2$$

EQUILIBRIUM CONFIGURATIONS
The stable equilibrium configurations correspond to local minimum points of the total potential energy function. It is advisable the use of a Quasi Newton type method to find these local minimums because it does not requires the evaluation of the stiffness matrix.

Considering x as the vector of unknown displacements and f as the vector of nodal forces, the total potential energy function and its gradient can be written as:

$$\pi(x) = \sum_{\text{elements}} \phi(x) - f^T x$$

$$\nabla\pi(x) = \sum_{\text{elements}} \nabla\phi(x) - f$$

EXAMPLE

A procedure to analyze a cable network in the presence of conservative forces and small deformations is summarized.

Figure 4 shows a cable network in its undeformed configuration. It is a cable beam, whose design is known as Zetlin.

Fig. 4. A cable network in its undeformed configuration.

A loading consisting of forces acting upward is applied on nodes of the top cable as a crude simulation of wind uplift action. The wind action is considered as distributed load acting along the span of the top cable. The self-weight of cables is considered in this analysis.

This loading results in compression of the top cable elements in this model. This should be interpreted as the elements becoming slack or flaccid.

This flaccidity is due the fact that the upward loading tends to increase stress at the bottom cable and decrease stress at the top cable. Since the structure was undeformed, this result is no surprise - the structure needs to be tensioned. The tensioning must be determined such that the upward loading produces no flaccidity.

Figure 5 shows the cable network in its undeformed configuration, where the first and last elements of the bottom cable are marked with dashed lines.

Fig. 5. The cable network in its undeformed configuration, where the first and last elements of the bottom cable are marked with dashed lines.

Imposing a constant tension to these elements may result in the required tensioning. Only the self-weight of cables is considered in this step.

It is important to notice that imposing a constant tension to selected elements of a cable network is an attempt to simulate what is accomplished through hydraulic jacks in practice.

The general problem is to choose a specific value for the constant tension that results in tension of all elements for all loading cases. A good trial value is to set the constant tension at a percentage of the breaking tension of the rope. The deformed structure can be said tensioned or pre-stressed.

Applying the constant cut value equivalent to the constant tension value found in the previous step, and the loading acting upward results in a deformed configuration shown in Figure 6, where all elements show tension.

Fig. 6.

It is important to notice that imposing a constant cut equivalent to the constant tension is an attempt to simulate what happens after the hydraulic jacks have been removed. The action of a hydraulic jack is pictured as to shorten the selected element where it is applied.

COMPUTATIONAL PERFORMANCE
Table 1 shows the computational performance on an ordinary Pentium machine (200 MHz). The Limited Memory BFGS method was used. The line search procedure used cubic interpolation.

Table 1.

	Loading	Tensioning	Tensioning Loading
Iterations	94	423	54
CPU time (s)	0	1	0

BIBLIOGRAPHY
1. GILL, P. E. AND MURRAY, W., Newton type methods for unconstrained and linearly constrained optimization, Mathematical Programming 7, 1974.
2. LASDON, L. S., Optimization theory for large systems, Macmillan, New York, 1970.
3. LUENBERGER, D. G., Linear and nonlinear programming, second edition, Addison Wesley, Reading, Massachusetts, 1989.
4. NOCEDAL, J. AND WRIGHT, S. J., Numerical Optimization, Springer-Verlag, 1999.

Fundamental Study on Cable Roof Structure with Variable-stiffness-spring and Damper

Hideo OKA, Research & Development Institute, Takenaka Corporation, Chiba, Japan,
Hidetoshi HAYASHIDA, R & D Institute, Takenaka Corp., Chiba, Japan,
Kiyoshi OKAMURA, R & D Institute, Takenaka Corp., Chiba, Japan,
Yasuzou FUKAO, R & D Institute, Takenaka Corp., Chiba, Japan,
Hideyuki NARITA, Office of Intellectual Property, Takenaka Corp., Tokyo, Japan,
Akihiro SUGIUCHI, Office of Engineering, Takenaka Corp., Tokyo, Japan, and
Xiaoguang LIN, Dept. of architecture, Osaka Institute of Technology, Osaka, Japan

1. INTRODUCTION

Since tension structures that consist of cable materials are very lightweight compared with frame or shell structures and have design uniqueness, they are used abundantly for the roof of large-span structures. Especially on the sports stadiums, to admit sunshine required for growing natural grass, a roof shape covering only upper part of the seats in the perimeter zone and having a big opening in the field central part is often used. One of the construction method of such a type roof with cable materials is to tense the ring-cable by pre-stressing the radial cables suspended from the boundary structure as shown in Figure 1 or [Ref 3 etc.].

Generally on the cable structure, quite large initial tensile force is introduced into each cable member so that cable tension will not be lost under all expected short-term loads, such as snow or wind. Therefore, the order of cable tension, especially in the hoop ring-cable, becomes very large and size-up of cable members or bundle of many cables are inevitably needed. These design methods also make the cable joint hardware or the anchor parts to be larger and more complicated. Moreover, since the cable roof structure shown in Figure 1 is originally unstable before pre-stressing, it is requested that the tension distribution of each cable member and the configuration of cable roof should be controlled prudently during the pre-stressing process, and this request causes an increasing of construction cost.

Owing to above reasons, although the cable structure is originally lightweight and rational, its merits are not realized sufficiently and many problems are still remained. In this study, in order to solve such problems, we try to develop a kind of cable-end fixing mechanism having the following two main functions.
The first function is to hold the cable tension level within the proper range in both cases of cable tension increasing and reducing, against the static loads such as an average component of wind load or snow load.
The second function is to transmit the excited motion of the cable end, caused by variable of wind load, to the damping mechanism, in order to reduce the tension amplitude quantity. In this case, the roof deformation is permitted to some extent.

Wind load can be divided into an average component and a variable component. If the ratio of these components can be considered 1:1 on the roof of large-span structure, the damping mechanism has a possibility to reduce initial tension order and cable section by half at the maximum. By giving additional functions to the above-mentioned devices, for example, initial tension installing mechanism or cable length adjustment mechanism due to construction error, the improvements for construction efficiency and quality can also be expected.

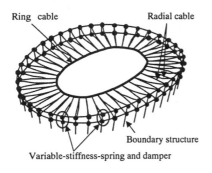

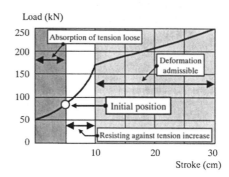

Figure 1. Structural system of roof   Figure 2. Variable-stiffness-spring character

2. THE DEVICE CONSISTS OF VARIABLE-STIFFNESS-SPRING AND DAMPER
2.1 Device mechanism
As described in the INTRODUCTION, since the cable members are determined by the sum of initial tension introduced and maximum tension caused by the short-term external loads, usually quite larger cable members are needed. Also, the boundary structure including foundation must be enlarged. Moreover, cable members setting work and initial tension introducing work will also become large-scale.

To solve these problems, we propose to install the devices having a spring and a damping mechanism to the radial cable-ends. This aims mainly at reducing the initial tension level and tension amplitude quantity caused by wind excitation. The functions of spring mechanism, under the static loads such as average component of usual level wind or snow load, are to make the roof deformation small when cable tension increase. On the other hand, when cable tension decreases, it will avoid losing tension. Namely, spring stiffness is comparatively strong when the cable tension is increasing, conversely when decreasing its stiffness is weak. In the case of high-level wind loads like a very large typhoon, the spring stiffness is weakened and the dynamic motion of the cable ends are transmitted to a damping mechanism to consume its vibration energy. This concept of spring character is shown in Figure 2.

Similar methods, that is, adding damping mechanisms to the cable or rod material can be seen in the [Ref 4, 5 etc.]. These methods try to stiffen and reduce the vibration of the frame structure. The spring elements used in these methods are the mechanical type, such as the coil or plate spring. These springs have linear stiffness character. However, the spring imaged in this study has high non-linearity giving the special character of spring-stiffness-weakening when the cable tension level becomes large.

Therefore, we examined the spring mechanism using gaseous and oil pressure. Compared with a mechanical spring, a hydraulic spring can be designed with the device-size compactly and also has the advantage of being easy to control the spring stiffness.

2.2 Spring character of proto-type device

Mechanism of variable-stiffness-spring and damper using gaseous (nitrogen gas) and oil pressure is shown in Figure 3. It consists of two oil cylinders and gaseous rooms. The variable-stiffness-spring character is realized by changing the first cylinder to the second one on a certain load level. Damping mechanism is obtained by moving resistance of oil. The proto-type model, with the spring-stroke length is set to about 1/5 against a full-size model, is shown in Photo 1. The spring with two cylinders should be applied to the lower radial cables for wind load, in the case of Figure 1 type structure. As for the device installed to the upper radial cables, because snow load is assumed to be the governing load, its spring is a one-cylinder type device shown in Figure 4 and Photo 2.

The spring characteristics under the condition that the damping mechanism is not made to act (state of shutting the damping valve) are shown in Figure 5 and Figure 6. This excitation is done by triangular wave load and its frequency is 0.25 Hz. Although some historical ropes are drawn by influence of cylinder friction etc., almost expected spring characteristics can be acquired.

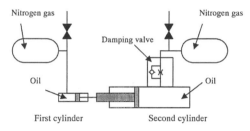

Figure 3. Lower cable device mechanism

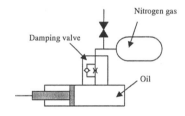

Figure 4. Upper cable device mechanism

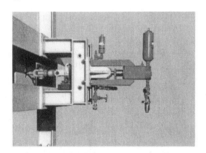

Photo 1. Lower cable proto-type device

Photo 2. Upper cable proto-type device

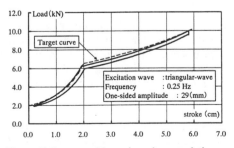

Figure 5. Lower cable spring characteristics

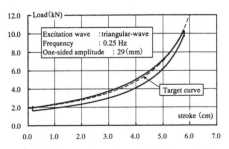

Figure 6. Upper cable spring characteristics

3. LOADING TESTS OF SMALL-SCALE CABLE-NET MODEL
In order to investigate the fundamental behavior and to collect basic data of cable structure with the devices, loading tests were performed using the small-scale cable-net model.

The experimental model is the circular cable-net structure with an opening in the central part, which consists of eight streets (A-H streets, Figure 7). The diameter of perimeter and ring cable in the central part is about 10m and 2m respectively, and distance between upper and lower radial cable-end is 1.6m. The 16-radial cables are stranded-rope (ϕ-9) and the ring cable is the same (ϕ-14). They are jointed by junction hardware and bolted to each other. The proto-type devices shown in the chapter 2 are connected to the upper and lower radial cable-ends through the sheave and fixed to the high-rigid steel frame around a perimeter part (Figure 8). Test model and measured data are shown in Photo 3, Photo 4 and Table 1. The initial cable tension is introduced by fastening the turnbuckles prepared on each radial cable and adjusted so that the 16-radial cable tensions is equal to about 3.5kN. The actuators for loading are set under the junction node position of the ring and the radial cables and concentrated loads are given to these points in the vertical direction (Figure 7 and Photo 4).

Since we considered that a chief aim of this experiment was to investigate the effect of the spring mechanism and also in order to make understanding of phenomena easy, the device damping valves were shut and its damping mechanism did not act in any experiment cases. Moreover, as the devices could be locked to the perimeter steel frame, both cases which installing the devices (*Spring ON*, as follow) and non-installing case (*Spring OFF*, that is same as the fixed conditions) were performed for comparison.

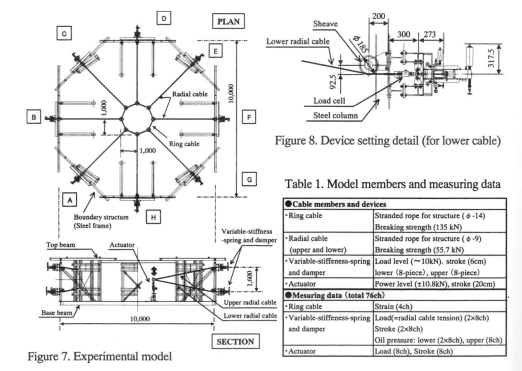

Figure 7. Experimental model

Figure 8. Device setting detail (for lower cable)

Table 1. Model members and measuring data

●Cable members and devices	
・Ring cable	Stranded rope for structure (ϕ-14)
	Breaking strength (135 kN)
・Radial cable	Stranded rope for structure (ϕ-9)
(upper and lower)	Breaking strength (55.7 kN)
・Variable-stiffness-spring	Load level (~10kN), stroke (6cm)
and damper	lower (8-piece), upper (8-piece)
・Actuator	Power level (±10.8kN), stroke (20cm)
●Mesuring data (total 76ch)	
・Ring cable	Strain (4ch)
・Variable-stiffness-spring	Load(=radial cable tension) (2×8ch)
and damper	Stroke (2×8ch)
	Oil pressure: lower (2×8ch), upper (8ch)
・Actuator	Load (8ch), Stroke (8ch)

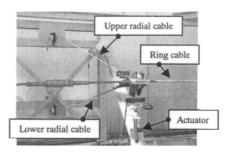

Photo 3. Overview of the experimental model   Photo 4. Cable joint in the experimental model

4. ANALYTICAL STUDY OF LOADING TESTS
4.1 Analytical method of the device

In analysis, cable elements are treated as elastic rod elements in consideration of slack and cable tension is lost when axial force is compressed. Load incremental analysis method, usually used, is employed to solve the geometrical non-linear equations of cable structure. As for the variable-stiffness-spring and damper devices, the damping force by the damper mechanism is not taken into consideration, the same as experiment conditions. If a spring is modeled as an element different from cable element, it may produce difficulty on numerical analysis when the length of the spring element becomes very short or 0. This may also cause an error to incremental procedure when the spring stroke reaches the limit and the structural stiffness changes suddenly. In order to avoid these problems, the cable and spring are combined to one element having equivalent stiffness. The spring characteristics are assumed to be non-linear curve, not a historical rope type, and approximated by the folding lines as shown in Figure 9.

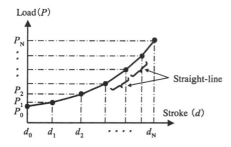

Figure 9. Modeling method of device non-linear spring character

Here, (N) means the number of the straight lines used for modeling, $(P_i: i = 0, \cdots, N)$ and $(d_i: i = 0, \cdots, N)$ means the load and the amount of strokes of each straight-line-end, respectively. Therefore the minimum and the maximum of a non-linear spring stroke can be represented as (d_0) and (d_N), respectively. If the initial length of cable, connected to the device, and corresponding initial tension are represented (L_{C0}) and (T_0), the composite element length (L_i) corresponding load (P_i) can be expressed by the equation (1).

$$L_i = d_i + L_{C0} + (P_i - T_0) L_{C0} / EA \quad , \quad (i = 0, \cdots, N) \qquad \cdots (1)$$

where (E) and (A) is cable young's modulus and section of area, respectively. An equivalent stiffness (K_T) and tension (P) of composite element can be represented by the following relations.

●When a spring is under the minimum stroke limit $(L < L_0)$
$$K_T = EA / (L - d_0), \quad P = P_0 + K_T (L - L_0) \qquad \cdots (2.a,b)$$

●When a spring is within the effective stroke $(L_0 \leq L \leq L_N)$
$$K_T = (P_{i+1} - P_i) / (L_{i+1} - L_i), \quad P = P_i + K_T (L - L_i) \qquad \cdots (3.a,b)$$

●When a spring is over the maximum stroke limit $(L_N < L)$
$$K_T = EA / (L - d_N), \quad P = P_N + K_T (L - L_N) \qquad \cdots (4.a,b)$$

If an equivalent stiffness (K_T) and load (P) can be determined from (2.a,b) to (4.a,b) equations, a composite element can be treated as same as a cable element.

4.2 Analytical model of loading tests

The analytical model of loading tests and cable elements data are shown in Figure 10 and Table 2. Cable members are not divided in the middle in both cases of (*Spring OFF*) and (*Spring ON*). Moreover, in the detail of the experiment model, each radial cable was connected to the variable-stiffness-spring device through the sheave and its spring stroke direction was always restrained to the horizontal (Figure 8). But in the analysis model, boundary nodes of a radial cable-ends are put on the touching points of the radial cable and the sheave. As for the variable-stiffness-spring characters, they have some historical friction, but as described before, this was not included here. Although some variations of the spring character were observed, they are approximated by the folding lines shown in Figure 11. They are the average values of 8 streets devices obtained in the experiment results and this same spring character is given to the all-street devices. Needless to say, shape analysis was carried out before structural analysis in order to determine the initial balanced system. This mostly fills the initial shape and tension distribution on the experiment model.

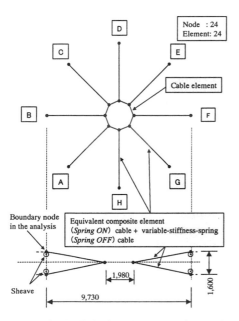

Figure 10. Analytical model of experimentation

Table 2. Cable elements data

	A (cm²)	w (N/m)
· Ring cable (ϕ-14)	0.955	92.3
· Radial cable (ϕ-9)	0.394	9.7
Young's modulas: E=137 GPa		

A: Section area, w: Unit weight (including cable jointing hardware)

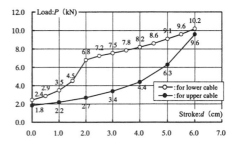

Figure 11. Modeling of device spring character
(for lower: N=12, for upper: N=6)

5. RESULTS OF EXPERIMENTATION AND ANALYSIS

5.1 Static loading tests

The representative results (A, B, E streets, and ring cable between B-C) of static loading test under the three load cases, that is all-area, half-area and 1-street(A) loading mode, are shown in Figure 12-14 (*Spring OFF*) and Figure 15-17 (*Spring ON*), respectively.

In the case of (*Spring OFF*), the structural stiffness is large and each cable tension also changes sharply. The phenomena of the radial cables losing tension can be seen in the each loading mode case. After losing radial cable tension, the structural stiffness is weakened and the ring cable tension begins to increase suddenly. In the half-area and 1-street loading mode case, the difference of the radial cable tension between each street is magnified and an imbalanced tension distribution is produced. It can be seen that the analysis results are well in agreement with these experiment results on each loading mode.

On the other hand, in the case of (*Spring ON*) where installing the variable-stiffness-spring devices, a little variation of spring character was seen in each street device. But on the whole, they operated well in the cable structure as expected. In comparison with (*Spring OFF*), however the structural stiffness is weakened and the tension amplitude of each cable is small. The tension-decreasing ratios of radial cables are loose and its losing phenomena do not occur. It means that the spring devices worked effectively to prevent loss of tension. Moreover, under the half-area or 1-street loading mode case, the effect that eases the tension unbalance between each street is seen. As for the analysis by the chapter (4) method, the experimental results are well simulated as well as in the case of (*Spring OFF*).

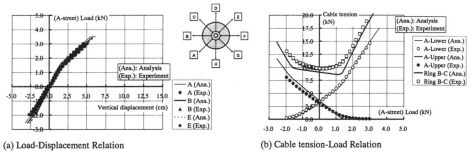

(a) Load-Displacement Relation   (b) Cable tension-Load Relation

Figure 12. Static loading, All-area loading mode, (*Spring OFF*)

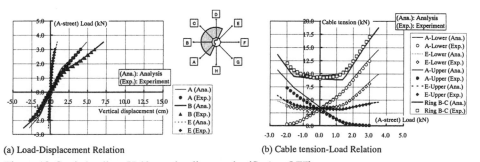

(a) Load-Displacement Relation   (b) Cable tension-Load Relation

Figure 13. Static loading, Half-area loading mode, (*Spring OFF*)

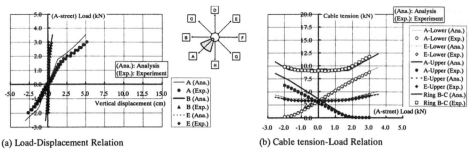

(a) Load-Displacement Relation

(b) Cable tension-Load Relation

Figure 14. Static loading, 1-street loading mode, (*Spring OFF*)

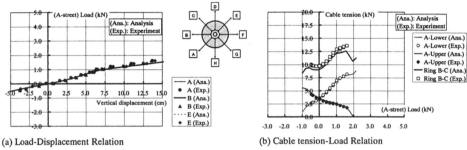

(a) Load-Displacement Relation

(b) Cable tension-Load Relation

Figure 15. Static loading, All-area loading mode, (*Spring ON*)

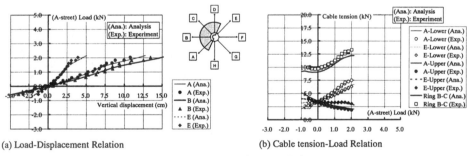

(a) Load-Displacement Relation

(b) Cable tension-Load Relation

Figure 16. Static loading, Half-area loading mode, (*Spring ON*)

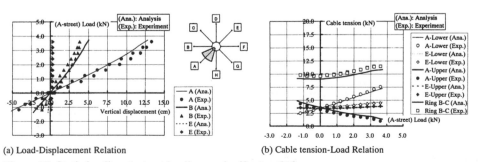

(a) Load-Displacement Relation

(b) Cable tension-Load Relation

Figure 17. Static loading, 1-street loading mode, (*Spring ON*)

5.2 Sine-wave excitation tests

Next, the results of dynamic problem are shown. Considering load case is all-area symmetrical loading mode with 0.5Hz sine-wave excitation.

Since the damping mechanism of device was not activated in the experiment, damping in the case of (*Spring ON*) is given as a 0.5% equivalent viscous damping against the eigenvalue, (5Hz; the vertical symmetry vibration mode), obtained by eigenvalue analysis. Analytical model mass is given to each node as concentrated mass and the masses of device movable part are not taken into consideration. The spring device parts connected to the radial cable-ends are modeled as the fixed boundary nodes here (Figure 10). The representative results (A street and ring cable between B-C) are shown in Figure 18-19.

In the case of (*Spring OFF*), upper and lower radial cable tension were lost and revive alternately. But analysis and experiment results are well in agreement.

In the case of (*Spring ON*), cable tension lost does not occur, but displacement amplitude of an experiment is small compared with analysis. Probably this result is caused by two main reasons: neglecting the inertia force of a device mass, which was relatively large compared with the cable members, and the under-estimation of total damping in this model. These points should be verified in future.

However, as a whole, good agreement can be seen in the experiment and the analysis results. The fundamental validity of employed analytical method can be confirmed.

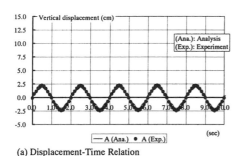

(a) Displacement-Time Relation

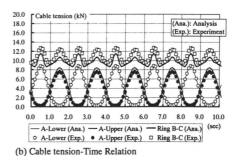

(b) Cable tension-Time Relation

Figure 18. 0.5Hz sine-wave excitation, All-area loading mode, (*Spring OFF*)

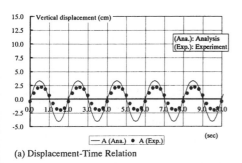

(a) Displacement-Time Relation

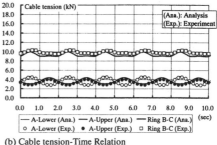

(b) Cable tension-Time Relation

Figure 19. 0.5Hz sine-wave excitation, All-area loading mode, (*Spring ON*)

6. CONCLUSION
In this paper, in order to solve the some existing problems of cable structure, we propose to install the devices consisting of variable-stiffness-spring and damper to the radial cable-ends. Proto-type devices were produced and some loading tests with circular cable-net model were performed. Moreover, these tests were examined by an analytical method, that is, non-linear spring and cable are combined to one element having an equivalent stiffness. In general, good agreement was found between the experiment and the analysis, and their fundamental validity was confirmed.

Although all studies are done with restricted conditions, that is, the damping mechanisms of device are not activated, the main results acquired under this condition are the following.
(1) A spring mechanism can delay the timing of cable tension lost. This effect means that the initial tension level introduced to cable members can be reduced.
(2) A spring mechanism has the effect of easing the imbalanced tension distribution between each street when partial loads are given.
(3) As for the dynamic problem on the (*Spring ON*) case, since the masses of device movable part are relatively large compared with the cable mass in this experiment case, it turned out that evaluation of its inertia force is needed. But in the real case, for example on the 100m- 200m span structures, their mass effects maybe become smaller and negligible.
(4) Although the analysis method is extremely simplified, say approximating the non-linear spring characteristic by the folding lines, the solutions obtained here are accurate enough for practical usage.

As for the proposing device, of course, optimal spring and damping character should be designed for the structure scale, roof shape or external loads level. But we think the following are important problems, which should be clarified on the fundamental research stage, and further studies are expected.
(1) Verification of the structural characteristics against wind excitation with the condition of device damping mechanism acted.
(2) Evaluation of wind and snow load as distributed load and its influence to the structure and the devices.
(3) Verification of device temperature dependability and reliability over a long period of time.
(4) Development of the device functions or utilization methods under the construction.

REFERENCES
1. Recommendations for Design of Cable Structures, Architectural Institute of Japan, 1994
2. Recommendations for Loads on Buildings, Architectural Institute of Japan, 1997
3. Edited by Kazuo Ishii, Gottlieb-Daimler Stadium (Schlaich Bergermann and Partners), Membrane Designs and Structures in the World, Shinkenchiku-sha, 1999, pp 98-103
4. T.Takeuchi, et al, Vibration control of tension structures by viscoelastic material with spring, J.Struct.Constr.Eng., AIJ, No.527, 2000.1, pp 117-124
5. T.Takeuchi, et al, Dynamic tests on viscoelastic dampers directly connected to tension strings, J.Struct.Constr.Eng., AIJ, No.534, 2000.8, pp 87-93

Retracting long span roof, Colonial Stadium, Melbourne Australia

M. L. SHELDON, B. K. DEAN
Connell Mott MacDonald, Melbourne

INTRODUCTION

Colonial Stadium, in Melbourne's Docklands, is a 52,000 seat multi-purpose venue featuring one of the largest retractable roofs in the world.

Colonial Stadium has been designed to function as an open stadium, with a roof which can be closed when required so that the venue can be used for a much wider range of functions and events than an open arena. Although the main sport to be played at the stadium will be AFL football, the venue will also host other sports such as soccer and rugby, which use a smaller pitch. To facilitate this, the lower level of the stand incorporates a retractable seating tier, which, when extended, will bring 12,500 spectators closer to the rugby/soccer field of play.

It is the exposed steel roof however which is the dominant feature of this stadium. A fundamental criterion of the roof geometry was the need to gain maximum sunlight onto the natural turf pitch. This requirement dictated that the roof opening should be as large as possible, and that the roof profile should be kept low to minimise shadows on the pitch, throughout the day at all times of the year. Conversely, the need to cover as many spectators as possible, and ensure each seat had world-standard sightlines tended to increase the shading on the turf.

Fig. 1. Colonial stadium alongside the City of Melbourne.

A number of alternative steel roof schemes were investigated with each being benchmarked for opening size, buildability, steel tonnage, cost, and most importantly, reliability of the moving roof system.

CONCEPT DESIGN

In order to make Colonial Stadium an "all-weather" facility, the design team reasoned that the revenue-earning potential of the stadium would be increased significantly if a moveable roof were to be provided. This provided the ability to fully enclose the stadium enabling sporting matches, concerts and other events to be held at any time, night and day, under any weather conditions. It was also realised that the closed roof could reduce the volume of crowd noise heard outside the stadium in nearby residential areas.

Key features of the Stadium include:
- covered seating for 52,000 spectators
- dining areas for 6,500 patrons
- 66 private spectator boxes
- carparking below the pitch and beneath the stands for 2,500 cars
- acoustically treated roof for concert mode
- a naturally ventilated stadium
- moveable seating tiers for approximately 12,500 spectators.

Figure 2 shows the plan of the stadium, its elliptical shape having a major axis dimension of 245m, and minor axis dimension of 215m.

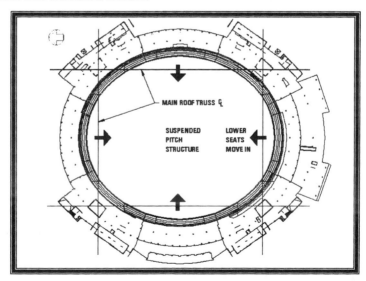

MAIN ROOF TRUSS ℄

SUSPENDED
PITCH
STRUCTURE

LOWER
SEATS
MOVE IN

Fig. 2. Plan of stadium.

Structural Form

The large perimeter surrounding a typical AFL pitch meant that the required seating capacity could be readily accommodated in a bowl in which even the uppermost and most rearward of seats would still be relatively close to the playing action. Whilst a large proportion of the seating could be positioned in a complete ring, it was decided early in the evolution of the architectural design, to "hold back" part of the upper tier at four "corners". This was to provide scope for scoreboards that must be highly visible, without obstructing the view of spectators. This concept also assisted in the structural economy of the great expanse of roof as the cores to be constructed at each of theses "corners" became the main supporting elements for the roof framing.

Figure 3 shows the typical cross section through the stand, with its three seating tiers, and suspended pitch with carparking below.

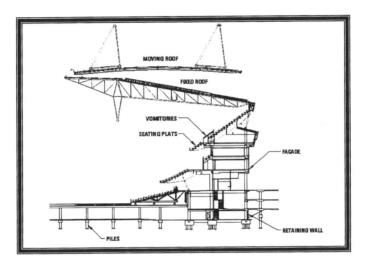

Fig. 3. Typical cross-section through stand.

Roof Concepts
Once the decision had been made that part of the roof should be retractable, the attention of the design team turned to devising alternative roof framing configurations and systems which would minimise these costs. Reducing the mass of the roof structure became a primary engineering objective, particularly because the poor ground conditions necessitated extensive piling to support the structure, which could be reduced if the mass of the superstructure were reduced.

The requirement to reduce roof mass, together with the possibility of creating a complete ring, which is an inherently efficient form, suggested that a shell or tensegrity structure might be appropriate.

In order for the stadium to operate in the closed concert mode, passive acoustic characteristics were preferred for the roof, which ultimately precluded the use of various fabric structures. The need to maximise the incidence of natural light on the pitch was considered vital to the facility's future operation, as was the avoidance of distinct shadows that might confuse players and/or spectators and diminish the quality of television coverage.

From these considerations, the following options were developed and reviewed. Acknowledgment is made of the key role Mr Steve Morley of Modus Consulting (now part of Sinclair Knight Merz) played in developing some of these schemes.

Option 1 – Toroidal Shell

This scheme used a segment of the surface of a torus making use of the elliptical plan of the stadium. Relatively efficient shell action was achieved for the fixed portion of the roof, with a compression ring at the inner boundary balanced by a tension ring on the outer boundary. Conventional trusses supported the moving roof panels.

The geometry of the bowl contributed to the aesthetic appeal of this roof configuration. However this geometry required a fairly complex motive system.

Option 2 – Shell Roof with Fabric Iris

This scheme investigated an innovative way of closing the roof over the infield using folding trusses and fabric, like the webbed wings of a bat.

The fixed roof was again a shell structure supported, in part, by an inner compression ring from which sprang a series of curved steel trusses, hinged at their ends about a vertical axis, as shown in Figure 4. In the closed position the trusses cantilevered up to 50m, with the end moments being resisted by the structure of the shell. Between the trusses, valley cables stressed by hydraulic pumps located at the truss tips tensioned the structural fabric. By rotating the trusses in plan until they folded against the inner compression ring, the roof could be opened with the tension in the fabric being maintained by varying the hydraulic pressures controlling the tension in, and position of, the valley cable.

More active acoustic control was required with this scheme to reduce noise breakout and this, coupled with the fact that such a scheme was untried on a scale of such magnitude, contributed to the decision not to pursue this innovative scheme.

Option 3 – Space Truss Scheme

The third scheme investigated was a space truss design. This roof scheme comprised four main trusses framing a 165m x 100m opening above the pitch, which supported the fixed roof secondary trusses spanning from the rear of the upper tier.

After reviewing the above mentioned options, plus others, this space truss scheme was ultimately adopted.

The main attributes of this scheme include the following:

- the flat roof profile maximises the incidence of natural light on the pitch
- the horizontal crane rail beam enables wheel driven bogies to be used, thus avoiding the use of cable-driven or rack-and-pinion driven system
- it is an economic and efficient design.

THE MOVING ROOF

The moveable roof consists of two panels, each 168 metres long and 52 metres wide, which run on rails and park over the top of the fixed east and west roof structures. Each of these panels is supported by two three-dimensional steel trusses, which are approximately 12 metres high with a clear span of 165 metres. Each of these roof panels weigh in excess of 1000 tonne, and during closing will travel the 50 metres in less than 10 minutes.

Each of the retractable roof panels is 168m x 52m and consists of two three-dimensional arch trusses spaced approximately 30m apart supporting the underslung rafters and roof. The curvature of the top chord is such that under gravity loads, the structure works more like a tied arch that a pure truss. The lack of space on the site precluded the assembly and jacking of a complete roof panel in one piece.

The methodology adopted was to split each panel longitudinally into two half-plates, each still 168m long, but 26m wide and formed by one of the three-dimensional trusses and half the underslung roof structure. Two temporary jacking towers approximately 50 m high on piled foundations were provided at each end and spaced at the same width as the truss bearing points on the mechanical drive bogie.

Temporary stillages were provided to allow the roof rafters, bracing and purlins to be erected at about 3.5 – 4.0m above ground level. This was to allow subsequent installation of the ceiling lining from below. These stillages were preset to allow for the vertical pre-camber of approximately 480mm in the truss.

Once the entire assembly was complete, including the drive bogies, jacking cables from a 600t capacity jack at the top of each tower were connected to the lifting eyes at the four corners of each truss.

The half plate was then de-propped (ie jacked up about 1.8m) and comprehensively surveyed to confirm the deformations with those predicted. As well as the vertical deflection and elastic elongation mentioned above, the trusses were also expected to twist and bow horizontally due to their asymmetry.

The jacking for each half plate took 16-20 hours to jack the 45 metres it had to travel in 450mm increments. At the top, horizontal needle beams were extended out from the towers, underneath the temporary rail and runway beam to take the load off the jacks.

Fig. 4. Raising one of the four half-plates.

After alignment, the roof plate was then traversed on temporary runway trusses onto the permanent structure using hand tirfers.

To minimise shadowing, the front plane of each truss in each roof panel is laid back away from the centre of the stadium. Also, each half plate is 26m wide, with the longitudinal centreline between bearing posts offset four metres from the longitudinal centreline of the roof panel. A consequence of this asymmetry is that each of the two half plates were predicted to deflect differentially when de-propped.

Initial calculations showed that the abutting rafter ends between each half plate would be out of alignment by approximately 900mm vertically. This problem was overcome by omitting various different areas of roof cladding and purlins on each half plate, so that the deflections of each half plate were similar. The infill roofing was then lifted and erected after the half plates had been spliced.

It was also necessary to ensure that the combination of stresses in the half plates due to erection, and in-service conditions in the combined panel were within the design limits. This required the reduction of residual stresses in one of the trusses in each panel prior to splicing. This was achieved by internally stressing the structure in a horizontal plane using a king post at midspan and cables to each end, similar to a giant Barrup truss. The king post was jacked until a predetermined horizontal displacement was achieved, and then the two half plates butted together. The twenty-three rafters along the 168m length of the two panels were then bolted together without the need for any site elongation of the bolt holes.

Comparison with the roof framing of other stadia has indicated this to be an extremely efficient and lightweight framing system. The largest roof-framing members in the project are the Grade 450 610mm diameter x 58mm thick CHS truss chords, with thin-walled 508mm diameter CHS chords being used in the moving roof trusses.

Wind Testing
Wind loading was critical to the structural design of a roof of this size, and so it was necessary to initiate a wind-tunnel test program during the design development phase in order to produce a cost-effective structure.

The design technique used a series of influence coefficients derived from the structural analysis to simulate actual wind pressure combinations acting on the roof. This process of iteration between the wind tunnel specialists and the structural engineer was time consuming, but it enabled lower design pressures to be justified for the design of the structure.

As the chord members for the main trusses are quite slender (168 dia spanning over 15metres), tuned dampers were installed in certain locations to control wind-driven oscillations.

Moving Roof Traction System
A number of roof traction systems were investigated, including electrically driven bogies, cable/winch drives, and rack-and-pinion systems.

The electrically driven bogie system was selected as the most economic and foolproof, the details of which are shown in Figure 5. Eight bogies travel along twin rail tracks supported by the Truss Type T3 support each moving roof panel. Bearings are provided at each bogie to provide for thermal expansion and rotation movements of the moving roof panels. The system has electronic controls to prevent skew of the moving roof.

Fig. 5. Moving roof bogies under fabrication.

THE FIXED ROOF

The fixed roof on the Colonial Stadium continues well forward of most of the seats in the stadium so that, even with the roof open, 98% of the seats are under cover behind the drip line. The concept for this fixed roof structure is simple, but elegant. Four major steel trusses form the main supports for the rectangular opening in the fixed roof, including bowstring trusses on the east and west sides which are 12 metres deep and span 150 metres. These trusses are underslung, to enable the moving roof panels to pass over in an east-west direction. Supporting the moving roof rails on the north and south ends are arched trusses rising 14 metres above the roof, and spanning 120 metres. Stability of the top chord member is a major consideration. Buckling is restrained by the fixed roof secondary trusses T1 which span back to bearings at the rear of the stand.

East & West Fixed Roofs

The east and west roofs consist of a major bow string truss which is formed by a top chord and two mid chords, approximately 4.5m below the top chord, and a parabolic shaped bottom chord, which is a further 8m below the mid chords at mid-span. The top chord and mid chords form a "triangular" shaped truss which also includes the final catwalk structure.

The adopted erection sequence was based upon the erection of this triangular truss over temporary support towers, followed by the installation of the parabolic bottom chord prior to de-propping, or loading the structure with roof cladding. In order to satisfy the residual stress limitations and deflection criteria, a maximum span between temporary towers of 37.5m was adopted, and each tower preset to height to allow for the truss precamber.

The underslung bottom chord (610mm diameter CHS) was butt-welded on site into one length of approximately 75m, prior to lifting into position for a further two major butt welds, to be made at the outer temporary tower locations at height. The temporary towers were designed to allow the de-propping to be then carried out, whilst still providing lateral restraint to the truss. As the precamber was significant, the de-propping was performed incrementally to a sequence agreed with Connell Mott MacDonald, over a period of approximately six hours. Upon completion, the secondary roof trusses, cantilever trusses, bearing and roof panels were erected.

North & South Fixed Roofs

The structure of the north and south roof is a consists of the arched truss above the roof cladding, the actual runway truss below rail level and the secondary roof trusses which span back to the stadium perimeter. The stability of the arched truss is dependent upon the secondary trusses below. It was decided to erect the runway truss section over the top of temporary towers, install the secondary roof trusses spanning back to the stadium perimeter and then assemble the main arch over the top. Once the arch truss was completed, the temporary towers were removed and roof panels lifted into place.

Again, detailed calculations were prepared to allow for the "erection" deformations between temporary towers, the preset of the height of the temporary towers and the limitations of the residual stresses in the final condition.

The adopted solution allowed for the runway truss to be propped at 30m spacing, with the arched roof truss assembled at the ground into two 60m halves prior to lifting. At 130 tonne each, these were the heaviest single lifts on the job. Fabrication accuracy was paramount as the arch truss had to match up to spigot connections on the runway truss at a total of 34 locations, with each typical spigot featuring 70mm thick plates.

Fig. 6. Construction of the runway trusses at north and south ends.

ROOF CONSTRUCTION ISSUES
The initial task for the design team was to establish the stadium geometry, and the Engineers and Shop Detailers created a three-dimensional wire frame model of the entire structure (bowl steelwork, fixed roof and retractable roof). After the member sizes were established, this wire frame model was built up into a solid model from which dimensions such as clearances between the retractable roof and the fixed roof, and their secondary members, could be readily confirmed. The model was then adjusted to allow for the necessary pre-camber of trusses, and, where applicable, pre-set of temporary support towers, and pre-camber of sub-assemblies between temporary support towers.

From this information the actual members and assemblies were detailed.

Each of the CHS nodes in the major three-dimensional trusses was of complex geometry. Some nodes were composed of six intersecting members and through-plate stiffeners to transfer local effects.

Tube profiles at the ends of truss webs/chords were generated by the shop detailers and given to the steel workshop in electronic format. The electronic input of tube profiles into the CNC controlled tube cutter allowed very quick, efficient and accurate cutting of the ends of all tubes, including bevelling for weld preparations. Some of the tubular sections were 58mm thick.

Also, prior to detailing, a critical review of tolerances and the provision of erection packers at bolted joints was undertaken. For example, the retractable roof trusses were essentially supported on a fixed bearing at the north end and a sliding bearing at the south end. Stringent fabrication tolerances were agreed to at the outset to allow criteria for the bearing design to be established.

The fixed roof was to be erected from within the stadium using a combination of large cranes including the latest Demag CC2800 crane (600t capacity).

MOVING TIERS
Another feature of the Colonial Stadium is the provision of moving tiers in the lower level of the stadium, as can be seen in Figure 2. Single steel frames on both the east and west sides of the pitch 115m long are able to be rolled forward 18 metres to bring the spectators close to the boundary of the soccer or rugby pitch. Similarly, at the north and south ends, structural steel frames 75 metres long are able to be rolled forward 11 metres to bring the seats closer to the pitch. In total, 12,500 seats are brought forward in the soccer/rugby mode. These massive frames consist of steel trusses, bracing, and folded steel plate.

SUMMARY
The Colonial Stadium, in Melbourne's Docklands, is a "latest generation" multipurpose facility designed to provide maximum flexibility for the operators. To provide such a facility, significant challenges were set for the designers and the construction team. Within a very tight program, features such as one of the largest moving roofs in the world, one of the largest moving seat arrangements in the world, Australia's first pitch suspended over a car park, and a stand designed to provide Colosseum-like atmosphere for spectators had to be incorporated on a site offering difficult ground conditions.

Numerous structural schemes were evaluated during the concept development of the stadium. These were then refined and finalised following input from the construction team, taking into account the progressive erection methodology developed by the team, and possible locked-in effects of partially completed elements experiencing load.

Fig. 7. Colonial Stadium.

COLONIAL STADIUM FACTS
- $A460 million (total project value)
- Completion – February 2000
- 52,000 seats
- One of the world's largest retractable roofs (168m x 102m)
- 12,500 seats in retractable tiers, capable of moving up to 18 metres
- 6,500 seats in dining facilities
- 66 corporate boxes
- 2,500 car spaces within stadium
- Natural turf playing surface, suspended over carpark
- Nightclubs, sports bars, studios and associated facilities
- Approximately 6000 tonnes of structural steelwork

Architects:	Daryl Jackson Bligh Lobb Sports ArchitectureJV
Structural Engineer:	Connell Mott MacDonald/ Modus Consulting Engineers (UK)
Building Services Engineer:	Connell Mott MacDonald
Civil Engineer:	Connell Mott MacDonald
Design and Construct Contractor:	Baulderstone Hornibrook
Main Steel Contractor:	Alfasi Constructions Pty Ltd

REFERENCES
1. DEAN, B K AND MORLEY, S. Victoria Docklands Stadium, Melbourne. LSA'98 International Congress, Sydney, October 1998.
2. SHELDON, M AND SHELDON, R. Colonial Stadium – A Highly Innovative Multipurpose Stadium Journal of the Australian Institute of Steel Construction, Vol 33, No 4.

Spatial Foldable Structures

P. FARRUGIA, Bezzina & Cole, Architects and Engineers, Malta and Space Structures Research Centre, University of Surrey, UK

SYNOPSIS

A spatial foldable structure may be defined as a structure that can change its shape in three-dimensions. One of the possible ways to generate spatial foldable systems is to use scissors-like units called duplets with multiple degrees of freedom at the joints. A novel type of scissors-like unit called a calix duplet is presented in the first part of the paper. The joint employed in a calix foldable structure is a modified Universal joint. This joint has two degrees of freedom. The geometric and kinematic properties of the calix duplet are presented. Different foldable shapes and forms may be obtained using the calix duplet including barrel vaults, grids and domes. The second part of the paper focuses on the matrix analysis of foldable systems and finite mechanisms in general. Based on an algebraic solution of interelemental constraint equations, the overall degree of freedom of a foldable system is determined. The proposed method identifies the different modes of displacement and the presence and location of bifurcation points.

INTRODUCTION

The analysis and design of deployable and foldable structures has been the subject of considerable research during the recent decades (Refs 1-3). The research in this area has been particularly useful in relation to the applications in outer Space. New concepts and innovations are continuously being developed in parallel with new mathematical and structural analysis solutions.

CALIX FOLDABLE STRUCTURES

Definitions

Scissors-like units called duplets may be used to generate foldable structures. A duplet is shown in Fig. 1. The duplet is composed of two members called uniplets. In general, duplets are pin-connected. Each joint in a foldable structure composed of duplets has one degree of freedom- a single axis of rotation.

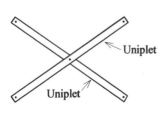

(a) Plan

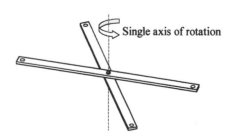

(b) Perspective view

Fig. 1 A pin connected duplet

Space Structures 5, Thomas Telford, London, 2002

The Calix Duplet

Fig. 2(a) shows a novel type of duplet called a calix duplet. The calix duplet is used to generate foldable structures similar to the example shown in Fig. 3. The calix duplet has three components, namely, two calix uniplets and an intermediate body consisting of a central cross-shaped connector. These are illustrated in Figs. 2(b), 2(c) and 2(d). A calix duplet is assembled as shown in Fig. 4 noting that the dimensions of the central ring of one of the uniplets is larger than the ring of the other uniplet such that the rings can fit loosely within one another. This is required so that the uniplets can rotate freely and independently. Each arm of the cross-shaped connector is the pivot for the calix uniplet that is attached to it. The arm determines an independent axis of rotation for each uniplet. A similar cross-shaped connector is used to connect calix duplets together as indicated in Fig. 3 to form a calix foldable structure.

The calix duplet differs from the duplet shown in Fig. 1 since each uniplet in a calix duplet has an independent axis of rotation determined by the cross-shaped connector that connects the calix uniplets. As a result, the joints in a calix foldable structure have two degrees of freedom.

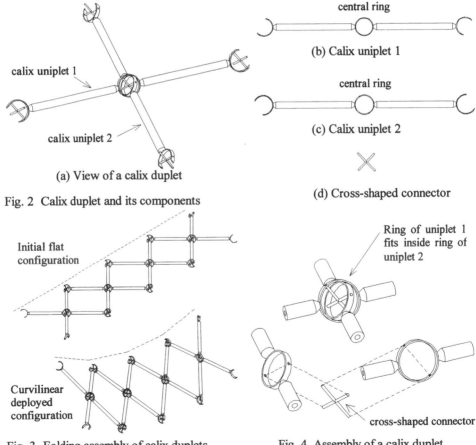

central ring

(b) Calix uniplet 1

calix uniplet 1

central ring

(c) Calix uniplet 2

calix uniplet 2

(a) View of a calix duplet

(d) Cross-shaped connector

Fig. 2 Calix duplet and its components

Initial flat configuration

Ring of uniplet 1 fits inside ring of uniplet 2

Curvilinear deployed configuration

cross-shaped connector

Fig. 3 Folding assembly of calix duplets

Fig. 4 Assembly of a calix duplet

The displacement characteristics of a calix duplet

Fig. 5 shows a schematic view of a calix duplet. Suppose that the calix duplet lies initially in the X-Y plane. If an equal and opposite displacement in the X-direction, Δ_X is applied at A and B, it can be proved that the central node rises by

$$\Delta_Z = \Delta_X \tan \theta \tag{1}$$

where θ is the angle subtended by the arms of the cross-shaped connector with the Y-Z plane as shown in Fig. 6.

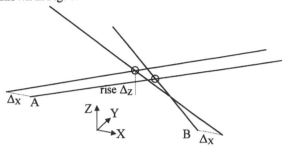

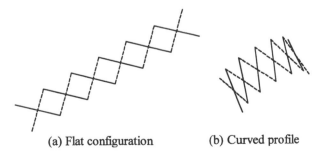

Fig. 5 Displacement of a calix duplet

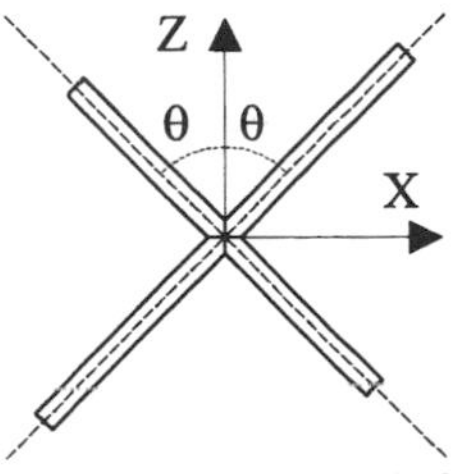

Fig. 6 Angle θ subtended by the arms of the cross-shaped connector with the Y-Z plane

Here the angle between the arms of the cross-shaped connector and the Y-Z plane is the same. If each arm subtends a different angle, say θ_1 and θ_2, it can be proved that Equation (1) becomes

$$\Delta_Z = \frac{2\Delta_X \tan\theta_1 \tan\theta_2}{(\tan\theta_1 + \tan\theta_2)} \tag{2}$$

Examples of Calix Foldable Structures

Example 1: Calix duplet chain

A chain of calix duplets is shown in Fig. 7. From an initial flat configuration, a curved profile is obtained in Fig. 7(b). Continuing the deployment process, the structure coils into a relatively thin cylindrical shape. A chain of calix duplets may be employed in different ways. For example, to obtain a curved profile from an initial flat position and perhaps more importantly, to obtain a flat or curved surface from a compact cylindrical assembly.

(a) Flat configuration   (b) Curved profile

Fig. 7 Chain of calix duplets at different foldable states

Example 2: Calix Foldable Grid
Fig. 8 shows a grid of calix duplets. In this example, the orientation of the calix duplets is vertical. From an initial rectangular form, the structure adopts a curvilinear form shown in the top view of Fig. 8(b).

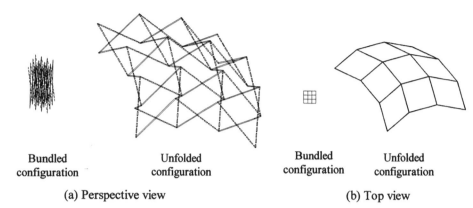

| Bundled configuration | Unfolded configuration | Bundled configuration | Unfolded configuration |

(a) Perspective view   (b) Top view

Fig. 8 Foldable grid using the calix duplet

Example 3: Calix Foldable Dome
The calix duplet employed in examples 1 and 2 is straight. The third example adopts a non-straight calix duplet. The free folding dome shown in Fig. 9 is composed of angulated calix duplets.

Fig. 9 Foldable dome assembly using angulated calix duplets

Calix, which is the Latin term for cup, reflects the curvilinear form of calix foldable structures and in particular the free folding dome obtained using the calix duplet.

MATRIX MODELLING AND KINEMATIC ANALYSIS OF FINITE MECHANISMS
One of the ways to perform the kinematic analysis of pin-connected frameworks is by manipulating the equilibrium matrix (Ref 4). However, this approach cannot be applied to foldable systems such as calix foldable structures since generally these are not pin-connected. A different formulation is necessary to model the geometric properties of foldable systems including topology, geometry and degrees of freedom. A matrix approach for the kinematic analysis and geometric design of foldable systems and finite mechanisms is proposed.

The proposed matrix method is based on the solution of interelemental constraint equations. Interelemental constraint equations are expressions that define the topology and geometry of interconnected members. In many texts concerned with the kinematic and dynamic modelling of multi-body systems, partial differentiation is applied to interelemental constraint equations to obtain differential equations (Refs 5-6). This is carried out to solve the dynamics of the problem. The solution is then obtained by using numerical techniques. The method presented here focuses on the fundamental vector spaces of interelemental constraint equations.

In kinematic theory, the overall degree of freedom, also called mobility, is generally addressed in topological terms where the number of freedoms and constraints are equated to obtain the overall degree of freedom, (Refs 7-8). However, this approach is not without problems. In certain instances, it may fail to give the correct degree of freedom. An example is a spatial four bar linkage known as a Bennett Linkage (Ref 9). The method outlined in this paper departs from the more traditional approach of computing mobility. Geometric and topological definitions are incorporated straight away in the interelemental constraint equations that model the structure, thereby obtaining an unequivocal picture of the kinematic behaviour of the structure.

Interelemental constraint equations

Fig. 10(a) shows member M_{12} in a two-dimensional coordinate system. The nodal coordinate of node 2 is given by the expression:

$$p_2 = p_1 + A_{12} \, p'_{2(12)} \tag{3}$$

where p_1 and p_2 are nodal coordinates for nodes 1 and 2 in the X-Y global coordinate system and $p'_{2(12)}$ is the nodal coordinate of node 2 in a local coordinate system, X'_{12}-Y'_{12}, associated with member M_{12}. The origin of the local coordinate sytem, X'_{12}-Y'_{12}, is located at node 2. A_{12} is a transformation matrix between the local co-ordinate system X'_{12}-Y'_{12} and the global coordinate system X-Y. In a way it is like going from node 1 to node 2.

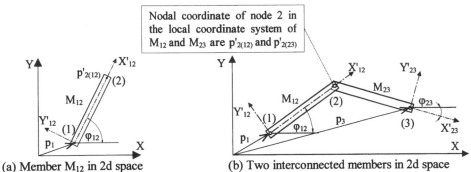

(a) Member M_{12} in 2d space   (b) Two interconnected members in 2d space
Fig. 10 Members in a two-dimensional space

In Fig. 10(b), two members M_{12} and M_{23}, of length L_{12} and L_{23}, respectively, are connected together at node 2. Note here that the local coordinate system X'_{23}-Y'_{23} associated with member M_{23} has its origin at node 3. The nodal coordinate of node 2 is obtained by going from node 1 to node 2 and from node 3 to node 2. In mathematical terms the nodal coordiante of node 2 is given by

$$p_2 = p_1 + A_{12} \, p'_{2(12)} = p_3 + A_{23} \, p'_{2(23)} \tag{4a}$$

This implies that

$$p_1 - p_3 = -A_{12} \, p'_{2(12)} + A_{23} \, p'_{2(23)} \tag{4b}$$

Expanding equation (4b)

$$\begin{bmatrix} x_1 \\ y_1 \end{bmatrix} - \begin{bmatrix} x_3 \\ y_3 \end{bmatrix} = -\begin{bmatrix} \cos(\varphi_{12}) & -\sin(\varphi_{12}) \\ \sin(\varphi_{12}) & \cos(\varphi_{12}) \end{bmatrix}\begin{bmatrix} L_{12} \\ 0 \end{bmatrix} + \begin{bmatrix} \cos(\varphi_{23}) & -\sin(\varphi_{23}) \\ \sin(\varphi_{23}) & \cos(\varphi_{23}) \end{bmatrix}\begin{bmatrix} -L_{23} \\ 0 \end{bmatrix} \qquad (4c)$$

Equation (4c) is an interelemental constraint equation. It establishes the following relationships:

    (i)   members M_{12} and M_{23} are interconnected at node 2 which has a pin joint.

    (ii)   node 2 allows one degree of freedom - rotation in the X-Y plane about node 2.

Using expressions similar to Equation (4c), it is possible to generate in a systematic manner a set of interelemental constraint equations that model the full geometry and topology of a structure. The members of the structure may be connected by pins, pivots and sliding joints. Furthermore, the members in the structure may be lines, duplets or a combination of both.

Setting up interelemental constraint equations

Reuleaux defined an individual joint in a mechanism, which he called a kinematic pair, as a joint connecting a pair of members (Ref 10). There may be more than one kinematic pair at a joint depending on the number of members that are connected at the joint. In a structure, it is possible to construct a set of interelemental constraint equations for each kinematic pair. The maximum number of kinematic pairs and hence the number of interelemental constraint equations that can be formulated for a system is:

$$J_K = \sum_{i=1}^{j}(M_i - 1) ! \qquad (5)$$

where J_K is the number of kinematic pairs, j is the number of joints in the system and M_i is the number of members that frame at each joint.

Fig. 11 shows a duplet composed of two uniplets 1-3-5 and 2-3-5 connected to two members 4-6 and 5-6.

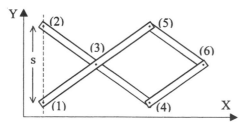

Member	Length
1-3-5, 2-3-4	2 units
5-6, 4-1	1 unit

Joints 1 and 2 are constrained to a line parallel with the Y axis

Fig. 11 Two-dimensional foldable assembly

In this example joints 1 and 2 are constrained to a line parallel with the Y-axis. The distance s between joints 1 and 2 may be regarded as a member having a variable length and subtending a constant angle with the X and Y axes. There are two kinematic pairs at joint 1 and hence two interelemental constraint equations are formulated for joint 1. One interelemental constraint equation for joint 1 is obtained by considering the vector going from joint 2 to joint 1 and the vector going from joint 3 to joint 1. Applying equation (4c):

$$\begin{bmatrix} x_2 \\ y_2 \end{bmatrix} - \begin{bmatrix} x_3 \\ y_3 \end{bmatrix} = -\begin{bmatrix} \cos(\varphi_{12}) & -\sin(\varphi_{12}) \\ \sin(\varphi_{12}) & \cos(\varphi_{12}) \end{bmatrix}\begin{bmatrix} -L_{12} \\ 0 \end{bmatrix} + \begin{bmatrix} \cos(\varphi_{13}) & -\sin(\varphi_{13}) \\ \sin(\varphi_{13}) & \cos(\varphi_{13}) \end{bmatrix}\begin{bmatrix} -L_{13} \\ 0 \end{bmatrix} \qquad (6a)$$

In this example $\varphi_{12} = 90°$, $L_{12} = s$ and $L_{13}=1$. Substituting these values in Equation (6a)

$$\begin{bmatrix} x_2 \\ y_2 \end{bmatrix} - \begin{bmatrix} x_3 \\ y_3 \end{bmatrix} = -\begin{bmatrix} 0 & -1 \\ 1 & 0 \end{bmatrix}\begin{bmatrix} -s \\ 0 \end{bmatrix} + \begin{bmatrix} \cos(\varphi_{13}) & -\sin(\varphi_{13}) \\ \sin(\varphi_{13}) & \cos(\varphi_{13}) \end{bmatrix}\begin{bmatrix} 1 \\ 0 \end{bmatrix} = \begin{bmatrix} -\cos(\varphi_{13}) \\ s - \sin(\varphi_{13}) \end{bmatrix} \tag{6b}$$

Equation (6b) may be rewritten in an algebraic form as follows:

$$\begin{bmatrix} 1 & -1 \end{bmatrix}\begin{bmatrix} x_2 & y_2 \\ x_3 & y_3 \end{bmatrix} = \begin{bmatrix} -\cos(\varphi_{13}) & s - \sin(\varphi_{13}) \end{bmatrix} \tag{6c}$$

The other interelemental constraint equation for joint 1 is perhaps not so obvious. It is obtained by considering the vector going from joint 3 to joint 1 and the vector going from joint 5 to joint 1. This defines the uniplet 1-3-5. Applying again Equation (4c), the second interelemental constraint equation for joint 1 is:

$$\begin{bmatrix} x_3 \\ y_3 \end{bmatrix} - \begin{bmatrix} x_5 \\ y_5 \end{bmatrix} = -\begin{bmatrix} \cos(\varphi_{13}) & -\sin(\varphi_{13}) \\ \sin(\varphi_{13}) & \cos(\varphi_{13}) \end{bmatrix}\begin{bmatrix} -L_{13} \\ 0 \end{bmatrix} + \begin{bmatrix} \cos(\varphi_{15}) & -\sin(\varphi_{15}) \\ \sin(\varphi_{15}) & \cos(\varphi_{15}) \end{bmatrix}\begin{bmatrix} -L_{15} \\ 0 \end{bmatrix} \tag{7a}$$

where $\varphi_{13} = \varphi_{15}$, $L_{13} = 1$ and $L_{15} = 2$. Substituting these values in Equation (7a)

$$\begin{bmatrix} x_3 \\ y_3 \end{bmatrix} - \begin{bmatrix} x_5 \\ y_5 \end{bmatrix} = -\begin{bmatrix} \cos(\varphi_{13}) & -\sin(\varphi_{13}) \\ \sin(\varphi_{13}) & \cos(\varphi_{13}) \end{bmatrix}\begin{bmatrix} -1 \\ 0 \end{bmatrix} + \begin{bmatrix} \cos(\varphi_{13}) & -\sin(\varphi_{13}) \\ \sin(\varphi_{13}) & \cos(\varphi_{13}) \end{bmatrix}\begin{bmatrix} -2 \\ 0 \end{bmatrix} = \begin{bmatrix} -\cos(\varphi_{13}) \\ -\sin(\varphi_{13}) \end{bmatrix} \tag{7b}$$

Equation (7b) is again rewritten in algebraic form as

$$\begin{bmatrix} 1 & -1 \end{bmatrix}\begin{bmatrix} x_3 & y_5 \\ x_3 & y_5 \end{bmatrix} = \begin{bmatrix} -\cos(\varphi_{13}) & -\sin(\varphi_{13}) \end{bmatrix} \tag{7c}$$

There are 15 kinematic pairs in the system of Fig. 11. The interelemental constraint equations are obtained by using the same approach adopted for joint 1. Table 1 summarizes the procedure for setting up the constraint equation for this example.

Joint reference number in Fig 11	M_i	interelemental constraint equations (2-1/3-1 means that the interelemental constraint equation is assembled by considering the vector from joint 2 to joint 1 and the vector from joint 3 to joint 1)						Number of interelemental constraint equations at each joint
1	3	2-1/3-1	3-1/5-1					2
2	3	1-2/3-2	3-2/4-2					2
3	4	1-3/2-3	1-3/4-3	1-3/5-3	2-3/4-3	2-3/5-3	4-3/5-3	6
4	3	2-4/3-4	3-4/6-4					2
5	3	1-5/3-5	3-5/6-5					2
6	2	5-6/4-6						1

$$J_K = 15$$

Table 1 Procedure for setting up the interelemental constraint equations for the foldable structure of Fig. 11

The value of J_K shown in Table 1 may be obtained directly by applying equation (5) to the foldable structure of Fig 11.

The assembly process results in a global interelemental constraint equation of the form

$$A_i [x, y] = [b_x, b_y] \tag{8}$$

Equation (8) has the same format like Equations (6c) and (7c) obtained for joint 1. The matrix A_i is a connectivity or topology matrix. A_i is called a "pair-joint" matrix. The order of A_i is J_K rows by j columns. The column vectors x and y in Equation (6) contain the joint coordinates of the system. The entries in the right hand side of Equations (6c) and (7c) and hence in $[b_X, b_Y]$ of Equation (8) are in terms of the angles subtended with one or more of the reference axes and member lengths. These vectors define the geometric properties of the members and the degrees of freedom of the joints. The vectors b_x and b_y thus contain member variables, which may be the angle between the member and the global coordinate system, the member length or both.

Equation (9) is the global interelemental constraint equation for the foldable system of Fig, 11. The incidence matrix A_i is of order 15 rows by 6 columns. Notice that the first two rows of A_i and $[b_X, b_Y]$ in Equation (9) are in fact Equations (6c) and (7c).

$$
\begin{bmatrix}
0 & 1 & -1 & 0 & 0 & 0 \\
0 & 0 & 1 & 0 & -1 & 0 \\
1 & 0 & -1 & 0 & 0 & 0 \\
0 & 0 & 1 & -1 & 0 & 0 \\
1 & -1 & 0 & 0 & 0 & 0 \\
1 & 0 & 0 & -1 & 0 & 0 \\
1 & 0 & 0 & 0 & -1 & 0 \\
0 & 1 & 0 & -1 & 0 & 0 \\
0 & 1 & 0 & 0 & -1 & 0 \\
0 & 0 & 0 & 1 & -1 & 0 \\
0 & 1 & -1 & 0 & 0 & 0 \\
0 & 0 & 1 & 0 & 0 & -1 \\
1 & 0 & -1 & 0 & 0 & 0 \\
0 & 0 & 1 & 0 & 0 & -1 \\
0 & 0 & 0 & 1 & -1 & 0
\end{bmatrix}
\begin{bmatrix}
x_1 & y_1 \\
x_2 & y_2 \\
x_3 & y_3 \\
x_4 & y_4 \\
x_5 & y_5 \\
x_6 & y_6
\end{bmatrix}
=
\begin{bmatrix}
-\cos(\varphi_{13}) & s-\sin(\varphi_{13}) \\
-\cos(\varphi_{13}) & -\sin(\varphi_{13}) \\
-\cos(\varphi_{23}) & -s-\sin(\varphi_{23}) \\
-\cos(\varphi_{23}) & -\sin(\varphi_{23}) \\
-\cos(\varphi_{13})+\cos(\varphi_{23}) & -\sin(\varphi_{13})+\sin(\varphi_{23}) \\
-\cos(\varphi_{13})-\cos(\varphi_{23}) & -\sin(\varphi_{13})-\sin(\varphi_{23}) \\
-2\cos(\varphi_{13}) & -2\sin(\varphi_{13}) \\
-2\cos(\varphi_{23}) & -2\sin(\varphi_{23}) \\
-\cos(\varphi_{13})-\cos(\varphi_{23}) & -\sin(\varphi_{13})-\sin(\varphi_{23}) \\
-\cos(\varphi_{13})+\cos(\varphi_{23}) & -\sin(\varphi_{13})+\sin(\varphi_{23}) \\
-\cos(\varphi_{23}) & -\sin(\varphi_{23}) \\
-\cos(\varphi_{23})-\cos(\varphi_{46}) & -\sin(\varphi_{23})-\sin(\varphi_{46}) \\
-\cos(\varphi_{13}) & -\sin(\varphi_{13}) \\
-\cos(\varphi_{13})-\cos(\varphi_{56}) & -\sin(\varphi_{13})-\sin(\varphi_{56}) \\
-\cos(\varphi_{46})-\cos(\varphi_{56}) & -\sin(\varphi_{46})-\sin(\varphi_{56})
\end{bmatrix}
\tag{9}
$$

This foldable system has 5 member variables, namely s, φ_{13}, φ_{23}, φ_{46} and φ_{56}. s refers to the varying length between joint 1 and 2, while the other four variables refer to the angles subtended by the members in the system with the X-axis.

Solution of global interelemental constraint equations
Equation (8) has a solution if $[b_X, b_Y]$ is in the column space of A_i. The column space of A_i, therefore is a basis for the system. In other words, the column space of A_i imposes conditions on the right hand side of Equation (8). The conditions are as follows:

For simplicity Equation (8) will be written as

$$A_i x = b \tag{10}$$

Let x_{ln} be the left null vector of A_i such that

$$A^T x_{ln} = 0 \tag{11}$$

Postmultiplying x_{ln}^T with b:

$$x_{ln}^T b = x_{ln}^T A x = (A^T x_{ln})^T x = 0 \; x = 0 \tag{12}$$

Equation (12) implies that the left null space is orthogonal to b. Hence the condition for b to be in the column space of A_i is given by:

$$x_{ln}^T b = 0 \qquad (13)$$

Equation (13) provides a method of obtaining a set of homogenous equations of the form

$$C\,m = 0 \qquad (14)$$

where C is called a dependency matrix for the system and m is a vector containing member variables. Equation (14) is generally non-linear and will therefore have more than one solution. Each solution reflects a mode of displacement of the foldable system. The solutions are expressed in terms of the free variables in m. Each solution of (14) has its own free variables, which may be different for each particular solution. The number of independent variables for each solution is a direct measure of the degree of freedom of the system.

It can be shown that the transpose of the left null vector of the matrix A_i in Equation (9) is:

$$\begin{bmatrix}
1 & 0 & -1 & 0 & 1 & 0 & 0 & 0 & 0 & 0 & 0 & 0 & 0 & 0 & 0 \\
0 & 0 & -1 & -1 & 0 & 1 & 0 & 0 & 0 & 0 & 0 & 0 & 0 & 0 & 0 \\
0 & -1 & -1 & 0 & 0 & 0 & 1 & 0 & 0 & 0 & 0 & 0 & 0 & 0 & 0 \\
-1 & 0 & 0 & -1 & 0 & 0 & 0 & 1 & 0 & 0 & 0 & 0 & 0 & 0 & 0 \\
-1 & -1 & 0 & 0 & 0 & 0 & 0 & 0 & 1 & 0 & 0 & 0 & 0 & 0 & 0 \\
0 & -1 & 0 & 1 & 0 & 0 & 0 & 0 & 1 & 0 & 0 & 0 & 0 & 0 & 0 \\
-1 & 0 & 0 & 0 & 0 & 0 & 0 & 0 & 0 & 1 & 0 & 0 & 0 & 0 & 0 \\
0 & 0 & -1 & 0 & 0 & 0 & 0 & 0 & 0 & 0 & 0 & 1 & 0 & 0 & 0 \\
0 & 0 & 0 & 0 & 0 & 0 & 0 & 0 & 0 & 0 & -1 & 0 & 1 & 0 & 0 \\
0 & -1 & 0 & 1 & 0 & 0 & 0 & 0 & 0 & 0 & 0 & 0 & 0 & 0 & 1
\end{bmatrix}$$

Postmultiplying the left null vector of A_i with [bx , by] generates the following equations in terms of the member variables:

$$\begin{bmatrix}
-2\cos(\varphi_{13}) + 2\cos(\varphi_{23}) & 2s - 2\sin(\varphi_{13}) + 2\cos(\varphi_{23}) \\
-\cos(\varphi_{13}) + \cos(\varphi_{23}) & s - \sin(\varphi_{13}) + \sin(\varphi_{23}) \\
-\cos(\varphi_{13}) + \cos(\varphi_{23}) & s - \sin(\varphi_{13}) + \sin(\varphi_{23}) \\
\cos(\varphi_{13}) - \cos(\varphi_{23}) & -s + \sin(\varphi_{13}) - \sin(\varphi_{23}) \\
\cos(\varphi_{13}) - \cos(\varphi_{23}) & -s + \sin(\varphi_{13}) - \sin(\varphi_{23}) \\
0 & 0 \\
\cos(\varphi_{13}) - \cos(\varphi_{23}) & -s + \sin(\varphi_{13}) - \sin(\varphi_{23}) \\
-\cos(\varphi_{13}) + \cos(\varphi_{23}) & s - \sin(\varphi_{13}) + \sin(\varphi_{23}) \\
\cos(\varphi_{23}) + \cos(\varphi_{46}) - \cos(\varphi_{13}) - \cos(\varphi_{56}) & \sin(\varphi_{23}) + \sin(\varphi_{46}) - \sin(\varphi_{13}) - \sin(\varphi_{56}) \\
\cos(\varphi_{13}) - \cos(\varphi_{23}) - \cos(\varphi_{46}) + \cos(\varphi_{56}) & \sin(\varphi_{13}) - \sin(\varphi_{23}) - \sin(\varphi_{46}) + \sin(\varphi_{56})
\end{bmatrix} = \begin{bmatrix}
0 & 0 \\
0 & 0 \\
0 & 0 \\
0 & 0 \\
0 & 0 \\
0 & 0 \\
0 & 0 \\
0 & 0 \\
0 & 0 \\
0 & 0
\end{bmatrix} \qquad (15)$$

Equation (15) contains only 4 different equations. These are

$$-\cos(\varphi_{13}) + \cos(\varphi_{23}) = 0, \quad \cos(\varphi_{13}) - \cos(\varphi_{23}) - \cos(\varphi_{46}) + \cos(\varphi_{56}) = 0$$
$$s - \sin(\varphi_{13}) + \sin(\varphi_{23}) = 0, \quad \sin(\varphi_{13}) - \sin(\varphi_{23}) - \sin(\varphi_{46}) + \sin(\varphi_{56}) = 0 \qquad (16)$$

Equations (16) are rearranged in the algebraic form C m = 0.

$$
\begin{bmatrix} 1 & -1 & 1 & 0 & 1 \\ 1 & -1 & 1 & -1 & 1 \end{bmatrix}
\begin{bmatrix} 0 & s \\ \cos(\varphi_{13}) & \sin(\varphi_{13}) \\ \cos(\varphi_{23}) & \sin(\varphi_{23}) \\ \cos(\varphi_{46}) & \sin(\varphi_{46}) \\ \cos((\varphi_{56}) & \sin((\varphi_{56}) \end{bmatrix}
= \begin{bmatrix} 0 & 0 \\ 0 & 0 \end{bmatrix}
\tag{17}
$$

Equation (17) has two solutions, namely:

$$
\begin{bmatrix} s \\ \varphi_{13} \\ \varphi_{23} \\ \varphi_{46} \\ \varphi_{56} \end{bmatrix}
= \begin{bmatrix} 0 & 2 \\ 1 & 0 \\ -1 & 0 \\ 1 & 0 \\ -1 & 0 \end{bmatrix}
\begin{bmatrix} \varphi_{13} \\ \sin(\varphi_{13}) \end{bmatrix}
\text{ and }
\begin{bmatrix} 0 & 0 \\ 1 & 0 \\ 1 & 0 \\ 0 & 1 \\ 0 & 1 \end{bmatrix}
\begin{bmatrix} \varphi_{13} \\ \varphi_{56} \end{bmatrix}
\tag{18}
$$

Interpretation of results : different modes of displacement and bifurcation points
The example of Fig. 11 is of particular interest. From Equation (18), for values of $s{\geq}0$, the member variables are expressed in terms of a single free variable φ_{13}. Hence, for values of $s{\geq}0$, the system has a constrained mobility. But, when $s = 0$, two free variables φ_{13} and φ_{56} are required to obtained the solution and hence define the general orientation of the members.

Consider the first solution, s attains a value of 0, when $\varphi_{13} = 0°$ or $180°$. Referring to Fig. 12, when $\varphi_{13} = 0°$ or $180°$, then $s = 0$. In this position, the foldable system behaves like two bar elements connected in sequence. The points where $\varphi_{13} = 0°$ or $180°$ and $s=0$ are called bifurcation points.

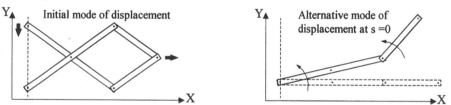

Fig. 12 Bifurcation point of foldable assembly used in the example

Restraints againt rigid body motion and computation of Cartesian coordinates
The position of restraints required to prevent rigid body motion is obtained by considering again Equations (8) and (9). A_i is an incidence matrix composed of ±1's in each row. All the columns add up to the zero column. Mathematically speaking, this implies that A_i has a single null vector composed of 1's. In physical terms, this means that the vectors x and y, each contain a single free nodal coordinate corresponding to the position of restraint necessary to prevent rigid body motion. Referring to Equations (8) or (9), the member variables in $[b_X, b_Y]$ are substituted with the solutions of Equations (17). The nodal coordinates are then obtained in terms of the free member variables and the free nodal coordinates.

Three-dimensional interelemental constraint equations
Setting up interelemental constraint equations in three dimensions follows a systematic approach similar to the two dimensional problem. They are in fact an extension to Equation (4c). An additional transformation matrix is incorporated in the interelemental constraint

equations. The transformation matrix models the axes of rotation and degrees of freedom of the joints.

CONCLUDING REMARKS

The calix duplet is a versatile building block suitable for the generation of spatial foldable structures of different forms. The displacement characteristics of calix foldable structures are directly related and dependent on the joint characteristics, which may have multiple degrees of freedom.

A systematic and general mathematical procedure suitable for the closed form kinematic analysis of foldable structures has been presented. The left null vectors of interelemental constraint equations are used to generate the dependency matrix. The solution or solutions of the dependency equation determine the modes of displacement and the corresponding bifurcation points of the system. The number of free variables in each solution of the dependency equation is a measure of the mobility in the system for the corresponding mode of displacement. The above mathematical technique is a general method for the modelling and understanding of the kinematic behaviour of more complex foldable systems and finite mechanisms.

Extension of the technique is possible by the inclusion of parameters such as loads, member stiffness and time dependent variables. These may be used to perform further analysis including static, dynamic and nonlinear analysis.

ACKNOWLEDGEMENTS
The author would like to thank Professor H Nooshin for his valuable guidance and contribution in developing the concepts and ideas outlined in this paper. Thanks must also go to Bezzina & Cole, a Maltese firm of Architects and Structural Engineers for the funding of this research project.

REFERENCES
1. ESCRIG F. Transformable Architecture in Journal of the International Association of Shell and Spatial Structures, 2000 Vol. 41 n.1, pp3-22
2. PELLEGRINO S, Foreword, Special Issue on Deployable Structures, The International Journal of Space Structures, Multi Science Publishing Co Ltd, 1993 Vol. 8 nos 1 & 2, p1.
3. GANTES C.J., Deployable Structures: Analysis and Design, WIT Press, 2001, pp6-16.
4. PELLEGRINO S and CALLADINE C.R., Matrix Analysis of Statically and Kinematically Indeterminate Frameworks, Int. J. Solids and Structures, 1986 Vol. 22, No 4, pp409-428.
5. HAUG E J, Computer Aided Kinematics and Dynamics of Mechanical Systems, Volume I: Basic Methods, Allyn and Bacon, 1989, pp48-71.
6. GERADIN M, "Finite Element Simulation of Deployable Structures" in Deployable Structures, ed. Pellegrino S, Springer Verlag, 2001, pp239-249.
7. HUNT K H, Kinematic Geometry of Mechanisms, Clarendon Press, 1978, pp33-37
8. HARTENBERG R and DENAVIT L, Kinematic Synthesis of Linkages, McGraw-Hill 1964, pp132-138.
9. Mc CARTHY J M, Geometric Design of Linkages, Springer-Verlag, 2000, p3.
10. REULEAUX F, The Kinematics of Machinery, 1876, Dover, pp86-92.

Snap-through-type deployable structures with arbitrary curvature

C J GANTES, Metal Structures Laboratory, National Technical University of Athens, Greece
E KONITOPOULOU, Pantechniki ATE, Athens, Greece

THE CONCEPT OF SNAP-THROUGH-TYPE DEPLOYABLE STRUCTURES

Deployable structures are prefabricated space frames consisting of straight bars linked together in the factory as a compact bundle, which can then be unfolded into large-span, load bearing structural shapes by simple articulation. Because of this feature they offer significant advantages in comparison to conventional, non-deployable structures for a wide spectrum of applications ranging from temporary structures to the aerospace industry, being mainly characterized by their feature of transforming and adapting to changing needs.

Because of their numerous advantages deployable structures have been investigated, designed and constructed by many engineers (Refs 1-5). The concept as well as the geometric and structural characteristics of the type of deployable structures considered here are the product of research work carried out since 1985 at the Massachusetts Institute of Technology, the Technion in Israel and the National Technical University of Athens by several investigators (Refs 6-10), who succeeded in converting the early ideas into a feasible type of structure. The main findings of this work are included in a recent book (Ref 11).

A fundamental design requirement of the structures investigated here is that they are self-standing and stress-free when fully closed or fully deployed. However, at intermediate geometric configurations during the deployment process incompatibilities between the member lengths lead to the occurrence of second-order strains and stresses resulting in a snap-through phenomenon that "locks" the structure in its deployed configuration (Ref 12). The response during deployment is, thus, characterized by geometric non-linearities, and simulation of the deployment process is, therefore, a difficult problem requiring sophisticated finite element modeling. The material behavior, however, must remain linearly elastic, so that no residual stresses reduce the load bearing capacity under service loads (Refs 8-11).

From a structural point of view, deployable structures have to be designed for two completely different loading conditions, under service loads in the deployed configuration, and during deployment. The structural design process is very complicated and requires successive iterations to achieve some balance between desired flexibility during deployment and desired stiffness in the deployed configuration (Refs 8-11).

From a geometric point of view, deployable structures of this type are based on the so-called scissor-like elements (SLEs), pairs of bars connected to each other at an intermediate point through a pivotal connection allowing them to rotate freely about an axis perpendicular to their common plane but restraining all other degrees of freedom, while their end points are hinged to the end points of other SLEs. Several SLEs are connected to each other in order to form units with regular polygonal plan views. The sides and radii of the polygons are SLEs. These polygons, in turn, are linked in appropriate arrangements constituting deployable structures, which are either flat or curved in their final deployed configuration.

Geometric design is performed according to a set of geometric constraints resulting from the requirement of zero stresses at the two extreme configurations (Refs 13-16). The stress-free condition is achieved by requiring straightness of the bars in the deployed configuration, thus obtaining a set of constraint equations, derived by looking at the development of adjacent SLEs on a common plane and applying basic geometric and trigonometric rules. The additional functional requirement that has to be satisfied through geometric design is a stress-free state in the folded configuration. By translating this also into a demand for straightness, one can obtain the so-called deployability constraint, requiring that the sums of the lengths between pivot and end node of the bars of SLEs that are connected to each other are equal.

The geometric constraint equations are derived by applying the above rules for all scissor-like elements of a unit, taking also symmetry or other special conditions into account. The formulation of a design procedure based on these constraint equations must be preceded by the choice of design parameters. Such parameters are usually certain external dimensions of the units, which are often imposed by architectural requirements. The other quantities that define the geometry, such as member lengths or angles between the members in the deployed configuration, are the unknown variables. Following this approach, one ends up with a system of simultaneous nonlinear equations that have to be solved numerically using an iterative algorithm such as the Newton-Raphson method.

This geometric design approach is initially followed at a polygonal unit level. Then, the additional constraints for deployment compatibility between adjacent units, and how this affects the overall geometric design process, must be accounted for. However, the snap-through-type deployable structures that had been designed so far according to this approach had a significant limitation. The geometric shapes that were possible in the deployed configuration were only flat or curved with constant curvature. Other shapes, which might be structurally more efficient or architecturally more desirable, could not be achieved.

In the present paper this limitation is addressed by unifying the two stages of the approach, namely design of individual units and then connectivity between adjacent units, into one (Ref 17). Thus, the desired final shape of the deployed structure is taken into account during geometric design of individual units. This increases the order of the resulting system of simultaneous equations, and thus the computational effort, but seems to be the only way to design snap-through-type deployable structures of arbitrary curvature. In addition, individual polygons are no longer necessarily regular. This methodology is applied successfully for the geometric design of a semi-elliptical arch. The arch is then modeled with finite elements, and a geometrically nonlinear analysis is performed in order to verify the deployability feature. Extension of the methodology to other geometric shapes does not appear to present any additional conceptual difficulties.

GEOMETRIC CONSTRAINTS FOR ELLIPTICAL DEPLOYABLE ARCH
As already mentioned, the proposed approach will be demonstrated for the case of a semi-elliptical arch. Consider an ellipse that will be the axis of the arch, shown in Fig 1 and described by the equation:

$$\frac{x^2}{a^2} + \frac{y^2}{b^2} = 1 \tag{1}$$

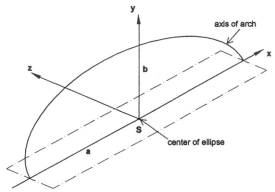

Figure 1. Axis of elliptical arch

Furthermore, consider two more ellipses, having the same axes lengths a, b as the previous one, that are placed on both sides of the original one (Fig 2) at a specific distance that will be defined later on. Thus, the upper surface of the elliptical arch is created. Then, this surface is divided into consecutive segments, having the same arch length (Fig 3). Because of the elliptical shape, the chords of the respective arch segments are of different length, called g_2. The distance g_1 between the ellipses placed on both sides of the axis is defined as equal to the average of all g_2. With the subdivision of the upper elliptical surface, an inscribed convex polyhedron is created. Each sub-plane constitutes the top view of a single structural unit.

So, the geometric design of the elliptical arch begins with the first structural unit. Some geometric parameters, such as the total size of the surface and the size of the individual unit are known, because they are imposed by architectural requirements. Assuming that the dimensions of the elliptical surface are known, we can choose the coordinates of points A_1 and B_1 (Fig 4), defined as the intersection of the two ellipses with the xz plane.

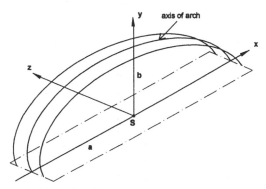

Figure 2. Upper surface of elliptical arch

If S is the center of the ellipse, then the points A_2 and B_2 belong to lines SA_1 and SB_1 and are at distance h_1 from A_1 and B_1, respectively, where h_1 is the thickness of the structural unit (Fig 4). Points C_1 and D_1 are on the elliptical surface and more specifically on the two edge ellipses. Their distance from B_1 and A_1, respectively, is g_2 (Fig 5). Similarly, points C_2 and D_2 are at distance h_2 from C_1 and D_1, respectively. The point O_1, which is the peak of the polygonal unit, is on the axis of the elliptical arch and more specifically in the middle of each unit's arch (Fig 6). The same holds for point O_2, and the distance between O_1 and O_2 is h_3. The dimensions h_2 and h_3 are not design parameters, but are derived from the geometric solution process of the structural unit. Thus, the solution of the structural unit starts with the coordinates of points A_1, A_2, B_1, B_2, C_1, C_2, O_1 and S as known parameters. The coordinates of points C_2, D_2, and O_2 are unknown. The complete first structural unit is shown in Fig 7.

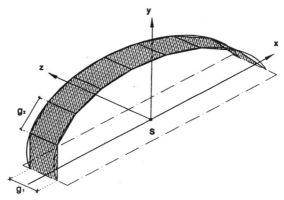

Figure 3. Sub-division of upper surface of elliptical arch in sub-planes

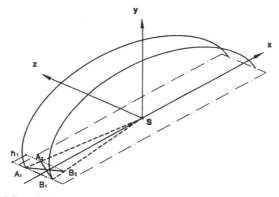

Figure 4. Location of first SLE

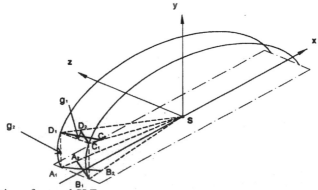

Figure 5. Location of second SLE

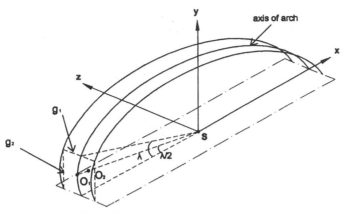

Figure 6. Location of points O_1 and O_2

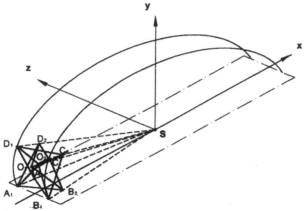

Figure 7. First basic unit

For the geometric design of the elliptical arch several different types of SLEs are used, contrary to the case of curved structures with constant curvature. In particular, the outer planes of the unit $A_1A_2B_1B_2$ and $C_1C_2D_1D_2$ are determined by symmetrical but different to each other SLEs, while the planes $A_1A_2D_1D_2$ and $B_1B_2C_1C_2$ are defined by non-symmetrical SLEs, which are the same for both of them (Fig 8). Similarly, the inner planes $O_1O_2A_1A_2$, $O_1O_2B_1B_2$, $O_1O_2C_1C_2$ and $O_1O_2D_1D_2$ are illustrated in Fig 9. Assuming that all the above are known, the angles φ_1, φ_2, φ_3, ω_1, ω_2, ψ_1, ψ_2, ψ_3, ψ_4, ψ_5, ψ_6 and the lengths L_1, L_2, which are defined in Figs 8 and 9, can be derived. L_1 and L_2 are found as the distances between two points with known coordinates, for example:

$$L_1 = \sqrt{(x_{O1} - x_{A1})^2 + (y_{O1} - y_{A1})^2 + (z_{O1} - z_{A1})^2} \tag{2}$$

For the angles the cosine law is used. For example angle φ_1 is derived by the equation:

$$(A_1B_1)^2 = (A_1S)^2 + (B_1S)^2 - 2(A_1S)(B_1S)\cos\varphi_1 \tag{3}$$

In order to derive the geometric constraints for the units, let us consider the development of two SLEs on a common plane. Two of them are shown indicatively in Fig 10. The following equations can be written for this problem:

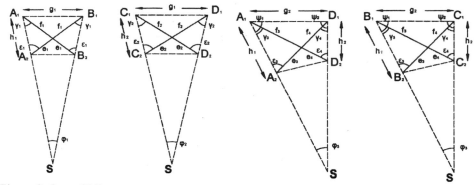

Figure 8. Outer SLEs

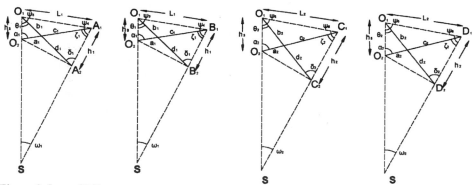

Figure 9. Inner SLEs

Foldabillity constraints

- Between outer SLEs:

$$e_1 + f_1 = e_3 + f_3, \; e_2 + f_2 = e_4 + f_4 \qquad (4)$$

- Between inner SLEs:

$$a_1 + b_1 = a_2 + b_2 \qquad (5)$$

- Between inner and outer SLEs:

$$e_1 + f_1 = c_1 + d_1, \; e_2 + f_2 = c_2 + d_2 \qquad (6)$$

Sine law:

- For the outer SLEs:

$$\frac{e_1}{f_1} = \frac{\sin\gamma_1}{\sin\varepsilon_1}, \; \frac{e_2}{f_2} = \frac{\sin\gamma_2}{\sin\varepsilon_2}, \; \frac{e_3}{f_3} = \frac{\sin\gamma_3}{\sin\varepsilon_3}, \; \frac{e_4}{f_4} = \frac{\sin\gamma_4}{\sin\varepsilon_4} \qquad (7)$$

- For the inner SLEs:

$$\frac{a_1}{b_1} = \frac{\sin\theta_1}{\sin\alpha_1}, \; \frac{c_1}{d_1} = \frac{\sin\delta_1}{\sin\zeta_1}, \; \frac{a_2}{b_2} = \frac{\sin\theta_2}{\sin\alpha_2}, \; \frac{c_2}{d_2} = \frac{\sin\delta_2}{\sin\zeta_2} \qquad (8)$$

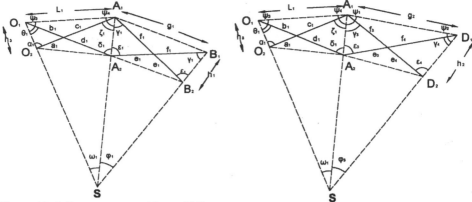

Figure 10. Adjacent outer and inner SLEs

Length projections in the radial direction:

- For the outer SLEs:

$$e_1 \cos\varepsilon_1 + f_1 \cos\gamma_1 = h_1, \ e_2 \cos\varepsilon_2 + f_2 \cos\gamma_2 = h_2,$$

$$e_3 \cos\varepsilon_3 + f_3 \cos\gamma_3 = h_1, \ e_4 \cos\varepsilon_4 + f_4 \cos\gamma_4 = h_2 \quad (9)$$

- For the inner SLEs:

$$a_1 \cos\alpha_1 + b_1 \cos\theta_1 = h_3, \ c_1 \cos\zeta_1 + d_1 \cos\delta_1 = h_1,$$

$$a_2 \cos\alpha_2 + b_2 \cos\theta_2 = h_3, \ c_2 \cos\zeta_2 + d_2 \cos\delta_2 = h_2 \quad (10)$$

Length projections in the tangential direction:

- For the outer SLEs:

$$2f_1 \sin\!\left(\frac{\varepsilon_1 + \gamma_1}{2}\right) = g_1, \ 2f_2 \sin\!\left(\frac{\varepsilon_2 + \gamma_2}{2}\right) = g_1, \ f_3 \cos(\psi_1 - \gamma_3) + f_4 \cos(\psi_2 - \gamma_4) = g_2 \quad (11)$$

- For the inner SLEs:

$$b_1 \cos(\psi_3 - \theta_1) + c_1 \cos(\psi_4 - \zeta_1) = L_1, \ b_2 \cos(\psi_5 - \theta_2) + c_2 \cos(\psi_6 - \zeta_2) = L_2 \quad (12)$$

Concurrency of O_1O_2, A_1A_2, B_1B_2:

$$\varepsilon_3 + \gamma_3 = \varepsilon_4 + \gamma_4, \ \alpha_1 + \theta_1 = \delta_1 + \zeta_1, \ \alpha_2 + \theta_2 = \delta_2 + \zeta_2, \ \gamma_1 + \varphi_1 = \varepsilon_1$$

$$\gamma_2 + \varphi_2 = \varepsilon_2, \ \gamma_4 + \varphi_3 = \varepsilon_3, \ \zeta_1 + \omega_1 = \alpha_1, \ \zeta_2 + \omega_2 = \alpha_2 \quad (13)$$

The above equations constitute a system of 34 equations for 34 unknowns, the 16 member lengths (e_1, f_1, e_2, f_2, e_3, f_3, e_4, f_4, a_1, b_1, c_1, d_1, a_2, b_2, c_2, d_2), the 16 angles (ε_1, γ_1, ε_2, γ_2, ε_3, γ_3, ε_4, γ_4, α_1, θ_1, δ_1, ζ_1, α_2, θ_2, δ_2, ζ_2) and the unit dimensions h_2, h_3. As already mentioned, the structural thickness h_1 is a known design parameter. As far as the solution of the above system is concerned, the equations (7a), (9a), (11a) and (13d) are linearly independent and they can be solved as a 4x4 system. Since all coordinates of the nodes of the first SLE ($A_1A_2B_1B_2$) are known, the lengths e_1, f_1 and the angles ε_1, γ_1 can be derived. Then, a system of 30 nonlinear equations for 30 unknowns remains to be solved numerically.

APPLICATION

The above process has been applied for the geometric design of the semi-elliptical arch shown in Fig 11. The geometric design methodology has been verified through the construction of a physical model in scale 1:20, shown in Fig 12 in some successive deployment stages. The members of the arch are made of straws, the outer hubs of circular wire rings and the pivotal connections of metal pins.

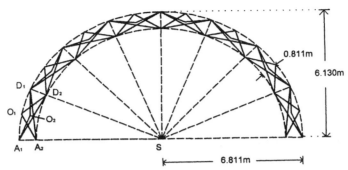

Figure 11. Semi-elliptical arch

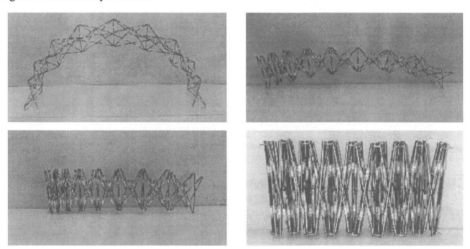

Figure 12. Successive deployment stages of the physical model

The design was further verified by finite element analysis of the arch, using rectangular hollow pipe 150mm x 60mm with a 15mm wall thickness made of low density polyethylene (E=150MPa, σ_y=20MPa) for the outer SLEs and the hubs, and rectangular hollow pipe 110mm x 50mm with a 15mm wall thickness made of acrylic (E=80MPa, σ_y=15MPa) for the inner SLEs. The general purpose finite element program MSC/NASTRAN 4.0 for Windows was used for the analysis, which consisted of two stages: (i) Linear analysis in the deployed configuration under service loads, and (ii) non-linear analysis of individual structural units during deployment. Fig 13 shows successive deployment stages as obtained from the above analysis, while Fig 14 illustrates the nonlinear load-displacement path, exhibiting clearly the snap-through nature of the response. The nonlinear analysis was displacement-controlled, consisting of 60 steps with 150 Newton-Raphson iterations each.

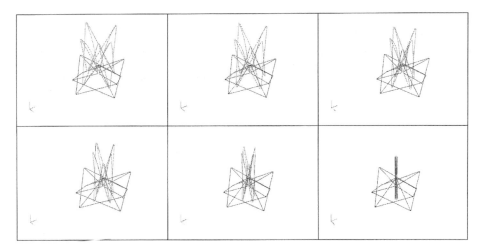

Figure 13. Successive deformed configurations of finite element model

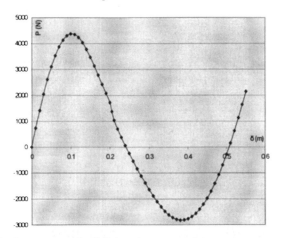

Figure 14. Load-displacement path of finite element model

SUMMARY AND CONCLUSIONS

Deployable structures that are self-standing and stress-free when fully closed or fully deployed but exhibit incompatibilities between the member lengths at intermediate geometric configurations during the deployment process, which lead to second-order strains and stresses and a snap-through phenomenon that "locks" the structures in their deployed configuration have been investigated. A geometric design methodology for deployable arches of arbitrary curvature has been proposed, thus overcoming a previous disadvantage of such structures that could be only flat or curved with constant curvature. The methodology has been applied successfully for the geometric design of a semi-elliptical arch. Verification of deployability has been achieved by the construction of a small-scale physical model as well as by deployment simulation with the finite element method.

REFERENCES

1.   PINERO E P, Expandable Space Framing, Progressive Architecture, vol 43, no 6, 1962, pp 154-155.

2.   MIURA K. and FURUYA H, Adaptive Structure Concept for Future Space Applications, AIAA Journal, vol 26, no 8, 1988, pp 995-1002.

3.   ZEIGLER T R, Collapsable Self-Supporting Structures, U.S. Patent No. 4.437.275, 1984.

4.   ESCRIG F, VALCARCEL J P and GIL DELGADO O, Design of Expandable Spherical Grids, Proceedings of the XXX IASS Congress, Madrid, Spain, 1989.

5.   KWAN A S K and PELLEGRINO S, The Pantographic Deployable Mast: Design, Structural Performance, and Deployment Tests, MARAS '91, International Conference on Mobile and Rapidly Assembled Structures, Southampton, U.K., April 9-11, 1991, pp 213-224.

6.   KRISHNAPILLAI S. and ZALEWSKI W P, The Design of Deployable Structures, Unpublished Research Report, Department of Architecture, M.I.T., Cambridge, Massachusetts, 1985.

7.   MERCHAN C H H, Deployable Structures, SM Thesis, M.I.T. Architecture Department, 1987.

8.   GANTES C J, A Design Methodology for Deployable Structures, PhD Thesis, available as Research Report No. R91-11, Department of Civil Engineering, MIT, Cambridge, Massachusetts, 1991.

9.   GANTES C J, CONNOR J J, and LOGCHER, R D , A Systematic Design Methodology for Deployable Structures, International Journal of Space Structures, vol 9, no 2, 1994, pp 67-86.

10. GANTES C J , Nonlinear Structural Behavior, Analysis and Design of Deployable Structures, NATO Advanced Research Workshop on Computational Aspects of Nonlinear Structural Systems with Large Rigid Body Motion, Pultusk, Poland, July 2-7, 2000.

11. GANTES C J, Deployable Structures – Analysis and Design, WIT Press, Southampton, England, 2001.

12. GANTES C J, Design Strategies for Controlling Structural Instabilities, International Journal of Space Structures, vol 15, nos 3&4, 2000, pp 167-188.

13. GANTES C J, Geometric Constraints in Assembling Polygonal Deployable Units to Form Multi-Unit Structural Systems, Fourth International Conference on Space Structures, Surrey, United Kingdom, Sep. 6-10, 1993, G A R Parke and C M Howard Editors, Thomas Telford, London, pp 793-803.

14. GANTES C J, LOGCHER R D, CONNOR J J, and ROSENFELD Y, Deployability Conditions for Curved and Flat, Polygonal and Trapezoidal Deployable Structures, International Journal of Space Structures, vol 8, nos 1/2, 1993, pp 97-106.

15. GANTES C J, LOGCHER R D, CONNOR J J, and ROSENFELD Y, Geometric Design of Deployable Structures with Discrete Joint Size, International Journal of Space Structures, vol 8, nos 1/2, 1993, pp 107-117.

16. GANTES C J, GIAKOUMAKIS A, and VOUSVOUNIS P, Symbolic Manipulation as a Tool for Design of Deployable Domes, Computers & Structures, vol 64, no 1-4, 1997, pp 865-878.

17. KONITOPOULOU E, Geometric Design of Deployable Arches with Arbitrary Curvature and their Structural Analysis Using the Finite Element Method, Diploma Thesis, Civil Engineering Department, National Technical University of Athens, February 2001 (in Greek).

Surface adjustment on thermal deformation of a large deployable space structure

S. ZHANG
Department of Building Structures, Tongji University, Shanghai, 200092, P. R. China
Q. ZHANG
Department of Building Structures, Tongji University, Shanghai, 200092, P. R. China
F. GUAN
Department of Civil Engineering, Zhejiang University, Hangzhou, 310027, P. R. China

INTRODUCTION

With the development of technology in aerospace and astronautics, deployable space antennas are becoming more and more larger. But the surface accuracy is required more stringent (especially within high frequency band). Flexible structures such as large deployable truss structures are required to have high sensibility and thermal stability. So when the thermal control method is not effective, active adjustment technology of thermal deformation is proposed to satisfy its surface accuracy requirement. Some researchers have studied the methods of surface adjustment of antenna structures, such as: (a) Assistant pulling plane method found in Ref 1; (b) Automatic surface adjustment technology associated with the length adjustment of the cables in the structure presented in Ref 2; (c) Automatic surface adjustment technology associated with the smart structure presented in Ref 3.

In space environment, there are following three types of disturbance on the shape of antenna structures:

(1) Transient disturbance. Structure recovers to its original shape once the disturbance damps out. Enhancing the damp of the structure can solve this disturbance.

(2) Fixed disturbance, such as manufacturing errors and assembling errors which were found in Ref 4, and so on. This type of disturbance can be adjusted by the force actuators which were presented in Refs 5-9, the actuators were placed on the points of the structures, or by the cable which can change the length of members. But these methods are complex and adaptive on the ground.

(3) Disturbance induced by thermal loads. In space environment, with the influence of sun radiation and shadowing effects of the earth and the structures, temperature varies from 140^0C to -140^0C, so the induced thermal deformation has much effect on the surface shape and surface accuracy of the antenna structure. In order to reduce the thermal deformation, composite materials with low thermal expansion coefficient were utilized, and other thermal control methods were taken. On the other hand, if both methods above can't satisfy

requirement either, active adjustment actuators are used to adjust the thermal deformation. That is the method proposed in this paper.

This method is based on the heating and cooling of the actuators which are set on the points of the structure. The temperature distribution in the whole structure can be changed, so that the thermal deformation can be reduced arbitrarily. Temperature adjustment is similar in its effect to the length adjustment of the members. And it's fit for structural elements, not only one-dimensional but also two- and three-dimensional. Adjustment of thermal deformation of a 5-m diameter cutting-parabolic antenna is analyzed efficiently by this method.

Temperature adjustment has the advantages as follows:
(1) Solar energy is available.
(2) It's based on the self-equilibrating thermal loads, so we needn't worry about inaccurate equilibrating forces which lead to drift in position or orientation.
(3) Stress associated with the heat smaller than those associate with applied forces.

TEMPERATURE ADJUSTMENT PRINCIPLE

It is assumed that ψ is the deformation vector at one moment, and m is the total points of the antenna structure.

$$\psi = [\psi_1 \quad \psi_2 \quad \cdots \quad \psi_m]^T \tag{1}$$

Thermal deformation of the structure is controlled by applied temperature at n $(n \leq m)$ control points, U is the vector used to correct the deformation.

$$U = [U_1 \quad U_2 \quad \cdots \quad U_m]^T \tag{2}$$

$$U_i = \sum_{j=1}^{n} U_{ji} \Delta T_j \tag{3}$$

where, U_{ji}— displacement vector at ith point due to the applied temperature at jth point.

ΔT_j— applied temperature at jth point.

After the structure is corrected, the total deformation at ith point can be expressed as U_{Ti}.

$$U_{Ti} = \psi_i + \sum_{j=1}^{n} U_{ji} \Delta T_j \tag{4}$$

Then the total deformation of the whole structure is denoted as follows:

$$U_T = \psi + U \cdot \Delta T \tag{5}$$

or

$$\begin{Bmatrix} U_{T1} \\ U_{T2} \\ \vdots \\ U_{Tn} \\ \vdots \\ U_{Tm} \end{Bmatrix} = \begin{Bmatrix} \psi_1 \\ \psi_2 \\ \vdots \\ \psi_n \\ \vdots \\ \psi_m \end{Bmatrix} + \begin{bmatrix} U_{11} & U_{21} & \cdots & U_{n1} \\ U_{12} & U_{22} & \cdots & U_{n2} \\ \vdots & \vdots & \cdots & \vdots \\ U_{1n} & U_{2n} & \cdots & U_{nn} \\ \vdots & \vdots & \cdots & \vdots \\ U_{1m} & U_{2m} & \cdots & U_{nm} \end{bmatrix} \begin{Bmatrix} \Delta T_1 \\ \Delta T_2 \\ \vdots \\ \Delta T_n \end{Bmatrix} \tag{6}$$

$$\begin{Bmatrix} U_{T1} \\ U_{T2} \\ \vdots \\ U_{Tn} \\ \vdots \\ U_{Tm} \end{Bmatrix} = \begin{Bmatrix} \psi_1 \\ \psi_2 \\ \vdots \\ \psi_n \\ \vdots \\ \psi_m \end{Bmatrix} + \begin{bmatrix} U_1 & U_2 & \cdots & U_n \end{bmatrix} \begin{Bmatrix} \Delta T_1 \\ \Delta T_2 \\ \vdots \\ \Delta T_n \end{Bmatrix} \tag{7}$$

where,
$$U_i = \begin{Bmatrix} U_{i1} \\ U_{i2} \\ \vdots \\ U_{in} \\ \vdots \\ U_{im} \end{Bmatrix} \tag{8}$$

The key problem is to determine those temperatures applied on the control points, which minimize the following formulation:

$$U_{rms}^2 = U_T^T U_T \tag{9}$$

where, U_T is defined in equation (5), substitute equation (5) into equation (9), and minimize U_{rms}^2 with respect to each ΔT_j, then we can get the following equations:

$$\frac{\partial U_{rms}^2}{\partial \Delta T_j} = \frac{\partial (U_T^T U_T)}{\partial \Delta T_j} = 0 \tag{10}$$

$$(\psi^T + \Delta T^T U^T) U_j = 0 \tag{11}$$

that is:
$$U_j^T U \Delta T = -U_j^T \psi \tag{12}$$

or
$$A \Delta T = R \tag{13}$$

where, the element in the *m*th row and *n*th column of A is:

$$A_{mn} = U_m^T U_n \tag{14}$$

and the element in the m th row of R is:

$$R_m = -U_m^T \psi \tag{15}$$

$$\Delta T = [\Delta T_1 \quad \Delta T_2 \quad \dots \quad \Delta T_n]^T \tag{16}$$

TEMPERATURE ADJUSTMENT OF ANTENNA STRUCTURE

A 5-m diameter large deployable cutting-parabolic antenna model is designed. It is composed of a supported backbone (deployable truss) and a reflector surface (flexible mesh), it's shown in Figure 1.   Fig.1a and Fig.1b denote the front view and side view of the antenna respectively.

Based on the finite element model of the antenna structure, adjustment of thermal deformation of this structure is carried out by this method. Simply, locations of three control points were considered, and they were shown in Figure 2. Determining the adjustment of temperature on the three control points, total thermal deformation of the structure will be calculated. The distortion curves of deformed and corrected along of the structure are shown in Figure 3.

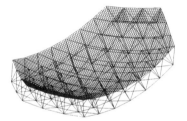

Fig. 1a.   Front view of the antenna.   Fig. 1b.   Side view of the antenna.

Fig. 1. Deployable cutting-parabolic antenna structure.

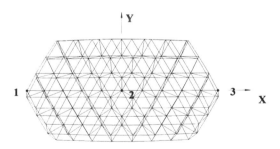

Fig. 2. Positions of three control points in the antenna.

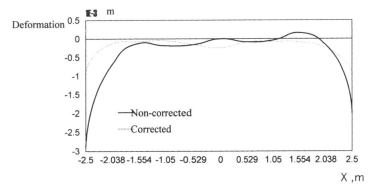

Fig. 3. Distortion curves of deformed and corrected along X axes.

CONCLUSIONS

This paper describes an approach of surface accuracy control for the large flexible space structures. Optimum numbers of control points are determined according to the required surface accuracy. Temperatures at the control points are determined to minimize the overall deformation of the structure. A matrix equation is derived, optimum control temperatures can be obtained by solving it. Deformation control results of the parabolic antenna demonstrates that temperature adjustment was an effective approach.

In a word, during the thermal design of space structures, when the passive control method can't satisfy the thermal deformation requirement, temperature actuators at control points can be used to adjust the temperatures and minimize the deformation of the whole structure.

REFERENCE

1. GEDFREY RD, The Evolution of the Tracking and Relay Satellite System (TDRESS), EASCON, DC, Sep. 29 Oct. 1, 1975.
2. Large Erect-able Antenna for Space Application. NASA CR-102522.
3. First-Fifth Joint U.S./Japan Conference on Adaptive Structures, 1990-1994.
4. GREENE, W. H., Effects of Random Member Length Errors on the Accuracy and Internal Loads of Truss Antenna, AIAA Paper 83-1019, May 1983.
5. BUSHNELL, D., Control of Surface Configuration by Application of Concentrated Loads, AIAA Journal, Vol. 17, Jan. 1979, pp. 71-77.
6. BUSHNELL, D., Control of Surface Configuration of Non-uniformly Heated Shells, AIAA Journal, Vol. 17, Jan. 1979, pp. 78-84.
7. WEEKS, C. J., Shape Determination and Control for Large Space Structures, JPL Pub. 81-91, Oct. 1981.
8. ASPINWALL, D. M. and KARR, T. J., Improved Figure Control with Edge Application of Forces and Moments, Proceedings of the SPIE Meeting, Vol. 228, 1980, pp. 26-33.
9. SCHAFER, W., Large 12 GHz Antennas in Advanced Technology, Proceedings of the International Astronautical Federation Meeting, Lisbon, Portugal, Sept. 1975.

Multi-reciprocal element (MRE) space structure systems

J. P. RIZZUTO, M. SAIDANI
School of Science and the Environment, Coventry University, UK
J. C. CHILTON
School of the Built Environment, University of Nottingham, UK

ABSTRACT
A Reciprocal Frame (RF) system is a three-dimensional grillage structure constructed of a closed circuit of mutually supporting beams. A number of RFs connected to each other at the outer end of each radiating beam results in the formation of a Multi-Reciprocal Element (MRE) space structure. The uniqueness of the MRE system is that only two elements contained within module circuits of mutually supporting beams require to be connected to one another. A module cell may theoretically consist of any number of elements however. This simplifies the connections considerably therefore giving this novel system potentially economic advantages over traditional space structure connection systems such as the Mero KK node joint with up to 18 threaded holes. MRE space structures can therefore provide an economic, aesthetic and convenient way of spanning small or very large areas without the need to provide intermediate vertical supports. The aim of this paper is to explore and discuss the historical development of these systems and to consider their present and future application.

Keywords: Reciprocal Frames, Multi-Reciprocal Element Space Structures

INTRODUCTION

Multi-Reciprocal Element (MRE) space structures are constructed from modules that are made up of elements reciprocally supporting one another. MRE space structures can be considered as an extension of the Reciprocal Frame (RF) [1]. The RF is a three-dimensional grillage structure constructed of a closed circuit of mutually supporting beams. This reciprocal support gives some MRE structures the appearance of being made from elements that are interwoven. Elements can be arranged so as to form two and three-dimensional space structures. At the joint locations within a module it is only necessary for two elements to be connected to each other. With each of the elements so connected a complete circuit can be formed and hence a module. This is a typical feature of a MRE joint system and has potentially economic advantages [2] over traditional space grid connections found for example in geodesic domes [3] that are very often complex and therefore expensive. The minimum number of elements required to create a module is three. There is no theoretical maximum number of elements that may be used however.

Alternatively, MREs have also been referred to as Nexorades, the elements as Nexors and modules as Fans [4]. Saidani *et al* [2] refers to MREs as multi-reciprocal grids (MRGs). Dean [5] and [6] refers to RFs as lamella beams. RFs are also known as Mandala Dach in Germany [7].

HISTORICAL REVIEW

The use of the RF method of construction dates back many centuries. Chilton *et al* [8] suggests that it is not possible to tell precisely from where the first idea for using a structure such as the RF came. It is likely, although there is no direct evidence, that the RF developed from simpler structures used by early human civilizations. Many societies in their early stages of evolution lived in covered pits or huts. The roofs of these primitive one-cell dwellings were constructed from woven tree branches that had similarities with the RF.

Chinese builders over 900 years ago used the principle of interwoven elements in the construction of relatively short span bridges. The principle they used relies on the fact that three elements are arranged so that the scissor action of any two traps a forth element that has been placed perpendicular to the original three. These typically make up a construction unit. When a series of these units are grouped together an arch form results [9]. The span and rise of the arch depends upon the diameter and length of the elements being used. This form of construction uses the principle of reciprocal support but is subtly different from the present RF closed circuit of beams.

The most likely country from which the RF originated in its present form is Japan. There is some evidence to suggest that in the late 12th century the Buddhist monk Chogen (1121–1206) established a technique of spiral layering wooden beams that were used in the construction of temples and shrines [8]. The technique used is identical to the structural principle of the RF.

Examples of elementary RF use are illustrated by Drew [10]. These include the North American Indian *Tipi* built of long poles that meet around a single point at the top of the structure and rely upon each other for support, and the Eskimo *Tupiq* tent. Faegre [11] shows that the conical tent originated from Siberia before spreading West into Lapp country and East into North America. North Siberian tribes such as the Evenk and the Yukaghir used the *Chorama-Dyu*, whereas the Koryak and Chukchi of the same region favoured the *Yaranga*. Both of these style of tents are examples of compound tents that use a cylindrical base and a RF style conical roof. The Afghan and Mongolian *Yurt* is also an example of a compound tent with a cylindrical base and a RF style conical roof. The Lapps used a forked-pole *Kata* tent that had a framework of mutually supporting tree branches similar in style to the *Tipi* tent. The first written reports of the North American conical tipis are believed to be from the Coronado expedition of 1540 [11].

Villard de Honnecourt, the medieval architect and engineer, explains "how to work on a house or tower even if the timbers are too short" [12]. Honnecourt illustrates his solution in sketches made sometime between 1225-1250. These sketches clearly indicate the construction of floors using timber beams that are all shorter than the required span. His grillage assembly uses the same principle as the RF. This suggests that the principle had been known for a lot longer than the date suggests as it is quite likely that it would have been passed down from previous generations as were other construction methods and techniques [13]. Leonardo da Vinci (1452–

1519) also appreciated how to span distances with short elements as illustrated in his 3-dimensional grillage structures and temporary timber bridges [14]. The principle used by Leonardo da Vinci is identical to that used by the Chinese several hundred years before.

Sebastiano Serlio (1475–1554) was also familiar with the need to span distances greater than the length of a single beam. He proposed a planar grillage very similar to Villard de Honnecourt, where typically, one beam relied upon another for support. If the beams were notched into each other, a two-dimensional flat surface resulted (ignoring the beam depth as the third dimension) that could be used for the construction of floors and ceilings [15].

Almost one hundred years after Serlio, John Wallis in 1652 started to build models of flat grillage assemblies with elements that were notched and fitted into one another. He also explored different planar morphologies in an attempt to study the geometry and load transfer mechanisms [16]. It is interesting to note that John Wallis' planar grid sketches are virtually identical to those of Leonardo da Vinci [14].

For over 260 years it appears that the RF system was lost with no apparent record of any RF structures being built. Chilton [17] describes some interesting examples of RF-like three-dimensional grillage structures built by the architect Jujol earlier this century. These include the roofs at Casa Negre, San Juan Despi, Barcelona, (1915) and Casa Bofarul, Pallaresos, Tarragona (1913 - 1918).

More recently in the early 1990's the roof of a salt storage building of 26m span was built in Lausanne, Switzerland by Natterer *et al* [18]. At about the same time, a similar roof of almost 7m diameter was used for a Puppet Theatre, near Kumamoto, on the island of Kyushu, Southern Japan [19]. Other recent RF structures include gazebos and whisky vat houses in Scotland and a Permaculture Centre in Bradford, UK [17].

PATENTS FILED ON MRE AND RELATED SYSTEMS
The Reciprocal Frame concept was patented in Britain in 1989 under the title of 'Three-dimensional structures' [1]. Here, sloping beams give three-dimensional structures. The basic geometrical relationships of the RF was described by Chilton and Choo [20]. Studies in the potential of RF use in Architecture were carried out by Popovic [21] at the University of Nottingham. The use of three-dimensional RF structures generally utilized as a complete buildings, or part of a building such as the roof were investigated. Popovic *et al* [22] considered a variety of RF arrangements that could be used for medium span buildings.

A multi-reciprocal element concept was first patented in Holland by Bijnen in 1976 [23] and used to construct a dome known as a *Nomadome*. Circular hollow tubes were used as the primary elements. These were connected together with scaffold couplings. Bijnen's patent also illustrates notched circular tubes resting upon un-notched supporting reciprocal tubes.

Gat [24] also introduced the self-supporting interlocking grids for multiple applications (SIGMA) system – a form of MRE in 1978 - where he explored the 'mutual support' idea for flat and later, curved surfaces. Gat *et al* [25] in 1992 attempted to study a similar problem for mutually restraining modular floating platforms. It is worth noting that Gat's system incorporated wedges between elements in order to create three-dimensional forms. The resulting structures built using Gat's method appear to be a re-evaluation of what Wallis [16] investigated before him. Wallis' structures in turn appear are a refashioning of Leonardo da Vinci's sketched planar grids illustrated in Pedretti's book [14].

Crooks [26] took out a Patent in 1980 in the USA under the title of 'Building construction of A-shaped elements'. In his system - similar in principle to the RF and MRG - triangular A-shaped basic building elements composed of bendable material were proposed and to be interconnected together to form the framework for a building such as a dome. A Rotegrity system [27] created in 1992 uses interwoven bent flexible elements in what appears to be a near identical system to that proposed by Crooks some twelve years earlier.

Some investigations on the potential of the MRE system using members of circular cross-section were initially carried out and discussed by Baverel and Saidani [28] that resulted in a Patent [29]. It is worth noting that there are similarities between this patent and Bijnen's patent [23]. The territorial time limit of 20 years had also elapsed since Bijnen's patent was filed in 1976, and the information contained in his patent on his MRE system had therefore already been added to the body of world knowledge.

MRE RELATIONSHIP TO OTHER SPACE STRUCTURES

MRE structures also have apparent similarities with tensegrity structures configured and Patented by Buckminster Fuller [30] in 1960. His tensegrity domes consist of diagonally orientated compression bars held apart by wires in tension. If the compression bars are collapsed onto each other by the removal of the wires, a structure very similar to an MRE would be produced.

Marks [31] illustrates Buckminster Fuller's 72-foot (21.95m) diameter 50-foot (15.24m) high, two-thirds spherical basket tensegrity dome built in 1962 at Southern Illinois University. This dome could (on inspection) be reduced to a number of large open cell multi-reciprocal element modules. The dome was constructed from lengths of interwoven bolted timber sections. Three orientations were used to align the sections. The first orientation followed a horizontal great circle, similar to the earth's equator. Smaller parallel circles were further provided in a horizontal direction approximating to the lines of latitude on the earth. For the second and third orientations, skewed smaller circles parallel to a great circle were aligned approximately to the lines of longitude, in as much that they followed a north north-west to south south-east direction one way, and a north north-east to south south-west direction the other way. Both were inclined at approximately 60 degrees from the horizontal but travelling in apparent opposite directions. The orientation of all three alignments resulted in the configuration of equilateral triangles attached to the edges of larger hexagons mapped on to the surface of the dome. A central pole with ties was used to stabilize the entire structure.

RECENT MRE INVESTIGATIONS
There has been some work on the general analysis of flat lamella grids (MREs) by Dean [5] in the United States. Dean obtained several new force and displacement formulas by deriving and solving their governing difference equations for arbitrarily loaded systems for regular hinged lamella beams. The effects of joint imperfections were also studied by formula modification in order to account for relative joint slippage and deformation. Bending moment and shear forces formulas for (RF) lamella elements were published from this earlier work in 1997 [6].

Some work on conceptual theories had been carried out on MRE geometry generation using circular tube sections by Baverel and Saidani, [2] and [28] in 1999. They also carried out some very limited work on retractable MRE structures [32] in 1998.

Use of direct formulations and genetic algorithms in the programming language 'Formian' [33] and [34] has been considered by Baverel [4] and Kuroiwa [35] at the University of Surrey. These studies use a powerful configuration processing tool utilising Formex algebra. Analytical geometry has been used by Baverel to study regular and semi-regular polyhedra. Both Kuroiwa [35] and Baverel [4] have looked at the use of the more versatile method of using a genetic algorithm that allows a wider range of configurations of Nexorades (MREs). Some success has been achieved to-date using this method and currently further work is required to fully develop and extend this method with new Formian implementation functions.

Some limited work has also been done in Germany [36] on what appears to be regular modules made from reciprocally supported elements. However, no complete structures are illustrated. Some architectural studies have also been carried out in Colombia by Ariza [37]. Some interesting architectural forms were investigated and many small-scale models made. No analytical modelling however was included in Ariza's study.

Research at Coventry University [38], [39] and [40] has been looking at MRE structures made of prismatic and circular cross-sectional elements. The objectives of this research include an attempt to simplify possible element connection systems such as the use of single bolted connections, and consider overall structural behaviour, as well as element orientation and numerical modelling. Further work using analytical geometry to study regular and semi-regular polyhedra is also being carried out.

PRESENT AND FUTURE APPLICATIONS
Applications of the RF and MRE systems have been limited. There are very few examples of full-scale MRE structures in use other than Bijnen's [23] Nomadome, Gat's [25] timber models, Baverel's [4] timber and steel models. The exception being the Toyoson Stonemason Museum roof structures in Japan [21]. Investigations using various models have also been carried out by Rizzuto *et al* [40]. See figures 1 and 2. There are by comparison many examples of RF roof structures in Europe and Japan that are known about and mentioned earlier in this paper.

For larger span structures the multi-reciprocal element system presents interesting opportunities for the construction of multi-layer grids. The advantages of the MRE system include ease of transport to site, and speed of construction provided the elements are pre-fabricated. This is ideal for the construction of emergency shelters. In addition, an MRE system could potentially also be suitable for use as temporary or permanent falsework in the construction of permanent structures. The MRE system could also be useful for military applications. These could be used to hide artillery, and provide temporary shelter for troops. The erected structure could be hidden with standard camouflage cover.

Fig. 1. Multi-Reciprocal Element module of cylindrical hollow steel tubes with bolted connections.

Fig. 2. Multi-Reciprocal Element cube Space structure using timber elements.

Fig. 3. Clad perimeter-truncated pentagonal Multi-Reciprocal Element space structure.

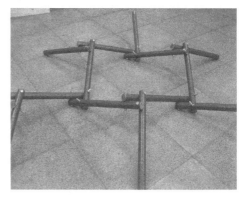

Fig. 4. Example of a planar grid made from 3-element modules with bolted joints.

Simple types of connection could be achieved using a bolting system, screws or nails depending on the material, size, type, and nature of the members used. See figure 3. The MRE system could be constructed using any type of material such as timber including bamboo, steel, aluminium, or suitable plastics. This may be an important factor in the construction of emergency shelters where it can be difficult to access an area.

One of many possible reasons why these space structure systems have not been used in the past is that generally they contain complex geometry. There is also a lack of understanding of their structural behaviour. With modern powerful computers and commercially available software, solving geometrical problems can be relatively straightforward. In order to improve fundamental understanding of their structural behaviour a full size MRE space structure is in the process of being designed and constructed at Coventry University in the UK. This will be used for load testing and monitoring of overall structural performance. See figure 4.

CONCLUSIONS
A historical review of RFs and MREs and has been presented together with their relationship with other space structures. Present and future applications of the MRE system has also been discussed.

MRE space frames provide an aesthetic and convenient way of spanning small or very large areas without the need to provide intermediate vertical supports.

They are potentially suitable for use as temporary emergency shelters or falsework in the construction of permanent structures. The MRE system could also be useful for military applications.

Previous work on the MRE system has been mainly concerned with geometry and has been generally devoid of analysis in terms of overall structural performance. Some limited experimental and numerical modelling has been carried out.

Further research is required to explore strength and stability problems in order to assess the engineering feasibility of MRE space structures and provide the basis for future design procedures.

REFERENCES
1.   Brown, G. *Three-dimensional structures*. United Kingdom Patent Office, 1989, Patent No. GB2235479B.
2.   Saidani, M., Baveral, O. and Cross-Rudkin, P. S. M. Investigation into a new type of multi-reciprocal grid. *Journal of the International Association for Shell and Spatial Structures,* 1998, **13**(4), 215-218.
3.   McHale, J. *R. Buckminster Fuller*. Prentice-Hall International, 1962.
4.   Baverel, O. *Nexorades: A family of interwoven space structures*. PhD thesis, University of Surrey, 2000.
5.   Dean, D. Lamella beams and grids. *Journal of the Engineering Mechanics Division, American Society of Civil Engineering,* 1964, **90** (EM2), 107 -129.

6.   Dean, D. (1997) One-dimensional circular arrays of Lamella beams. *Journal of the International Association for Shell and Spatial Structures,* 1997, **38**(1), 19 -21.

7.   Chilton, J.C., Choo, B. S. and Yu, J. Morphology of Reciprocal Frame three-dimensional grillage structures. In: *Proceedings of the IASS-ASCE International Symposium,* Atlanta, USA, 1994, 1065-1074.

8.   Chilton, J.C., Choo, B. S. and Popovic, O. Reciprocal Frames – Past, Present and Future. In: *Proceedings Lightweight structures in Civil Engineering,* (Ed. Jan B. Obrebski), **1**, Warsaw, Poland, 1995, 26 - 29.

9.   NOVA Online|*Secrets of Lost Empire*|China Bridge. URL: http://www.pbs.org/wgbh/nova/lostempire/china/buildsthumbs.html [03 March 2002]

10.   Drew, P. *Tensile Architecture.* Granada Publishing Ltd, 1979.

11.   Faegre, T. *Tents: Architecture of the Nomads.* John Murray Ltd. London,1979.

12.   Bowie, T *The sketchbook of Villard de Honnecourt,* Indiana University Press, Bloomington and London, 1959.

13.   Heyman, J, How to design a cathedral: some fragments of the history of structural engineering. In: *Proceedings Institution of Civil Engineers. Civil Engineering,* 1992, **92**, 24 - 29.

14.   Pedretti, C. *Leonardo: Architect.* Thames and Hudson Ltd, London, 1986.

15.   Serlio, S. *First Book of Architecture by Sebastiano Serlio,* Benjamin Bloom Inc. Publishers, New York,1970. (*First published 1611*).

16.   Wallis, J. *Opera Mathematica,* Georg Olms Verlag Hildesheim, New York, 1972. (*First published 1695*).

17.   Chilton, J. C. Polygonal living: some environment-friendly buildings with reciprocal frame roofs. *Journal of the International Association for Shell and Spatial Structures,* 1995, **36**(2), 83 -89.

18.   Natterer, J., Herzog, T. and Volz, M. Holzbau Atlas Zwei, Institut fur International Architektur, Munich, 1991, p.179.

19.   Japan Architect Annual, 1993-1, p.67.

20.   Chilton, J. C. and Choo, B. S. Reciprocal Frames Long Span Structures. In: *Proceedings Innovative Large Span Structures,* (Ed. Srivastava N. K., Sherbourne A. N., and Roorda, J.), **2**,Toronto, Canada, 1992, 100 -109.

21.   Popovic, O. (1996) *The architectural potential of the reciprocal frame.* PhD thesis, University of Nottingham, 1996.

22.   Popovic, O., Chilton, J. C. and Choo, B. S. The variety of reciprocal frame (RF) morphologies developed for medium span assembly building: a case study. *Journal of the International Association for Shell and Spatial Structures,* 1998, **39** (1), 29-35.

23.   Bijnen, A. *Een geodetiese knoopkonstruktie.* Octrooiraad Nederland. Terinzagelegging. Aanvrage, 1976, Nr. 7603046.

24.   Gat, D. Self-supporting interlocking grids for multiple applications (SIGMA). *Architectural Science Review,* 1978, 105 -110.

25.   Gat, D., Eisenberger, M. and David, I. Mutually restraining modular floating platforms. In: *Proceedings of the International Association for Shell and Spatial Structures,* Toronto, Canada, 1992, 859-868.

26.   Crooks, M. *Building Construction of A-Shaped Elements.* United States Patent Office, 1980, US Patent No 4182086.

27. Rotegrity. Leaflet Parity Products, Simple Science Toys, Eugene, OR 97401 USA, 1972.
28. Baverel, O. and Saidani, M. The multi-reciprocal grid system. *Journal of the International Association for Shell and Spatial Structures,* 1999, **40**(1), 33-41.
29. Coventry University. *Module for a space structure.* United Kingdom Patent Office, 1998, Patent No GB9817785.0
30. Buckminster Fuller, R. *A Framework Structure for Buildings and the like.* United States Patent Office, 1960, US Patent No 963259.
31. Marks, R. W. *The dymaxion world of Buckminster Fuller.* Southern Illinois University Press, 1960.
32. Baverel, O. and Saidani, M. Retractable Multi-Reciprocal Grid Structures. *Journal of the International Association for Shell and Spatial Structures,* 1998, **39**(3), 141-146.
33. Nooshin, H. and Disney, P. Formex configuration processing I. *International Journal of Space Structures,* 2000, **15**(1), 1 - 52.
34. Nooshin, H. and Disney, P. Formex configuration processing II. *International Journal of Space Structures,* 2001, **16**(1), 1 - 56.
35. Kuroiwa, Y. *Regularisation of structural forms using genetic algorithms.* PhD thesis, University of Surrey, 2000.
36. Wörgberger, R. Stabflechtwerkschalen – ein Beitrag fur schlichte und exponierte Bauaugaben. *Bauen mit Textilien,* **1.** Jahrgang Heft 2. Ernst & Sohn, Wiley, 1998.
37. Ariza Ruiz, J. O. *Estructuras Reciprocas bases para su aplicacian arquitectonica.* Monografia para optar al titulo de Arquitecto. Universidad Nacional De Colombia, 2000.
38. Rizzuto, J., Saidani, M. and Chilton, J. C. The Self-supporting multi-reciprocal Grid (MRG) system using notched elements. *Journal of the International Association for Shell and Spatial Structures,* 2000, **41**(2),125-131.
39. Rizzuto, J.P., Saidani, M. and Chilton, J. C. Joints and Orientation of Module Elements in Multi-Reciprocal Grid (MRG) systems. In: *Proceedings IASS International Symposium,* Nagoya, Japan, 2001, (Extended Abstracts) 308-309. (Full paper) TP138, 1-8, on CD-ROM included with Proceedings.
40. Rizzuto, J., Saidani, M. and Chilton, J. C. Polyhedric space structures using reciprocally supported elements of various cross-section. *Journal of the International Association for Shell and Spatial Structures,* 2001, **42**(3), 149-159.

Visible Transparency

E.E.S. FRIJTERS MSc, Department of Building Engineering, University of Delft, NL.
A. BORGART MSc, Department of Building Engineering, University of Delft, NL.

INTRODUCTION

The development of transparency in buildings moves parallel with the development of glass as a construction material. Glass is a strong and transparent material with disadvantages like fragility and unpredictability in behavior.[Fig1] These contradicting properties of glass make it a challenge to use as a construction material. The safe use of glass in a main construction forms the subject of this design. The subject contains the design and development of a maximal transparent glass roof system based on a tensegrity- structure.

Figure 1: Fragility and unpredictability of glass

CONSTRUCTION

The substance of a tensegrity-structure is that there is a totality of compression and tension elements, by which the tension is contributed by slim continuous elements, meanwhile independent short thick elements take the compression forces for their account. Since in tensegrity-structures the compression and tension elements are used in an additional manner, they can carry forces which lay on a higher level than traditional structure principles.[Fig 2]

Figure 2: Tensegrity-structure in a glass façade construction [1]

MATERIAL

Glass has the same molecular structure as a liquid. Liquids are in general transparent. The cause of this can be found by the fact that the molecular structure is irregular.[Fig 3] Thereby the microscopic cross section of the molecules are smaller than the wave length of visible light. This means that there are no obstructions for light waves. The internal homogeneity of glass, that is typical for liquids, forms the basis for transparency.

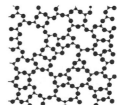

Figure 3: Irregular molecular structure of glass [2]

A material is therefore transparent if there are no obstructions that hinder light. This principle is used to optimise the transparency of this construction. By using a tensegrity-structure you anticipate on the physical source of transparency. The combination of an optimum of construction elements with the use of glass as a construction material will lead to a minimum of obstructions.[Fig 4]

Figure 4: Optimum of construction elements in combination with glass products

ELEMENTS

The construction of the roof system consists of linear elements and plates. These elements are coupled to each other with hinges. The linear elements can be divided in compression and tension elements. The main idea of the construction is the use of existing glass products for the compression rods [Fig 5] and the use of glass plates as a major construction-element.[Fig 6]

Figure 5: Glass rod in a roof system [3]

SAFETY ASPECTS

A glass roof system has to fulfil high safety requirements. These requirements can be reached on four different levels, on material-, elemental-, component- or construction level.

The safety requirements on construction level [Fig 8] in combination with those on material level [Fig 7] forms the basis for the development of the main design.

Figure 6: Glass plates in a hanging roof system [4]

Figure 7: Composition of the glass panel in the design:
1. Thermally toughened glass plate 6 mm
2. Combination coating and PET-foil
3. Air 12 mm
4. 2 x 10 mm glass plates with PVB-foil
5. Strengthened glass strokes

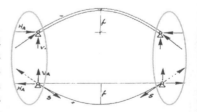

Figure 8: Distribution of forces in two directions.

DESIGN AND DEVELOPMENT

The design can be described as a closed lens system. The system is a composition of an arch of laminated glass panels on top and a cable of steel elements below.[Fig 9] Between these components there are glass rods situated which will function as hinged compression elements. The use of the lens system in two directions will lead to a shell construction with a double curve in the upper and under scale.

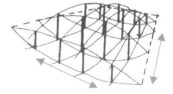

Figure 9: Closed symmetric system

APPLICATION

The principle of this design can be implicated by two different systems: a single curved construction and a double curved shell construction. Both applications are simulated with computer programs.

Single curved construction

The single curved construction is imported in the computer (MATRIX 2D) as a linear construction. The construction is calculated and observed with several load combinations.

In the load combination where the downward load is higher than the upward load , the construction act as followed:
- the upper curve is in compression
- the lower curve is in tension
- the vertical rods are in compression
- the crossed rods are in tension
- [Fig 10, Tab1]

The sum of the support reactions is equal to the sum of the loads. So the construction is evenly balanced.

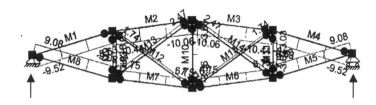

Figure 10: Member forces N_{max} (kN)

	F_{appear} (kN)	$F_{tadmisible}$ (kN)
Glass panel	17.1	38.0
Glass rod	6.0	6.3

Table 1: Member forces N_{max} (kN) in case of extreme snow load

However, when the upward load is higher than the downward load (in case of extreme wind suction), all the elements are loaded in opposite position.[Tab 2]

	F_{appear} (kN)	$F_{admissible}$ (kN)
Glass panel	0.70	116.0
Glass rod	0.20	2.7
Steal rod	-0.29	-0.6
Connections	10.50	26.0

Table 2: Member forces N_{max} (kN) in case of extreme wind suction

Double curved shell construction

The double curved shell construction is imported in DIANA (Finite Element Method). [Fig 11] The construction is calculated and observed with several load combinations.

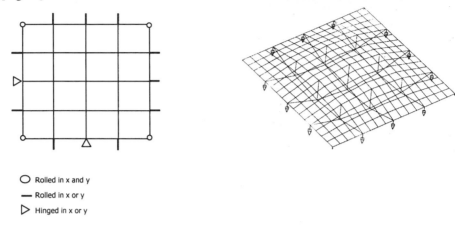

○ Rolled in x and y
— Rolled in x or y
▷ Hinged in x or y

Figure 11: Double curved shell construction and its border connections principles.

In the load combination, when the downward load is higher than the upward load, the global and local working of this application of the system act as showed in figure 12 and 13.

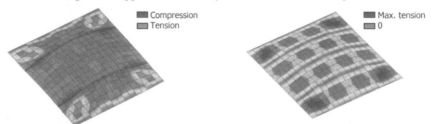

Figure 12: Global working: σ = N/A   Figure 13: Local working: σ = M/W

The global working in combination with the local working of the system results in a totality of tension and compression in the construction, showed in figure 14.

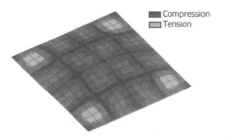

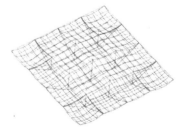

Figure 14: Compression in the strengthened glass strokes and tension in the surface of the plates.

Figure 15: Minimum and equally divided Transformation.

The sum of the support reactions is equal to the sum of the loads. So the construction is evenly balanced.

However, when the upward load is higher than the downward load (in case of extreme wind suction), all the elements are loaded in opposite position. [Tab 3]

	σ_{appear} (N/mm^2)	$\sigma_{admissible}$ (N/mm^2)
Glass panel	0.1	20.0
Glass rod	0.2	7.0
Steal rod	- 4.8	- 5.4

Table 3: Member tensions σ (N/mm^2) in case of extreme wind suction

CONCLUSION

Since the design is a closed symmetric system it only requires vertical reaction forces at the borders, in every allowable stress level. The closed and symmetric shape also leads to less tension in the vertical glass rods and the horizontal glass panels. This leads to a minimum of transformation.[Fig 15] The design can therefore be described as a system which condition is guaranteed by a mechanism of equal distributed tension and compression, and will lead to using glass products safely in a primary construction. This effective system allows minimal construction elements and can therefore be described as maximal transparent.

Figure 16: shopping gallery

The single curved version as a linear construction is especially suitable in cases where the length/width ratio of the space is large. Applications areas for example are: shopping galleries or railway stations.[Fig 16]

The double curved version as a shell construction is especially suitable in cases where the length/width ratio is one. Applications areas for example are: courtyards [Fig 17] or as an 'eyecather' in a building.

Figure 17: Courtyard

REFERENCES
[1]   Glass Façade, Tel Aviv, Mick Eekhout, Octatube Space Structures
[2]   Handbook of glass in construction, Joseph S. Amstock,1997, Mc.Graw-hill
[3]   Roof system Zwitserleven-Building, Amsterdam, ABT and Octatube Space Structures
[4]   Application Sentry Glass Plus in roof system, Court of Justice, Phoenix, R .Meier

Space frames and classical architecture

J. F. GABRIEL
School of Architecture, Syracuse University, New York

INTRODUCTION

Despite the obvious contrast in the appearance of space frames and classical buildings, these two architectural languages are alike in significant ways. The forms of classical architecture were derived from masonry construction over 20 centuries. The French architect C.N. Ledoux (1735-1806) declared that the circle and the square are the letters of the alphabet that creators use in their best works. Columns, barrel vaults and hemispherical domes are generated from the circle. The plans of most rooms and the shape of doors and windows are based on a square, a square and a half, or a double square.

Space frames use triangles to form octahedra and tetrahedra, where the square is only incidental and the circle totally absent. The octahedron and the tetrahedron were familiar to the Ancients, but they came to the attention of designers when they were able to be exploited for their structural potential. This could only occur after steel had become truly affordable and when serviceable joints began to be produced after World War Two.

The intention of this paper is to show that analogies exist between space frames and classical architecture and that the analogies are more significant than the differences.

PATTERNS IN CLASSICAL ARCHITECTURE

Mention classical buildings and your interlocutor will immediately form a mental picture of columns with elaborate capitals, pediments and multiple moldings with obscure Latin names. These elements do indeed characterize classical architecture but they are superficial. Underlying classical designs are canons, which are more substantive:

• The use of simple geometric shapes.

• Two-fold symmetry in plan and in elevation.

• Rhythmic regularity, most evident in colonnades and fenestration.

• The recurrence of triads or sets of three parts.

Let us observe how these canons find expression in a typical classical design (Fig 1). Our example is one of the pavilions that punctuate the lateral façade of the Mint, *La Monnaie*, in Paris. It was designed by J.D. Antoine and completed in 1775. Countless variations of this archetypal design are found in the United Kingdom, Italy, the United States and France, all countries with a strong classical tradition.

The formal organization of the façade consists of three tiers vertically and three bays horizontally. A symmetrical design, the vertical axis is given emphasis by a series of unique elements: the door with an arch and an oculus at ground level, a projecting balcony and a pediment on the *piano nobile*. All the elements on one side of the axis are repeated on the other. Already divided in three storeys, the façade is stratified in three zones: a rusticated base, a muscular cornice at the top and, between those two, an area that includes two storeys. This "double reading" is a device frequently used by classical designers to reiterate a tripartite division.

The square figure plays a critical role in the composition: not only is the largest portion of the façade a near square but the six major openings, including the door, are double squares. A more detailed analysis would reveal the presence of other, implied squares in the façade.

Lastly, rhythmic regularity can be observed in the rigorous alignment of the windows and their even spacing, in the equal height of the masonry courses of the ground floor, in the identical balusters of the balcony and the brackets of the cornice.

Of the four canons, the most important is that of organizing the design in triads. Aristotle had recommended that a story be centered on one unified action and that the telling consist of three parts: the beginning, the middle and the end in order to form a whole. Plato advocated the same strategy, using the metaphor of the human body complete with head, body and legs. In architecture, nearly all the components of the classical vocabulary are formed of three parts. For example, a column has a base, a shaft and a capital; an entablature is made of architrave, frieze and cornice; a balustrade has a cap and a base with balusters between them. It is this insistence on triads that gives classicism its other name: humanism.

Fig. 1. A typical facade for a classical building. There is a two-fold symmetry with emphasis on the centre; there is rhythmic regularity; it makes use of simple geometric shapes; the facade is organized in three parts horizontally and vertically.

PATTERNS IN SPACE FRAMES

Space frames, made of indeformable, triangular frames, are structurally redundant. According to Buckminster Fuller, they produce the single most powerful structural system in the universe. Like classical buildings they use simple geometric forms. The octahedron and the tetrahedron both allow for two-fold symmetry. In space they are ordered into regular, rhythmic patterns.

Triads are more in evidence in space frames as triangular figures than they are in any classical building. Even more significant is the fact that, in multi-layer, *three-way* space frames, a three-part cycle is complete when a chord is exactly superposed to another, node for node: this occurs with every fourth chord, *in every set of three consecutive storeys*. Triads are integral to the topology of space frames.

The Star Beam

Space frames are such redundant structures that many struts can be dispensed with at the design stage and a number of minimal, repetitive patterns can be identified. There will be no loss of rigidity within certain limits concerning the height of the structure, spans, acceptable cross-section of struts and other criteria.

The star beam is one such pattern. It owes its name to the six-pointed star figure appearing in the chords as the result of the systematic elimination of certain braces (leaning or oblique struts) in the search for larger spaces within the framework of a three-way multi-storey space frame.

The star pattern was introduced by the author in 1978 (Ref 1) and described at some length in subsequent publications (Refs 2-6). To avoid repetition only essentials are sketched out here.

The module is a geometrically rigid, habitable frame obtained from two star-shaped floor elements connected by ten braces (Fig 2). Lined up lengthwise, modules open onto one another through square bays four metres wide, forming by increments a through-truss. Through-trusses, i.e., star beams can be joined laterally in one of two ways, each producing a different spatial pattern (Fig 3). Four operational openings, two on each side of every module, permit lateral circulation from one star beam to another.

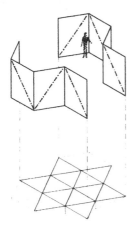

Fig. 2. The essential elements of a module: a star-shaped floor element and eight braces with vertical enclosures placed against them. The upper star-shaped floor element and the last two braces are not shown because, in context, they are normally provided by adjoining modules.

Fig. 3. A schematic model of one star beam resting on two others. Gaps in the floor allow us to see the star shapes.

When star beams converge at a 180° angle, three different patterns can be observed on consecutive storeys. Joined vertically, star modules create leaning towers (Fig 4). Any and all star modules can belong simultaneously to through-trusses and towers, and the space within is unaffected (Fig 5).

The overall shape of the proposed design is a megatetrahedron generated by three towers leaning against each other. Additional star beams are attached in saddlebag fashion on either side (Fig 6).

The ground plane remains mostly unobstructed. Self-imposed walk-up restrictions determine the height of the structure. Large spaces such as auditoriums, exhibition spaces and open terraces are found on the fourth and fifth levels. The upper storeys, much smaller in area, do not contain public spaces.

Clustering three megatetrahedra causes a large octahedral space to materialize in the middle (Fig 7). In topological terms, the close-packing system octahedra-tetrahedra is reproduced at a large scale. In terms of architecture, the potential for varied spaces and beautiful forms is great.

Fig. 4. Three leaning towers made of star modules. Any and all modules can be integrated in both a truss and a tower.

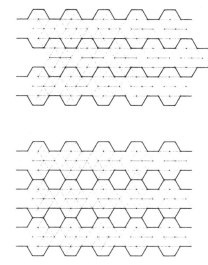

Fig. 5. Star trusses can be connected laterally in two ways, each creating a different spatial pattern. Architectural requirements determine the choice between the two.

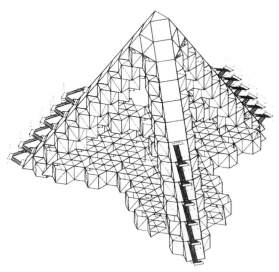

Fig. 6. The star beam geometry suggests a triangle-based pyramid, i.e., a megatetrahedron. Three stairs at the periphery are accessible from all parts of every storey.

CONCLUSION

The patterns of classical architecture and space frames are governed by the same canons, which include: Simple geometric shapes, two-fold symmetry, rhythmic regularity and recurring sets of tree parts.

The great architectural historian, Sir John Summerson, argued that the classical language of architecture, inherited from Rome, was the common architectural language of nearly the whole civilized world in the five centuries between the Renaissance and our own time. The ongoing development of an architecture of space frames can be understood as a continuation and an enrichment of the classical tradition.

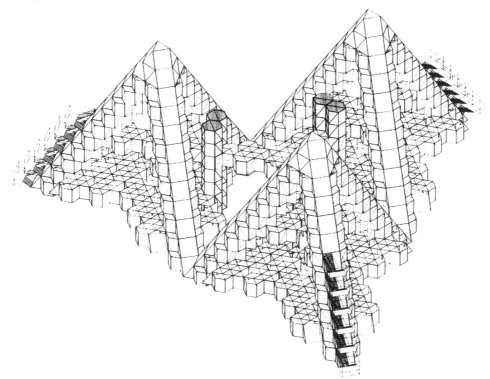

Fig. 7. The octahedral gap between the three megatetrahedra could be spanned by additional star beams to increase the useable space of the structural framework. Three sets of lifts are housed in the hexagonal towers.

REFERENCES

1. Gabriel JF, La Poutre-Etoile, Techniques et Architecture, No 320, June-July 1978, pp 70-71.

2. Gabriel JF, From Space Lattice to Architecture, Bulletin of the International Association for Shell and Spatial Structures, No 70, 1978, pp 19-23.

3. Gabriel JF, The Star Beam, Proceedings of the Symposium on Shell and Spatial Structures, September 1983, pp13-24.

4. Gabriel JF, Space Frames: The Space Within – A Guided Tour, International Journal of Space Structures, Vol 1 No 1, 1985, pp3-12.

5. Gabriel JF, Polyhedra and Folded Plate Structures, Proceedings of the International Colloquium on Structural Morphology, August 1997, pp 30-35.

6. Gabriel JF, Beyond the Cube: The Architecture of Space Frames and Polyhedra, John Wiley and Sons, New York, 1997, pp 459-465.

7. Summerson Sir J, The Classical Language of Architecture, The MIT Press, Cambridge, Massachusetts, and London, 1963, p 7.

ACKNOWLEDGEMENTS
The author is indebted to the work of the late R. Buckminster Fuller and to his kind personal advice. Professor Bruce Abbey, Dean of the School of Architecture at Syracuse University, generously provided the funds necessary to present this paper. Brian Ackerman and Jeremy Munn did the computer graphic part under the author's supervision.

Ecohouse: implementing a space structures *ecotecture*

P. J. PEARCE, Pearce Research and Design, 6160 Periwinkle Way, Woodland Hills CA 91367 USA

WHAT IS ECOTECTURE

Ecotecture can be characterized as an approach to environmental design that is driven by guidelines implied by the concept of sustainability. Such guidelines, inevitably, have their roots in ecologically based sensibilities. Fundamentally, this means that the built environment would be designed to achieve its programmatic goals with the lowest possible energy costs over time, and to minimize or eliminate the need to use depletable sources of energy for heating, cooling, and lighting.

Since these programmatic goals are found to be consistent with sound ecological principles, we can characterize such an approach for building design as *ecotecture*.

At the same time, the programmatic goals of this *ecotecture* would embrace classical architectural attributes, which contribute to human well being. These attributes include access to daylight, natural ventilation, temperateness, open space, and views. In this *ecotecture* there is a complementary relationship between these classical attributes and the minimization of long term and short-term energy costs.

Also, implied by the concept of sustainability is the concept of long life solutions requiring low maintenance. Long life, in turn, requires high performance with respect to the ability of a structure to sustain assault from dramatic events of nature such as earthquakes, high winds, heavy precipitation, and fire storms. In short, then, *ecotecture* comprises a high-performance design ethic in which form is considered an agent of performance.

CLIMATE MANAGEMENT CANOPIES

Ecotecture can be thought of as a building designed to efficiently and effectively manage the effects of regional climates, in order to provide *people-friendly* environments. The concept of a *Climate Management Canopy* is seen as an effective strategy towards achieving such an *ecotecture*. The *Climate Management Canopy* comprises an interface between the natural elements and human habitats. These natural elements include solar radiation, wind, precipitation, and ambient temperature. The regional characteristics of these natural elements give meaning to the concept of local climate. The management of these natural elements enables *people-friendly* environments to be built. Such environments can be large or small, and can comprise fully enclosed or open-air spaces.

Climate Management Canopies comprise three-dimensional structures conceived as a means of optimizing the effects of climate on human environments with minimal or no use of mechanical or electronic systems. In temperate latitudes, the management of solar radiation (insolation) is critical to the achievement of comfortable environments. The *Climate*

Management Canopy can provide an effective method for mitigating overheating caused by solar radiation.

The example of *ecotecture* presented in the following is a residence designed to serve the needs of this author and his wife, and to manifest a vision of building design that has been percolating for many decades. In addition to serving as a residence, the building must function as a facility for professional activity. At this point in my career, I am able to work on product design and architectural projects of my own initiative, which often involve physical models and prototypes. Therefore, the building must provide appropriate workspace for both design activity (most of which is computer based) and a modest model and prototype shop. In addition, my wife Susan is a consultant who works on feasibility studies and strategic planning for non-profit organizations. She requires appropriate office space for this activity.

This *Ecohouse* is located in Southern California, in which the regional climate creates an array of physical issues, which must be addressed. Although the management of the regional daily microclimate is the primary focus of the design, issues of wind, seismic activity, and fire ecologies must also drive the design. The actual site is a 2.5-acre plot in the Santa Moncia Mountains, at 2200 feet above sea level. The site offers a remarkable view of the Pacific Ocean, which includes Santa Moncia Bay, the City of Santa Monica, Pales Verdes Peninsula, and Santa Catalina Island. The bay creates a shoreline that is south facing, enabling the orientation of the house to optimized for both view and earth/sun geometry. The site is also geologically challenging, and is found in a classic Southern California fire zone.

INTERCEPTING SOLAR RADIATION

Solar radiation can be intercepted, and dissipated. In the process, convection currents are formed which create natural ventilation and cooling effects. This *solar cooling* is implemented through an aluminum louver system of high surface to volume, in which heat is conducted and dissipated rapidly (much like the fins of an air-cooled engine). This louver system is contained in a three-dimensional exoskeletal space truss, under which a space enclosure envelope is hung. The louver-equipped exoskeleton becomes a *Climate Management Canopy*.

 Openable zones (e.g. skylights) that are strategically located overhead in the space enclosure can be provided. When this is done, the heated up canopy creates a low-pressure condition in which the cooler air from the interior is drawn upwardly, resulting in natural cooling effects. This, in turn, creates a desirable microclimate,

Even without the *Climate Management Canopy*, there is the *stack- effect* that results from the cross-ventilation from lower walls through the skylights. The convection currents emanating from the canopy will enhance such stack-effects. Passive configurations can be created whose geometry is defined by seasonal sun angles (for any given latitude). In such schemes the summer sun is intercepted and dissipated, with a concomitant summer cooling effect, while winter sun is used as a source of heat gain to warm an environment. This can be accomplished without mechanical systems of any kind. In equatorial latitudes, year round protection from insolation can be accomplished.

In the case of canopies which support envelopes for building enclosures, the *Climate Management Canopy*, which forms the roof of the building, is designed so that summer insolation is interrupted before it actually reaches the enclosing roof surface. This strategy is equivalent to erecting a building entirely in the shade, under a very large tree. The canopy becomes a kind of synthetic tree. It takes the form of a large three-dimensional louver configuration with high surface area (like the fins on an air-cooled engine or leaves of a tree) to intercept and dissipate solar heat gain. The enclosing surface is suspended under the canopy in such a way that the roof (and wall) surfaces are never exposed to solar radiation (during the hot times of the year). In this way heat gain on the building enclosure from solar radiation is mitigated. Since the primary source of heat gain is solar radiation, not ambient temperature, very effective passive climate management can be achieved with this approach. (See Figure 1.)

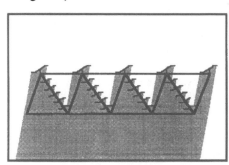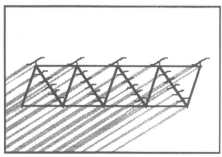

Fig. 1. *Climate Management Canopy* intercepts solar radiation during the summer months, and allows radiation into the interior during the winter months. The summer and winter solstices are shown for the Southern California latitude, which is 34° north. The left image shows the sun angle at 80°, at noon on June 21. The right image shows the sun angle at 31.5°, at noon on December 21. The shading indicates the zones in shadow. The drawings show a view from the west of a space truss segment with integral sun control louvers. Note that the louvers do not change position between the two conditions. In addition to the sunlight studies shown in Figure 1, more sophisticated sunlight studies have been undertaken using three-dimensional computer software, which validate the performance of the *Climate Management Canopy* of the *Ecohouse* as a manager of solar radiation. Inclusion of this material is beyond the scope of the present document.

Natural Light

The *Climate Management Canopy* is designed so that natural indirect light from the sun, and ambient daylight are always provided. In this way the need for artificial light is minimized or eliminated, during daylight hours. In the case of enclosed spaces the opportunity exists for the use of large expanses of overhead glazing without the problems of enormous heat gain and large amounts of glare. Ambient and reflected natural light, managed by the canopy as solar interface, can reach into the interior of a building space in abundance. Thus, the need for artificial light is minimized. Not only does this create a more people friendly environment; it reduces energy costs in two ways. First, the energy cost of electric lights is minimized or eliminated, and second, there is no heat generated by the use of electric lights during the daylight hours. Heat from electric lights can add significantly to cooling loads.

THREE-DIMENSIONAL SPACE TRUSSES

My personal experience has included considerable work in the design and construction of three-dimensional space truss projects, for architectural applications. For approximately 15 years (1980 to 1995) I operated a company called *Pearce Structures*. During this period the Company completed some 80 architectural scale projects. Many of these built architectural projects are still considered to be *state-of-the-art* examples.

The most well know of these projects is probably the Biosphere 2, facility near Tucson, Arizona. For this project we designed and built the space truss architectures, and developed and built an airtight glazing system, which encloses seven million cubic feet of volume. Five different, but linked, airtight glazed structures were designed and built of unprecedented performance and size. Never before in the history of building has an enclosure of this size been airtight. The project was completed in 1991.

Fig. 2. Biosphere 2, from the air, showing a diverse collection of building form, all designed around space truss technology and airtight cladding developed by this author.

The last major project completed by Pearce Structures was a 1400-foot long vaulted canopy at Fremont Street in Las Vegas. This project links a four block long segment of downtown Las Vegas. This overhead canopy is 100 feet wide and extends 1400 feet in a series of 180 and 200 foot spans in the longitudinal direction. These spans are supported with branching tree-shaped column capitals. The space truss vault supports a state-of-the-art graphic display system incorporating approximately two million light pixels. This project was completed in 1995.

Fig. 3. Fremont Street, showing on the left the space truss vault prior to the installation of the lattice cladding which creates shade and support for the light pixels. On the right is shown the vault at night after the lattice cladding has been installed.

In addition to the fifteen years of practical experience, intimate familiarity with the geometry of these structure types was gleaned in my early theoretical work reported in my book: *Structure in Nature is a Strategy for Design, The MIT Press, 1978, 1990.*

With this background of experience, it occurred to me sometime ago that there was an opportunity inherent in the nature of spatial structures to address the concept of managing the climate of environmental spaces. As is well known, space trusses are typically comprised of two parallel layers of structural members, separated by spatial diagonals. The outer members are called chord struts and the spatial diagonals are called web struts. Space trusses can be planar, multi-directional, or curved (e.g. vaults and domes).

In some special cases there can be only a single layer grid but these are usually convex domes, and often of relatively short spans. Note that large span domes of apparent single layer configuration are not true space trusses because they must introduce bending resistance into the strut members to avoid collapse. Because it comprises a fully triangulated framework, a true space truss has only axial loads in its struts, with no bending moments. Such space trusses provide structures of very high strength to material weight, and have the ability to create very large unsupported areas. Because of these attributes they are very effective for the creation of large column-free spaces. Experience has shown that sophisticated space trusses can be adapted to a myriad of structural configurations, and building form.

It occurred to me that when the attributes of large span and configuration adaptability are combined with the three-dimensional grid of the space truss, that a very effective passive sun control device could be fashioned. It also occurred to me that such a structure could be adapted to both open-air canopy applications and entire buildings comprising integral climate management envelopes.

EMBODIMENT OF ECOTECTURE
Perhaps the most complete embodiment of an *ecotecture* is found in the aforementioned concept for the new residence that I am designing for my wife, Susan, and myself. It is a concept that develops an entire architecture around the principle of maintaining a temperate, user friendly environment, through passive management of the natural climate. It is also an embodiment of the classical environmental attributes of access to daylight, natural ventilation, temperateness, open space, and views. We call this the *Ecohouse*, for reasons outlined in the foregoing.

A key element to the design of this residence is the *Climate Management Canopy*, which forms the basic architecture of the building. The project is currently in design development, and does not have a completion schedule at the present time. I am the architect, engineer, product designer, and contractor, clearly a labor of love.   Documentation for permit applications should be completed prior to the end of 2002, with groundbreaking as soon thereafter as the building permit is issued.

HIGH-PERFORMANCE DESIGN ETHIC
The house is an innovative design that will strive to synthesize what I have learned in nearly 50 years of attempting to practice a high-performance design ethic. It addresses energy efficiency through the incorporation of passive climate management principles. It incorporates new space structures and cladding/glazing technology of manufacturing ease, material

efficiency, and convenience of assembly. It is also designed to minimize site intervention with respect to structural interface on a topographically and geologically challenging site. The design will establish high standards for residential seismic safety, and resistance to high winds. Fire protection will be addressed through material selection, configuration, and, landscaping.

The house will be built from high-performance materials and finishes (steel, aluminum, precast concrete, and glass) of extremely long life, and minimal maintenance. Most of these materials are recyclable and produced from partially recycled materials (all but concrete). However important this may be from an ecological perspective, what is even more important is that these materials will have a very long life. It is also important that the configuration of the structure and building envelope use materials efficiently. Another attribute of the building system is that it can be disassembled and the components reassembled in different configurations at other locations. All of this is achievable within the parameters of surprisingly favorable economics.

INTEGRATED PRODUCT SYSTEM
As a design strategy I am approaching the project as an integrated product system. All elements from basic structural and space enclosure components, to floor systems, electrical distribution, communication networks, and mechanical services are designed as an integrated total system. Although this project is modest in scale, it is comprehensive in scope.

SPATIAL PARADIGM
From an architectural perspective the house, for which schematic design is essentially completed, is based upon the spatial paradigm of a loft or barn, though in a very technically advanced form. The large open spaces of flexible use implied by this spatial paradigm are deemed preferable. In this case, the inspiration of the barn is mostly metaphorical. There is no interest in imitating the appearance or materials of the barn.

The house comprises an open-plan column-free space of approximately 2700 square feet on one level; centrally placed upon a deck of 4900 square feet (inclusive of the interior floor area) which surrounds the house enclosure. In plan, the structure has the shape of a hexagon, slightly elongated along the east/west axis of the building. The deck is supported from 6 reinforced concrete piers, which reach 60 feet below grade to bedrock. The deck, a space truss on a triangular grid in plan, spans between the piers. Floor cladding will be pre-cast concrete panels, which match the grid, serve as thermal mass, and include integrated radiant heating capability. The only truly enclosed *rooms* in this "one-room" building space will be two bathrooms. Otherwise, spatial differentiation will be achieved with a modular storage wall system, which will be designed specifically for this project. Below the main deck (floor) level is a 2000 square foot enclosed garage, workshop, storage, and service core space.

ATTRIBUTES OF ECOHOUSE
Springing from three symmetrically displaced zones at the boundary of the deck is an overarching *Climate Management Canopy*. This canopy is formed by a three-dimensional space truss, which comprises an exoskeleton. This exoskeleton forms the basic structure from which a 100% glass (transparent) space enclosure is hung (stepped roof and vertical walls), and to which, overhead, is located passive sun control panels (large-scale louvers) which manage the radiant energy from the sun. Winter sun can penetrate the enclosed space and summer sun is interrupted and it's heat dissipated. (See Figure 1.)

Three hundred and sixty-degree natural ventilation (and view) is provided, as well as overhead ventilation (and view). As explained in the foregoing, natural cooling effects are precipitated by the "stack effect", as well as the heat dissipation from the overhead shading panels, that generate natural convection currents, which pull air through the building. In effect, the house is *solar cooled* by the *Climate Management Canopy*.

The building enclosure is comprised entirely of insulated glass, including all exterior walls, doors, and roof. There are 88 openable windows around the entire perimeter of the enclosure, and 41 openable windows in the overhead (roof) glazing. A large (2700 sq. ft.) open plan is formed with 360 degrees of view, including the ocean on the south, and the sky towards the north (through the glass roof), with an interior washed in natural reflected (indirect) light.

Because of the glass enclosure, the primary source of light is natural light, which is reflected and controlled by the outboard louvers. Energy is saved by reducing electricity demand, and the heat gain caused by artificial lighting is eliminated, thereby reducing cooling loads in the summer months. In combination with cellular drapes, insulated glass minimizes thermal transfer. In addition, the cellular drapes, which are installed on the interior of the vertical glass wall, enable privacy to be controlled and add further refinement to the ability for daylight to be managed.

Surface area of the building envelope is minimal relative to enclosed volume, to optimize thermal performance (and to minimize the amount of building material required). Based on the hexagon derived plan form in which exterior walls meet at corners of 120° interior angles, and the inherent simplicity of the building envelope shape, surface area of the building enclosure is minimized. Surface area is 5444 square feet, and volume is 40,000 cubic feet, a ratio of .136 square feet to cubic feet. This low surface area geometry, helps to mitigate thermal transfer between inside and out. This is particularly important during the lower outside temperatures of winter. Sprinklers will be located outboard in the overhead structural system for washing, fire protection, and added cooling effects.

Although there will be a solar powered (hot water) radiant heating system built into the floor, it is anticipated that solar gain in the concrete floor will meet most heating needs in the winter. A utility corral will be located remotely from the house. It will be enclosed with an opaque wall and spanned by an open-air space truss, which will support a photovoltaic array to provide electricity for the house (and shade within the utility corral).

As mentioned above, the building is anchored to bedrock via six cast-in-place concrete piers that descend approximately 60 feet below grade. These piers are tied laterally with reinforced concrete grade beams. The piers terminate at column capitals, which are in the form of inverted hexagonal pyramids. These capitals support the space truss, which forms the deck/floor structure. This strategy in combination with the high performance space truss insures maximum effectiveness in resisting large seismic and wind forces. It also minimizes the invasiveness of the structural interfaces with the site.

3D COMPUTER RENDERINGS

A series of computer renderings have been generated which show various aspects of the building design as described in the foregoing. Most of these renderings show shadow patterns from sunlight for specific months of the year, and at very specific times of the day. These drawings are as follows:

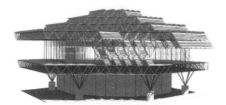

Fig. 4. Views from the southeast, at 9:00am, June 21.

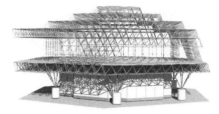

Fig. 5. View of space truss exoskeleton, and deck/floor.

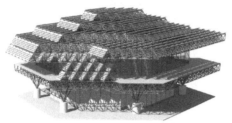

Fig. 6. View from southwest, at 4:00pm, June 21.

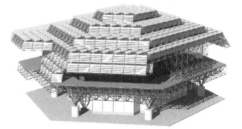

Fig. 7. View from the northeast, at 7:00am, June 21.

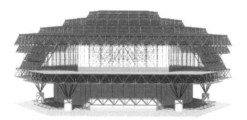

Fig. 8. View from the south, at 12:00pm, December 21.

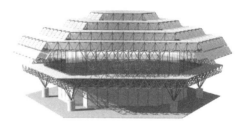

Fig. 9. View from the north, at 7:00am, June 21.

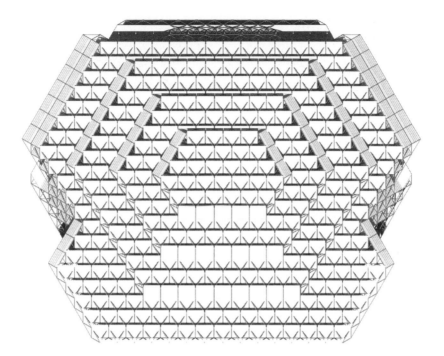

Fig. 10.   Roof plan, showing the array of aluminum louvers integrated into the three dimensional exoskeleton.

CONCLUSION

The *Ecohouse*, as described above, develops the concept of the *Climate Management Canopy* into a fully formed *ecotecture*. In this example an new paradigm for building form is created in which the classical environmental design attributes of access to daylight, natural ventilation, temperateness, open space, and view, are fully integrated with building performance. Passive climate management and adaptation to site are given high priority as fundamental elements of a sustainable *ecotecture*. At the same time, a *people-friendly* environment is provided with an abundance of natural light and ventilation, open space and views, which nurtures a sense of well being.

Methods of virtual architecture used for the design of geodesic domes and multi-petal shells

G. N. PAVLOV
Nizhny Novgorod, University of Architecture and Civil Engineering, Russian Federation

ABSTRACT

This paper is devoted to the modern method of virtual architectural used for designing domes and shells. The methods are completely based on the using of the automated architectural designing systems. Method of getting of library elements for CAD systems and several examples of the library elements using in the process of virtual designing of geodetic domes and multi-petal shells are given.

INTRODUCTION

Revolutionary changes in architectural designing have occurred during the last years, now designing is completely based on the computer equipment, from the rough draft stage to carrying out of the working drawings. In design firms nowadays traditional for architects instruments, such as drawing boards, pencils, pens, watercolour, paints, gouache, mascara and brushes are not absolutely used. In the interior of architectural company there are only several computers, connected with each other by a computer network, scanners and printers. Manual performing of drawings and demonstration plane-tables have become an anachronism.

Presently computer programs allow to get new products of architectural designing, which were impossible earlier. One of the fantastic possibilities of the programs is getting of films with virtual showing of a designing object. The customer now can see a designing object from all the sides, the customer can rise on the helicopter and see the building from above, then he can enter inside the building and see its interiors (Table 1).

All mentioned above is concerned both ordinary buildings and buildings with spatial structures. One of the peculiarities of the process of modern computer designing is necessity of creating of library elements for multiple use. CAD systems usually offer a big library element list, but hitherto there is no one system which contains data on geodesic structures and some other types of shells.

LIBRARY ELEMENTS

The author of the article has studied five optimized geodesic systems and one of the multi-petal system. Information on them is published in different sources (Ref 1-6). For all the systems were created programs, allowing to get the framework drawings of the network breakdown of the surface. These framework drawings made easier architectural designing greatly, however for getting a general view of buildings it was necessary to use traditional labour-consuming methods of architectural design of the projects - the performing of the projects manually, in watercolour, paints and modeling. But in the automatic systems of architectural designing the single-line drawings are little suitable. Besides, the automatic designing supposes to use library elements with visualized surfaces in the process of designing, the colour and texture of the library elements should satisfy conditions of natural sunlight in different climatic areas.

Table 1. Examples of the modern virtual architectural design of buildings.

PROJECT OF A TEMPLE
Architect Pavlov K..G.

PROJECT OF A SCHOOL
Architects: Pavlov O.G.
Pavlenkov V.A.

In geodetic systems the library elements are a visualized surface of the whole sphere, added by possibility of singling out of its separate   fragments, chosen for designing a dome. Standard parts can be library elements for other shell systems.

Fig.1 the upper row illustrates library elements of three geodetic systems (on the left an example of the library element of the six-petal shell is shown). In the lower row - there are examples of singling out of shell fragments necessary for domes designing by the architect.

In the pictures we can see that the library elements are not just framework sketches, but the colour of their ribs and surfaces were given by the architect with taking into account of sunlight conditions in given climatic or geographical area.

Fig. 1. Examples of library elements and their fragments.

In double-contour schemes of shells it is possible to fix a top of pyramids in any height. By fixing different values of height we can get fantastic virtual images of geodesic shells (Fig. 2).

Table 2, (a) and (b), present a full list of library elements for geodesic system «C». Table of classification of types network of this system is in (Ref 1, 4).

Fig. 2. Virtual geodetic structure images.

Table 2. Library elements of the system « C ».

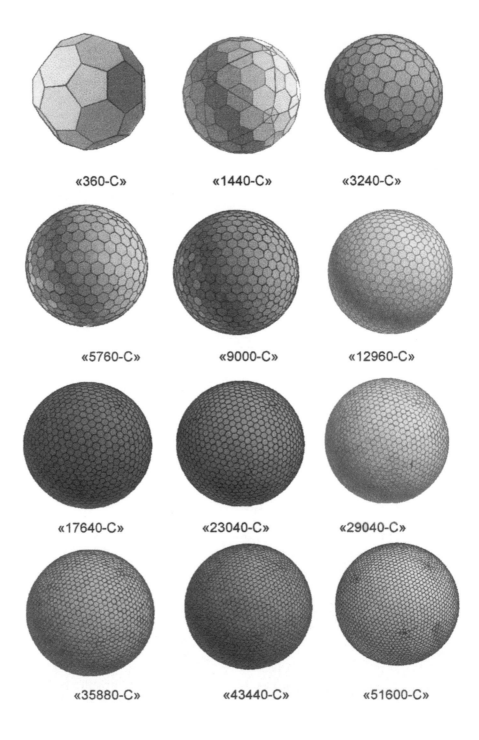

«360-C»   «1440-C»   «3240-C»

«5760-C»   «9000-C»   «12960-C»

«17640-C»   «23040-C»   «29040-C»

«35880-C»   «43440-C»   «51600-C»

Table 3. Methods of virtual architectural design of the dome based on the breakdown « 12960-C ».

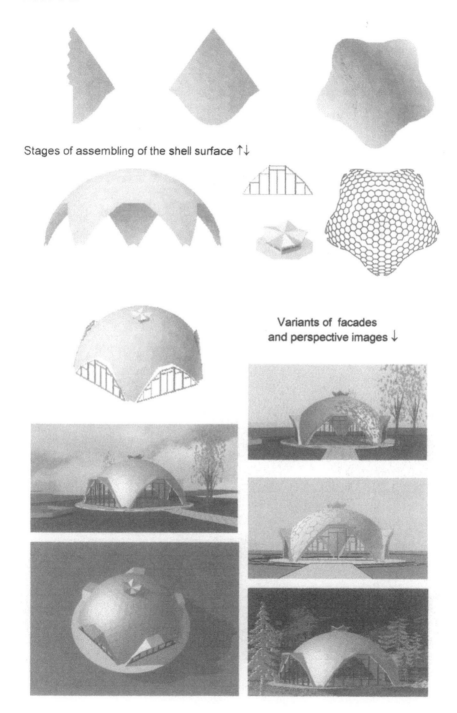

Stages of assembling of the shell surface ↑↓

Variants of facades
and perspective images ↓

Table 4. Examples of the virtual architectural design of the geodesic dome based on the breakdown « 17640-A''(''8820-A1''+''2940-A2 »)

Definition of geometric parameters of the dome

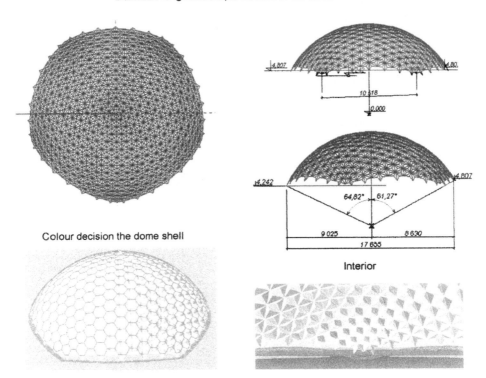

Colour decision the dome shell

Interior

Perspective image of the dome

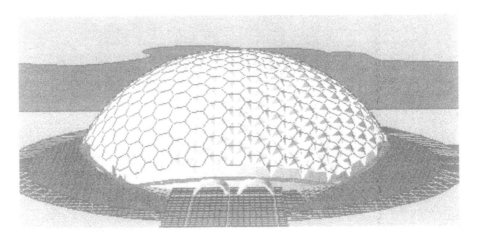

Table 5. Example of virtual architectural design of the geodesic dome based on the breakdown '' 17280-D » (« 2880-D1 ''+'' 8640-D2»).

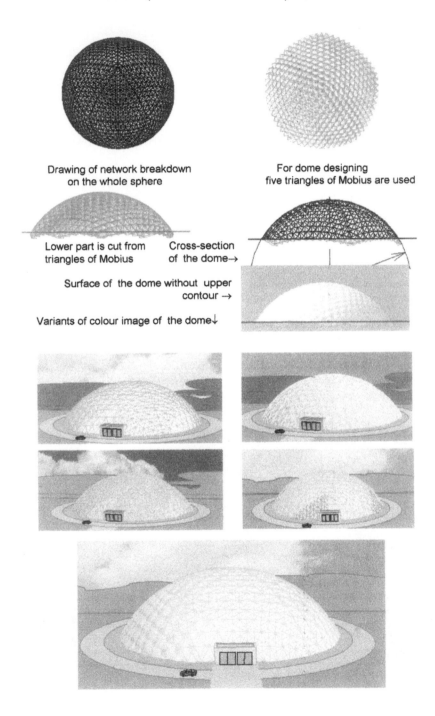

Drawing of network breakdown
on the whole sphere

For dome designing
five triangles of Mobius are used

Lower part is cut from
triangles of Mobius

Cross-section
of the dome→

Surface of the dome without upper
contour →

Variants of colour image of the dome↓

Table 6. Examples of the creation of architectural virtual variants of six -, eight -, and twelve – petal shells.

Stages of getting the library element and shaping of shells

VIRTUAL DESIGN
In tables 3, 4, 5 and 6, shown previously, a few examples of shell virtual design are given.

Table 3 illustrates the process of virtual designing of the pavilion, built in N. Novgorod. The dome is based on breakdown «12960-C». The library element of this type is shown above in Table 2,(a). In Table 3 the process of forming of the virtual image of the pavilion, from cutting out of a shell fragment to creating of variants of architectural decisions is shown.

In tables 4, 5 and 6 there are some examples of dome virtual design on the basis of the systems A, B and D.

CONCLUSIONS
Architectural designing of the geodesic shells has always been a difficult process, but after the computer library elements of shells appear this difficulty will disappear. Nowadays it can be said, that modern professional architectural CAD-designing of shells is impossible without computer library elements.

Methods of getting of computer library elements for CAD systems allow to fill a missing section in the problem of shell architectural designing. Architects and students of architectural schools will get a facilities for designing of buildings.

REFERENCES
1. Pavlov G N, Compositional form-shaping of crystal domes and shells, Spherical grid structures, Hungarian Institute for Building Science, Budapest, 1987, p. 9-124.
2. Pavlov G N, Determination of parameters of crystal latticed surfaces composed of hexagonal plane facets. International Journal of Space Structures. Vol. 5, No. 3-4, 1990, p. 69-185.
3. Pavlov G N, Geodesic Domes Bounded by Symmetrical Mainly Hexagonal Elements, International Journal of Space Structures. Vol. 9, No. 2, 1994, p. 53-66.
4 Pavlov G N, Domes and shells from flat hexagonal panels (in Russian), Nizhny Novgorod, 1995, 92 p.
5. Pavlov G N, Geodesic domes and shells (in Russian). Nizhny Novgorod, 1997, 197 p.
6. Pavlov G N, Multy-petal shells (in Russian), Nizhny Novgorod, 1995, 24 p.

ALOSS - album of space structures

D. YAMATO
Architecture and Civil Engineering, Graduate School of Engineering, Fukui University, Japan
K. ISHIKAWA
Department of Architecture and Civil Engineering, Fukui University, Japan
Y. ISONO
Chuuouku 4-8-4 Sayama-shi 350-1308, Japan
T. MURAKAMI
Nesty, Usui 2-402 Fukui-shi 918-8114, Japan
H. NOOSHIN
Space Structures Research Centre, University of Surrey, Guildford, Surrey, UK

ABSTRACT
Named by the last author Professor Nooshin, ALOSS is an abbreviation for "Album of Space Structures", which is a database of space structures throughout the world, containing building names, photographs, addresses, architects, structural engineers, completion years and so on. ALOSS is available on the Internet to be able to access the information easily.

The URL is "http://www.anc-d.fukui-u.ac.jp/~ishikawa/Aloss.htm". So far, ALOSS has data on 697 structures, which have been collected by the third author Mr. Isono. Now, ALOSS is being extended with new information and its the content is being edited.

INTRODUCTION
Many space structures have been built all over the world. These structures, techniques, and designs are useful for students, structural engineers, architects and researchers to learn space structures of the world on the Internet freely. The third author, Mr. Isono, is collecting more than 1000 photographs of shell and space structures around the world, and is reviewing information such as buildings name, photographs, addresses, structural engineers, completion years and so on. Therefore, we are providing this information in an electronic version for readers and users of ALOSS.

Fundamental information frames of ALOSS are composed of the outside and inside photographs with information. ALOSS includes both the photographs and valuable information. The purpose of this paper is to describe what ALOSS is and how to use it to find information.

CONCEPT OF DATABASE OF ALOSS

The outline of ALOSS is described in Fig. 1. ALOSS mainly classifies information into three types which are construction region, structural forms and structural engineers. Administrators of ALOSS are looking for volunteers, to help to expand it in the future. Moreover, we will continue to increase the information and improve the database.

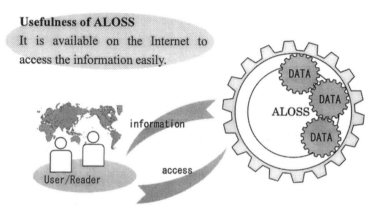

Fig. 1. Outline of ALOSS

The elements of ALOSS are shown in Fig. 2. A space structural database of the world, ALOSS offers information which include addresses, architects, structural engineers, traffic access information, notes, the building names and the completion year as exemplified in Fig.5. The most important feature of ALOSS is the provision of information, which has been collected by Mr. Isono on the Internet in electronic form. Fig. 5 shows a sample entry from the database.

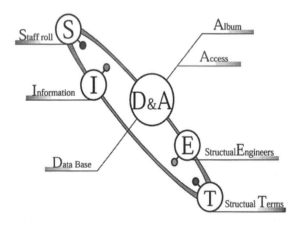

Fig. 2. Elements of ALOSS

An index constitution of ALOSS is shown in Fig. 3. ALOSS is separated into 9 sections which are available to access information of your choice. With "structural system", "countries" and "structural engineers", we can access information which are shown in Fig. 5. Other indices include profiles of Mr. Isono, explanations of architectural and structural technical terms and so on. Details of these indices are described in later pages.

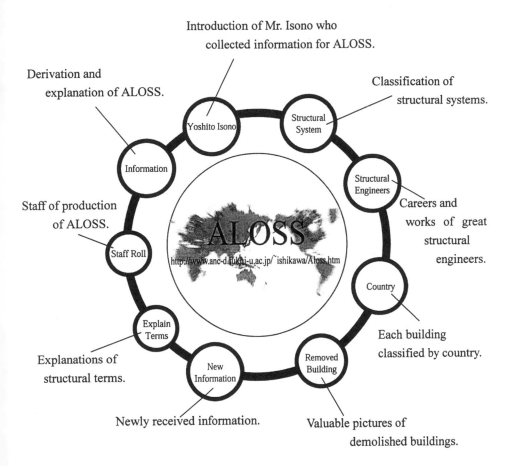

Introduction of Mr. Isono who collected information for ALOSS.

Derivation and explanation of ALOSS.

Classification of structural systems.

Staff of production of ALOSS.

Careers and works of great structural engineers.

Explanations of structural terms.

Each building classified by country.

Newly received information.

Valuable pictures of demolished buildings.

Fig. 3. Index constitution

A special feature of the database is "Removed Building information". Moreover, one can see all the pictures which have been gathered together so far. One also can access the information from each picture freely. When the administrators of ALOSS have improved the database, it will become a more impressive database in the future. That is why we hope to have the collaboration of volunteers all over the world.

CLASSIFICATION OF SPACE STRUCTURES INFORMATION

ALOSS is composed of a fundamental information frame as shown in Fig. 4. This is an example of a frame which includes name of building, structural system, country, construction year, address, architect, structural engineer, usage, admission, requirement for admission, traffic access, reference page of proceedings, notes and so on.

ALOSS also classifies the building information frames into the country, the structural engineers and main structural system. The information has been inputted since 1997.

1. Country

Inputted number of countries are 16. As shown in Fig. 4, United Kingdom is one of the inputted countries. The other countries are Australia, Canada, China, France, Germany, Italy, Malaysia, Mexico, Poland, Singapore, Spain, Switzerland, Taiwan, Turkey and USA. Although the number of total data by construction region is 375, total information numbers of structural type identifications are 697. 322 of the remainder were not classified into the construction region because of the shortage of information when we inputted newly received information into structural forms. Thus, when these 322 new items of information are classified by construction region, new countries will be added.

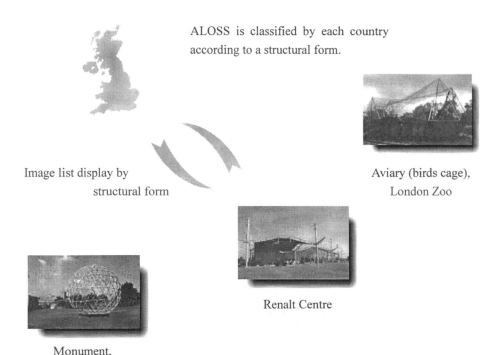

ALOSS is classified by each country according to a structural form.

Image list display by
structural form

Aviary (birds cage),
London Zoo

Renalt Centre

Monument,
University of Surrey

Fig. 4. Sample pictures in United Kingdom

Name of Building	:	Solone Nervi, Parco Acqua Santa
Structural System	:	shell, RC
Country	:	ITALY
Construction year	:	1952
No.	:	39TO01 [5-106]
Address	:	Viale Guido Baccelli/Viale delle Terme, Chianciano Terme, Toscana.
Architect	:	Mario Loreti/Mario Marchi
Structural Engineer	:	Pier Luigi Nervi
Use	:	assembly hall
Demolished or not	:	No
Admission	:	OK
Requirement for Admission	:	everyday 07:00-12:00/17:00-19:00
Traffic Access	:	20 km west from Stazione Chiusi, bus to Montepulciano., get off at Parco Argua Santa. 1 time/hr.taxi avail at station.
Reference	:	Structures, Pier Luigi Nervi, F.W.Dodge Corp.,NY,1956, p.75,81.
Page of Proc. of "Tsuboi Memorial Anual Seminar"	:	Space Structures vol.5 , p.106
Notes	:	PLN-fig.5-29. tel:0578-6-8111/8410. fax:0578-6-0622.

Fig. 5. Sample of information

2. Structural systems

ALOSS classifies the structural systems by the categories shown in Fig. 6. Structural systems may be divided into multi-layer structural system and single-layer structural system (ref. 2). A multi-layer structural system includes flat, plate and beam structures.

A single-layer structural system may be divided into three categories:

● 'compressive force system' such as a folded plate, arch(bridge) or shell structure
● 'axial force system' such as a space frame structure
● 'Tensile force system' such as a cable, membrane or hybrid structure

Classified by each structural system, the catalog of all inputted pictures are displayed on the top page of each structural system. It looks like an album.

3. Structural Engineers

A summary of inputted information for each structural engineer and structural system is shown in Table 1. Particularly, ALOSS classifies each structural system by structural engineers Robert Maillart, Eduardo Torroja, Pier Luigi Nervi, Felix Candela and Heinz Isler. ALOSS has data on 168 structures about the structural engineers who are shown in Table 1. The structural systems are divided into 8 sections which are shell, arch, beam, flat plate, folded plate, hanging roof, steel and miscellaneous. Classified by each structural engineer, the catalog of all inputted pictures are displayed on the top page of each structural engineer as well as by the classification method of the construction region and structural system.

Table 1. Classified Structural Engineers

Structural Engineers	Shell	Arch (Bridge)	Beam	Flat plate	Folded plate	Hanging Roof	Steel	etc	Total
Robert Maillart	2	22	2	—	—	—	—	1	27
Eduardo Torroja	10	3	—	—	2	—	1	—	16
Pier Luigi Nervi	12	1	—	1	4	1	1	3	23
Felix Candela	56	—	—	—	—	—	1	—	57
Heinz Isler	45	—	—	—	—	—	—	—	45
Total	125	26	2	1	6	1	3	4	168

ALOSS has not been completed yet; inputted information may include errors. Hence, we have been modifying and improving the database. Henceforth, users and readers opinions will be reflected in ALOSS.

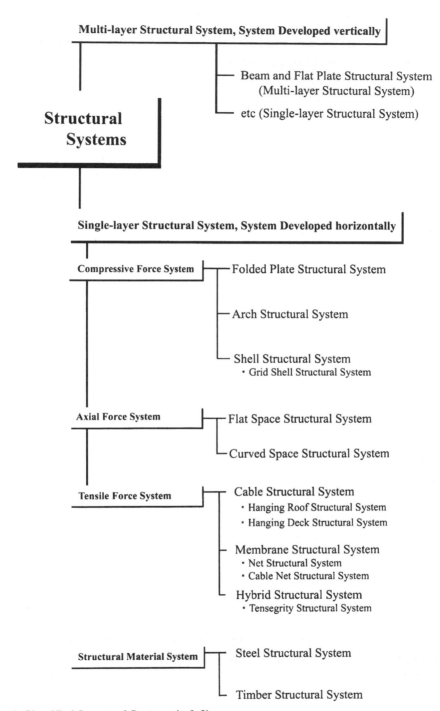

Fig. 6. Classified Structural Systems (ref. 2)

Robert Maillart

Viaduc de Grandfey
(1927)

Salginatobel Brucke
(1930)

Spital Brucke
(1931)

Schwandbach Brucke
(1933)

Eduardo Torroja

Viaducto del Aire
(1933)

Puente de la Tordera
(1939)

Iglesia Nuestra Senora
de la Ascencion
(1952)

Templo Parroquial
San Nicolas
(1961)

Pier Luigi Nervi

Assembly Hall,
UNESCO
(1956)

Secretariat Building,
UNESCO
(1956)

CNIT
(1958)

Palazzo dello Sport
(1960)

Felix Candela

Pabellones de Rayos
de Cosmicos
(1951)

Schwandbach
Brucke
(1958)

Auditorio,
Hotel Casino de la Selva
(1960)

Parroquia del Senor
del Campo Florido
(1963)

Heinz Isler

Wyss Garten
Haus Zuchwil
(1962)

Migross Mercato
(super market)
(1964)

Gips Union SA
(1968)

Tennishalle Burgdorf
(1980)

CONCLUSIONS

The database of space structures from around the world, ALOSS was constructed on the Internet as an electronic version of information which has been collected by Mr. Isono. ALOSS has also a simple information retrieval system which displays the catalog of all inputted pictures on the top page of each index for accessing detailed information. It looks as if it was an album. From our point of view, we think this information retrieval system is quite useful. The visual effects are attractive and appealing. Thus, we think that ALOSS is better than the conventional word-based retrieval systems.

ALOSS IN THE FUTURE

We really hope to involve collaborating volunteers all over the world. Improved by volunteers and ALOSS production staff, ALOSS will become a impressive and large database. If you have any interesting information please do not hesitate to contact us. Administrators of ALOSS plan to expand ALOSS in the following ways:

1. A frame which explains structural material, transition of the construction technology and the influence of the relation of a structural form will be added.

2. We will increase the visual effects of the database. For instance, ALOSS will be equipped with a clickable-map as one of its web functions.

3. We will make an English version of ALOSS, and this information will be sent all over the world.

4. Information of ALOSS will be improved and expanded by joint work with volunteers. We hope to exchange information with volunteers to further expand ALOSS.

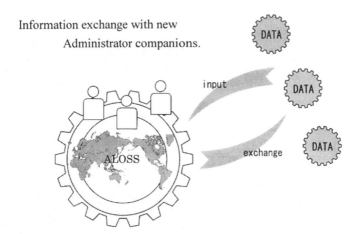

Fig. 7. The planned expansion of ALOSS

ACKNOWLEDGEMENT

We wish to acknowledge the help of the following students in the Fukui University, Ishikawa laboratory, who have been working hard for the creation of ALOSS since 1997:

  1997 T. Ishii, K. Tamura, K. Suzaki, K. Nishimura, K. Ikehata, S. Okubo

  1998 Y. Ando, Y. Tanaka, Y. Mae, K. Muranaka

  1999 T. Atagi, N. Ikehata, S. Takeda, T. Tanimura, T. Noda

  2000 T. Itou, T. Oshima, Y. Nakatani, T. Nogami, Y. Hashimoto

  2001 T. Igarashi, N. Takeda, C. Tadokoro, T. Murakami, D. Yamato

REFERENCES

(1) Y. Tuboi : HISTRICAL OUTLINE OF SPACE STRUCTURES IN JAPAN, IASS Journal col42,Recent Spatial Structures In Japan p9-14, 2001,Oct.

(2) Architectural, structural system society(Y. Tuboi, T. Saitou, K. Hayashida, T. Watanabe, Another 14) : Architectural structure -A structural system is understood- Syoukoku Campany,1997

Double – layer structures in design of flat covered roofs

W. BOBER
Division of Structures and Building Engineering, Department of Architecture, Wroclaw University of Technology, Poland

DEFINITIONS

Double-layer flat structures are mostly used as a solid system for large span roofs. Because of high damage stability these systems are especially useful to cover such buildings as hangars, chemical factories, etc.

The space structures, denoted: B{ T-T}A and B{ D-O}A, belong to the numerous group of structures proposed by J. Rebielak for structural roofs of large span. The B{ T-T}A structure, formed on the basis of a classical double-layer {T-T} truss, consists of regular tetrahedrons and regular octahedrons. To get the B{T-T}A form, each edge of tetrahedron is divided into two equal parts. Created joints are connected with new short bars of the same length. In this method each of tetrahedron consists of four equilateral triangles. The B{D-O}A structure was built on the basis of classical {D-O} truss as a result of section each edge of half octahedron modulus.

In both types of structures the bars of lower layer are longer than the bars of other ones. These elements join the lower nodes of modular solids. The paper presents the results of analysis made for different configurations of skylights and support systems for proposed double-layer structures.

RESEARCH MODEL

The linear static analyses of the research structures were made by using the computer program Robot V14.1. All investigated shapes of the space structures were supported in the perimeter nodes of lower bar layer. The vertical load was applied to every node of upper bar layer. All bars in the structure have the same size of square-tube cross section: 40 x 40 x 5 mm, and are made of steel of strength f=215 Mpa and E=205 Gpa. For all the cases one may assume that these structures are composed of weightless bars.

B{T-T}A structure

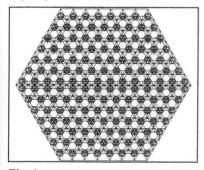
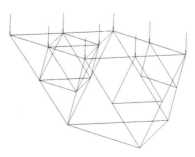

Fig. 1.

The structure B{T-T}A, shown in Figure 1., was shaped on a hexagon plan of a diagonal of 28 m. Bar lengths are equal: 1,0 m for upper and middle layer and 2,0 m only for lower layer. The first part of analyses was made for different systems of support, as follow:

- S1: pivot bearing in every perimeter node of lower layer
- S2: pivot bearing in every second node of lower layer
- S3: pivot bearing in every quoin node of lower layer
- S4: pivot bearing in four pillars with a triangular cap supporting three nodes of lower layer
- S5: pivot bearing in eight pillars with a single cap supporting one node of lower layer

In Figure 2a-f there are shown configurations of skylights in the area of shaped covers. In Figure 2a. it is shown a central hexagon skylight. For calculations it was taken in three dimensions:

H1 = 4 %   H2 = 10 %   H3 = 20 %

The perimeter of skylights is shaped with long bars surrounding the central hole.

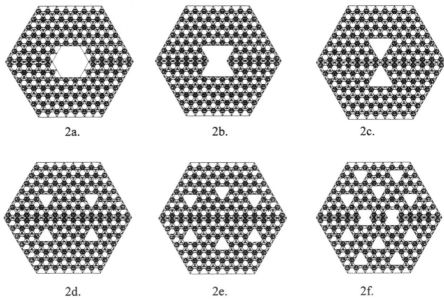

2a.   2b.   2c.

2d.   2e.   2f.

Fig. 2.

The next suggested skylight has a form of concave hexagon situated in the central part of the structure (Fig. 2b). It was calculated for the dimension

X1 = 7 %   X2 = 16 %   X3 = 28 %

In Fig.2c are shown two triangular holes, connected by the central node in upper layer of the structure. These skylights were investigated in the followed areas:

2T1 = 5 %   2T2 = 12 %   2T3 = 22 %

In Fig. 2d-f the skylights have the same triangular area. Only the holes in the middle of structure in Fig. 2f are proposed in concave hexagon form. Total sum of hole areas in each type is as follow:

4T = 4 %   6T = 6 %   10T2X = 12 %

Loading

In every node in upper layer of the models act concentrated forces of the value 1,50 kN. Loading of nodes situated in the edges of skylights was calculated in two alternative cases : the first model with weightless skylight and the second one with the reactions of skylight acting in perimeter nodes. The value of these forces is equal to the area of a skylight.

B{D-O}A structure

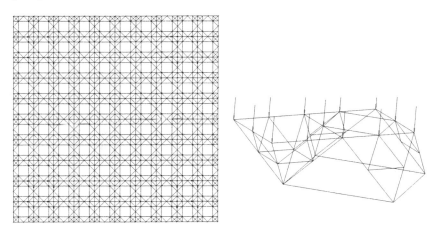

Fig. 3.

Double-layer structure denoted B{D-O}A, shown in Fig. 3 was shaped on a square plan of dimension 40 m. Construction depth is 2 m. Static calculation was made for different systems of support, as follow:

S1: every perimeter node of lower layer
S2: every second node of lower layer
S3: every quoin node of lower layer
S4: four pillars with a square cap supporting four nodes of lower layer.

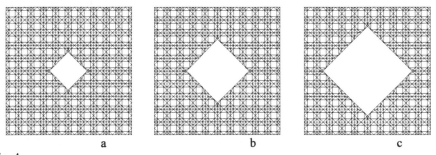

a   b   c

Fig. 4.

In Fig. 4a-c there are shown shapes of skylights in three dimensions:
O1 = 5 %   O2 = 12 %   O3 = 24 %
The perimeter nodes of holes among half-octahedron modules, supported by extra bars, connect them with nodes in the middle layer of the structure.

Loading
In all nodes in upper layer there were applied forces of value 2,00 kN. Loading of edges of skylights was calculated for two cases: one with weightless skylights and another one with the loading equal to the area of skylights.

RESULTS OF CALCULATION
B{T-T}A structure
As a result of calculations there were found extrema inner forces as compression axial forces as well as deformations of the structures.

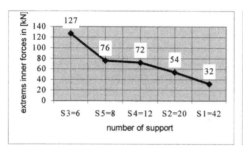
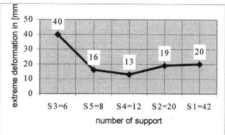

Fig. 5.

In Fig. 5 it is shown dependence between the support system and the value of forces as well as deflections in each case of search-truss. The values of extrema compression forces and extrema deflections when support is in six nodes, are three times as big as when support is in every node in perimeter of the structure. Relatively small deflections for eight- and twelve-nodes support may be explained by the position of support situated in some distance from the perimeter.

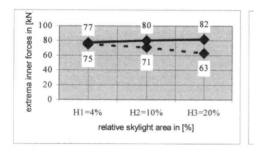
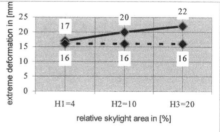

Fig. 6.

The calculation for the B{T-T}A structures with skylights was made for eight-nodes support. The full line in diagram presents results for the loading edges of skylight. Dash-line presents results for loading less edges of skylight.
It is worth noticing that the deflection for an open skylight made into the shape of hexagon is constant but for H3 skylight is smaller than for H2 one. This can be explained due to shorter distance from support line in H3.

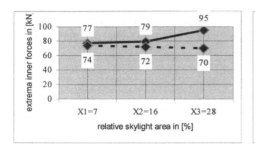
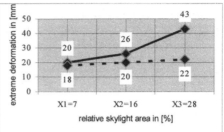

Fig. 7.

Diagrams for convex and concave hexagon skylights are similar for forces, and only the difference may be noticed for extrema deflections in case of the biggest holes. Much greater value for X3 can be seen because the edge node of extrema displacement is situated near the centre of the truss.

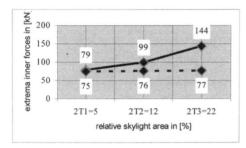

Fig. 8.

 Graph relations between forces, as well as deflections and areas of skylights are proportional. The value of extrema forces in 2T3 case is twice as big as in S5 one of support the whole skylight structure. Extrema forces act in bars located in the narrowing central part in two triangles .

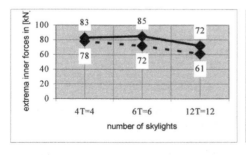
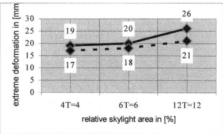

Fig. 9.

Figure 9 presents the results of calculations for different numbers of similar triangular skylights. It is worth noticing that in case of the biggest amount of skylights, values of extrema forces are smaller than in less number of skylights, while deflection is ascending.

B{D-O}A structure
The edges of skylights in the B{D-O}A structure are paralleled to up-layer net. The calculations for forces and deflections are made for S2 case of support every second node of the perimeter. The surface loading for the B{D-O}A structure is similar to that one for the B{T-T}A structure.

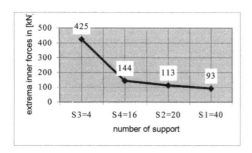
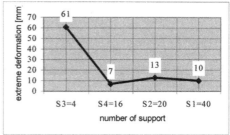

Fig. 10.

Boundary conditions of supported system influence the values of extrema forces. Especially, the bearing in only four quoin nodes gives much bigger extrema forces than in another one. Similar to the previous results supporting system on four pillars with four-point caps is the best solution.

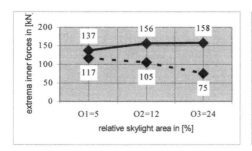
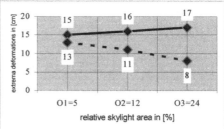

Fig. 11.

The central skylight with a square hole in a diagonal position of different areas gives slightly bigger inner extrema forces and deformation. The most vertically displaced nodes are situated in the middle of edges of skylights holes, because the distance between support points and the nodes is relatively long.

CONCLUSIONS
Suggested structures: the B{T-T}A and B{D-O}A have relatively small amount of nodes in the space of the structure referring to the number of bars; the factor is approximately 4.

1. Formation in the B{T-T}A structure of the external edges and the edges of skylights in triangular and hexagonal systems give the possibility of using the existing modulus in the structure. The corresponding problem for the B{D-O}A how to shape the outer and inner edges of the structure requires to add some bars to create a stable system of the structure.

2. In case of greater number of support nodes in perimeter of the structures, the values of extrema inner forces decrease. However, better solution is to design only a few pillars with branched caps supporting nodes near the perimeter line. This configuration causes the decrease of values of vertical displacement of the structure.

3. Extrema inner forces increase when the edges of the skylight holes are loading by concentrated forces. The values of these forces are relatively fewer for the central holes of a sub circular shape than the oblong-shaped holes, which give greater values of extrema forces and much bigger vertical displacement.

4. Application of many small skylights on the surface of the flat structure slightly increases the extrema inner forces, but considerably influences on the vertical displacement of the central part of the structure.

5. To compare the results mentioned above to the results of previous researches for typical {T-T} and {D-O} structures it can be claimed that multiplication of bars in upper layer of the B{T-T}A and B{D-O}A structures cause greater rigidity of the truss.

REFERENCES
1. BOBER W, REBIELAK J, Initial Static Analysis of B{T-T}A Space Structure, Proceeding of the Local Seminar of IASS Polish Chapter, Warsaw, Poland, December 1996, pp 12-15.
2. BOBER W, REBIELAK J, Space Structures {T-T}, B{T-T}A and B{T-T}B as Constructions of Flat Structural Covers of Great Spans, Proceeding of the International Conference : Challenges to Civil and Mechanical Engineering in 2000 and Beyond, vol 2, Wroclaw, Poland, June 1997, pp 159-164.
3. BOBER W, REBIELAK J, Estimation of Space Structure B2{D-O}B as Flat Structural Cover of Great Span, Proceeding of the International Colloquium on Computation of Shell and Spatial Structures, Taipei, Taiwan, November 1997,
4. REBIELAK J, space Structures – Proposal for Shaping, International Journal of Space Structures, 3(7), 1992, pp 175-190.
5. BOBER W, Double-Layer Structure {D-D} with Central Hole, Lightweight Structures in Civil Engineering, Proceeding of the Local Seminar of IASS Polish Chapter, Warsaw – Wroclaw, Poland, December 2001, pp 13-14.
6. BOBER W, Distribution of Forces in Space of Double-Layer Structure {D-O} Design on the Circular Plane, Lightweight Structures in Civil Engineering, Proceeding of the Local Seminar of IASS Polish Chapter, Warsaw – Cracow, Poland, December 2000, pp 22-23.

The space frame of Elephant Tower

PIYAWAT CHAISERI
Arun Chaiseri Consulting Engineers Co., Ltd.Bangkok, THAILAND
ARUN CHAISERI
Arun Chaiseri Consulting Engineers Co., Ltd.Bangkok, THAILAND
TAYUTI ISARIYARUTTHANON
Arun Chaiseri Consulting Engineers Co., Ltd.

INTRODUCTION

Elephant Tower (Figure 1) is one of Bangkok's distinguished skyscrapers. The complex consists of three towers of thirty-two stories each and 32 metres apart. These towers are linked by seven stories office-cum-residential condominium. The total construction area is 140,000 sq.m. The linking structure has total area of 13,500 sq.m. Vierendeel space frames were considered the most appropriate for the project in view of both construction and harmony with architecture.

Considering the structure at the planning stage helped all designers to create the most suitable architectural and engineering works. Furthermore, this proper design yielded comparatively low construction cost even though the 32 metre vierendeel truss structure cost more than a normal short span structure. The project was completed in 1997 and the building is satisfactory and thus, became one of the most renowned structures in Thailand.

Fig. 1. The Elephant Tower.

SELECTION OF THE STRUCTURE

The most crucial part of the project was the 32 metre - linking structures at the top of the building. This would have effects on many aspects especially the aesthetics of the project.

Consideration for the most appropriate structure was affected by constraints of:

a) Building Program: The program of the 25th to the 32nd floor at the top of the building was determined to be office space in Tower A and residential condominium in Tower B and Tower C. Banquet facilities were planned to be on the 25th floor.

b) External façade: Because of the need to minimise transmission of heat energy from sunlight into the building for energy conservation, the concept of a solid façade was introduced. Together with the method of construction, concrete precast panels of 3.0 m x 3.7 m with window opening of 1. 5 m x 1.8 m were implemented all over the building. Figure 2 shows detail of the facade.

Fig. 2. External facade.

c) Shape of the premise: The construction site is rather tedious as shown in Figure 3. The shape is quadrilateral with a length of about 200 meter. The width is 45 m in front and is 50 m at the back. The building fits tightly in the premise with clearance of only 6 m.

Therefore, a vierendeel space frame was chosen to be the structure linking these portions. The frame structure has dimension of 32 m long and 27 m wide. The frame consists of four sets of vierendeel trusses (Figure 3). Each set is a combination of 7 layers of truss, each of which is equal to one story; 3.7 m. The lower most truss is 4.5 m high. It is determined that spacing of the vertical member of the frame was in accordance with the external facade. Center-to-center of the vertical member is set to be 2.725 m.

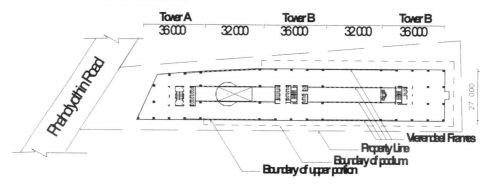

Fig. 3. Site plan and positions of vierendeel frames.

DESIGN OF THE SPACE FRAME

A major concern in designing the frame is the sequence of construction. This is due to the fact that there are many physical constraints that effect the construction method. Construction of the higher level must utilize the lower floor as a platform for all construction facilities. The design of the frame, therefore, has to be conducted at every step of construction by considering all of the loads, including dead load and construction load, and all conditions. Final design is to have every member that is capable to withstand all loads from the very first stage of construction to the last one.

Stage of construction

Construction of the vierendeel space frame is performed in sixteen stages as depicted in Figure 4a – Figure 4e. Details are listed below:

a) Stage I : Construction of the linking structure is planned to start with the fabrication of two steel vierendeel trusses of two stories high together. They are linked by horizontal steel floor beams and a metal deck. The frame is to be lifted up for installation at the 25th floor. The lower chords stay on the 25th floor and the upper ones are on the 27th floor.

b) Stage II : Reinforcing bars are placed around the steel members of the 25th floored frame and concrete is poured to form a composite structure. Reinforced concrete slabs are cast on top of the metal deck as a topping.

c) Stage III : Reinforcing and concrete pouring of vertical and horizontal members of the frame on the 26th floor are completed. Reinforced concrete floor beams and slabs of certain portions, which are just sufficient to be a platform for further construction, is cast. This procedure, therefore, minimizes load burden to the frame.

d) Stage IV : The 27th floored horizontal steel members are covered with reinforced concrete, whilst vertical members of reinforced concrete are added.

e) Stage V : Addition of the 28th floored frame.

f) Stage VI : Addition of the 29th floored frame.

g) Stage VII : Completion of the 26th storied reinforced concrete floor beams and slabs.

h) Stage VIII : Addition of the 30th floored frame.

i) Stage IX : Completion of the 27th storied reinforced concrete floor beams and slabs.

j) Stage X : Addition of the 31st floored frame.

k) Stage XI : Completion of the 28th storied reinforced concrete floor beams and slabs.

l) Stage XII : Addition of the 32nd floored frame. Only horizontal members exist. Temporary steel diagonals at 25th and 26th were removed.

m) Stage XIII : Completion of the 29th storied reinforced concrete floor beams and slabs.

n) Stage XIV : Completion of the 30th storied reinforced concrete floor beams and slabs.

o) Stage XV : Completion of the 31st storied reinforced concrete floor beams and slabs.

p) Stage XVI : Completion of the 32nd. storied reinforced concrete floor beams and slabs.

Analysis

Analysis of the vierendeel space frame is conducted based on the sequence of construction as aforementioned. One important consideration is the capacity of the existing frame structure and respective loading. It is clear that at each stage of construction, the incomplete frame has limited strength whilst load of the newly cast frame acts as dead load to the existing structure. Hence, designing each member of the frame has to take account of every stage of construction.

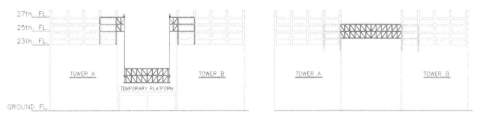

(Lifting the steel frame) (Installation the steel frame)

Fig. 4a Stage I.

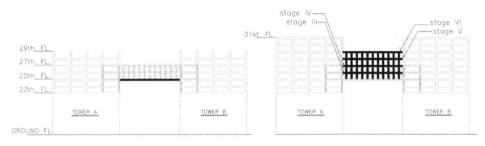

Fig. 4b Stage II. Fig. 4c Stage III-VI.

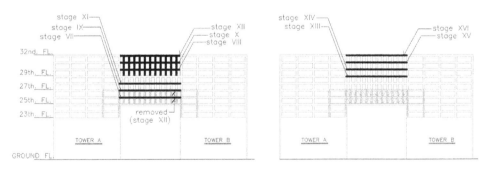

Fig. 4d Stage VII-XII. Fig. 4e Stage XIII-XVI.

Fig. 4. Stages of construction.

Modeling of the frame is as shown in Figure 5. Supports of the frame are modeled as hinges, which are at the second column from the rim of the building. Fixed supports are on the 24th floor at the rim. The frame is modeled to have rigid joints. Although a real condition of the structure is different especially the supports, which are rather fixed supports than hinges, the structure was modeled in such way. This helps practical analysis become less tedious and yield a conservative result.

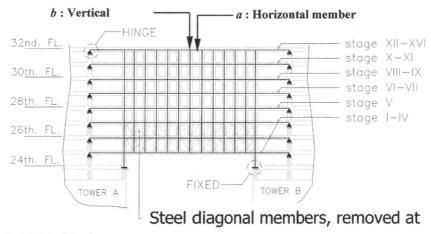

Fig. 5. Model of the frame.

Analysis of the frame is conducted step-by-step. Stress at each member is a summation of stress of each stage, which the structure of that stage is subject to loads of construction and fresh structure above it. Result of the step-by-step analysis is shown below. Figure 6a shows moment in the horizontal member at point *a*. Figure 6b shows moment in vertical member at point *b*. Axial force in horizontal member at point *a* is displayed in Figure 6c whilst Figure 6d shows the force in vertical member at *b*. The results are compared with calculations of stresses and forces of the whole frame. It could be seen that stress and axial force obtained from the step-by-step analysis are greater than that of the whole frame analysis especially at the members of the 25th and the 26th floor. Results of the upper floor are not much different. The larger value of stress and force is to be considered in the detailed design.

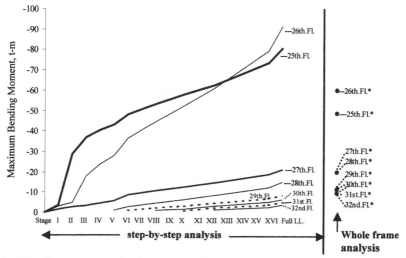

Fig. 6a. Bending moment at horizontal member (point *a*).

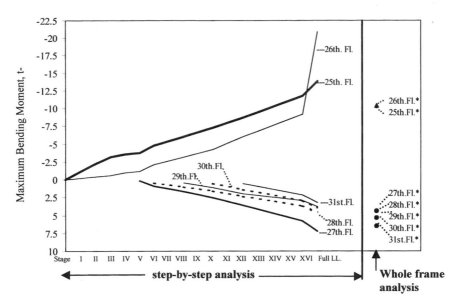

Fig. 6b. Bending moment at vertical member (point *b*).

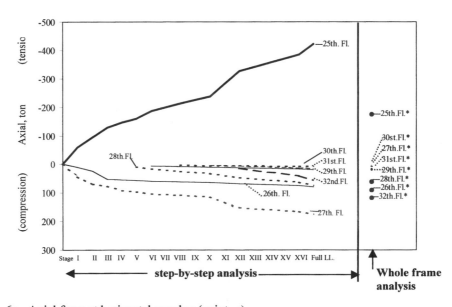

Fig. 6c. Axial force at horizontal member (point *a*).

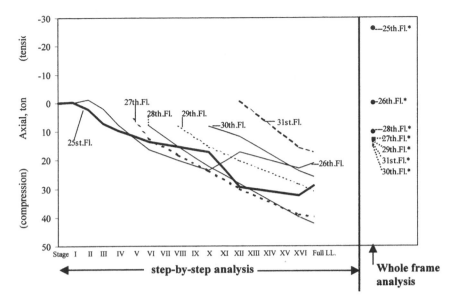

Fig. 6d. Axial force at vertical member (point *b*).

Thermal expansion and shrinkage of concrete of the upper floors of the Elephant Tower is taken care of. Because the upper floors are 172 m in length, effect of thermal expansion is calculated to be significant especially on the 29th floor onward. The 25th floor to the 27th floor had less displacement due to the fact that the steel trusses were tightly connected with the shear cores. The 28th floor had little effect. However, due to stability of the structure and architectural requirements, expansion joint of any type is not acceptable. Solution to the problem is to have the whole structure of the upper floors confined. Therefore, reinforcement to endure displacement caused by thermal effect is implemented. Moreover, prestressing tendons are placed on the 29th floor to the 32nd floor to hold two towers together. There were two groups of tendon on each floor (Figure 7). One group of strands linked main structure of Tower B with half of the floor that was on the trusses. The other group ran from Tower A to Tower C continuously. Stress in the tendons is applied so that stress in the slabs is 0,7 MPa as required for minimum stressing by Thai Building Code.

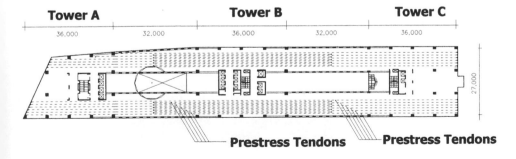

Fig. 7. Pre-stressing strand layout.

During pre-stressing, column hinges at the tower's columns of one floor below were made in order that the pre-stressing force was not restraint by stiff columns.

CONSTRUCTION OF THE FRAME
As aforementioned, stage of construction of the frame is in the sequence that is considered the most suitable in terms of strength of the frames and convenience for construction.

Another critical parameter that would affect strength of the frame is differential vertical displacement between two towers, which are supports of the frame. This is mainly caused by settlement of the towers in very soft Bangkok clay-and-sand soil layers. To minimize differential settlement effect, construction of the two towers must reach approximately equal level of loading before connecting with the linking structure. The settlement of each tower would, therefore, be the same. After the two towers were joined, construction of the two towers must continue at the same rate.

Construction of initial frame
The initial frames were fabricated at ground level (Figure 8). There were two sets of initial frame for connecting Tower A and Tower B and another two for Tower B and Tower C. The frames were two stories high. They are linked by horizontal steel floor beams and metal deck. Horizontal diagonal bracings were attached in order to counter lateral sway during lifting up and installation. Diagonal members are temporarily installed to strengthen the frames during construction.

Fig. 8. Initial frame.

Pre-founded columns and beams
The vertical members at the end of the initial frame are to be connected into pre-founded columns at the rim of each tower. The steel trussed column embedded two floors in the middle of concrete columns of the main structure down to the 23rd floor. The horizontal members are joined to pre-founded beams that link into columns of the second row or the shear core (Figure 9).

Fig. 9. Pre-founded column and beam.

Lifting up and installation

The initial frame was lifted up using four hydraulic jacks. The jacks pulled cables that hung four corners of the frame (Figure 10) at the rate of 150 mm. per minute. The frame was lifted up 90 m. and took 10 hours to reach its final position. The lifting process was very delicate that all four cables must be pulled at the same rate so that the initial frame remained horizontal. The frame was connected to the pre-founded beams and columns by high tensile strength bolts.

Fig. 10. Lifting the frame.

Further construction

Construction continued after the initial frames were all installed into their position. Stages of construction were as aforementioned. Members of the vierendeel frames from the 27th floor onward were of reinforced concrete. Cast-in-place method was implemented using lower members as platform for construction. Until the entire frame was completed, floor structures were seldom cast due to the fact that it could increase loading burden to the unfinished frame.

Construction of the middle bay structure, which was in between the two initial frames, was started from the 25th floor. Steel beams and metal deck were placed between the two frames. Reinforced concrete slabs were cast as topping. The floor was used as a platform for construction of the 26th floor.

CONCEALING THE FRAME

Working closely with architect and interior designer, the vierendeel space frame is hardly observed. The frame at the rim of the building hides behind the precast facade. Spacing of the frame is in accordance with spacing of the precast panel and the window opening.

The middle frames, which run along the middle bay of the building, are treated as colonnade of the atrium as shown in Figure 11. This helps relaxing a feeling of being confined with walls of huge column. Whilst the 25th floor is used as banquet facilities, interior decoration of colonial style windows and verandas does veil the vierendeel frame as shown in Figure 12.

Fig. 11. Atrium at 26th floor.   Fig. 12. Banquet facility at 25th floor.

CONCLUSION

The Elephant Tower with vierendeel space frame at the top 7 stories is in use at present. The summary of construction cost showed that this long span frame costs only about 125% of normal structure of the same material and dimension. This is owing to the fact that every detail, both architecture and engineering work was considered and planned as early as the conceptual design stage. Construction methodology was also determined roughly in an early stage before the contractor proposed the details in accordance with the planned one.

Engineering-wise, the space frame yields satisfactory performance. The structure could attain strength as designed and is perfectly functional. At the 25th floor, where floor slabs of the tower joins the reinforced concrete topping of the metal deck of the frame, small cracks along the joints are observed. Nevertheless, the main structures; beams, columns and members of the frame are completely intact. The upper floor slabs, with prestressed tendons that stay the two towers and the frame together, have no crack either caused by thermal expansion or vibration of the towers.

In view of construction, these vierendeel space frames were no hardship to the contractor. Fabrication and the lifting went smoothly. Construction of the rest of the frame was as scheduled.

Architecturally, the upper floor of the Elephant Tower with the space frame structures could fulfill the requirement program. The frames could be blended to the architectural concept and do not have any obstacle for aesthetic. On the other hand, it yields exciting aspect to the architectural concept and, thus, makes the Elephant Tower one of the most fascinating buildings in Thailand.

Form finding and experiments for three types of space trusses

KIM JIN-WOO, Department of Civil & Environmental Engineering, Gyeongsang National University, Korea, and
KIM JONG-JU, Construction Environmental Subdivision, Kyungnam College of Information & Technology, Korea, and
LEE JUNG-WAN, Department of Civil & Environmental Engineering, Gyeongsang National University, Korea

DEFINITIONS

This paper is concerned with the form finding of dome-shaped space truss achieved by post-tensioning. These dome-shaped space structures are assembled initially in a planar layout on the ground and post-tensioned by cables placed inside the short bottom chords. As a result, the top chords become curved and the space truss is shaped. Various types of space structures have under gone small-scale tests with the shaping and theoretical analysis both performed in the laboratory. One model was post-tensioned along the middle line, and the other models were post-tensioned along the circumference. The feasibility of the proposed shape formation method and the reliability of the established geometric models were confirmed through both experimental investigation and the finite element analysis of the models. In this paper the authors suggest the possibility of shaping formation in accordance with various types space trusses, and we discuss the trends of shaping behaviour for dome-shaped space structures. Potentially this procedure can be useful to eliminate the need of scaffolding providing structures which are both safe and fast to erect.

INTRODUCTION

Since the beginning of the commercial use of space trusses, space trusses have now become widely used throughout the world. They are usually used to provide long span roof coverings for buildings such as stadiums, public halls, exhibition centers, aeroplane hangers, offshore drilling platform, power transmission towers and many other structures, where there is a need to avoid columns. Because the space truss consists of tension and compression member that can be maximized the capacity of load resistance, the space truss is an efficient and light structure type compare to conventional beam-column typed structure. In general, because a space truss consists of many joints that are sensitive to stress and strain, and the characteristic of this joint effect to total behavior of structure, it is necessary to care in design and construction. During the post decades, extensive research has been performed on the behaviour of dome shaped space truss formed by post-tensioning methods. Although the post-tensioning method has been used principally in concrete structures, its application to steel structures has gained acceptance over recent decades. Traditionally although the post-tensioning method has been used in concrete structures, recently the post-tensioning method has been applied to steel structures similar to arch shaped structures or planar frame steel structures. These structures were investigated by the theoretical analysis and experiments in laboratory for the post-tensioned and shaped barrel vaults, domes and hypar structures.(Refs 2-13) The main advantage of post-tensioned and shaped space truss is their fast and economical construction and fabrication. Because the planar layouts are assembled at ground level with a single-layer of top chords and pyramid units of web members, work efficiency and safety are increased. In this study, for the three kinds of dome shaped space trusses, the shaping formation tests have been performed with applying the post-tensioning to the strand located in the bottom chords of space trusses, these results are compared to the results computed by linear and nonlinear analysis with MSC/NASTRAN, and the characteristics of the behaviour for each models are analysed. As a result the authors can suggest the possibility and feasibility of shaping formation for dome-shaped space truss. And because the shape formation procedures are integral with the erection process, this method can save the construction costs by eliminating or minimizing the need for scaffolding and heavy cranes.

PRINCIPLE OF SHAPING AND PLANAR LAYOUTS OF DOME-SHAPED SPACE TRUSSES

In general, the principle of a post-tensioned and shaped space truss is shown in Fig. 1. The shape formation principle of space structure is based on two basic conditions; the mechanism condition and the geometric compatibility condition. In the case of three-dimensional space, the mechanism condition of a post-tensioned and shaped space truss can be expressed by a general Maxwell criterion (Ref 1,14).

$$R-S+M=0; \text{ where } R=b-(3j-r) \tag{1}$$

where R = degree of statical indeterminacy; S = number of independent prestress states that exist; M = number of independent mechanisms; b = total number of members; j = total number of joints; and r = number of restraints on the structure. Using this criterion, a mechanism condition of a post-tensioned and shaped dome can be expressed as: $M > 0$ ($R < 0$,

$S = 0$) in its initial planar layout, and $M = 0$ ($R \geq 0$, $S \geq 0$) in its final space shape. Mechanical properties of the each members used in these space trusses are shown in Table 1. The top chords with hollowed square sectioned tubes are connected with the web member by bolted gusset plates. The lower joints of web members are connected with welding and these members are composed of hollowed circular sectioned tubes. The devices for the guide socket of strand are fastened in the lower joints of the web members with the type of pyramid unit, the bottom chords which are used in post-tensioning are composed of hollowed circular sectioned tubes. The top-chord members and web members are left at their true length, and the bottom chord members are given gaps in proportion to the desired final shape.

Table 1. Properties of members for each model

Model Size	Dome(A): 2,946×2,946 *mm*
	Dome(B): 2,488.25×2,488.25 *mm*
	Dome(C): 4,729.1×4,729.1 *mm*
Top Chord	Dome(A): SHS 13×13×1.8 *mm*
	Dome(B): SHS 13×13×1.5 *mm*
	Dome(C):
	Outer: SHS 15×15×2 *mm*
	Inner: SHS 13×13×1.6 *mm*
Web	CHS 13×2.5 *mm*
Bottom Chord	Dome(A): CHS 13.5×2.3 *mm*
	Dome(B): CHS 13×2.5 *mm*
	Dome(C): CHS 13×2 *mm*
Strand Area	24.6 *mm*^{2}
Poisson's Ratio (ν)	0.3
Yield Stress (σy)	Top Chord : 450 *MPa*
	Web Chord : 440 *MPa*
	Bottom Chord : 440 *MPa*
	Strand : 935 *MPa*
Young's Modulus (E)	200,000 *Mpa*

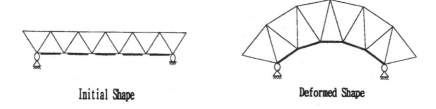

Initial Shape   Deformed Shape

Fig. 1. Shape formation of planar structure after post-tensioning in lower chord.

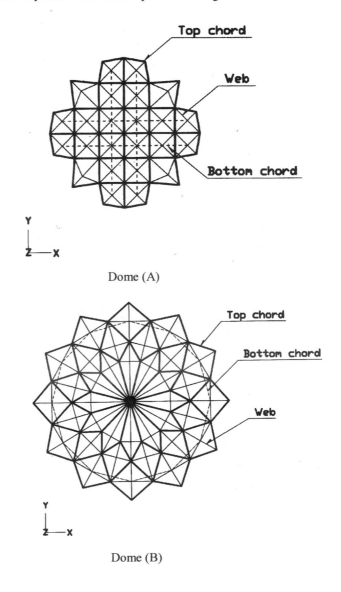

Dome (A)

Dome (B)

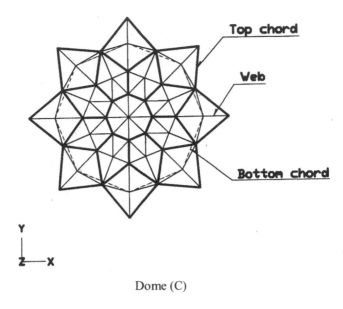

Dome (C)

Fig. 2. Layouts of dome-shaped space trusses.

Before the post-tensioning, the space trusses are flat, their planar layouts of dome shaped space trusses are the same as shown in Fig. 2, the depths of dome(A) and (B) are 250*mm*, the depth of dome(C) is 350*mm*. Generally the amount of gaps in space trusses are determined by the shaping of space trusses and the type of dome. In Fig. 2, the bottom chords are shown with the dotted lines, the post-tensioning forces are applied in the strands that are located inside of the bottom chords. When the gap is closed in the each bottom chords, the post-tensioning is finished, and the final space shape is obtained.

FINITE ELEMENT ANALYSIS OF SHAPING FORMATION

In this study, the finite element method is used for the following purposes; to predict the final space shape, to investigate the feasibility of the proposed the post-tensioning method, and to predict the member force and post-tensioning force during the shaping formation. The commercial program (MSC/NASTRAN) is used in the theoretical analysis of these test models. This program is known to be an appropriate program in the nonlinear finite element analysis for the types of barrel vault, dome and hypar structure investigated by previous researches. In the post-tensioning of space truss, because the large displacements of joints must be considered, analytical modeling of such a procedures should be based on non-linear analysis. Consequently, the nonlinear finite element analysis should be used in these models. To simulate the procedures of shaping formation, it is analysed by the closing method of the bottom chords gaps with applying the negative thermal loads. In the theoretical analysis the top chords are modeled with beam elements, web chords are modeled with rod elements. Through these procedures, we can find the final shape and the characteristic of behaviour for space trusses. Also we can predict the member force and the post-tensioning force during the shaping formation in practice. From the flat condition of initial planar structures, the deformed space trusses by nonlinear finite element analysis are illustrated in Fig. 3. (Before the post-tensioning, the flat condition of space trusses are shown with thick line)

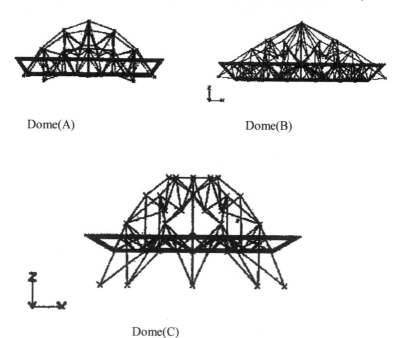

Dome(A)   Dome(B)

Dome(C)

Fig. 3. Deformed shape by finite element analysis.

Dome(A)

Dome(B)

Dome(C)

Photo 1. Post-tensioned and shaped space trusses.

DISCUSSION

The final shapes of the dome-shaped space trusses were obtained after closing the gaps in each bottom chords. Post-tensioned and shaped space trusses of three types are shown in Photo1. In these shape formation the experimental model had an out-of-plane deflection in top chord and in plane deflection of top-chord-grid. In the test models the forces used for shape formation were very small. During the post-tensioning, the top chords are not stressed but undergo finite inextensible displacements. Hence, the top chords remain the same length before and after. The average post-tensioning force for dome(A) is 13*kN* for closing the gaps, the average post-tensioning force for dome(B) is 3.25*kN* for closing the gaps, the average post-tensioning force for dome(C) is 2.7*kN* for closing the gaps. With the comparative small post-tensioning force, the feasibility of the proposed method for dome-shaped space trusses has been presented by these results. The applied post-tensioning force for shaping of dome (A) is greater than those of dome (B) and dome (C). In these results, the circular shaped space domes as like dome (B) and dome (C) could be a good types to construct with respect to be applied post-tensioning force. The results of finite element analysis and the shaping formation test are shown in the Figs. 4, 5 and 6. As the Figs. 4, 5 and 6, we can predict the general trends for shaping behavior of space truss. In these results, for the case of dome (A), in terms of, the deflection for top joint, the difference between the theory and experiment is small compare to the dome (B) and dome (C). Though the imperfections included the geometric imperfections of members and assembly, in shaping formation, nonlinear finite element analysis for shaping is the comparative accurate solution technique compare to linear analysis, and then in practice we recommend the use of nonlinear analysis for he prediction of shaping formation.

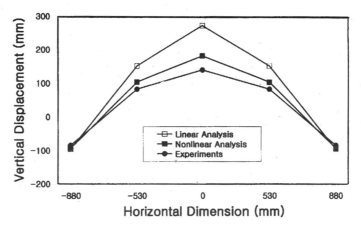

Fig. 4. Shape and deflection of top joints in dome (A).

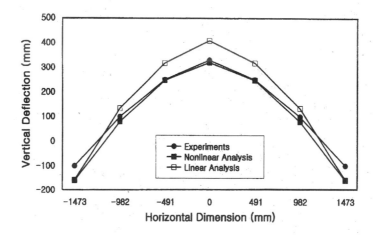

Fig. 5. Shape and deflection of top joints in dome (B).

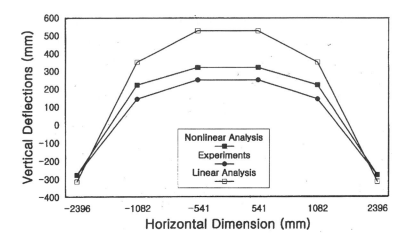

Fig. 6. Shape and deflection of top joints in dome (C).

Though the post-tensioned and shaped space trusses have the characteristics of both mechanism and structures, a finite element method considers the structural characteristics, and then it may lack efficiency in these structures. For the further study to minimize the difference between theoretical analysis and experiments, this finite element program should be approved in the simulation for the each steps in deployment for space trusses.

CONCLUSIONS

Based on the results of test and analysis of a post-tensioned and shaped space trusses, the following conclusions are drawn.

1) By comparative small post-tensioning force, the feasibility of the proposed method for dome-shaped space trusses has been presented.

2) The nonlinear finite element analysis method can be used in the prediction of shaping formation and the post-tensioning force for dome-shaped space trusses.

3) There is some discrepancy in some of the results between theory and experiment, due to differences between test model and modeling techniques. The shape formation behaviour needs further investigation.

4) The shape formation process is integral with the erection process, and then these methods over conventional space trusses are economical by eliminating or minimizing the need for scaffolding and the use of high capacity cranes.

ACKNOWLEDGMENTS

The authors are grateful to Emeritus Professor Lewis Schmidt of University of Wollongong, Australia, for the recommendation and the support of laboratory facility.

REFERENCES

1. Calladine, C. R., Buckminster Fuller's "tensegrity" structures and Clerk Maxwell's rules for the construction of stiff frames, Int. J. Solids Struct. 14(3), 1978, pp. 161-172.

2. Clark, M. J. and Hancock, G. J., Test and nonlinear analyses of small-scale stressed-arch frames. J. of Str. Eng., ASCE, Vol. 121, No. 2, 1995, pp. 187-200.

3. Dehdashti, G. and Schmidt, L. C., Dome shaped space trusses formed by means of post tensioning. J. of Str. Eng., ASCE, Vol. 122, No. 10, 1996, pp. 1240-1245.

4. Kim Jin-Woo, Nonlinear Behaviour of Dome Shaped Space Truss. Journal of Ocean Engineering and Technology, Vol. 12, No. 4, 1998, pp. 1-7.

5. Kim Jin-Woo and L. C. Schmidt, Shape Formation and Behaviour in Dome-Shaped Space Truss. The Conference of the Korean Society of Civil Engineers Proc. Vol. 1, Korea, 1998, pp. 85-88.

6. Kim Jin-Woo, Shape Formation and Analysis of Diamond-Shaped Grid Dome. The Conference of the Korean Society of Civil Engineers Proc. Vol. 1, Korea, 1999, pp. 185-188.

7. Kim Jin-Woo, Analysis and Test of Dome-Shaped Space Truss. Journal of the Korean Society of Civil Engineers, Vol.20, No.1-A, 2000, pp. 39-46.

8. Kim J. W. and L. C. Schmidt, Test of Deployable Dome-Shaped Space Truss. International Conference on Computing in Civil and Building Engineering (ICCCBE-□) Proc. Vol. 1, California, USA, 2000, pp. 66-73.

9. Kim Jin-Woo, Shape Creation and Ultimate Load Test of Space Truss by Means of Post-tensioning. Journal of the Architectural Institute of Korea. Vol. 17, No. 5, 2001, pp.51-57.

10. Kim Jin-Woo, Jiping Hao and Lee Kang-Woon, New Attachment Device for Post-tensioning of Full Size Scale Space Truss. International Conference(EASEC-8), Singapore, December, 2001, paper No.1034.

11. L. C. Schmidt, H. Li and Manuel Chua, Posttensioned and Shaped Hexagonal Grid Dome: Test and Analysis. J. of Str. Eng., ASCE, Vol. 124, No. 6, 1998, pp. 696-703.

12. L. C. Schmidt and S. Selby, Domical Space Trusses and Braced Domes: Shaping, Ultimate Strength and Stiffness. International Journal of Space Structures, Vol. 14, No. 1, 1999, pp. 17-23.

13. Li, H. and Schmidt, L. C., Posttensioned and Shaped Hypar Space Trusses. J. of Str. Eng., ASCE, Vol.123, No. 2, 1997, pp.130-137.

14. S. Pellegrino and C. R. Calladine, Matrix Analysis of Statically and Kinematically Indeterminate Frameworks. Int. J. Solids Struct. Vol. 22, No. 4, 1986, pp. 409-428.

Parametric evaluation to induce a modified design in double layer grid space frames

J. VASEGHI AMIRI
Assistant Professor, University of Mazandaran, Iran
M. H.ALIBEYGI
Assistant Professor, University of Mazandaran, Iran

ABSTRCT
Double-layer space frames are one of the conventional kinds of space structures, which have a wide range of application in the construction industry. This type of space frame can be designed and constructed in various textural patterns, and the evaluation of their efficiency is essential and necessary.

In this research seven kinds of conventional double-layer space frames with different grid patterns have been investigated. The influence of some of the important parameters such as: span to depth ratio, the number and location of peripheral support systems and type of connections (whether they are rigid or hinged) on the behaviour, weight and the stiffness of different grid patterns were determined and compared. In order to obtain acceptable results, in addition to providing the required subroutines, 116 type of double-layer space frames have been analysed and designed.

INTRODUCTION
Nowadays, space structures have gained increasing popularity throughout the world and this is because of it's significant characteristics, such as it's ability to cover large spans, revealing elegance, light weight, together with the easiness of fabrication transportation, speedy erection and etc. Space structures have large variation of patterns which will be induced due to changes in the pattern of elements of upper, lower and middle layers. In this research we will imply the most conventional double layer patterns (figure 1) which are as follow:
- Square on square grids. (SOS)
- Square on large square grid. (SOLS)
- Square on diagonal grid. (SOD)
- Diagonal on square grid. (DOS)
- Diagonal on diagonal grid. (DOD)
- Truss grid.
- Three-way grid. (3-way)
There has been a lot of research on the double layer system of space structure in the past. Mr. West carried out the first study in 1967. In the years 1981 and 1998, Makowski and El-sheikh carried out the same studies respectively (Refs. 1, 2). With regard to the variation of this kind of structure it is quiet essential to carry out more studies, because with changes in the patterns of one system, the distribution of stress, amount of required materials, the stiffness of structures, degree of indeterminacy and etc will vary.
Therefore, the aim of this study is the parametric evaluation of conventional double layer grids. In order to determine the optimum system, the performance of each of these grids must

be studied investigating different parameters. It is important to keep in mind the criteria for the optimisation of design which is, it's economy and its final cost. Although this cost depends on different parameters such as weight, cost of joint construction, number, length and the variation of nodes, length and number of elements and corresponding parameters. With regard to factors mentioned above it is difficult, to consider all these factors in a parametric evaluation and perhaps it is impossible investigate them all in just one article.

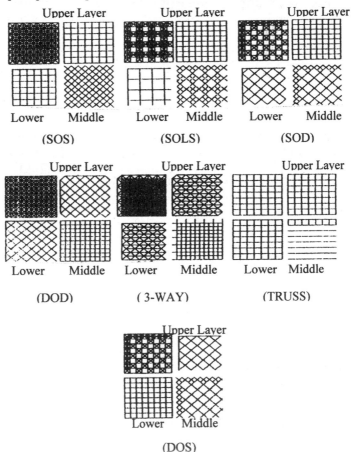

Fig. 1. Conventional double layer grids.

Because the weight of the structure is the most influential factor in the design and the final cost of structures, therefore in this study weight has been considered as the major criteria for the comparison of various patters.

PARAMETERS UNDER EVALUATION
There are different parameters have been considered, which have influence on the behaviour of space structures; these are as follow:
- Span to depth ratio, which in this study have three different values of 15, 20 and 25.
- Aspect ratio which has been considered are for two values of 1:1 and 2:1.

- The type of element connections which have been investigated are for pinned and rigid joints.
- The support conditions, which have been investigated cover three different conditions and are illustrated in figure (2).

In this study structures with spans of 30 x 30 and 30 x 60 meter dimensions have been selected. The dead load of 30 kg / m^2, live load of 200 kg / m^2 and basic wind pressure of 75 kg / m^2 have been considered for all pattern and structures. The all structures have been calculated according to the Iranian code of practice 2800, assuming that the structures were located in a zone with greatest seismic risk [3]. It is essential to mention that the analysis of the structure was linear and performed with the well-known computer program SAP90. The design of structures carried out according to the AISC code.

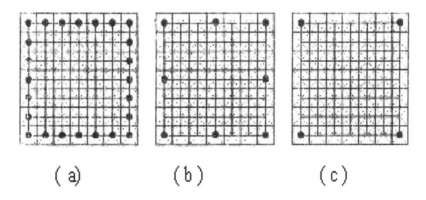

Fig. 2. a) Peripheral support, b) Middle support, c) Corner support.

In this parametric study, 116 models of space structure have been designed and analysed. In order to evaluate the effects of different parameter, computer programs have been prepared using the Fortran language to define geometry in a more simple approach and to create an easy access for the performance of design and analysis. The output of the mentioned programs was used as input data in SAP 90.

EVALUATION OF CHANGES IN THE WEIGHT

One of the most important criteria for the economical evaluation of one pattern is the structure weight. Of course the other parameters also have influence on design (such as length and number of elements, amount and type of nodes, cost of cover and etc). But we should note that it is a very difficult task or nearly impossible to consider all the parameters which effect the economical evaluation of the structure. In this research all types of different patterns were analysed and designed under loading conditions that were mentioned above. Table 1 has been provided giving the results of cross sections, which will shows the final weight of the patterns.

Table 1. Comparison of weight for various patterns.

Pinned Connections			Weight (kg/m^2)		
Pattern	Aspect ratio	Span to Depth	Peripheral support	Middle support	Corner support
SOLS	1:1	15	15.9	17.9	27.5
		21	17.1	19.4	31.7
		26	20.4	23.4	42.6
	2:1	15	17.7	-	-
		21	19.7	-	-
		26	29.18	-	-
DOS	1:1	15	20.1	20.8	31.7
		21	21.3	22.8	42.6
		26	23.8	25.0	44.5
	2:1	15	24.0	-	-
		21	27.0	-	-
		26	31.2	-	-

Pinned connections			Weight (kg/m^2)		
Pattern	Aspect ratio	Span to depth	Peripheral support	Middle Support	Corner support
TRUSS	1:1	15	20.4	-	-
		21	26.0	-	-
		26	27.9	-	-
	2:1	15	21.2	-	-
		21	27.6	-	-
		26	28.7	-	-
3-WAY	1:1	15	25	26.9	36.0
		21	23.21	26.0	38.8
		26	25.5	28.7	45.2
	2:1	15	26.8	-	-
		21	25.9	-	-
		26	29.2	-	-
SOS	1:1	15	17.9	19.8	29.3
		21	18.7	20.9	35.0
		26	22.5	24.8	43.0
	2:1	15	19.9	-	-
		21	21.3	-	-
		26	25.6	-	-
SOD	1:1	15	16.9	19.6	26.0
		21	17.2	19.2	22.0
		26	20.4	22.7	39.7
	2:1	15	19.8	-	-
		21	22.1	-	-
		26	26.5	-	-
DOD	1:1	15	17.2	20.2	29.5
		21	17.9	20.1	34.2
		26	20.3	22.36	40.6
	2:1	15	20.9	-	-
		21	24.1	-	-
		26	27.6	-	-

As it has been determined in table 1 the weight of different patterns for various spans, different aspect ratio, and three kinds of support conditions have been evaluated. From table 1 we can evaluate the efficiency of different patterns in comparison to each other. This table shows that the behaviour of the entire pattern considering various supports is different. It has minimum effect on the 3-way pattern. It also indicates that (using the SOLS and DOD patterns) the weight of the structures in the peripheral support case is approximately half of the corner supports case. According these results it's not wise to assume a space structure without any columns in any condition. Therefore we should not attempt to decrease the number of columns. These results are still correct even if we take into account the weight of columns. The lightest pattern is the SOLS and the heaviest is the 3-way pattern.

EVALUATION OF THE STIFFNESS OF MODELS

Stiffness or deflection of the structure versus the load, is another parameter for the evaluation of efficiency of a pattern. Because one of the limiting parameters in the design of most structures is deflection, which has direct relationship with the stiffness.

In order to compare the different patterns, all the patterns were analysed and designed under constant loading, the ratio of imposed concentrated load in the centre of grid to the deflection of that point was calculated and defined as the structural stiffness. The percent of increase stiffness of various patterns that causes by type of support, aspect ratio and ratio of span to depth is shown in table 2. This definition of stiffness has been obtained from ref 2.

Table 2. Percent of increase stiffness for various patterns.

Pattern	Peripheral S. /Middle S.	Peripheral S. /Corner S.	Aspect ratio 1:1/2:1	Span to Depth 15/21	Span to Depth 15/26
SOS	28 %	170 %	18 %	51 %	112 %
SOLS	42 %	160 %	21 %	47 %	102 %
SOD	13%	117 %	50 %	54 %	115 %
DOS	30 %	187 %	87 %	35 %	80 %
DOD	5.8 %	179 %	83 %	58 %	90 %
TRUSS	-	-	3 %	92 %	102 %
3-WAY	18 %	238 %	24 %	73 %	111 %

We can see from table 2 that increase of stiffness is very dependable on support conditions. The increase of stiffness in the case of peripheral support in comparison with corner support is considerable. This increase is even larger than the increase of stiffness due to the increase in the depth of the grid. The analytical results show that the effect of increase in the depth on the stiffness of a structure is less than the effect of support condition. In such a way that the decreases of ratio of span to the depth from 25 to 15 causes increase in the stiffness to amount of 115%. Where as the effects of support conditions for example in the 3-way pattern in the case of peripheral and corner supports were about 238%. This increase in other patterns is also considerable and it will differ between 117% to 180% whilst no sensitivity is show to the other parameters.

When concentrating only on the stiffness we can not obtain useful data about the behaviour of structure. Therefore new criteria as stiffness to weight ratio can be defined, it compares the increase or decrease in stiffness to the weight parameter. These criteria, however, will evaluate the changes in stiffness with relation to the cost. The analytical results for four patterns have been shown in table 3. It is necessary to mention that this type of study has been carried out considering rigid connections. But we should keep in mind that the double layer grids space structure in the case of linear analysis and within the range of allowable stress has no considerable sensitivity to the rigidity of the connections. Perhaps, in this issue, using non-linear analysis or in the other forms of space structures such as single layer, domes and other similar structures could have a considerable effect which in beyond the scope of this study.

THE EFFECT OF INTIAL ESTIMATION OF CROSS SECTIONS ON THE FINAL RESPONSE OF THE STRUCTURE.

The initial estimation of cross section of components of a structure is the first stage of design and analysis of the structure, which seems to have influence on the final selection of cross sections. It is supposed that the initial inconvenient estimation might change the path of load transformation, because the load bearing of on member is a function of its stiffness. As a consequence the larger cross-sections may become stronger and the smaller cross sections may become weaker. The fact, which is mentioned above, seems to be more sensitive in the space structures that have a greater numbers of members.

Table 3. Comparison of stiffness to weight ratio for various patterns.

Pin Connection			Stiffness/Weight (1/m)		
Pattern	Aspect ratio	Span to Depth	Peripheral Support	Middle Support	Corner Support
SOLS	1:1	15	372	228	80
		21	233	142	49
		26	136	89	27
	2:1	15	266	-	-
		21	171	-	-
		26	104	-	-
DOD	1:1	15	419	334	83
		21	251	209	48
		26	180	159	35
	2:1	15	185	-	-
		21	101	-	-
		26	77	-	-
SOS	1:1	15	385	257	82.37
		21	233	168	48.4
		26	136	101	27.6
	2:1	15	283	-	-
		21	178	-	-
		26	105	-	-
SOD	1:1	15	349	267	92
		21	219	170	57
		26	132	105	32
	2:1	15	195	-	-
		21	112	-	-
		26	72	-	-

In order to investigate this case the structures with space of 30 x 30 and 30 x 60 meter with different support conditions, two different connection namely rigid and pin connection and different aspect ratio with three initial estimations of cross sections have been analysed and designed. In the first estimation all of cross sections were identical, in the second estimation axial stiffness of members in one direction were twenty times greater than axial stiffness in the perpendicular direction and in the third case, members have been selected randomly with four different size.

In table 4 the final weight of 3-WAY and DOD pattern having dimensions of 30 × 30 meter with peripheral supports with rigid and pin connections in three different cases have been illustrated. As indicated in the table the initial estimation has no effect on the final sections and existing negligible difference is due to design restrictions. A comparison between individual members in three cases will confirm the conditions mentioned above.

Table 4. Final weight of 3-WAY and DOD (Kg/m^2).

	3-WAY		DOD	
	Pin Connection	Rigid Connection	Pin Connection	Rigid Connection
First Estimation	25.5	24.0	20.3	19.0
Second Estimation	26.0	24.8	21.5	20.5
Third Estimation	26.4	25.2	21.0	20.0

The final sections are the results of several analytical and design process. In fact the final section is referred to the stage of analytical and designs process in which two sequential analysis process yield identical values for cross sections. In the analysis carried out in this research the minimum and maximum number of trial and error for obtaining the final cross section is four and six respectively. The influence of various pattern, type of element connection, ratio of span to depth and aspect ratio is not significant at number of trail and error. From this investigation we came to the conclusion that the final cross sections of the double layer grid space structure are independent from initial estimation and this is one of the advantages of this kind of structures..

CONCLUSIONS
It is well known that space structures are able to cover the large spans with limited number of supports. But this does not mean we should reduce the number of supports without any reason, because it will have a major influence on the reduction of the final weight as well as on the uniform distribution of stress and also on the increase of stiffness. Incorporating linear analysis it has been shown that:
- The influence of rigidity of nodes on the structural behaviour in the case of double layer is not significant, and therefore the semi-rigid connections in the double layer grids are not very important.
-    The initial estimation of cross sections has no influence on their final values.

- The effect of the type of support on the stiffness of all kind of double layer grids was relatively large and it is more than the effect of the increase in the depth to the span ratio and this will indicate the importance of choosing a proper support.

REFERENCES
1. EL-SHEIKH A, Configurations of Double Layer Space Trusses, vol. 6, No5, 1998, pp. 543-554.
2. MAKOWSKI, Z. S, Analysis Design And Construction Of Double Layer Grids, Elsevier Applied Science Publishers LTD, 1985.
3. Iranian Code For Earthquake Loading Of Structures 2800.

India's first stainless steel space frame roof

N. SUBRAMANIAN
Computer Design Consultants, Chennai, India.
C. C. SAMPATH
Sreevatsa Stainless Steel Fabricators (P) Ltd., Chennai, India.

ABSTRACT
Corrosion is a big problem in steel structures, especially in coastal areas. To avoid corrosion, steel structures have to be painted at periodic intervals, thus increasing their life cycle cost. Due to these factors, many owners and architects do not specify steel space frames. To solve this problem, stainless steel is being increasingly used. This paper describes India's first stainless steel space frame roof. This double layer grid roof was erected at Hissar, Haryana, and has a saucer shape in the longitudinal direction. This roof of size 11m x 18m is supported by two circular reinforced concrete columns of 500mm diameter. The roof was modeled as a space truss and analysed using STAAD III program. It was fabricated and erected using MERO type joints in just 8 weeks.

INTRODUCTION
Though structural steel offers many advantages such as high strength to weight ratio, easy fabrication and erection, prefabrication, demountability and recycling, uniform properties, ductility, etc., its use in coastal areas is very much limited due to fear of corrosion. Hence stainless steel was developed at the beginning of 20th century as a rust-resistant alloy steel. Since it was expensive, it was used in small quantities, for limited applications, in military supplies use and in the chemical industry. During 1950's it was possible to mass produce stainless steel at low cost due to the advancement in refining processes (Ref. 1). This lead to the greater use of stainless steel in kitchen products and other consumer durable, especially in Japan. Recent years saw the growth of stainless steel applications in the Japanese Construction Industry, in roofing, building interior and exterior and structural applications. As seen in Fig.1, the production of raw stainless steel in the Western world has increased considerably from 1960 until today.

In India, stainless steel is produced in various grades (SS 301, SS 304, SS 304L, SS 310S, SS 316, SS 316L, SS 321, SS 409, SS 409M, SS 410S, SS 420 and SS 430) and forms but has found limited use in structural applications. Its use has been limited to those applications where aesthetics rather than structural performance is the sole criteria. However, though the stainless steel production in India was only 20,000 tones in 1978, it has grown to 625,000 tons in 1997 (Ref. 2).

STAINLESS STEEL
Stainless steel is essentially a low carbon steel to which chromium has been added. It is this addition of chromium, in amounts greater than 10.5% by weight, that gives the steel its unique 'stainless' corrosion resistant properties.

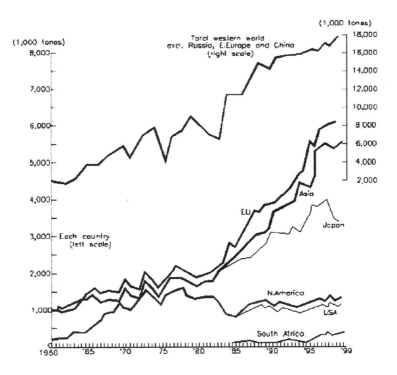

Fig. 1. Western World Production of Stainless Steel (Ref. 1)

Like other steels, stainless steel has a high strength to weight ratio, is weldable and environment friendly and its main advantage over the other steels are its aesthetic appearance, corrosion resistance, high tensile strength, high toughness, impact and heat resistance. The most important difference from the structural designer's point of view is that it has a non-linear stress-strain relationship (Fig. 2). Thus as compared to carbon steel, which exhibits linear elastic behaviour up to the yield stress and a plateau before strain hardening is encountered, stainless steel has a more rounded response with no-well defined yield stress. Therefore, yield strength of stainless steel is generally quoted in terms of a proof strength defined for a particular offset permanent strain (conventionally the 0.2% strain) as indicated in Fig. 2. The initial modulus is approximately the same as that for structural steel, but the proportional limit is generally quiet low. For some stainless steels, the stress-strain curve may be different in tension and compression and the results may also be different in the longitudinal and transverse direction, depending upon the method by which they are produced. Therefore, the guidelines used in designing conventional steel members are not applicable for stainless steel members. Hence several countries have developed separate guidelines for the design of stainless steel structural members (Ref. 3-5). But Indian Code provisions for the design of stainless steel members are not available. This is one of the reasons in India for the non application of stainless steel members to engineering structures.

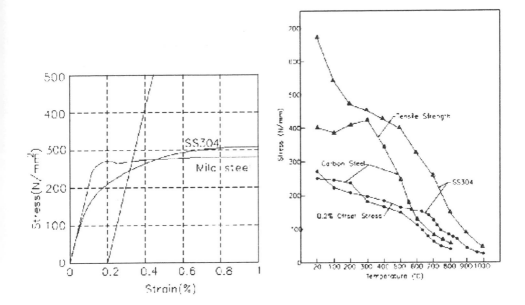

Fig. 2. Stress-strain curves of stainless
steel and carbon steel

Fig. 3. Stress-temperature relation of
less steel and carbon steel (Ref.1)

Apart from corrosion resistance, stainless steel is superior to carbon steel in such qualities as elongation and fire resistance. Fig. 3 shows how yield strength and tensile strength depend on temperature. While the proof strength of ordinary steel begins to decline at a steel temperature of 300° to 500°C, that of stainless steel has a small rate of decrease up to about 700°C. This means that stainless steel has far superior fire resistance, making possible the construction of buildings using stainless steel without fire - proof insulation.

Out of the various available grades, SS 304, SS 304L, SS 306, SS 409 & SS 430 are suitable for structural applications. The chemical, mechanical and physical properties of these grades are given in Tables 1, 2 and 3 (The properties for mild steel are also given for comparison).

Most stainless steels fall into one of three main classes, namely martensitic, ferritic and austenitic, titles taken from their metallurgical structures. Martensitic types are rarely used in buildings, only when high strength or hardness are required. The ferritic steels, which are magnetic, contain chromium alone most commonly to the extent of 12-30 per cent. They are identified in the AISI 400 Series (SS 409 and SS 430 grade). They are low priced stainless steel useful in less aggressive environment, where surface condition is of less importance (Ref. 2).

Table 1 Chemical composition of stainless steels (Ref. 6)

Grade	C Max	Si Max	Mn Max	Cr	Ni	P Max	S Max
SS 316	0.08	0.75	2.00	16-18	10-14	0.045	0.03
SS 304	0.08	0.75	2.00	18-20	8-10.5	0.045	0.03
SS 304L	0.03	0.75	2.00	18-20	8-12	0.045	0.03
SS 409	0.08	1.00	1.00	10.50-11.75	0.50 max	0.045	0.03
SS 430	0.12	1.00	1.00	16-18	0.75 max	0.040	0.03
Mild Steel (IS:2062)	0.22	0.40	1.50	-	-	0.045	0.045

Note: C- Carbon, Si- Silicon, Mn- Manganese, Cr-Chromium, Ni-Nickel, P-Phosphorus, S-Sulphur

Table 2. Mechanical properties (annealed condition) of cold rolled stainless steel (Ref. 6 and 7).

Grade	UTS N/mm^2 Min	0.2% Proof Stress N/mm^2 Min	% Elongation on 50 mm GL Min	Hardness RB Max
SS 316	515	205	40	95
SS 304	515	205	40	92
SS 304 L	485	170	40	92
SS 409	380	205	20	88
SS 430	450	205	22	89
Mild Steel	410 min	250 min	23	80-105

Table 3 Physical properties of CRSS (Ref. 6 and 7).

Details	SS 316	SS 304 / 304 L	SS 409	SS 430	Mild Steel
Density (kN /m³)	77.5	77.5	75.5	75.5	78.5
Modulus of Elasticity (N/mm²)	193000	193000	200000	200000	210000
Coefficient of Thermal Expansion (°C x 10⁻⁶, 0-500°C)	19.8	18.4	11.52	11.34	12.0
Melting Range (°C)	1400-1420	1400-1455	1430-1510	1430-1510	1350-1450

They possess good resistance to corrosion and have good weldability (However, their corrosion resistance is inferior to that of the 300 series). They are highly recommended for components which need a paint coating such as commercial garage, canopies and road signs.

The Austenitic steels are many in number and varied in composition. They contain 16-26 percent Chromium and 6-22% Nickel. They are in annealed condition, non-magnetic and have excellent corrosion resistance. They are not hardenable on cold working. However they develop high strength on cold working. They have excellent weldability and identified in the

AISI 300 series (Ref. 3). Normally used grades in structural applications are SS 304, SS 304L and SS 316 grades. Grade 304 should not be used for external elements near coastal areas. Grade 316 is the best grade for all external applications (SS 316 is basically a 304 grade with the addition of 2-3% molybdenum).

The outstanding plastic deformation capacity of stainless steel makes it an ideal material for space trusses, bracings, column bases and base-isolated structures for earthquake – response control. Life - cycle cost analysis shows that stainless steel structures will be economical after 20 years. Current design philosophy of designing structures which should have a minimum life span of 100 years with materials which could be recycled makes stainless steel the ideal structural material of the future. This paper deals with the first stainless steel space frame construction in India.

Surface Finish
Colled rolled (thin gauge up to 3mm) stainless steels are available with the following finishes (Ref.2):

No.BA - A bright, smooth mirror finish.
No.2B - A bright, smooth, silvery grey, moderately reflective finish.(most common finish for sheet material)
N.2D - A matt, dull, silvery grey, non reflective finish.(gives better paint adhesion)
No.3 - A directional, uniform, polished finish, using 100 to 120 grit abrasive.
No.4 - A bright, polished finish with a visible directional grain from a 120 to 180 grit abrasive.

The above finishes are mill finishes. Other finishes such as various polished finishes, coloured BA finishes, regidised patterns and decorative steel are also commercially available in stainless steel.

THE SPACE FRAME ROOF
M/s. Jindal Strips Limited, who manufactures stainless steel products in India, wanted to show the advantage of stainless steel to the general public by erecting an elegant stainless steel roof at the entrance of their factory at Hissar, Haryana. The plan and side elevations of this space frame roof are shown in Fig. 4,5 and 6. This double layer grid has a saucer shape in the longitudinal direction and has a dimension of 11m x 18m. At the support points, the roof has a pyramidal shape and supported by two circular reinforced concrete columns of 500mm diameter. These columns are spaced at 5m as shown in Fig. 5. The foundation for these columns are placed at 2m from the ground level. The size of the foundation is 3.5m x 3.5m and is reinforced with 10mm ϕ rebars at 125mm c/c at bottom and 10mm ϕ rebars at 200mm c/c at top.

This space frame roof was modeled as a space truss and analysed using the STAAD III program, developed by Research Engineers, Inc., New Jersy, U.S.A. Though the roof is not going to be covered right now, just to show the excellent corrosion resistant properties of stainless steel, in the design a light weight covering was considered. The structure was analysed for dead, live, wind load and their combinations as specified in the Indian Code of Practice(Ref. 8). It has 520 joints and 2150 members.

As already discussed, Indian standard provisions are not available for the design of stainless steel members. Hence the design was done based on Ref. 3. It is interesting to note that

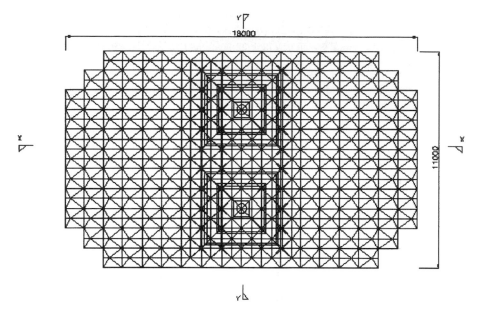

Fig. 4. Top plan of the structure.

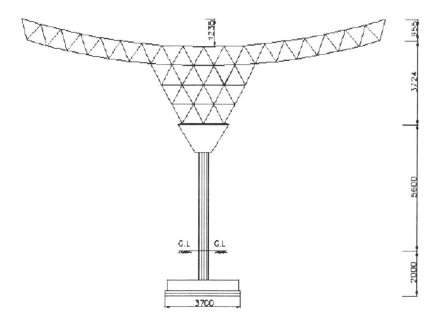

Fig. 5. Section X-X of the frame.

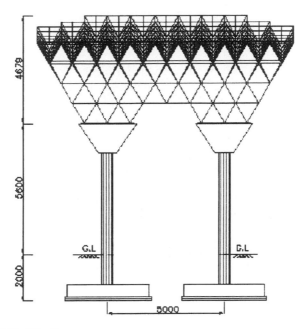

Fig. 6. Section Y-Y of the frame.

Satish Kumar and Audisesha Reddy have recently proposed the following non-dimensional column strength curve for stainless steel tubular members (Ref. 9)

$$\psi = \frac{1}{\phi + (\phi^2 - \lambda^2)^{0.5}} \qquad\qquad --- \qquad (1)$$

$$\phi = 0.5\,(1 + \eta + \lambda^2) \qquad\qquad --- \qquad (2)$$

Where the imperfection parameter η is a non-linear function of the slenderness ratio parameter

$$\eta = \alpha\,(\lambda - \lambda_1)^\beta - \lambda_0 \qquad\qquad --- \qquad (3)$$

where $\alpha = 1.31$, $\beta = \lambda_0 = 0.67$ and $\lambda_1 = 0.37$

$$\Psi = \frac{\sigma_u}{\sigma_{0.2}} \qquad\qquad --- \qquad (4)$$

$$\lambda = \frac{L}{r\,\pi}\sqrt{\frac{\sigma_{0.2}}{E_0}} \qquad\qquad --- \qquad (5)$$

Where σ_u, L and r are the ultimate stress, effective length and radius of gyration respectively.

Fig. 7. Two views of the completed structure.

The above column strength curve has been obtained by approximating the non-linear stress-strain curve by the Ramberg – Osgood relationship

$$\varepsilon = \frac{\sigma}{E_0} + 0.002 \left(\frac{\sigma_n}{\sigma_{0.2}}\right) \qquad\qquad \text{---} \qquad (6)$$

Where E_0 = initial Young's modulus

$\sigma_{0.2}$ = 0.2% proof strength (see Fig. 2)

and   n   = exponent which controls the sharpness of the knee of the stress-stain curve.

Tubes of size 50mm OD and 2mm thick are used for the longitudinal members of the top and bottom chord members of the longitudinal direction and size 50mm OD and 1.6mm thick mm tubes for the top and bottom chord members of the shorter side and also for the bracing members. The corners of the pyramids were designed to be of size 50mm OD and 4mm thick. This frame was built using stainless steel tubes of SS 304 grade and with No. 4 finish. MERO type joints were used. Proper care was taken while welding the conical ends of the members. Stainless steel bolts of size 16mm were used. The lengths of different members were calculated using the AUTOCAD program. Due to the complex geometry of the roof, a typical frame with single row of module was first fabricated at the factory and assembled to check the dimensions, angles, etc. After this test fabrication, all the members and nodes were taken to the site. The erection was done with minimum scaffolding using the cantilevering method. The whole frame was fabricated and erected within 8 weeks. The completed structure is shown in Fig. 7.

CONCLUSION
The trend in these days is towards constructions designed to last 100 years, which requires materials of high durability. Stainless steel with its excellent performance is ideally suited for such highly durable constructions and is readily recyclable. Life cycle cost calculations have found that stainless steel pays for itself in about 20 years. In spite of the obvious advantages of stainless steel, it has not been adopted in structural engineering applications in India. This may be due to the absence of codal provisions. For the first time, stainless steel tubular members were used in a double layer space frame roof constructed at the entrance of a factory building in Haryana, India. The details of this space frame are given along with the details of different grades and finishes of stainless steel which can be used in structural engineering applications. In the exposed condition, the erected stainless steel space frame roof provides elegance and aesthetic appearance to the Jindal Steel Strips factory

ACKNOWLEDGEMENTS
The author wishes to thank the help they received from Mr. B. Madeshkumar, Ms. S. Mangalam and Mr. C. Anbu while preparing this paper.

REFERENCES
1. HOSHINO,K., Market Development for Stainless Steel, Steel Today and Tomorrow, No.155, Sept.2001, pp. 5-8.
2. MATHUR, N.C., Stainless Steel in Architecture, Building and Construction, Steel in Construction 2000, Joint Plant Committee, Govt. of India, First edition, Feb.2000, pp.179-193.
3. EURO INOX, European Stainless Steel Development Group, Design Manual for Structural Stainless Steel, Nickel Development Institute, Toronto, Canada, 1994
4. ASCE /ANSI – 8 – 90, Specification for the design of Cold – formed Stainless Steel Structural Members, ASCE, 1991.
5. ENV 1993 – 1 – 4, Euro code 3 Design of Steel Structures, Part 1.4, General Rules and supplementary rules for stainless steels, CEN, 1992.
6. SALEM STEEL – Users Guide, Steel Authority of India, Salem Steel Plant, Salem, India, 1998, 48pp.
7. IS 6603 – 1972, Specification for Stainless steel bars and flats, Bureau of Indian Standards, New Delhi, 1972.
8. IS 875 – 1987, Code of Practice for Design Loads (other than Earthquake) for Building and Structures, Part 2 – Imposed Loads, Part 3 – Wind Loads and Part 5 – Special Loads and Load combinations, Bureau of Indian Standards, New Delhi, March 1989.
9. SATISH KUMAR, S.R., and AUDISESHA REDDY, I., "Design of Stainless Steel Tubular Members", Seminar on Modern Trends in Steel Structures, Nagpur, Feb.2002, pp.7-12.

Adjustable and non-adjustable connecting systems

N. BĂLUȚ
Building Research Institute INCERC, Timişoara, Romania
V. GIONCU
Department of Architecture, Politehnica University, Timişoara, Romania

INTRODUCTION

The impressive development of space structures (especially since the second half of the 20th century) is due to their undeniable advantages as compared to the traditional systems. A great variety of geometrical forms and joint systems are successfully used nowadays all over the world. However, the behaviour of such structures exhibits some important peculiarities. The designer is expected to be aware of them, in order to make sure that the analysis model he intends to use can be qualified as satisfactory.

Among the above mentioned peculiarities, imperfection sensitivity can be essential in certain cases. In the following, the discussion is restricted to two categories of triangulated space structures, namely single-layer curved (shell-like) structures, and respectively double-layer planar ones.

Imperfections are geometrical or mechanical ones. Geometrical imperfections can be classified into three main categories:
    (a) deviations from the nominal member lengths;
    (b) deviations from the nominal dimensions of the member cross sections;
    (c) initial deformations of the component members and initial eccentricities.

The assembling of members which deviate from their nominal lengths has two possible consequences. On the one hand, they result in deviations of the nodes from their theoretical position. On the other hand, additional member forces may occur in certain circumstances, in order to compensate the lack of fit, giving rise to a state of undesirable prestressing. Which of those two effects prevails is a matter that depends on the type of structure. In a given case, the effect of deviations from the nominal lengths depends essentially on the adopted connecting system.

Imperfections of the second category are usually covered by a partial safety factor, as in EC3 (Ref 1), that also allows for variations in the mechanical characteristics of the material (which belong to mechanical imperfections).

Initial deformations can be classified into plate deformations and member deformations. Plate deformations are initial deformations of the component walls. They influence local buckling and, in the case of reticulated structures, are only relevant for thin-walled members. Member deformations are initial bows and twists. They give rise to bending, torsion and (in the case of open sections) warping, and those effects are currently taken into account by means of the buckling reduction factors. But they also cause an apparent reduction in the member axial stiffness. As for joint eccentricities, they can be assimilated with initial bows. However, they are negligible for most of the modern connecting systems. The discussion will be limited herein to the influence of the imperfections belonging to the above categories 'a' (which represent the main topic of this paper) and 'c'.

MEMBER LENGTH INACCURACIES AND THE INFLUENCE OF THE CONNECTING SYSTEM
Adjustable and non-adjustable systems
The joint connecting systems used for space structures can be classified from many points of view. One possibility is to classify them into 'adjustable' and 'non-adjustable' systems.

In the first case, the member lengths can easily be_modified (e.g. by means of a threaded device) in order to respect the nominal values, and the term 'geometrical tolerance' is not relevant. Such modifications are not possible in the second case, and therefore the member lengths entirely depend on the manufacturing accuracy. Strictly speaking, non-adjustable systems do not exist, but the term will be extended herein, to include the cases where the distance between the end connections can be modified, but only within very small limits. Only the particular example shown in Fig 1 will be discussed in the following, but the conclusions can be extended to other non-adjustable systems.

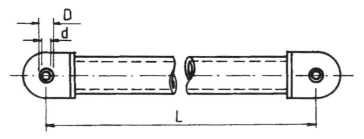

Fig. 1. Example of non-adjustable connecting system.

The length L can be modified within the limits $L \pm \Delta_c$ without needing the application of an axial load. Here

$$\Delta_c = D - d \tag{1}$$

where D and d are respectively the hole and the bolt diameter.
Note: In the theoretical case of a proper 'non-adjustable' system, $\Delta_c = 0$.

Denoting by Δ_m the allowable deviation from the nominal length due to manufacturing inaccuracies, the distance between the hole centres can vary between $L - \Delta_m$ and $L + \Delta_m$. Considering both Δ_c and Δ_m, the limits of the member length after erection can be $L \pm \Delta L$, where

$$\Delta L = \Delta_m + \Delta_c \tag{2}$$

While Eqn (2) is relevant for the variation limits of the member lengths, the values to be considered for estimating the member forces arising from the lack of fit are different and will be denoted by. ΔL_f. Assuming that $\Delta_m > \Delta_c$, one obtains:

$$\Delta L_f = \Delta_m - \Delta_c \tag{3}$$

The difference between Eqns (2) and (3) is due to the fact that in the first case the total deviation from the nominal value L being significant, both Δ_m and Δ_c are involved. In the second case, the significant phenomenon is the lack of fit, which occurs only when the gap Δ_c has been consumed. The only practical possibility to eliminate this effect is to have $\Delta_m \le \Delta_c$ by diminishing Δ_m, which demands a high manufacturing accuracy. The alternative of increasing Δ_c is generally not acceptable, because solutions like oversize holes (or similar) are not recommendable, especially for statically indeterminate structures. Normally, for bolted connections, Δ_c = 2 mm and Δ_m= 5 mm (Refs 1-2). Hence the upper limits of deviations are ΔL = 7 mm and ΔL_f = 3 mm.

The analysis for lack of fit is very similar to that for temperature variations.

The case of single-layer structures
For such structures, the main effect of member length inaccuracies is the deviation of the nodes from their theoretical position. This kind of geometrical imperfection is known to flavour instability phenomena (see e.g. Refs 3-9). As for the additional member forces due to the lack of fit, those are less important because the structure is less rigid and can accommodate more easily to variations in member lengths without producing a significant prestressing.

Considering an isolated cell of a structure, if the member initial shortening is ΔL, the theoretical rise f is diminished by a quantity f_i (Fig 2):

$$f_i = f - [f^2 - 2L\Delta L + (\Delta L)^2]^{0.5} \approx f - (f^2 - 2L\Delta L)^{0.5} \tag{4}$$

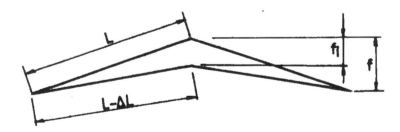

Fig. 2. The decrease in rise due to inaccuracies in member lengths.

If $\Delta L = \Delta L_{lim} = f^2/2L$, the imperfection f_i equals the theoretical rise. If ΔL exceeds that limit, the assembling of the component members simply cannot be achieved in the case of a single cell, and is either impossible or leads to very large geometrical imperfections in the case of an actual (multi-cell) structure.

The geometry of the shallow dome shown in Fig 3 is very similar to the one presented in Ref 3, with the only difference that all dimensions are 10 times greater. All members are 51x2.6 mm circular hollow sections. Their cross section area is 395 mm². The supports are in the nodes 8, 9, 10,11,12 and 13 and are assumed to be hinged, and so are all the joint connections. The elasticity modulus E = 210 000 N/mm². All dimensions in the figure are in mm.

In order to investigate the effect of the lack of fit, the structure is analysed in an unloaded state, neglecting even the influence of the permanent load. If $\Delta_m = 5$ mm and $\Delta_c = 2$ mm, one

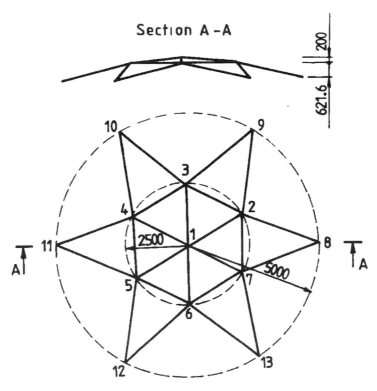

Fig 3. Single-layer dome.

obtains $\Delta L = 7$ mm and $\Delta L_f = 3$ mm. In case that only the member 1-2 is 5 mm shorter, and the lengths of all the other members equal their nominal values, then the initial stresses σ_o range between 1.8 and 14.7 N/mm^2. Although a geometrically nonlinear analysis was carried out, the differences in comparison with the linear analysis are insignificant when no external loads are present. From the point of view of geometrical imperfections, the most unfavourable situation occurs if all the members connected in the central node 1 are 7mm shorter than their nominal length ($\Delta L = 7$ mm). It should be noticed, however, that this assumption is rather conservative. Eqn (4) gives $f_i = 130.0$ mm (i.e. 65.0 % of the initial local rise of the central cell, $f = 200$ mm), a value that is simply unacceptable. Reducing ΔL results in the following values of the initial imperfection: $f_i = 77.9$ mm (38.9 % of the initial rise) for $\Delta L = 5$ mm; 42.0 mm (21.0 %) for $\Delta L = 3$ mm, and 13.0 mm (6.5 %) for $\Delta L = 1$ mm (which corresponds to a substantially improved accuracy). If ΔL is the same for all the members connected in the central node, and all connections are hinged, there is no lack of fit and no initial state of stresses.

An approximate solution for estimating the effect of the possible deviation of the nodes from their theoretical position (due to manufacturing and/or erection inaccuracies) was suggested in Ref 4. Both a linear and a geometrically nonlinear analysis are performed for the structure, which is assumed to be perfect and subjected to the loads considered in the ultimate state design. Then the member axial forces are amplified by a factor depending on the ratios f_i/f and w_l/w. Here w is the decrease in the local rise f, resulting from the nonlinear analysis. The symbol w_l has the same meaning as w, but corresponds to the linear analysis. The above method can provide no more than a rough approximation, but gives the possibility to anticipate the effect of geometrical imperfections for preliminary design. For structures that are very important and/or particularly sensitive to imperfections, this procedure should be completed by a geometrical survey of all the nodes' position. It does not seem feasible to perform such measurements in every current case. For a usual structure, the designer is expected to make a realistic estimate of the possible imperfections.

The case of double-layer structures

Such structures are much more rigid than single-layer ones. Therefore, the most important effect of member length inaccuracies is the generation of an initial state of stresses due to the lack of fit. Disregarding this aspect may lead to dramatic consequences. The research work carried out by several authors (see e.g. Refs 10-13) has led to the conclusion that double-layer structures do not possess significant post-critical reserves. One of the authors of the present paper once witnessed a case when a diagonal buckled during erection as a consequence of the enforced assembling of geometrically inaccurate elements. On the contrary, the deviation of the nodes from their theoretical position is less important (except from the aesthetic view point).

It follows that adjustable connecting systems are preferable for such structures. But if a non-adjustable system is chosen, one should allow for the influence of a possible lack of fit. Since ΔL is a random variable, an analysis model could be achieved by numerical simulation, assuming a distribution law based on reliable statistical data. In the absence of such data, a simplified procedure is suggested herein. From each group of members (e.g. those belonging to the top layer, the bottom layer or the diagonals) one should select separately the one with the greatest tension load and/or the one with the greatest compression load from permanent + variable actions. In case that a group contains members which differ by their cross sections and/or effective lengths, the group should be partitioned into subgroups and the selection be operated within each subgroup. Then, in order to allow for the lack of fit, a separate 'loading case' should be considered for each of the selected members. It will be assumed that the initial length of the member under discussion is $L - \Delta L$ if the axial load from permanent + variable actions is tensile, and $L + \Delta L$ if compressive. The lengths of all the other members are supposed to take their nominal values.

The above procedure is applied to the double-layer square mesh grid presented in Fig 4. All the joint connections are hinged. The structure is supported along the top layer contour. The supports in the corner nodes 1, 5, 37 and 41 are spherical hinges, and the other nodes on the contour are simply supported in the vertical direction. The member cross sections are the same as in the previous example. Three separate cases are examined:

   (a) the top layer member 20-21 is 3 mm longer than its nominal length;
   (b) the bottom layer member 16-17 is 3 mm shorter;
   (c) the diagonal 6-11 is 3 mm longer.

The variation of the initial stresses σ_0 in the imperfect member and its vicinity in each of the three cases is shown in Fig 4. It may be noticed that the stresses in the neighbouring members not affected directly by imperfections are substantially smaller than in the imperfect member itself. Moreover, it should be added (although this fact is not illustrated in the figure) that the

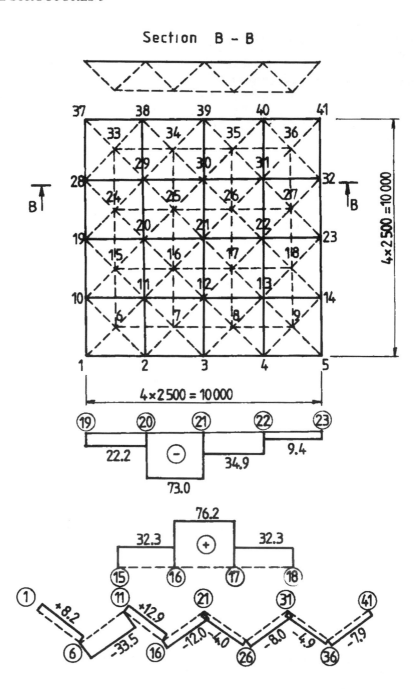

Fig. 4. Double-layer grid.

influence on any member not belonging to the same layer as the imperfect member is even smaller.

THE EFFECT OF INITIAL BOWS
The effect of initial bows on the buckling load of compression members is common to all metal structures. But there is also another effect which is specific to reticulated structures, i.e. the apparent reduction in the axial stiffness of the members. (The reduction is apparent because the relative displacement of the member ends in the longitudinal direction is not caused exclusively by the axial load, but also by the bending moments from transverse deflections). The latter effect is important for compression members only, and is negligible for tension members (see Ref 3). Since the structures are statically indeterminate, the omputed axial loads in the more compressed members result higher than in reality. For double-layer structures, disregarding this effect seems at first sight to be on the safe side. However, it is possible that in certain cases, some of the members and their connections are overstressed owing to the redistribution of the member forces due to the modified stiffnesses. But the effect of reduction in axial stiffness is much more important for single-layer structures. In that case, the buckling loads related to different instability modes (node instability, line instability – which is specific to braced barrel vaults – and overall instability) are affected in a negative sense.

Both effects are due to the conjugate influence of two categories of imperfections: geometrical (initial bows) and mechanical ones (residual stresses). Those two categories are different by their nature. While geometrical imperfections can be replaced by equivalent transverse loads, mechanical imperfections cause a reduction in the bending stiffness. In the case of an isolated compression member, both influences act in the same sense. This is not always true for statically indeterminate structures.

A procedure for the separate consideration of the two categories of imperfections was suggested in Ref 14. If a member is subjected to a relatively low compression load, the stresses arising from eccentric compression added to the residual stresses do not reach the yield strength, and the member behaves elastically. On the contrary, if the compression load value is higher, the stresses attain the yield strength in certain zones, and the partial plasticizing causes a decrease in the bending stiffness. Since the bending stiffness in the postelastic range is variable along the member, an average value can be assumed. Then a pseudoelastic analysis is performed. This is obtained if the elastic stiffness is multiplied by a reduction factor allowing for the inelastic behaviour. One of the main factors the reduction factor depends on is the magnitude of the member compression load. Because in a statically indeterminate structure the values of the member forces are influenced by their stiffness, the analysis is necessarily iterative.

In the case of an isolated member, the simplified approach based on the elastic behaviour hypothesis and the equivalent geometrical imperfection concept is always conservative in comparison with that based on the actual initial bow and the reduced bending stiffness. The above mentioned concept is meant to incorporate the effect of the actual geometrical imperfection and that of residual stresses. The equivalent imperfection e_0 can be expressed (see Ref 15) as:

$$e_0 = \alpha \, (\, \bar{\lambda} - 0.2) \, W/A \tag{5}$$

where α, $\bar{\lambda}$, W and A are respectively the imperfection parameter, the nondimensional slenderness, the section modulus and the cross section area. Assuming that both member ends are hinged and that the deflected shape is a sine half-wave, the axial shortening is:

$$\delta = \frac{N}{A} + \frac{\pi^2(e^2 - e_o^2)}{4L} \tag{6}$$

where N is the axial compression load and e is the final bow. Conventionally, one can take into account this phenomenon by means of a fictitious reduced cross section area A_r:

$$A_r = A/(1 + \beta) \tag{7}$$

$$\beta = 0.25\alpha^2 \frac{(\bar{\lambda} - 0.2)^2 (2 - v\chi\ddot{\lambda}^2)}{(1 - v\chi\ddot{\lambda}^2)^2} (\frac{i}{v})^2 \tag{8}$$

$$v = \frac{N}{\chi N_{pl}} \tag{9}$$

where i is the radius of gyration, v is the distance from the centroid to the extreme fibre, N_{pl} is the plastic axial load, and χ is the buckling reduction factor.

Since β is a function of the axial load N, the analysis must be iterative even for this simplified model. Applying the equivalent geometrical imperfection concept leads to an underestimate of the axial stiffness of the members with relatively low compression loads, which are actually in the elastic range. In the case of a single-layer structure, this underestimate is conservative because it overestimates the geometrical nonlinearity.

CONCLUSIONS
(a) Deviations of the member lengths from their nominal values depend to a great extent on the connecting systems, which can be classified into 'adjustable' and 'non-adjustable' ones. The first category is better suited to double-layer grids, where deviations of the nodes from their theoretical position are less important than the additional stresses due to the lack of fit. If such systems are used for single-layer structures, special devices for the control of the correct geometry (namely the local rise of each cell) are recommended. It must be emphasized that inaccuracies in both senses are unfavourable (too long or too short members).

(b) If a non-adjustable system is adopted for a single-layer structure, the acceptable deviations from the nominal lengths should be much smaller than for usual structures. The designer should anticipate the effect of such deviations. One possibility is to amplify the axial loads (obtained from a geometrically nonlinear analysis carried out for the perfect structure) by a factor taking into account the effect of imperfections, as suggested in Ref 4. A more accurate model (see Ref 6) can be obtained if the geometrical imperfection are assumed to be proportional with the vertical displacements of the nodes resulting from a geometrically nonlinear analysis of the perfect structure and then a new nonlinear analysis is performed for the imperfect structure. A survey of the nodes' position after erection and a subsequent analysis based on the actual imperfections is advisable for important structures. The stresses arising from the lack of fit are not significant in the case of single-layer structures (because of their flexibility) and may be neglected. The opposite is true for double-layer structures (owing to their much greater stiffness), where this effect should be considered. In the absence of reliable statistical data concerning the distribution of the imperfections, it is suggested to assume that only one or two members of each group (e.g. the top layer members) is affected by such inaccuracies. The effect of this undesirable prestressing is more important for steel structures than for aluminium alloys, because of the much lower value of E in the latter case. In order to diminish the additional stresses due to the lack of fit, the only realistic possibility is to impose very strict tolerances. Moreover, it is necessary to prevent any subsequent slip of the connections after erection, which could lead to an uncontrolled redistribution of the member forces. For instance, if bolted connections are used, they should belong to Category C (slip-resistant at ultimate limit state) as defined in Ref 1.

(c) The initial bows reduce both the bearing capacity and the apparent axial stiffness of the component members subjected to compression axial loads. The latter effect is more substantial for slender members subjected to high compression loads (see Ref 16). It is recommendable to take it into account for single-layer structures, where it affects the buckling loads related to different instability modes. A simplified approach (although theoretically incorrect) based on the equivalent geometrical imperfection concept seems acceptable from the practical point of view.

The question whether it is necessary or not to allow for the influence of the decrease in member axial stiffness in the case of double-layer structures is still open to discussion, but it seems that in many cases the effect is not important.

In the above, all the joint connections were assumed to be hinged. In reality, they are semi-rigid and they provide a partial restraint that varies from one system to another, but has always a favourable effect by reducing both effects of initial bows. However, considering the effect of partial restraint must be based on detailed information, including laboratory tests on joints, in order to determine the moment-rotation characteristics. In the absence of such data, it is suggested to accept the hinged connection hypothesis, which is always conservative.

The only possibility to keep this category of imperfections within the specified limits is the quality control during the manufacturing process. A survey of the initial bow of each member after erection does not seem to be feasible.

As a general conclusion, the influence of imperfections is more complex and more important for reticulated space structures than in many other cases. Certain aspects are essential for single-layer structures and other aspects for double-layer ones. A correct analysis must take into account the consequences of those inaccuracies which are relevant for the structure under examination.

REFERENCES
1. CEN – European Committee for Standardisation, Eurocode 3: Design of Steel Structures – Part 1.1: General Rules and Rules for Buildings, ENV 1993 1 – 1.
2. CEN – European Committee for Standardisation, Execution of Steel Structures – Part 1: General Rules and Rules for Buildings, pr ENV 1090 – 1.

3. ROTHERT H and GEBBEKEN N, On Numerical Results of Reticulated Shell Buckling, International Journal of Space Structures, Vol. 7, No 4, 1992, pp 299-319.

4. BĂLUŢ N, PORUMB D and GIONCU V, Some Aspects Concerning the Behaviour and the Analysis of Single-Layer Latticed Roof Structures, Stability of Metal Structures, International Colloquium, Paris, 16-17 November 1983, Preliminary Report, pp 133-144.

5. HATZIS D, The Influence of Imperfections on the Behaviour of Shallow Single-Layer Lattice Domes, Ph D Thesis, University of Cambridge, 1987.

6. GIONCU V and BĂLUŢ N, Instability Behaviour of Single Layer Reticulated Shells, International Journal of Space Structures, Vol. 7, No 4, 1992, pp 243-252.

7. SUZUKI T, OGAWA T and IKARASHI K, Elastic Buckling Analysis of Rigidly Jointed Single Layer Reticulated Domes with Random Initial Imperfection, International Journal of Space Structures, Vol 7, No 4, 1992, pp 265-273.

8. BORRI C and CHIOSTRINI S, Numerical Approaches to the Nonlinear Analysis of Single Layer Reticulated and Grid-Shell Structures, International Journal of Space Structures, Vol 7, No 4, 1992, pp 285-297.

9. IVAN A, Contributions to the Stability Analysis of Reticulated Structures (in Romanian), Ph D Thesis, Politehnica University, Timişoara, 1999.

10. SUPPLE W J and COLLINS I M, Limit Analysis of Double-Layer Grids, in Analysis and Construction of Double-Layer Grids (Ed. Makowski Z S), Applied Science Publications, London, 1981, pp 93-117.

11. COLLINS I M, An Investigation into the Collapse Behaviour of Double-Layer Grids, Proceedings of the Third International Conference on Space Structures, (Ed. Nooshin H), Elsevier Applied Science Publishers, London, 1985, pp 400-405.

12. HANAOR A, Analysis of Double Layer Grids with Material Non-linearities a Practical Approach, Space Structures, Vol. 1, No 1, 1985, pp 33-40.

13. TADO M and WAKIYAMA K, Load Carrying of Space Structures, Space Structures (Eds Parke and Howard C M), Thomas Telford, London, 1993, pp 205-201.

14. BĂLUŢ N, A Suggestion for the Separate Consideration of Geometrical and Mechanical Imperfections, International Colloquium, Stability of Steel Structures, Budapest, 1990, pp 237-244.

15. RONDAL J and MAQUOI R, Formulations d'Ayrton-Perry pour le flambement des barres métalliques, Construction Métallique, No 4, 1979, pp 41-53.

16.BĂLUŢ N and GIONCU V, The Influence of Geometrical Tolerances on the Behaviour of Space Structures, International Journal of Space Structures, Vol. 15, Nos. 3$ 4, 2000, pp 189-194.

Structural Behaviours of Various Joint Types and Twisted H-shaped Steels for Single-layer Lattice Shells

Y KUROIWA, N MASAOKA, I KUBODERA - Tomoe Corporation, Tokyo, Japan,
T OGAWA, Y KIMURA - Tokyo Institute of Technology, Tokyo, Japan, and
S KATO - Toyohashi University of Technology, Aichi, Japan

INTRODUCTION

When diagonal truss members are used for a space frame, torsional discrepancy usually occurs at joints, giving rise to complicated joints in case of H-shaped section (see Photo 1). Compared with general steel structures, standardisation of the joints is significant since the number of joints and members is considerable. One measure that the members are twisted around the axes makes it possible to connect H-shaped members at joints in a common manner (see photo 2) and this idea has been realised [Refs 1-3]. Another idea of standardisation is to eliminate some splice plates of flanges or webs, nevertheless, the existence of bending moment and shear force must be taken in consideration and they should be transferred safely even though axial forces are dominant in lattice shells.

It is reported that in case of no web splice plates, the flange splice plates are bended by shear forces and this causes the deterioration of slip strength of the joint. This secondary stress has been studied and a number of methods of evaluation have been proposed by analysing the deformed region [Ref 4]. On the other hand, the type of web joint without flange splice plates has been also studied. In this case, the thickness of web splice plates tends to be great since such a joint is required to be a rigid one. The case of semi-rigidity has been also evaluated.

The purpose of this paper is to elucidate the structural behaviors of H-shaped twisted members with various joint types. In this study, a number of simplified joints are tested under bending moment and shear force (defined as 'bending shear') to obtain the allowable strength and its mechanism of transferring stresses. Also, twisted members with various joint types are loaded under axial compressive force to obtain the relationship between the structural behaviour of each joint type and the restricting condition, which is relating to joint type.

Photo 1 Complicated Joint

Photo 2 Joint with Twisted Members

Space Structures 5, Thomas Telford, London, 2002

BASIC STRUCTURAL TESTS ON SINGLE-LAYER LATTICE SHELLS OF TWISTED H-SHAPED STEEL MEMBERS

Single-layer lattice shell

The truss system studied in this paper is for single-layer lattice shells and adopts twisted H-shaped steel members if needed. An example of the lattice shells is shown in Photo 3. Fig. 1 shows the details of the joint of the current system. The 'core' is the node of the lattice shells, where truss members are rigidly connected and is designed so that all or most of them will be simply connected. To do this, the truss members are twisted around their axes in the shaded zone. The friction joint of high strength bolts is adopted to connect the truss members.

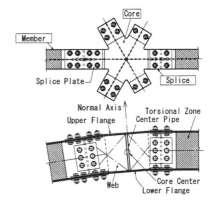

Photo 3 Shell with Twisted Members   **Fig. 1 Joint for the System**

Torsional operation

Photo 4 shows a scene of twisting an H-shaped steel member using a special device. This device grabs the both ends of the member and twists it around the axis so that the member will have the designed torsional angle (see Fig. 2). However, the member twisted tends to revert to its original form and thus, more than designed angle is required to be installed. This structural behaviour has been studied experimentally and analytically in terms of a number of parameters, such as the section size [Refs 1-2]. Using the data accumulated, the torsional operation can be carried out for required lattice shells. It is found that H-shaped steel members can be twisted uniformly since the restricting condition at the ends is a type of roller that does not confine their warping deformations.

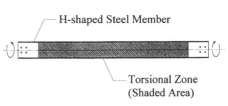

Fig. 2 Torsional Zone

Photo 4 Torsional Operation

BENDING SHEAR TEST ON VARIOUS JOINT TYPES

The structural behaviours of various joints are evaluated in this section, regarding a single-layer lattice shell using H-shaped steel members. To do this, bending shear tests are carried out in terms of conventional joints and simplified joints.

Outline of bending shear test

Fig. 3 illustrates an actual single-layer lattice shell with H-shaped steel members and Fig. 4 shows the relationship between the bending moment, M, and the shear force, Q, existing in the shell. The notations in the graph, DL, SL, EQ, indicate the dead load, snow load and earthquake load, respectively. The shear span ratio of the specimens is determined as 0.35 for the bending shear tests such that the greatest shear force will occur, compared with the bending moment. Each specimen is simply supported at the ends and is loaded at the middle point that is 350 mm away from the loading point. The inclining straight line in Fig. 4 shows the ratio of M/Q that is adopted for the current tests. The vertical displacement and the rotational angle are measured at both the loading point and the joint. Also, the histories of the stress transferring at the joint and the axial forces of the bolts are observed using strain gages.

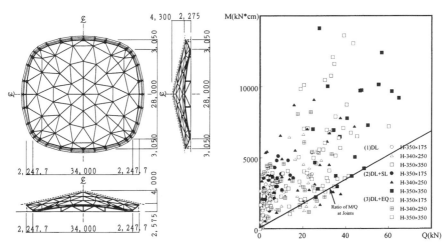

| Fig. 3 An Example of Single-layer Lattice Shell | Fig. 4 Relationships between M and Q in the Shell |

Specimens with various joint types

The joints of the specimens for the current tests are depicted in Fig. 5 together with their properties in Table 1. The size of the section is H-300x150x6.5x9 for all the specimens and their length is chosen to be 1200 mm. The number of bolts and the thickness of the splice plates of No. 1 to No. 3 are designed so that the tensile full plastic strength of the effective area can be transferred. No. 1 has a full strength joint against the axial force, bending moment and shear force. This specimen is the prototype and is modified to No. 4 to No. 6 by reducing the number of bolts or the thickness of the splice plates. No. 7 satisfies the full strength design conditions in the section with no boltholes. The steel grade of the specimens is SS400, which is standardised to have the ultimate strength of 400 N/mm^2 in the Japanese Industrial Standards.

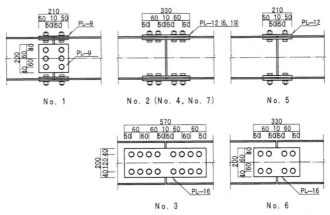

Fig. 5 Details of Various Joint Types

Table 1 List of Specimens

Specimen	Section H×B×t_w×t_f	Splice Plate web	Splice Plate flange	Bolts web	Bolts flange	Joint Type
No.1		6	9	6	4	Normal Joint
No.2			12		8	Flange Joint
No.3	H-300×150	16		16		Web Joint
No.4	×6.5×9		6		8	Flange Joint
No.5	*l*=1200mm		12		4	Flange Joint
No.6		16		8		Web Joint
No.7			19		8	Flange Joint

Influence of joint types on bending shear behaviours

Fig. 6 shows the load-displacement curves of the specimens from No. 1 to No. 3. The vertical axis indicates the non-dimensional load with reference to the yield load, P_y. The horizontal axis shows the non-dimensional displacement at the loading point with reference to δ_y, which is the displacement corresponding to the load of the yield strength.

As shown in the figure, the normal joint of No. 1 has main slides of bolts in the elastic region and then, the magnitude of the load increases slowly. The local buckling occurs in the compressive flange close to the loading point when the displacement is around $12\delta_y$. The actual slip strength is found to be approximately the same as that of the numerical one. This load is more than the yield strength of the effective sectional area when the nominal value is used for the material strength. Because of the difference between the nominal and actual material strengths, the actual slip strength is found to be less than the yield strength and thus, the slides occur earlier than expected.

The specimen of No. 2, a flange joint type, has the splice plates yielded at the clearance when the load increases to be around $1.2P_y$. The shear force and the bending moment, which are transferred through the splice plates, give rise to the secondary bending stress in the plates. After the yield of the splice plates, the slides of the bolts occur and the sharp movements of load can be seen. Consequently, the bending shear deformation proceeded in the plates. The increase of the load can also be seen after the movements. This is because the bearing effect occurs between the compressive flanges.

The notations $_bP_s$ and $_{pl}P_y$ in Fig. 6 indicate the slip strength of bolts and the bending shear yield strength of splice plates, respectively. Although the slip strength is more than both the yield strength, P_y, and the bending shear yield strength, $_{pl}P_y$, the actual value of $_bP_s$ is found to be 87% of the theoretical one because the splice plates yield in the bending shear manner. Even when a flange joint type is adopted, the strength will be sufficient if all the shear forces can be transferred through the splice plates.

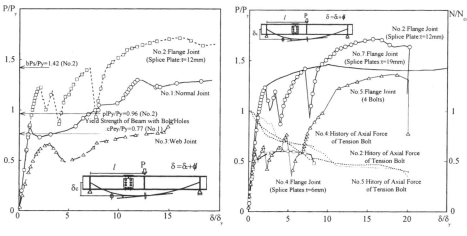

Fig. 6 P-δ Curves of Specimens with Various Joint Types (No. 1, 2, 3)

Fig. 7 P-δ Curves of Specimens with Various Joint Types (No. 2, 4, 5, 7)

Since the specimen of No. 3 has relatively lower bending rigidity at the joint clearance, the initial rigidity is lower than the others and the maximum strength is found to be around $0.6P_y$, as can be seen from Fig. 6. In the process of loading, the splice plates yield before the bolts slip and thus, any sharp change of the load cannot be recognised. The displacement due to one slide of the bolts is greater than those of the other cases, such as the normal type of No. 1 and the flange joint type of No. 2. After a rotation of the joint, a bearing state occurs between the compressive flanges. The actual slip strength agrees with the theoretical value of the bolt closest to the loading point.

Fig. 7 shows the load-displacement curves regarding the specimens of the flange joint type, namely No. 2, 4, 5 and 7. It is seen from the figure that the rigidities of these four specimens are approximately the same in the elastic region. In the case of No. 4, the splice plates yield under the bending shear force before the load reaches the slip strength of the bolts because of the secondary bending stress. It is found that No. 5 has half strength of No. 2, which is about 60% of the yield strength. After the beginning of the bolt slip, the load increases sharply in a similar manner to No. 2. In the case of No. 7, the local buckling occurs in the compressive flange close to the loading point and the strength steeply decreases. However, neither the yield of the splice plates nor the slip of the bolts can be seen.

It is concluded from the results that the flange joint type shows stable structural behaivours even in the ultimate state when enough thickness of splice plates and sufficient number of bolts are adopted.

Evaluation of various joint types

When a bending moment, M, and a shear force, Q, apply simultaneously, a flange joint type deforms in the manner of Fig. 8a. The splice plates at the clearance of the joint show the local bending deformation caused by the secondary stress. The stress in the out-of-plane direction of each splice plate is measured at the most outer edge of the flange splice plate. The total stress of the splice plates under consideration is the combination of the axial stress due to the bending moment, the secondary bending stress and the shear stress, which are caused by the shear force, as illustrated in Figs. 8b and 8c.

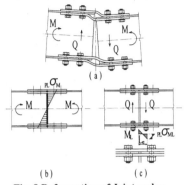

(a)

(b)   (c)

Fig. 8 Deformation of Joint under Bending Moment and Shear Force

M :	Applying Bending Moment
Q :	Applying Shear Force
$_{PL}\sigma_M$:	Axial Stress by M
$_{PL}\sigma_{ML}$:	Secondary Bending Stress
τ :	Shear Stress
$_{PL}Z$:	Section Modulus of Splice Plate
$_{PL}A_a$:	Sectional Area of Outer Splice Plate
$_{PL}A_b$:	Sectional Area of Inner Splice Plates
L :	Length of Specimen
L_c :	Length of Torsional Zone
L_e :	Span for Secondary Moment

The axial stress at the most outer edge of the splice plates (see Fig.8b) can be calculated as follows:

$$_{PL}\sigma_M = \frac{M}{_{PL}Z} \tag{3.1}$$

When the web splices are eliminated, the flange splice plate can be evaluated as a fixed beam by considering that the end exists at the most inner bolt in the flange and therefore, a corresponding distribution of bending moment will be obtained. The distance between the ends of the beam is assumed as depicted in Fig. 8c (L_e), which is determined from the experimental results of deformation. In this case, the secondary bending stress will be

$$_{PL}\sigma_{ML} = \frac{\frac{_{PL}A_a}{_{PL}A_a + 2_{PL}A_b} \cdot \frac{Q}{2} \cdot \frac{L_e}{2}}{_{PL}Z_a} \tag{3.2}$$

The shear stress, which normally passes through web splice plates, is transferred to the flange splice plates around the clearance of the joint. This value can be obtained as

$$\tau = \frac{\frac{_{PL}A_a}{_{PL}A_a + 2_{PL}A_b} \cdot \frac{Q}{2}}{_{PL}A_a} \tag{3.3}$$

Formula 3.4 can evaluate the yield strength of the flange splice plates under bending moment and shear force. This formula represents the effective stress obtained from the three stresses described above:

$$_{PL}\sigma_y \geq \sqrt{\left(_{PL}\sigma_M + _{PL}\sigma_{ML}\right)^2 + 3\tau^2} \tag{3.4}$$

The material strength of the splice plates is used as the yield stress for the tensile side and the buckling stress of the splice plate is used for the compressive side. In this study, the half of the distance between the most inner bolts is used as the buckling length of the plates [Ref 4].

The strengths of the current specimens are tabulated in Table 2. In the cases of No. 3 and No. 6, which are type of web joints, the theoretical slip strengths of bolts, $_bP_s$'s, agree with the experimental ones, $_{eb}P_s$'s. The strengths of No. 2 and No. 4 can be defined by the yield strength of the splice plates at the clearance of each joint, $_pP_{gy}$'s. Nevertheless, the slip strength of No. 4 can not be determined and therefore, the load at the limit of elasticity is taken for its $_{eb}P_s$. The slip strength of No. 5 shows a good coincidence with the experimental one.

Table 2 List of Strengths of Specimens with Various Joint Types

Specimen	P_y (kN)	P_{ey} (kN)	$_pP_y$ (kN)	$_bP_s$ (kN)	$_pP_{gy}$ (kN)	$_{eb}P_s$ (kN)
No.1	245.75	261.60	481.11	247.03		205.94
No.2	242.62	263.28	534.85	385.99	267.30	286.84
No.3	257.42	356.94	140.33	153.28		170.15
No.4	253.80	275.18	329.01	395.89	97.35	100.52*
No.5	252.42	274.40	549.17	204.61	270.21	171.62
No.6	251.05	352.32	140.33	81.59		102.97
No.7	220.50	238.86	899.64	396.35	693.28	269.68*

P_y : Yield Strength of Full Section P_{ey} : Yield Strength of Effective Sectional Area $2M_{ey}/85$
$_pP_y$: Yield Strength of Joint (Splice Only) $_bP_s$: Slip Strength of Bolt
$_pP_{gy}$: Yield Strength at Clearance of Joint Considering Effects of Splice Plates
$_{eb}P_s$: Experimental Slip Strength of Bolt * Load at Limit of Elastisity of P-d Relationship

AXIAL COMPRESSIVE TEST ON ELEMENT MEMBERS FOR SINGLE-LAYER LATTICE SHELLS

In this section, axial compressive test is performed regarding twisted H-shaped steel members with various joint types for single-layer lattice shells. The buckling strength is evaluated using the parameters of the torsional angle and the joint type to elucidate the buckling behaivours.

Axial compressive test and specimens

Using an amsler type testing machine, each specimen is axially loaded with the ends simply supported. Section size of H-200x100x5.5x8 is adopted for all the specimens. A whole specimen and the details of the joints are illustrated in Fig. 9. The hatched part of the specimen is twisted around the axis by the torsional operation. Two kinds of member lengths, 2900 mm (Type A) and 2400 mm (Type B), are chosen and the corresponding slenderness ratios are 130 and 108, respectively.

Table 3 shows the properties of the specimens: the first alphabet of each name indicates the member length and the following figure indicates the joint type and the last figure shows the torsional angle. The joint types are as follows: normal joint (Type 2), flange joint (Types 3, 4), web joint (Type 5). Also, specimens with no joints are tested (Type 1). The difference between Types 3 and 4 is the thickness of the splice plates.

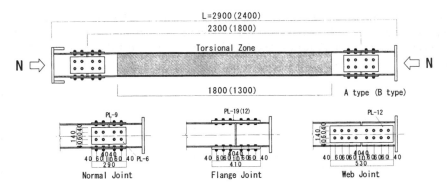

Fig. 9 Specimen of Axial Compressive Test and Joint Types

Table 3 List of Specimen for Axial Compressive Test

Size of Section HxBxt$_w$xt$_f$ (mm)	200x100x5.5x8 (SS400)					
Slenderness Ratio l	130.4			107.9		
Length of Specimen (mm)	2900			1800		
Length of Torsional Zone (mm)	1800			1300		
Torsional Angle (degree)	0	5	10	0	5	10
Joint Type 1. No Joint	A1-0	A1-5	A1-10	B1-0	B1-5	B1-10
2. Normal Joint		A2-5			B2-5	
3. Flange Joint (PL-19)		A3-5			B3-5	
4. Flange Joint (PL-12)	A4-0	A4-5	A4-10	B4-0	B4-5	B4-10
5. Web Joint	A5-0	A5-5		B5-0	B5-5	

Type 2 has normal joints of elastic design and is designed so that it has the full strength of the member. The simplified joints of Types 3 and 5 are designed so that each has the full strength for the axial force only. In contrast, the splice plates of Type 4 are designed so that they will be the thinnest but will not buckle between the most inner bolts. As shown in the table, three torsional angles are used, namely 0, 5 and 10 degrees, which are ordinarily used for the actual diagonal trusses (1.5-2.0 degrees/m). Steel grade of SS400 is chosen for the test.

Structural behaviours obtained from the compressive test
Fig. 10 shows the relationships between the axial force and the axial deformation regarding Type B. The axial force N indicated by the vertical axis is non-dimensioned by N_y, which is the yield axial force of the corresponding H-shaped steel. The axial deformation, δ, is also non-dimensioned by δ_y, which is the axial deformation when N_y is applied to the specimen.

B1-5 and B2-5 behave similarly in terms of the elastic rigidity and the inclination of deterioration. In contrast, the deterioration of B4-5 does not proceed quickly but steadily. This is because the rigidity at the joint of B4-5 is greater than that of the member and the splice plates restrict the out-of-plane deformation of buckling in contrast that the bending rigidity of B2-5 is the same as that of the member. The elastic rigidity of B5-5 is smaller than the others and its structural behaviour is relatively steady. The reason is that the web splice plates are loaded with eccentricities from the beginning and then the bending moment is added. This behaviour has been confirmed by the experimental results of the strains measured on the web splice plates. It is found that the deterioration of B5-5 is rather abrupt after the reach to the maximum strength. The reason of this is the yield in the web splice plates accompanied with the buckling deformation, which causes the decrease of the bending rigidity.

Fig. 11 shows the relationships between the axial and lateral deformations at the middle of each specimen of Type B. All the specimens except B3-5 show a similar behaivour, that is, the lateral deformation increases steeply after the flexural buckling and the maximum strength. On the other hand, the strength of B3-5 decreases owing to the local buckling in the flange.

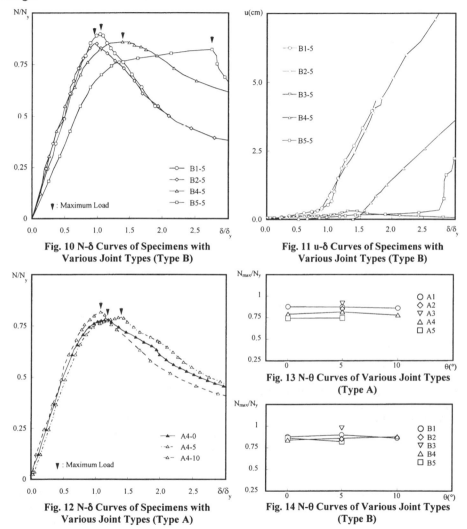

Fig. 10 N-δ Curves of Specimens with Various Joint Types (Type B)

Fig. 11 u-δ Curves of Specimens with Various Joint Types (Type B)

Fig. 12 N-δ Curves of Specimens with Various Joint Types (Type A)

Fig. 13 N-θ Curves of Various Joint Types (Type A)

Fig. 14 N-θ Curves of Various Joint Types (Type B)

Fig. 12 shows the relationships between the axial force and the axial deformation of A4-0, A4-5 and A4-10, each of which has different torsional angles: 0, 5, and 10 degrees, respectively. It is seen from the graph that all specimens behave similarly in terms of the elastic rigidity, the deformation under the maximum axial force, the maximum strength and the inclination of deterioration. Accordingly, it can be said that the torsional angle manufactured around the axis does not seriously affect the buckling behaivour of H-shaped steel members under the compressive axial force.

Figs. 13,14 show the correlations between the buckling strength and the torsional angle, regarding Types A and B, respectively. From these graphs, the strengths of A3-5 and B3-5 are found to be relatively high. As far as B3-5 is concerned, its strength is defined as the local buckling strength of the flange rather than the flexural buckling one since the bending rigidity at the joints is the greatest in the specimens. For this reason, a plastic hinge appears at the middle of the specimen but not at the joints of the ends.

Judging from the study above, any prime difference in structural strength cannot be recognised among the specimens with different torsional angles. Because a member with a torsional angle has two contradictory factors in terms of the mechanical property: the plastic parts formed by the torsional operation and the ambiguous direction of the weak axis. Because of these, it can be said that the member with a torsional angle does not cause the deterioration of strength.

CONCLUSIONS
1) When the flange joint type with the splice plates of sufficient thickness and enough number of bolts are adopted, the bending moment and the shear force can be transferred together with the axial force. For this reason, it is expected that this joint type can be used for rigidly jointed single-layer lattice shells.
2) The strength at the elastic limit can be evaluated as the minimum strength between the slip strength and the yield strength of the splice plates. This thought has been experimentally confirmed regarding various joint types.
3) In the case of simplifying joints, the buckling strength and rigidity will be sufficient when the bending rigidity of the joint is in the same level of the base metal. However, the initial rigidity of web joint type can be low and this gives rise to a large deformation before the load reaches the buckling strength due to the eccentricity at the joint.
4) It is found that H-shaped steel members with different torsional angles behave similarly under compressive loads in terms of the buckling strength and rigidity. This is because a twisted member has two contradictory properties: the residual stress and the ambiguous direction of the weak axis. Therefore, the influence of the torsional angles is considered to be negligible.

REFERENCES
1. Imagawa T, Kimura Y, Kato S, Ueki T, Ogawa T and Masaoka N, Structural Behavior of Single Layer Truss System Organized H-shaped Steel Part 1, Part 2 (in Japanese), AIJ Annual Meeting 1999, pp953-956

2. Kuroiwa Y, Masaoka N, Ueki T, Ogawa T and Kato S, High Structural Performance of Twisted Wide Flange Members Applied to Node Connections for Single Layer Lattice Shells - H Diamond Shell -, Proc. of International Symposium on Theory, Design and Realization of Shell and Spatial Structures, 2001, Nagoya, TP095

3. Masaoka N, Kuroiwa Y, Ueki T, Kato S and Ogawa T, A New Node Connection System with Twisted Wide Flange Members for Single Layer Lattice Shells - H Diamond Shell -, Proc. of International Symposium on Theory, Design and Realization of Shell and Spatial Structures, 2001, Nagoya, TP096

4. Uno N, Inoue K, Takeuchi I, Azuma S and Kita T, Experimental Study on High Strength Bolted Friction Joints with Hf-splice-plates (in Japanese), Journal of Structural and Construction Engineering, No.502, pp.127-133, Dec.1997

A general method for the design of bolted connections for space frames

S. STEPHAN, C. STUTZKI
MERO GmbH & Co. KG, D-97064 Wurzburg, Germany

1.   INTRODUCTION

The key problem of the design of space frames is the layout of the connections. With regard to the space frame concept, bolted connections are the preferred choice. For single layer structures, which are the most appropriate for glazed roofs and facades, the members often have to be connected by means of more than one bolt in order to increase the bending capacity of the connection. The design procedure of single bolt connections is well described in [2] [3] [4] [5], however, a general method for the design of multi-bolt connections is not available.

This paper will present a general design method for single or multi-bolt connections of beams with arbitrary thin-walled cross sections, suitable for application in computer programs. The design method is based on the classical strain iteration algorithm for cross sections, which is described in [6]. In this method, the ultimate capacity of bolted connections will be obtained using an iterative numerical determination of the elastic-plastic stress distribution in the connection elements. The numerical method will be derived in two steps – the numerical determination of the stress distribution in the connection for a given combination of internal forces and – the determination of the ultimate capacity of the connection. Furthermore analytical design formulas for a multi-bolt tube connection will be derived. Finally results of numerical and analytical calculations will be compared with corresponding test results.

2.   BASES AND ASSUMPTIONS

For the derivation of both, numerical and analytical design methods, some general restrictive assumptions have to be made. Basis of the discussion is the bolted connection of a prismatic beam with a longitudinal axis x and cross section axes y and z. The cross section axes of the beam do not need to be central axes. Only normal forces and bending moments will be considered, lateral forces and torsion moment will be ignored.

The connection profile, i.e. the cross section of the beam at the bolted connection, must be thin-walled. The outline of the connection profile can be open or closed. The wall of the connection profile is modeled as a set of line elements and curve elements. In the contact zone of the connection profile only compression forces can be transferred. The bolts are modeled as a set of point elements. Only tensile forces can be transferred through the bolts. The connection profile together with the bolts remains planar under full load (hypothesis of planar cross sections).

Thus the strain equation for any point (y, z) of the planar connection can be written as follows (for explanation of symbols refer to the notation list at the end of this paper):

$$\varepsilon(y,z) = \kappa_z \cdot y - \kappa_y \cdot z + \varepsilon 0 \tag{1}$$

Hence the compression stress (negative) for any point (y,z) of the connection profile can be obtained using the following formula:

$$\sigma(y,z) = \begin{cases} \sigma_{lim} ; & E \cdot \varepsilon(y,z) \leq \sigma_{lim} \\ E \cdot \varepsilon(y,z); & (E \cdot \varepsilon(y,z) > \sigma_{lim}) \wedge (E \cdot \varepsilon(y,z) \leq 0) \\ 0 ; & E \cdot \varepsilon(y,z) > 0 \end{cases} \tag{2}$$

Further the tensile stress (positive) for any bolt can be calculated from the following formula:

$$\sigma b(y,z) = \begin{cases} \sigma b_{lim} ; & E \cdot \varepsilon(y,z) > \sigma b_{lim} \\ E \cdot \varepsilon(y,z) ; & (E \cdot \varepsilon(y,z) > 0) \wedge (E \cdot \varepsilon(y,z) \leq \sigma b_{lim}) \\ 0 ; & E \cdot \varepsilon(y,z) \leq 0 \end{cases} \tag{3}$$

Finally the normal force and the bending moments at the connection can be determined from:

$$N = \sum_{v} Ab_v \cdot \sigma b(y_v, z_v) + \int_A \sigma(y,z) \, dA \tag{4a}$$

$$My = -\sum_{v} Ab_v \cdot \sigma b(y_v, z_v) \cdot z_v - \int_A \sigma(y,z) \cdot z \, dA \tag{4b}$$

$$Mz = \sum_{v} Ab_v \cdot \sigma b(y_v, z_v) \cdot y_v + \int_A \sigma(y,z) \cdot y \, dA \tag{4c}$$

Thus a certain combination of strain parameters κ_y, κ_z and ε_0 corresponds with a combination of internal forces and moments. This interdependence will be used further.

3.   NUMERICAL CALCULATION OF MULTI-BOLT CONNECTIONS

For the numerical design method some additional restrictive assumptions should be made. The wall of the connection profile is modelled as a set of only line elements. Curved sections must be adequately approximated by polygonal line elements. Any line element (index J) is defined by two nodes (index i, k) and a particular thickness t_J which is constant over the length of the element (Fig. 1). Any bolt (index v) has a particular thread diameter d_v which determines the accompanying stress area Ab_v.

3.1. Iterative Determination of Stress Distribution in the Bolted Connection

First of all it is necessary to derive an algorithm for the determination of the stress distribution in the connection elements for an arbitrary combination of internal forces and moments N_{orig}, My_{orig}, Mz_{orig}.

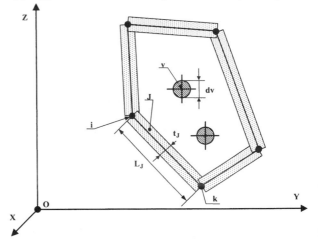

Fig. 1: General scheme of a multi-bolt connection

This combination of normal force and bending moments will be used as initial values for the iteration parameters (iteration index it):

$$it = 0 , \quad n^{\langle it \rangle} = N_{orig} , \quad my^{\langle it \rangle} = My_{orig} , \quad mz^{\langle it \rangle} = Mz_{orig} \tag{5}$$

With those iteration parameters the strain parameters can be calculated from:

$$\kappa_z^{\langle it \rangle} = \frac{mz^{\langle it \rangle}}{E \cdot Jz} , \qquad \kappa_y^{\langle it \rangle} = \frac{my^{\langle it \rangle}}{E \cdot Jy} , \qquad \varepsilon_0^{\langle it \rangle} = \frac{n^{\langle it \rangle}}{E \cdot A} \tag{6}$$

Hence the strain distribution of the planar connection (Fig. 2) for the current iteration is determinable as follows:

$$\varepsilon(y,z)^{\langle it \rangle} = \kappa_z^{\langle it \rangle} \cdot y - \kappa_y^{\langle it \rangle} \cdot z + \varepsilon_0^{\langle it \rangle} \tag{7}$$

Figure 2 shows the strain distribution in the connection for an arbitrary combination of internal forces and moments.

The surfaces of the adjacent connection profiles will contact each other only within areas with negative strain ($\varepsilon < 0$). The contact zone is limited by the zero strain line ($\varepsilon = 0$) Compression forces will be transferred only within the contact zone. Therefore only line elements with negative strain at least in one node must be regarded for the determination of compression stress in the connection profile. The bolts will be activated only within areas with positive strain ($\varepsilon > 0$). Within those areas there is a gap between the surfaces of the adjacent connection profiles.

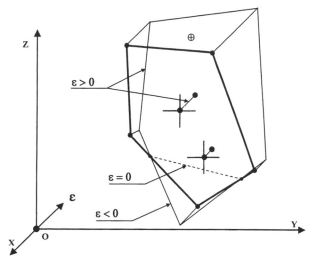

Fig. 2: Strain distribution in the connection

3.1.1.   *Determination of Compression Stress in the Connection Profile*

Figure 3 shows the strain and the stress distribution for a selected line element within the contact zone of the connection profile. First the element length must be ascertained:

$$L_J = \sqrt{\left(y_{i,J} - y_{k,J}\right)^2 + \left(z_{i,J} - z_{k,J}\right)^2} \tag{8}$$

Additionally the node strains are needed for the further calculation:

$$\varepsilon_i = \varepsilon\left(y_{i,J}, z_{i,J}\right)^{\langle it \rangle} , \qquad \varepsilon_k = \varepsilon\left(y_{k,J}, z_{k,J}\right)^{\langle it \rangle} \tag{9}$$

If there is a point with zero strain along the line element, the distance from node i must be determined:

$$u0_J = \frac{E \cdot \varepsilon_i}{E \cdot \varepsilon_i - E \cdot \varepsilon_k} \cdot L_J , \qquad u0_J \in \left[0 \ldots L_J\right] \tag{10a}$$

If the limit stress is reached in one point along the line element, the distance from node i can be obtained in a similar way:

$$u_{lim_J} = \frac{E \cdot \varepsilon_i - \sigma_{lim}}{E \cdot \varepsilon_i - E \cdot \varepsilon_k} \cdot L_J , \qquad u_{lim_J} \in \left[0 \ldots L_J\right] \tag{10b}$$

Thus the coordinates of the point with zero strain can be calculated:

$$y0_J = y_{i,J} + \frac{u0_J}{L_J} \cdot \left(y_{k,J} - y_{i,J}\right) , \quad z0_J = z_{i,J} + \frac{u0_J}{L_J} \cdot \left(z_{k,J} - z_{i,J}\right) \tag{11a}$$

In a similar way the coordinates of the point with limit stress will be determined:

$$y_{\lim_J} = y_{i,J} + \frac{u_{\lim_J}}{L_J} \cdot \left(y_{k,J} - y_{i,J}\right) , \quad z_{\lim_J} = z_{i,J} + \frac{u_{\lim_J}}{L_J} \cdot \left(z_{k,J} - z_{i,J}\right) \tag{11b}$$

Elements with at least one of those intermediate points (with zero strain or with limit stress) must be split into sub-elements to simplify the calculation of the resulting normal force and bending moments. During this the position of the intermediate points must be considered.

The arising sub-elements are line elements (index j) with two nodes and a constant thickness t_j (Fig. 4).

For further calculations only the sub-elements will be used.

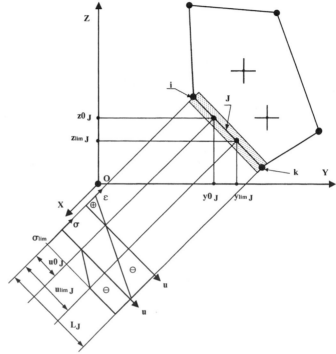

Fig. 3: Strain and stress distribution for a selected line element

The splitting operation into sub-elements has to be executed as follows:

$$Y_J = \begin{cases} \begin{bmatrix} y_{i,J} & y0_J & y_{\lim_J} & y_{k,J} \end{bmatrix} ; & u0_J \leq u_{\lim_J} \\ \begin{bmatrix} y_{i,J} & y_{\lim_J} & y0_J & y_{k,J} \end{bmatrix} ; & u0_J > u_{\lim_J} \end{cases} \tag{12a}$$

$$Z_J = \begin{cases} \begin{bmatrix} z_{i,J} & z0_J & z_{\lim_J} & z_{k,J} \end{bmatrix} ; & u0_J \leq u_{\lim_J} \\ \begin{bmatrix} z_{i,J} & z_{\lim_J} & z0_J & z_{k,J} \end{bmatrix} ; & u0_J > u_{\lim_J} \end{cases} \tag{12b}$$

The length of a sub-element (Fig. 4) with its two nodes (index i, k) must be calculated from:

$$L_{sub_j} = \sqrt{\left(Y_{i,j} - Y_{k,j}\right)^2 + \left(Z_{i,j} - Z_{k,j}\right)^2} \tag{13}$$

The strain calculation for nodes of sub-elements is similar to (9). On that basis the node stress can be obtained using the following formula:

$$\sigma_{i,j} = \begin{cases} \sigma_{\lim} ; & E \cdot \varepsilon_{i,j} \leq \sigma_{\lim} \\ E \cdot \varepsilon_{i,j} ; & \left(E \cdot \varepsilon_{i,j} > \sigma_{\lim}\right) \wedge \left(E \cdot \varepsilon_{i,j} \leq 0\right) \\ 0 ; & E \cdot \varepsilon_{i,j} > 0 \end{cases} \tag{14}$$

The same formula can be applied for $\sigma_{k,j}$ accordingly.

Thus the resulting normal force and bending moments of a sub-element for the current iteration can be determined. This will be done using a linear stress distribution between $\sigma_{i,j}$ at node i and $\sigma_{k,j}$ at node k.

Hence the arising normal force will be obtained from the product of the stress trapezoid area and the element thickness. With that the bending moments can be determined as the product of the normal force and the corresponding coordinate of the centre of gravity of the stress trapezoid:

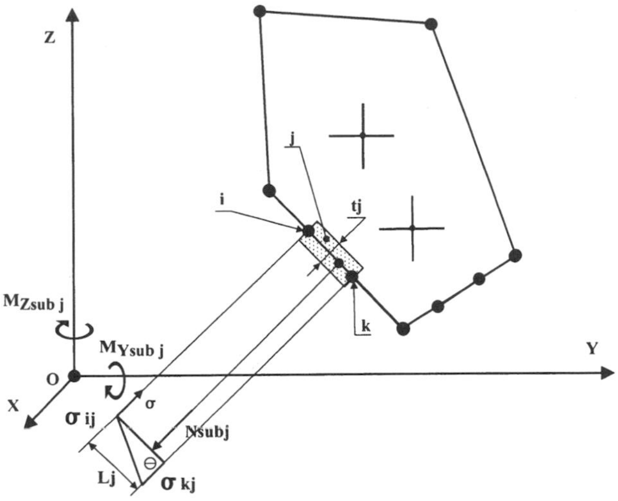

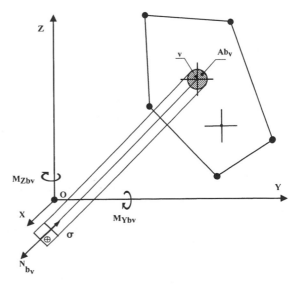

Fig. 4: Stress distribution for a sub-element

$$N_{sub_j} = \frac{1}{2} \cdot \left(\sigma_{i,j} + \sigma_{k,j} \right) \cdot t_j \cdot L_{sub_j} \tag{15a}$$

$$My_{sub_j} = N_{sub_j} \cdot \left[Z_{i,j} + \frac{\sigma_{i,j} + 2 \cdot \sigma_{k,j}}{3 \cdot \left(\sigma_{i,j} + \sigma_{k,j} \right)} \cdot \left(Z_{k,j} - Z_{i,j} \right) \right] \tag{15b}$$

$$Mz_{sub_j} = N_{sub_j} \cdot \left[Y_{i,j} + \frac{\sigma_{i,j} + 2 \cdot \sigma_{k,j}}{3 \cdot \left(\sigma_{i,j} + \sigma_{k,j} \right)} \cdot \left(Y_{k,j} - Y_{i,j} \right) \right] \tag{15c}$$

3.1.2.   Determination of Tensile Stress in the Bolts

Figure 5 shows a selected bolt (index v) with its thread diameter d_v and its stress area Ab_v. That bolt is located at the position (y_v, z_v) within the tension zone of the connection. The bolt strain for the current iteration will be calculated as follows:

$$\varepsilon_v = \varepsilon \left(y_v, z_v \right)^{\langle it \rangle} \tag{16}$$

With that the tension stress in the bolt is determinable from:

$$\sigma b_v = \begin{cases} \sigma b_{lim} ; & E \cdot \varepsilon_v > \sigma b_{lim} \\ & \left(E \cdot \varepsilon_v > 0 \right) \wedge ... \\ E \cdot \varepsilon_v ; & \left(E \cdot \varepsilon_v \leq \sigma b_{lim} \right) \\ 0 ; & E \cdot \varepsilon_v \leq 0 \end{cases} \tag{17}$$

Fig. 5: Internal forces and moments of a bolt

3.1.3.   Determination of Internal Forces and Moments

The resulting normal force in the bolts must be obtained from the product of the tensile stress and the stress area of the bolt. The bending moments will then be determined as the product of the normal force and the corresponding coordinate of the bolt position.

Now it is possible to calculate the normal force and bending moments in the connection, which results from the combination of strain parameters (6) for the current iteration step $\langle it \rangle$:

$$N^{\langle it \rangle} = \sum_v Ab_v \cdot \sigma b_v + \sum_j N_{subj} \tag{18a}$$

$$My^{\langle it \rangle} = -\sum_v Ab_v \cdot \sigma b_v \cdot z_v - \sum_j My_{subj} \tag{18b}$$

$$Mz^{\langle it \rangle} = \sum_v Ab_v \cdot \sigma b_v \cdot y_v + \sum_j Mz_{subj} \tag{18c}$$

3.1.4.   Check of Convergence

To verify whether convergence is reached in the current iteration, it is necessary to determine the deviation of the normal force and bending moments from the given original combination of internal forces and moments N_{orig}, My_{orig}, Mz_{orig}:

$$\Delta N^{\langle it \rangle} = N_{orig} - N^{\langle it \rangle} \tag{19a}$$

$$\Delta My^{\langle it \rangle} = My_{orig} - My^{\langle it \rangle} \tag{19b}$$

$$\Delta Mz^{\langle it \rangle} = Mz_{orig} - Mz^{\langle it \rangle} \tag{19c}$$

Convergence will be reached, if the following condition is fulfilled:

$$\left(\Delta N^{\langle it \rangle} \cong 0 \right) \wedge \left(\Delta My^{\langle it \rangle} \cong 0 \right) \wedge \left(\Delta Mz^{\langle it \rangle} \cong 0 \right) \tag{20}$$

If condition (20) is not fulfilled, then it will be necessary to execute a new iteration step. To begin with the new iteration the following iteration parameters must be adjusted:

$$it = it + 1, \quad n^{\langle it \rangle} = n^{\langle it-1 \rangle} + \Delta N, \quad my^{\langle it \rangle} = my^{\langle it-1 \rangle} + \Delta My, \quad mz^{\langle it \rangle} = mz^{\langle it-1 \rangle} + \Delta Mz \tag{21}$$

Afterwards all equations from (6) to (19) has to be calculated again and condition (20) has to be checked. The iteration must be repeated until condition (20) is fulfilled.

The stress and strain distribution in the connection elements, which is caused by the given combination of internal forces and moments N_{orig}, My_{orig}, Mz_{orig}, is thus determined.

3.2.   Iterative Determination of the Ultimate Connection Capacity

The above described determination of the stress distribution in the connection elements for an arbitrary combination of normal force and bending moments will be used now for the calculation of the ultimate connection capacity.

Two of three parameters N, My and Mz (normal force and bending moments) must be given, the limit value of the third parameter will be determined. Beginning with zero, the third parameter has to be increased in steady steps. For each step the procedure described in pragraph 3.1 has to be carried out. If the convergence condition (20) cannot be fulfilled within a reasonable number of iterations, then the limit value of the third parameter was obviously exceeded.

Hence, the previous increase of the third parameter has to be cancelled, the parameter step size has to be halved and the third parameter will be increased with reduced step size.

After that procedure 3.1 has to be repeated. If the convergence condition (20) can be fulfilled now, then the third parameter will be continually increased with reduced step size and procedure 3.1 will be carried out for each step. If, however, the convergence condition (20) cannot be fulfilled within a reasonable number of iterations, then the previous increase of the third parameter has to be cancelled once more, the parameter step size has to be halved again and the third parameter should be increased with the smaller step size. Then procedure 3.1. has to be carried out once more.

This process must be repeated until the parameter step size is lower than a small predefined value. If that condition is met, the limit value of the third parameter is found. The current combination of all three parameters N, My and Mz is the ultimate capacity of the connection.

The advantage of the presented numerical method is that it can be easily adapted to arbitrary thin-walled connection profiles with any bolt scheme.

4.   ANALYTICAL CALCULATION OF MULTI-BOLT TUBE CONNECTIONS

In this paragraph an analytical calculation method will be presented for a bolted tube connection as shown in Fig. 6, which has a simple and continuous connection profile. This analytical method will be used to check the results of the above described numerical method.

The idea of the above numerical method was the iterative modification of strain parameters until the resulting internal forces are equal to the given combination of internal forces (refer to [6]).
The numerical calculation of the ultimate connection capacity was finally a stepwise increase of a given internal force or moment until the equality of the resulting internal forces with the given

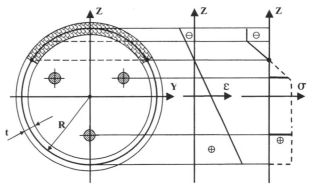

Fig. 6: Multi-bolt tube connection

combination cannot be achieved anymore. Thus, the ultimate capacity of a connection can be determined only as a result of a double iteration process.

In contrast, the analytical method does not require any iteration. Ultimate strain conditions have to be defined for the connection profile and for the bolts. Together with assumed coordinates of the zero strain line this is sufficient to calculate all strain parameters and hence the resulting combination of normal force and bending moments That combination represents simultaneously the ultimate capacity of the connection. Modifying the coordinates of the zero strain line, a limit state interaction of normal force and bending moments can be determined.

Obviously this is an implicit calculation method, since it is impossible to determine the combination of normal force and bending moments directly - knowing two of three parameters N, My and Mz. The combination of normal force and bending moments can be ascertained only from a set of parametric equations based on the strain parameters.

It is assumed again that the connection profile together with the bolts remains planar under full load (hypothesis of planar cross sections). In order to simplify the derivation it is assumed further that the bending moment Mz is zero.

Hence it follows that only two strain parameters are necessary: the coordinate of the zero strain line z0 and a certain strain value $\varepsilon 1$ with its corresponding coordinate z1. Strain $\varepsilon 1$ is either the ultimate strain of the outermost tube wall in the compression zone or the ultimate strain of the outermost bolt in the tension zone, depending on which of the two is the decisive criterion. This procedure is described below in paragraph 4.4.

Consequently, the strain distribution for the tube connection (Fig. 6) can be determined:

$$\varepsilon(z) = \varepsilon 1 \cdot \frac{z - z0}{z1 - z0} \tag{22}$$

4.1. Determination of Compression Stress in the Tube Wall

Considering equations (2) and (22), the stress distribution in the tube wall can be obtained from:

$$\sigma(z) = \begin{cases} \sigma_{\lim} \; ; & E \cdot \varepsilon(z) \le \sigma_{\lim} \\ E \cdot \varepsilon(z); & \begin{array}{l}(E \cdot \varepsilon(z) \le 0) \wedge ... \\ (E \cdot \varepsilon(z) > \sigma_{\lim})\end{array} \\ 0 \; ; & E \cdot \varepsilon(z) > 0 \end{cases} \tag{23}$$

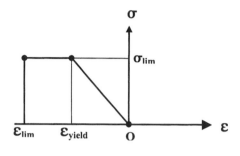

Fig. 7: Elastic-plastic tube material

The corresponding stress distribution is shown in figure 8. Figure 7 shows the elastic behaviour of the tube material until the yield strain ε_{yield} is reached and the plastic behaviour until ultimate strain $\varepsilon_{\lim}$ is reached. According to Hooke's law the yield strain can be determined from the following equation:

$$\varepsilon_{yield} = \frac{\sigma_{\lim}}{E} \tag{24}$$

Provided that $z_{\lim}$ is the coordinate with the yield strain, equation (22) can be transformed to:

$$\varepsilon_{yield} = \varepsilon 1 \cdot \frac{z_{\lim} - z0}{z1 - z0} \tag{25}$$

Considering equations (24) and (25), the coordinate with the yield strain $z_{\lim}$ can be determined for $(-R \le z_{\lim} \le R)$ from:

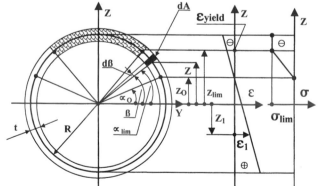

Fig. 8: Stress and strain distribution for the tube wall

$$z_{\lim} = \frac{\sigma_{\lim}}{E \cdot \varepsilon 1} \cdot (z1 - z0) + z0 \tag{26}$$

The geometrical characteristics in figure 8 will be obtained from:

$$\beta = \arcsin\left(\frac{z}{R}\right) , \quad \alpha 0 = \arcsin\left(\frac{z0}{R}\right) , \quad \alpha 1 = \arcsin\left(\frac{z1}{R}\right) , \quad \alpha_{\lim} = \arcsin\left(\frac{z_{\lim}}{R}\right) \tag{27}$$

As can be seen from figure 8, the area differential of the tube wall is defined as follows:

$$dA = R \cdot t \cdot d\beta \tag{28}$$

The tube wall can be divided in a section with no contact and hence no stress ($-R \le z < z0$), a section with elastic stress ($z0 \le z < z_{lim}$) and a section with plastic stress ($z_{lim} \le z \le R$).

Thus the normal force of the section with only elastic stress will be obtained from:

$$N_e = 2 \cdot \int_{z0}^{z_{lim}} \sigma(z) \cdot dA = 2 \cdot \int_{\alpha0}^{\alpha_{lim}} E \cdot \epsilon 1 \frac{\sin(\beta) - \sin(\alpha0)}{\sin(\alpha1) - \sin(\alpha0)} \cdot R \cdot t \cdot d\beta \tag{29a}$$

$$N_e = 2 \cdot E \cdot \epsilon 1 \cdot R \cdot t \cdot \frac{\cos(\alpha0) - \cos(\alpha_{lim}) + (\alpha0 - \alpha_{lim}) \cdot \sin(\alpha0)}{\sin(\alpha1) - \sin(\alpha0)} \tag{29b}$$

The normal force of the section with only plastic stress will be derived accordingly:

$$N_p = 2 \cdot \int_{z_{lim}}^{R} \sigma(z) \cdot dA = 2 \cdot \int_{\alpha_{lim}}^{\pi/2} \sigma_{lim} \cdot R \cdot t \cdot d\beta \tag{30a}$$

$$N_p = (\pi - 2 \cdot \alpha_{lim}) \cdot \sigma_{lim} \cdot R \cdot t \tag{30b}$$

The bending moment of the section with elastic stress will be derived in a similar way:

$$My_e = 2 \cdot \int_{z0}^{z_{lim}} \sigma(z) \cdot z \cdot dA \tag{31a}$$

$$My_e = 2 \cdot \int_{\alpha0}^{\alpha_{lim}} E \cdot \epsilon 1 \frac{\sin(\beta) - \sin(\alpha0)}{\sin(\alpha1) - \sin(\alpha0)} \cdot R^2 \cdot t \cdot \sin(\beta) \cdot d\beta \tag{31b}$$

$$My_e = E \cdot \epsilon 1 \cdot R^2 \cdot t \cdot \frac{2 \cdot \sin(\alpha0) \cdot \cos(\alpha_{lim}) + (\alpha_{lim} - \alpha0) - \dfrac{\sin(2 \cdot \alpha_{lim})}{2} - \dfrac{\sin(2 \cdot \alpha0)}{2}}{\sin(\alpha1) - \sin(\alpha0)} \tag{31c}$$

The bending moment of the section with plastic stress will be obtained from:

$$My_p = 2 \cdot \int_{z_{lim}}^{R} \sigma(z) \cdot z \cdot dA = 2 \cdot \int_{\alpha_{lim}}^{\pi/2} \sigma_{lim} \cdot R^2 \cdot t \cdot \sin(\beta) \cdot d\beta \tag{32a}$$

$$My_p = 2 \cdot \cos(\cdot \alpha_{lim}) \cdot \sigma_{lim} \cdot R^2 \cdot t \tag{32b}$$

4.2.   Determination of Tensile Stress in the Bolts

Considering equations (2) and (22), the stress distribution in the bolts can be calculated as follows:

$$\sigma b(z) = \begin{cases} \sigma b_{lim} \; ; & E \cdot \epsilon(z) > \sigma b_{lim} \\ E \cdot \epsilon(z) \; ; & \begin{array}{l}(E \cdot \epsilon(z) > 0) \wedge ... \\ (E \cdot \epsilon(z) > \sigma b_{lim})\end{array} \\ 0 \; ; & E \cdot \epsilon(z) \le 0 \end{cases} \tag{33}$$

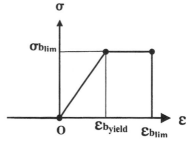

Fig. 9: Elastic-plastic bolt material

The corresponding stress distribution is shown in figure 10. Figure 9 shows the elastic behaviour of the tube material until the yield strain ϵb_{yield} is reached and further the ideal plastic behaviour until ultimate strain ϵb_{lim} is reached.

According to Hooke's law the yield strain can be determined from the following equation:

$$\epsilon b_{yield} = \frac{\sigma b_{lim}}{E} \tag{34}$$

Provided that zb_{lim} is the coordinate with the yield strain, equation (22) can be transformed to:

$$\epsilon b_{yield} = \epsilon 1 \cdot \frac{zb_{lim} - z0}{z1 - z0} \tag{35}$$

Considering equations (34) and (35), the coordinate with the yield strain zb_{lim} can be calculated from the following equation ($-R \leq zb_{lim} \leq R$):

$$zb_{lim} = \frac{\sigma b_{lim}}{E \cdot \varepsilon 1} \cdot (z1 - z0) + z0 \tag{36}$$

The bolt area can be divided in a section with no stress due to contact in the tube wall ($z0 \leq z < R$), a section with elastic stress ($zb_{lim} \leq z < z0$) and a section with plastic stress ($-R \leq z \leq zb_{lim}$).

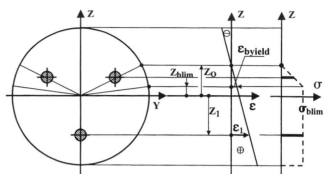

Fig. 10: Stress and strain distribution for the bolts

Thus the normal force of all bolts (index v) will be determined as follows:

$$N_b = \sum_v \begin{cases} \sigma b_{lim} \cdot Ab_v \; ; & z_v < zb_{lim} \\ E \cdot \varepsilon 1 \cdot \dfrac{z_v - z0}{z1 - z0} \cdot Ab_v \; ; & (z_v \geq zb_{lim}) \wedge (z_v \leq z0) \\ 0 \; ; & z_v > z0 \end{cases} \tag{37}$$

The bending moment of all bolts will be calculated accordingly:

$$My_b = \sum_v \begin{cases} \sigma b_{lim} \cdot Ab_v \cdot z_v \; ; & z_v < zb_{lim} \\ E \cdot \varepsilon 1 \cdot \dfrac{z_v - z0}{z1 - z0} \cdot Ab_v \cdot z_v \; ; & (z_v \geq zb_{lim}) \wedge (z_v \leq z0) \\ 0 \; ; & z_v > z0 \end{cases} \tag{38}$$

4.3.   Determination of Ultimate Normal Force and Bending Moment

The normal force and bending moment represent the ultimate capacity of the connection, since the calculation is based on ultimate strain conditions. The ultimate normal force and ultimate bending moment of the multi-bolt tube connection will be determined from the following equations:

$$N = N_e + N_p + N_b \tag{39a}$$

$$My = -My_e - My_p - My_b \tag{39b}$$

The limit state interaction of the normal force and the bending moment can be determined by a stepwise increase of the zero strain coordinate z0 from $-R$ to R.

4.4.   Determination of Ultimate Strain Conditions

For the above described analytical method, a certain strain value $\varepsilon 1$ with its corresponding coordinate z1 has to be assumed. The strain value $\varepsilon 1$ is either the ultimate strain of the outermost tube wall in the compression zone or the ultimate strain of the outermost bolt in the tension zone, depending on the critical strain conditions. According to [1], the ultimate strain for mild steel can be assumed as:

$$\varepsilon_{lim} = 0.1 \tag{40a}$$

Obviously, the corresponding coordinate of the outermost tube wall is:

$$z_{\cdot R} = R \tag{40b}$$

Likewise, according to [1] the ultimate strain of high strength bolts (grade 8.8 and 10.9) can be calculated as follows:

$$\varepsilon b_{lim} = \frac{\pi \cdot do \cdot \sigma b_{lim}}{4 \cdot E \cdot Lb} \cdot \left(\frac{1.6}{\pi \cdot d} + \frac{4 \cdot Ls}{\pi \cdot d^2} + 4 \cdot \frac{ko \cdot (Lb - Ls) + kn \cdot hn}{\pi \cdot do^2} \right) \tag{41a}$$

The corresponding coordinate of the outermost bolt in the tension zone can be obtained from:

$$z_{\cdot B} = \min(z_v) \tag{41b}$$

The thread strain coefficient has to be determined from the following formula:

$$ko = \begin{cases} 0.8 + \dfrac{0.2}{0.021} & ; \quad \text{grade } 8.8 \\ 0.9 + \dfrac{0.1}{0.013} & ; \quad \text{grade } 10.9 \end{cases} \tag{42}$$

The nut strain coefficient has to be calculated as follows:

$$kn = \begin{cases} 0.48 + \dfrac{0.12}{0.021} & ; \quad \text{grade } 8.8 \\ 0.54 + \dfrac{0.06}{0.013} & ; \quad \text{grade } 10.9 \end{cases} \tag{43}$$

The resulting ultimate strain of the bolts usually lies between 0.02 and 0.06, depending on the geometric parameters of the bolts (diameter, shaft length and thread length).

Providing that the specific ultimate strain appears at the outermost bolt in the tension zone, the resulting strain at the outermost tube wall in compression zone has to be calculated as follows:

$$\varepsilon_R = \frac{z_R - z0}{z_B - z0} \cdot \varepsilon b_{lim} \tag{44}$$

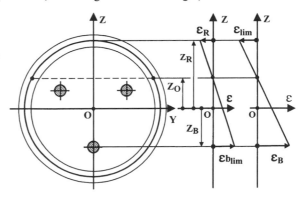

Providing, however, that the specific ultimate strain appears at the outermost tube wall in the compression zone,

Fig. 11: Two possibilities of ultimate strain distribution

the resulting strain at the outermost bolt in the tension zone has to be calculated as follows:

$$\varepsilon_B = \frac{z_B - z0}{z_R - z0} \cdot \varepsilon_{lim} \tag{45}$$

The strain value $\varepsilon 1$ depends on which of the two strain distributions is permissible:

$$\varepsilon 1 = \begin{cases} \varepsilon_{lim} & ; \quad (\varepsilon_B \leq \varepsilon b_{lim}) \wedge (\varepsilon_R > \varepsilon_{lim}) \\ \varepsilon b_{lim} & ; \quad (\varepsilon_B > \varepsilon b_{lim}) \wedge (\varepsilon_R \leq \varepsilon_{lim}) \end{cases} \tag{46a}$$

Hence the corresponding coordinate z1 has to be calculated accordingly:

$$z1 = \begin{cases} z_R & ; \quad (\varepsilon_B \leq \varepsilon b_{lim}) \wedge (\varepsilon_R > \varepsilon_{lim}) \\ z_B & ; \quad (\varepsilon_B > \varepsilon b_{lim}) \wedge (\varepsilon_R \leq \varepsilon_{lim}) \end{cases} \tag{46b}$$

Those values are sufficient for the analytical calculation according to 4.1, 4.2 and 4.3.

5. TENSILE AND BENDING TESTS

The calculated ultimate connection capacity can be verified only by experimental tests. Usually, these tests will be carried out as single axial load tests either with pure tensile load or with pure bending load. Multiaxial load tests, i.e. with interaction of normal force and bending moments, are difficult to realize and very expensive. Every specific test has to be carried out at least three times. The test results must be evaluated according to relevant codes or regulations, e.g. BS 5950-1. The objective of the tests is the determination of the connection capacity for a given load. Another objective is the determination of the corresponding connection stiffness, which has to be used in the structural model to reflect the real semi-rigid behaviour of the bolted connection. In order to determine the connection stiffness, the applied load and the corresponding deflection must be continuously recorded. The applied load will be gradually increased from zero to the ultimate load (connection failure or extreme deflections).

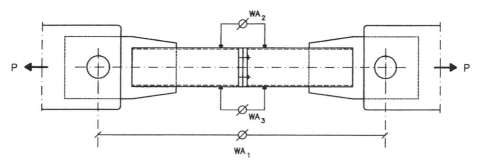

Fig. 12: Typical tensile test arrangement

Figure 12 shows a typical tensile test of a multi-bolt tube connection. The tensile force will be measured directly at the testing machine. The tensile deflection of the connection will be measured with inductive path-measuring instruments (WA_2, WA_3 in Fig. 12).

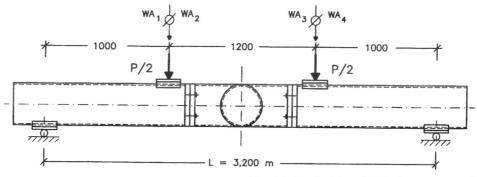

Fig. 13: Typical bending test arrangement

Figure 13 shows a typical bending test of a multi-bolt tube connection. The bending force will be measured with load cells underneath the crossbeam of the testing machine. The bending deflection of the connection will be measured with inductive path-measuring instruments (WA_1, WA_2, WA_3, WA_4 in Fig. 13). The applied bending moment is constant between the two points of force introduction and can be easily calculated due to the symmetric four-point test arrangement.

6. COMPARISON OF NUMERICAL, ANALYTICAL AND TEST RESULTS

The above described numerical calculation method was converted to a computer program by means of the programming language FORTRAN 90. The data model of the connection profile and the bolt scheme is very flexible and can be adopted to any configuration. The analytical calculation method for tube connections was programmed by means of the mathematical software MATHCAD 2000.

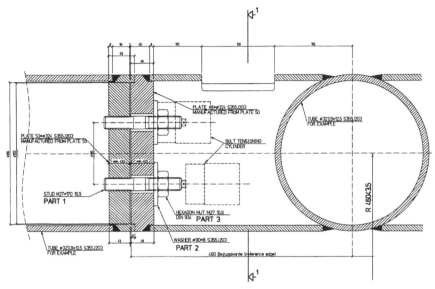

Fig. 14: Longitudinal section of the tested 4-bolt connection for Glasgow Science Center

Figure 14 shows a longitudinal section through the 4-bolt connection of the Exploratorium roof structure of the Glasgow Science Center. Figure 15 shows the corresponding cross section. This connection was designed according to the paragraphs 3, 4 and 5.

The 4-bolt connection of the above structure has the following parameters:

tube diameter	323.9 mm
tube wall thickness	12 mm
tube material	S355 J2H
bolt size, grade	4 x M27 – 10.9
pitch diameter	180 mm

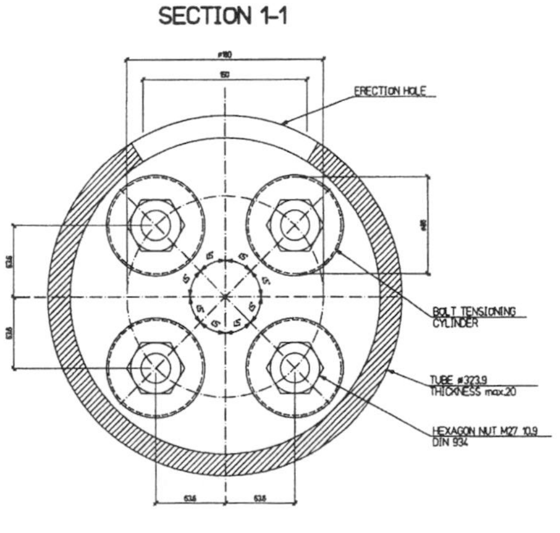

The results of the numerical and analytical calculations as well as the test results are shown in figure 16. As can be seen from the diagram the analytical and numerical results are close together. These calculations have been confirmed by test results for the relevant load combinations.

The evaluation of the test results has been done according to BS 5950-1.

SECTION 1-1

Fig. 15: Cross section of the tested 4-bolt connection

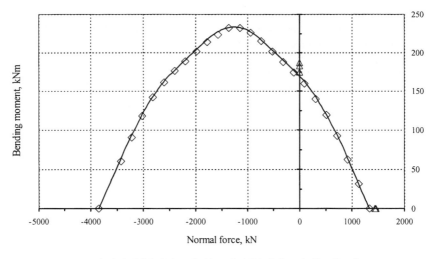

—— Analytical Calculation ◇ Numerical Calculation △ Test Results

Fig. 16: Analytical & numerical calculation and test results for Glasgow Science Center

Although the node connection for the roof structure of the Eden Project is a single bolt connection, it can be designed according to the above described methods.

The results of calculations and tests are shown in figure 17. The connection parameters are:

tube diameter	193.7 mm	tube wall thickness	10 mm
tube material	S355 J2H	bolt size, grade	1 x M33 – 10.9

As can be seen from the diagram the analytical and numerical results are close together and have been confirmed by the test results for the relevant load combinations.

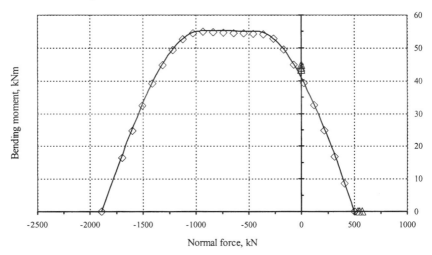

—— Analytical Calculation ◇ Numerical Calculation △ Test Results

Fig. 17: Analytical & numerical calculation and test results for Eden Project

7. REFERENCES

[1] STEURER A., Das Tragverhalten und Rotationsvermögen geschraubter Stirnplatten-verbindungen, ETH Zürich: IBK Bericht Nr. 247, Birkhäuser Verlag, Basel 1999, pp 242 – 244 & 255 – 256
[2] Allgemeine bauaufsichtliche Zulassung Z-14.4-10: MERO Raumfachwerk, DIBt Berlin
[3] KLIMKE H., Developing a Space Frame System, IASS Singapore 1997, pp 439 – 446
[4] STUTZKI C., MERO Plus – Handbuch, Würzburg 1990, pp II.10.34 – II.10.48
[5] KLIMKE H., How Space Frames Are Connected, IASS Madrid 1999, pp B4.13 – B4.19
[6] KINDMANN R. & FRICKEL J., Elastische und plastische Querschnittstragfähigkeit, Ernst & Sohn, Berlin 2002, pp 411 – 412 & 485 – 499

8. NOTATION LIST

8.1. Symbols

$\alpha 0$	angle with zero strain (z0)	kn	nut strain coefficient
$\alpha 1$	angle with given strain (z1)	ko	thread strain coefficent
α_{lim}	angle with limit stress (z_{lim})	L, L_{sub}	element length
ß	current angle (z)	Lb	clamping length of bolt
σ	compression stress (profile)	Ls	shaft length of bolt
σ_{lim}	ultimate compression stress	My, Mz	bending moments
σb	tensile stress (bolt)	My_{sub}, My_{sub}	bending moments
σb_{lim}	ultimate tensile stress (bolt)	my, mz	iteration parameters
ΔM_y, ΔM_z	deviation of bending moment	My_b	bending moment of bolts
ΔN	deviation of normal force	My_e, Mz_e	bend. moments of elastic area
κ_y, κ_z	strain parameters	My_p, Mz_p	bend. moments of plastic area
ε	strain in x-direction	My_{orig}, Mz_{orig}	given bending moments
ε_{lim}	ultimate strain	N, N_{sub}	normal force
ε_{yield}	yield strain	n	iteration parameter
εb_{lim}	ultimate strain of bolt	N_b	normal force of bolts
εb_{yield}	yield strain of bolt	N_e	normal force of elastic area
ε_B	strain at outermost bolt	N_p	normal force of plastic area
ε_R	strain at outermost tube wall	N_{orig}	given normal force
$\varepsilon 0$	strain parameter	R	tube radius
$\varepsilon 1$	given strain	t	wall thickness
A	cross section area	u0	offset of zero strain point
Ab	stress area of bolt	u_{lim}	offset of limit stress point
d	thread diameter of bolt	y, z	node coordinates
do	diameter of stress area of bolt	Y, Z	node coordinates
dA	area differential	y0, z0	coordinates of zero strain point
dß	angle differential	y_{lim}, z_{lim}	coordinates of limit stress point
E	modulus of elasticity	z_B	coordinate of outermost bolt
hn	height of nut	z_R	coord. of outermost tube wall
Jy, Jz	moments of inertia	z1	coordinate with given strain

8.2. Indices

i, k	node index	it	iteration index
J, j	element index	v	bolt index

TekCAD – A new system for creating space structures

D. J. ANDERSON
TekStar International Corporation, Longwood FL, USA
W. R. WENDEL
Starnet International Corporation, Longwood FL, USA
V. GANE HOK
Aviation Division, Orlando FL, USA

TekStar International Corporation[1] has published TekCAD[2], a new software system for creating space structures. Under development since February 2000 and released in March 2002, this computer-aided design system leverages modern software application technology to streamline the space structure design process. The software is fast, easy to use, has a shallow learning curve with a rich set of functions and features, and can interoperate with existing CAD programs. TekCAD users can directly interact with designs consisting of hubs and struts creating truly visionary geometric structures. These structures may be graphically, numerically or symbolically manipulated in multiple views simultaneously. Operations focusing on creating, editing, transforming, reporting and rendering are all provided, as well as operations for importing and exporting structures using industry-standard formats. TekCAD also includes unique features, such as built-in mathematical expression evaluation, overlaid geometric displays and an extensive web-based context-sensitive help system. When paired with TekStar's TekKit[3] model-building system, scale models of designs may be built quickly to aid in the visualization and understanding of space structures.

THE TEKCAD PHILOSOPHY: OPERATIONS, SELECTIONS AND VIEWS

TekCAD is used to create and manipulate 3D structures made of hubs and struts. Hubs may be located at any position in space, and struts are connections between any two hubs. These objects have properties such as color and layer, and each hub and strut in a structure has a unique identifier. Within the program, networks of these hubs and struts represent geometric vertices and edges. For example, in Figure 1, an octet-truss framework is represented as a network of interconnected hubs and struts. While there are few limits to the types of frameworks that can be created from hubs and struts, design software must offer a

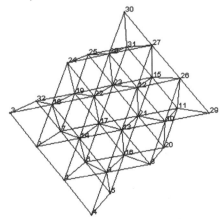

Figure 1. A simple network of hubs and struts
(the hub identifiers are shown)

reasonable way to interact with them. TekCAD does this by letting the user select these objects and execute operations upon them.

Effectively, TekCAD is an operating system for design. It functions in the same way a computer's operating system does: when an operation is requested, it is loaded into the system and executed. The system provides the resources and environment that the operation needs to perform its work. Some operations execute silently, performing their tasks and finishing. Other operations require interactive user input. Several even manage the efforts of subordinate operations, serving to coordinate multiple tasks and to divide work between them. All of the mechanisms that affect structures in TekCAD exist in the form of loadable operations.

TekCAD operations are grouped into several broad categories:

> **File** operations deal with structures as a whole
> **Create** operations add new hubs, struts and structures
> **Edit** operations adjust the properties of selected hubs and struts
> **Transform** operations modify the arrangements of selections
> **Model** operations aid in building scale models of the structure
> **Information** operations access tools and build reports
> **Help** operations display documentation

Some of these operations need to access selected groups of hubs and struts. A designer can select objects either by using the mouse (selecting individual objects or larger groups) or by executing specialized selection operations. Once a selection is made, an operation is chosen and executed. This noun-verb interaction style is the essence of event-driven computer programs. TekCAD brings this comfortable mechanism to space structure design.

When working in TekCAD, the user may open multiple structures and multiple views of those structures. The different views of a structure are displayed simultaneously, which is especially valuable when manipulating it. Manipulative operations may be started in one view and continued in another when the vantage in the different view becomes more appropriate. There are six types of views available in TekCAD, and all may be active at once:

> **Perspective** Views (Figure 2) show a perspective rendering of the structure from any point (the viewpoint) to any point (the lookpoint).
>
> **Plan** Views (Figure 3a) show the structure from directly above. The height of the structure is effectively collapsed.
>
> **Front** Elevation Views (Figure 3b) show the structure directly from the front. The depth of the structure is effectively collapsed.
>
> **Side** Elevation Views (Figure 3c) show the structure directly from the side. The breadth of the structure is effectively collapsed.
>
> **Isometric** Views show the structure from a vantage that is at an equal angle to each of the Cartesian coordinate axes.

Oblique Views show the structure directly from the front with its components shifted according to their depths.

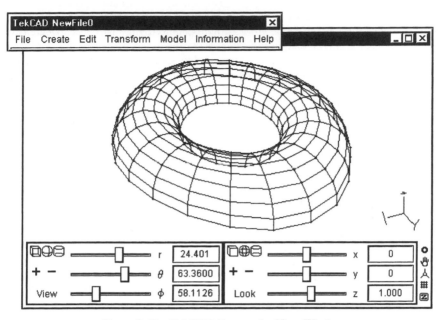

Figure 2. The TekCAD Perspective View Window

These views all reflect the current state of the structure whatever it may be, however only their content is synchronized. Views may be independently scaled, zoomed or adjusted at any time in TekCAD. The view windows each contain controls that adjust the view within. The perspective view allows independent adjustment of the viewpoint and the lookpoint, while the other view windows contain sliders and buttons that can be used to adjust their views. View windows also contain facilities for labeling, panning and zooming.

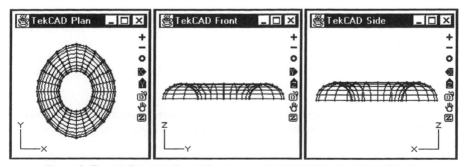

Figure 3. The (a) Plan, (b) Front Elevation, and (c) Side Elevation View Windows

TEKCAD OPERATIONS

At the top of figure 2 is a menu bar. It may be moved freely within the windowing environment but is logically attached to the perspective view window. Loadable operations are accessed from this menu bar by navigating popup menus. Clicking on a heading will bring up a hierarchical popup menu from which operations may be chosen. Alternatively, many operations have keyboard equivalents that allow the user to access an operation directly from the keyboard. There is also a freely moveable shortcut tool (Figure 4) that allows popular operations to be started by clicking a small icon. All of these ways to navigate to an operation are equivalent; the user may choose his preferred method of interaction.

When an operation requires input, an interaction window appears on the screen that contains a series of controls that affect that operation. Each interactive operation has its own set of controls that are specific to its actions. As the values of these controls are changed, the effects of the operation are changed, and many times this will affect the contents of the view windows. In this way, changes to operational parameters translate directly into changes in the views, bringing a unique feeling of interactivity to the design process.

Figure 4. TekCAD Iconic Shortcuts

File Operations

The File operations generally pertain to either the whole product or the whole structure.

The first group of operations adjusts the product. A new structure may be created, an exiting one may be opened, preferences may be set, the color standards and environment may be adjusted, windows and views may be made visible, and TekCAD may be exited.

The second group of operations concerns the entire structure. The current structure may be saved or closed, its units may be changed, comments and information may be added, duplicate hubs and struts may be eliminated, objects may be labeled and renumbered, other structures may be imported into the structure, and the structure may be exported in a variety of formats.

Different export formats promote interoperability between TekCAD and other programs. Images of the structure can be created in GIF and JPG formats for printing or inclusion in electronic media. Interactive renderings can be created to work within web browsers using facilities such as the Live3D Java applet[4] or VRML[5]. Structures can also be exported in wireframe or solid DXF[6], the preferred import formats for most traditional CAD programs.

Create Operations

There are three types of Create operations in TekCAD: operations that create hubs and struts, operations that generate networks of these objects parametrically and operations that load predefined structures.

The first type of operation is used to create small numbers of hubs and struts. Designers can build sets of hubs and sets of struts and complete modules in grid frameworks.

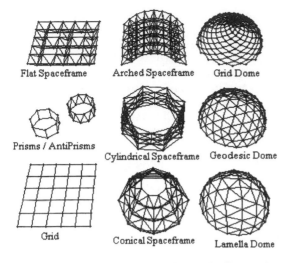

Flat Spaceframe   Arched Spaceframe   Grid Dome

Prisms / AntiPrisms

Cylindrical Spaceframe   Geodesic Dome

Grid

Conical Spaceframe   Lamella Dome

Figure 5. Some of TekCAD's Generable Geometries

The second type of operation generates networks of hubs and struts. Some of these geometries are illustrated in Figure 5. Besides spaceframes with a wide variety of modular forms and structures, there are different types of domes and grids with various tessellations, toroidal forms, polygons, etc. Simply changing values in interaction windows will automatically generate these complex structures in TekCAD.

The third type of operation loads standard forms. Standard forms are predesigned structures that are not yet generated automatically. These include towers, pyramids and all of the Platonic, Archimedean and Johnson polyhedra. The standard forms are simply TekCAD structures placed in a specified location in the computer's file system. TekCAD users can create their own forms and add them to the standard form directories. They are then reflected in the standard form popup menus.

Unlike language-driven production systems, TekCAD does not build structures from a script. This level of specification may be included through synthesis operations added to the system in the future, but for now it is felt that non-graphical structure development is difficult for untrained designers to understand. TekCAD is oriented toward designers that work on their structures visually, building them up interactively.

Edit Operations
Like a word processor or a graphics tool, TekCAD is an editor – but an editor that edits structures instead of documents or images. The Edit operations modify the networks of objects, select sets of objects, and change the characteristics of those objects.

The operations that modify the networks are the simple Cut, Copy, Paste, Undo and Redo that are present in most editing environments. Selected parts of structures may be cut or copied and then pasted into different structures. Any operation that modifies a structure may be undone, or subsequently redone. The visibility and participation of objects in the different views can also be controlled through an Edit operation.

The operations that select objects are provided in addition to the mouse selection facility. Using the mouse, a designer can select and toggle individual objects and groups of objects by clicking and dragging. The select operations modify selections using Boolean logic and object properties based on everything from color and layer to proximity, angle and length. Very complex selections can be made using these operations – selections that would be difficult to make using only the mouse. These selection operations are also useful for locating objects that are difficult to find through visual inspection alone. Once selections are made, they may be operated on by manipulative operations.

The remaining Edit operations deal with coloring the selection, setting the selection's layers, and splitting, merging and locking objects.

Transform Operations
The Transform operations comprise some of the most powerful operations in TekCAD. They operate on selections or (in absence of a selection) the entire structure. Generally, there are simple transformations, advanced transformations and symbolic transformations.

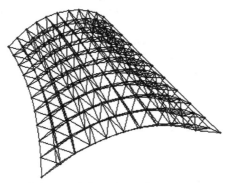

Figure 6. A curved, flaired spaceframe arch created using the Shape Transform Operation

The simple transformations are Move, which moves a selection from one location to another; Scale, which linearly scales a selection from a reference point; Rotate, which rotates a selection around a reference axis; and Mirror, which mirrors a selection through a reference point, line or plane.

The advanced transformations are Shape, which uses a Bezier patch to shape the selection (an example of a structure created using the Shape operation is shown in Figure 6 and was produced in less than 30 seconds); the Align Line and Align Plane operations, which align a selection to a line or plane; Stretch, which rotates and scales as a single operation; and Skew, which linearly adjusts a structure based on a skewing direction and a distance from a reference line.

The symbolic transformations bring real mathematical power to the design process. In these operations formulae with symbolic variables are specified to transform the selections. Coordinate transformations are done using the Set operation in which complicated formulae may be applied to existing structures to quickly derive new structures that would be difficult (at best) to create in traditional CAD systems. An example of a structure created using the Set operation is shown in Figure 7, produced in less than 20 seconds. The Duplicate operation distributes copies of a selection according to a formula specified in Cartesian, Cylindrical or Spherical coordinates. The Twist operation, which is a superset of Rotate, allows a structure to be twisted based on a formula containing radial and axial components. The Transfer operation reprojects structures between points, lines and planes. These four transformations make sophisticated structures relatively easy to produce.

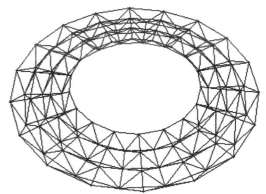

Figure 7. A spaceframe ring created using the Set Transform Operation

Model Operations

Building scale models is important for both understanding the structures being created and communicating them to others. TekCAD contains operations to create Bill of Materials and Strut Cutting Schedule reports for scale modeling. These reports are produced in HTML[7] and are generic, but provide precisely the information needed to build models using TekKit, TekStar's model building system.

Producing reports in HTML is an important consideration. In a world where web browsing has become widespread, using the browser to display and print reports is a powerful synergy. Rather than spend time rewriting existing applications, it was decided that leveraging universally available technology is a better alternative. When reports and images that are rendered in universal formats may be displayed and printed by anyone with a browser - which today *is* everyone - the sensibility of the choice is obvious.

Information Operations

Getting information about the structure while it is being manipulated is critical during design. TekCAD includes operations that provide real time feedback, create reports and perform conversions and calculations with specialized software tools to help the user make better design decisions.

The feedback operations include getting the overall dimensions of selections, measuring the angles between struts and measuring distances within the structure. The reporting operations include generating full-scale Bill of Materials, Geometric Summary and Component Detail reports in HTML.

The calculations and conversions are handled by Tools, persistent calculators that display positions, lengths and angles in different units, continuously display current dimensions, determine angles and lengths using trigonometry and determine distances from planes and lines. In many situations, using these tools can eliminate the need for electronic calculators.

Help Operations

TekCAD includes extensive local and online documentation including a complete User's Reference Guide, Index of Operations, and step-by-step examples that include tips and tricks for building complex structures and using the product more effectively. There is also access to TekStar's TekNotes, a compilation of useful mathematic, design and construction information. The documentation is written in HTML and is extensively hyperlinked.

UNIQUE FEATURES

Although TekCAD has the advanced features needed to very quickly create and manipulate space structures, it also has several features that set it apart from existing tools. The three features that stand out are Mathematical Expression Evaluation that lets any numeric (and in some cases, formulaic) expression be solved directly; Overlaid Geometric Display that shows you what the structure is and what it will be as it is modified; and the Web-Based Context-Sensitive Help System that, from within TekCAD, directly navigates to appropriate documentation in a browser when help is requested during the execution of an operation.

Mathematical Expression Evaluation

The advent of electronic calculators has had a dramatic influence on the design, architectural and engineering professions. Being able to carry a calculator in one's pocket has fostered a generation that has never seen slide rules and has not experienced the dependence on rules-of-thumb that once were the only ways to complete large designs. TekCAD goes one step further in this progression by building a calculator into every input field in its interface that accepts a number.

Instead of going to a calculator, in TekCAD the designer can enter mathematical expressions directly into numeric fields. A full complement of arithmetic operators is provided, as well as physical constants and over 30 standard mathematical functions - including conversions, algebraics, trigonometrics and hyperbolics. If the expressions are not parenthesized, standard precedence rules will be used to find their results.

In some of the operations (most notably the symbolic transformation operations) more than just numerical expressions may be entered. In these operations, mathematical formulae are

Figure 8. Grid (a) initially, (b) with a sinusoid added, and (c) an exponential added

used. In the symbolic transformation operations, these formulae are evaluated at each hub location (as was done in Figure 8). Variables for each coordinate in each of the Cartesian, Cylindrical and Spherical coordinate systems are available for use as variables in the formulae. In other such operations, different variables are available. The context of the operation determines which variables are allowed.

By using these formulaic manipulations, remarkable structures may be created very quickly. For instance, the structure in Figure 9 was created by generating a grid, distorting it using the Set operation in cylindrical coordinates, mirroring it using the Mirror operation and finally curving it with an exponential, again in the Set operation. The entire design was made interactively and took less than two minutes. These sorts of manipulations can take much more time to complete in traditional CAD programs. In TekCAD, by specifying a few simple equations in a handful of operations, elegant space structures may be developed through creative experimentation.

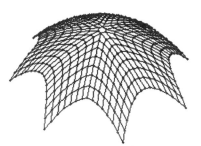

Figure 9. The outer hull of an exponential enclosure

Overlaid Geometric Display
When modifying structures in traditional CAD systems, the designer must typically apply changes in a vacuum. As the change proceeds, the structure is either modified immediately to reflect the current state of the change, or, worse, the change does not take place until the command is completed. TekCAD opts for a different strategy. When modifying a structure in TekCAD, the designer sees the existing selection and an overlaid version of what the selection will become. This overlay gives the user the context needed to work more quickly and with more assurance.

For example, in Figure 10 a Shape operation is being done: a flat spaceframe is being curved in the vertical direction. As the operation proceeds, the new form is overlaid on the original structure. When values are changed in the operation's interaction window or when control points are moved graphically in a view, the overlay is updated to indicate where the results will be placed when the operation is complete.

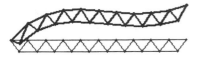

Figure 10. A side view of a flat
spaceframe is being shaped.
The new form is overlaid on the
original as the user interacts with
the structure.

This immediate visual feedback, shown in all of the View Windows, gives the user a precise indication of what is taking place. With this feedback, the user may decide to commit the change, cancel it or continue to change values in the interaction window to achieve the desired result.

Web-Based Context-Sensitive Help System

Today, online help is available in many software packages. But, by using the web browser as the vehicle for displaying help, TekStar has taken the help system to the next level. When help is requested from within TekCAD, a user-assigned preference value is checked; this value specifies whether the documentation on the local disk should be displayed or if the documentation should be accessed through the Internet. Since TekStar continuously updates TekCAD's documentation, the latest revisions are always online. Users that opt to go to the Internet automatically get the most up-to-date documentation, with hyperlinks to new examples, techniques and information. Users may also periodically download the new documentation to their local disk to replace the outdated information there.

Help is not just available through the Help operations. Each operation's interaction window has a Help button that, when pressed, navigates directly to the documentation for that operation, either on the local disk or the Internet, depending on the setting of the preference value. In addition to the Help button, each control is configured for context-sensitive help retrieval. In an interaction window, clicking the mouse on a control's label while holding the keyboard's shift key opens the documentation for the control (again, either from the local disk or the Internet), which explains exactly what the control is for and what values are acceptable.

TEKCAD AND TEKKIT

By itself, the TekCAD software is very powerful and can significantly reduce the time and effort needed to generate space structures. With a full set of creation, manipulation and rendering operations, the software allows structures to be built quickly through a modern interface in a logical, straightforward way. Yet it faces the limitations all software faces: the structures it produces are virtual. Although you can examine structures within TekCAD quite easily, you cannot hold those structures in your hand.

TekStar's companion TekKit model building system complements the software nicely by enabling users to quickly build scale models of the structures that they have created in TekCAD. These models aid in verifying, visualizing and presenting these structures. TekKits consist of reusable plastic hubs and struts that can be joined together to match the geometries created in the program. The patented hubs come in four varieties: five-, six-, eight- and nine-legged. They snap into the struts, which the user cuts to length according to the scaled model reports generated by TekCAD's Model operations. The struts are available in thirteen colors (including glow-in-the dark) and may be painted if desired. To quickly cut struts to required lengths, TekStar has also developed *Le Guillotine*[8], a high precision cutting tool.

Starnet International Corporation[9] of Longwood, Florida, a specialty structures fabricator and early adopter of TekCAD and TekKit, has successfully used the software to design space structures and has created models of these structures for architectural presentations. The ability to quickly conceive both traditional and contemporary architectures using TekCAD has expanded the design capabilities of the company. Being able to take a physical model of a complex structure to its clients has smoothed the way for receiving engineering, fabrication and construction contracts.

At the Aviation Division of HOK[10], a leading worldwide architectural firm, TekCAD was used in the schematic design phase to explore and generate very large structures. These designs are quite elegant expressions of organic structural forms.

As space structures increasingly become part of the designer's collection of solutions, TekCAD and TekKit bridge the gap opened by deficiencies in traditional computer-aided design and modeling tools. TekStar has anticipated the needs of this emerging marketplace, offering a significant step forward on the path to bring visionary structures into the mainstream. More improvements are being planned that will fit directly into TekCAD's operational framework, allowing the system to grow as the works of architects become ever-more sophisticated. The authors sincerely hope that with TekCAD, the real advent of truly elegant structures is finally arriving.

REFERENCES
1. TekStar International Corporation – Where Visions Take Shape, http://www.tekstaronline.com, TekStar International Corporation, Longwood FL, 2002.
2. TekCAD 2002, http://www.tekstaronline.com/products/tekcad.html, TekStar International Corporation, Longwood FL, 2002.
3. TekKit Products, http://www.tekstaronline/products/TekKit.html, TekStar International Corporation, Longwood FL, 2002.
4. LiveGraphics3D Homepage, http://wwwvis.informatik.uni-stuttgart.de/~kraus, Martin Kraus, Stuttgart, Germany, 2001.
5. Virtual Reality Modeling Language - VRML 97 Specification, http://www.web3d.org/technicalinfo/specifications/vrml97, The VRML Consortium Incorporated, 1997.
6. AutoCAD 2000 DXF Reference, htttp://www.autodesk.com/techpubs/autocad/acad2000/dxf, Autodesk Corporation, San Rafael CA, 2000.
7. Hypertext Markup Language – HTML Home Page, http://www.w3.org/MarkUp, W3C, 2002.
8. Le Guillotine, http://www.tekstaronline.com/products/guillotine.html, TekStar International Corporation, Longwood FL, 2002.
9. Starnet – Covering Space Creatively, http://www.starnetint.com, Starnet International Corporation, Longwood FL, 2002.
10. HOK, http://www.hok.com, Hellmuth, Obata & Kassabaum, Inc., Saint Louis MO, 2002.

Compliance of Hollow Nodes

B.K RAGHU PRASAD, Professor, Dept. of Civil Engineering, Indian Institute of Science, Bangalore, India, and
M. PRADYUMNA, Associate Director, Geodesic Techniques, Bangalore, India (former Research Scholar, Dept. of Civil Engineering, Indian Institute of Science, Bangalore, India)

Abstract

A 3D finite element analysis of hollow spherical and octahedral nodes is presented. ·The stiffness matrix of the nodes has been derived numerically for different sizes. A new method of incorporating the stiffness matrix in the regular analysis of the space truss has been developed. The new method proposed yields realistic values for the forces in the members and takes into account the elastic deflections or in other words the compliance of the node. The implementation of the proposed method has been carried out by writing the custom program using the state of the art object oriented programming technique. A sample problem has been analysed and the influence of the flexibility or compliance of the joint or node has been obtained.

Introduction

Space frames of the most widely used configurations of double layer grids (square on square and square on diagonal) typically have eight members attached to each node. There are four members in the chord plane, which occupy the right angle slots around a circle. Each of the other four members is inclined to the chord plane at 45°. The projections of these members on the chord plane are again along four directions, which are similarly displaced around in a circle. The chord members and projections of the web members are offset by 45° in plan. Two types of hollow nodes have been studied viz. truncated octahedral nodes (referred as "Octa" node) and hollow spherical nodes. Figs. 1 & 2. show the finite element models of a typical Octa node and Spherical node with holes for this configuration.

Each FE model is made of several hexahedron elements. At each corner of these elements are present, what are generally called as nodes. However, in this paper these nodes shall be referred to as FE nodes, to avoid confusion with the space frame node, which is the subject of study. The models, in the present case, refer to *space frame nodes* and these are always referred to as 'nodes' throughout this report.

If the node were to be assumed to behave in a totally flexible manner as if it is a hinge there would be no moment transmitted through the joint. On the other hand, moments would be fully transmitted through the joint if it were to be infinitely rigid. In the former case, the structure would deflect more than that in the latter case. The joint, in practice, is actually in between these two extremities and can be classified as being semi-rigid. However, it is necessary to determine how semi-rigid it is or in other words, it is required to assess the flexibility of the node.

The maximum forces are transmitted through a node when it is considered rigid. Hence, for the design of members this assumption would lead to a safe and conservative design.

However, assumption of rigid joints would lead to an underestimate of the deflections in the structure.   From serviceability point of view, it may become essential to estimate the deflections more precisely.   Further, it also overestimates the natural frequencies.   It also leads to erroneous conclusions regarding the fracture behaviour.   Thus, it becomes essential to evaluate the degree of flexibility offered by the joints in a structure. The literature pertaining to the above is reviewed (Refs 1 – 9) and it is observed that a simple method of including the compliance of a node in the behaviour of the space structure is not available.

Therefore an elegant new method is proposed to obtain the flexibility of a space frame joint and apply it as part of the analysis of a space frame.   While the method proposed is quite general in nature, it is illustrated using a node in the square on square configuration of double layer grids.

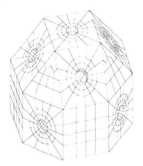
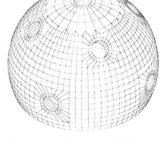

Fig. 1 Octa Node – FEM model   **Fig. 2 Spherical Node: FEM model**

A typical node of the square on square configuration, as already mentioned, has eight members attached.   These members are symbolically shown as arrows in the following Figs. 3 & Fig. 4 Member local axes and global axes (Typical)

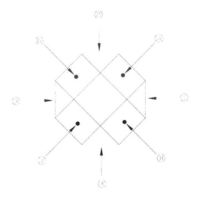

Fig 3 Directions of Positive axial forces

Fig. 4 Member local axes and global axes (Typical)

Stiffness Matrix of the Node

The stiffness equation of the node with the sign conventions as shown in above figure and directions can be written in the matrix form as

$$\{P\} = [J]\{\Delta\}$$

------------------------ (1)

Where, $\{P\}$ and $\{\Delta\}$ are column matrices of forces and displacements respectively. The stiffness matrix of the joint is represented by $[J]$. Since there are 8 members at each node and considering 6 forces (3 translational and 3 rotational) from each member there are 48 forces acting on the node. Hence, the column matrices have 48 rows and the stiffness matrix is a square matrix of order 48.

The stiffness matrix of the joint can be obtained by modeling and analyzing the joint in 3D to obtain the reactions when the joint is subjected to unit displacement in each of the 48 directions. In practice, symmetry reduces the actual number of analyses required. The reactions from each analysis yield one column of the stiffness matrix. The total stiffness matrix can thus be assembled easily.

In any double layer grid, it is well established that axial forces are dominant in comparison to the shears and moments (Ref 8). Hence, it is normal practice to analyze double layer grids as space trusses rather than space frames. Keeping this in mind, the 48 directions for the forces on the joint are renumbered as follows.

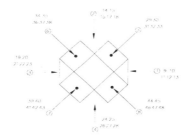

Fig. 5 Numbering of Forces on a joint

The axial force directions of all the members are numbered first from 1 to 8. Then, at each member end the y-shear, z-shear, torsion, y-moment and z-moment are numbered in that order taken one after the other. The full numbering is shown in Fig. 5 above. With this numbering scheme, the stiffness matrix will contain the effects due to the axial forces in the top left 8 x 8 sub-matrix.

The joint when modeled for analysis actually has a bolthole corresponding to each member direction. The model is made of 8-noded brick elements, which have 3 degrees of freedom at each of the corner nodes. All the nodes around the boltholes are held in all the three co-ordinate directions. Local coordinate directions are defined at each of the web member locations to correspond with the local co-ordinate axes of the corresponding member already adopted. The reactions obtained from any analysis will thus correspond with the local axes defined earlier. Unit displacements in the three linear directions at each member end can be simulated by applying the same to each of the finite element nodes around the corresponding bolthole. Since there are only 3 degrees of freedom at each finite element node, reactions are obtained in only these three directions. The moments (torsion and moments about Y and Z axes) can be obtained by multiplying the reactions by the corresponding lever arms at each FE node from the center of the bolthole and summing.

Observing symmetry it is necessary to analyze for the following displacement directions only:

1, 9, 10, 11, 12, 13 => for the chord directions

5, 29, 30, 31, 32, 33 => for the web directions.

This leads to a set of 12 static linear analyses. With this set of analyses results, the entire stiffness matrix of size 48 x 48 can be assembled by inspection.

This stiffness matrix represents the total behaviour of the joint when subjected to the 48 forces applied at the boltholes corresponding to the eight member axes. This is not the complete stiffness matrix of the joint since the effects at other FE nodes in the model are not listed. This stiffness matrix is a condensed form of the total stiffness matrix.

However, since these joints are used in truss analyses, there are only eight axial forces to contend. The stiffness matrix is to be condensed to an 8x8 matrix using the method of static condensation as described below (Ref 7).

Static condensation
The Stiffness method of analysis can be represented in its most general form as

$$[P] = [K][\Delta]$$

------------------------- (2)

Where [P] represents a column matrix of the applied forces, [K] represents the global stiffness matrix of the structure and [Δ] the column matrix of displacements caused by the loads [P]. The above equation can be partitioned into two sets of equations in the displacements. The method of static condensation is a numerical way of representing all the displacements in terms of only the first set of displacements. This is accomplished as follows:

$$\begin{bmatrix} P_1 \\ \hline P_2 \end{bmatrix} = \begin{bmatrix} K_1 & K_2 \\ \hline K_3 & K_4 \end{bmatrix} \begin{bmatrix} \Delta_1 \\ \hline \Delta_2 \end{bmatrix}$$

------------------------- (3)

The above partitioned equations can be written as

$$P_1 = K_1\Delta_1 + K_2\Delta_2$$
$$P_2 = K_3\Delta_1 + K_4\Delta_2$$

------------------------- (4)

The second equation can be rewritten as

$$\Delta_2 = K_4^{-1}\left(P_2 - K_3\Delta_1\right)$$
-------------------------- (5)

Substituting this in the first equation of (4), we have

$$K_1\Delta_1 + K_2K_4^{-1}\left(P_2 - K_3\Delta_1\right) = P_1$$
-------------------------- (6)

Collecting similar terms the above can be rewritten as

$$\left[K_1 - K_2K_4^{-1}K_3\right]\Delta_1 = P_1 - K_2K_4^{-1}P_2$$
-------------------------- (7)

The terms enclosed in square brackets in the above equation is called as the equivalent condensed stiffness matrix and the equation itself represents a condensed form of the stiffness equation (2). Thus, equation (7) can be rewritten as

$$\left[K_{eq}\right]\left[\Delta_1\right] = \left[P_1\right] - \left[K_2\right]\left[K_4\right]^{-1}\left[P_2\right]$$
-------------------------- (8)

In the present case P_2 is equal to zero since only axial forces are being considered and Δ_1 represents the 8 degrees of freedom in the 1 to 8 axial directions of the members meeting at a node. Thus, equation (8) is written as

$$\left[K_{eq}\right]\left[\Delta_1\right] = \left[P_1\right]$$
-------------------------- (9)

Matrix K_{eq} is designated as matrix **J**, which represents the condensed stiffness matrix of the joint. Matrix Δ_1 is designated as Δ_j denoting that this is the response of the joint. In addition, matrix P_1 is designated as **P**. Thus, equation (9) takes the following general form

$$P = J.\Delta_j$$
--------------------------(10)

Here, **J** is the condensed stiffness matrix of size 8x8 and

$$P = \begin{bmatrix} P_1 \\ P_2 \\ P_3 \\ P_4 \\ P_5 \\ P_6 \\ P_7 \\ P_8 \end{bmatrix} ; and \quad \Delta_j = \begin{bmatrix} \Delta_1 \\ \Delta_2 \\ \Delta_3 \\ \Delta_4 \\ \Delta_5 \\ \Delta_6 \\ \Delta_7 \\ \Delta_8 \end{bmatrix}$$
--------------------------(11)

Where P_1 to P_8 represent the axial forces from the members with the sign convention already mentioned and Δ_1 to Δ_8 are the corresponding displacements in the node.

In any space truss analysis, the members exert axial forces only on the nodes. Thus, any node is subjected to axial forces from the members connected to it. The response of the node is according to the stiffness equation given by equation (10). In this equation [P] and [J] are known and [Δ_j] can be solved for easily using the following relationship.

$$\left[\Delta_j\right] = \left[J^{-1}\right]\left[P\right]$$
--------------------------(12)

The values obtained in [Δ_j] are dependent on the axial forces from the members, which are in turn evaluated from the displacements at the joints of the structure. Hence, there is no direct way of obtaining the true displacement response of the nodes and the corresponding member forces. In addition, the response of the node in the axial direction of any member is dependent on all the forces acting on the node and not only on the force from that member. Hence, an iterative approach is necessary to obtain the actual nodal responses and member forces causing these.

Member Stiffness including Node Flexibility

To assimilate the response of the node into the structure, each member can be visualized as having a spring at either end. The stiffness values of these springs are equivalent to the stiffness of the nodes at each end of the member.

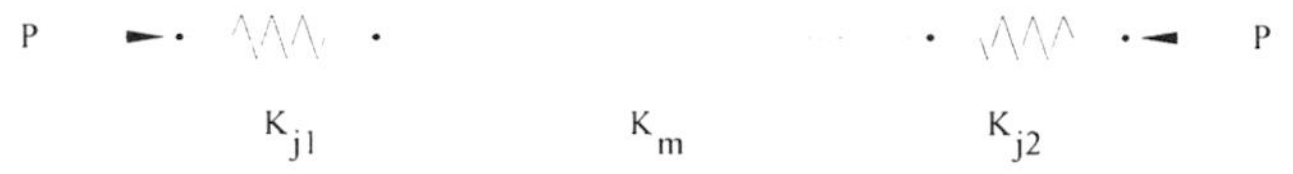

Fig. 6 Member idealization

Referring to Fig 6 the stiffness of the member is designated as K_m and the spring stiffness as K_{j1} and K_{j2} respectively. The Δ_j obtained earlier give the responses of each of the springs in Fig. 6. Under the action of the member force P, the stiffness of the two springs can be found as

$$K_{j1} = \frac{P}{\Delta_{j1}} \quad and \quad K_{j2} = \frac{P}{\Delta_{j2}} \quad\quad\quad\text{------------------------(13)}$$

The stiffness of the member is already known as $K_m = E.A/L$. The equivalent stiffness of the member in the presence of the two springs can be derived as follows.

The stiffness of the two springs and member stiffness are in series and the equivalent stiffness for this is given by

$$\frac{1}{K_{eq}} = \frac{1}{K_{j1}} + \frac{1}{K_m} + \frac{1}{K_{j2}} = \frac{K_m.K_{j2} + K_{j1}.K_{j2} + K_{j1}.K_m}{K_{j1}.K_m.K_{j2}} \quad\text{------------------(14)}$$

This can be rewritten as

$$K_{eq} = \frac{K_{j1}.K_m.K_{j2}}{K_m.K_{j2} + K_{j1}.K_{j2} + K_{j1}.K_m} = \frac{K_m.(K_{j1}.K_{j2})}{K_m(K_{j1} + K_{j2}) + (K_{j1}.K_{j2})} \quad\text{------(15)}$$

Taking $(K_{j1}.K_{j2})$ to the denominator equation (15) can be rewritten as

$$K_{eq} = \frac{K_m}{K_m.\left(\dfrac{K_{j1} + K_{j2}}{K_{j1}.K_{j2}}\right) + 1} \quad\quad\quad\text{------------------------(16)}$$

Equation (16) can be rearranged and simplified as

$$K_{eq} = \frac{K_m}{1 + K_m\left(\dfrac{1}{K_{j1}} + \dfrac{1}{K_{j2}}\right)} \quad\quad\quad\text{------------------------(17)}$$

The stiffness of the member can now be modified as an equivalent stiffness as given in equation (17). This can now be substituted back into the analysis routine to obtain fresh values for member forces. With the new member forces, Δ_j, the responses of each of the nodes are recalculated from equation (12). These values are then used to recalculate the stiffness contributions to each member using relations in (13) and the equivalent stiffness of each member is obtained using equation (17). These steps are carried out until the differences in deflections in the structure at two consecutive iterations are all within specified tolerances.

Summary of Method to Incorporate flexibility of Hollow nodes

The method of incorporating the flexibilities (or stiffness!) of the nodes into the structural analysis described above can be summarized as:

1. The space truss analysis is carried out at the outset assuming rigid joints and the member forces are obtained.

2. At each node of the structure, the member forces acting on it are resolved as joint forces using the sign convention that forces acting towards the joint are positive and forces away are negative.

3. Using the stiffness matrix of the node **J** (which is already known from separate detailed analyses) the nodal responses, Δ_j are found from the expression $\Delta_j = J^{-1}.P$. Here **P** is the matrix of forces on the node obtained in step 2 above.

4. Each member is connected to two nodes. The nodal responses of these two nodes in the direction of the member have been obtained separately in step 3 above. These nodal responses are converted into stiffness values using the expressions

   $K_{j1} = \dfrac{P}{\Delta_{j1}}$ and $K_{j2} = \dfrac{P}{\Delta_{j2}}$. K_{j1} and K_{j2} represent the stiffness of the two nodes at either end of a member and P is the axial force in the member. Δ_{j1} is the nodal response of the node at one end in the direction of the member and Δ_{j2} is the corresponding nodal response at the other end of the member.

5. The stiffness of each member is now replaced by an equivalent stiffness to include the effect of the nodal flexibility (or stiffness!). The equivalent stiffness is obtained from equation $(2 - 17)$.

6. The global stiffness matrix is now reassembled with the revised stiffness values of the members, and analysis of the structure is carried out as before.

7. The deflections obtained in the structure at the end of this analysis are compared with the deflections in the previous analysis run.

8. If the differences in the deflections are within specified tolerances, the iteration is stopped. Otherwise steps 2 to 7 are repeated as another iteration until the above conditions are satisfied.

9. After obtaining the correct global stiffness matrix, when the deflections from two successive iterations are close (within specified errors), the analysis proceeds to calculate the reactions and forces in the members like in any other regular analysis.

Standard Nodes and Sizes

The stiffness matrix for several standard nodes have been evaluated. The Octa and spherical nodes selected are given below in Table 1

Table 1 Geometry of Hollow Nodes

Sl. No	Outer Dia	Thickness	Hole Dia	Bolt Size
1	110	10	13	M-12
2	120	10	13	M-12
3	135	15	17	M-16
4	150	15	17	M-16
5	170	25	21	M-20
6	210	25	25	M-24

The diameters of the holes shown in the table are the same for all four chords and four webs in the particular node.

Condensed Stiffness Matrices of nodes:

The stiffness matrix for each of the nodes selected above has been obtained by conducting a series of static linear analyses using a commercially available software, NISA developed and marketed by EMRC, USA.

Each of the stiffness matrices has been condensed to a size of 8x8 using the method of static condensation as described. The condensed stiffness matrices for each of the nodes are given below for Octa nodes in Table and for Spherical nodes in Table 3. All the nodes are for the square on square configuration.

Table 2 Octa 110/10 – Condensed Stiffness Matrix

	1	2	3	4	5	6	7	8
1	300384.99	154810.03	102757.31	154810.03	-98700 .62	98598 .84	98587.79	-98458 .32
2	154810.03	300384.99	154810.03	102757.31	-98458 .32	-98700 .62	98598.84	98587 .79
3	102757.31	154810.03	300384.99	154810.03	98587 .79	-98458 .32	-98700 .62	98598 .84
4	154810.03	102757.31	154810.03	300384.99	98598 .84	98587 .79	-98458 .32	-98700 .62
5	-98700 .62	-98458 .32	98587 .79	98598.84	324004 .08	-128274 .15	-66858 .32	-128274 .15
6	98598.84	-98700 .62	-98458 .32	98587.79	-128274 .15	324004 .08	-128274 .15	-66858 .32
7	98587 .79	98598 .84	-98700 .62	-98458.32	-66858 .32	-128274 .15	324004 .08	-128274 .15
8	-98458 .32	98587 .79	98598 .84	-98700.62	- 128274 .15	-66858 .32	-128274 .15	324004.08

Table 3 Spherical 110/10 – Condensed Stiffness Matrix

	1	2	3	4	5	6	7	8
1	276574	145591.6	81234.49	145591.6	-97732	97892.13	97763.06	-97768.9
2	145591.6	276574	145591.6	81234.49	-97768.9	-97732	97892.13	97763.06
3	81234.49	145591.6	276574	145591.6	97763.06	-97768.9	-97732	97892.13
4	145591.6	81234.49	145591.6	276574	97892.13	97763.06	-97768.9	-97732
5	-97732	-97768.9	97763.06	97892.13	316687.7	-122841	-73433.7	-122841
6	97892.13	-97732	-97768.9	97763.06	-122841	316687.7	-122841	-73433.7
7	97763.06	97892.13	-97732	-97768.9	-73433.7	-122841	316687.7	-122841
8	-97768.9	97763.06	97892.13	-97732	-122841	-73433.7	-122841	316687.7

An object-oriented program has been written using Delphi's Object Pascal to analyze space trusses and the proposed method of incorporating joint flexibility has been implemented. (Delphi is a proprietary software tool available from Inprise/Borland for development of software on the Windows platform. Delphi is the successor of the popular Turbo Pascal.)

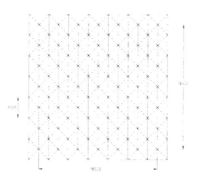

Fig. 7 Sample space frame

Table 4 Nomenclature for Space frame supports

Legend	Description
Sq 4	4 sides Supported
Sq 4 J	4 sides Supported with Joint flexibility
Sq 2	2 sides Supported
Sq 2 J	2 sides Supported with Joint flexibility
Sq C	Corner supports
Sq C J	Corner supports with Joint flexibility

A sample space frame of the square on square configuration has been selected to check the effect of including the joint flexibility in the space truss analysis. The sample space frame in Fig 7 has a span of 18 m both ways and three support conditions were selected with identical loadings in all cases. The first load case had all four bottom-edges supported, the second had two opposite edges supported and the third had only the four corners of the bottom grid supported. The Table 5 below shows the results of analyses with and without joint flexibility. Table 4 shows the explanation of the nomenclature used in Table 5.

Table 5 shows the maximum deflections obtained from analysis of the sample space frame using rigid joints and flexible joints separately for several support conditions. The percentage change in maximum deflections on using flexible joints is also shown.

Table 5 Maximum Deflections in mm in Space frames

	Sq 4	Sq 4 J	% Change	Sq 2	Sq 2 J	% Change	Sq C	Sq C J	% Change
Max +ve X	0.24	0.31	25.60	0.43	0.59	37.16	0.71	0.94	31.90
Max +ve Y	0.36	0.42	16.68	0.27	0.37	34.93	1.49	1.70	14.01
Max +ve Z	0.15	0.18	19.19	0.21	0.31	47.97	1.32	1.55	17.69
Max –ve X	-0.24	-0.30	25.76	-0.43	-0.60	37.08	-0.67	-0.87	29.26
Max –ve Y	-0.36	-0.41	16.52	-0.27	-0.37	34.14	-1.49	-1.70	13.99
Max –ve Z	-1.35	-1.59	17.61	-2.63	-3.45	31.25	-6.95	-8.18	17.76

Conclusions:

A new method of incorporating joint flexibility or compliance in to the space frame analysis is presented. The important feature of the method proposed is that the size of the structure stiffness matrix is not changed due to the incorporation of flexibilities of the nodes. From the sample analysis done on a chosen space frame it is observed that there is considerable increase in the deflections of the space frames due to joint flexibility. The percentage increase could be nearly 50% depending on the overall size of the frame and boundary conditions.

References:

1. Abdalla, K.M. and Chen, Wai-Fah., Expanded Database of Semi-rigid Steel Connections, Comp. & Structures, Vol 56, No 4, 1995.

2. Abu-yasein, O.A. and Frederick, G.R., Analysis of Frames with Semi-rigid Joints, Computers & Structures, Vol 52, No 6, 1994.

3. Analysis, Design and Realization of Space Frames-A state-of-the-Art Report, IASS working group on Spatial Steel structures, Bulletin of IASS, Vol XXV – 1/2, 1984

4. Anantharaman. S and Pradyumna, M., Design and construction management features of the 'Barrel Dome' for the indoor stadium at Koramangala, The Indian Concrete Journal, Vol. 71, 6, 1997.

5. Archer, G.C., Fenves, G. and Thewalt, C., A new object-oriented finite element analysis program architecture, Computers & Structures, Vol 70, 63-75, 1999.

6. Attiogbe, Emmanuel and Morris, Glenn., Moment Rotation Functions for Steel connections, J of Str Engg, ASCE, Vol 117, No 6, 1991.

7. Cook, Robert D., Concepts and Applications of Finte Element Analysis, Second Edition, John Wiley & sons, 1981.

8. Makowski, Z.S., Shaping the future in space structures and their impact on Architectural and Structural Engineering, Proceedings of IASS symposium on Innovative Applications of Shells and Spatial Forms, Bangalore, India. 1988.

9. Murtha-Smith, Erling, Hwang, Sun-Hee and Bean, John E., Load Transfer in a Space Frame Connection, Int J of Space Structures, Vol 7, No 3, 1992.

The envelopes of the arts centre in Singapore

H. KLIMKE, J. SANCHEZ, M. VASILIU, W. STÜHLER, C. KASPAR.
MERO GmbH & Co. KG, D-97064 Würzburg, Germany

SUMMARY
This paper describes the planning and realisation of the envelopes for the Lyric Theatre and Concert Hall of the Arts Centre in Singapore, called The Esplanade Theatres. Starting with the geometry of the free form surfaces, the computer aided design process and the realisation by means of a customized building system is described.

The large variety of structural elements (members and nodes, glass panes and shadow panels) could be reduced by the application of complex, but rational geometric rules, to make the logistic of planning, fabrication, transport and erection feasible.

INTRODUCTION
At a site between Marina Centre and Marina Bay in Singapore, the arts Centre is progressing towards completion. The design is based on an earlier concept of Architects Michael Wilford and Partners in London and is executed by DP Architects (Vikas Gore and Pietro Stallon in charge) in Singapore. Significant characteristics are the envelopes of the Lyric Theatre and the Concert Hall (fig. 1).

The structures behind the envelopes are custom designed space trusses, providing triangular top chord grids for adjustment to the free form surfaces.

The cladding system comprises triangular panes of insulated glass and an arrangement of aluminium shading elements, giving the observer varying impressions of transparency and opaqueness [1].

In contrast, the roof cladding is opaque, consisting of a foil water barrier, which is covered by aluminium panels with open joints, similar to the façade shadow panels, but nearly flat. These panels also cover the deep gutters, which are positioned between each facade and roof structure.

Fig. 1.

The major challenges of the realisation have been

- the iterative adjustment process of the shading panels to meet the aesthetical and functional demands,
- the design, production, transport and erection of a very great number of similar, but different building components, and
- the coordination of an unusually large number of consultants, institutes and subcontractors.

THE GEOMETRY

Geometry on its own does not create a building, but without geometry the Arts Centre envelopes would never have become reality. Although dome and membrane structures are extreme examples for applied geometry, the practicability of many different structures depends on geometrical knowledge. Geometry played a major role from the design to the fabrication of the Arts Centre [7].

'Formfinding'

The surfaces of both envelopes are NURBS, which stands for 'Non Uniform Beta Splines', a mathematical description for free form surfaces. Stimulation for the development of NURBS came from the ship building, automobile and airplane industries. Coons and Bezier developed the theoretical basis for the implementation into CAD programs, which simplifiy the application. Parameters and equations were substituted by 'weights' and 'control points' and their influence on the form of the surface can be controlled by graphic representation.

With this technology available, the Arts Centre's envelopes were designed with the CAD program Microstation by 'Atelier One' in London, which were consultants to DP Architects. Four sided surface areas (Coons patches) were generated in a mesh of spline curves, which could be modified individually without changing the whole mesh.

Members and Nodes

After deciding on the surface form, a net of members and nodes could be generated. A method was applied, that is known from the generation of cable nets. A square net of constant 1,5 m module was 'laid' on the surface in such a way, that only the nodal points meet the surface and the connecting members remain straight.

The difference between Concert Hall and Lyric Theatre layouts is that the net of the former is orthogonal, while it is diagonal for the latter (fig. 2).

Fig. 2.

Splitting each rhombic configuration resulted in a net of plane triangles, which was required for the support of plane glass panes as well as for the structural stability. However, a square net of members was added 90 cm below the triangular net. Together with diagonal members between both nets, space trusses were generated that enhanced the stiffness without affecting the lightness of the structures (fig. 3).

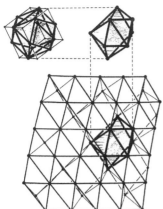

Fig. 3.

Colour Management

For the management of the great number of building components (8.300 nodes, 34.500 members, 10.500 triangular glass panes, 4.900 shading elements and 2.230 roof panels), colour identification marks for the nodes and members were introduced, which were related to their position. By means of this colour-coding it was possible to identify each group of components throughout the design, fabrication and erection.

DESIGN AND CALCULATION

The design and calculation is based on the following steps:
- the bearing concept
- the evaluation of loads
- the calculation of deflections, member forces and reactions
- the dimensioning.

The Support Concept

The space trusses are supported at each second or third top chord node on the concrete edge girders. The upper edges of the space trusses are only supported at some bottom chord nodes on top of concrete columns (fig. 4).

The upper edge bearings differ from the lower edge bearings by
- the realisation of horizontal movements
- the vibration insulation.

Static Bearings

For the stabilisation of the facade structures it was necessary to support the lower and upper edges of the space frames. This was achieved at the lower edges by a full restraint of the bearing points.

However, at the upper edges it was necessary to choose a layout of bearings that allowed for the thermal expansion of the structures; each structure is rigidly fixed to the non resilient stair towers. All other bearing points are supported horizontally only where statically required.

Fig. 4.

The flexibility of the stair towers was considered in the analytical model, as well as the flexibility of the concrete girders at the lower edges and the concrete columns.

Vibration Insulation of the upper Bearings

The purpose of the vibration insulation is to prevent the transfer of body sound from the cladding into the concert and theatre auditorium. A possible solution had to solve the problems of unrestrained support, minimum size and sustainability. Taking the experience with similar applications into consideration, natural rubber was the appropriate solution for the problem. A low creeping value and a moderate increase of stiffness for dynamic loading can be achieved by special rubber compositions.

For the layout of a vibration isolation with rubber cushions, ARTEC Consultants Inc. from New York N.Y. / USA proposed the following design parameters:
- eigenfrequency   < 10 Hz
- maximum strain   < 15 % for permanent and < 45% for temporary loading
- statical deflection ≈ 4 mm.

These parameters were considered in a design concept, that was developed together with Wölfel Consultants from Würzburg / Germany and coordinated with Wilson, Ihrig & Associates from Oakland, CA / USA, specialists for vibration insulation, working for ARTEC.

The biggest problem of the layout of the rubber bearings was the strong interaction between statical and dynamical requirements. Two values of support stiffness were considered to meet the stress depending Young's modulus of the rubber cushions:
- soft support for permanent loads
- stiff support for wind peak loads.

The horizontal bearings are structurally separated from the vertical bearings. As the rubber cushions can only transfer compression, two rubber cushions are required for each bearing. Furthermore, the rubber cushions are prestressed to avoid gaping and to provide an almost constant stiffness under varying loads.

The required surface area of the chosen 2 x 26 mm thick rubber cushions for the vertical bearings was iteratively evaluated on the basis of approx. 4 mm statical deflection, thus keeping the eigenfrequency of the bearings below 10 Hz.

For the horizontal bearings, the required surface of the rubber cushions is derived from the prestress force, which is equivalent to the maximum wind load, equally keeping the eigenfrecency of the bearings below 10 Hz.

Loading
The following loads had to be considered:
- dead loads
- wind loads
- live loads and installations
- thermal loads

Dead Loads (DL)
The dead load of the space trusses was considered automatically by the calculation program (Meroprog). The cladding dead weight is 0,85 kN/m².

Live Load (LL)
Local regulations require the consideration of 0,5 kN/m² live load. Furthermore, an installation load of 0,1 kN/m² was considered.

Wind loads (WL)
Wind tunnel tests were performed at the City University of London in the initial design phase to obtain the wind pressure coefficients for the envelopes.

The pressure tabs were applied to the smooth model skin, equivalent to the real building's glass skin. The shading caps were applied with a separate wire model. The tests were performed for 'caps off' and 'caps on'. For the design calculation, the average values of pressure coefficients from both measurements were applied. The drag forces were calculated from the pressure values.

As this approach left some questions open, the client requested additional tests for clarification. For this reason, the resulting reactions of the envelopes were measured in the wind tunnel of the steel building department at the RWTH Aachen / Germany [4]. Furthermore, the pressure coefficients were measured again. The models from the previous tests could be used again by courtesy of the City University London.

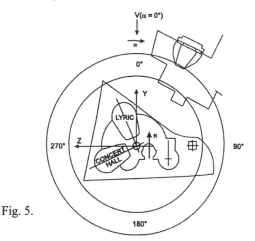

Fig. 5.

Test set up
Fig. 5 illustrates the set up on the turntable, which could be rotated up to 360°. The boundary layer and the natural wind turbulence were simulated.

Measurement of Pressure Values
The measure points, 1 mm bore holes with pressure tabs, were applied to the inside of the model. The position of the tabs is indicated on figures 6 for the theatre and concert hall respectively. The frequency of the measured pressure coefficients is shown in fig.7 for the Lyric Theatre, tab 5 / 150° wind direction.

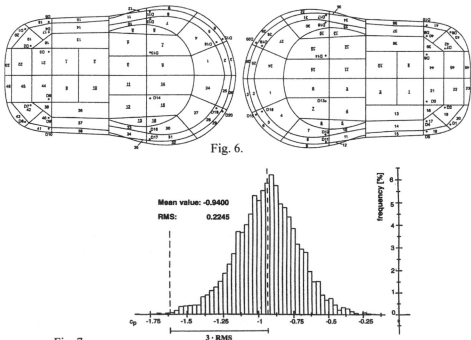

Fig. 6.

Fig. 7.

The foundation loads were measured for wind directions from 0° to 360° in steps of 30°. For comparison, the resulting forces were calculated from the measured pressure values. Measured and calculated results are listed in table 1 for 'caps on', indicating that the calculated values, with consideration of drag forces, are significantly greater than the measured values.

Table 1	Measurement 'caps on'		Calculation			
			no drag		drag	
angle α	force [kN]	direction [°]	force [kN]	direction [°]	force [kN]	direction [°]
0°	924	170	536	178	1332	177
30°	616	262	504	252	1124	222
60°	594	275	550	274	1178	249
90°	1062	293	1024	286	1882	283
120°	581	326	784	301	1700	304
150°	206	302	228	282	1079	321
180°	398	8	194	54	923	6
210°	1127	22	387	37	1051	28
240°	827	41	567	34	1259	44
270°	840	53	598	52	1294	80
300°	1156	78	1005	79	1737	101
330°	568	108	836	117	1628	133

Evaluation of Nodal Forces for the Analysis

The nodal forces were calculated by a program from the surface loads, related to
- the loading zones with constant pressures (fig. 6) and
- the effective areas for the nodal loading.

As discussed above (chapter 2.3), the nodes in each loading area were assigned individual colours. Depending on the wind direction (load case), the program then related each colour to a certain value of pressure.

Thermal Loading (TL)
Temperature changes in Singapore are very small. The biggest temperature change would result from the air conditioning and was considered with a temperature difference of $\pm 20°C$.

Static Calculation

Load Combinations

In principle, British Standards have to be applied in Singapore. BS 5950: Part 1: 1985 describes the combination of basic load cases, considering partial safety factors:

1.4 DL + 1.6 LL + 1.2 TL
1.0 DL + 1.4 W + 1.2 TL
1.2 DL + 1.2 LL + 1.2 WL + 1.2 TL

A design wind speed of 34,5 m/sec was calculated from a basic wind speed of 33 m/sec with CP3: Chapter V, Part 2, leading to a design wind pressure of 0,75 kN/m².

The evaluation of wind loads was performed as follows: the wind components orthogonal to the building surfaces were considered for 'caps off' and 'caps on'. Additionally, drag components were calculated for 'caps on'. This resulted in a conservative structural layout as indicated already in table 1.

Twelve wind directions were considered (0° – 330° in steps of 30°) for 'caps on' and 'caps off'. Together with three dead load cases, one live load and two temperature load cases, a total of fifty and two load case combinations were considered. Furthermore, eight erection conditions were calculated.

Dimensioning Concept

The basic calculations and the dimensioning were performed by the above mentioned Mero-prog. This program was developed by Mero for the design of space trusses, covering the steps from the geometry to the numerical controlled manufacturing.

The lateral bending of the top chord members from dead weight and direct support of the glass panes was not considered for the analysis, but for the following dimensioning of the members.

The related procedure is covered by a general approval from the German Building Institute [5]; the concept is adjusted to the German Standard (DIN 18 800), which again is generally equivalent to BS 5950 Part 1 (ultimate limit state design) and, therefore, was accepted in Singapore. The member sizes could be minimized by the iterative dimensioning procedure.

Stability checks were performed for selected load combinations with the optimised structures. The nonlinear analyses were performed with the program STAAD. This program was also used by the professional engineer in Singapore (who prepared all calculations and drawings for the accredited checker), so that only the input data had to be transferred to Singapore.

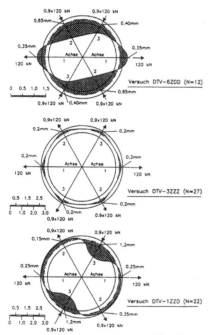

A speciality of any space truss design is the lay-out of the nodes. The main objective is, avoiding collisions between connecting members. A stress check is not possible, however, empiric rules from tests are available for the dimensioning [6].

Two different node types were used for the Arts Centre:

- spherical nodes for the bottom chord

and

- bowl nodes for the top chord.

Tests were performed at the University of Karlsruhe / Germany with bowl nodes. They revealed that the bolts of the connected members failed before the nodes reached the limit state of serviceability (fig. 8). With these results, the check of the nodes could be substituted by stress checks of the connecting bolts.

Results of the Dimensioning

The relevant results from the dimensioning are listed in table 2.

Fig. 8.

Table 2	Results of the Dimensioning (mm)	
	Lyric Theatre	Concert Hall
Circular tubes for bottom	d = 48,3 – 88,9	d = 48,3 – 88,9
Chord and diagonals	95% d ≤ 60,3	98% d ≤ 60,3
Square hollow sections	60 x 60 – 90 x 90	60 x 60 – 90 x 90
for the top chord	98% h = 60 x 60	99% h = 60 x 60
MERO – spherical nodes	d = 110 – 155	d = 110 – 155
	90% d = 110	92% d = 110
MERO – bowl nodes	d = 160; 200	d = 160; 200
	98% d = 160	96% d = 160

CONSTRUCTION

The Steel Structure
DP architects and Atelier One had planned the structures as single layer tubular frames with welded connections. As an alternative, MERO suggested double layer space trusses with bolted connections, which provide high accuracy through machined fabrication and, therefore, could be easily adjusted to the free form surfaces.

This meant that MERO had to take the full responsibility for the cladding design and build.

The Space Trusses
To avoid a secondary support system for the glass panes, the top chord members of the space trusses had to be square hollow sections, which could directly support the glass panes.

Bowl type nodes have been selected for the connection of the top chord members and the tubular diagonal members. To keep the nodes small, special head plates were used for the square diagonal members, which divide the rhombic configurations into triangles (fig. 9).

The space diagonal members and the bottom chord members are made from round tubes and the nodes are forged spheres, corresponding to the design rules of the Mero approval.

All components were produced at the MERO workshops in Würzburg / Germany.

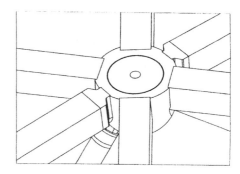

Fig. 9.

The bearings
Corresponding to the support concept the bearings were realised as follows:

- the fixed bearings on the concrete edge girders support the space trusses in the top chord nodes. The bowl nodes were welded to rectangular interfaces, that were guided within rectangular hollow sections, which eventually transfer the support loads to the bearing plates. The chosen layout allowed the compensation of vertical and horizontal tolerances (fig. 10).

- the bearing nodes of the upper edges were realised by cylindrical stubs with hemispherical heads. The cylindrical stubs are guided in tubes for vertical adjustment. The horizontal adjustment was done by moving the tubes on the head plates of the rubber bearings prior to welding (see fig. 4).

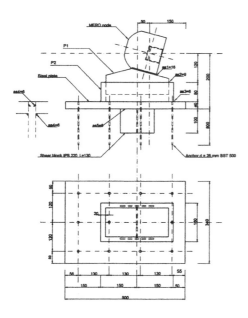

Fig. 10.

The Glazing

Theoretically, only every second of the 10.500 glass panes is identical. By allowing a tolerance of ± 2,5 mm, which could be adjusted within the joints, the number of pane types could be reduced to approximately 1.500.

The layout of the insulated glass is equal for all panes: outside 6 mm fully toughened green glass, followed by 12 mm airspace and inside 2 x 5 mm laminated heat strengthened glass. A low E-layer is positioned on the inside of each laminated pane.

The glass panes are fixed against wind suction by means of aluminium discs at the top chord nodes and by two additional clamps at each top chord member (fig. 11). The clamps are positioned on top of the glass, as the durability of the glass compound could not be guaranteed for clamps applied within the edge sealing of the insulating glass panes.

Special attention was given to the layout of the joints between the glass panes. A drainage system of primary and secondary EPDM gutter profiles guide the water, that may occur from condensation, to EPDM discs, which are positioned on top of the bowl nodes, and further down to the main gutter of the concrete edge girders (fig. 12).

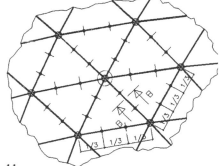

Fig. 11.

The outside sealing was realised by silicon joints with a constant width of 20 mm.

The drainage system can only work if the joints are ventilated. Because of the possible condensation of humid air inside the joints, a ventilation to the outside seemed risky. It could be shown however, that condensation can only appear for very short periods of time.

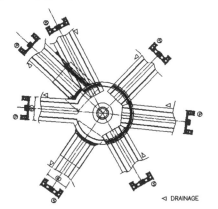

⊲ DRAINAGE

Fig. 12.

Fig. 13.

Shading Panels

The facades of the Lyric Theatre and the Concert Hall are covered by 4.900 panels and the roofs by 2.230 panels. The design proceeded in two steps:

- first step was the iterative optimisation of the rise of each panel, which required the intense cooperation with the architect and the cladding consultant (fig 14),

- second step was the optimisation of panel cutting types, similar to the glass panes. This step produced approximately thirty basic panel patterns and thirty more special patterns for the edge panels.

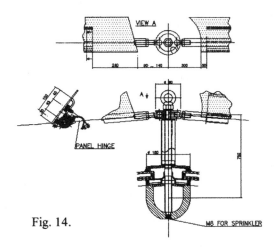

Fig. 14.

The panels consist of 4 mm thick aluminium sheets, which are supported approximately 300 mm above the top chord of the space trusses by means of aluminium tubular frame structures with special connectors (fig.14).

The panels are curved along the short diagonal and are fixed to the tubular frames by means of hinges, so that each panel can be opened in two directions for maintenance.

A certain worry was the possible wind vibration of the shading panels. However, a calculation revealed, that the lowest eigenfrequency of the folded panes was well above 10 Hz, so that a gust excitation of vibrations could be excluded. A high frequency vortex vibration, however, could not be observed during tests performed in Singapore.

THE ERECTION
It was intended to base the erection planning on the sections introduced for the representation of the wind loads and used for the colour the management. The erection would follow the sections from top to bottom and proceed to the horizontally adjacent sections.

Erection Planning
Two independent working groups would start erecting from one end of the symmetry lines. This meant that the scaffolding had to be provided in due time prior to starting the erection in the consecutive sections.

A scaffolding, similar to the locally common bamboo-scaffolding, was chosen to allow adjustment to the hardly predictable requirements of the three dimensional space truss geometry. For that reason, the loads from supporting the space trusses had to be calculated 'real time' with the erection progress and considered for the layout of the scaffolding.

The erection of the cladding system (drainage, glazing and shading) was to follow the same procedure by means of temporary platforms. They were to be provided in horizontal sections consecutively from the upper to the lower edge. For the practical realisation however, it proved favourable to provide the platforms for all levels and several sections simultaneously.

Erection Performance
The erection started with the installation of the bearing plates, including anchor bolts and shear stubs. After fixing the plates with grout and after a final measurement, the bearing components were adjusted and welded to the bearing plates.

Next, some space truss units of 4,5 m x 4,5 m up to a maximum of 9 m were preassembled on ground and lifted on top of the concrete edge girder. This units were stable after connection to the bearings and required no additional support.

After completion of these basic units, the erection proceeded with single members and nodes, first in the horizontal and then in the vertical direction of each section. However, the machined faces of the nodes and members fabricated with intentionally small minus tolerances, together with permanent measurements of the node positions, enabled the erected structure to meet the prescribed geometry.

The erection of the glazing and of the shading panels followed after the completion of two space truss sections. The erection of the drainage system for the façade glazing started from the upper edges with two working groups and proceeded downwards in each section. The EPDM gutter profiles were temporary fixed on top of the RHP members by self taping strips and finally fixed by screws.

The erection of the glass panes followed immediately after the gutter profiles were fixed. The glass panes were delivered in frames, each containing six panes. They were lifted to the erection platforms and distributed to their individual positions on top of the EPDM gutter profiles. After adjustment and horizontal fixation, the joints between the panes were sealed with silicon, before the panes were fixed vertically by means of discs at the nodes and two clamps at each supporting beam.

Fig. 15.

The erection of the shading panels followed the same procedure (fig. 15) and started with the erection of the clamp nodes for the fixation of the tubular frame members. These provide spherical end pieces, which fit into the clamp nodes and allow the adjustment to any angle prior to fixing the clamps. Finally, the shading panels were fixed to the tubular frame members by means of hinges.

CONCLUDING REMARK
The envelopes of the Arts Centre in Singapore have proved, that modular systems together with the application of CAD tools allow the realisation of complex structures without unreasonable restrictions to the architects conceptual design.

PARTICIPANTS

Client:	The Esplanade Company Ltd. Singapore
Architect:	DP Architects Pte. Ltd., Singapore
General Contractor:	Penta-Ocean Construction Co. Ltd., Singapore
Project Manager:	Public Works Department (PWD), Ministry of National Development;

Arts Centre Development Division (ACDD), Singapore

Cladding Consultant:	Atelier One, London / U.K.
Cladding Contractor:	MERO GmbH & Co. KG, Würzburg and
	MERO Asia Pacific Pte. Ltd., Singapore

BIBLIOGRAPHY

[1] GORE, V., "My take on the design", Web page: http://www.vikas-gore.com/design.htm

[2] Engineering design with natural rubber, NR Technical Bulletin, The Malaysian Rubber Producer Research Association, ISSN 0956-3856, 1992

[3] Wind tunnel tests on a Model of the New Arts Centre, Singapore, Centre of Aeronautics, City University London, March, 1996

[4] Windkanaluntersuchungen zur Bestimmung der windinduzierten Fundamentlasten der Hallen des 'New Arts Centre', Singapore, Bericht W 694/1197, RWTH Lehrstuhl für Stahlbau, Windingenieurtechnik, Aachen, November, 1997

[5] Zulassung Z-14.4-10 für das MERO- Raumfachwerk, Deutsches Institut für Bautechnik, Berlin, 1999

[6] Tragfähigkeitsuntersuchungen von MERO- Napfknoten, Bericht Nr. 983008, Versuchsanstalt für Stahl, Holz und Steine, Universität Karlsruhe (TH), 1998

[7] "The geometrical processing of the free-formed envelopes for The Esplanade Theatres in Singapore" by Sanchez,J., Proceedings of the 5th International Conference on Space Structures, Surrey, GB; 2002

Construction of the roof steel structure of the Saitama Super Arena, Japan

Y. MATSUOKA
Nippon Steel Corporation Ltd., Tokyo, Japan

OVERVIEW OF SAITAMA SUPER ARENA

Saitama Super Arena, Saitama Prefecture, Japan, which opened in September 2000, is a multi-purpose facility. It can hold various events, including sports, such as American football and basketball, music events, such as concerts and musicals, and exhibitions, such as trade shows. It is one of the largest facilities in Japan of the kind, with the site area of 45,007.22 m², the building area of 43,730.25 m² and the total floor area of 132,397.75 m². The feature of the building is the "Moving Block," which includes covered seating areas, washrooms and air-conditioning systems. The Moving Block can move 70 m in 20 minutes to convert the facility from the arena mode for boxing and basketball, to the stadium mode for athletic sports and American football. The semi-circular Moving Block is 41.5 m high and weighs roughly 15,000 tons. The whole facility, including the Moving Block, is covered with main roof of 130 m × 120 m, which is supported by steel space structure.

In this paper, I describe several problems, and their solutions that occurred during the construction of the roof steel structure of Saitama Super Arena.

Photo 1. General view of Saitama Super Arena.

CHARACTERISTICS OF THE ROOF STEEL STRUCTURE

As shown in Figure 1, the Saitama Super Arena's main roof is shaped as a part of an inclining conical surface. At first, a pair of huge truss beams, called the End girder (hereinafter EG) and the Arched keel girder (hereinafter KG), was placed along the arch-direction of the inclining cone. The EG is a Warren truss which consists of H-shaped steel

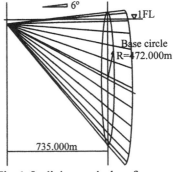

Fig. 1. Inclining conical surface.

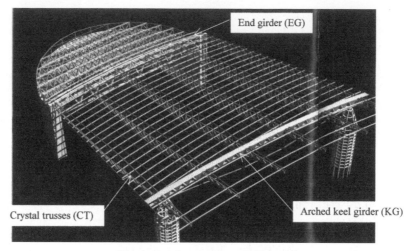

Fig. 2. Roof structure model.

members with a span of approximately 115 m and maximum truss height of 14 m. The KG has a span of approximately 140 m; the depth of the string beam is 10 m; and the upper member is, itself, a 5 m Warren truss and is made of steel members whose sectional view resembles three H-shaped steel members positioned side by side (ꓝꓮꓝ). The string members are three parallel wire cables, and each cable takes approximately 10,000 kN of initial tensile force. KG has approximately 5 degrees of angle in relation to the circle of the inclining conical surface; therefore, in a strict sense, KG's upper member is not a circle.

These two huge trusses are arranged approximately 120 m apart. Between them, six truss beams, called crystal truss (hereinafter CT), are placed along the generatrices of the inclining cone with intervals of approximately 25 m. The section of CT is an inverted-triangle. Since CT's are arranged on the inclining conical surface, each inverted-triangle section is slightly tilted so that the topside follows the curve of the inclining conical surface. In addition, since CT's are simple beams, in order to bear the large bending moment at the center of the trusses, the depth of the trusses increases from 4 m at the edges to approximately 10 m at the center. In relation to the depth, the sectional shape also changes gradually in proportion. Thus, CT has a complex three-dimensional shape with several inter-related factors. As for the members, the upper and lower chord members are H-shaped steel, and the diagonal members are circular pipes.

Between CT's, there are H-shaped steel tie beams to support roof materials, and square pipe braces maintain the horizontal rigidity of overall roof structure.

In the course of constructing such a complex steel structure, there were many challenges. I would like to discuss these problems and their solutions below.

LINEAR SHAPING OF STEEL MEMBERS
Since the shape of the main roof is a part of inclining conical surface, arranging steel members to follow the surface exactly will require complex curved members. For example, in a strict sense, the upper members of CT are not in the direction of the inclining cone's generatrices. Therefore, in order to make these members follow the inclining cone shape,

they must have a two-dimensional curve not in a circle.   In addition, since the H-shaped steel members themselves have directivity, the relationship between the two-dimensional curve and the direction of H-shaped members had to be determined.

Trying to regenerate the inclining cone-shape strictly, a vast amount of work and time would be required in fabricating steel members.   Steel members are made based on linear members and arch-shaped bending members.   Therefore, in order to fabricate the steel members efficiently, the key is to find out how to combine the linear and arch-shaped members in order to assimilate a part of inclining conical surface as an architectural expression.

The basic rules for determining the shape of crystal trusses

In order to realize the architectural design and optimize the fabrication of steel members at the same time, the shape of CT is the key factor.

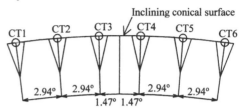

Fig. 3. Crystal trusses X axis section.

The shape of CT is defined by the inclining conical surface.   If the cone is cut square to the center of the cone, in the section, the six crystal trusses are angled evenly along the roof surface arch.   For each truss, the direction towards the center of the circle is defined as the height direction of the inverted triangle of the truss.   Then, the height of the inverted triangle is determined to be 4 m at EG side and 5 m at KG side.   The bottom member of the crystal truss (the bottom peak of the inverted triangle) is horizontal from EG to the midway to KG.

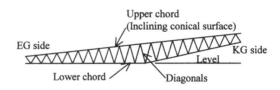

Fig. 4. Crystal truss Y axis section.

The rest of the triangular section is determined by drawing a straight line from the midway point to KG end, with the height of 5 m.   The shape of the inverted triangle is determined with the topside as the half of its height.

Chord members are arranged at each peak of inverted triangles to form trusses.   On these chord members, the connections for lattices are placed evenly in a horizontal direction.   Thus, the basic shape of the crystal trusses is determined.

Designing the shape of steel members

Based on the basic shape discussed above, we go on to design the shapes of actual steel members.   First, the tie beams are placed on sections that are square to the center of the cone (Fig.5); therefore, if the cone surface is to be followed exactly, the tie beams should form a single circular line.   If they do, we could arrange the upper members of the crystal trusses

Fig. 5. CT on cone surface.

Fig. 6. CT on multi-sided pyramid surface.

exactly on the cone surface. However, since the sectional shape of inverted-triangle changes gradually, the upper chord members would have three-dimensional curves.

Instead, we define the tie beams as straight members that are tangent to the cone arches (Fig.6). By doing so, the upper chord members become straight members, though they deviate slightly from the cone surface.

Thus, the basic inclining conical surface is replaced by an inclining multi-sided pyramid surface. By placing the two upper chord members of a CT in a flat plane, the shapes of steel members are drastically simplified. By introducing the local coordinates to the plane, we do not have to deal with complex three-dimensional coordinates anymore.

CONNECTION DETAIL

Since the roof has a three-dimensional curve, the steel members are connected three-dimensionally. In the case of CT's, several diagonal members are connected to a H-shaped chord member at the same point; for all connections, the connection detail must satisfy various criteria, such as smooth and definite transfer of stress, sufficient weld length, the easy assembly of steel plates to maintain accuracy, and enough work space to perform welding without difficulty.

Photo 2. Upper chord connection.

CT's consist of H-shaped steel chord members and diagonal steel-pipe members. Two diagonal members meet at each upper chord member connection, and four diagonal members meet at each lower chord member connection. Both ends of steel pipes have cross-shaped plates welded in place, which will be bolted to the connection with high-tension bolts. In the course of designing the weld connection detail, one of the key points was to smoothly transfer the stress on the cross-shaped plates to other members.

Constraints on connection details

In order to determine the details, we have to consider the following factors, as well as the structural constraints discussed above.

Photo 3. Lower chord connection.

1. Architectural design intent to minimize size of connection.
2. Sufficient working clearance to perform sound welding at shop.
3. Sufficient working clearance to facilitate bolting at site.
4. Consistent connection detail to accommodate varied connection angle.

Based on these conditions, the connection detail is determined for the upper and lower chord members, as described below.

Connections at the upper-chord members

At first, PL- (1) (see Fig.7) is welded to the upper-chord member flange. This plate ensures a sufficient welding length with the diagonal member plate.

Next, PL- (2) is attached vertically. This plate has two functions: it transfers stress between two diagonal members, and resists the diagonal member's cross-plate buckling. In addition, PL- (2) also transfers the shear force from the tie beam webs, and the plate's angle is the same as that of the tie beam web.

Then, PL- (3) is attached. PL- (3) will be welded to the cross-plate of the diagonal member later. This plate transfers part of the diagonal member's stress (the stress that is square to the upper-chord member) to the adjacent diagonal member's plate through the welding to PL- (2). The stress that is parallel to the upper-chord member is directly transferred to the upper-chord member, through the welding to the upper-chord member web. At this stage, PL- (2) and PL- (3) are welded together. This welding is done with Double-bevel-groove partial penetration welding.

Lastly, the diagonal member's cross-plate, PL- (4) is attached. The stress within PL- (4) is transferred through the weld between PL- (4) and PL- (1). The stress transferred to PL- (1) is transferred to the adjacent diagonal member, while the force component that is parallel to the upper-chord member is transferred to the upper-chord member.

Since the PL- (4) is the last plate to be joined, the working clearance is very limited. Because it is impossible to perform welding from the inside at the connection, Double-bevel-groove welding is not suitable. Therefore, Single-bevel-groove welding is adopted within the connection so that the welding is done from one side only. However, Double-bevel-groove welding is performed for the outside of the connection in order to minimize the amount of welding and possible distortion caused by welding heat.

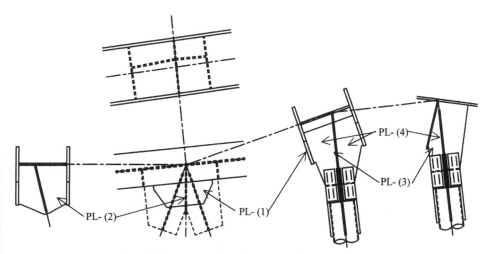

Fig. 7. Connection detail at upper-chord member.

Connections at the lower-chord members

Since four diagonal members connect to the lower-chord member at the same point, the connection detail is even more complex than that of the upper-chord member.

The first step is the same as the upper-chord member; PL- (1) (see Fig.8) is welded to the lower-chord member flange.

Next, PL- (2) is attached to the center of the lower-chord member web so that PL- (2) is parallel to the flange. As a stiffener against buckling for PL- (2), PL- (3) is attached vertically. PL- (2) has important functions; it transfer stress between the diagonal members, and also transfers a portion of stress from four diagonal members (the stress component that is parallel to the lower-chord member) to the lower-chord member. Since the welding up to this stage is done in advance, there is sufficient work clearance, and it is possible to perform Double-bevel-groove partial penetration welding from both sides of plates.

The cross-plate attached to the diagonal member joins to PL- (4) and PL- (5). PL- (5) is attached to PL- (4) only; thus, all stress from the diagonal member is received by PL- (4). As PL- (5) acts as PL- (4)'s buckling stopper, PL- (5) is designed to be the maximum possible length. The welding between PL- (4) and PL- (5) is performed in advance.

Lastly, four sets of PL- (4) and PL- (5) welded together are attached to the lower-chord member. PL- (4) is welded to PL- (1) and PL- (2), and each plate transfers stress to the adjacent diagonal members and the lower-chord member. Since PL- (4) is welded in a limited space within a connection, Double-bevel-groove and Single-bevel-groove welding are combined to achieve a satisfactory result.

Thus, by clarifying the stress flow and determining the steps of assembly and welding based on the stress flow, we can come up with reasonable connection detail that satisfies various factors described above.

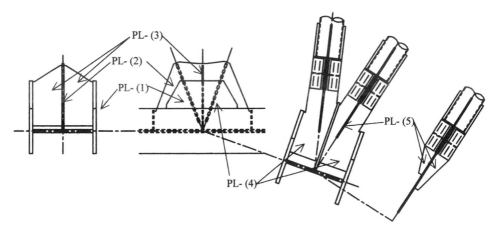

Fig. 8. Connection detail at lower-chord member.

CONSTRUCTION METHOD

CT has planar trusses (the upper-chords) as compression members, and a single member (the lower-chord) as a tension member; the truss has an inverted-triangular section. This shape is very stable once the structure is complete. However, since the top is heavier than the bottom, it is unstable until set in place. A long-span structure generally requires accurate assembly; therefore, we have to seek a simple strategy to achieve an accurate assembly.

In the construction, 4 out of 6 CT's (except the ones on both ends) are assembled at the ground level, then lifted up by 40 m to be connected to EG and KG, which are assembled at their respective height. At these connections, H-shaped members are welded together; it is almost impossible to achieve the same level and angle, and a suitable distance, between two components. Therefore, the assembly method must have some flexibility to allow for some variances.

Maintaining accuracy in the ground assembly

For CT, 3 chord members and diagonal members are delivered to the site as separate components.   They   are   bolted   together   on   the   construction   site   to   form   their inverted-triangular shapes.   When the trusses are assembled, the chord members have three-dimensional angles. To realize the truss shape in a strict sense, each connection must be positioned exactly according to three-dimensional coordinates; the assembly requires a vast amount of jigs.

At this point, we come back to the idea of linear shaping of CT's. A pair of upper-chords is defined as a flat plane. If we can assemble the plane accurately, we can ensure the accuracy of the whole structure. Based on this idea, the following construction method is adopted.

The basic rule is to assemble the trusses upside-down, with the upper-chord at the bottom, and the lower-chord at the top. On the ground level, a completely flat surface is prepared for assembly of the upper-chords. Once the surface is ready, the assembly of the upper-chord members can be managed two-dimensionally on the surface. After the upper-chords are assembled accurately, the diagonal and the lower chord members are assembled. They

Photo 4. Ground assembly (upside-down).

don't have to be as accurate as that of the upper-chord. In addition, because the heavier upper members are placed at the bottom, the structure is stable even during the assembly. In this condition, the assembly is performed up to the final tightening stage of high-tension bolts.

When the assembly is done, 2 large cranes to flip and set it into proper position lift a block of CT up. The size of the block is determined by the capacity of the cranes. Since the triangular shape is already established, there is no major problem in maintaining accuracy at this stage.

Coping with the variance of lift-up
When CT's assembled at ground level is lifted up and set in place, the position of connections may deviate from their proper position. This deviation can be expressed with 6 factors in the three-dimensional space: the distance in the direction of the chord, the vertical distance, the horizontal distance, the rotation angle along the chord, the vertical rotation angle, and the horizontal rotation angle. Among these factors, the vertical distance is not a problem, since it can be controlled when CT is lifted up. The following discusses the remaining 5 factors.

Photo 5. Ground assembly (proper position).

Photo 6. Lifting-up of CT's.

First, by performing the welding on site, 3 factors (the distance in the direction of the chord, the vertical rotation angle, and the horizontal rotation angle) can be construed as a matter of welding root gap, although there is

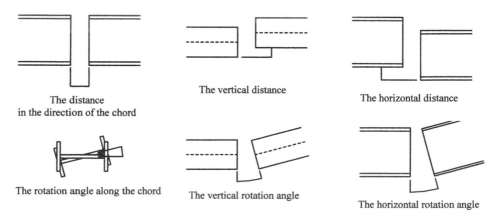

The distance
in the direction of the chord

The vertical distance

The horizontal distance

The rotation angle along the chord

The vertical rotation angle

The horizontal rotation angle

Fig. 9. Six deviations in the three-dimensional space.

still a problem of allowing a large gap at the root. The remaining factors (the horizontal distance and the rotation angle along the chord) can be construed as that of misalignment of plates. A thick component inserted between the chord members can solve this problem.

In order to prevent a large gap at the welding root and to position a thick component, we adopt the following method. First, when CT's are lifted up, a clearance of 300 mm is left at each connection. After the lift-up is done, the clearance is measured. Based on this measurement, built-up H section of approximately 300-mm length is

Photo 7. Built-up H section by very thick plate.

fabricated with very thick material. By having the steel material prepared in advance, the construction time can be reduced. Thus, by fabricating built-up H section on the actual measurement, the gap at the welding point is minimized. At the same time, by utilizing very thick plates for the built-up H section, misalignment between the chord members can be absorbed within the thickness of the steel. The actual thickness of the built-up H section is 80 mm, while regular members are made of 22 to 32 mm thickness steel plates.

CONCLUSIONS
The Saitama Super Arena's main roof is a gigantic structure and has complex three-dimensional shape. There were many challenges during the course of engineering and construction of this structure, and we solved these problems as discussed in this paper. As a result, we were able to construct a high-quality steel structure complying with stringent technical requirement, while maintaining cost efficiency. Our solution should be utilized in the future projects where similar large complex steel structures are involved.

ACKNOWLEDGEMENTS
Construction Entity: Saitama Prefecture, Japan
Design: MAS·2000 Design Team (Representative: Nikken Sekkei)
General contractor: Special joint venture consisting of Taisei Corporation, Mitsubishi Heavy Industries, UDK company group

Simultaneous Engineering in construction

I ORTEGA
Ortega & Kanoussi Technologies, Mexico City, Mexico

INTRODUCTION

Simultaneous Engineering is a systematic method for improving the product development process in industrial mass production. It is based on parallelisation and integration of the product development activities and on multidisciplinary teamwork. This paper discusses the transfer of Simultaneous Engineering to building construction. The resulting parallelisation and integration of architecture, engineering, and construction can lead to substantial benefits such as shorter duration of the construction process, lower construction costs, higher construction quality, increased owner satisfaction, and innovations. The potential benefits of Simultaneous Engineering in construction are verified by four case studies.

Before discussing the use of Simultaneous Engineering in construction, an overview of Simultaneous Engineering in industrial mass production is given.

SIMULTANEOUS ENGINEERING IN INDUSTRIAL MASS PRODUCTION

In industrial mass production, Simultaneous Engineering stands for a method employed to achieve the following objectives: reduced product development and production costs, shortened total product development and production time, improved quality, increased customer satisfaction, and innovations.

Simultaneous Engineering in industrial mass production is based on *parallelised activities* and *multidisciplinary teamwork* and includes some of the following techniques or elements:

1. Computer Aided Design
2. Rapid Prototyping
3. Design for Manufacture, Design for Assembly, Quality Function Deployment, Failure Mode and Effects Analysis, etc.

The following pages present some of the techniques of Simultaneous Engineering in industrial mass production.

Parallelised Activities

Simultaneous Engineering requires parallel product and production processes design. By overlapping several phases of industrial production, starting with the product idea, up to the market introduction of a new product, total production time is shortened. The duration of individual phases of a Simultaneous Engineering project can turn out to be longer than the corresponding duration of the individual phases of a project carried out sequentially. However, as the example from car manufacturing in *Fig. 1* shows, the *total* duration time of a Simultaneous Engineering project is shorter than the *total* duration of a conventional, sequential project.

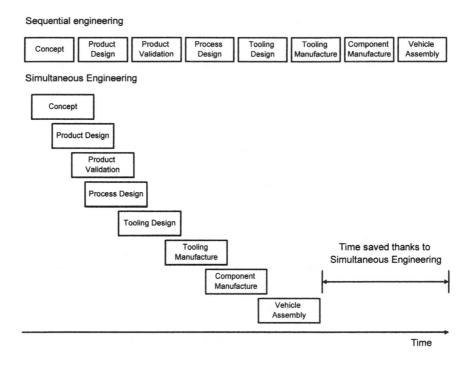

Fig. 1. Main activities of a sequential, conventional vs. a Simultaneous Engineering project in car manufacturing [1].

Multidisciplinary Teams

A characteristic feature of Simultaneous Engineering is multidisciplinary teams that carry out industrial projects [2]. The teams consist of representatives from all departments involved in the product life cycle such as design, manufacture, assembly, purchasing, and marketing.

The team-oriented approach adopted in Simultaneous Engineering offers substantial advantages compared to conventional, "over the fence" or "over the wall" engineering. In "over the fence" engineering, the different departments involved in the product life cycle work separately. Product Design engineers work out a product design and throw it "over the fence" to Production Engineering. There, production engineers struggle to plan production processes. Production engineers may suggest product changes to improve production. However, because the product has been completely designed, any improvements suggested by production engineers imply at least a partial product redesign. Therefore, changes suggested by "outsiders," such as production engineers, are often resented by product designers [2]. When production engineers complete the design of production processes, they throw their designs "over the fence" to Manufacturing Design. Because production design is considered to be complete, manufacturing engineers cannot propose changes to improve production processes without triggering at least a partial redesign of production processes. In "over the fence" engineering, problems start to emerge shortly before production begins. Changes to improve product quality and lower production costs become necessary. These changes require major revisions to product and process design and therefore lead to substantial production delays, cost overruns, and ultimately, to unsatisfied customers.

In "over the fence" engineering, specialists from the different design phases work in isolation. As specialists, they often lack the necessary, detailed knowledge to anticipate problems that may emerge in the design phases following the one they are working on. The various specialists required to work on a conventional engineering project are available in a sequence that is determined *a priori*. Therefore, due to the absence of the required specialists on a particular project phase, problems are often discovered during a later project phase, when the required specialists come into play. This happens, for example, when manufacturing engineers are confronted with product designs that cannot be manufactured unless they are substantially modified. Due to the fragmentation of conventional engineering projects, problems discovered during an advanced project phase trigger changes on the results of several early project phases. The further a project has progressed, the more the early project phases may have to be revised. Therefore, late design changes are usually more costly to make than early design changes. The multidisciplinary teams of Simultaneous Engineering projects bring together specialized knowledge from all departments involved in the product life cycle. As a result, Simultaneous Engineering projects lead to fewer design changes than conventional engineering projects. Thus, reduced costs and shortened lead times are achieved.

A large portion of the product costs is committed during the early engineering project phases, when changes are more easily accommodated. Hartley [2] estimates that about 60 to 80 percent of product costs are committed during the product design phase, while Syan [3] provides an estimate of 60 to 95 percent. Therefore, improvements accomplished during the design phase have the largest cost-cutting potential [3]. The earlier the improvements are achieved, the larger the savings are. For example, the design costs of a new automobile model make up only about 5% of its total cost [2]. Hence, an increase of 20% in design costs would lead to an increase of only 1% of the total costs. In automobile design, the design costs of Simultaneous Engineering projects are usually larger than those of conventional engineering projects. However, due to the improved design achieved during the extended design phase, the overall costs of a Simultaneous Engineering project turn out to be lower than those of a conventional project [2].

Sequential engineering projects tend to inhibit innovation because, as projects progress, conceptual freedom decreases [4]. The first broad step of a sequential engineering project is to optimize the product characteristics. During the second broad step, an attempt is made to optimize the process characteristics. The process characteristics, however, cannot be optimized beyond the boundaries laid down by the previously defined product characteristics. Process innovations offering substantial benefits are not introduced if they require major modifications to the product characteristics. Therefore, sequential engineering projects optimize the results of every individual project phase strictly within the boundaries laid down by the previous phases. To go beyond these boundaries often requires modifications to the results of previous phases and can therefore lead to cost overruns and delays.

Simultaneous Engineering projects are carried out by teams made up of representatives from all departments involved in the engineering project. Therefore, the team members can propose improvements at any stage of the engineering project. For example, process engineers can offer innovative processing solutions early in the product design phase without waiting for the process design phase to begin. Such innovative processing solutions can be integrated into the product design before it has been completed. Since innovative solutions, belonging to later project phases, can be incorporated during early project phases, conceptual freedom is not restricted as the project progresses. Innovative process improvements proposed or at least considered during product design help avoid costly and time-consuming modifications to the original product design. Therefore, Simultaneous Engineering is a *total-system* approach that promotes *innovations* and optimizes the overall results of the engineering project. On the other hand, sequential engineering focuses on sequential optimization of the individual phases of the engineering projects and therefore leads to sub-optimal project results.

Rapid Prototyping
Rapid Prototyping stands for the rapid (within minutes or hours), automatic generation of solid models based on virtual models created on CAD systems [1]. Rapid prototypes are used to assess the overall design, appearance, function, manufacture, and assembly of products [1]. Rapid prototypes created early in the design phase assist in anticipating problems that may emerge in later design and production phases.

Other Simultaneous Engineering Techniques
The following techniques are often applied in Simultaneous Engineering projects:
1. *Design For Manufacture.* Design for Manufacture (DFM) aims to minimize manufacturing costs by taking into account the manufacturing process early in the design phase. This is often done by minimizing the number of steps and setups to convert a part or assembly into a product [1]
2. *Design for Assembly.* The objective of Design for Assembly (DFA) is to minimize the costs of assembly operations by taking into account assembly operations early in the design phase. This is usually accomplished by simplifying the assembly process
3. *Design for Engineering Analysis.* Designs that are difficult to analyze may unnecessarily delay project completion. The project lead time maybe shortened by carrying out only designs that can be easily analyzed [5]

Now that the application and the benefits of Simultaneous Engineering in industrial mass production have been discussed, the transfer of Simultaneous Engineering to construction will be analyzed.

SIMULTANEOUS ENGINEERING IN CONSTRUCTION
From what has been said so far, it can be concluded that Simultaneous Engineering in construction implies to simultaneously carry out architectural design, engineering design, and construction planning and to form multidisciplinary project teams. Parallelised activities in construction require, as in industrial mass production, the formation of multidisciplinary teams and intense communication between those involved in the construction project. The individual activities - architectural design, engineering design, and construction planning - may last longer than in a construction project carried out sequentially. However, the total duration of the parallelised construction project can be shorter than the total duration of a conventional, sequential construction project. Since construction is a sequential and highly fragmented activity, construction projects can benefit from parallelised activities and the formation of multidisciplinary teams.

Construction project teams made up of architects, engineers, and contractors may help overcome the problems caused by "over the fence" construction projects. A major problem in construction is design changes. Multidisciplinary teams made up of architects, engineers, and contractors may help reduce the number of design changes by anticipating problems that appear in late project phases. In construction, as in industrial mass production, a large portion of the total project costs is committed during the early project phases. Therefore, improvements achieved during the early phases of a construction project offer the largest potential benefits. The design costs in construction are estimated at about 10% [6]. Therefore, an increase in design costs of about 20% amounts to only 2% additional total project costs. Therefore, it appears worthwhile to invest in architectural and engineering design and in planning the construction process.

The main hypothesis of this paper is that Simultaneous Engineering in construction can lead to substantial benefits such as shorter duration of the construction process, lower construction costs, higher construction quality, increased owner satisfaction, and innovations. The plausibility of the hypothesis is supported by empirical evidence from industrial mass production and theoretical arguments. However, to determine the potential benefits of Simultaneous Engineering in construction, empirical evidence from *construction* is necessary. The required evidence is provided by the following case studies on four major designers: Heinz Isler, Fazlur Khan, Félix Candela, and Pier Luigi Nervi.

CASE STUDIES
The case studies presented herein concern designers that have been mainly active in building construction, particularly shells and high-rise buildings. The designers selected for the case studies have practiced in Switzerland (Isler), the U.S. (Khan), Mexico (Candela), and Italy (Nervi). It is hoped that the international selection emphasizes the universality of the concepts discussed.

Heinz Isler
The Swiss engineer Heinz Isler (1926) is the owner of the engineering office *Ingenieur- und Studienbüro Heinz Isler* in Burgdorf, Berne (Switzerland). He has become next to Félix Candela and Pier Luigi Nervi, one of the major designers of shell structures and is one of the most enthusiastic proponents of shell construction.

Isler has been constantly searching for organic, harmonious forms for shells. His preferred method for finding shell forms is the experimental method. He came up with this method of Rapid Prototyping accidentally. By observing the shape of his pillow, he discovered the dome shape with a rectangular ground plan [7]. He determined the minimal shape of the pillow in his laboratory with a rubber membrane clamped in a timber frame. Air pressure applied from below the rubber membrane curved its shape. Precise measurements of the model in the laboratory helped transfer the shape of the membrane to a structure on the construction site.

Isler has applied the experimental method of finding forms to structures with different kinds of borders: squares, triangles, circles, ellipses and irregular shapes, and even to structures situated at different levels. Shapes determined this way are minimal, i.e. for a given crown level, they provide the smallest possible surface. The practical advantages of structures with minimal surfaces are [7]: minimal building material expenditures, minimal weight, and minimal radiating surface, i.e. minimal loss of energy.

The observation of a hanging wet jute cloth led Isler to a further experimental method for finding shell forms: the hanging form method. This method not only generates shapes with closed edges, it also produces shapes with point supports. The procedure requires a cloth to be soaked in a hardening liquid. The wet form is then suspended until it hardens. The stiffened form is then turned upside down and behaves as a membrane under compression [8]. Inverted hanging forms carry their own weight only by compressive stresses. After the inverted form has been measured using a special procedure ensuring the highest accuracy (*Fig. 2*), the shell coordinates can be input into a computer program to calculate the stresses. More refined experiments on scale models may be necessary to confirm the computer results.

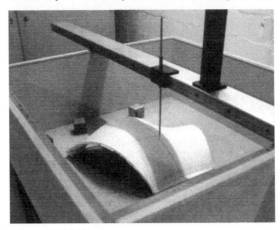

Fig. 2. Installation to measure the shape of a shell model generated from an inverted hanging form in Isler laboratory.

Since shells are three-dimensional structures, they are difficult to describe in words or with the aid of mathematical models. Therefore, Isler prefers to use scale models [7]. Full-sized, 1:1 models are rarely affordable in construction; therefore, scale models offer the opportunity to apply Rapid Prototyping in shell design. Models enable the engineer to identify the advantages and shortcomings of a proposed design. Moreover, communication with project participants is simpler with models.

In shell design, parallelised activities and multidisciplinary teamwork, i.e. close cooperation between architect, engineer, and contractor are essential. The architect, the engineer, and the contractor have to carefully balance form, structure, and economy. To reduce construction costs, the formwork design and the construction schedule should be defined early and in close cooperation with the building contractor. Even though shells consist of doubly curved surfaces, which (theoretically) cannot be covered with rectangular plates, in practice it is nevertheless possible to build shells out of rectangular elements [7]. Thus, by parallelising design and construction, shells can be more economical than other structural systems. By the re-use of curved board elements (assuming careful handling) further savings are possible. Isler built many of his designs in close cooperation with the building contractor *W. Bösiger AG* from Langenthal. Thus, he achieved additional cost reductions.

Isler's design method involves the application of the following Simultaneous Engineering methods in construction: Parallelised activities, multidisciplinary teamwork, Rapid Prototyping, Design for Construction (in analogy to Design for Manufacture), and Design for Engineering Analysis.

Fazlur Khan
Fazlur Khan (1929-1982) worked as an engineer at the architectural and engineering office *Skidmore, Owings & Merrill (SOM)* in Chicago. He vigorously promoted the multidisciplinary and collaborative character of design. He rejected the idea that architects are

solely responsible for esthetic concerns, while engineers should be limited to providing the necessary technical solutions. In almost all his designs, Khan took the initiative, working from the early design phases in close collaboration with his architectural partners (mostly the architects Bruce Graham and Myron Goldsmith). Khan introduced the concept of *overall system thinking* [9] to architectural and engineering design. According to this concept, for each height or span of a building there is an optimal structural system. Crucial for an architectural and engineering design is the selection of the overall system. The optimization and refinement of the subsystems play a secondary role.

Khan's main contribution to civil engineering was to develop optimal structural systems for steel and concrete buildings of growing heights. Khan also optimized the construction processes by applying prefabrication. The common objective of Khan's innovations was to provide column-free interiors to the occupants of a building. For this purpose, he eliminated the interior columns and transferred their load-carrying function to the building façade. Thus, the façade became an integrated system providing enclosure and load-carrying resistance. Khan's innovations consisted of different structural systems for integrated, load-carrying building façades. Today, the majority of skyscrapers built are designed according to the concepts and principles developed by Khan. Khan's innovations simultaneously increased the stiffness of very tall buildings while reducing their structural costs. Khan not only developed several new types of efficient structural systems, he also laid the foundation for a new architectural vocabulary. Interestingly, many of Khan's designs proved to be starting points for further scientific research - and not the other way around [10]. Close cooperation with his architectural partners and his interest in the construction process led to Khan's many groundbreaking innovations (*Fig. 3*).

Fig. 3. John Hancock Center, Chicago: the innovative result of early and intensive collaboration between architect and engineer [11].

As this case study shows, Khan applied the Simultaneous Engineering principles of parallelised activities, multidisciplinary teamwork, and Design for Construction.

Félix Candela
The Spanish architect Félix Candela (1910-1997), who stayed in Mexico during his professional career, built mostly shells in the shape of hyperbolic paraboloids, a geometrical

form with remarkable properties. The surface of a hyperbolic paraboloid can be determined with a surprisingly simple equation. Additionally, the surface of a hyperbolic paraboloid can be decomposed into straight lines. Thus, a hyperbolic paraboloid can be built using formwork made up of straight boards, having a significant cost advantage over shell surfaces requiring expensive curved formwork. In addition, in sharp contrast to other shell shapes, the stresses in a hyperbolic paraboloid shell can be determined by direct integration of the differential equations for static equilibrium, thus simplifying the design and analysis process. Candela systematically exploited all the advantages of the hyperbolical paraboloid to build approximately 900 shells. Even though most of Candela's shells were hyperbolic paraboloids, the originality and variety of his designs proved the versatility and potential of the hyperbolic paraboloid as the basis for shell shapes (*Fig. 4*). Often, Candela was simultaneously architect, engineer, and contractor and could therefore make full use of the potential of the hyperbolic paraboloid. Even in those cases when Candela practiced simply as an architect, he took the construction process carefully into account. In his designs, Candela parallelised architectural and engineering design and construction planning. Thus, he built many innovative, economically competitive structures with a lasting architectural impact.

Fig. 4. Sales office in Guadalajara [12].

As this case study shows, Candela systematically applied the Simultaneous Engineering techniques of parallelised activities, multidisciplinary teams, Design for Construction, and Design for Engineering Analysis.

Pier Luigi Nervi
Even though the Italian Pier Luigi Nervi (1891-1979) was educated as an engineer, he had comprehensive knowledge of architecture and construction. Most of his structures were built as a result of competitive design-and-construction bids [13]. In competitive design-and-construction bidding, contractors are invited to submit tenders for the design and construction of a structure at a binding price. To win competitive design-and-construction bids, it is necessary to carry out and simultaneously optimize the design and the construction phase of a building project. This requires comprehensive knowledge of design and construction and close collaboration between design engineers and construction engineers. To further optimize the construction process, Nervi built many of his structures from prefabricated elements.

Nervi was a pioneer in experimenting with models. He used scale models to systematically analyze the stresses in large and complex structures [13]. In 1935, Nervi built a 110 x 8 m hangar for the Italian Air Force. The design of the hangar was largely based on the analysis of experiments with a celluloid model. The loaded and instrumented model is shown in *Fig. 5*.

Fig. 5. Stress analysis of a hangar, undertaken on a celluloid scale model [13].

The present case study shows that Nervi applied the Simultaneous Engineering principles of parallelised activities, multidisciplinary teamwork, Rapid Prototyping, and Design for Construction.

CONCLUSIONS

As the preceding case studies have shown, the benefits of Simultaneous Engineering in construction are similar to those provided by Simultaneous Engineering in industrial mass production: reduced construction costs, shortened total construction time, improved quality, increased owner satisfaction, and innovations. The case studies share an interesting pattern: the achievement of innovations by integrating different systems of a building. For example, Isler integrates the space-enclosing system with the load-carrying system in his shell structures; and so did Khan, Candela, and Nervi. Khan integrated the space-enclosing system with the load-carrying system in his tubular systems. It is not the innovative integration *per se* that is beneficial, but the elimination of obstructing load-carrying systems such as interior columns through physical integration of the load-carrying system with the space-enclosing system. Owners prefer ground plans without interior columns, because unobstructed ground plans provide more flexibility. However, integrated systems do not offer the mere advantage of absent interior columns, they are also more cost efficient than separated systems. For example, Isler's shells are cost efficient because their shapes are optimal surfaces, carrying only compression loads. Candela's hypar shells are economical because of their ease of construction. Nervi's structures are cost efficient because of their efficient shape and the economical benefits of prefabrication. Khan's tubular structures are also cost efficient because, by moving the interior columns to the perimeter, their leverage increases, thus improving the resistance of a building to wind loads. Thus, Simultaneous Engineering in construction offers the opportunity to improve the overall results of all activities involved in construction and to generate innovations achieved through the techniques of Simultaneous Engineering and through the integration of systems of buildings.

SUMMARY
The present paper discussed the merits of Simultaneous Engineering in industrial mass production. Then, the potential benefits of applying Simultaneous Engineering to construction were determined. Case studies served to empirically validate the potential benefits of Simultaneous Engineering in construction: shorter duration of the construction process, lower construction costs, higher construction quality, increased owner satisfaction, and innovations.

ACKNOWLEDGEMENTS
The work on this paper was supported by the Basic Research Commission and the Institute for Technology Management, University of St. Gallen, Switzerland.

REFERENCES
1.   RANKY P G, Concurrent / Simultaneous Engineering (Methods, Tools & Case Studies), Biddles Limited, Guildford, Surrey, 1994.
2.   HARTLEY J R, Concurrent Engineering: Shortening Lead Times, Raising Quality, and Lowering Costs, Productivity Press, Cambridge, Massachusetts, 1992.
3.   SYAN CH S, Introduction to Concurrent Engineering, Concurrent Engineering: Concepts, Implementation, and Practice, Syan Ch S and Menon U (eds.), Chapman & Hall, London, 1994, pp 3-23.
4.   KROTTMAIER J, Leitfaden Simultaneous Engineering, Springer-Verlag, Berlin, 1995 (in German).
5.   JO H H, PARSAEI H R, AND SULLIVAN W G, Principles of Concurrent Engineering, Concurrent Engineering: Contemporary Issues and Modern Design Tools, Parsaei H R (ed.), Chapman and Hall, London, 1993.
6.   SCHNEIDER J, Introduction to Safety and Reliability of Structures, International Association for Bridge and Structural Engineering, Zurich, 1997.
7.   ISLER H, Moderner Schalenbau, Gestalten in Beton: Zum Werk von Felix Candela. Die Kunst der leichten Schalen, Verlagsgesellschaft Rudolf Müller GmbH, Colony, 1992, p 50-66 (in German).
8.   Author's interview with Prof. Dr. h.c. H. Isler on December 9 1999.
9.   IYENGAR H, Structural and Steel Systems, Technique and Aesthetics in the Design of Tall Buildings / Fazlur Khan Memorial Session, Billington D P and Goldsmith M (eds.), Institute for the Study of the High-Rise Habitat / Lehigh University, Bethlehem, PA., 1986, pp 57-69.
10.   BILLINGTON D P AND GOLDSMITH M (eds.), Technique and Aesthetics in the Design of Tall Buildings / Fazlur Khan Memorial Session, Institute for the Study of the High-Rise Habitat / Lehigh University, Bethlehem, PA., 1986, pp ix-x.
11.   FRANCONE M, Bruce Graham of SOM, Electa, Milano, 1989.
12.   FABER C, Candela: The Shell Builder, Rheinhold Publishing Corporation, New York, 1963.
13.   NERVI P L, Structures, F. W. Dodge Corporation, New York, 1956.

The analysis and design of the City of Manchester stadium roof.

M. SIMPSON
Arup, Manchester, UK

SYNOPSIS
The City of Manchester Stadium will form the centrepiece for the Commonwealth Games 2002. One of the most spectacular aspects of this stadium is the simplicity and graceful lines of the lightweight steel roof, supported by a cable net, in turn supported by 12 slender masts. A key feature of this roof is the innovative grounded tension ring system, used to prestress the cable net and combat the effects of wind uplift on the lightweight roof.

This paper will highlight the major steps taken during the analysis and design of the cable stayed roof.

INTRODUCTION
The city of Manchester has always had the ambition to stage an international sporting event. The Commonwealth Games provides an excellent opportunity and the City of Manchester Stadium provides the centrepiece and athletics venue with a capacity of 41,000 seats, see Figure 1. However, the long term legacy provided by the City of Manchester Stadium is as a world class football venue. Therefore the stadium shown in Figure 1, is in fact only Phase 1, though the roof and roof support structure are permanent.

© Jim Mackintosh Photography

Fig. 1. The City of Manchester Stadium – Phase I, Athletics.

After the Commonwealth Games work will begin on site to convert the stadium into a 48,000 seat football stadium. This will involve removal of the athletics track and excavating 5m to place the football pitch at a lower level. At the same time, the final end stand will be built and additional terraces will be built down to the new football pitch level, see Figure 2.

Fig. 2. The City of Manchester Stadium – Phase II, Football.

ROOF CONCEPT

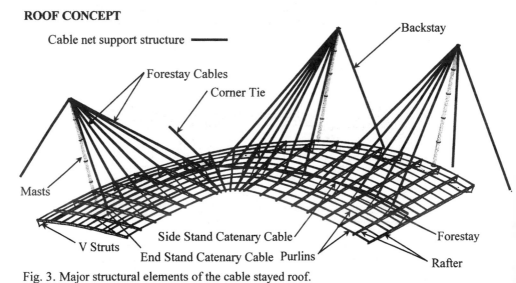

Cable net support structure ▬▬

Forestay Cables

Corner Tie

Backstay

Masts

V Struts

Side Stand Catenary Cable

End Stand Catenary Cable Purlins

Forestay

Rafter

Fig. 3. Major structural elements of the cable stayed roof.

Roof under downward load.

The downward load path is intuitive and apart from gravity, most other loads are applied directly to the cladding. The cladding comprises of standing seam metal cladding (Kalzip) to the rear portion of the roof and polycarbonate to the leading edge. The cladding is supported by purlins which in turn span onto the rafters. There are 76 rafters that run in a radial pattern and are formed from a fabricated box section 900mm deep x 300mm wide, tapering to

450mm deep at the leading end and rear end. Each rafter is supported from the bowl by two inclined struts; the "V-strut". The forward support point is achieved by hanging the rafter from the cable net by means of the "Forestay-strut". The maximum free span of a rafter is 37m together with a 15m cantilever. This ratio was derived for the longest rafter to equate the cantilever moment and backspan moments induced in the rafter under various load cases, thereby optimising the rafter design. The geometry of backspan to cantilever for other rafters was then dictated by the geometry of the cable net.

Each rafter is supported by a single spiral strand cable. These forestays are grouped together in fans of either 5 or 7 cables which are supported by a single mast. The mast is supported at the base by either a concrete plant tower linked back to the bowl structure, or at ground level. Under downward load conditions, the tension in the forestay cable is resolved by compression in the mast and tension in the backstay cables.

Roof under uplift
As the roof is of lightweight construction, uplift due to wind is a considerable factor in the design. Existing solutions for maintaining tension in cables include increasing the mass of the roof by additional ballast or using opposing cables with tension and compression rings.

For the City of Manchester Stadium, neither of these existing solutions was appropriate due to program/budget restraints and the two stage construction required for the venue to perform first as an athletic stadium for the commonwealth games and then as football stadium.

The solution adopted was an innovative adaptation of the opposing cable solution called "the grounded tension ring", see figure 4. The forestay cables were linked together by a "catenary cable". This cable was then tied back to ground at the four corners by means of the four "corner tie cables". Therefore a pretension could be induced into the entire cable net by pulling down at the four corner tie positions. The geometry of the cable net was derived so that exactly the right amount of tension was induced in each forestay cable preventing it from going slack under uplift.

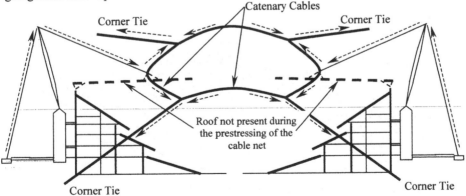

Fig. 4. Diagram showing how the grounded tension ring works.

Another advantage that this solution possesses is that after stressing the cable net, the system is statically determinate and can be erected independently of the rafters. Because it is erected separately to the roof structure, none of the prestress forces in the cables are transmitted into the roof structure creating an efficient solution. Therefore, under uplift the rafter is again supported at the front by the forestay strut. This strut is in compression and is restrained against buckling by the forestay-strut brace in the vertical direction and the catenary cable in the transverse direction. The compressive load in the forestay strut decreases the residual pre-tension built into the forestay cables during the erection of the cable net, thereby retaining the stability of the structure.

ANALYSIS AND DESIGN OF MAJOR ELEMENTS
The major structural elements of the roof consist of:
- The Cable Net
- The Rafters
- The Masts

Cable Net
Due to the nature of the cable net and the philosophy adopted the first task was to analyse a simple static 3D model of the stadium under wind uplift. By using individual bar elements for the cables (bar elements can only take axial load and cannot buckle), and placing pinned restraints at the top of the masts it was possible to apply a net upward force to the model. This analysis gave a compressive force in the forestay cables. The prestress force applied to the cable net by the grounded tension ring must be equal to, or exceed this pseudo-compressive force if the cable net is to always remain in tension, and therefore stable, through all loading conditions.

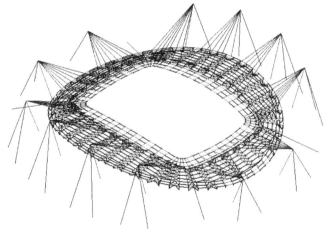

Fig. 5. Finite Element model of stadium – OASYS GSA.

The level of tension in each of the forestay cables is governed by two factors:
- The geometry of the catenary cable.
- The tensile force in the corner tie.

A funicular catenary geometry would provide the optimum geometry for the cable net so that just the right amount of tension would be produced in each forestay cable. The other option was to impose a set geometry on the catenary cable and then increase the forces until all the forestay cables had at least the correct tension. This option was rejected on the grounds of efficiency and therefore cost.

The method adopted to create the funicular geometry was an iterative 'soap-film' form-finding routine. Though the catenary shape was funicular, there were several architectural restraints imposed on the solution, these included:

- The positions of the tops of the masts were fixed in space.
- The fans on forestays on the side stands and those on the end stands were required to lie in the same plane.
- The intersection of the forestay cable and catenary cable had to lie in a vertical plane above the corresponding rafter.
- The intersection of the side and end catenary cable and the corner tie cable was fixed in space.
- The corner tie was to lie in a particular orientation on plan, at half grid, and clear the rear edge of the roof.

The formfinding process was iterative and is best highlighted from the flow chart shown in figure 6:

The first loop concentrates on the geometry of the side and end catenary in turn.

1. The first stage uses the soap-film formfinding facility of OASYS Fablon in which the actual member properties such as axial stiffness are ignored. By using the desired forestay cable tensile forces from the uplift case and the geometry constraints mentioned above an initial catenary geometry is created. The new geometry is saved as a result of the formfinding.
2. Stage 2 involves a non-linear analysis of the new geometry from stage 1, but with the geometric restraints on the catenary temporarily removed and the correct material properties. This run allows the forces between the catenary cables and forestay cables to re-distribute until equilibrium is reached.
3. Stage 3 involves another formfinding run, but with the new geometry from stage 1, the geometric constraints re-applied and the redistributed catenary forces from the non-linear analysis in stage 2. This stage modifies the geometry further and stage 2 is then repeated with this new geometry.

By repeating stages 2 and 3 it is possible to quickly create a catenary geometry that satisfies the geometric constraints and the force constraints imposed by the wind uplift model.

However, the geometry of the corner tie cable is a resultant of the side catenary and end catenary. The geometry on plan is a result of the ratio in catenary forces, whilst the geometry in elevation is a result of the magnitude of both catenary cables taken together, ie the greater the force, the shallower the angle in elevation. The second loop involves adjustments to the side and end catenary forces and re-running the first loop. With experience it became a quick process to obtain the correct funicular shape that satisfied all the geometric constraints as well as producing the correct tension in each of the forestay cables.

Each of the analysis loops involved subtle changes to the geometry of the cable net. Therefore a final check was always carried out using the static analysis model to ensure that the uplift forces applied to the cable net were still correct and that the geometric changes to the cable net had not increased the pseudo-compressive force that a forestay attracted.

The design of the cable elements was carried out in accordance to the manufacturer's data (Bridon Ropes) and the relevant British Standards. The ultimate load combination governing strength was dead (permanent) + imposed (snow and services) + downward wind + cable net pre-tension. The cable sizing was based on the worst case of 0.66 maximum breaking load (MBL) at ULS or 0.45 MBL at SLS.

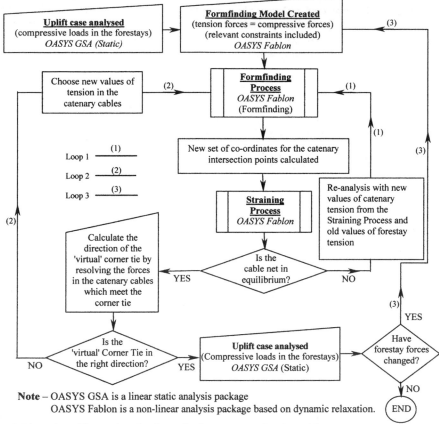

Fig. 6. Flowchart illustrating the form-finding process for the cable net.

Rafters
The rafters are Grade 50 steel fabricated box sections 300mm wide that vary in depth from 900mm for main span and taper to 450mm at the rear edge and tip. The webs of the rafters are formed from 6mm plate and the flanges vary from 12mm to 55mm.

As the design progressed from the concept stage, one of the key problems was the occurrence of arching in the roof system. The curvature of the roof was producing a secondary load path through the purlins in the side stands. This secondary load path was a problem for both downward loads and thermal expansion and would require a significant increase in the size of the purlins to facilitate it. The standard solution would be to incorporate movement joints at the four corners to coincide with the major thermal joints in the concrete bowl at the end of the side stands. However, waterproofing details at movement joints are notoriously difficult and expensive, so the design team adopted a solution of regular but much smaller movement joints throughout the roof. These joints were incorporated every other bay through simple slotted holes in the purlins and shoulder bolts. The movements at each joint were in the order of ⁺/. 25mm, which the standard cladding and flashing details were able to accommodate.

The lateral stability of the rafter was provided by cross bracing every other bay as shown in Figure 7. Lateral torsional stability and minor axis buckling of the rafter was also a key factor in the design, especially under wind uplift conditions where effective restraint from the purlins could not be provided. The solution adopted was to use U-frame stability, more commonly associated with bridge design. By using internal diaphragms at the purlin positions, we were able to use the bending resistance of the purlin to restrain the bottom flange of the rafter. This principle is shown in Figure 8.

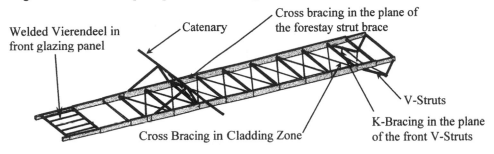

Fig. 7. Lateral stability system for rafters.

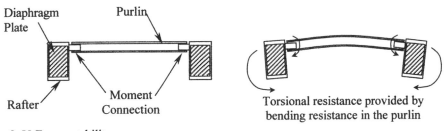

Fig. 8. U-Frame stability.

Masts
The roof is supported by 12 masts that vary in length from 35m to 40m. The masts are non-uniform in cross section and vary from 750mm diameter at either end to 1500mm at the centre as shown in figure 9.

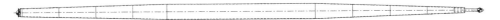

Fig. 9. Elevation of a typical mast.

In order to optimise the design of the masts it was necessary for the design team to adopt a method based on first principle moment magnification. This method allows the first order moments due to wind loads and the second order moments due to eccentricity and construction tolerances to be used in order to derive the buckling characteristics of the mast. Therefore, the wall thickness and diameter of the sections used were adjusted along the length of the mast in order to reduce self weight and maximise efficiency.

REQUIREMENTS OF ITERATIVE DESIGN

All good design is an iterative process, however complex analytical and design work usually requires special procedures to allow design loops to take place quickly and effectively. Throughout the design stages of the stadium roof there were many design loops, which could be attributed to the following reasons:

- Normal design procedure.
- New data becoming available.
- Design changes due to value engineering.

Normal design

Some iterative design procedures have already been mentioned, such as the cable net design. Others included changes to the roof geometry or the position of key points such as the mast tops or catenary cable/corner tie cable intersection points. These changes occurred due to the design development process within the design team.

New data becoming available

The most important piece of information for lightweight roofs is the wind uplift calculations. Most of the initial scheme design was carried out using data from the British wind codes and from experience built up from other stadia around the world. However, a wind tunnel study was carried out by Rowan, Williams, Davis and Irwin (RWDI) of Canada. The results from the wind tunnel study became available late in the design and had to be rapidly incorporated into the design of the elements.

Design changes due to value engineering

The method of procuring and funding the stadium was such that value engineering was a critical and frequent source of design modifications. An example of how value engineering can dramatically affect the design of the roof was in the cladding packages. The original concept for cladding the roof was that the transparent area along the leading edge was to consist of 13m of laminated glass. This was reduced to 10m to facilitate a cost saving. The relative weight difference between the glazing and the standing seam cladding was approximately half. Therefore the overall weight of the roof substantially decreased leading to an increase in the required prestress force in the cable net and subsequent foundations.

Systems in place to allow rapid redesign

The key to flexible iterative design for the stadium was attributed to the inter-relationship between spreadsheets and the finite element packages used. The geometry of the stadium was such that it could be defined in a logical manner, in fact the most complex equations used were that of a line and circle intersecting. The use of simple repetitive numbering sequences for the nodes and elements allowed spreadsheets to be used firstly to define geometry, then to attribute properties and loadings and finally to design the elements themselves. The system was so successful that the full re-design of the roof due to the reduced glazing zone took place in only two weeks.

DESIGN OF THE STADIUM ROOF UNDER SERVICE CONDITIONS

The assessment of the roof in service was in many ways more complex than the design of the roof elements for strength. This was further complicated by the strict limits imposed by the cladding. At the time the assessment into behaviour during service was carried out, it was still the intention to use laminated glass as the cladding to the leading edge. The behaviour of the roof was assessed for the following criteria:

- Drainage Slope – The only drainage to the roof was from the gutter at the rear edge of the stadium. Therefore each rafter and cladding panel was assessed to ensure that the minimum drainage slope of 1.5° was always maintained and that the potential occurrence of ponding was removed.
- Visual Deformation – The visual deformation of the rafters was maintained at 1:100 for the cantilever and 1:200 for the backspans, however the overall total deflection was controlled by the deflection of the cable net.
- Shear Strain and Warping – The shear strain, measured as the change in angle at the four corners of a cladding panel, and the warping, measured as the mean change in distance out of plane at each of the four corners, was assessed for every cladding panel.
- Movement – The length of slots cut in the purlins to allow movement was assessed for various load cases and studies were carried out to determine the effect of the movement joints seizing up.

Though each of the rafters are supported by a unique forestay they are grouped together in 'fans' supported by unique masts. These fans act as load sharing devices with respect to relative displacement between rafters. If a rafter is heavily loaded compared to its neighbour, then the increase in force, extension of backstays and rotation of the mast, largely controls subsequent deformation of that rafter. This rotation of the mast also has the effect of lowering all the other rafters in the same fan of cables. Therefore, one of the key areas for concern was the relative displacement between groups of cables. This philosophy cut down the amount of loadcases required, a selection of which are seen below.

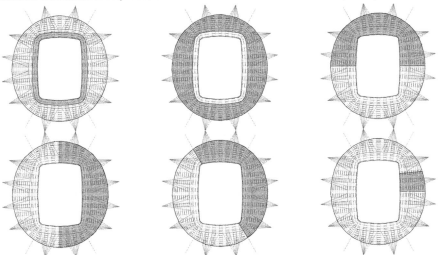

Fig. 10. Pattern load cases required for deformation study.

The relative displacement between rafters in adjacent fans was acceptable, except for the corners between the corner and end masts. In this location, the usual format of alternate braced bays was modified so that the three bays across the fan could be linked together.

DESIGN OF THE STADIUM ROOF FOR ROBUSTNESS
The events of September 11th 2001 reaffirmed the requirements for designers to take into account the effects of disproportionate collapse. This is especially true for stadia where the roofs cover thousands of spectators.

The robustness requirements of the stadium were investigated thoroughly leading to the use of multiple cables for highly sensitive loadpaths.

- Each backstay is made from four elements, which are in turn attached to two discrete foundation plinths.
- Each catenary cable and corner tie is made from four cables.
- Non-linear studies were carried out to investigate the results of losses of elements ranging from forestay cables, to entire towers.

Under emergency loadcase, stability of the four cable elements is maintained with any three cables from the group of four. In addition the primary connections involving multiple cables, i.e. the mast heads, were designed taking into account the eccentricities caused by loosing one or more forestay or backstay cable. The conclusion was that there is a sufficient degree of redundancy in the load paths which should prevent disproportionate collapse.

Fig. 11. Primary elements; (a) Mast Head; (b) Backstay (c) Catenary; (d) Corner Tie.

CONCLUSIONS

The successful design of the cable stayed roof for the City of Manchester Stadium can largely be attributed to a holistic approach taken to the design incorporating the structural engineers, architects and steelwork fabricators. The roof is an interesting structure incorporating many innovative features that are sure to be repeated on future stadia projects around the world, especially the first use of the grounded tension ring philosophy. The recognition that the design would be highly iterative at the start of the project allowed systems to be devised to allow rapid re-design and optimisation routines to be carried out efficiently leading to an elegant, yet very cost effective solution. The main vehicle for the iterative analysis and design work took place using the interface between analysis software (OASYS GSA/Fablon) and spreadsheets (Excel). To conclude, the cable stayed roof of the City of Manchester Stadium roof would not be possible in the short time scale available, without the use of such analysis techniques and the skills to use them effectively and appropriately.

Assembly and deployment of large precision space reflectors

V. I. BUJAKAS
Astro Space Center of P.N.Lebedev Physical Institute, Russian Academy of Sciences,
Moscow, Russia

ABSTRACT

To meet technical requirements of advanced astronomical projects large high-precision space mirrors are needed. In the paper solid-state statically determinate multi-mirror structures are considered as a possible candidate for such mirrors development. Both manually assembled and automatically deployable designs are studied. To avoid reflecting surface adjustment in orbit, self-adjusting locks are introduced in the design. To check the repeatability of unfolding, physical models were made and tested.

INTRODUCTION

Large high-precision space mirrors are needed in advanced astronomical projects (Next Generation Space Telescope, USA, (Ref 1) Millimetron Project, Russia, (Ref 2)). Large precision panels form the reflecting surface of such reflectors. During the last decades a considerable progress was achieved in the technology of large, solid-state, high-precision, thermostable panels (Ref 3.). However the unfolding of large precision space reflectors made from panels is still a serious problem, because technical requirements to reflecting surface accuracy and repeatability of unfolding are very high (Ref 4.).

According to the Millimetron Project a 12-meter space telescope operating at 22.2, 43, 110 300 GHz is to be built. If the requirement to the surface accuracy is $1/16\lambda$, then for the wave length $\lambda=1$mm the errors of unfolding must not exceed 60 microns under the assumption, that there are no other causes of shape deformation. Technical requirements for large far infrared or sub-millimeter instruments are significantly more stringent.

To meet technical requirements of the Projects various designs of reflectors, both assembled and automatically deployable, are considered. We propose that multi-mirror statically determinate structures be used for large precision reflector designs and consider assembled and deployable large reflectors. Within this approach a large unfolded mirror should be a statically determinate structure formed by solid-state high-precision reflecting panels. The number of links in the unfolded reflector is chosen to meet Maxwell condition. In the design links are distributed so as to ensure geometric invariability of the structure and are made as self-adjusting locks. Such locks have to provide high repeatability of assemblage (or deployment) and will permit, we hope, avoiding reflector surface adjustment in orbit.

Two types of designs – petal-type structures and mult-mirror reflectors formed by hexagonal panels - are considered.

REFLECTOR FORMED BY HEXAGONAL PANELS.

In this section we consider the telescope that has a large spherical primary mirror and uses an aspherical secondary mirror for wave front correction. The primary mirror is formed by 7 high-precision (hexagonal in plan) reflecting panels (modules) of the same curvature. The reflecting panels 3.5 m in diagonal are envisaged.

The spherical shape of primary mirror is chosen:
- to simplify the technology of panel fabrication
- to reduce the number of adjusting mechanisms for composed mirror alignment (3 mechanism per panel for spherical panels instead of 6 for parabolic one),
- to simplify the system for adjustment and unfolding repeatability control.

Both manually assembled and automatically deployable versions of primary mirror are considered. The secondary mirror is automatically deployed after the primary mirror is unfolded.

In working state the primary mirror is a statically determinate structure and each side panel is connected with the central one by three locks. The adjustment of the composed mirror may be realized by mutual movement of panels. To perform these movements three adjusting mechanisms for each side panel are foreseen. The adjustment mechanisms are incorporated in the locks. The usage of a statically determinate structure in primary mirror design makes it possible to avoid stress in the design during assemblage, deployment or shape adjustment.

The alignment of a composed mirror in orbit, we hope, may be avoided. That is, the reflector adjusted on the ground and unfolded in orbit has to obtain, with a high degree of accuracy, the desirable shape without additional alignment. To achieve this property the locks are planned as self-settled connections. In this case the adjusted position of the composed mirror becomes the equilibrium state of the design.

Physical model of assembled mirror

To check the concept, the model of a statically determinate multi-mirror reflector equipped with self-settled locks and adjusting mechanisms was made and tested. The model consists of 7 spherical modules (mirrors hexagonal in plan) of optical quality. The curvature radius of the modules is 1041 mm, the dimension – 170 mm in diagonal. The model in assembled and decomposed positions is given in figs.1,2.

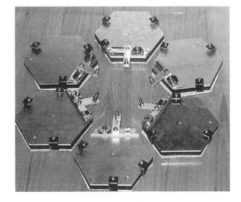

Fig. 1.   Fig. 2.

The locks

used to connect the mirror contain
-   supporting element (figs.3a, 3b, 3c)
-   spherical tip mounted at the end of precision screw (adjustment mechanism) (fig.3c),
-   spring capture that keeps the spherical tip in the supporting element.

Three types of supporting elements and therefore three types of locks are used in the model:
-   a cone socket (lock type I),
-   a wedge-like groove, (lock type II),
-   a flat support (lock type III).

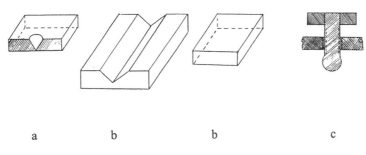

a   b   b   c

Fig. 3.

A spherical tip kept in the cone socket operates as a spherical hinge and adds three kinematic constraints in to the design,
-   a lock with a wedge-like groove introduces two kinematic restrictions in the system,
-   a lock with a flat support introduces one kinematic constrain into the system.

Each side mirror is connected with the central one by three locks (type I, II, and III). These locks introduce 6 kinematic restrictions in the configuration and form statically determinate connection between the mirrors. Therefore the assemblage of the model is stress free.

Adjustment mechanisms

The precision screws of adjustment mechanisms are mounted on the brackets protruded from under the mirrors (fig.6). The supporting elements and spring captures of the locks are placed on the back side of the mirrors (fig.7). Three adjustment mechanisms of the side mirror make the curvature centers of the side and central mirrors coincide. The shape adjustment is also stress free.

Control device

To check the multi-mirror spherical reflector adjustment and to control the repeatability of model assemblage, a simple and convenient mechanical device was proposed, made and used. Let us describe the control device development as a step-by-step modernization of standard spherometer.
1. A standard spherometer is a tripod equipped by a micrometer. The supports of the device form an equilateral triangle, and a micrometer is installed in its center. When the device is mounted on a perfect sphere and the surface is touched by the tip of

the micrometer, the surface bending depth is measured and a curvature radius of the sphere may be calculated.

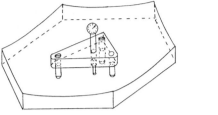

Fig. 4.   Fig. 5.

2. Let us replace the screw micrometer by a digital spring indicator (fig. 4). The scale of the indicator is mobile and its origin may be arbitrarily chosen. Let us place the device on a hexagonal mirror and set the scale in the zero position, then the tip of the indicator touches the surface. *The important property of the device* had to be emphasized: whatever the position of the device on the perfect sphere the reading of the indicator remains zero.

3. We now remove the indicator outside the triangle (see fig. 5), place the device of new configuration on the hexagonal mirror and set the scale of the indicator in the zero position, then the indicator tip touches the surface. Again in the arbitrary position of the device on the perfect sphere the reading of the indicator remains zero.

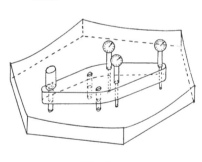 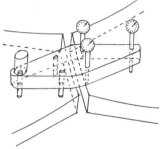

Fig. 6.   Fig. 7.

4. Let us install three indicators on the device as is shown in fig.6, put the device on the hexagonal mirror and set the scales of the indicators in the zero positions.

5. Now we put the device on the model, as in fig.7. The tripod is placed on the central mirror, three tips of indicators are put on the same side mirror. At the beginning the readings of the indicators will be arbitrary.

Adjustment

6. Three adjustment mechanisms located between the side mirror and the central one are used to achieve zero readings of indicators. The zero indexes of indicators mean that the curvature centers of the side mirror and central mirrors coincide. The

accuracy of adjustment is determined by the accuracy of the indicators (10 microns in our case).

Control of assemblage repeatability
7. . After all side mirrors are aligned the design is decomposed and assembled again. Then the device is again used to check the surface accuracy of the composed mirror.

Simulation results
The assemblage repeatability better when 20 microns was achieved in the model, this accuracy may be improved. Simple assemblage of a large composed mirrors with a high accuracy seems possible.

Deployable version of reflector formed by hexagonal panels
The deployable version of composed reflector with hexagonal mirrors is now under consideration. In the folded position the hexagonal mirrors are stacked in column. General idea of deployment is

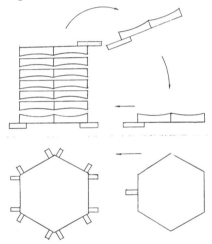

Fig. 8.

presented in fig.8. The mirrors are deployed one by one and each mirror is unfolded in two stages. Preliminary deployment is realized with a low accuracy. After that the hexagonal mirror is put by self - settled locks in the preset position with high accuracy.

PETAL TYPE REFLECTOR
10-meter cm band the petal-type antenna recently was fabricated for Radioastron telescope (Ref 5). The reflector of the telescope is a deployable structure that contain the central mirror and a number of solid-state petals. The petals are deployed simultaneously by synchronous rotation around proper chosen axes. Computer simulation of petal-type antenna opening is shown in fig 9.

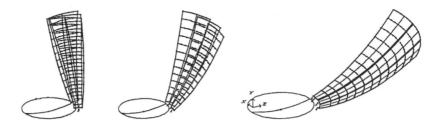

Fig. 9.

To increase accuracy and repeatability of such reflectors opening the unit for high precision petals deployment was developed and tested (Ref 6). The moment of physical model testing is presented in fig.10.

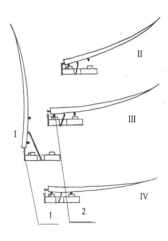

Fig.10.

Fig.11.

1 – the axis of petal rotation during low-precsion deployment

2 – the axis of petal rotation during high-precision fixation.

Within this approach high-precision deployment is realized in two stage. At first the petal is moved from the initial state to a position close to the final operating position; this is low-precision opening. In the second stage a high-precision fixing of the petal is performed.

A self-settled statically determinate was used to achieve the required high precision position of the petal in the deployed state. Three spherical metal supports were placed on the frame of the petal; they form an equilateral triangle. There are also three wedge-like grooves on the foundation of the central reflector, whose axes are oriented at 120° from each other. While the antenna is being deployed, the petal rotates around the axis of the deployment mechanism. Supporting spheres then enter the wedge-like grooves, and a pressing device locks the petal into its final prescribed position.

Three wedge-shape grooves introduce 6 kinematic restrictions to the configuration, and the resulting statically determinate petal support system has the property that the ultimate position of spherical supports in the grooves also is the equilibrium state of the system. Any movement of the spheres does work against the fixing forces, which will return the system to its equilibrium state.

The tests showed that the errors in the final position of the petal model in the direction of the normal to the reflecting surface did not exceed 10 microns.

REFERENCES

1.   http://www.NGST.nasa.gov;
2.   http://www.asc.rssi.ru/millimetron
3.   **HINKLE J. D. PETERSON L. D., HARDAWAY L. M. R., HACHKOWSKI M. R.,** Microdynamics of Lightweight Precision Deployable Structures, Proceedings of IUTAM Symposium on Deployable Structures: Theory and Application, Kluwer A. P., 2000, pp 143 – 153.
4.   **HELWIG G., DEYERLER M., BRUNER O.,** Composite Technology for Precision Reflectors, 20th ESTEC Antenna Workshop on Antenna Technology and Antenna Measurement, Noordwijk, 1997.
5.   **http://www.asc.rssi.ru/radioastron/Description/progress_eng.htm.**
6.   **BUJAKAS V. I., RYBAKOVA A. G.,** Millimetron Project: a Unit for High-Precision Deployment of Antenna Petals, Proceedings of P.N. Lebedev Physical Institute, 2000, v.228, pp 129 – 142, (in Russian).

Product development of the glued aluminium delta trusses with customized cast joints for the Dutch aluminium centre

M. EEKHOUT
Faculty of Architecture, Delft University of Technology, the Netherlands

INTRODUCTION

Architect Micha de Haas, as the result of winning a design competition, has designed the Dutch Aluminium Centre in Houten, the Netherlands. The concept was a rectangular all aluminium single storey building with an aluminium structure, floors and claddings, resting on more than 600 aluminium piles. This contribution deals with the design and development of a special aluminium space frame system with customized cast aluminium joints and glued tubular member ends. The eight main features of interest in this project are:

- The all aluminium 'Streamline' space frame, fitting in an all aluminium building;
- The curved form of the space frame trusses and the individualization of the joints;
- The essential form of the designed components in a 'Streamline' system;
- The problems of an axial screwing method with diagonal bracings in assembly;
- The use of glued connections between tubes and tube ends;
- The individualized, yet industrialized aluminium cast joints;
- The transfer of the aeronautical/car industry technology to the building industry;
- The different processes speeds of experiments in an actual building project.

Fig. 1. Aluminium Centre under construction.   Fig. 2. Delta trusses.

In the first phase of the building design, a standard space frame was intended. Then, the actual building started with a further commitment of Dutch industrial aluminium parties. Instead of the Tuball space truss system, already developed in 1984, Octatube proposed to add a new development to the building process. Whereas the Aluminium Centre, being the client, was more interested in an entire building and all its commitments to that, Octatube thought it more interesting to take up new technical challenges. From the very beginning, it was agreed that Octatube would maintain the leading position in the project. The choice for a new development consisted of the introduction of customized cast joints and the aeronautical gluing technology. Parties in the Netherlands were willing to contribute to both techniques. Furthermore, the aluminium would be supplied. By the nature of the development process, the chair Product Development of the Delft University of Technology would support and publish the process, so that the general message of product development would become public knowledge.

PRODUCT DEVELOPMENT IN GENERAL
During the product development process of the space aluminium roof structure for the Aluminium Centre in Houten, it became clear that fundamental research, directed at product development, could only be fitted into a current building project with the greatest caution. Basically, the product development process and the building process are two independent routines. They are not compatible in the aspect of time, but they often need one another in the sense of stimulation, or for source and cause. It is a good thing to deal with this contradictory dependence.

Aspects of time
The following aspects are of importance. When a development process is started up, there is a rough indication of the direction in which the process will move, but only with the making of the design it becomes clear how this will actually look and how it will be assembled. This applies to the design of an entire building, as well as to the design of a building component, like this project. The designers give a very personal interpretation to the design by their choice of composition. Insight, concerning the possible realization of the solutions, arises gradually during the product development process. In the initial stage, relatively little knowledge and experience of the actual experimental material application will be available, but the most important design decisions must be made. At that time, no feedback has been possible and consequently no experience yet. (Therefore, it is tempting to bring a composition of well-known techniques to the task). It often occurs that the design process manifests itself iteratively. First of all a concept has to be made, after that a prototype has to be tested and subsequently improvements to the original concept are always next. This cycle may have to be passed through for several times to come to a permanent useable product that complies with all the requirements of the basic principles. These are considerations of unfamiliarities and insecurities that accompany each new process of product development.

On the other hand is the process of the building project. Working against the clock, as is usual with current larger building projects, there is only a limited length of time for each part. This means that for the engineering, the sub productions and assembly at the factory, as well as for the assembly of the various components of the building, limited passing through times are inevitable. These time limits are usually specified on time sheets of the project's planning scheme. The architect must identify the components that will require much development work in their processing, and assess length of time and complexity. It is sensible to set the starting point of development for the most crucial components as early as possible. For example, particular aluminium project façades are often supplied before the architectural tendering. Particularly experimental building components, like the delta trusses in this case, require sufficient time and quiet to consider broad frame decisions. Some aspects need to pass through one or more iterative rounds before the final consequences for building can be established, especially for the other building parts (i.e. in the form of reactive forces, fastenings or connections of a building component). Especially for the sake of reliable processes of product development, continuous feedbacks and reconsiderations are a necessity to acquire high quality projects. Furthermore, new prototypes are often needed after the design has been changed because of testing results, which means new production and assembly cycle for the prototype repeatedly.

Unfortunately, in the case of the Aluminium Centre, the gaining of time as a result of delays in the preparation and production of other building components and their assembly at the building site, could not or hardly be used for the important design and development process of the delta trusses. The process already had a kind of **'just-in-time'** character, due to its free of charge nature. This 'just-in-time' character in particular, takes away all possible elbowroom for setbacks in experiments. Therefore, the building preparation and realization ran parallel with the development process of the newly glued aluminium construction with cast joints. This increasing arrears in the preparation, production and realization of the remaining building components had nothing to do with the product development of these components, but with the ordering of materials and the starting up of their production, after the dimensions were established and the detail drawings were made. This could be called the engineering of building parts, building engineering or building development.

Aspects of budget

The Aluminium Centre took on various parties from the aluminium industry, to collectively bring about a basically gratuitous realization of their new accommodation. Led by the **'pilot'** company Octatube Space Structures, each of the parties bore its own expenses and ran its own risks in return for a share in the intellectual ownership of the final product. This principle had both a bright side and a drawback for the development process.

A positive result was that the companies were receptive to acquire new knowledge, expertise and insights. Octatube stated only to be interested in a contribution for the Aluminium Centre if actual experimental steps forward could be taken. The casting company had developed a casting method that, in principle, could be suitable for very small series. They would gladly examine and test this method. Sergem, an engineering office, manned with former Fokker employees, was specialized in aircraft construction and interested to take the step to the building industry by giving advice on gluing technologies.

Because a development budget in this **'free of charge'** project was lacking, the participants wanted to try to realize their own share in the strived for new industrial product only with minimal means, within a set period. This restricted effort caused parties to wait for the other parties to make their contribution and subsequently react, rather than participate in a collaborative engineering process and anticipate on the results of team members. Only to discover later that much energy was spent in vain, because successive activities could not be realized. Therefore, the consequence was that this 'free of charge' project was not carried out with full labour force. Waiting for each other's results extended the total development process in such a way, that the profits from building delays could not be cashed.

To make the entire process run effectively and efficiently, a cooperation of the five participating parties was set up. It was agreed to respect each other's contribution in terms of copyrights, but also to communicate openly with one another. The proceeds of each participating company had to be from the offspring of the company products, and/or the company services in a later stage and the copyrights for every own contribution by means of an individual or mutual patent application.

In general, it can be stated that many models and variants need to be made for the development of modernizing products and their industrial constructing processes. Like the products, time to think and do is expensive too, but is necessary if a good product is the result.

Project independent product development is an expensive undertaking for a company, but it has its function within the framework of the continuation of the company's future, as an investment in the long run and in the quest for more benefit to the company. This also shows the difference between the functioning of contracting or producing companies. In particular, now that the customer, as compared to some decades ago, dominates the current market it is important for companies to strive for a considerable amount of individuality, identity or possible benefit, by which they distinguish themselves on the market and so obtain commissions. However, many companies mainly follow the route of short-term investments of applications, tendering and cashing the commissions.

Basically, the present process has passed through only one cycle of design, engineering, production, assembly and tests. The series of tests was restricted to the testing of individual components. There was no time left for the full scale mechanical testing of an entire truss, as was planned, due to the fitting into the assembly of the building schedule. One last attempt to full scale testing in the building itself at level, failed after calculation of the necessary water load and its consequences for the floor and foundation of the building.

DESIGN
Objective
The design of the lenticular aluminium delta trusses should lead to a new aluminium building system (the development of a new type of aluminium truss girder with a triangular diameter for medium-sized free spans). Furthermore, the introduction of the transfer of the gluing technology from aeronautics to commercial and industrial building, as well as the introduction of the technology of singular aluminium casting. In addition, the experimental product development team was a prototype of the cooperation of companies in co-makership for a building component.

Configurative delta trusses
Initially, the design of the architect showed one delta truss and two half-delta trusses. Because of the triple division in the functional programme of the exhibition room, this shortly changed to two large delta trusses with two half trusses (so, three interspaces). Subsequently, Octatube turned the truss around, so that the basis became the top and a nicer, logical architectonical arrangement of the space was established. The roofing sheet span was reduced by that (from 3600 mm. to 1800 mm.) and so became realizable in simple aluminium profiled sheeting.

Fig. 3. Delta trusses under construction.

The delta girders would each be built from two plane trusses, with bent upper members and straight lower members, consisting of aluminium tubes, aluminium cast joints and structural glued joints. The roof girders were used as stabilizing cross bracings for the 'spatial girder'. Octatube drastically reduced the number of diagonals by choosing a larger modulus size (from 1200 mm. to 1800 mm.). The advantage was the reduction of the number of joints and therefore time and money. All connections had to be designed in a 'Streamline' shape. Yet, this basic principle should be set aside if the multiplicity of the joint would become an issue.

ENGINEERING
Configuration of joints and bars
Industrial production considerations resulted in an important change. Initially, Octatube thought of a fixed joint distance, through which a variable triangulation would occur. From this, a fixed triangulation (all equilateral triangles with 60° angles) with more different joint distances was developed. The advantage was that a more regular joint geometry in the lower member was achieved. Placing all diagonals in 60° angles reduced the tension in the bars. The reduction of the tension proved to be an additional advantage. Initially, the upper and lower members were separately divided in equal bar lengths (therefore, two types of lengths). The change to a fixed triangulation resulted in the small disadvantage of an increasing number of bar lengths, but this was nothing compared to the advantages of a reduction in the number of cast joint types.

Introduction of screwed joint
A direct structural glued joint between aluminium bar and cast joint resulted in too much tension concentrations in the glued joint. This was caused by the wall thickness of the bar, which was too thin. In addition to this problem, the tolerances of the cast joints, the dimensions of the truss and the chosen triangulation turned out to be so big that they could not be measured up to the much smaller tolerances (0,2-0,3mm.) which were required for the glued joint.

The solution was a screwed joint at the ends of the cast joints. This way, the tolerances of the triangulation could be absorbed. Subsequently, the development group chose for the direct absorption of screw threads in the lost foam model and not for the making of threads afterwards. This was done to do as much justice as possible to the industrial production. Only a minimum number of acts resulted in an optimal qualitatively good product. One consequence of this choice was that two types of foam models would be necessary, one with left-hand screw thread and one with right-hand thread. Tensile tests showed that the screw thread, which was cast along (left and right), would meet the required strength to a large extent.

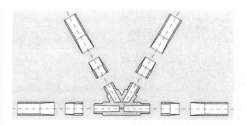
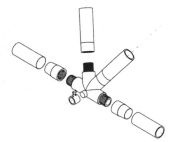

Figs. 4, 5. Cross section and exploded view of castnode & glued connections.

Introduction of glue-ring

The above resulted in the introduction of aluminium 'glue-ring' on both sides of the aluminium bars. The structural glued joint shifted to the bar glue-ring connection. Initially, a glue-ring, with an internal screw thread, was entirely glued in on the inside of the bar. Thus, the glue-ring remained invisible and only one seam between bar and cast joint could be seen.

Tensile tests showed that the aluminium member wall was too thin at the joint. Consequently, a second visible seam occurred and the glue-ring became not just the physical, but also the optical connecting element. The wall thickness of the aluminium member could be reduced from 10mm to 5mm, while the bar ends got an internal narrowing to keep the tension concentrations in the glue line minimal. The bar was pushed over the glue-ring. For the sake of safeguarding the glue thickness, the member butts against the glue-ring. Therefore, the narrowing is not entirely carried through, but stops at a thickness of 1 mm. This prevented the glue from squeezing out of the joint.

PRODUCTION
Members

The quality of the aluminium was 6060 T6 AIMgSi-0,5. The total number of members was 186. The outsides were anodized with a matt surface. Calculations showed that all occurring (normal) forces should be absorbed by tubes with an external diameter of 60mm and a wall thickness of 5mm.

The accuracy of the internal diameter at the point of narrowing was important, because the narrowing at the outside of the glue-ring should not entirely be completed, so that the member could butt against an edge of 1 mm. height. All this was necessary for the safeguarding of the thickness of the glue line. The narrowed surfaces had to be blasted thoroughly with aluminium oxide, to improve the bond of the glue. The tolerance for the linear measurement of the members was set to approximately 0,1mm, but was basically of marginal importance. It was safe, because the screwed assembly could absorb the tolerance for the entire truss length. At this point, a not exactly fitting connection with a margin of one turn per member, was already considered.

Cast joints

The casting process passed off according to the principle of the lost foam model. This technique is often employed for the production of small series, in this case 102 pieces, divided over 6 types. The choice for the type expanded polystyrene affected the surface condition of the joint (globule appearance). The assembly of foam model components into a complete model was done by hand. The resulting tolerances could not be united into a glued joint. The introduction of a screw thread proved to be necessary.

Figs. 6, 7, 8. Polystyrene models, casting and rough cast products.

By employing a high quality aluminium cast alloy (A356-T6), the required strength of the screw thread could be guaranteed. The metric screw thread M48 was chosen to go along with the foam model, instead of the making of screw threads afterwards. On the one hand, two moulds would have to be made (right- and left-handed screw thread), which would involve considerable extra costs. On the other hand, an additional treatment of the product would not help the total industrial character of the product. Tensile tests of the screw thread which was cast along, showed that the required strength could be required to a large extent.

By the triangulation of the delta girders, during the assembly phase, it showed that more tolerance in the screw thread was needed. Therefore, the thread needed additional rotation to make a good fit.

Glue-rings
Firstly, the glue-ring as a whole was inserted into the member. The sudden material transition from 5mm to 10mm caused too much tension in the glued connection. As mentioned earlier, the glue-ring now inevitably became visible (so, two seams), to keep the material thickness at the point of the glued connection constant. One half was externally narrowed under an angle of 5,7° with an edge against which the bar would butt. The other half was internally provided with metrical screw thread M48. The accuracy of the external diameter at the point of the narrowing, again was of great importance (see 'Members'). The narrowed surfaces had to be blasted thoroughly with aluminium oxide to improve the bond of the glue.

Glued connection
Cold setting glue was chosen over warm-tempered glue, so that no additional production step of heating was needed. It could be applied to large elements that do not fit into an autoclave. It will be fully tempered after seven days at room temperature.

Subsequently, nine tensile tests were carried out with Epoxy glue 'Scotch-Weld 2216 B/A' (3x 12,5, 25 and 50mm) with singular lap joint test pieces and six with bevelled lap joints (50mm; 5,7°). All test pieces were 5mm thick and 30mm wide. The average shear strain was 12,5N/mm² ± 1,7. With this, approximately four times the required (calculated) force can be resisted. Greater length of overlap over 15-20mm would hardly have an effect, because tension concentrations at the edges are determining. However, by employing bevelled components, an increasing length of overlap proves to be sensible, because tension concentrations are then partly levelled. Surface treatment also proves to be of great influence, datasheets show an average shear strain of 17,0N/mm².

In a phase, already too late, it was discovered that a roof strip of 2 metres width had to be supported by the 'half' delta girders. The glue joints would collapse by this. To this purpose, additional structural arrangements had to be made outside the roof construction.

ASSEMBLY
The theory
Beforehand, it was established that the assembly of the triangular trusses by means of screw joints would not be possible in theory, and in practice should be based upon broad tolerances. The assembly should be carried out as follows: after spreading out all cast joints and members with glue-rings, all bars would be tightened simultaneously. In theory, it was recognized that it would be impossible to form the triangular trusses from axially turning tubes. Yet, it was expected that the members would yield sufficiently. It was also recognized

that the maximum 'tightening difference' between the various screw connections could amount to one rotation per joint.

To minimize this tolerance, the production process would have to become far more accurate, more time-consuming and cost increasing. Some examples are that both glue-rings at the ends of the bars should be better geared to one another. Furthermore, during the composition of the separate foam models to a joint, the screw thread on each expanded polystyrene foam model should be precisely determined. Because this composing of the foam models was done by hand, this proposed quality-increasing measure would be impossible.

Practical problems

Yet, some greater problems arose. The simultaneous tightening of the members failed. They soon got stuck. This was caused by defects in the angles of the cast joints that could not entirely be absorbed by tolerance. This could have made the members bend slightly. However, due to their tremendous rigidity, they did not bend and because of this, they caused immense tensile forces perpendicular to the glued connections which, in their turn (three of them), caused delaminating.

Time was short. New joints could not be cast and the bar heads could not be finish-turned wider anymore. It was an unexpected detail that the glue line apparently could not bear the tensions. Subsequently, it was doubted to what extent the glue lines in the not collapsed joints could still be trusted.

Firstly, a design was drawn up in which only glued connections were used. By the triangulation of the truss, the members and joints would all have to be connected in one go. Since there are no autoclaves in which a truss of such dimensions would fit and also because it would not come to a general structural system, it seemed necessary to apply a cold setting glue epoxy glue type 'Scotch-Weld 2216 B/A' of the firm 3M (seven days at room temperature). A disadvantage of cold setting glue types, as opposed to warm setting ones, is the lower quality.

Practical solutions

The screw thread of all cast joints were cut once again. This was a lot of unnecessary work. The assembly, however, was much easier and the tensions in the glue connections were minimized during the assembly process.

A safety locking system was installed in the members, because of the suspicion regarding the non-delaminated glued connections. A stainless steel cable with screw-threaded end (M20) with a diameter of 10mm. was pulled trough the lower bar and anchored this in the ends of the castings. To this purpose, the joints had to be bored! For the upper side, a lead-through was made for the bearing cams of the girders. Here as well, the stainless steel cables were anchored in the in the ends of the castings.

Because the tolerances, with regard to casting and glued connection, were not in proportion, the glued connection was shifted. A glue-ring was introduced by which the structural glue line was relocated between the bar end and the glue-rings. These smaller elements were suitable for a warm setting epoxy glue type.

Delta girder as a whole

In their factory, Octatube assembled the cast joints and the members with glue-rings to six flat lenticular truss girders. Subsequently, two whole delta girders were formed from these, by the connection of the straight lower members by means of a rotated aluminium connecting ring at the point of the cast joints, and an M6 securing bolt. It was important for this connection that the joints were accurately put next to each other in the linear direction of both truss halves, to make the aluminium connection ring fitting. To realize this, the bars were tightened equally as much as possible.

Fig. 9. Assembly

Fig. 10. Connecting the trusses.

ASSEMBLY OF COMPONENTS

To guarantee the 60° positions of the two whole delta girders during transportation, strips served temporarily as the third side. These two whole delta girders, together with the 'halves' were transported to the building site at Houten. The following types of connection for these elements were used.

Connection delta girder – I– edge beam

For the endings of the delta girders, special cast joints were made, to which an L-profile was cast. To absorb changes in the length of the truss because of temperature differences, a roller support at one side of the delta girder was realized by a slotted hole in the mentioned L-profile in the linear direction of the girder. This connection was made at the point of the bottom flange of the I-edge beam. A hinging connection was fitted at the other side, by means of one single bolt connection in the linear direction of the truss. At first, it was considered to connect the delta girders to the body, but then the truss would not be capable of moving freely and unwanted moments in the body of the I-edge beam might occur. Fastening them crosswise with two bolts in the mentioned L-profiles prevented toppling over of the whole delta girders. This crosswise stability idea proved to be sufficient during the assembly.

Connection 'half' delta girder – C-edge beam

To prevent buckling of the lenticular 'half' delta girder, it was hinged-connected to the body of a C-profile at every joint in the lower member. Again, this was done by a (longer) turned aluminium distance block, locked with an M6 securing bolt and fastened with an M12 bolt.

Changes in the length of the truss, as a result of temperature fluctuations were absorbed in the C-profile by applied horizontal slotted holes at the point of the cast joint C-girder connection.

Connection delta girder – roof purlin

The temporarily applied strips at the basis of the delta girders were replaced by roof purlins. The cast joints of the upper members have an aluminium block that was cast along, the upper surface is 30 x 50mm., in it is a hole with a tapped screw thread. The aluminium extrusion C-profiles (50 x 100 x 50 x 2mm.) were already equipped with holes at the point of connection with the cast joint in Octatube's factory. Only at the building site, delta girders and roof purlins were connected to each other. For this connection as well, it was of importance that the cast joints were nicely placed on one surface, crosswise in the separate (half) delta girders. This was done to stop transformations in the roof purlins as much as possible, casu qou to prevent them.

Fig. 11. Construction view.

CONCLUSION

It is possible to design and develop an aluminium space frame system with customized cast aluminium joints if an appropriate node design is maintained. In theory, all joints can be different, with regard to the development of digital baroque architecture with its irregular a-systematic composition. It is also possible to use glued tubular endings. Due to the nature of the triangulated space frame geometry, it is not advisable to design axially screwed connections in a triangulated space frame. It would be better to develop cross pin connections. The development of such structural systems is best done independently from ongoing building projects.

Structures of high-rise buildings designed by means of prismatic space frames

J. REBIELAK, Department of Architecture, Wroclaw University of Technology, Poland

INTRODUCTION

The shaping of the structural systems for the high-rise buildings, by means of certain types of the space structures, belongs to the two main branches of the research activity of the author. The systems proposed earlier by himself are described e.g. in Refs 3-14. The terrorist attack against USA in September 11th, 2001 and the collapse of the twins of the WTC in New York caused the new perception of the form and the tasks of the tall building structures. The paper contains brief description of the structural systems developed by the author, shapes of which seemed to him earlier as slightly complex, but in these new circumstances they can be considered as some of the many possible answers for the new challenge of the design of the tall buildings. The main goal in the processes of shaping of these systems was the endeavour to give to the structure as much as possible great spatial rigidity by means of relatively simple means. At the same time it was aimed to make possible the evacuation of the people, in cases of emergency, not only by the vertical transportation means, which at present are located usually only in the central cores of these tall buildings.

SIMPLE FORMS OF PRISMATIC STRUCTURES

It is assumed that when the bar length equals approximately 4 meters, like the crystal structure model of which is shown in Fig. 1a, then the space structure can be habitable, Ref 1. The double-layer space frames, in the propositions given previously by the author, were vertically placed around the perimeter of the building. The spatial rigidity of them is high but this arrangement may cause the huge increasing of the force sizes acting in some of its component parts arranged particularly in the border areas of each structural segment. Even the subdivision of the perimeter space frame into some will be not the sufficient solution in order to decrease the unjustified too high level of the force sizes acting in members of the main support space structure. On the other hand the space of the perimeter frame could be used as a space for the another option of the evacuation ways in the unusual situations. Figure 1b shows

Space Structures 5, Thomas Telford, London, 2002

the view of the physical model of a skew belt of a crystal structure, which may be applied in the designing of the alternatively types of the support structures. In this particular case the inner space of it is devoted the staircase of a multi-storey building. Figure 1c shows the result of the transformation of another shape of the space frame "square on square" being at the beginning the basic form of the circumferential space structure of a tall building. The number of the structure members has been reduced in the way that the remainders of the initial arrangement create the skew spatial belts. The diagonal members, placed onto the main façade plane of the building, will transfer the forces to the all columns located along the perimeter of the building.

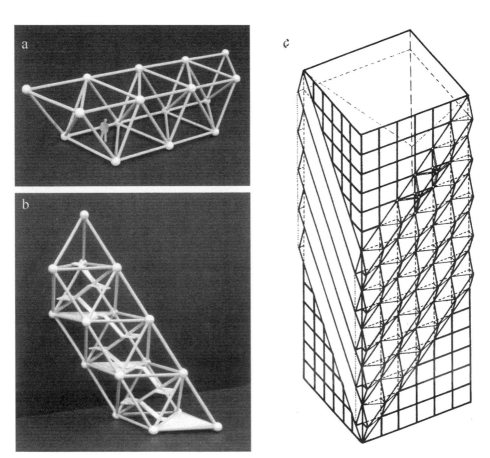

Fig. 1. a) Example of habitable form of a space frame, b) physical model of a prismatic structure, c) simple form of a prismatic structure.

The outer diagonal members will transfer the horizontal forces in the same way except the members located close to the corners. The spatial skew belts are not connected on the surfaces of the outer diagonals in order to focus the forces exactly along the corner columns. Figure 1c shows only an example of endings of these skew belts spatial belts. In the vertical cross section of this part of the façade one may notice the triangular, prismatic shapes of these belts. It is obvious that the space of all these skew belts and between them may belong to the inner space of the building by appropriate arrangement of the curtain walls, see the left top part of Fig. 1c. The dimensions of these skew belts should make possible to design in these areas the additional means of the transportation between particular storeys, Ref 14.

The skew prismatic space structure, shown in Fig. 1b, may be arranged onto the vertical facade of a tall building in the way presented in Fig. 2a. It is dictated by the endeavour to keep the advantages of application of the double-layer space frame and to play the jet forces along the main and single directions of edges of the huge triangular area of a façade and in order to avoid the disadvantages mentioned previously. The strips of the prismatic structure are placed inside area of the chosen triangle along its borders. The height of a single floor (h) may equal the height of a modular triangle of that grid.

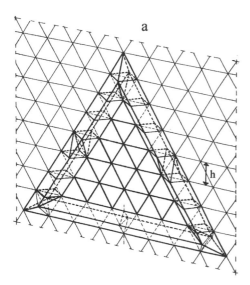
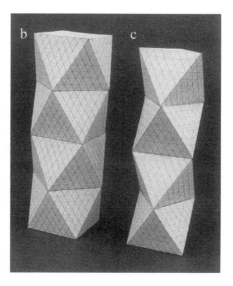

Fig. 2. a) Triangular form of the prismatic structure, b) shape of a tall building consisting of square antiprisms c) form of a tall building built by means of regular triangular antiprisms.

These form of the proposed prismatic structures can be placed onto each triangular face of the polyhedral forms of the example shapes of the multi-storey buildings shown in Fig. 2b and in Fig. 2c. It could be more economic when they will be arranged appropriately onto every second face of them. The spaces of the boundary trusses may belong simultaneously to the outer and to the inner space of the designed building. Figure 8 presents the model of the same part of that system, which boundary bands of the trusses are only outside the main space of the multi-storey building. The case when the space of the prismatic structure is arranged outside the surface of the main support columns or elements could be easily applied for the reinforcement of the existing high-rise buildings.

STAR SHAPED PRISMATIC STRUCTURES

The last presented shape of the prismatic structure is built by means of the space truss bands, which are placed along the edges of a huge triangle. This triangle, as itself, can be the starting form for the creating a gigantic prismatic space frame. Figure 3a shows the basic shape of the huge prismatic structure. This shape consists of the three tetrahedral forms, which are joined together by means of two triangular faces. The entire set of the three tetrahedrons is inscribed inside e.g. the equilateral triangle. The entire form of such a structure is supplemented by means of additional short and horizontal members, which will be placed on levels of the designed floors. On the example of the this form of the prismatic space frame it is easy to notice that the resultants of the forces from its side faces will be focused along the side edges of this huge spatial structures. The streams of forces will be directed along the designed lines what will make possible the simple and relatively easy way of the force transmissions in the entire structural system.

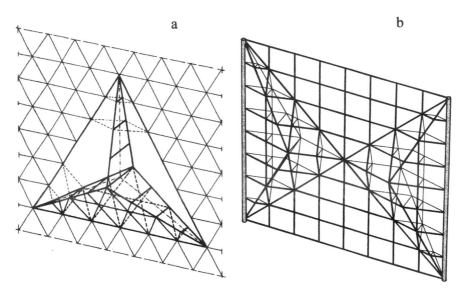

a b

Fig. 3 a) Basic form of the star prismatic structure, b) example of the transformed shape of the star prismatic structure.

The proposed form of the prismatic space frame is very rigid and it has enough inner space in order to contain the installations necessary to arrange the alternative ways of the evacuation in chosen parts of the perimeter spaces of the building. The alternative staircases can be for instance bounded by means of the transparent glass walls, which may be fireproof and which may allow almost the unobstructed daylight of the inner floor. It can be expected that this form could in the very good way fulfil the requirements of the safe and the rigid structural system for the high-rise buildings.

The outlines of the star shaped prismatic space frames may obtain many forms of different triangles and they may be located in various ways onto the facades of a tall building. The another way of the application of the proposed types of the prismatic space frames in structural systems of the high-rise buildings shows Fig. 3b. In this case the two prismatic frames, which have the forms of rectangular triangles, are spaced between two main columns of a high-rise building, they are opposite directed and are joined in the central node. In this way it is built a kind of the spatial girder which could be considered as the main huge component part of the lacing system of the whole vertical space structure. The main columns are located in the corners of the tall building. In this case the height of the single prismatic structure, along the vertical direction, equals the height of a set of the eight floors.

The basic form of the prismatic space frame can be adjusted to many various shapes what enables to design the tall buildings of many various and interesting architectonic views, Ref 14. For instance a single set of the triangular prismatic space frames can be located only onto the second square area of each façade. The typical arrangement of these structures onto the every square field of each façade is shown in Fig. 4a. The whole perimeter structure of a building consists of five gigantic segments vertically located each on other. Figure 4a is obtained as a kind of visualisation of an execution of the programme written in Formian, Ref 2, which programmes defines the numerical model of the proposed structural system.

The structural system designed in that way may prove as the very efficient technical solution but because of some architectonic reasons it sometimes could be considered as not enough satisfactory proposition. Therefore it is proposed another form of the structural system which also resembles the pattern of the simple vertical located truss with the X-shaped bracing system. The general scheme of the system presents Fig. 4b. The height of the unit will equal the height of several storeys. The structure will be made by means of these units vertically arranged along every side of the building. In that way it is built the elongated version of the prismatic space frame. In the middle between the main columns are located simple forms of the vertical frames. In area close to the middle distance between the two main columns of the building it is plan to arrange the additional means of the vertical transportation. All the remarks regarding these additional means remain the same like they were spoken before. The endeavour to locate there the additional means of the vertical transport is caused by some

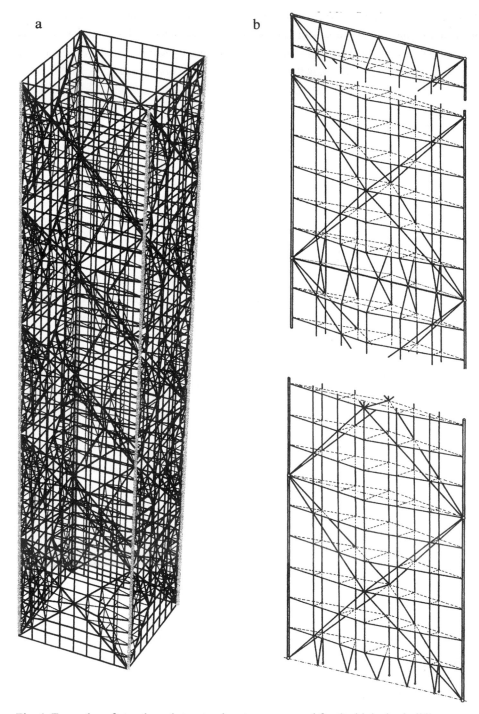

Fig. 4. Examples of star shaped structural systems proposed for the high-rise buildings.

reasons. One of the most important is that the central areas of each side edge of the horizontal projection of the floors are usually subjected to act the smallest forces and which cause relatively small deformations of the structural system of the entire tall building. The prismatic structure is located along the perimeter of the designed tall building, Ref 14.

The proper application of the offered structure requires the suitable arrangement of other component parts along the perimeter and the whole height of the tall building. Owing to this the system can be applied in the design of the real building but simultaneously the additional component parts make the structure somewhat complex. The necessity of the special arrangement of some areas of that system is caused by the reasons like for instance the influence of the strains of the entire structure on the sizes of strains and forces appearing in its component parts. The thermal load, caused by the temperature differences between the outside and inner space of the building, can be important for the high-rise building structure. But the most important factor in case of this shape of the support structure is the unsymmetrical way of the load of the intermediate columns spaced between the main columns placed in the corners of the tall building. The inner intermediate columns will be loaded by the forces of several times greater values than the forces falling on columns designed along the perimeter.

In order to decrease the influence of all these elements on the strain values and on the force sizes acting in the particular component parts of the whole system is divided into some segments along the vertical direction. The height of a single segment will be equal the height of e.g. a set of three square units with the X-shaped system of the bracing. The two particular segments are connected together on levels where they have the common nodes placed onto the central vertical plane of the elongated prismatic structure. Because the inner columns are subjected to the greater sizes of the load, than the perimeter columns, that is why they run in the uninterrupted way to the fundaments, being especially massive along the perimeter of the building. The eccentric shape of the prismatic space structure will cause the increasing of the force values acting in the intermediate columns but it should allow to decrease more greater force sizes in many other component parts of the system, which may be caused by the bending of the whole building. It could be also a good technical solution regarding the influence of the thermal load. The huge X-shaped bracing system can be also located onto a single central and vertical plane of the prismatic structure but it this case it is necessary to arrange an additional system of short members in the chosen parts of its inner space.

CLOSING REMARKS

All the proposed types of the structural systems should be subjected to many comprehensive analyses in order to estimate their practical suitability to the designed purposes. The high-rise buildings, due to the application of the proposed structural system, could be of the great space rigidity , their safety in many emergency cases could be much greater than systems applied today and finally these buildings can obtain the interesting and the individual architectonic views.

ACKNOWLEDGEMENTS

The material of this paper was prepared by the author recently when he was involved in several research programmes. The numerical models of the proposed structural systems are prepared in the programming language Formian. The author wishes to thank Prof. Hoshyar Nooshin and Dr. Peter Disney from the Space Structures Research Centre of the University of Surrey in the United Kingdom for their kind support in the use of Formian.

Many numerical models of the proposed structural systems were prepared during the realisation of the program No 8 TO7F 019 20 entitled "Numerical methods in the design of the architectonic space structures" and supported by Komitet Badan Naukowych (KBN) in Warsaw. The author would like to thank the KBN for the support in this matter.

Basic models of the proposed structures were prepared during the author's stay in the Kawaguchi Lab in the Institute of the Industrial Science at the University of Tokyo in Japan. His stay there was possible due to the support of the Japan Society for the Promotion of Science (JSPS). The author wishes to extend his thanks to Prof. Ken'ichi Kawaguchi and to the Japan Society for the Promotion of Science for the opportunity to carry out the research in Japan.

REFERENCES

1. GABRIEL, J. F., Are Space Frames Habitable?, in: Beyond the Cube. The Architecture of Space Frames and Polyhedra, ed. Gabriel J. F., John Wiley & Sons, Inc. ISBN 0-471-12261-0, New York/Chichester/ Weinheim/Brisbane /Singapore/ Toronto, 1997, pp. 439-494.

2. NOOSHIN H, DISNEY P, YAMAMOTO C, Formian, Multi-Science Publishing, Brentwood, 1993.

3. REBIELAK J, Space structures Used in the Construction of Large Span Roofs and Tall Buildings, in: Space Structures 4, Proceedings of the Fourth International Conference on Space Structures, eds. G.A.R. Parke, C.M. Howard, Guildford, UK, 5-10 September 1993, Vol. 2, London: Thomas Telford, 1993, pp. 1581-1590.

4. REBIELAK J, Proposals of Shaping Multi-Layer and Vertical Space Structures, in: Proceedings of the International Conference on Lightweight Structures in Civil Engineering, ed. J.B. Obrebski, Warsaw, Poland, 25-29 September 1995, Vol. I, pp. 77-82.

5. REBIELAK, Proposals of Application of Space Structures in Shaping of Constructional

Systems for Large Span Roofs and Tall Buildings, in: Proceedings of the 7th International Conference on Engineering Computer Graphics and Descriptive Geometry, Cracow, Poland, 18-22 July 1996, Vol. 1, pp. 57-61.

6. REBIELAK J, Proposal of Structural System of a High-Rise Building Formed by Means of Chosen Types of Space Structures, in: Lightweight Structures in Civil Engineering - Local Seminar of IASS Polish Chapter, ed. J.B. Obrebski, Warsaw, 6th December 1996, pp. 148-151.

7. REBIELAK J, Examples of Shaping for Large Span Roofs and for High-Rise Buildings, International Journal of Space Structures, special issue on Morphology and Architecture, Volume 11, Nos 1&2, 1996, pp. 241-250.

8. REBIELAK J, Construction Systems for Tall Buildings Shaped by Means of Space Structures, Proceedings of the International Conference: Challenges to Civil and Mechanical Engineering in 2000 and Beyond, June 2-5, 1997, Wroclaw, Poland, Vol. III, pp. 389-395.

9. REBIELAK J, Some Proposals of Space Structures Shaping, Proceedings of the International Colloquium: Structural Morphology - Towards the New Millennium, Eds: J.C. Chilton, B.S. Choo, W.J. Lewis & O. Popovic, University of Nottingham, August 15-17, 1997, Nottingham, United Kingdom, pp. 144-151.

10. REBIELAK J, Structural Proposals for Long Span Roofs and High-Rise Buildings, ed. J.B. Obrebski, Proceedings of International Conference on Lightweight Structures in Civil Engineering, Warsaw, 1998, pp. 108-117.

11. REBIELAK J, Some Proposals for Shaping of High-Rise Buildings, Proceedings of the International Conference "Engineering a New Architecture", Aarhus, Denmark, May 26-28, 1998, pp. 169-176.

12. REBIELAK J, Some Proposals of Structural Systems for Long Span Roofs and High-Rise Buildings, Journal of the International Association for Shell and Spatial Structures, Vol. 40, No 1, 1999, pp. 65-75.

13. REBIELAK J, Some examples of space structures shaping by means of Formian, in: Bridge between civil engineering and architecture, Proceedings of 4th International Colloquium on Structural Morphology, IASS, Working Group No 15, August 17-19, 2000, Delft University of Technology, The Netherlands, pp. 131-137.

14. REBIELAK J, Prismatic space frames as the main support structures for high-rise buildings, Lightweight Structures in Civil Engineering, Local Seminar of IASS Polish Chapter, Micro-Publisher Jan B. Obrebski Wydawnictwo Naukowe, Warsaw-Wroclaw, 7th December, 2001, pp. 81-89.

Study of edge restraints and localized weakness in the double layer grids of square and rectangular layouts

S. A. RIZWAN
Civil Engineering Department, University of Engineering & Technology (UET), Lahore 54890, Pakistan.
H. M. AHMED
Government College of Technology, Raiwind Road, Lahore, Pakistan.

ABSTRACT

This research addresses the effect of edge restraints and localized weakness on the structural response of Double Layer Square and rectangular grids in terms of axial forces and deflection. This work is an extension of previous theoretical study done by the authors. It contains the analysis of two double layer grid model structures of 8'x8' and 8'x4' in plan which were modeled and analyzed by using standard SAP90 software under a pressure of 1 psi (145 Mpa) to investigate the stated parameters. Localized weakness problems can occur due to poor connections, heavy concentrated loads, aerial attacks, blasts or due to fire. The grids were braced and loaded at the top chord and supported at supports at the bottom chord. Twelve different edge restraint conditions were studied in terms of deflections and maximum chord forces with a view to suggest some suitable arrangement for temperature and lateral loads which impose opposing nature of support constraints. Three weaknesses in square layouts were considered in maximum moment and maximum shear zones. In rectangular grids two weaknesses were considered for analysis purposes. Dropping the axial stiffness of members in the concerned zone by 60000 times during modeling stages has simulated the localized weakness in both types of layouts analyzed.

Generally by increasing number of supports axial forces and deflections decrease. By removing horizontal restraint at supports for avoiding temperature stresses, vertical deflection increases. For square layouts if the weakness occurs in peripheral zone it becomes critical which might be taken care of by central stiffness arrangements .For rectangular layouts, weakness at the centre is critical for which again stiffening of central portion may be useful.

INTRODUCTION

It is very rare that an actual structure is analysed. Invariably an idealized centre line linear elastic analysis of structures is accomplished for idealized loads and support conditions which makes further discrepancies in analysis results. The effect of variation between theoretical (considered in analysis) and actual edge restraints (obtaining at site) may produce a marked variation in the response of double layer grids when measured in terms of member forces and deflections. Very limited research has been reported in the literature in this very important area.

Similarly temperature and seismic loads are also not considered and reported in the literature very often. The supports of a double layer grid can be columns in steel or in reinforced concrete. The connection between the double-layer grids and columns is termed as supports (edge restraints) in this paper. The supporting columns may be flexible or stiff depending on whether they are slender or massive. Moreover they may or may not be vertically braced.

For ideal frictionless hinge supports, columns seem to provide lateral support to grids so that temperature stresses are developed in grid members. These forces will be small for unbraced or flexible supporting columns. In vertically braced columns, resistance to lateral loads is provided by bracing.

In addition to dead, live and wind loads, other loads like temperature and earthquake must be considered and evaluated in suitable load combinations for the members of grids. For supporting steel columns, the consideration of ratio of width to thickness ratio especially on compression flange side is also very important to avoid local bucking due to lateral loads.

The design of connections (supports) between space grids and substructure is significant and stiffness of sub-structures may vary the structural response. Karamanos(6) states that resisting temperature and seismic loads at connections (supports) is often a compromise between two opposing requirements. In general the connections (supports) should be arranged such that they permit free movement due to temperature changes to avoid temperature member stresses and at the same time provide resistance to seismic forces (supports not allowing translations).

Makowski(1) presents stress distribution for square and rectangular plan layouts. Salajegheh(2) has analyzed square on square grids for various types of supports. West (3) has also worked in this area. Rizwan and Bhatty (4, 5) have also addressed this area for square on square grids. With the above information in mind 12 different support conditions at periphery (edge restraints) have been assumed in the present analysis of square-on-square model grids having tubular member of external dia ½" and a wall thickness of 1/16" and a 9" structural depth for a 8'x8' grid for 1 psi pressure. Later after applying stress checks according to AISC specifications, the allowable pressure for the models investigated appears to be around ⅓ psi.

The paper examines two areas in detail. Firstly for fixed type of gravity loads, the structural response has been studied for various types of connections (supports) between grid and columns. Secondly the creation of localized weakness and its effect on structural response for a given edge restraint (support conditions). Sequential failures of individual members (progressive weakness of steel) can lead to collapse of double layer grids occurring due to the excessive amounts of heat transfer from hot gases to the grid members in case of fire (7). For avoiding fire limit state special fire and corrosion resistant FCR steel may be used (8).

STUDY OF EDGE RESTRAINTS

Provision of suitable number of supports and their effective orientation while in position can reduce the internal number stresses arising due to already mentioned loads and their combinations. In addition to loads, the internal member stress distribution also depends upon length to width ratio of structure in plan.

This theoretical study has been carried out by using SAP90 linear elastic analysis for a 8'x8' square on square latticed grid having ½" x 1/16" hollow circular section of members. The nodal loads correspond to 1 psi pressure. The supports (edge restraints) were at the lower chord while loads were applied to the braced upper chord. The structure was supported in 12 different arrangements. Fig. 1 gives the general arrangement of structure and node numbing. Fig. 2 gives the member numbering.

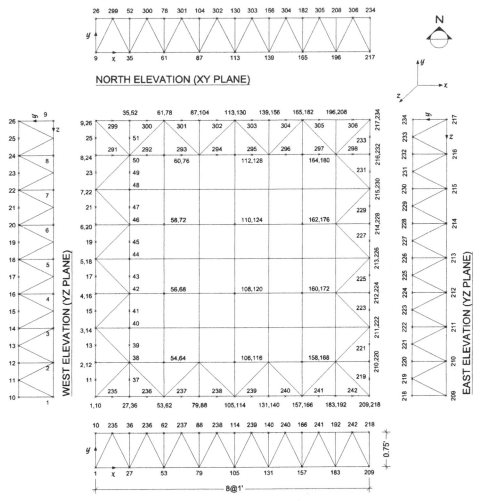

Fig. 1:   Plan (ZX Plane) and Node Numbering Scheme and General Arrangement of Structure.

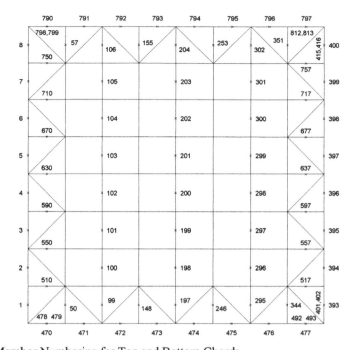

Fig. 2: Member Numbering for Top and Bottom Chords

The detailed description of types of supports (restraint conditions) is shown in table 1.

Table 1: SAP90 Analysis Cases For Edge Restraints Effect.

Case #	CONDITION OF SUPPORT
S1	Supports at corners, only which prevent displacement in all directions and allow rotations.
S2	Supports at corners only which allow displacement in x-directions at one, support (9) and allow rotations at all supports.
S3	Displacement is allowed in x-direction at two supports (9&217) and rotation is allowed at all ones.
S4	Displacement is allowed in x-direction at three supports (9,209&217) and rotation is allowed at all ones.
S5	Displacement is allowed in x & z direction at one support (9) and rotation is allowed at all ones.
S6	Displacement is allowed in x & z direction at two supports (9&217) and rotation is allowed at all ones.
S7	Supports at corners only, which prevent displacement & rotation in all directions.
S8	Displacement is prevented in all directions at supports (1,5,9,105,113,209,213 & 217) and rotation is allowed at all ones.
S9	Hinge supports at corners (1,9,209 & 217) and roller supports (vertical restraint only) at mid edge nodes (5,105,113 & 213).
S10	Displacement is prevented in all directions at supports (all nodes located at edges) and rotation is allowed at all ones.
S11	Displacement and rotation is prevented in all directions at supports (all nodes located at edges).
S12	Hinge supports at corners (1,9,209 & 217) and roller supports (only vertical restraints) at remaining edge nodes 2 to 8,27,53,79,105,131,157,183,209,210 to 216,35,61,87,113,139, 165 & 191.

The structural response in terms of maximum displacement is shown in table 2.

Table 2: Study of Edge Restraints (type of supports) in terms of maximum displacement in columns.

Case No.	Maximum Displacement (Inches)					
	X-Displace	At Joint	Y-Displace	At Joint	Z-Displace	At Joint
S1	0.026	11,25, 219,233	-0.214	122	0.0256	235,242, 299,306
S2	0.07 ▲	9	-0.233	96	0.030	235
S3	0.07 ▲	9,217	-0.246	122	0.032	235,242, 299,306
S4	0.07 ▲	9,217	-0.246	122	0.032	235,242, 299,306
S5	0.07 ▲	9	-0.252	98	0.0699 ▲	9
S6	0.07 ▲	9,217	-0.271 ▲	124	0.04	9,217
S7	0.03	11,25, 219,233	-0.21	122	0.03	235,242, 299,306
S8	0.01 ■	115,129	-0.08	122	0.01 ■	267,274
S9	0.02	105,113	-0.10	122	0.02	5,213
S10	0.01 ■	115,129	-0.05 ■	122	0.01 ■	267,274
S11	0.01 ■	115,129	-0.05 ■	122	0.01 ■	267,274
S12	0.01 ■	105,113	-0.08	122	0.01 ■	5,213

▲ Maximum displacement   ■ Minimum displacement

Table 3 shows structural response in terms of axial member forces.

Table 3: Study of Edge Restraints in terms of Maximum Axial Forces.

Case No.	Max. Tension (Lower Chord)		Max. Tension (Bracing)		Max. Compression (Top Chord)	
	Value	Member #	Value	Member #	Value	Member #
S1	1019.57	4,5, 396,397, 473,474, 793,794	1368.6	41,42, 433,434, 501,502, 821,822	2003.9	16,17, 408,409, 485,486, 805, 806
S2	2538	5	1630.9	502	2562.9	481
S3	2474.86	4,5, 396,397	1668.12 ▲	501,502, 821,822	2523.68	481,490, 801,810
S4	2474.86	4,5, 396,397	1668.12 ▲	501,502, 821,822	2523.68	481,490, 801,810
S5	2698.87	5,793	1602.63	502	2579.8 ▲	413,481
S6	2922.12 ▲	793,794	1632.44	501,502	2417.96	481,490
S7	1018.76	4,5, 396,397, 473,474, 793, 794	1363.15	41,42, 433,434, 501,502, 821,822	1993.39	16,17, 408,409, 485,486, 805, 806
S8	585.31	151,152, 249,250, 593,594, 673,674	898.42	237,238, 661,662	1127.18	212,213, 645,646
S9	1041.65	200,201, 633,634	606.24	237,238, 661,662	844.346	163,164, 261,262, 605,606, 685,686
S10	240.99 ■	151,152, 249,250, 593,594, 673,674	396.40	237,238, 661,662	787.229	212,213, 645,646
S11	285.48	200,201, 633,634	394.05	237,238, 661,662	778.751 ■	212,213, 645,646
S12	915.07	200,201, 633,634	392.37 ■	237,662	797.223	212,213, 645,646

▲ Maximum force   ■ Minimum force

STUDY OF LOCALIZED WEAKNESS
Localized weakness may be defined as a situation in which a member or a group of members fails to act as a structural component of the system and is no longer contributing to the strength of structure. Such situation may arise due to fire, earthquakes, terrorist and aerial attacks, poor joint connections and accidental overloads.

STRUCTURAL MODELING
To study the effect of localized weakness, on same previous 8'x8' square on square and 8'x4' rectangular grid shown in figs 1 and 4 with same member properties and loading were analyzed by SAP90 program wherein the axial member stiffness was reduced by 60,000 times in the concerned regions W_1, W_2 and W_3 of double-layer grid. The members at the periphery were assumed to be connected as true frictionless hinges (translational restraint conditions) at four corners and then at all peripheral nodes. For 8'x8' square grid same node and member numbering has been used as was shown in figure 1 & 2 respectively. Three areas W_1, W_2 and W_3 subjected to localized weakness were considered as shown in fig 3 below. The results of displacement and axial forces pertaining to the square-on-square grid as presented in tables 2 and 3 would be termed as zero-weakness case (w_0).

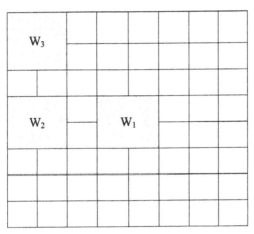

Fig. 3:   Location of localized weakness in plan in moment and shear zones.
Total area = 64 Sft. area of a weakness = 4 Sft., % weakness area = 6.25%

Table 4 presents the maximum displacements for square-on-square grids for three different types of weaknesses shown in fig 3 for two different support conditions i.e. corner supports and all peripheral lower nodes supported by frictionless pins (3D translations prevented while 3D rotations allowed).

Table 4: Maximum displacement in inches for square-on-square grids.

Case No.	Corners Supported			All Nodes Supported		
	X	Y	Z	X	Y	Z
W_0	0.0256	-0.2145	0.0256	0.0096	-0.0505	0.0096
W_1	0.0244	-0.2266	0.0244	0.01356	-0.0688	0.01356
W_2	0.0267	-0.2410	0.038	0.0111	-0.0509	0.00990
W_3	0.0450	-0.2360	0.0459	0.0111	-0.0529	0.01110

Table 5 gives maximum axial forces for two external supports conditions and three localized weaknesses.

Table 5: Maximum axial forces (lbs) for square-on-square grids.

Case No.	Only Corners Supported		All Nodes Supported	
	Tension	Compression	Tension	Compression
W_0	1019.574	2003.929	240.9	787.23
W_1	1114.025	1850.151	299.36	700.90
W_2	1055.5	2207.02	388.2	1040
W_3	1121	2533	305.4	828

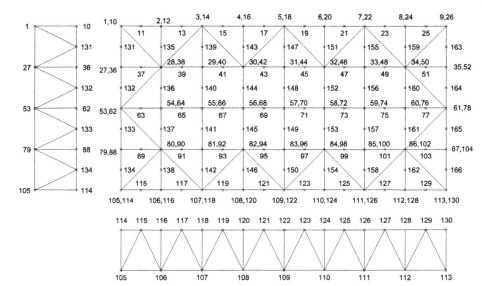

Fig. 4:   Node numbering and general arrangement of structure

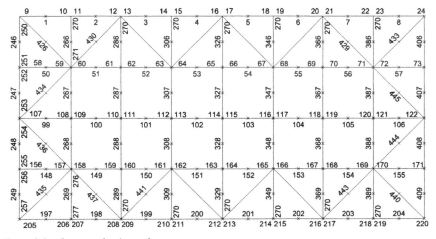

Fig. 5:   Member numbering scheme

Fig 6 shows localized weakness in rectangular plans.

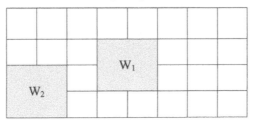

Fig. 6: Rectangular double layer grid showing two weakened regions.
Total area = 32 Sft., area of each weakness = 4 Sft. % weakness area = 12.5%

Table 6 shows the results of maximum displacements along the three axis for various weaknesses.

Table 6: Maximum displacement for 2 supports conditions and two weaknesses in rectangular grid.

Case No.	Corners Supported			All Nodes Supported		
	X	Y	Z	X	Y	Z
W_0	0.01768	-0.0972	0.0296	0.0014	-0.0088	0.0022
W_1	0.0193	-0.1108	0.0148	0.00159	-0.00918	0.0032
W_2	0.022	-0.1092	0.00852	0.00157	-0.00899	0.00237

Table 7 gives maximum axial forces for various weaknesses in rectangular grid.

Table 7: Maximum axial forces (lbs) in Rectangular grids.

Case No.	Only Corners Supported		All Nodes Supported	
	Tension	Compression	Tension	Compression
W_0	859.4	1368.90	126.70	372.16
W_1	584.63	1488.70	147.41	268.59
W_2	840.19	1441.07	129.23	400.25

CONCLUSIONS

1. In general by eliminating transnational restraints in X-direction (allowing translation), vertical deflection increases. This effect is more pronounced in case of four corner supports compared with all peripheral nodes supported.

2. By eliminating restraint in horizontal direction at one support only tensile force increases by 149% and maximum compression increases by 28%. By introducing the elimination of horizontal restraints at other supports, rate of increase in maximum tensile and compressive forces slightly reduces.

3. By eliminating horizontal restraints (allowing translations at more supports) in various directions maximum tensile force at bottom chord increases while maximum compression at upper chord does not increase significantly.

4. S_{10} and S_{12} arrangements may be good for releasing temperature induced stresses while S_{11} may be recommended for lateral load resistance.

5. Tensile and compressive chord forces decrease significantly by increasing number of peripheral supports.

6. For square grids W1 seems to be critical weakness in term of displacement while W2 is critical in terms of chord forces.

7. For rectangular grids, W1 is critical for displacements for both corner column supports and all peripheral columns supported.

8. Increase in axial compression in diagonal grids for the two peripheral support conditions remains almost same.

9. From the analysis it is tremendously clear that role of edge restraints is very significant for overall structural response of double layer grids. Localized weakness may not cause collapse but definitely alters the stress distributions, deflections and axial forces.

ACKNOWLEDGEMENTS

The Principal author is grateful to Professor Z.S Makowski and to Professor H. Nooshin for their valuable guidance in the subject during his post-graduate studies at University of Surrey, UK. Thanks are also due to the vice-chancellor UET Lahore, Pakistan for patronizing the research activities.

REFERENCES

1. Lee, H.G. and Makowski, Z.S "Study of factors effecting the stress distribution in double layer grids on square diagonal type", Architectural Science Review, Volume 20, No. 4 1977.

2. Salajegheh, E "An analytical study of double layer square on square grids". M.Sc Thesis, University of Surrey, 1977.

3. West, F.E.S "A study of efficiency of double layer grid structures" M. Phil Thesis, University of Surrey, UK 1967.

4. Rizwan, S.A. and Bhatty, A.A. "On the performance of double layer square on square grids", paper presented at 4th International Conference on Space Structures, September 1993, University of Surrey, UK. Conf. Proc Vol. 1, pp 614-621.

5. Rizwan, S.A. and Bhatty, A.A. "Analysis and design of double layer skeletal latticed grids subjected to local weakness", EASEC-4 September 27-29, Seoul, Korea. Conf. Proc. Vol. 1 pp 269-274. 1993.

6. A.S. Karamanos and S.A. Karamanos "Seismic Design of double layer space grids and their supports", 4th International conference on Space Structures Proc. Vol. 1 pp 476-484 University of Surrey, September 1993.

7. A. Yarza, P. Cavia and G.A.R. Parke "An Introduction to the Fire analysis of double layer grids" 4th International Conference on Space Structures, University of Surrey, September 1993, Proc Vol. 1, pp 683- 692.

8. T. Ono *et al* "Wide-span structures using fire and corrosion resistant steel", 4th International Conference on Space Structures, University of Surrey, September 1993, Proc Vol. Conf. Proc. Vol. 1, pp 862 – 870.

A 40 metre diameter aluminum dome used for a tank

I. L. ROUJANSKI
Department of Space and Lightweight Structures, The Melnikov Central Research and
Design Institute of Steel Structures, Moscow, RF

DEFINITIONS

The paper considers the structural conception of the 40-m diameter aluminum dome for
20000 m^3 capacity steel tanks designed by the Melnikov Institute in the year 2000. The dome
is spherical having the radius 29 m and the rise of about 8 m. The load-bearing framework of
the dome was designed in the grid scheme and it differs from the "ULTRAFLOT" structure
both in the design of the grid itself and in the design of the dome joints and the bearing joints
on the tank shell. The project of load-bearing and cladding structures of the aluminum dome
for the 20000 m^3 capacity tank with a pontoon for oil storage in Nizhny Novgorod was
developed in the year 2001.

The specific technological requirements for the dome:

- ventilation – is naturally ventilated;
- gas impermeability – is not required;
- water impermeability – is required;
- drainage system – is natural runoff.

The requirements for material of the dome: home-made aluminum alloys meeting the
requirements for weldability, strength, thermal stability and corrosion resistance.

TECHNICAL CHARACTERISTICS OF THE DOME LOAD-BEARING FRAMEWORK AND CLADDING

The aluminum dome with 40.34 m in diameter along the axes of the supports has a spherical
shape with the radius of 29.0 m and the rise of 8.25 m. The cladding structure is a 1.5-mm
thick aluminum sheeting. The load-bearing framework is designed as a grid variant (Fig. 1).

Unlike the "ULTRAFLOT" system the bar grid of the dome load-bearing framework was
designed in the star-shaped scheme and has 64 supports which are attached to the steel tank
supporting stiffening ring using hinged joints. This ring is at the same time the dome lower
thrust ring. Bearing joints prevent the dome supports from displacing in relation to the tank
shell. All the dome framework members, without exception, represent rectilinear bars made
of 240-mm deep extruded I-beams. Here two types of section are used: one type of section is
for all the dome elements, and the other, with a larger cross-section area, is for the dome
supporting bars.

All the dome framework members are connected with bolts. The joint for connection of the
dome framework members was designed with the use of M 16 high-strength bolts (shear-
resistant joint), see (Fig. 2) which requires an order of magnitude smaller accuracy of
manufacturing as compared to the joint with shear bolts of the "ULTRAFLOT" system.

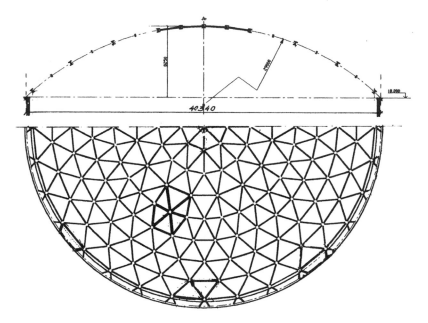

Fig. 1.

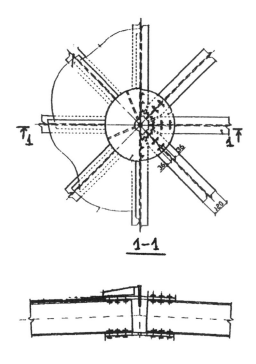

Fig. 2.

The overall stability of the dome is provided by the framework design in combination with the design of joints. The attachment of the assembly elements of cladding to the framework members is carried out as a lap joint using self-tapping screws through cover plates. In case the width of cladding sheets is smaller than the size of the assembly element of cladding, the sheets are to be butt welded using full-strength welds at the manufacturing plant. The framework members and joint elements (cover plates taking force actions) shall be made with account for the following requirements for accuracy of manufacturing. The length of the framework members shall be provided with the accuracy stipulated in the Russian Building Rules and Regulations. While connecting the joint elements to the framework members in the joints with M 16 high-strength bolts for 18-mm diameter holes the tolerances shall be as follows:
– distance between groups of holes is within the accuracy 0 +0.5 mm;
– distance between holes in a group is within the accuracy 0 ± 0.2 mm;
– 18-mm diameter hole is within the accuracy 0 ± 0.2 mm.

Marking and drilling of holes shall be done using a jig. The joint cover plates taking force actions were designed as cone-shaped with the angle of cone opening being a design value that made it possible to have the same type and size of cover plates for a whole dome. This joint was patented. The authorship priority is effective from 01. 2001 (Fig. 3).

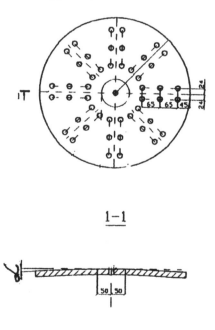

1–1

Fig. 3.

DESIGN CALCULATIONS OF THE DOME
The design calculations of the dome were made for the following main loads and their combinations: dead weight of the structure, snow load, wind load, 80° C temperature differential. As the bearing joints of the dome were realized as undisplacing in relation to the

tank shell, combined work of the dome and the tank shell was taken into account in the calculations. In this case the design combinations are the following loading combinations: dead weight + symmetrical snow load + asymmetrical snow load. For this structure variations of annual maximum values of snow cover weight in the construction region were analysed. The analysis was based on meteorological data for a multi-year period of observations provided by four local weather stations which are the nearest to construction site. The approximation of these variations was calculated by means of Gumbel first limit distribution. Considering the data provided, the design values of snow loads were taken as 280 kgf/m^2 (the average value for the four weather stations, with the recurrence once during 25 years).

The design diagrams of snow loading.
Symmetrical snow loading of the tank shell was adopted according to Fig. 4.
where $\mu1 = \cos 1.6\ \alpha$
α – the roof slope in degrees (for all angles β)
Asymmetrical loading. When making calculations of the dome roof, for a half of it the zero value load was taken, and for the other one load is determined by the following formula:
$\mu2 = (\alpha2\ /700 + \sin 48\alpha) * \sin \beta$
where α – the angle of roof slope;
β – the angle counted from the fixed radius of the dome roof on the plan to the radius passing through the projection of the considered point of the roof. For example, for section II –II coefficient $\mu2 = 0$. For section I-I on the diameter $\mu2 = 2$ (see Fig. 4).

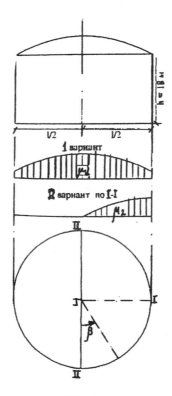

Fig. 4.

The final calculations were made using finite element method according to the CATRAN program. The dome-tank structure is a space system consisting of plates and bars. The position of each joint is characterised by six degrees of freedom. Two types of finite elements were used for assembly of the design diagram: space bar and rectangular plate. Numerical experiments showed that below the third course of the tank shell (about 6 m in height) disturbances in the stressed and strained state due to joint work of the dome and the tank disappeared. As a result of the calculations two upper courses of the tank shell with thickness 8 mm and total height of about 4 m were strengthened by stiffening ribs. The structure fully meets the domestic requirements for deformability, strength and also those for manufacturing and the system erection procedure.

DOME ERECTION PROCEDURE

The first dome was erected in two stages: at the first stage a check assembly of the dome load-bearing framework was carried out on open areas of a manufacturing plant, see Fig. 5. Then the assembled framework was dismantled and transported to erection site, where the dome was fully reassembled and after that it was lifted to the design elevation. A check assembly of the second dome was no longer carried out at the manufacturing plant. The sequence of the dome erection procedure is described below.

The dome was assembled in the tank, on its bottom (Fig. 6). The dome was erected starting from the dome central joint to the periphery by adding structural members from below. Each time after closing the successive ring, the assembled part of the dome was lifted to a given height by means of standard manual hoists (Fig. 6) and then the framework members of the next tier were installed. In this process the framework members of each tier were assembled using joint lower round cover plates and after all the joint elements had been placed and attached to the joint lower round cover plate, the upper round cover plates were installed (Fig. 7). In each joint high-strength bolts were tightened only after the assembly completion of all the adjacent joints. Contact surfaces were sandblasted. The friction coefficient of joined elements was adopted as 0.45.

As the assembly works of the framework progressed, after tightening bolted joints of the successive tier assembly elements of cladding (Fig. 6) were installed. They represent plate flanged triangular elements a part of which has technological holes and welded nozzles. The assembly elements of cladding were made at the manufacturing plant and delivered as packaged to erection site. The attachment of cladding to the framework bars was carried out by means of self-tapping screws through cover plates.

After all the assembly elements of cladding being adjacent to the joint had been installed, the joint was tightly covered with an aluminum cap preventing water penetration. All structural joints were sealed. After the assembly completion the entire dome was lifted to the design elevation using sixteen hand-operated chain pulley blocks which at the top were caught on hooked assembly elements installed peripherally on the tank stiffening ring (Fig. 8), and in sixteen outside joints of the dome – on special erection cover plates. The dome was hung over the tank and after that the supporting bars were placed into the dome joints (Fig. 9). After tightening bolts in these joints the dome was installed in the design position. The general view of the structure is shown in Fig. 10. Presently 2 similar domes for the 20000 m^3 capacity tanks have been installed.

GENERAL CONCLUSIONS

It should be noted in conclusion that, for the first time, the design, fabrication and installation works of the domestic aluminum domes for tanks considered in this report, have been rather

Fig. 5.

Fig. 6.

Fig. 7.

Fig. 8.

Fig. 9.

Fig. 10.

successfully realized in our country by domestic designers, manufacturers and erectors. All the structural members were fabricated from aluminum alloys of the national grades and using the national production range. A guaranteed service life of this aluminum dome for a medium-aggressive environment is 25 years without an overhaul that exceeds several times the service life of steel tank roofs of the considered type. Moreover, the weight of the aluminum dome is about 5 times smaller than that of the steel dome with the same parameters, thus facilitating greatly work of the tank shell. Actually the conditions formed are favourable to a worldwide application of domestic aluminum domes as fixed roofs for tanks of various capacities and designation.

The general theory of the Volume and Displacement Indicators

Ph SAMYN, Department of Civil Engineering, Vrije Universiteit Brussel (Brussels Free University), Brussels, Belgium, and
Th VILQUIN, Samyn and Partners, 1537 chaussée de Waterloo, 1180 Brussels, Belgium.

1. INTRODUCTION

The dimensionless VOLUME and DISPLACEMENT INDICATORS arc first defined. As an example, their expression is given for various structural morphologies concerning an horizontal span.

A closer look is then taken at their accuracy and at the conditions to their practical use. The influence of various parameters is studied: first the buckling for bars in compression, and the volume of end fittings for rods or cables in traction; then the necessary approximation induced by the limited number of industrial steel L, U or I sections; and finally dynamic loading, through the relation between the first eigen frequency and these indicators.

2. VOLUME AND DISPLACEMENT INDICATORS

2.1. Definition

It has been shown [Ref 1] that the volume V $[m^3]$ of material needed for a fully stressed isostatic structure is proportional to ρ/σ (with ρ $[N/m^3]$ the specific weight and σ $[Pa]$ the allowable stress of the composing material), and to a dimensionless number, the VOLUME INDICATOR $W = V\sigma/(FL)$, with F $[N]$ the total intensity of the loads applied to the structure and L $[m]$ its main dimension.

Similarly, the corresponding maximal (absolute) displacement at mid-span δ $[m]$ is proportional to L, σ/E (with E $[Pa]$ the Young modulus of the composing material) and a dimensionless number, the DISPLACEMENT INDICATOR $\Delta = E\delta/(\sigma L)$.

An optimisation can then be achieved by minimising these factors, among others these two indicators, what makes them useful tools for the preliminary design.

They depend only on the slenderness ratio of the structure, L/H, where H $[m]$ is the main dimension perpendicular to L.

These definitions apply for fully stressed structures when not subjected to buckling nor to dynamic loading. The influence of such points is discussed further, and the case of buckling in great details by P. Latteur in Ref 2. But it should be noted that various other reasons could require to reduce the allowable stress taken into account, like inaccuracies on dimensions or loads (a.o. dead weight), fire resistance, fatigue and ductility. Generally speaking, second order effects exert few influence on W, while they can be significant on Δ. The connection design might also substantially influence the weight of the structure.

Although based on simple mathematical concepts, their determination asks for lengthy calculations.

Samyn and Partners and the department of Civil Engineering at Brussels Free University are working since 1998 on the determination of these indicators and, since 2000, with a grant from the Brussels Ministry for Economic Affairs.

The indicators are now well known for:
- the horizontal isostatic span vertically loaded;
- masts under any load cases;
- a great number of porticoes under vertical and horizontal loads;
- various membranes of revolution.

The cases of beams with a variable cross-section and guyed and suspended structures under a uniformly distributed load, as well as the use of the indicators for the design at Samyn and Partners, have already been presented in former papers (see Refs 3, 4 and 5).

The present developments concern, among others, horizontal multiple spans, asymmetric solicitations of arches, guyed columns and structures with variable longitudinal profile. The results of these researches in progress will influence in an important way structures that will be conceived by our team in the future.

The actual knowledge has however already allowed to clarify the thought and to conceive structures with a greater objectivity.

2.2. Example of the horizontal span, transversally loaded

Various configurations of rectilinear beams, with symmetrical and constant section or truss beams, and parabolic arches and cables are considered (see Fig 1).

Four types of trusses are considered (see Fig 2): Howe-Pratt, Warren, multitruss (i.e. a truss whose mesh bases are trusses as well, and so on) and multistrings (patented [Ref 6]).

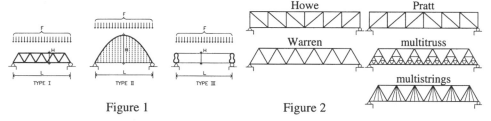

Figure 1   Figure 2

These structures are internally and externally isostatic and subjected to a downwards load F uniformly distributed on their whole length L, or concentrated on the nodes for the trusses. The bars of the trusses are hinged at each end and of constant section between nodes.

Moreover, the structures are fully-stressed designed, i.e.:
- trusses: all bars work at the same allowable stress;
- beams and cables: their constant section is determined by the maximum stress, i.e. the one
                          respectively at mid-span and on supports;
- arches: exhibit a variable section, in order to work at the same allowable stress
                          as the tie, columns or strings.

It should nevertheless be noted that this last assumption is not usual in practice, as, for instance, cables in steel structures work at an allowable stress about 5 to 10 times higher than the one of the compressed or bent elements.

The expressions of the volume and displacement indicators for all these structures present the form:

$W = k_1\, L/H + k_2\, H/L$ and $\Delta = k_3\, L/H + k_4\, H/L$.

They are represented as functions of L/H on figures 3 and 4. The envelope curves, corresponding to the best mesh proportion, are considered for the trusses.

It should be noticed that, under a mobile point load F, the W values of beams and trusses are approximately twice as high as the ones for the uniformly distributed load.

These figures illustrate that:
- arches without tie are always lighter than trusses;
- Warren trusses are always lighter and stiffer than Howe-Pratt ones;
- Warren multistrings and optimum variable multitrusses are hardly heavier than Warren ones, while able to stand an uniformly distributed load on their lower chord;
- trusses are lighter than arches with ties when the slenderness ratio is larger than 8;
- beams of constant continuous section are heavier than all other structures, but are also the stiffest, followed by arches without tie, trusses and finally arches with tie.

3. CORRECTION OF THE VOLUME INDICATOR OF ELEMENTS IN COMPRESSION DUE TO BUCKLING

3.1. Introduction

The allowable stress is limited by the buckling stress σ_{cr}, so that:
- if $\sigma < \sigma_{cr}$, then $W = 1$;
- if $\sigma \geq \sigma_{cr}$, then $W = \sigma/\sigma_{cr}$.

According to Euler, σ_{cr} is a function of the reduced slenderness $\Lambda = \dfrac{1}{\pi}\dfrac{\mu L}{H}\sqrt{\dfrac{\Omega H^2}{I}\dfrac{\sigma}{E}}$ $[/]$,

with μ $[/]$ the end-fixity constant, as: $W = \sigma/\sigma_{cr} = \Lambda^2$. The actual relations for the various building materials, given in literature and standards, diverge from this theoretical expression. For example, for steel and $\Lambda \geq 1$: $\Lambda^2 + 1 > W > \Lambda^2 + 0.5$; for concrete, $W = \Lambda^2 + 1$, and for wood and $1.152 \geq \Lambda \geq 0.72$: $W = 3.5\Lambda^2$ (see Fig 5). Nevertheless, W and Λ^2 vary proportionally, and the study of the influence of the various components of Λ keeps meaningful.

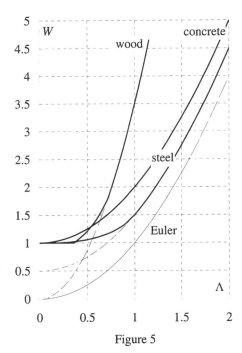

Figure 5

3.2. The material: σ/E

σ/E varies few for usual materials ($4 \cdot 10^{-4}$ for reinforced concrete, $7 \cdot 10^{-4}$ for S235 steel, $9 \cdot 10^{-4}$ for wood, 10^{-3} for S235 steel).

For wood and concrete, one has $E = \sqrt{\sigma}/k$ with k a constant, so that $\sigma/E = k\sqrt{\sigma}$.

But, while keeping the same σ/E, it is possible to vary σ, and E in the same proportion then, by modifying the material microstructure.

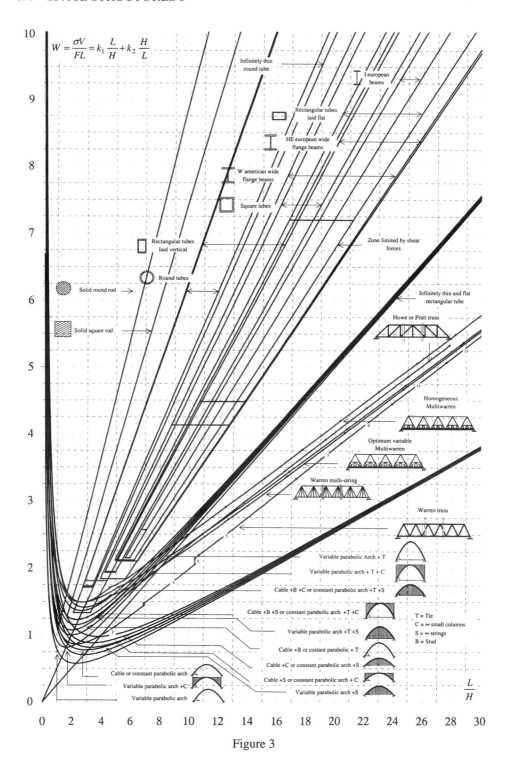

Figure 3

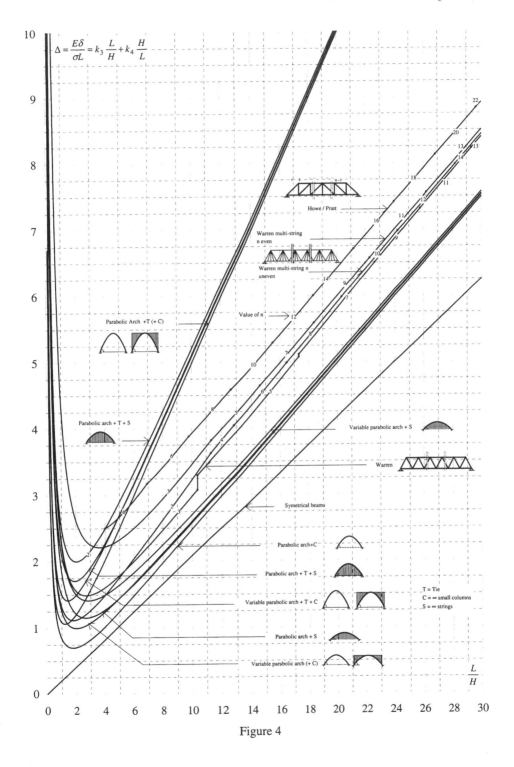

Figure 4

For instance, a solid rod can be replaced by a large amount of small tubes, presenting the same total solid section Ω but an apparent section Ω_a, an apparent allowable stress σ_a and an apparent Young modulus E_a such as $\sigma_a/\sigma = E_a/E = 4k(1-k)$ where $k = e/H$, with H the width and e $[m]$ the wall thickness of a tube. For the same bearing capacity, the reduced slenderness ratio of the tubes set is $2\sqrt{k-k^2}$ times smaller than the one of the solid section (see Fig 6).

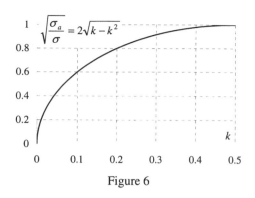

Figure 6

3.3. Transverse geometry: $\Omega H^2/I$

Typical values of $\Omega H^2/I = (H/r)^2$ and steel profiles (following Ref 7) are:

16	round-section rod
12	rectangular-section rod
8.2 to 10.1	usual round tubes
6.2 to 7.3	usual square tubes

6.61 to 7.55	L profiles
5.18 to 6.71	HE or WF profiles
5.78 to 6.10	IPE profiles

All values for the larger I .

It can be observed that, for the same efficiency, a element of square section (whatever full or hollow) viewed along its diagonal is bulkier than the corresponding round-section one. But the difference keeps small (~ 6%), while the simplicity of connections for square-section elements can still make them preferable to round-section ones.

3.4. Longitudinal geometry: $\mu L/H$

Various values of μ are:

1	0.7	0.5
element hinged at each end	one end is fixed, one is hinged	both ends are fixed

In the peculiar case of trusses, this advantage is lessen by the effect of the parasite bending moments induced in the structure by making some nodes fixed.

3.5. General influence of buckling

It can be shown that the increase of W due to the limitation of the allowable stress to σ_{cr} is limited when the drawing of the structure is carefully done. For example, in the case of S355 steel tubular or I profiles, this increase is limited to 20 % when the structural slenderness $\mu L/H$ is limited to 20 (see Fig 7).

4. CORRECTION OF THE VOLUME INDICATOR OF ELEMENTS IN TRACTION DUE TO VOLUME OF END FITTINGS

The volume indicator of a bar in traction, without taking into account the volume of end fittings, equals 1.

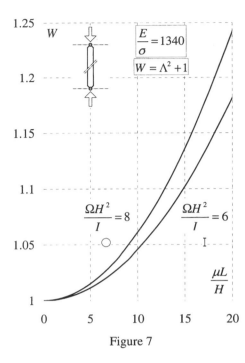

Figure 7

Let us now consider a solid round section bar or cable equipped with open spelter sockets at each end. Let $L\,[m]$ be its length, $H\,[m]$ its constant diameter, $\sigma_s\,[Pa]$ the allowable stress of sockets' material and $V_s\,[m^3]$ the volume of a socket.

Then $V = \Omega L + 2V_s$ and $W = \dfrac{\sigma}{F}\Omega + 2\dfrac{\sigma}{FL}V_s$.

As $\Omega = \dfrac{F}{\sigma}$, $W = 1 + 2\dfrac{V_s}{\Omega L}$.

Further, $\Omega = \dfrac{1}{4}\pi H^2$, and $W = 1 + 8\dfrac{1}{\pi}\dfrac{V_s}{H^3}\dfrac{H}{L}$.

In practice, V_s is approximately proportional to H^3. For instance, considering open spelter socket in stainless steel with $\sigma_s = 630\,MPa$ from the German company Pfeifer (Ref 8), for cable whose breaking load is $1450\,MPa$:

$$P_{s[N]} \cong 0.00489 H^3_{[mm]}.$$

With $\rho = 76.52\,kN/m^3$,

$$W = 1 + 8\dfrac{1}{\pi}\dfrac{P_s}{\rho H^3}\dfrac{H}{L} \approx 1 + 162.7\dfrac{H}{L}.$$

This means that the bar or cable should present at least a length of 813 times its diameter to limit the increase of W to 20 %, and that this length has to be doubled if one considers the same minimum amount of material for the fixing on the supports.

It should however be noted that, instead of open spelter sockets, a cable can also be equipped at its ends with cylindrical sockets (usually with a diameter equal to $3H$ and of a length equal to $5H$). This halves the weight of the fittings.

Another typical case is the one of concrete reinforcement bars. Their anchoring is achieved by adhesion, and an anchoring length of at least 30 times the diameter must be considered. The total length amounts then to $L + 60H$ for a usable length L, and its volume indicator becomes $W = 1 + 60H/L$.

In order to limit this increase to 20 % of the theoretical value, L/H must be higher than 300, what is not always feasible. This is illustrated on Figure 8.

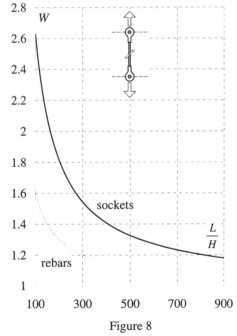

Figure 8

5. CORRECTION OF THE VOLUME INDICATOR DUE TO PRACTICAL INACCURACY

As a complement to what precedes, it should be noted that it may not be possible in practice to match precisely the theoretically needed characteristic for structural elements in laminated or extruded materials, like steel or aluminium; this characteristic is Ω for an element in simple compression without buckling, I for an element in simple compression with buckling and I/H for an element in simple bending.

Series should ideally be produced to keep constant, and as small as possible, the relative difference c between two successive values k_n and k_{n+1} ($k_{n+1} = (c+1)^n k_0$). These geometrical series are known as Renard series (see Ref 9).

If the necessary values are just slightly higher than the one of a serie, c represents the maximal increase and $c/2$ the mean increase of W resulting from this inaccuracy.

As an example, let us consider the specific situation of steel profiles, and in particular:
- 18 IPE, 92 HE, 65 L and 16 U profiles (Refs 7 and 10);
- 184 hot-finished square tubes, 292 cold-finished square tubes and 1599 round tubes (Ref 11).

The corresponding c values, expressed in %, are given in the following table.

type	Ω			I			I/H		
	c_{min}	c_{mean}	c_{max}	c_{min}	c_{mean}	c_{max}	c_{min}	c_{mean}	c_{max}
HEA	7	16	24	20	46	72	13	30	47
HEB	6	19	31	18	55	92	11	36	60
HEM	1	18	34	11	44	77	9	30	51
HEAA+	6	19	32	18	48	78	11	33	54
All HE	0.08	9	17	0.2	24	48	0.1	20	40
IPE	16	26	35	37	54	70	26	39	71
square tubes	0.02	5	9	0.001	0.1	0.2	0.02	10	20
round tubes	0.01	4	8	0.001	0.05	0.1	0.01	5	10

This shows that c varies widely. Moreover, it is observed that it is larger for the smaller I or H sections, while being very low for the tubes.

The half of c_{mean} constitutes an approximation of the increase of W for an optimised structure composed of many different profiles.

This effect gets stronger if the number of available profiles is limited, what could lead to a morphology which is not theoretically optimal, but that exploit them at maximal allowable stress σ.

6. DETERMINATION OF THE FIRST EIGEN FREQUENCY AND INFLUENCE OF DYNAMIC LOADING FOR A HORIZONTAL SPAN, VERTICALLY LOADED

Let us now consider a horizontal span, vertically loaded, and distinguish the various components of the total force F_T to which the structure is subjected:

$F_T = F_p + F_m + F_0$ where F_p $[N]$ is the permanent load, $F_m = (1/x - 1)F_p$ $[m]$ the mobile or temporary load (with $0 \le x \le 1$) and $F_0 = \rho V$ $[N]$ the dead weight.

$F_P = F_p + F_0$ is the part of it that is exerted permanently on the structure, that is the one present when it is subjected to dynamic loading.

Let δ_p $[m]$ be the corresponding maximal displacement, and $K_P = F_P/\delta_P$ $[N/m]$ the structure stiffness.

The first eigen frequency amounts then to (see for example Ref 12):

$$\omega = \frac{1}{2\pi}\sqrt{g\frac{K_P}{F_P}} = \frac{1}{2\pi}\sqrt{g\frac{1}{\delta_P}} \; [Hz], \text{ with } g = 9.81 \; m/s^2 \; .$$

Further, $\dfrac{\delta_P}{\delta_T} = \dfrac{F_P}{F_T} = x + (1-x)\rho\dfrac{V}{F_T}$.

As $\Delta = \dfrac{E\delta_T}{\sigma L}$ and $W = \dfrac{\sigma V}{F_T L}$, $\omega = \dfrac{1}{2\pi}\sqrt{\dfrac{gE}{\Delta L[\sigma x + (1-x)\rho L W]}}$.

If $x = 1$, all loads are permanent: $\omega_p = \dfrac{1}{2\pi}\sqrt{g\dfrac{E}{\sigma}\dfrac{1}{\Delta L}}$;

if $x = 0$, all loads are mobile: $\omega_0 = \dfrac{1}{2\pi}\sqrt{g\dfrac{E}{\rho}\dfrac{1}{\Delta W L^2}}$.

Figure 9 represents, on one hand, $1/\sqrt{\Delta}$ as a function of L/H for main optimal structures, and, on the other hand, ω_p for a span L in a given material. One should be aware that the Young modulus to be considered here is the instant one, which can amounts to up to twice the permanent one.

It can be observed that these optimised structures present all a rather low first eigen frequency under total load. As a consequence and in order to avoid any resonance problem, they should be equipped with dampers or the allowable stress should be strongly reduced.

ω_0 represents the upper limit of the first eigen frequency that can be reached by reducing the loads that are present during dynamic loading.

7. CONCLUSION

The volume and displacement indicators allow a relatively accurate evaluation of the volume of material of a structure and its maximal deflection under a given load case; the influence of buckling for elements in compression, of connecting pieces for elements in traction, and of the availability of commercial profiles can easily be determined in order to improve further this accuracy.

However, the method can up to now only be used for a restricted number of load cases. The research in progress will allow to extend its use progressively, for additional load cases as well as other morphologies.

8. NOTATIONS

c	/	relative difference between two successive values in series
$E_{(a)}$	Pa	Young modulus of the material of the (lightweight) structure
$F_{(T)}$	N	total intensity of the loads applied to the structure
F_m	N	mobile or temporary component of $F_{(T)}$
F_p	N	permanent component of $F_{(T)}$, except dead weight
F_P	N	permanent component of $F_{(T)}$, including dead weight
F_0	N	dead weight
g	m/s^2	gravity acceleration
H	m	main dimension of a structure, perpendicular to L
I	m^4	inertia of a structure
$k_{(i)}$	/	real number; relative wall thickness of a tube

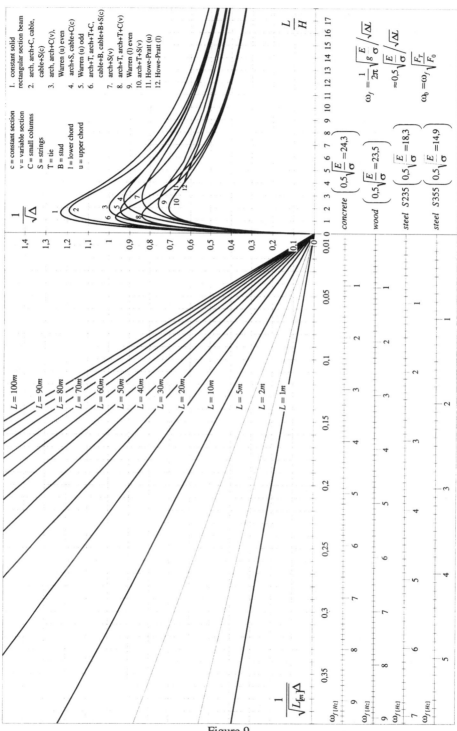

Figure 9

K_P	N/m	structure stiffness under F_P
L	m	main dimension of a structure
L/H	$/$	slenderness ratio of a structure
P_s	N	weight of a socket
V	m^3	volume of a structure
V_s	m^3	volume of a socket
W	$/$	volume indicator of a structure
x	$/$	real number between 0 and 1
$\delta_{(i)}$	m	(absolute) displacement at mid-span of a structure (under F_i)
Δ	$/$	displacement indicator of a structure
Λ	$/$	reduced slenderness of a structure
μ	$/$	end-fixity constant of a compressed bar
ρ	N/m^3	specific weight of the material
$\sigma_{(a)}$	Pa	allowable stress of the material of the (lightweight) structure
σ_{cr}	Pa	buckling stress of a structure
σ_s	Pa	allowable stress of the composing material of a socket
ω	Hz	first eigen frequency of a structure under $F_{(T)}$
ω_p	Hz	first eigen frequency of a structure under F_p
ω_0	Hz	first eigen frequency of a structure under F_m
$\Omega_{(a)}$	m^2	cross-section of a (lightweight) structure

9. REFERENCES

[1] SAMYN Ph, Étude comparée du volume et du déplacement de structures bidimensionnelles sous charges verticales entre deux appuis - Vers un outil d'évaluation et de prédimensionnement des structures, final work for a Ph.D. in Applied Sciences, Liège University, 1999.

[2] LATTEUR P, Optimisation des treillis, arcs, poutres et câbles sur base d'indicateurs morphologiques – Application aux structures soumises en partie ou en totalité au flambement, final work for a Ph.D. in Applied Sciences, Free University Brussels, 2000.

[3] VAN STEIRTEGHEM J, SAMYN Ph and DE WILDE W P, The use of morphological indicators in the optimisation of structures, proceedings of the 1st international conference on High Performance Structures and Composites, Sevilla, 2002.

[4] SAMYN Ph, VILQUIN Th and LAUWERYS S, Study of guyed and suspended structures under a uniformly distributed load using volume and displacement indicators, proceedings of the International IASS (International Association for Shell and Spatial Structures) Symposium on Lightweight Structures in Civil Engineering – Contemporary Problems, Warsaw, 2002.

[5] SAMYN Ph, BODARWE É and VILQUIN Th, Shaping the Structures at Samyn and Partners, proceedings of the International IASS (International Association for Shell and Spatial Structures) Symposium on Lightweight Structures in Civil Engineering – Contemporary Problems, Warsaw, 24-28 June 2002.

[6] SAMYN Ph, Lattice girder and bridge comprising such a girder, European Patent EP0953 684 B1, 5 April 2000, Bulletin of the European Patent Agency 2000/14, Den Haag.

[7] EUROPROFIL, Programme de vente – poutrelles, profilés U et cornières ARBED, Esch-sur-Alzette (Luxemburg), 1995.

[8] PFEIFER, Cable Structures (technical catalogue), Memmingen (Germany), 2000.

[9] Association Française de Normalisation (AFNOR), NFX 01-002 - Guide pour le choix des séries de nombres normaux et des séries comportant des valeurs plus arrondies de nombres normaux, Paris, December 1967.

[10] Centre belgo-luxembourgeois d'information de l'acier (CBLIA), Produits sidérurgiques pour la construction en acier, Brussels, 1973.

[11] VAN LEEUWEN BUIZEN, Technical information booklets n°1+2 - Dimensions and sectional properties of square, rectangular and flat-oval hollow sections, and n°3 - Dimensions and sectional properties of round (circular) hollow sections, Zwijndrecht (Netherlands), ca 1996.

[12] Institut Belge de Normalisation, NBN B03-002-2 - Actions du vent sur les constructions, effets dynamiques du vent sur les structures flexibles, Brussels, December 1988.

Index

Volume 1: pages 1 to 898
Volume 2: pages 899 to 1614